Seventh Edition

Intermediate Algebra
An Applied Approach

With Math Study Skills Workbook, Second Edition
by Paul D. Nolting

Richard N. Aufmann
Palomar College, California

Vernon C. Barker
Palomar College, California

Joanne S. Lockwood
Plymouth State University, New Hampshire

HOUGHTON MIFFLIN COMPANY BOSTON NEW YORK

INTERMEDIATE ALGEBRA: AN APPLIED APPROACH, SEVENTH EDITION
by Richard N. Aufmann, Vernon C. Barker and Joanne S. Lockwood
Copyright © 2006 by Houghton Mifflin Company. All rights reserved.
Editor-in-Chief: Jack Shira
Senior Sponsoring Editor: Lynn Cox
Associate Editor: Melissa Parkin
Editorial Assistant: Noel Kamm
Senior Project Editor: Tamela Ambush
Editorial Assistant: Sage Anderson
Manufacturing Manager: Karen Banks
Senior Marketing Manager: Ben Rivera
Marketing Assistant: Lisa Lawler
MATH STUDY SKILLS WORKBOOK, SECOND EDITION
by Paul D. Nolting
Copyright © 2005 by Houghton Mifflin Company. All rights reserved.
Publisher: Jack Shira
Senior Sponsoring Editor: Lynn Cox
Assistant Editor: Melissa Parkin
Associate Project Editor: Kristin Penta
Manufacturing Manager: Florence Cadran
Senior Marketing Manager: Ben Rivera
Marketing Assistant: Lisa Lawler

Custom Publishing Editor: Dan Luciano
Custom Publishing Production Manager: Kathleen McCourt
Custom Publishing Project Coordinator: Andrea Wagner
Cover photo: © Getty Images, Inc./Comstock Images
Photo Credits from Intermediate Algebra: Chapter 1: p. 1, Stock Works / CORBIS; p. 44 Sky Bonilla/ PhotoEdit, Inc.; p. 52 David Stoecklein/ CORBIS; p. 34 Stephen Mark Needham / Foodpix / Getty Images. Chapter 2: p. 55 Michael Newman/ PhotoEdit, Inc.; p. 72 Leonard de Selva / CORBIS; p. 82 Michael S. Yamashita / CORBIS; p. 93 Richard Cummins/ CORBIS; p. 106 Stephen Chernin / Getty Images; p. 113 Alan Oddie/PhotoEdit, Inc. Chapter 3: p. 119 Jeff Greenberg/PhotoEdit, Inc.; p. 125 Craig Tuttle / CORBIS; p. 131 Ulrike Welsch / PhotoEdit, Inc.; p. 141 Robert W. Ginn/PhotoEdit, Inc.; p. 162 Eric Fowke/PhotoEdit, Inc.; p. 172 Royalty-Free /CORBIS; p. 173 David Keaton / CORBIS; p. 185 The Granger Collection; p. 188 Stan Honda/Getty Images. Chapter 4: p. 199 David Stoecklein / CORBIS; p. 232 Spencer Grant /PhotoEdit, Inc.; p. 237 Jose Carillo / PhotoEdit, Inc; p. 238 Royalty-Free / CORBIS; p. 239 Michael Newman / PhotoEdit, Inc.; p. 245 AP/Wide World Photos; p. 247 Michael Newman / PhotoEdit, Inc.; p. 254 Susan Van Etten/PhotoEdit, Inc. Chapter 5: p. 257 AP/Wide World Photos; p. 269 Stocktrek / CORBIS; p. 270 NASA/ JPL Handout/ Reuters Newmedia Inc. / CORBIS; p. 270 John Neubauer /PhotoEdit, Inc.; p. 325 David Young-Wolff / PhotoEdit, Inc.; p. 327 Roger Ressmeyer/ CORBIS; p. 334 Susan Van Etten/ PhotoEdit, Inc.; Chapter 6: p. 339 Rachel Epstein / PhotoEdit, Inc.; p. 365 Allan Morgan; p. 366 Michael Newman / PhotoEdit, Inc.; p. 369 James Marshall / CORBIS; p. 374 Joel W. Rogers / CORBIS; p. 374 HMS Group/ CORBIS; p. 379 Tony Freeman /PhotoEdit,Inc; p. 384 Jonathan Nourak/ PhotoEdit, Inc.; p. 392 Royalty-Free/ CORBIS. Chapter 7: p. 395 Benjamin Shearn/TAXI / Getty Images; p. 419 Bettmann/CORBIS; p. 421 Sandor Szabo / EPA / Landov; p. 433 Royalty-Free/ CORBIS; p. 442 Sky Bonillo / PhotoEdit, Inc. Chapter 8: p. 443 David Young-Wolff / PhotoEdit, Inc.; p. 458 Photex/ CORBIS; p. 477 Bill Aron / PhotoEdit, Inc.; p. 478 Reuters/CORBIS; p. 480 Dale C. Spartas/ CORBIS. Chapter 9: p. 491 Tim Boyle/ Getty Image; p. 492 Lon C. Diehl / PhotoEdit, Inc.; p. 500 Jim Craigmyle / CORBIS; p. 507 Rich Clarkson /Getty Images; p. 508 Vic Bider/ PhotoEdit, Inc.; p. 511 Nick Wheeler /CORBIS; p. 517 Jose Fuste Raga / CORBIS; p. 521 Joel W. Rogers / CORBIS; p. 522 Robert Brenner / PhotoEdit, Inc.; p. 533 Michael Newman / PhotoEdit, Inc.; p. 542 Kim Sayer / CORBIS. Chapter 10: p. 543 Rudi Von Briel / PhotoEdit, Inc.; p. 556 The Granger Collection; p. 573 Express Newspapers/Getty Images; p. 574 Richard T. Nowitz/CORBIS; p. 574 Courtesy of the Edgar Fahs Smith Image Collection /University of Pennsylvania Library, Philadelphia, PA 19104-6206; p. 575 Bettmann/CORBIS; p. 577 Mark Harmel / STOKE/ Getty Images; p. 578 Myrleen Fergusun Cate / PhotoEdit, Inc.; p. 579 Michael Johnson - www.earthwindowcom; p. 580 Roger Ressmeyer / CORBIS; p. 583 David Young-Wolff /PhotoEdit, Inc.; p. 588 Macduff Everton/ CORBIS; p. 592 Frank Siteman / PhotoEdit, Inc. Chapter 11: p. 593 Roger Ressmeyer / CORBIS; p. 594 Jeff Greenberg / PhotoEdit, Inc.; p. 595 Jennifer Waddell / Houghton Mifflin Company; p. 596 Galen Rowell/ CORBIS; p. 597 Bettmann /CORBIS; p. 627 Photodisc Green / Getty Images; p. 628 Joseph Sohm; Visions of America / CORBIS; p. 638 Mark Cooper/ CORBIS. Chapter 12: p. 639 Colin Young-Wolff / PhotoEdit, Inc; p. 646 The Granger Collection; p.652 Ariel Skelley / CORBIS; p. 662 William James Warren/ CORBIS; p. 663 The Granger Collection; p. 664 The Granger Collection; p. 669 AP / Wide World Photos; p. 678 David Young-Wolff/PhotoEdit, Inc.; p. 680 Tony Freeman / PhotoEdit, Inc.

Printed in the United States of America.
ISBN: 0-618-58681-4
N-04493

2 3 4 5 6 7 8 9 – DS – 07 06 05

222 Berkeley Street • Boston, MA 02116
Address all correspondence and order information to the above address.

Contents

Preface

The seventh edition of *Intermediate Algebra: An Applied Approach* provides comprehensive, mathematically sound coverage of the topics considered essential in an intermediate algebra course. The text has been designed not only to meet the needs of the traditional college student, but also to serve the needs of returning students whose mathematical proficiency may have declined during years away from formal education.

In this new edition of *Intermediate Algebra: An Applied Approach*, we have continued to integrate some of the approaches suggested by AMATYC. Each chapter opens with an illustration and a reference to a mathematical application within the chapter. At the end of each section there are Applying the Concepts exercises, which include writing, synthesis, critical thinking, and challenge problems. At the end of each chapter there is a "Focus on Problem Solving," which introduces students to various problem-solving strategies. This is followed by "Projects and Group Activities," which can be used for cooperative-learning activities.

NEW! Changes to This Edition

In response to user requests, in this edition of the text students are asked to write solution sets of inequalities in both set-builder notation and in interval notation, as well as to graph solution sets of inequalities in one variable on the number line. See, for example, pages 89 and 90 in Section 2.4.

Section 3 of Chapter 6 now presents two methods of simplifying a complex fraction: (1) multiplying the numerator and denominator of the complex fraction by the least common multiple of the denominators and (2) multiplying the numerator by the reciprocal of the denominator of the complex fraction.

We have found that students who are taught division of a polynomial by a monomial as a separate topic are subsequently more successful in factoring a monomial from a polynomial. Therefore, we have added a new objective, "To divide a polynomial by a monomial," to Section 4 of Chapter 5.

The concept of function is given greater emphasis in this edition of the text. For example, Chapter 6 begins with a new objective titled "To find the domain of a rational function." The next objective is on simplifying rational functions. In Section 9.3, the material on composition of functions has been expanded, and students are given more opportunities to apply the concept to applications.

In the previous edition, complex numbers were presented in Section 7.3. In this edition, complex numbers have been moved to the last section of the chapter. This provides for a better flow of the material in Chapter 7 and places complex numbers immediately before Chapter 8, *Quadratic Equations*, where it is used extensively.

In Section 2 of Chapter 10, the introduction to logarithms has been rewritten. Motivation for the need for logarithms is developed within the context of an application. The slower pace of the presentation of this topic will help students to better understand and apply the concept of logarithm.

The in-text examples are now highlighted by a prominent HOW TO bar. Students looking for a worked-out example can easily locate one of these problems.

As another aid for students, more annotations have been added to the Examples provided in the paired Example/You Try It boxes. This will assist

students in understanding what is happening in key steps of the solution to an exercise.

Throughout the text, data problems have been updated to reflect current data and trends. Also, titles have been added to the application exercises in the exercise sets. These changes emphasize the relevance of mathematics and the variety of problems in real life that require mathematical analysis.

The Chapter Summaries have been remodeled and expanded. Students are provided with definitions, rules, and procedures, along with examples of each. An objective reference and a page reference accompany each entry. We are confident that these will be valuable aids as students review material and study for exams.

In many chapters, the number of exercises in the Chapter Review Exercises has been increased. This will provide students with more practice on the concepts presented in the chapter.

The calculator appendix has been expanded to include instruction on more functions of the graphing calculator. Notes entitled Integrating Technology appear throughout the book and many refer the student to this appendix. Annotated illustrations of both a scientific calculator and a graphing calculator appear on the inside back cover of this text.

<div style="float:left">

chapter

11 Conic Sections

OBJECTIVES

Section 11.1
A To graph a parabola

Section 11.2
A To find the equation of a circle and then graph the circle
B To write the equation of a circle in standard form

Section 11.3
A To graph an ellipse with center at the origin
B To graph a hyperbola with center at the origin

Section 11.4
A To solve a nonlinear system of equations

Section 11.5
A To graph the solution set of a quadratic inequality in two variables
B To graph the solution set of a nonlinear system of inequalities

When creating a parabolic-shaped mirror for use in a telescope, such as the one shown in this photo, there is little room for error. The multimillion dollar Hubble Telescope made international headlines when it was discovered that the outer edge of its primary mirror was gound too flat by one fiftieth the thickness of a human hair! This flaw (which was later corrected) caused Hubble to transmit fuzzy images because the light from viewed objects was scattering instead of reflecting to a single point. This single point is called the *focus*, as explained in **Exercises 25 and 26 on page 600.**

Need help? For online student resources, such as section quizzes, visit this textbook's website at **math.college.hmco.com/students.**

</div>

Page 593

Chapter Opening Features

NEW! Chapter Opener

New, motivating chapter opener photos and captions have been added, illustrating and referencing a specific application from the chapter.

The at the bottom of the page lets students know of additional online resources at math.college.hmco.com/students.

Objective-Specific Approach

Each chapter begins with a list of learning objectives that form the framework for a complete learning system. The objectives are woven throughout the text (i.e., Exercises, Prep Tests, Chapter Review Exercises, Chapter Tests, Cumulative Review Exercises) as well as through the print and multimedia ancillaries. This results in a seamless learning system delivered in one consistent voice.

Page 594

Prep Test and Go Figure

Prep Tests occur at the beginning of each chapter and test students on previously covered concepts that are required in the coming chapter. Answers are provided in the Answer Section. Objective references are also provided if a student needs to review specific concepts.

The **Go Figure** problem that follows the *Prep Test* is a playful puzzle problem designed to engage students in problem solving.

PREP TEST • • •

Do these exercises to prepare for Chapter 11.

1. Find the distance between the points whose coordinates are $(-2, 3)$ and $(4, -1)$. Round to the nearest hundredth.

2. Complete the square on $x^2 - 8x$. Write the resulting perfect-square trinomial as the square of a binomial.

3. Solve $\frac{x^2}{16} + \frac{y^2}{9} = 1$ for x when $y = 3$ and when $y = 0$.

4. Solve by the substitution method:
$7x + 4y = 3$
$y = x - 2$

5. Solve by the addition method:
$4x - y = 9$
$2x + 3y = -13$

6. Find the axis of symmetry and the vertex of the graph of $y = x^2 - 4x + 2$.

7. Graph: $f(x) = -2x^2 + 4x$

8. Graph the solution set of $5x - 2y > 10$.

9. Graph the solution set:
$x + 2y \le 4$
$x - y \le 2$

GO FIGURE • • •

What is the minimum number of times a paper must be folded in half so that its thickness is at least the height of the Empire State Building? Assume that the paper can be folded in half indefinitely and that the paper is 0.01 in. thick. The height of the Empire State Building is 1250 ft.

Aufmann Interactive Method (AIM)

Page 419

You Try It 3

Strategy To find the diagonal, use the Pythagorean Theorem. One leg is the length of the rectangle. The second leg is the width of the rectangle. The hypotenuse is the diagonal of the rectangle.

Solution

$c^2 = a^2 + b^2$ • Pythagorean Theorem
$c^2 = (6)^2 + (3)^2$ • Replace a by 6 and b by 3.
$c^2 = 36 + 9$ • Solve for c.
$c^2 = 45$
$(c^2)^{1/2} = (45)^{1/2}$ • Raise each side to the $\frac{1}{2}$ power.
$c = \sqrt{45}$ • $a^{1/2} = \sqrt{a}$
$c \approx 6.7$

The diagonal is 6.7 cm.

Page S23

An Interactive Approach

Intermediate Algebra: An Applied Approach uses an interactive style that provides a student with an opportunity to try a skill as it is presented. Each section is divided into objectives, and every objective contains one or more sets of matched-pair examples. The first example in each set is worked out; the second example, called "You Try It," is for the student to work. By solving this problem, the student actively practices concepts as they are presented in the text.

There are complete worked-out solutions to these examples in an appendix. By comparing their solution to the solution in the appendix, students obtain immediate feedback on, and reinforcement of, the concept.

Page xxv

AIM for Success Student Preface

This student 'how to use this book' preface explains what is required of a student to be successful and how this text has been designed to foster student success, including the Aufmann Interactive Method (AIM). *AIM for Success* can be used as a lesson on the first day of class or as a project for students to complete to strengthen their study skills. There are suggestions for teaching this lesson in the *Instructor's Resource Manual*.

AIM for Success

Welcome to *Intermediate Algebra: An Applied Approach*. As you begin this course, we know two important facts: (1) We want you to succeed. (2) You want to succeed. To do that requires an effort from each of us. For the next few pages, we are going to show you what is required of you to achieve that success and how you can use the features of this text to be successful.

Motivation One of the most important keys to success is motivation. We can try to motivate you by offering interesting or important ways mathematics can benefit you. But, in the end, the motivation must come from you. On the first day of class, it is easy to be motivated. Eight weeks into the term, it is harder to keep that motivation.

Problem Solving

Focus on Problem Solving

At the end of each chapter is a Focus on Problem Solving feature which introduces the student to various successful problem-solving strategies. Strategies such as drawing a diagram, applying solutions to other problems, working backwards, inductive reasoning, and trial and error are some of the techniques that are demonstrated.

Focus on Problem Solving

Solve an Easier Problem

We begin this problem-solving feature with a restatement of the four steps of Polya's recommended problem-solving model.

1. Understand the problem.
2. Devise a plan.
3. Carry out the plan.
4. Review your solution.

One of the several methods of devising a plan is to try to solve an easier problem. Suppose you are in charge of your softball league, which consists of 15 teams. You must devise a schedule in which each team plays every other team once. How many games must be scheduled?

To solve this problem, we will attempt an easier problem first. Suppose that your league contains only a small number of teams. For instance, if there were only 1 team, you would schedule 0 games. If there were 2 teams, you would schedule 1 game. If there were 3 teams, you would schedule 3 games. The diagram at the left shows that 6 games must be scheduled when there are 4 teams in the league.

Here is a table of our results so far. (Remember that making a table is another strategy to be used in problem solving.)

Number of Terms	Number of Games	Possible Pattern
1	0	0
2	1	1
3	3	1 + 2
4	6	1 + 2 + 3

1. Draw a diagram with 5 dots to represent the teams. Draw lines from each dot to a second dot, and determine the number of games required.
2. What is the apparent pattern for the number of games required?
3. Assuming that the pattern continues, how many games must be scheduled for the 15 teams of the original problem?

After solving a problem, good problem solvers ask whether it is possible to solve the problem in a different manner. Here is a possible alternative method of solving the scheduling problem.

Begin with one of the 15 teams (say team A) and ask, "How many games must this team play?" Because there are 14 teams left to play, you must schedule 14 games. Now move to team B. It is already scheduled to play team A, and it does not play itself, so there are 13 teams left for it to play. Consequently, you must schedule 14 + 13 games.

4. Continue this reasoning for the remaining teams and determine the number of games that must be scheduled. Does this answer correspond to the answer you obtained using the first method?
5. Making connections to other problems you have solved is an important step toward becoming an excellent problem solver. How does the answer to the scheduling of teams relate to the triangular numbers discussed in the Focus on Problem Solving in the chapter titled Linear Functions and Inequalities in Two Variables?

Page 245

Problem-Solving Strategies

The text features a carefully developed approach to problem solving that emphasizes the importance of *strategy* when solving problems. Students are encouraged to develop their own strategies—to draw diagrams, to write out the solution steps in words—as part of their solution to a problem. In each case, model strategies are presented as guides for students to follow as they attempt the "You Try It" problem. Having students provide strategies is a natural way to incorporate writing into the math curriculum.

Page 78

Example 3

A long-distance runner started a course running at an average speed of 6 mph. One and one-half hours later, a cyclist traveled the same course at an average speed of 12 mph. How long after the runner started did the cyclist overtake the runner?

You Try It 3

Two small planes start from the same point and fly in opposite directions. The first plane is flying 30 mph faster than the second plane. In 4 h the planes are 1160 mi apart. Find the rate of each plane.

Strategy

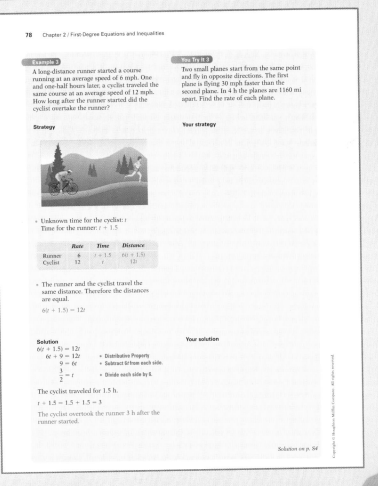

- Unknown time for the cyclist: t
 Time for the runner: $t + 1.5$

	Rate	Time	Distance
Runner	6	$t + 1.5$	$6(t + 1.5)$
Cyclist	12	t	$12t$

- The runner and the cyclist travel the same distance. Therefore the distances are equal.

 $6(t + 1.5) = 12t$

Your strategy

Solution

$$6(t + 1.5) = 12t$$
$$6t + 9 = 12t \qquad \bullet \text{ Distributive Property}$$
$$9 = 6t \qquad \bullet \text{ Subtract } 6t \text{ from each side.}$$
$$\frac{3}{2} = t \qquad \bullet \text{ Divide each side by 6.}$$

The cyclist traveled for 1.5 h.

$$t + 1.5 = 1.5 + 1.5 = 3$$

The cyclist overtook the runner 3 h after the runner started.

Your solution

Solution on p. S4

xvii

Real Data and Applications

Applications

One way to motivate an interest in mathematics is through applications. Wherever appropriate, the last objective of a section presents applications that require the student to use problem-solving strategies, along with the skills covered in that section, to solve practical problems. This carefully integrated applied approach generates student awareness of the value of algebra as a real-life tool.

Applications are taken from many disciplines including astronomy, business, carpentry, chemistry, construction, Earth science, education, manufacturing, nutrition, real estate, and telecommunications.

Real Data

Real data examples and exercises, identified by

, ask students to analyze and solve

problems taken from actual situations. Students are often required to work with tables, graphs, and charts drawn from a variety of disciplines.

Page 126

Integrating Technology

See the Keystroke Guide: *Scatter Diagrams*, for instructions on using a graphing calculator to create a scatter diagram.

A researcher may investigate the relationship between two variables by means of *regression analysis*, which is a branch of statistics. The study of the relationship between the two variables may begin with a **scatter diagram,** which is a graph of the ordered pairs of the known data.

The following table shows randomly selected data from the participants 40 years old and older and their times (in minutes) for a recent Boston Marathon.

Age (x)	55	46	53	40	40	44	54	44	41	50
Time (y)	254	204	243	194	281	197	238	300	232	216

TAKE NOTE

The jagged portion of the horizontal axis in the figure at the right indicates that the numbers between 0 and 40 are missing.

The scatter diagram for these data is shown at the right. Each ordered pair represents the age and time for a participant. For instance, the ordered pair $(53, 243)$ indicates that a 53-year-old participant ran the marathon in 243 min.

Example 5

The grams of sugar and the grams of fiber in a 1-ounce serving of six breakfast cereals are shown in the table below. Draw a scatter diagram of these data.

	Sugar (x)	Fiber (y)
Wheaties	4	3
Rice Krispies	3	0
Total	5	3
Life	6	2
Kix	3	1
Grape-Nuts	7	5

Strategy

To draw a scatter diagram:

- Draw a coordinate grid with the horizontal axis representing the grams of sugar and the vertical axis the grams of fiber.
- Graph the ordered pairs (4, 3), (3, 0), (5, 3), (6, 2), (3, 1), and (7, 5).

Solution

You Try It 5

According to the National Interagency Fire Center, the number of deaths in U.S. wildland fires is as shown in the table below. Draw a scatter diagram of these data.

Year	Number of Deaths
1998	14
1999	28
2000	17
2001	18
2002	23

Your strategy

Your solution

Solution on p. S7

Page 270

117. Astronomy The distance to Saturn is 8.86×10^8 mi. A satellite leaves Earth traveling at a constant rate of 1×10^5 mph. How long does it take for the satellite to reach Saturn?

118. The Federal Government In 2004, the gross national debt was approximately 7×10^{12} dollars. How much would each American have to pay in order to pay off the debt? Use 3×10^8 as the number of citizens.

119. Physics The mass of an electron is 9.109×10^{-31} kg. The mass of a proton is 1.673×10^{-27} kg. How many times heavier is a proton than an electron?

120. Geology The mass of Earth is 5.9×10^{24} kg. The mass of the sun is 2×10^{30} kg. How many times heavier is the sun than Earth?

121. Physics How many meters does light travel in 8 h? The speed of light is 3×10^8 m/s.

122. Physics How many meters does light travel in 1 day? The speed of light is 3×10^8 m/s.

123. Astronomy It took 11 min for the commands from a computer on Earth to travel to the rover Sojourner on Mars, a distance of 119 million miles. How fast did the signals from Earth to Mars travel?

Sojourner

124. Measurement The weight of 31 million orchid seeds is 1 oz. Find the weight of one orchid seed.

125. Physics A high-speed centrifuge makes 4×10^8 revolutions each minute. Find the time in seconds for the centrifuge to make one revolution.

Centrifuge

APPLYING THE CONCEPTS

126. Correct the error in the following expressions. Explain which rule or property was used incorrectly.
 a. $x^0 = 0$
 b. $(x^4)^5 = x^9$
 c. $x^2 \cdot x^3 = x^6$

127. Simplify. **a.** $1 + [1 + (1 + 2^{-1})^{-1}]^{-1}$
 b. $2 - [2 - (2 - 2^{-1})^{-1}]^{-1}$

Student Pedagogy

Icons

The at each objective head remind students that both a video and a tutorial lesson are available for that objective.

Key Terms and Concepts

Key terms, in bold, emphasize important terms. The key terms are also provided in a **Glossary** at the back of the text.

Key Concepts are presented in orange boxes in order to highlight these important concepts and to provide for easy reference.

Point of Interest

These margin notes contain interesting sidelights about mathematics, its history, or its application.

Take Note

These margin notes alert students to a point requiring special attention or are used to amplify the concept under discussion.

Page 424

Study Tips

These margin notes remind students of study skills presented in the *AIM for Success*; some notes provide page references to the original descriptions. They also provide students with reminders of how to practice good study habits.

HOW TO Examples

HOW TO examples use annotations to explain what is happening in key steps of the complete, worked-out solutions.

Integrating Technology

These margin notes provide suggestions for using a calculator or refer the student to an appendix for more complete instructions on using a calculator.

Exercises and Projects

Exercises

The exercise sets of *Intermediate Algebra: An Applied Approach* emphasize skill building, skill maintenance, and applications. Concept-based writing or developmental exercises have been integrated with the exercise sets. Icons identify appropriate writing , data analysis, and calculator exercises.

Included in each exercise set are **Applying the Concepts** that present extensions of topics, require analysis, or offer challenge problems. The writing exercises ask students to explain answers, write about a topic in the section, or research and report on a related topic.

Page 327

Page 566

14. $f(x) = \log_3(2 - x)$

15. $f(x) = -\log_2(x - 1)$

16. $f(x) = -\log_2(1 - x)$

APPLYING THE CONCEPTS

 Use a graphing calculator to graph the functions in Exercises 17 to 22.

17. $f(x) = x - \log_2(1 - x)$

18. $f(x) = -\frac{1}{2}\log_2 x - 1$

19. $f(x) = \frac{x}{2} - 2\log_2(x + 1)$

20. $f(x) = x + \log_3(2 - x)$

21. $f(x) = x^2 - 10\ln(x - 1)$

22. $f(x) = \frac{x}{3} - 3\log_2(x + 3)$

23. **Astronomy** Astronomers use the *distance modulus* of a star as a method of determining the star's distance from Earth. The formula is $M = 5\log s - 5$, where M is the distance modulus and s is the star's distance from the Earth in parsecs. (One parsec $\approx 1.9 \times 10^{13}$ mi)
 a. Graph the equation.
 b. The point with coordinates (25.1, 2) is on the graph. Write a sentence that describes the meaning of this ordered pair.

24. **Typing** Without practice, the proficiency of a typist decreases. The equation $s = 60 - 7\ln(t + 1)$, where s is the typing speed in words per minute and t is the number of months without typing, approximates this decrease.
 a. Graph the equation.
 b. The point with coordinates (4, 49) is on the graph. Write a sentence that describes the meaning of this ordered pair.

Projects and Group Activities

The Projects and Group Activities featured at the end of each chapter can be used as extra credit or for cooperative learning activities. The projects cover various aspects of mathematics, including the use of calculators, collecting data from the Internet, data analysis, and extended applications.

Page 327

Projects and Group Activities

Astronomical Distances and Scientific Notation

Astronomers have units of measurement that are useful for measuring vast distances in space. Two of these units are the astronomical unit and the light-year. An **astronomical unit** is the average distance between Earth and the sun. A **light-year** is the distance a ray of light travels in 1 year.

1. Light travels at a speed of 1.86×10^5 mi/s. Find the measure of 1 light-year in miles. Use a 365-day year.

2. The distance between Earth and the star Alpha Centauri is approximately 25 trillion miles. Find the distance between Earth and Alpha Centauri in light-years. Round to the nearest hundredth.

3. The Coma cluster of galaxies is approximately 2.8×10^8 light-years from Earth. Find the distance, in miles, from the Coma Cluster to Earth. Write the answer in scientific notation.

4. One astronomical unit (A.U.) is 9.3×10^7 mi. The star Pollux in the constellation Gemini is 1.8228×10^{12} mi from Earth. Find the distance from Pollux to Earth in astronomical units.

5. One light-year is equal to approximately how many astronomical units? Round to the nearest thousand.

Gemini

Shown below are data on the planets in our solar system. The planets are listed in alphabetical order.

Point of Interest

In November 2001, the Hubble Space Telescope took photos of the atmosphere of a planet orbiting a star 150 light-years from Earth in the constellation Pegasus. The planet is about the size of Jupiter and orbits close to the star HD209458. It was the first discovery of an atmosphere around a planet outside our solar system.

Planet	Distance from the Sun (in kilometers)	Mass (in kilograms)
Earth	1.50×10^8	5.97×10^{24}
Jupiter	7.79×10^8	1.90×10^{27}
Mars	2.28×10^8	6.42×10^{23}
Mercury	5.79×10^7	3.30×10^{23}
Neptune	4.50×10^9	1.02×10^{26}
Pluto	5.87×10^9	1.25×10^{22}
Saturn	1.43×10^9	5.68×10^{26}
Uranus	2.87×10^9	8.68×10^{25}
Venus	1.08×10^8	4.87×10^{24}

Jupiter

6. Arrange the planets in order from closest to the sun to farthest from the sun.

7. Arrange the planets in order from the one with the greatest mass to the one with the least mass.

8. Write a rule for ordering numbers written in scientific notation.

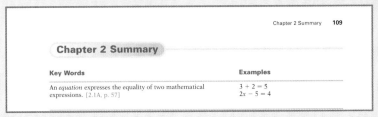

Page 109

End of Chapter

Chapter Summary

At the end of each chapter there is a Chapter Summary that includes Key Words, Essential Rules and Procedures, and an example of each. Each entry includes an objective reference and a page reference indicating where the concept is introduced. These chapter summaries provide a single point of reference as the student prepares for a test.

Essential Rules and Procedures	Examples
Addition Property of Equations [2.1A, p. 57] If $a = b$, then $a + c = b + c$.	$x + 5 = -3$ $x + 5 - 5 = -3 - 5$ $x = -8$

Page 110

Page 436

Chapter Review Exercises

Chapter Review Exercises are found at the end of each chapter. These exercises are selected to help the student integrate all of the topics presented in the chapter.

Page 439

Chapter Test

Each Chapter Test is designed to simulate a possible test of the material in the chapter.

Chapter 7 Test 439

Chapter 7 Test

1. Write $\frac{1}{2}\sqrt[3]{x^3}$ as an exponential expression.

2. Simplify:
$\sqrt[3]{54x^7y^3} - x\sqrt[3]{128x^4y^3} - x^2\sqrt[3]{2xy^3}$

3. Write $3y^{2/5}$ as a radical expression.

4. Simplify: $(2 + 5i)(4 - 2i)$

Page 441

Cumulative Review Exercises

Cumulative Review Exercises, which appear at the end of each chapter (beginning with Chapter 2), help students maintain skills learned in previous chapters.

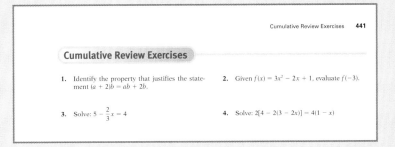

The answers to all Chapter Review Exercises, all Chapter Test exercises, and all Cumulative Review Exercises are given in the Answer Section. Along with the answer, there is a reference to the objective that pertains to each exercise.

Page A23

CUMULATIVE REVIEW EXERCISES

1. The Distributive Property [1.3A] 2. $f(-3) = 34$ [3.2A] 3. $\frac{3}{2}$ [2.1B] 4. $\frac{2}{3}$ [2.1C]

5. $\frac{1}{3}$ and $\frac{7}{3}$ [2.5A] 6. $\{x \mid x > 1\}$ [2.4A] 7. $\{x \mid -6 \le x \le 3\}$ [2.5B] 8. $(9x + y)(9x - y)$ [5.6A]

9. $x(x^2 + 3)(x + 1)(x - 1)$ [5.6D] 10. $y = \frac{1}{3}x + \frac{7}{3}$ [3.5B] 11. 3 [4.3A] 12. $C = R - nP$ [2.1D]

Instructor Resources

Intermediate Algebra: An Applied Approach has a complete set of support materials for the instructor.

Instructor's Annotated Edition This edition contains a replica of the student text and additional resources just for the instructor. These include: *Instructor Notes, New Vocabulary/Symbols, etc., Vocabulary/Symbols, etc. to Review, In-Class Examples, Discuss the Concepts, Concept Checks, Optional Student Activities, Suggested Assignments, Quick Quizzes, Answers to Writing Exercises/Focus on Problem Solving/Projects and Group Activities,* and *PowerPoints.* Answers to all exercises are also provided.

Instructor's Solutions Manual The *Instructor's Solutions Manual* contains worked-out solutions for all exercises in the text.

Instructor's Resource Manual with Testing This resource includes eight ready-to-use printed *Chapter Tests* per chapter, suggested *Course Sequences,* and a printout of the *AIM for Success* PowerPoint slide show. All resources are also available on the Instructor website and *Class Prep* CD.

HM ClassPrep with HM Testing CD-ROM *HM ClassPrep* contains a multitude of text-specific resources for instructors to use to enhance the classroom experience. These resources can be easily accessed by chapter or resource type and can also link you to the text's website. *HM Testing* is our computerized test generator and contains a database of algorithmic test items as well as providing **online testing** and **gradebook** functions.

Instructor Text-Specific Website The resources available on the *Class Prep* CD are also available on the instructor website at math.college.hmco.com/instructors. Appropriate items are password protected. Instructors also have access to the student part of the text's website.

WebCT ePacks *WebCT ePacks* provide instructors with a flexible, Internet-based education platform providing multiple ways to present learning materials. The *WebCT ePacks* come with a full array of features to enrich the online learning experience.

Blackboard Cartridges The *Houghton Mifflin Blackboard cartridge* allows flexible, efficient, and creative ways to present learning materials and opportunities. In addition to course management benefits, instructors may make use of an electronic grade book, receive papers from students enrolled in the course via the Internet, and track student use of the communication and collaboration functions.

NEW! HM Eduspace® is a powerful course management system powered by Blackboard that makes preparing, presenting, and managing courses easier. You can use this distance-learning platform to customize, create, and deliver course materials and tests online, and easily maintain student portfolios using the grade book, where grades for all assignments are automatically scored, averaged, and saved.

Student Resources

Student Solutions Manual The *Student Solutions Manual* contains complete solutions to all odd-numbered exercises in the text.

Math Study Skills Workbook by Paul D. Nolting This workbook is designed to reinforce skills and minimize frustration for students in any math class, lab, or study skills course. It offers a wealth of study tips and sound advice on note

taking, time management, and reducing math anxiety. In addition, numerous opportunities for self-assessment enable students to track their own progress.

HM Eduspace® Online Learning Environment *Eduspace* is a text-specific, web-based learning environment which combines an algorithmic tutorial program with homework capabilities. Specific content is available 24 hours a day to help you further understand your textbook.

HM mathSpace® Tutorial CD-ROM This tutorial CD-ROM allows students to practice skills and review concepts as many times as necessary by providing algorithmically-generated exercises and step-by-step solutions for practice.

SMARTHINKING™ Live, Online Tutoring Houghton Mifflin has partnered with SMARTHINKING to provide an easy-to-use and effective online tutorial service. **Whiteboard Simulations** and **Practice Area** promote real-time visual interaction.

Three levels of service are offered.

- **Text-specific Tutoring** provides real-time, one-on-one instruction with a specially qualified 'e-structor.'

- **Questions Any Time** allows students to submit questions to the tutor outside the scheduled hours and receive a reply within 24 hours.

- **Independent Study Resources** connects students with around-the-clock access to additional educational services, including interactive websites, diagnostic tests, and Frequently Asked Questions posed to SMARTHINKING e-structors.

Houghton Mifflin Instructional Videos and DVDs Text-specific videos and DVDs, hosted by Dana Mosely, cover all sections of the text and provide a valuable resource for further instruction and review.

Student Text-Specific Website Online student resources can be found at this text's website at math.college.hmco.com/students.

Acknowledgments

The authors would like to thank the people who have reviewed this manuscript and provided many valuable suggestions.

Bernadette Antkoviak, *Harrisburg Area Community College, PA*
Fred E. Brown, *University of Maine at Augusta, ME*
Greg Dietrich, *Florida Community College at Jacksonville, FL*
Lori S. Holdren, *Manatee Community College, FL*
John W. Horton Sr., *St. Petersburg College, FL*
Frederick W. Keene, *Pasadena City College, CA*
Helen Medley
Lauri Semarne
John Searcy

With special thanks to Dawn Nuttall for her contributions in the developmental editing of three editions of this textbook series.

AIM for Success

Welcome to *Intermediate Algebra: An Applied Approach*. As you begin this course, we know two important facts: (1) We want you to succeed. (2) You want to succeed. To do that requires an effort from each of us. For the next few pages, we are going to show you what is required of you to achieve that success and how you can use the features of this text to be successful.

Motivation

One of the most important keys to success is motivation. We can try to motivate you by offering interesting or important ways mathematics can benefit you. But, in the end, the motivation must come from you. On the first day of class, it is easy to be motivated. Eight weeks into the term, it is harder to keep that motivation.

To stay motivated, there must be outcomes from this course that are worth your time, money, and energy.

List some reasons you are taking this course.

TAKE NOTE

Motivation alone will not lead to success. For instance, suppose a person who cannot swim is placed in a boat, taken out to the middle of a lake, and then thrown overboard. That person has a lot of motivation but there is a high likelihood the person will drown without some help. Motivation gives us the desire to learn but is not the same as learning.

Although we hope that one of the reasons you listed was an interest in mathematics, we know that many of you are taking this course because it is required to graduate, it is a prerequisite for a course you must take, or because it is required for your major. Although you may not agree that this course is necessary, it is! If you are motivated to graduate or complete the requirements for your major, then use that motivation to succeed in this course. Do not become distracted from your goal to complete your education!

Commitment

To be successful, you must make a commitment to succeed. This means devoting time to math so that you achieve a better understanding of the subject.

List some activities (sports, hobbies, talents such as dance, art, or music) that you enjoy and at which you would like to become better.

ACTIVITY	TIME SPENT	TIME WISHED SPENT

Thinking about these activities, put the number of hours that you spend each week practicing these activities next to the activity. Next to that number, indicate the number of hours per week you would like to spend on these activities.

Whether you listed surfing or sailing, aerobics or restoring cars, or any other activity you enjoy, note how many hours a week you spend doing it. To succeed in math, you must be willing to commit the same amount of time. Success requires some sacrifice.

The "I Can't Do Math" Syndrome

There may be things you cannot do, such as lift a two-ton boulder. You can, however, do math. It is much easier than lifting the two-ton boulder. When you first

learned the activities you listed above, you probably could not do them well. With practice, you got better. With practice, you will be better at math. Stay focused, motivated, and committed to success.

It is difficult for us to emphasize how important it is to overcome the "I Can't Do Math" Syndrome. If you listen to interviews of very successful atheletes after a particularly bad performance, you will note that they focus on the positive aspect of what they did, not the negative. Sports psychologists encourage athletes to always be positive—to have a "Can Do" attitude. Develop this attitude toward math.

Strategies for Success

Textbook Review Right now, do a 15-minute "textbook review" of this book. Here's how:

First, read the table of contents. Do it in three minutes or less. Next, look through the entire book, page by page. Move quickly. Scan titles, look at pictures, notice diagrams.

A textbook reconnaissance shows you where a course is going. It gives you the big picture. That's useful because brains work best when going from the general to the specific. Getting the big picture before you start makes details easier to recall and understand later on.

Your textbook reconnaissance will work even better if, as you scan, you look for ideas or topics that are interesting to you. List three facts, topics, or problems that you found interesting during your textbook reconnaissance.

The idea behind this technique is simple: It's easier to work at learning material if you know it's going to be useful to you.

Not all the topics in this book will be "interesting" to you. But that is true of any subject. Surfers find that on some days the waves are better than others, musicians find some music more appealing than other music, computer gamers find some computer games more interesting than others, car enthusiasts find some cars more exciting than others. Some car enthusiasts would rather have a completely restored 1957 Chevrolet than a new Ferrari.

Know the Course Requirements To do your best in this course, you must know exactly what your instructor requires. Course requirements may be stated in a *syllabus*, which is a printed outline of the main topics of the course, or they may be presented orally. When they are listed in a syllabus or on other printed pages, keep them in a safe place. When they are presented orally, make sure to take complete notes. In either case, it is important that you understand them completely and follow them exactly. Be sure you know the answer to each of the following questions.

1. What is your instructor's name?

2. Where is your instructor's office?

3. At what times does your instructor hold office hours?

4. Besides the textbook, what other materials does your instructor require?

5. What is your instructor's attendance policy?

6. If you must be absent from a class meeting, what should you do before returning to class? What should you do when you return to class?

7. What is the instructor's policy regarding collection or grading of homework assignments?

8. What options are available if you are having difficulty with an assignment? Is there a math tutoring center?

9. If there is a math lab at your school, Where is it located? What hours is it open?

10. What is the instructor's policy if you miss a quiz?

11. What is the instructor's policy if you miss an exam?

12. Where can you get help when studying for an exam?

Remember: Your instructor wants to see you succeed. If you need help, ask! Do not fall behind. If you are running a race and fall behind by 100 yards, you may be able to catch up but it will require more effort than had you not fallen behind.

TAKE NOTE

Besides time management, there must be realistic ideas of how much time is available. There are very few people who can *successfully* work full-time and go to school full-time. If you work 40 hours a week, take 15 units, spend the recommended study time given at the right, and sleep 8 hours a day, you will use over 80% of the available hours in a week. That leaves less than 20% of the hours in a week for family, friends, eating, recreation, and other activities.

Time Management We know that there are demands on your time. Family, work, friends, and entertainment all compete for your time. We do not want to see you receive poor job evaluations because you are studying math. However, it is also true that we do not want to see you receive poor math test scores because you devoted too much time to work. When several competing and important tasks require your time and energy, the only way to manage the stress of being successful at both is to manage your time efficiently.

Instructors often advise students to spend twice the amount of time outside of class studying as they spend in the classroom. Time management is important if you are to accomplish this goal and succeed in school. The following activity is intended to help you structure your time more efficiently.

List the name of each course you are taking this term, the number of class hours each course meets, and the number of hours you should spend studying each subject outside of class. Then fill in a weekly schedule like the one printed below. Begin by writing in the hours spent in your classes, the hours spent at work (if you have a job), and any other commitments that are not flexible with respect to the time that you do them. Then begin to write down commitments that are more flexible, including hours spent studying. Remember to reserve time for activities such as meals and exercise. You should also schedule free time.

	Monday	Tuesday	Wednesday	Thursday	Friday	Saturday	Sunday
7–8 a.m.							
8–9 a.m.							
9–10 a.m.							
10–11 a.m.							
11–12 p.m.							
12–1 p.m.							
1–2 p.m.							
2–3 p.m.							
3–4 p.m.							
4–5 p.m.							
5–6 p.m.							
6–7 p.m.							
7–8 p.m.							
8–9 p.m.							
9–10 p.m.							
10–11 p.m.							
11–12 a.m.							

We know that many of you must work. If that is the case, realize that working 10 hours a week at a part-time job is equivalent to taking a three-unit class. If you must work, consider letting your education progress at a slower rate to allow you to be successful at both work and school. There is no rule that says you must finish school in a certain time frame.

Schedule Study Time As we encouraged you to do by filling out the time management form above, schedule a certain time to study. You should think of this time the way you would the time for work or class—that is, reasons for missing study time should be as compelling as reasons for missing work or class. "I just didn't feel like it" is not a good reason to miss your scheduled study time.

Although this may seem like an obvious exercise, list a few reasons you might want to study.

Of course we have no way of knowing the reasons you listed, but from our experience one reason given quite frequently is "To pass the course." There is nothing wrong with that reason. If that is the most important reason for you to study, then use it to stay focused.

One method of keeping to a study schedule is to form a ***study group***. Look for people who are committed to learning, who pay attention in class, and who are punctual. Ask them to join your group. Choose people with similar educational goals but different methods of learning. You can gain insight from seeing the material from a new perspective. Limit groups to four or five people; larger groups are unwieldy.

There are many ways to conduct a study group. Begin with the following suggestions and see what works best for your group.

1. Test each other by asking questions. Each group member might bring two or three sample test questions to each meeting.
2. Practice teaching each other. Many of us who are teachers learned a lot about our subject when we had to explain it to someone else.
3. Compare class notes. You might ask other students about material in your notes that is difficult for you to understand.
4. Brainstorm test questions.
5. Set an agenda for each meeting. Set approximate time limits for each agenda item and determine a quitting time.

And finally, probably the most important aspect of studying is that it should be done in relatively small chunks. If you can only study three hours a week for this course (probably not enough for most people), do it in blocks of one hour on three separate days, preferably after class. Three hours of studying on a Sunday is not as productive as three hours of paced study.

Text Features That Promote Success There are 12 chapters in this text. Each chapter is divided into sections, and each section is subdivided into learning objectives. Each learning objective is labeled with a letter from A to D.

Preparing for a Chapter Before you begin a new chapter, you should take some time to review previously learned skills. There are two ways to do this. The first is to complete the ***Cumulative Review Exercises***, which occurs after every chapter (except Chapter 1). For instance, turn to page 255. The questions in this review are taken from the previous chapters. The answers for all these exercises can be found on pages A14 and A15. Turn to page A14 now and locate the answers for the Chapter 4 Cumulative Review Exercises. After the answer to the first exercise, which is $-\frac{11}{28}$, you will see the objective reference [2.1C]. This means that this question was taken from Chapter 2, Section 1, Objective C. If you missed this question, you should return to that objective and restudy the material.

A second way of preparing for a new chapter is to complete the ***Prep Test***. This test focuses on the particular skills that will be required for the new chapter. Turn to page 200 to see a Prep Test. The answers for the Prep Test are the first set of answers in the answer section for a chapter. Turn to page A12 to see the answers for the Chapter 4 Prep Test. Note that an objective reference is given for each question. If you answer a question incorrectly, restudy the objective from which the question was taken.

Before the class meeting in which your professor begins a new section, you should read each objective statement for that section. Next, browse through the objective material, being sure to note each word in bold type. These words indicate important concepts that you must know in order to learn the material. Do not worry about trying to understand all the material. Your professor is there to assist you with that endeavor. The purpose of browsing through the material is so that your brain will be prepared to accept and organize the new information when it is presented to you.

Turn to page 3. Write down the title of the first objective in Section 1.1. Under the title of the objective, write down the words in the objective that are in bold print. It is not necessary for you to understand the meaning of these words. You are in this class to learn their meaning.

_____ _____ _____ _____

_____ _____ _____ _____

_____ _____ _____ _____

_____ _____ _____ _____

_____ _____ _____ _____

Math Is Not a Spectator Sport To learn mathematics you must be an active participant. Listening and watching your professor do mathematics is not enough. Mathematics requires that you interact with the lesson you are studying. If you filled in the blanks above, you were being interactive. There are other ways this textbook has been designed to help you be an active learner.

Annotated Examples The HOW TO feature indicates an example with explanatory remarks to the right of the work. Using paper and pencil, you should work along as you go through the example.

$$3x + 2 < -4$$
$$3x < -6$$
$$\frac{3x}{3} < \frac{-6}{3}$$
$$x < -2$$

The solution set is $\{x \mid x < -2\}$.

HOW TO Solve: $3x + 2 < -4$
Write the solution set in set-builder notation.

$$3x + 2 < -4$$
$$3x < -6$$
$$\frac{3x}{3} < \frac{-6}{3}$$
$$x < -2$$

- Subtract 2 from each side of the inequality.
- Divide each side of the inequality by the coefficient 3.

The solution set is $\{x \mid x < -2\}$.

When you complete the example, get a clean sheet of paper. Write down the problem and then try to complete the solution without referring to your notes or the book. When you can do that, move on to the next part of the objective.

Leaf through the book now and write down the page numbers of two other occurrences of a HOW TO example.

You Try Its One of the key instructional features of this text is the paired examples. Notice that in each example box, the example on the left is completely worked out and the "You Try It" example on the right is not. Study the worked-out example carefully by working through each step. Then work the You Try It. If you get stuck, refer to the page number at the end of the example, which directs you to the place where the You Try It is solved—a complete worked-out solution is provided. Try to use the given solution to get a hint for the step you are stuck on. Then try to complete your solution.

Example 2

Solve: $3x - 5 \le 3 - 2(3x + 1)$
Write the solution set in interval notation.

Solution
$$3x - 5 \le 3 - 2(3x + 1)$$
$$3x - 5 \le 3 - 6x - 2$$
$$3x - 5 \le 1 - 6x$$
$$9x - 5 \le 1$$
$$9x \le 6$$
$$\frac{9x}{9} \le \frac{6}{9}$$
$$x \le \frac{2}{3}$$
$$\left(-\infty, \frac{2}{3}\right]$$

You Try It 2

Solve: $5x - 2 \le 4 - 3(x - 2)$
Write the solution set in interval notation.

Your solution
$$5x - 2 \le 4 - 3(x - 2)$$
$$5x - 2 \le 4 - 3x + 6$$
$$5x - 2 \le 10 - 3x$$
$$8x - 2 \le 10$$
$$8x \le 12$$
$$\frac{8x}{8} \le \frac{12}{8}$$
$$x \le \frac{3}{2}$$
$$\left(-\infty, \frac{3}{2}\right]$$

Solution on p. S5

When you have completed your solution, check your work against the solution we provided. (Turn to page S5 to see the solution of You Try It 2.) Be aware that frequently there is more than one way to solve a problem. Your answer, however, should be the same as the given answer. If you have any question as to whether your method will "always work," check with your instructor or with someone in the math center.

Browse through the textbook and write down the page numbers where two other paired example features occur.

Remember: Be an active participant in your learning process. When you are sitting in class watching and listening to an explanation, you may think that you understand. However, until you actually try to do it, you will have no confirmation of the new knowledge or skill. Most of us have had the experience of sitting in class thinking we knew how to do something only to get home and realize that we didn't.

TAKE NOTE

There is a strong connection between reading and being a successful student in math or any other subject. If you have difficulty reading, consider taking a reading course. Reading is much like other skills. There are certain things you can learn that will make you a better reader.

Word Problems Word problems are difficult because we must read the problem, determine the quantity we must find, think of a method to do that, and then actually solve the problem. In short, we must formulate a *strategy* to solve the problem and then devise a *solution*.

Note in the paired example below that part of every word problem is a strategy and part is a solution. The strategy is a written description of how we will solve the problem. In the corresponding You Try It, you are asked to formulate a strategy. Do not skip this step, and be sure to write it out.

Example 6

Find three consecutive positive odd integers whose sum is between 27 and 51.

Strategy

To find the three integers, write and solve a compound inequality using n to represent the first odd integer.

Solution

$$\underset{\text{of the sum}}{\text{Lower limit}} < \text{sum} < \underset{\text{of the sum}}{\text{upper limit}}$$

$$27 < n + (n + 2) + (n + 4) < 51$$
$$27 < 3n + 6 < 51$$
$$27 - 6 < 3n + 6 - 6 < 51 - 6$$
$$21 < 3n < 45$$
$$\frac{21}{3} < \frac{3n}{3} < \frac{45}{3}$$
$$7 < n < 15$$

The three odd integers are 9, 11, and 13; or 11, 13, and 15; or 13, 15, and 17.

You Try It 6

An average score of 80 to 89 in a history course receives a B. Luisa Montez has grades of 72, 94, 83, and 70 on four exams. Find the range of scores on the fifth exam that will give Luisa a B for the course.

Your strategy

To find the scores, write and solve an inequality. Let N be the score on the last test.

Your solution

$$80 \le \frac{72 + 94 + 83 + 70 + N}{5} \le 89$$
$$80 \le \frac{319 + N}{5} \le 89$$
$$5(80) \le 5\left(\frac{319 + N}{5}\right) \le 5(89)$$
$$400 \le 319 + N \le 445$$
$$400 - 319 \le 319 + N - 319 \le 445 - 319$$
$$81 \le N \le 126$$

The range of scores to get a B is $81 \le N \le 100$.

Solutions on p. S5

Page 88

Rule Boxes Pay special attention to rules placed in boxes. These rules give you the reasons certain types of problems are solved the way they are. When you see a rule, try to rewrite the rule in your own words.

> *TAKE NOTE*
>
> If a rule has more than one part, be sure to make a notation to that effect.

I can add the same number to both sides of an inequality and not change the solution set.

> *TAKE NOTE*
>
> If you are working at home and need assistance, there is online help available at math.college.hmco.com/ students, at this text's website.

When solving an inequality, we use the **Addition and Multiplication Properties of Inequalities** to rewrite the inequality in the form *variable* < *constant* or in the form *variable* > *constant*.

> **The Addition Property of Inequalities**
>
> If $a > b$, then $a + c > b + c$.
> If $a < b$, then $a + c < b + c$.

Page 83

Chapter Exercises When you have completed studying an objective, do the exercises in the exercise set that correspond with that objective. The exercises are labeled with the same letter as the objective. Math is a subject that needs to be learned in small sections and practiced continually in order to be mastered. Doing all of the exercises in each exercise set will help you master the problem-solving techniques necessary for success. As you work through the exercises for an objective, check your answers to the odd-numbered exercises with those in the back of the book.

Preparing for a Test There are important features of this text that can be used to prepare for a test.

- Chapter Summary
- Chapter Review Exercises
- Chapter Test

After completing a chapter, read the Chapter Summary. (See page 248 for the Chapter 4 Summary.) This summary highlights the important topics covered in the chapter. The page number following each topic refers you to the page in the text on which you can find more information about the concept.

Following the Chapter Summary are Chapter Review Exercises (see page 251) and a Chapter Test (see page 253). Doing the review exercises is an important way of testing your understanding of the chapter. The answer to each review exercise is given at the back of the book, along with its objective reference. After checking your answers, restudy any objective from which a question you missed was taken. It may be helpful to retry some of the exercises for that objective to reinforce your problem-solving techniques.

The Chapter Test should be used to prepare for an exam. We suggest that you try the Chapter Test a few days before your actual exam. Take the test in a quiet place and try to complete the test in the same amount of time you will be allowed for your exam. When taking the Chapter Test, practice the strategies of successful test takers: (1) scan the entire test to get a feel for the questions; (2) read the directions carefully; (3) work the problems that are easiest for you first; and perhaps most importantly, (4) try to stay calm.

When you have completed the Chapter Test, check your answers. If you missed a question, review the material in that objective and rework some of the exercises from that objective. This will strengthen your ability to perform the skills in that objective.

Is it difficult to be successful? YES! Successful music groups, artists, professional athletes, chefs, and *Write your major here* have to work very hard to achieve their goals. They focus on their goals and ignore distractions. The things we ask you to do to achieve success take time and commitment. We are confident that if you follow our suggestions, you will succeed.

1 Review of Real Numbers

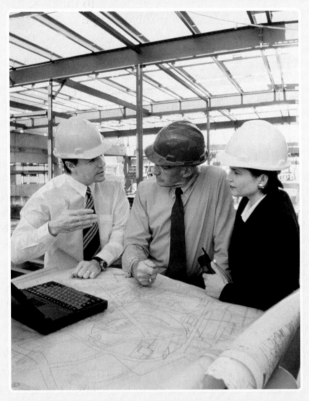

The architects in this photo are discussing and analyzing blueprints. Thorough analysis and effective collaboration involve problem solving, as does the job you will hold after you graduate. The basic steps involved in problem solving are outlined in the **Focus on Problem Solving on page 43** and then integrated throughout the text. Your success in the workplace is heavily dependent on your ability to solve problems. It is a skill that improves with study and practice.

OBJECTIVES

Section 1.1

A To use inequality and absolute value symbols with real numbers

B To write sets using the roster method and set-builder notation

C To perform operations on sets and write sets in interval notation

Section 1.2

A To add, subtract, multiply, and divide integers

B To add, subtract, multiply, and divide rational numbers

C To evaluate exponential expressions

D To use the Order of Operations Agreement

Section 1.3

A To use and identify the properties of the real numbers

B To evaluate a variable expression

C To simplify a variable expression

Section 1.4

A To translate a verbal expression into a variable expression

B To solve application problems

Need help? For online student resources, such as section quizzes, visit this textbook's Website at **math.college.hmco.com/students.**

PREP TEST • • •

Do these exercises to prepare for Chapter 1.

For Exercises 1 to 8, add, subtract, multiply, or divide.

1. $\dfrac{5}{12} + \dfrac{7}{30}$

2. $\dfrac{8}{15} - \dfrac{7}{20}$

3. $\dfrac{5}{6} \cdot \dfrac{4}{15}$

4. $\dfrac{4}{15} \div \dfrac{2}{5}$

5. $8 + 29.34 + 7.065$

6. $92 - 18.37$

7. $2.19(3.4)$

8. $32.436 \div 0.6$

9. Which of the following numbers are greater than -8?
 a. -6 **b.** -10 **c.** 0 **d.** 8

10. Match the fraction with its decimal equivalent.
 a. $\dfrac{1}{2}$ **A.** 0.75
 b. $\dfrac{7}{10}$ **B.** 0.89
 c. $\dfrac{3}{4}$ **C.** 0.5
 d. $\dfrac{89}{100}$ **D.** 0.7

GO FIGURE • • •

A pair of perpendicular lines are drawn through the interior of a rectangle, dividing it into four smaller rectangles. The areas of the smaller rectangles are x, 2, 3, and 6. Find the possible values of x.

1.1 Introduction to Real Numbers

Objective A To use inequality and absolute value symbols with real numbers

It seems to be a human characteristic to put similar items in the same place. For instance, an astronomer places stars in *constellations*, and a geologist divides the history of Earth into *eras*.

Mathematicians likewise place objects with similar properties in *sets*. A **set** is a collection of objects. The objects are called **elements** of the set. Sets are denoted by placing braces around the elements in the set.

The numbers that we use to count things, such as the number of books in a library or the number of CDs sold by a record store, have similar characteristics. These numbers are called the *natural numbers*.

Natural numbers = {1, 2, 3, 4, 5, 6, 7, 8, 9, 10, 11, . . .}

Each natural number greater than 1 is a *prime* number or a *composite* number. A **prime number** is a natural number greater than 1 that is divisible (evenly) only by itself and 1. For example, 2, 3, 5, 7, 11, and 13 are the first six prime numbers. A natural number that is not a prime number is a **composite number.** The numbers 4, 6, 8, and 9 are the first four composite numbers.

The natural numbers do not have a symbol to denote the concept of none—for instance, the number of trees taller than 1000 feet. The *whole numbers* include zero and the natural numbers.

Whole numbers = {0, 1, 2, 3, 4, 5, 6, 7, 8, . . .}

The whole numbers alone do not provide all the numbers that are useful in applications. For instance, a meteorologist needs numbers below zero and above zero.

Integers = {. . . , −5, −4, −3, −2, −1, 0, 1, 2, 3, 4, 5, . . .}

The integers . . . , −5, −4, −3, −2, −1 are **negative integers.** The integers 1, 2, 3, 4, 5, . . . are **positive integers.** Note that the natural numbers and the positive integers are the same set of numbers. The integer zero is neither a positive nor a negative integer.

Still other numbers are necessary to solve the variety of application problems that exist. For instance, a landscape architect may need to purchase irrigation pipe that has a diameter of $\frac{5}{8}$ in. The numbers that include fractions are called *rational numbers.*

Rational numbers = $\left\{ \dfrac{p}{q}, \text{ where } p \text{ and } q \text{ are integers and } q \neq 0 \right\}$

The numbers $\frac{2}{3}$, $-\frac{9}{2}$, and $\frac{5}{1}$ are examples of rational numbers . Note that $\frac{5}{1} = 5$, so all integers are rational numbers. The number $\frac{4}{\pi}$ is not a rational number because π is not an integer.

A rational number written as a fraction can be written in decimal notation by dividing the numerator by the denominator.

HOW TO Write $\frac{3}{8}$ as a decimal.

Divide 3 by 8.

$$\begin{array}{r} 0.375 \\ 8\overline{)3.000} \\ -2\,4 \\ \hline 60 \\ -56 \\ \hline 40 \\ -40 \\ \hline 0 \end{array}$$

← This is a terminating decimal.

← The remainder is zero.

$\frac{3}{8} = 0.375$

HOW TO Write $\frac{2}{15}$ as a decimal.

Divide 2 by 15.

$$\begin{array}{r} 0.133 \\ 15\overline{)2.000} \\ -1\,5 \\ \hline 50 \\ -45 \\ \hline 50 \\ -45 \\ \hline 5 \end{array}$$

← This is a repeating decimal.

← The remainder is never zero.

● The bar over 3 indicates that this digit repeats.

$\frac{2}{15} = 0.1\overline{3}$

Some numbers cannot be written as terminating or repeating decimals—for example, $0.01001000100001\ldots$, $\sqrt{7} \approx 2.6457513$, and $\pi \approx 3.1415927$. These numbers have decimal representations that neither terminate nor repeat. They are called **irrational numbers.** The rational numbers and the irrational numbers taken together are the **real numbers.**

TAKE NOTE

The real numbers are the rational numbers and the irrational numbers. The relationship among sets of numbers is shown in the figure at the right, along with examples of elements in each set.

Positive Integers
(Natural numbers)
7 1 103

Zero
0

Negative Integers
−201 −8 −5

Integers
−201 7 0 −5

Rational Numbers
$\frac{3}{4}$ $3.12\overline{12}$ −1.34 −5

Irrational Numbers
−0.101101110...
$\sqrt{7}$ π

Real Numbers
$\frac{3}{4}$ $3.12\overline{12}$
−1.34 7
0 −5 1
103 −201
−0.101101110...
$\sqrt{7}$ π

The **graph of a real number** is made by placing a heavy dot directly above the number on a number line. The graphs of some real numbers follow.

Consider the following sentences:

A restaurant's chef prepared a dinner and served *it* to the customers.
A maple tree was planted and *it* grew two feet in one year.

In the first sentence, "it" means dinner; in the second sentence, "it" means tree. In language, this word can stand for many different objects. Similarly, in mathematics, a letter of the alphabet can be used to stand for some number. A letter used in this way is called a **variable.**

It is convenient to use a variable to represent or stand for any one of the elements of a set. For instance, the statement "*x* is an *element of* the set {0, 2, 4, 6}" means that *x* can be replaced by 0, 2, 4, or 6. The set {0, 2, 4, 6} is called the *domain* of the variable.

The symbol for "is an element of" is \in; the symbol for "is not an element of" is \notin. For example,

$$2 \in \{0, 2, 4, 6\} \qquad 6 \in \{0, 2, 4, 6\} \qquad 7 \notin \{0, 2, 4, 6\}$$

Variables are used in the next definition.

> **Definition of Inequality Symbols**
>
> If a and b are two real numbers and a is to the left of b on the number line, then a **is less than** b. This is written $a < b$.
>
> If a and b are two real numbers and a is to the right of b on the number line, then a **is greater than** b. This is written $a > b$.

Here are some examples.

$$5 < 9 \qquad -4 > -10 \qquad \pi < \sqrt{17} \qquad 0 > -\frac{2}{3}$$

The inequality symbols \leq (is less than or equal to) and \geq (is greater than or equal to) are also important. Note the examples below.

$$4 \leq 5 \quad \text{is a true statement because } 4 < 5.$$
$$5 \leq 5 \quad \text{is a true statement because } 5 = 5.$$

The numbers 5 and -5 are the same distance from zero but on opposite sides of zero. The numbers 5 and -5 are called **additive inverses** or **opposites** of each other.

The additive inverse (or opposite) of 5 is -5. The additive inverse of -5 is 5. The symbol for additive inverse is $-$.

$-(2)$ means the additive inverse of *positive* 2. $-(2) = -2$

$-(-5)$ means the additive inverse of *negative* 5. $-(-5) = 5$

The **absolute value** of a number is its distance from zero on the number line. The symbol for absolute value is $|\ \ |$.

Note from the figure above that the distance from 0 to 5 is 5. Therefore, $|5| = 5$. That figure also shows that the distance from 0 to -5 is 5. Therefore, $|-5| = 5$.

> **Absolute Value**
>
> The absolute value of a positive number is the number itself.
> The absolute value of a negative number is the opposite of the negative number.
> The absolute value of zero is zero.

HOW TO Evaluate: $-|-12|$

$-|-12| = -12$ • The absolute value symbol does not affect the
 negative sign *in front of* the absolute value symbol.

Integrating Technology

See the Keystroke Guide: *Math* for instructions on using a graphing calculator to evaluate absolute value expressions.

Example 1

Let $y \in \{-7, 0, 6\}$. For which values of y is the inequality $y < 4$ a true statement?

Solution

Replace y by each of the elements of the set and determine whether the inequality is true.

$$y < 4$$
$$-7 < 4 \quad \text{True}$$
$$0 < 4 \quad \text{True}$$
$$6 < 4 \quad \text{False}$$

The inequality is true for -7 and 0.

You Try It 1

Let $z \in \{-10, -5, 6\}$. For which values of z is the inequality $z > -5$ a true statement?

Your solution

Example 2

Let $y \in \{-12, 0, 4\}$.
a. Determine $-y$, the additive inverse of y, for each element of the set.
b. Evaluate $|y|$ for each element of the set.

Solution

a. Replace y in $-y$ by each element of the set and determine the value of the expression.

$$-y$$
$$-(-12) = 12$$
$$-(0) = 0 \quad \bullet \text{ 0 is neither positive}$$
$$\qquad\qquad\qquad \text{nor negative.}$$
$$-(4) = -4$$

b. Replace y in $|y|$ by each element of the set and determine the value of the expression.

$$|y|$$
$$|-12| = 12$$
$$|0| = 0$$
$$|4| = 4$$

You Try It 2

Let $d \in \{-11, 0, 8\}$.
a. Determine $-d$, the additive inverse of d, for each element of the set.
b. Evaluate $|d|$ for each element of the set.

Your solution

Solutions on p. S1

Objective B **To write sets using the roster method and set-builder notation**

The **roster method** of writing a set encloses the list of the elements of the set in braces. The set of whole numbers, written $\{0, 1, 2, 3, 4, \ldots\}$, and the set of natural numbers, written $\{1, 2, 3, 4, \ldots\}$, are **infinite sets.** The pattern of numbers continues without end. It is impossible to list all the elements of an infinite set.

The set of even natural numbers less than 10 is written $\{2, 4, 6, 8\}$. This is an example of a **finite set;** all the elements of the set can be listed.

The set that contains no elements is called the **empty set,** or **null set,** and is symbolized by \varnothing or $\{\ \}$. The set of trees over 1000 feet tall is the empty set.

HOW TO Use the roster method to write the set of whole numbers less than 5.

$\{0, 1, 2, 3, 4\}$ • Recall that the whole numbers include 0.

A second method of representing a set is **set-builder notation.** Set-builder notation can be used to describe almost any set, but it is especially useful when writing infinite sets. In set-builder notation, the set of integers greater than -3 is written

$\{x | x > -3,\ x \in \text{integers}\}$

The set of all x such that $x > -3$ and x is an element of the integers.

This is an infinite set. It is impossible to list all the elements of the set, but the set can be described using set-builder notation.

The set of real numbers less than 5 is written

$$\{x | x < 5, x \in \text{real numbers}\}$$

and is read "the set of all x such that x is less than 5 and x is an element of the real numbers."

HOW TO Use set-builder notation to write the set of integers greater than 20.

$\{x | x > 20, x \in \text{integers}\}$

Example 3

Use the roster method to write the set of positive integers less than or equal to 7.

Solution
$\{1, 2, 3, 4, 5, 6, 7\}$

You Try It 3

Use the roster method to write the set of negative integers greater than -6.

Your solution

Example 4

Use set-builder notation to write the set of integers less than -9.

Solution
$\{x | x < -9, x \in \text{integers}\}$

You Try It 4

Use set-builder notation to write the set of whole numbers greater than or equal to 15.

Your solution

Solutions on p. S1

Objective C **To perform operations on sets and write sets in interval notation**

Just as operations such as addition and multiplication are performed on real numbers, operations are performed on sets. Two operations performed on sets are *union* and *intersection*.

Union of Two Sets

The **union of two sets**, written $A \cup B$, is the set of all elements that belong to either set A **or** set B. In set-builder notation, this is written

$$A \cup B = \{x \mid x \in A \quad \text{or} \quad x \in B\}$$

Given $A = \{2, 3, 4\}$ and $B = \{0, 1, 2, 3\}$, the union of A and B contains all the elements that belong to either A or B. Any elements that belong to both sets are listed only once.

$$A \cup B = \{0, 1, 2, 3, 4\}$$

Intersection of Two Sets

The **intersection of two sets**, written $A \cap B$, is the set of all elements that are common to both set A **and** set B. In set-builder notation, this is written

$$A \cap B = \{x \mid x \in A \quad \text{and} \quad x \in B\}$$

Given $A = \{2, 3, 4\}$ and $B = \{0, 1, 2, 3\}$, the intersection of A and B contains all the elements that are common to both A and B.

$$A \cap B = \{2, 3\}$$

Point of Interest

The symbols \in, \cup, and \cap were first used by Giuseppe Peano in *Arithmetices Principia, Nova Exposita* (The Principle of Mathematics, a New Method of Exposition), published in 1889. The purpose of this book was to deduce the principles of mathematics from pure logic.

HOW TO Given $A = \{2, 3, 5, 7\}$ and $B = \{0, 1, 2, 3, 4\}$, find $A \cup B$ and $A \cap B$.

$A \cup B = \{0, 1, 2, 3, 4, 5, 7\}$ • List the elements of each set. The elements that belong to both sets are listed only once.

$A \cap B = \{2, 3\}$ • List the elements that are common to both *A* and *B*.

Set-builder notation and the inequality symbols $>$, $<$, \geq, and \leq are used to describe infinite sets of real numbers. These sets can also be graphed on the real number line.

The graph of $\{x | x > -2, x \in \text{real numbers}\}$ is shown below. The set is the real numbers greater than -2. The parenthesis on the graph indicates that -2 is not included in the set.

The graph of $\{x | x \geq -2, x \in \text{real numbers}\}$ is shown below. The set is the real numbers greater than or equal to -2. The bracket at -2 indicates that -2 is included in the set.

In many cases, we will assume that real numbers are being used and omit "$x \in \text{real numbers}$" from set-builder notation. For instance, the above set is written $\{x | x \geq -2\}$.

HOW TO Graph $\{x | x \leq 3\}$.

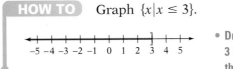 • Draw a bracket at 3 to indicate that 3 is in the set. Draw a solid line to the left of 3.

The union of two sets is the set of all elements belonging to either one or the other of the two sets. The set $\{x | x \leq -1\} \cup \{x | x > 3\}$ is the set of real numbers that are either less than or equal to -1 or greater than 3.

The set is written $\{x | x \leq -1 \text{ or } x > 3\}$.

The set $\{x | x > 2\} \cup \{x | x > 4\}$ is the set of real numbers that are either greater than 2 or greater than 4.

The set is written $\{x | x > 2\}$.

HOW TO Graph $\{x | x > 1\} \cup \{x | x < -4\}$.

 • The graph includes all the numbers that are either greater than 1 or less than -4.

The intersection of two sets is the set that contains the elements common to both sets. The set $\{x | x > -2\} \cap \{x | x < 5\}$ is the set of real numbers that are greater than -2 and less than 5.

The set can be written $\{x | x > -2 \text{ and } x < 5\}$. However, it is more commonly written $\{x | -2 < x < 5\}$, which is read "the set of all x such that x is greater than -2 and less than 5."

The set $\{x|x < 4\} \cap \{x|x < 5\}$ is the set of real numbers that are less than 4 and less than 5.

The set is written $\{x|x < 4\}$.

HOW TO Graph $\{x|x < 0\} \cap \{x|x > -3\}$.

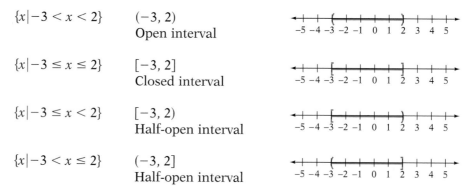

• The set is $\{x|-3 < x < 0\}$.

Some sets can also be expressed using **interval notation.** For example, the interval notation $(-3, 2]$ indicates the interval of all real numbers greater than -3 and less than or equal to 2. As on the graph of a set, the left parenthesis indicates that -3 is not included in the set. The right bracket indicates that 2 is included in the set.

An interval is said to be a **closed interval** if it includes both endpoints; it is an **open interval** if it does not include either endpoint. An interval is a **half-open interval** if one endpoint is included and the other is not. In each example given below, -3 and 2 are the **endpoints of the interval.** In each case, the set notation, the interval notation, and the graph of the set are shown.

$\{x	-3 < x < 2\}$	$(-3, 2)$ Open interval	
$\{x	-3 \le x \le 2\}$	$[-3, 2]$ Closed interval	
$\{x	-3 \le x < 2\}$	$[-3, 2)$ Half-open interval	
$\{x	-3 < x \le 2\}$	$(-3, 2]$ Half-open interval	

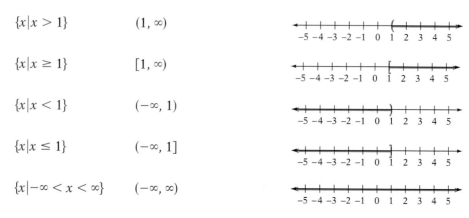

To indicate an interval that extends forever in one or both directions using interval notation, we use the **infinity symbol** ∞ or the **negative infinity symbol** $-\infty$. The infinity symbol is not a number; it is simply a notation to indicate that the interval is unlimited. In interval notation, a parenthesis is always used to the right of an infinity symbol or to the left of a negative infinity symbol, as shown in the following examples.

$\{x	x > 1\}$	$(1, \infty)$	
$\{x	x \ge 1\}$	$[1, \infty)$	
$\{x	x < 1\}$	$(-\infty, 1)$	
$\{x	x \le 1\}$	$(-\infty, 1]$	
$\{x	-\infty < x < \infty\}$	$(-\infty, \infty)$	

HOW TO Graph $(-2, 5)$.

<<----+---+---+--(--+---+---+---+---+---+--)-->>
−5 −4 −3 −2 −1 0 1 2 3 4 5

• The graph is all numbers that are greater than -2 and less than 5.

Example 5

Given $A = \{0, 2, 4, 6, 8, 10\}$ and $B = \{0, 3, 6, 9\}$, find $A \cup B$.

Solution

$A \cup B = \{0, 2, 3, 4, 6, 8, 9, 10\}$

You Try It 5

Given $C = \{1, 5, 9, 13, 17\}$ and $D = \{3, 5, 7, 9, 11\}$, find $C \cup D$.

Your solution

Example 6

Given $A = \{x | x \in \text{natural numbers}\}$ and $B = \{x | x \in \text{negative integers}\}$, find $A \cap B$.

Solution

$A \cap B = \varnothing$ • There are no natural numbers that are also negative integers.

You Try It 6

Given $E = \{x | x \in \text{odd integers}\}$ and $F = \{x | x \in \text{even integers}\}$, find $E \cap F$.

Your solution

Example 7

Graph $\{x | x \geq 3\}$.

Solution

Draw a bracket at 3 to indicate that 3 is in the set. Draw a solid line to the right of 3.

<<----+---+---+---+---+---+---+---[===+===+===>>
−5 −4 −3 −2 −1 0 1 2 3 4 5

You Try It 7

Graph $\{x | x < 0\}$.

Your solution

<<----+---+---+---+---+---+---+---+---+---+--->>
−5 −4 −3 −2 −1 0 1 2 3 4 5

Example 8

Graph $\{x | x > -1\} \cup \{x | x < 2\}$.

Solution

This is the set of real numbers greater than -1 *or* less than 2. Any real number satisfies this condition. The graph is the entire real number line.

<<====+===+===+===+===+===+===+===+===+===+===>>
−5 −4 −3 −2 −1 0 1 2 3 4 5

You Try It 8

Graph $\{x | x < -2\} \cup \{x | x > -1\}$.

Your solution

<<----+---+---+---+---+---+---+---+---+---+--->>
−5 −4 −3 −2 −1 0 1 2 3 4 5

Solutions on p. S1

Graph $\{x|x < 3\} \cap \{x|x \geq -1\}$.

Solution

This is the set of real numbers greater than or equal to -1 *and* less than 3.

Example 10

Write each set in interval notation.
a. $\{x|x > 3\}$
b. $\{x|-2 < x \leq 4\}$

Solution
a. The set $\{x|x > 3\}$ is the numbers greater than 3. In interval notation, this is written $(3, \infty)$.
b. The set $\{x|-2 < x \leq 4\}$ is the numbers greater than -2 and less than or equal to 4. In interval notation, this is written $(-2, 4]$.

Example 11

Write each set in set-builder notation.
a. $(-\infty, 4]$
b. $[-3, 0]$

Solution
a. $(-\infty, 4]$ is the numbers less than or equal to 4. In set-builder notation, this is written $\{x|x \leq 4\}$.
b. $[-3, 0]$ is the numbers greater than or equal to -3 and less than or equal to 0. In set-builder notation, this is written $\{x|-3 \leq x \leq 0\}$.

Example 12

Graph $(-2, 2]$.

Solution
Draw a parenthesis at -2 to show that it is not in the set. Draw a bracket at 2 to show it is in the set. Draw a solid line between -2 and 2.

You Try It 9

Graph $\{x|x < 1\} \cap \{x|x > -3\}$.

Your solution

You Try It 10

Write each set in interval notation.
a. $\{x|x < -1\}$
b. $\{x|-2 \leq x < 4\}$

Your solution

You Try It 11

Write each set in set-builder notation.
a. $(3, \infty)$
b. $(-4, 1]$

Your solution

You Try It 12

Graph $[-2, \infty)$.

Your solution

Solutions on p. S1

1.1 Exercises

Objective A **To use inequality and absolute value symbols with real numbers**

For Exercises 1 and 2, determine which of the numbers are **a.** integers, **b.** rational numbers, **c.** irrational numbers, **d.** real numbers. List all that apply.

1. $-\dfrac{15}{2}, 0, -3, \pi, 2.\overline{33}, 4.232232223\ldots, \dfrac{\sqrt{5}}{4}, \sqrt{7}$

2. $-17, 0.3412, \dfrac{3}{\pi}, -1.010010001\ldots, \dfrac{27}{91}, 6.1\overline{2}$

For Exercises 3 to 12, find the additive inverse of the number.

3. 27

4. -3

5. $\dfrac{3}{4}$

6. $\sqrt{17}$

7. 0

8. $-\pi$

9. $-\sqrt{33}$

10. -1.23

11. -91

12. $-\dfrac{2}{3}$

For Exercises 13 to 20, solve.

13. Let $y \in \{-6, -4, 7\}$. For which values of y is $y > -4$ true?

14. Let $x \in \{-6, -3, 3\}$. For which values of x is $x < -3$ true?

15. Let $w \in \{-2, -1, 0, 1\}$. For which values of w is $w \leq -1$ true?

16. Let $p \in \{-10, -5, 0, 5\}$. For which values of p is $p \geq 0$ true?

17. Let $b \in \{-9, 0, 9\}$. Evaluate $-b$ for each element of the set.

18. Let $a \in \{-3, -2, 0\}$. Evaluate $-a$ for each element of the set.

19. Let $c \in \{-4, 0, 4\}$. Evaluate $|c|$ for each element of the set.

20. Let $q \in \{-3, 0, 7\}$. Evaluate $|q|$ for each element of the set.

Objective B **To write sets using the roster method and set-builder notation**

For Exercises 21 to 26, use the roster method to write the set.

21. the integers between -3 and 5

22. the integers between -4 and 0

23. the even natural numbers less than 14

24. the odd natural numbers less than 14

25. the positive-integer multiples of 3 that are less than or equal to 30

26. the negative-integer multiples of 4 that are greater than or equal to −20

For Exercises 27 to 34, use set-builder notation to write the set.

27. the integers greater than 4

28. the integers less than −2

29. the real numbers greater than or equal to −2

30. the real numbers less than or equal to 2

31. the real numbers between 0 and 1

32. the real numbers between −2 and 5

33. the real numbers between 1 and 4, inclusive

34. the real numbers between 0 and 2, inclusive

Objective C **To perform operations on sets and write sets in interval notation**

For Exercises 35 to 42, find $A \cup B$.

35. $A = \{1, 4, 9\}, B = \{2, 4, 6\}$

36. $A = \{-1, 0, 1\}, B = \{0, 1, 2\}$

37. $A = \{2, 3, 5, 8\}, B = \{9, 10\}$

38. $A = \{1, 3, 5, 7\}, B = \{2, 4, 6, 8\}$

39. $A = \{-4, -2, 0, 2, 4\}, B = \{0, 4, 8\}$

40. $A = \{-3, -2, -1\}, B = \{-2, -1, 0, 1\}$

41. $A = \{1, 2, 3, 4, 5\}, B = \{3, 4, 5\}$

42. $A = \{2, 4\}, B = \{0, 1, 2, 3, 4, 5\}$

For Exercises 43 to 50, find $A \cap B$.

43. $A = \{6, 12, 18\}, B = \{3, 6, 9\}$

44. $A = \{-4, 0, 4\}, B = \{-2, 0, 2\}$

45. $A = \{1, 5, 10, 20\}, B = \{5, 10, 15, 20\}$

46. $A = \{1, 3, 5, 7, 9\}, B = \{1, 9\}$

47. $A = \{1, 2, 4, 8\}, B = \{3, 5, 6, 7\}$

48. $A = \{-3, -2, -1, 0\}, B = \{1, 2, 3, 4\}$

49. $A = \{2, 4, 6, 8, 10\}, B = \{4, 6\}$

50. $A = \{-9, -5, 0, 7\}, B = \{-7, -5, 0, 5, 7\}$

For Exercises 51 to 66, graph.

51. $\{x \,|\, x < 2\}$

$-5\ -4\ -3\ -2\ -1\ \ 0\ \ 1\ \ 2\ \ 3\ \ 4\ \ 5$

52. $\{x \,|\, x < -1\}$

$-5\ -4\ -3\ -2\ -1\ \ 0\ \ 1\ \ 2\ \ 3\ \ 4\ \ 5$

53. $\{x \,|\, x \geq 1\}$

$-5\ -4\ -3\ -2\ -1\ \ 0\ \ 1\ \ 2\ \ 3\ \ 4\ \ 5$

54. $\{x \,|\, x \leq -2\}$

$-5\ -4\ -3\ -2\ -1\ \ 0\ \ 1\ \ 2\ \ 3\ \ 4\ \ 5$

55. $\{x \,|\, -1 < x < 5\}$

$-5\ -4\ -3\ -2\ -1\ \ 0\ \ 1\ \ 2\ \ 3\ \ 4\ \ 5$

56. $\{x \,|\, 1 < x < 3\}$

$-5\ -4\ -3\ -2\ -1\ \ 0\ \ 1\ \ 2\ \ 3\ \ 4\ \ 5$

57. $\{x \,|\, 0 \leq x \leq 3\}$

$-5\ -4\ -3\ -2\ -1\ \ 0\ \ 1\ \ 2\ \ 3\ \ 4\ \ 5$

58. $\{x \,|\, -1 \leq x \leq 1\}$

$-5\ -4\ -3\ -2\ -1\ \ 0\ \ 1\ \ 2\ \ 3\ \ 4\ \ 5$

59. $\{x \,|\, x > 1\} \cup \{x \,|\, x < -1\}$

$-5\ -4\ -3\ -2\ -1\ \ 0\ \ 1\ \ 2\ \ 3\ \ 4\ \ 5$

60. $\{x \,|\, x \leq 2\} \cup \{x \,|\, x > 4\}$

$-5\ -4\ -3\ -2\ -1\ \ 0\ \ 1\ \ 2\ \ 3\ \ 4\ \ 5$

61. $\{x \,|\, x \leq 2\} \cap \{x \,|\, x \geq 0\}$

$-5\ -4\ -3\ -2\ -1\ \ 0\ \ 1\ \ 2\ \ 3\ \ 4\ \ 5$

62. $\{x \,|\, x > -1\} \cap \{x \,|\, x \leq 4\}$

$-5\ -4\ -3\ -2\ -1\ \ 0\ \ 1\ \ 2\ \ 3\ \ 4\ \ 5$

63. $\{x \,|\, x > 1\} \cap \{x \,|\, x \geq -2\}$

$-5\ -4\ -3\ -2\ -1\ \ 0\ \ 1\ \ 2\ \ 3\ \ 4\ \ 5$

64. $\{x \,|\, x < 4\} \cap \{x \,|\, x \leq 0\}$

$-5\ -4\ -3\ -2\ -1\ \ 0\ \ 1\ \ 2\ \ 3\ \ 4\ \ 5$

65. $\{x \,|\, x > 2\} \cup \{x \,|\, x > 1\}$

$-5\ -4\ -3\ -2\ -1\ \ 0\ \ 1\ \ 2\ \ 3\ \ 4\ \ 5$

66. $\{x \,|\, x < -2\} \cup \{x \,|\, x < -4\}$

$-5\ -4\ -3\ -2\ -1\ \ 0\ \ 1\ \ 2\ \ 3\ \ 4\ \ 5$

For Exercises 67 to 76, write each interval in set-builder notation.

67. $(0, 8)$ **68.** $(-2, 4)$ **69.** $[-5, 7]$ **70.** $[3, 4]$ **71.** $[-3, 6)$

72. $(4, 5]$ **73.** $(-\infty, 4]$ **74.** $(-\infty, -2)$ **75.** $(5, \infty)$ **76.** $[-2, \infty)$

For Exercises 77 to 84, write each set of real numbers in interval notation.

77. $\{x \mid -2 < x < 4\}$ **78.** $\{x \mid 0 < x < 3\}$ **79.** $\{x \mid -1 \leq x \leq 5\}$ **80.** $\{x \mid 0 \leq x \leq 3\}$

81. $\{x \mid -\infty < x < 1\}$ **82.** $\{x \mid -\infty < x \leq 6\}$ **83.** $\{x \mid -2 \leq x < \infty\}$ **84.** $\{x \mid 3 \leq x < \infty\}$

For Exercises 85 to 96, graph.

85. $(-2, 5)$

86. $(0, 3)$

87. $[-1, 2]$

88. $[-3, 2]$

89. $(-\infty, 3]$

90. $(-\infty, -1)$

91. $[3, \infty)$

92. $[-2, \infty)$

93. $[-5, 0) \cup (1, 4]$

94. $(-\infty, -2] \cup [3, \infty)$

95. $[-3, 3] \cap [0, 5]$

96. $(-\infty, 1) \cap (-4, \infty)$

APPLYING THE CONCEPTS

Let $R = \{\text{real numbers}\}$, $A = \{x \mid -1 \leq x \leq 1\}$, $B = \{x \mid 0 \leq x \leq 1\}$, $C = \{x \mid -1 \leq x \leq 0\}$, and \varnothing be the empty set. Answer Exercises 97 to 106 using R, A, B, C, or \varnothing.

97. $A \cup B$ **98.** $A \cup A$ **99.** $B \cap B$ **100.** $A \cup C$ **101.** $A \cap R$

102. $C \cap R$ **103.** $B \cup R$ **104.** $A \cup R$ **105.** $R \cup R$ **106.** $R \cap \varnothing$

107. The set $B \cap C$ cannot be expressed using R, A, B, C, or \varnothing. What real number is represented by $B \cap C$?

108. A student wrote $-3 > x > 5$ as the inequality that represents the real numbers less than -3 or greater than 5. Explain why this is incorrect.

1.2 Operations on Rational Numbers

Objective A To add, subtract, multiply, and divide integers

An understanding of the operations on integers is necessary to succeed in algebra. Let's review those properties, beginning with the sign rules for addition.

Point of Interest

Rules for operating with positive and negative numbers have existed for a long time. Although there are older records of these rules (from the 3rd century A.D.), one of the most complete records is contained in *The Correct Astronomical System of Brahma,* written by the Indian mathematician Brahmagupta around A.D. 600.

> **Rules for Addition of Real Numbers**
>
> **To add numbers with the same sign,** add the absolute values of the numbers. Then attach the sign of the addends.
>
> **To add numbers with different signs,** find the absolute value of each number. Subtract the smaller of these two numbers from the larger. Then attach the sign of the number with the larger absolute value.

HOW TO Add: **a.** $-65 + (-48)$ **b.** $27 + (-53)$

a. $-65 + (-48) = -113$

- The signs are the same. Add the absolute values of the numbers. Then attach the sign of the addends.

b. $27 + (-53)$
$|27| = 27$ $|-53| = 53$
$53 - 27 = 26$

$27 + (-53) = -26$

- The signs are different. Find the absolute value of each number.
- Subtract the smaller number from the larger.
- Because $|-53| > |27|$, attach the sign of -53.

Subtraction is defined as addition of the additive inverse.

> **Rule for Subtraction of Real Numbers**
>
> If a and b are real numbers, then $a - b = a + (-b)$.

HOW TO Subtract: **a.** $48 - (-22)$ **b.** $-31 - 18$

Change to +

a. $48 - (-22) = 48 + 22 = 70$

Opposite of -22

Change to +

b. $-31 - 18 = -31 + (-18) = -49$

Opposite of 18

HOW TO Simplify: $-3 - (-16) + (-12)$

$-3 - (-16) + (-12) = -3 + 16 + (-12)$

- Write subtraction as addition of the opposite.

$= 13 + (-12) = 1$

- Add from left to right.

The sign rules for multiplying real numbers are given below.

Sign Rules for Multiplication of Real Numbers

The product of two numbers with the same sign is positive.

The product of two numbers with different signs is negative.

HOW TO Multiply: **a.** $-4(-9)$ **b.** $84(-4)$

a. $-4(-9) = 36$

● The product of two numbers with the same sign is positive.

b. $84(-4) = -336$

● The product of two numbers with different signs is negative.

The **multiplicative inverse** of a nonzero real number a is $\frac{1}{a}$. This number is also called the **reciprocal** of a. For instance, the reciprocal of 2 is $\frac{1}{2}$ and the reciprocal of $-\frac{3}{4}$ is $-\frac{4}{3}$. Division of real numbers is defined in terms of multiplication by the multiplicative inverse.

Rule for Division of Real Numbers

If a and b are real numbers and $b \neq 0$, then $a \div b = a \cdot \frac{1}{b}$.

Because division is defined in terms of multiplication, the sign rules for dividing real numbers are the same as the sign rules for multiplying.

HOW TO Divide: **a.** $\dfrac{-54}{9}$ **b.** $(-21) \div (-7)$ **c.** $-\dfrac{-63}{-9}$

a. $\dfrac{-54}{9} = -6$

● The quotient of two numbers with different signs is negative.

b. $(-21) \div (-7) = 3$

● The quotient of two numbers with the same sign is positive.

c. $-\dfrac{-63}{-9} = -7$

Note that $\dfrac{-12}{3} = -4$, $\dfrac{12}{-3} = -4$, and $-\dfrac{12}{3} = -4$. This suggests: If a and b are real numbers and $b \neq 0$, then $\dfrac{-a}{b} = \dfrac{a}{-b} = -\dfrac{a}{b}$.

Properties of Zero and One in Division

● Zero divided by any number other than zero is zero.

$\dfrac{0}{a} = 0, a \neq 0$

TAKE NOTE

Suppose $\frac{4}{0} = n$. The related multiplication problem is $n \cdot 0 = 4$. But $n \cdot 0 = 4$ is impossible because any number times 0 is 0. Therefore, division by 0 is not defined.

- Division by zero is not defined.

 $\dfrac{a}{0}$ is undefined.

- Any number other than zero divided by itself is 1.

 $\dfrac{a}{a} = 1, a \neq 0$

- Any number divided by 1 is the number.

 $\dfrac{a}{1} = a$

Example 1 Simplify: $-3 - (-5) - 9$

Solution
$-3 - (-5) - 9$
$= -3 + 5 + (-9)$
$= 2 + (-9)$
$= -7$

- Rewrite subtraction as addition of the opposite. Then add.

You Try It 1 Simplify: $6 - (-8) - 10$

Your solution

Example 2 Simplify: $-6|-5|(-15)$

Solution
$-6|-5|(-15)$
$= -6(5)(-15)$
$= -30(-15)$
$= 450$

- Find the absolute value of -5. Then multiply.

You Try It 2 Simplify: $12(-3)|-6|$

Your solution

Example 3 Simplify: $-\dfrac{36}{-3}$

Solution $-\dfrac{36}{-3} = -(-12) = 12$

You Try It 3 Simplify: $-\dfrac{28}{|-14|}$

Your solution

Solutions on p. S1

Objective B **To add, subtract, multiply, and divide rational numbers**

Recall that a rational number is one that can be written in the form $\frac{p}{q}$, where p and q are integers and $q \neq 0$. Examples of rational numbers are $-\frac{5}{9}$ and $\frac{12}{5}$. The number $\frac{9}{\sqrt{7}}$ is not a rational number because $\sqrt{7}$ is not an integer. All integers are rational numbers. Terminating and repeating decimals are also rational numbers.

HOW TO Add: $-12.34 + 9.059$

$12.340 - 9.059 = 3.281$

- The signs are different. Subtract the absolute values of the numbers.

$-12.34 + 9.059 = -3.281$

- Attach the sign of the number with the larger absolute value.

HOW TO Multiply: $(-0.23)(0.04)$

$(-0.23)(0.04) = -0.0092$

- The signs are different. The product is negative.

HOW TO Divide: $(-4.0764) \div (-1.72)$

$(-4.0764) \div (-1.72) = 2.37$

- The signs are the same. The quotient is positive.

To add or subtract rational numbers written as fractions, first rewrite the fractions as equivalent fractions with a common denominator. The common denominator is the **least common multiple (LCM) of the denominators.**

HOW TO Add: $\dfrac{5}{6} + \left(-\dfrac{7}{8}\right)$

$$\dfrac{5}{6} + \left(-\dfrac{7}{8}\right) = \dfrac{5}{6} \cdot \dfrac{4}{4} + \left(\dfrac{-7}{8} \cdot \dfrac{3}{3}\right)$$

- The common denominator is 24. Write each fraction in terms of the common denominator.

$$= \dfrac{20}{24} + \left(\dfrac{-21}{24}\right) = \dfrac{20 + (-21)}{24}$$

- Add the numerators. Place the sum over the common denominator.

$$= \dfrac{-1}{24} = -\dfrac{1}{24}$$

TAKE NOTE

Although the sum could have been left as $\dfrac{-1}{24}$, all answers in this text are written with the negative sign in front of the fraction.

HOW TO Subtract: $-\dfrac{7}{15} - \left(-\dfrac{1}{6}\right)$

$$-\dfrac{7}{15} - \left(-\dfrac{1}{6}\right) = -\dfrac{7}{15} + \dfrac{1}{6}$$

- Write subtraction as addition of the opposite.

$$= \dfrac{-7}{15} \cdot \dfrac{2}{2} + \dfrac{1}{6} \cdot \dfrac{5}{5} = \dfrac{-14}{30} + \dfrac{5}{30}$$

- Write each fraction in terms of the common denominator, 30.

$$= \dfrac{-9}{30} = -\dfrac{3}{10}$$

- Add the numerators. Write the answer in simplest form.

The product of two fractions is the product of the numerators over the product of the denominators.

HOW TO Multiply: $-\dfrac{5}{12} \cdot \dfrac{8}{15}$

$$-\dfrac{5}{12} \cdot \dfrac{8}{15} = -\dfrac{5 \cdot 8}{12 \cdot 15} = -\dfrac{40}{180}$$

- The signs are different. The product is negative. Multiply the numerators and multiply the denominators.

$$= -\dfrac{20 \cdot 2}{20 \cdot 9} = -\dfrac{2}{9}$$

- Write the answer in simplest form.

In the last problem, the fraction was written in simplest form by dividing the numerator and denominator by 20, which is the largest integer that divides evenly into both 40 and 180. The number 20 is called the **greatest common factor (GCF)** of 40 and 180. To write a fraction in simplest form, divide the numerator and denominator by the GCF. If you have difficulty finding the GCF, try finding the prime factorization of the numerator and denominator and then divide by the common prime factors. For instance,

$$-\dfrac{5}{12} \cdot \dfrac{8}{15} = -\dfrac{\overset{1}{\cancel{5}} \cdot (\overset{1}{\cancel{2}} \cdot \overset{1}{\cancel{2}} \cdot 2)}{(\underset{1}{\cancel{2}} \cdot \underset{1}{\cancel{2}} \cdot 3) \cdot (3 \cdot \underset{1}{\cancel{5}})} = -\dfrac{2}{9}$$

> **Division of Fractions**
>
> To divide two fractions, multiply by the reciprocal of the divisor.
>
> $$\frac{a}{b} \div \frac{c}{d} = \frac{a}{b} \cdot \frac{d}{c}$$

HOW TO Divide: $\dfrac{3}{8} \div \dfrac{9}{16}$

$$\frac{3}{8} \div \frac{9}{16} = \frac{3}{8} \cdot \frac{16}{9} = \frac{48}{72} = \frac{2}{3}$$

- Multiply by the reciprocal of the divisor. Write the answer in simplest form.

Example 4 Simplify: $-\dfrac{3}{8} - \dfrac{5}{12} - \left(-\dfrac{9}{16}\right)$

Solution

$$-\frac{3}{8} - \frac{5}{12} - \left(-\frac{9}{16}\right) = \frac{-3}{8} + \frac{-5}{12} + \frac{9}{16}$$

$$= \frac{-3}{8} \cdot \frac{6}{6} + \frac{-5}{12} \cdot \frac{4}{4} + \frac{9}{16} \cdot \frac{3}{3}$$

$$= \frac{-18 + (-20) + 27}{48}$$

$$= \frac{-11}{48} = -\frac{11}{48}$$

You Try It 4 Simplify: $\dfrac{5}{6} - \dfrac{3}{8} - \dfrac{7}{9}$

Your solution

Example 5 Simplify: $\left(-\dfrac{5}{12}\right)\left(\dfrac{4}{25}\right)$

Solution $\left(-\dfrac{5}{12}\right)\left(\dfrac{4}{25}\right) = \dfrac{-5 \cdot 4}{12 \cdot 25} = -\dfrac{1}{15}$

You Try It 5 Simplify: $\dfrac{5}{8} \div \left(-\dfrac{15}{40}\right)$

Your solution

Solutions on pp. S1–S2

Objective C **To evaluate exponential expressions**

Integrating
Technology

The caret key on a graphing calculator is used to enter an exponent. For example, to evaluate the expression on the right, press 2 6 [ENTER]. The display reads 64.

Repeated multiplication of the same factor can be written using an exponent.

$$2 \cdot 2 \cdot 2 \cdot 2 \cdot 2 \cdot 2 = 2^6 \leftarrow \textbf{Exponent}$$
$$\uparrow\text{—— Base}$$

$$b \cdot b \cdot b \cdot b \cdot b = b^5 \leftarrow \textbf{Exponent}$$
$$\uparrow\text{—— Base}$$

The **exponent** indicates how many times the factor, called the **base,** occurs in the multiplication. The multiplication $2 \cdot 2 \cdot 2 \cdot 2 \cdot 2 \cdot 2$ is in **factored form.** The exponential expression 2^6 is in **exponential form.** The exponent is also said to indicate the **power** of the base.

2^1 is read "the first power of two" or just "two."

2^2 is read "the second power of two" or "two squared."

2^3 is read "the third power of two" or "two cubed."

2^4 is read "the fourth power of two."

2^5 is read "the fifth power of two."

b^5 is read "the fifth power of b."

- Usually the exponent 1 is not written.

> **nth Power of a**
>
> If a is a real number and n is a positive integer, the nth power of a is the product of n factors of a.
>
> $$a^n = \underbrace{a \cdot a \cdot a \cdot \cdots \cdot a}_{a \text{ as a factor } n \text{ times}}$$

$5^3 = 5 \cdot 5 \cdot 5 = 125$

$(-3)^4 = (-3)(-3)(-3)(-3) = 81$

$-3^4 = -(3)^4 = -(3 \cdot 3 \cdot 3 \cdot 3) = -81$

Note the difference between $(-3)^4$ and -3^4. The placement of the parentheses is very important.

Example 6 Evaluate $(-2)^5$ and -5^2.

Solution
$(-2)^5 = (-2)(-2)(-2)(-2)(-2)$
$\qquad = -32$
$-5^2 = -(5 \cdot 5) = -25$

You Try It 6 Evaluate **a.** -2^5 and **b.** $(-5)^2$.

Your solution

Example 7 Evaluate $\left(-\dfrac{3}{4}\right)^3$.

Solution
$\left(-\dfrac{3}{4}\right)^3 = \left(-\dfrac{3}{4}\right)\left(-\dfrac{3}{4}\right)\left(-\dfrac{3}{4}\right)$
$\qquad = -\dfrac{3 \cdot 3 \cdot 3}{4 \cdot 4 \cdot 4} = -\dfrac{27}{64}$

You Try It 7 Evaluate $\left(\dfrac{2}{3}\right)^4$.

Your solution

Solutions on p. S2

Objective D **To use the Order of Operations Agreement**

Suppose we wish to evaluate $16 + 4 \cdot 2$. There are two operations, addition and multiplication. The operations could be performed in different orders.

Multiply first. $16 + \underbrace{4 \cdot 2}$

Then add. $\underbrace{16 + 8}$
$\qquad\qquad 24$

Add first. $\underbrace{16 + 4} \cdot 2$

Then multiply. $\underbrace{20 \cdot 2}$
$\qquad\qquad\quad 40$

Note that the answers are different. To avoid possibly getting more than one answer to the same problem, an *Order of Operations Agreement* is followed.

Integrating Technology

A graphing calculator uses the Order of Operations Agreement. Press 16 [+] 4 [×] 2 [ENTER]. The display reads 24.

> **Order of Operations Agreement**
>
> **Step 1.** Perform operations inside grouping symbols. **Grouping symbols** include parentheses (), brackets [], braces { }, the absolute value symbol, and the fraction bar.
>
> **Step 2.** Simplify exponential expressions.
>
> **Step 3.** Do multiplication and division as they occur from left to right.
>
> **Step 4.** Do addition and subtraction as they occur from left to right.

Study Tip

The HOW TO feature indicates an example with explanatory remarks. Using paper and pencil, you should work through the example. See *AIM for Success,* pages xxix and xxx.

HOW TO Simplify: $8 - \dfrac{2 - 22}{4 + 1} \cdot 2^2$

$$8 - \dfrac{2 - 22}{4 + 1} \cdot 2^2 = 8 - \dfrac{-20}{5} \cdot 2^2$$

- The fraction bar is a grouping symbol. Perform the operations above and below the fraction bar.

$$= 8 - \dfrac{-20}{5} \cdot 4$$

- Simplify exponential expressions.

$$= 8 - (-4) \cdot 4$$

- Do multiplication and division as they occur from left to right.

$$= 8 - (-16)$$

$$= 24$$

- Do addition and subtraction as they occur from left to right.

One or more of the steps in the Order of Operations Agreement may not be needed. In that case, just proceed to the next step.

HOW TO Simplify: $14 - [(25 - 9) \div 2]^2$

$$14 - [(25 - 9) \div 2]^2 = 14 - [16 \div 2]^2$$

- Perform the operations inside grouping symbols.

$$= 14 - [8]^2$$

$$= 14 - 64$$

- Simplify exponential expressions.

$$= -50$$

- Do addition and subtraction as they occur from left to right.

A **complex fraction** is a fraction whose numerator or denominator contains one or more fractions. The fraction bar that is placed between the numerator and denominator of a complex fraction is called the **main fraction bar.** Examples of complex fractions are shown at the right.

$$\dfrac{\dfrac{2}{3}}{-\dfrac{5}{2}}, \quad \dfrac{\dfrac{1}{7} + 5}{\dfrac{3}{4} - \dfrac{7}{8}} \quad \leftarrow \text{Main fraction bar}$$

When simplifying complex fractions, recall that $\dfrac{\dfrac{a}{b}}{\dfrac{c}{d}} = \dfrac{a}{b} \div \dfrac{c}{d} = \dfrac{a}{b} \cdot \dfrac{d}{c}$.

HOW TO Simplify: $\dfrac{\dfrac{3}{4} - \dfrac{1}{3}}{\dfrac{1}{5} - 2}$

$$\dfrac{\dfrac{3}{4} - \dfrac{1}{3}}{\dfrac{1}{5} - 2} = \dfrac{\dfrac{9}{12} - \dfrac{4}{12}}{\dfrac{1}{5} - \dfrac{10}{5}} = \dfrac{\dfrac{5}{12}}{-\dfrac{9}{5}}$$

- Perform the operations above and below the main fraction bar.

$$= \dfrac{5}{12}\left(-\dfrac{5}{9}\right)$$

- Multiply the numerator of the complex fraction by the reciprocal of the denominator of the complex fraction.

$$= -\dfrac{25}{108}$$

Example 8

Simplify: $-3 \cdot 4^2 + [2 - 5(8 \div 2)]$

Solution

$-3 \cdot 4^2 + [2 - 5(8 \div 2)]$

$= -3 \cdot 4^2 + [2 - 5(4)]$ • Inside parentheses

$= -3 \cdot 4^2 + [2 - 20]$ • Inside brackets

$= -3 \cdot 4^2 + [-18]$

$= -3 \cdot 16 + [-18]$ • Exponents

$= -48 + [-18]$ • Multiplication

$= -66$ • Addition

You Try It 8

Simplify: $-2\left[3 - 3^3 + \dfrac{16 + 2}{1 - 10}\right]$

Your solution

Example 9

Simplify: $\left(\dfrac{1}{2}\right)^3 - \left[\left(\dfrac{2}{3} + \dfrac{1}{4}\right) \div \dfrac{5}{6}\right]$

Solution

$\left(\dfrac{1}{2}\right)^3 - \left[\left(\dfrac{2}{3} + \dfrac{1}{4}\right) \div \dfrac{5}{6}\right]$

$= \left(\dfrac{1}{2}\right)^3 - \left[\dfrac{11}{12} \div \dfrac{5}{6}\right]$ • Inside parentheses

$= \left(\dfrac{1}{2}\right)^3 - \left[\dfrac{11}{12} \cdot \dfrac{6}{5}\right]$ • Inside brackets

$= \left(\dfrac{1}{2}\right)^3 - \dfrac{11}{10}$

$= \dfrac{1}{8} - \dfrac{11}{10}$ • Exponents

$= -\dfrac{39}{40}$ • Subtraction

You Try It 9

Simplify: $\dfrac{1}{3} + \dfrac{5}{8} \div \dfrac{15}{16} - \dfrac{7}{12}$

Your solution

Example 10

Simplify: $9 \cdot \dfrac{\dfrac{5}{6} - 2}{\dfrac{3}{8}} \div \dfrac{7}{6}$

Solution

$9 \cdot \dfrac{\dfrac{5}{6} - 2}{\dfrac{3}{8}} \div \dfrac{7}{6} = 9 \cdot \dfrac{-\dfrac{7}{6}}{\dfrac{3}{8}} \div \dfrac{7}{6}$

$= 9 \cdot \left(-\dfrac{7}{6} \cdot \dfrac{8}{3}\right) \div \dfrac{7}{6}$

$= 9 \cdot \left(-\dfrac{28}{9}\right) \div \dfrac{7}{6}$

$= -28 \div \dfrac{7}{6}$ • Multiplication

$= -28 \cdot \dfrac{6}{7} = -24$ • Division

You Try It 10

Simplify: $\dfrac{11}{12} - \dfrac{\dfrac{5}{4}}{2 - \dfrac{7}{2}} \cdot \dfrac{3}{4}$

Your solution

Solutions on p. S2

1.2 Exercises

Objective A To add, subtract, multiply, and divide integers

1. **a.** Explain how to add two integers with the same sign. **b.** Explain how to add two integers with different signs.

2. Explain how to rewrite $8 - (-12)$ as addition of the opposite.

For Exercises 3 to 38, simplify.

3. $-18 + (-12)$

4. $-18 - 7$

5. $5 - 22$

6. $16(-60)$

7. $3 \cdot 4 \cdot (-8)$

8. $18 \cdot 0 \cdot (-7)$

9. $18 \div (-3)$

10. $25 \div (-5)$

11. $-60 \div (-12)$

12. $(-9)(-2)(-3)(10)$

13. $-20(35)(-16)$

14. $54(19)(-82)$

15. $8 - (-12)$

16. $-6 - (-3)$

17. $|12(-8)|$

18. $|7 - 18|$

19. $|15 - (-8)|$

20. $|-16 - (-20)|$

21. $|-56 \div 8|$

22. $|81 \div (-9)|$

23. $|-153 \div (-9)|$

24. $|-4| - |-2|$

25. $-|-8| + |-4|$

26. $-|-16| - |24|$

27. $-30 + (-16) - 14 - 2$

28. $3 - (-2) + (-8) - 11$

29. $-2 + (-19) - 16 + 12$

30. $-6 + (-9) - 18 + 32$

31. $13 - |6 - 12|$

32. $-9 - |-7 - (-15)|$

33. $738 - 46 + (-105) - 219$

34. $-871 - (-387) - 132 - 46$

35. $-442 \div (-17)$

36. $621 \div (-23)$

37. $-4897 \div 59$

38. $-17|-5|$

> **Objective B** **To add, subtract, multiply, and divide rational numbers**

39. **a.** Describe the least common multiple of two numbers. **b.** Describe the greatest common factor of two numbers.

40. Explain how to divide two fractions.

For Exercises 41 to 68, simplify.

41. $\dfrac{7}{12} - \left(-\dfrac{5}{16}\right)$ **42.** $\dfrac{3}{8} - \dfrac{5}{12}$ **43.** $-\dfrac{5}{9} - \dfrac{14}{15}$ **44.** $\dfrac{1}{2} + \dfrac{1}{7} - \dfrac{5}{8}$

45. $-\dfrac{1}{3} + \dfrac{5}{9} - \dfrac{7}{12}$ **46.** $\dfrac{1}{3} + \dfrac{19}{24} - \dfrac{7}{8}$ **47.** $\dfrac{2}{3} - \dfrac{5}{12} + \dfrac{5}{24}$ **48.** $-\dfrac{7}{10} + \dfrac{4}{5} + \dfrac{5}{6}$

49. $\dfrac{5}{8} - \dfrac{7}{12} + \dfrac{1}{2}$ **50.** $-\dfrac{1}{3} \cdot \dfrac{5}{8}$ **51.** $\left(\dfrac{6}{35}\right)\left(-\dfrac{5}{16}\right)$ **52.** $\dfrac{2}{3}\left(-\dfrac{9}{20}\right) \cdot \dfrac{5}{12}$

53. $-\dfrac{8}{15} \div \dfrac{4}{5}$ **54.** $-\dfrac{2}{3} \div \left(-\dfrac{6}{7}\right)$ **55.** $-\dfrac{11}{24} \div \dfrac{7}{12}$ **56.** $\dfrac{7}{9} \div \left(-\dfrac{14}{27}\right)$

57. $\left(-\dfrac{5}{12}\right)\left(\dfrac{4}{35}\right)\left(\dfrac{7}{8}\right)$ **58.** $\dfrac{6}{35}\left(-\dfrac{7}{40}\right)\left(-\dfrac{8}{21}\right)$ **59.** $-14.27 + 1.296$ **60.** $-0.4355 + 172.5$

61. $1.832 - 7.84$ **62.** $-3.52 - (-4.7)$ **63.** $(0.03)(10.5)(6.1)$ **64.** $(1.2)(3.1)(-6.4)$

65. $5.418 \div (-0.9)$ **66.** $-0.2645 \div (-0.023)$ **67.** $-0.4355 \div 0.065$ **68.** $-6.58 - 3.97 + 0.875$

For Exercises 69 to 72, simplify. Round to the nearest hundredth.

69. $38.241 \div [-(-6.027)] - 7.453$ **70.** $-9.0508 - (-3.177) + 24.77$

71. $-287.3069 \div 0.1415$ **72.** $6472.3018 \div (-3.59)$

Objective C **To evaluate exponential expressions**

For Exercises 73 to 92, simplify.

73. 5^3

74. 3^4

75. -2^3

76. -4^3

77. $(-5)^3$

78. $(-8)^2$

79. $2^2 \cdot 3^4$

80. $4^2 \cdot 3^3$

81. $-2^2 \cdot 3^2$

82. $-3^2 \cdot 5^3$

83. $(-2)^3(-3)^2$

84. $(-4)^3(-2)^3$

85. $4 \cdot 2^3 \cdot 3^3$

86. $-4(-3)^2(4^2)$

87. $2^2(-10)(-2)^2$

88. $-3(-2)^2(-5)$

89. $\left(-\dfrac{2}{3}\right)^2 \cdot 3^3$

90. $-\left(\dfrac{2}{5}\right)^3 \cdot 5^2$

91. $2^5(-3)^4 \cdot 4^5$

92. $4^4(-3)^5(6)^2$

Objective D **To use the Order of Operations Agreement**

93. Why do we need an Order of Operations Agreement?

94. Describe each step in the Order of Operations Agreement.

For Exercises 95 to 118, simplify.

95. $5 - 3(8 \div 4)^2$

96. $4^2 - (5 - 2)^2 \cdot 3$

97. $16 - \dfrac{2^2 - 5}{3^2 + 2}$

98. $\dfrac{4(5 - 2)}{4^2 - 2^2} \div 4$

99. $\dfrac{3 + \dfrac{2}{3}}{\dfrac{11}{16}}$

100. $\dfrac{\dfrac{11}{14}}{4 - \dfrac{6}{7}}$

101. $5[(2 - 4) \cdot 3 - 2]$

102. $2[(16 \div 8) - (-2)] + 4$

103. $16 - 4\left(\dfrac{8 - 2}{3 - 6}\right) \div \dfrac{1}{2}$

104. $25 \div 5\left(\dfrac{16 + 8}{-2^2 + 8}\right) - 5$

105. $6[3 - (-4 + 2) \div 2]$

106. $12 - 4[2 - (-3 + 5) - 8]$

107. $\dfrac{1}{2} - \left(\dfrac{2}{3} \div \dfrac{5}{9}\right) + \dfrac{5}{6}$

108. $\left(-\dfrac{3}{5}\right)^2 - \dfrac{3}{5} \cdot \dfrac{5}{9} + \dfrac{7}{10}$

109. $\dfrac{1}{2} - \dfrac{\frac{17}{25}}{4 - \frac{3}{5}} \div \dfrac{1}{5}$

110. $\dfrac{3}{4} + \dfrac{3 - \frac{7}{9}}{\frac{5}{6}} \cdot \dfrac{2}{3}$

111. $\dfrac{2}{3} - \left(\dfrac{3}{8} + \dfrac{5}{6}\right) \div \dfrac{3}{5}$

112. $\dfrac{3}{4} \div \left(\dfrac{5}{8} - \dfrac{5}{12}\right) + 2$

113. $0.4(1.2 - 2.3)^2 + 5.8$

114. $5.4 - (0.3)^2 \div 0.09$

115. $1.75 \div 0.25 - (1.25)^2$

116. $(3.5 - 4.2)^2 - 3.50 \div 2.5$

117. $25.76 \div (6.96 - 3.27)^2$

118. $(3.09 - 4.77)^3 - 4.07 \cdot 3.66$

APPLYING THE CONCEPTS

119. A number that is its own additive inverse is _____ .

120. Which two numbers are their own multiplicative inverse?

121. Do all real numbers have a multiplicative inverse? If not, which ones do not have a multiplicative inverse?

122. What is the tens digit of 11^{22}?

123. What is the ones digit of 7^{18}?

124. What are the last two digits of 5^{33}?

125. What are the last three digits of 5^{234}?

126. **a.** Does $(2^3)^4 = 2^{(3^4)}$? **b.** If not, which expression is larger?

127. What is the Order of Operations Agreement for a^{b^c}? *Note:* Even calculators that normally follow the Order of Operations Agreement may not do so for this expression.

1.3 Variable Expressions

To use and identify the properties of the real numbers

The properties of the real numbers describe the way operations on numbers can be performed. Following is a list of some of the real-number properties and an example of each property.

PROPERTIES OF THE REAL NUMBERS

The Commutative Property of Addition

$$a + b = b + a$$

$3 + 2 = 2 + 3$
$5 = 5$

The Commutative Property of Multiplication

$$a \cdot b = b \cdot a$$

$(3)(-2) = (-2)(3)$
$-6 = -6$

The Associative Property of Addition

$$(a + b) + c = a + (b + c)$$

$(3 + 4) + 5 = 3 + (4 + 5)$
$7 + 5 = 3 + 9$
$12 = 12$

The Associative Property of Multiplication

$$(a \cdot b) \cdot c = a \cdot (b \cdot c)$$

$(3 \cdot 4) \cdot 5 = 3 \cdot (4 \cdot 5)$
$12 \cdot 5 = 3 \cdot 20$
$60 = 60$

The Addition Property of Zero

$$a + 0 = 0 + a = a$$

$3 + 0 = 0 + 3 = 3$

The Multiplication Property of Zero

$$a \cdot 0 = 0 \cdot a = 0$$

$3 \cdot 0 = 0 \cdot 3 = 0$

The Multiplication Property of One

$$a \cdot 1 = 1 \cdot a = a$$

$$5 \cdot 1 = 1 \cdot 5 = 5$$

The Inverse Property of Addition

$$a + (-a) = (-a) + a = 0$$

$$4 + (-4) = (-4) + 4 = 0$$

$-a$ is called the *additive inverse* of a. a is the additive inverse of $-a$. The sum of a number and its additive inverse is 0.

The Inverse Property of Multiplication

$$a \cdot \frac{1}{a} = \frac{1}{a} \cdot a = 1, \quad a \neq 0$$

$$(4)\left(\frac{1}{4}\right) = \left(\frac{1}{4}\right)(4) = 1$$

$\frac{1}{a}$ is called the *multiplicative inverse* of a. $\frac{1}{a}$ is also called the *reciprocal* of a. The product of a number and its multiplicative inverse is 1.

The Distributive Property

$$a(b + c) = ab + ac$$

$$3(4 + 5) = 3 \cdot 4 + 3 \cdot 5$$
$$3 \cdot 9 = 12 + 15$$
$$27 = 27$$

Example 1

Complete the statement by using the Commutative Property of Multiplication.

$$(x)\left(\frac{1}{4}\right) = (?)(x)$$

Solution

$$(x)\left(\frac{1}{4}\right) = \left(\frac{1}{4}\right)(x)$$

You Try It 1

Complete the statement by using the Inverse Property of Addition.

$$3x + ? = 0$$

Your solution

Example 2

Identify the property that justifies the statement: $3(x + 4) = 3x + 12$

Solution
The Distributive Property

You Try It 2

Identify the property that justifies the statement: $(a + 3b) + c = a + (3b + c)$

Your solution

Solutions on p. S2

Objective B **To evaluate a variable expression**

An expression that contains one or more variables is a **variable expression.** The variable expression $6x^2y - 7x - z + 2$ contains four **terms:** $6x^2y$, $-7x$, $-z$, and 2. The first three terms are **variable terms.** The 2 is a **constant term.**

Each variable term is composed of a **numerical coefficient** and a **variable part.**

Variable Term	Numerical Coefficient	Variable Part
$6x^2y$	6	x^2y
$-7x$	-7	x
$-z$	-1	z

- When the coefficient is 1 or -1, the 1 is usually not written.

Replacing the variables in a variable expression by a numerical value and then simplifying the resulting expression is called **evaluating the variable expression.**

Integrating Technology

See the Keystroke Guide: *Evaluating Variable Expressions* for instructions on using a graphing calculator to evaluate variable expressions.

HOW TO Evaluate $a^2 - (a - b^2c)$ when $a = -2$, $b = 3$, and $c = -4$.

$a^2 - (a - b^2c)$

$$(-2)^2 - [(-2) - 3^2(-4)] = (-2)^2 - [(-2) - 9(-4)]$$
$$= (-2)^2 - [(-2) - (-36)]$$
$$= (-2)^2 - [34]$$
$$= 4 - 34$$
$$= -30$$

- Replace each variable with its value: $a = -2$, $b = 3$, $c = -4$. Use the Order of Operations Agreement to simplify the resulting variable expression.

Example 3

Evaluate $2x^3 - 4(2y - 3z)$ when $x = 2$, $y = 3$, and $z = -2$.

Solution

$2x^3 - 4(2y - 3z)$
$2(2)^3 - 4[2(3) - 3(-2)] = 2(2)^3 - 4[6 - (-6)]$
$$= 2(2)^3 - 4[12]$$
$$= 2(8) - 4[12]$$
$$= 16 - 48 = -32$$

You Try It 3

Evaluate $-2x^2 + 3(4xy - z)$ when $x = -3$, $y = -1$, and $z = 2$.

Your solution

Example 4

Evaluate $3 - 2|3x - 2y^2|$ when $x = -1$ and $y = -2$.

Solution

$3 - 2|3x - 2y^2|$
$3 - 2|3(-1) - 2(-2)^2| = 3 - 2|3(-1) - 2(4)|$
$$= 3 - 2|(-3) - 8|$$
$$= 3 - 2|-11| = 3 - 2(11)$$
$$= 3 - 22 = -19$$

You Try It 4

Evaluate $2x - y|4x^2 - y^2|$ when $x = -2$ and $y = -6$.

Your solution

Solutions on p. S2

Objective C **To simplify a variable expression**

Like terms of a variable expression are terms with the same variable part.

Constant terms are like terms.

$$4x \;-\; 5 \;+\; 7x^2 \;+\; 3x \;-\; 9$$

(like terms: $4x$ and $3x$; like terms: -5 and -9)

To simplify a variable expression, combine the like terms by using the Distributive Property. For instance,

$$7x + 4x = (7 + 4)x = 11x$$

Adding the coefficients of like terms is called **combining like terms.**

The Distributive Property is also used to remove parentheses from a variable expression so that like terms can be combined.

HOW TO Simplify: $3(x + 2y) + 2(4x - 3y)$

$3(x + 2y) + 2(4x - 3y)$

$= 3x + 6y + 8x - 6y$ • Use the Distributive Property.

$= (3x + 8x) + (6y - 6y)$ • Use the Commutative and Associative Properties of Addition to rearrange and group like terms.

$= 11x$ • Combine like terms.

HOW TO Simplify: $2 - 4[3x - 2(5x - 3)]$

$2 - 4[3x - 2(5x - 3)] = 2 - 4[3x - 10x + 6]$ • Use the Distributive Property to remove the inner parentheses.

$= 2 - 4[-7x + 6]$ • Combine like terms.

$= 2 + 28x - 24$ • Use the Distributive Property to remove the brackets.

$= 28x - 22$ • Combine like terms.

Study Tip

One of the key instructional features of this text is the Example/You Try It pairs. Each Example is completely worked. You are to solve the You Try It problems. When you are ready, check your solution against the one in the Solution Section. The solution for You Try Its 5 and 6 below are on page S2 (see the reference at the bottom right of the You Try It box). See *AIM for Success,* page xxx.

Example 5

Simplify: $7 + 3(4x - 7)$

Solution
$7 + 3(4x - 7)$
$= 7 + 12x - 21$ • The Order of Operations Agreement requires multiplication before addition.
$= 12x - 14$

You Try It 5

Simplify: $9 - 2(3y - 4) + 5y$

Your solution

Example 6

Simplify: $5 - 2(3x + y) - (x - 4)$

Solution
$5 - 2(3x + y) - (x - 4)$
$= 5 - 6x - 2y - x + 4$ • Distributive Property
$= -7x - 2y + 9$ • Combine like terms.

You Try It 6

Simplify: $6z - 3(5 - 3z) + 5(2z - 3)$

Your solution

Solutions on p. S2

1.3 Exercises

Objective A **To use and identify the properties of the real numbers**

For Exercises 1 to 14, use the given property of the real numbers to complete the statement.

1. The Commutative Property of Multiplication
$3 \cdot 4 = 4 \cdot ?$

2. The Commutative Property of Addition
$7 + 15 = ? + 7$

3. The Associative Property of Addition
$(3 + 4) + 5 = ? + (4 + 5)$

4. The Associative Property of Multiplication
$(3 \cdot 4) \cdot 5 = 3 \cdot (? \cdot 5)$

5. The Division Property of Zero
$\frac{5}{?}$ is undefined.

6. The Multiplication Property of Zero
$5 \cdot ? = 0$

7. The Distributive Property
$3(x + 2) = 3x + ?$

8. The Distributive Property
$5(y + 4) = ? \cdot y + 20$

9. The Division Property of Zero
$\frac{?}{-6} = 0$

10. The Inverse Property of Addition
$(x + y) + ? = 0$

11. The Inverse Property of Multiplication
$\frac{1}{mn}(mn) = ?$

12. The Multiplication Property of One
$? \cdot 1 = x$

13. The Associative Property of Multiplication
$2(3x) = ? \cdot x$

14. The Commutative Property of Addition
$ab + bc = bc + ?$

For Exercises 15 to 26, identify the property that justifies the statement.

15. $\frac{0}{-5} = 0$

16. $-8 + 8 = 0$

17. $(-12)\left(-\frac{1}{12}\right) = 1$

18. $(3 \cdot 4) \cdot 2 = 2 \cdot (3 \cdot 4)$

19. $y + 0 = y$

20. $2x + (5y + 8) = (2x + 5y) + 8$

21. $\dfrac{-9}{0}$ is undefined.

22. $(x + y)z = xz + yz$

23. $6(x + y) = 6x + 6y$

24. $(-12y)(0) = 0$

25. $(ab)c = a(bc)$

26. $(x + y) + z = (y + x) + z$

Objective B **To evaluate a variable expression**

For Exercises 27 to 56, evaluate the variable expression when $a = 2$, $b = 3$, $c = -1$, and $d = -4$.

27. $ab + dc$

28. $2ab - 3dc$

29. $4cd \div a^2$

30. $b^2 - (d - c)^2$

31. $(b - 2a)^2 + c$

32. $(b - d)^2 \div (b - d)$

33. $(bc + a)^2 \div (d - b)$

34. $\dfrac{1}{3}b^3 - \dfrac{1}{4}d^3$

35. $\dfrac{1}{4}a^4 - \dfrac{1}{6}bc$

36. $2b^2 \div \dfrac{ad}{2}$

37. $\dfrac{3ac}{-4} - c^2$

38. $\dfrac{2d - 2a}{2bc}$

39. $\dfrac{3b - 5c}{3a - c}$

40. $\dfrac{2d - a}{b - 2c}$

41. $\dfrac{a - d}{b + c}$

42. $|a^2 + d|$

43. $-a|a + 2d|$

44. $d|b - 2d|$

45. $\dfrac{2a - 4d}{3b - c}$

46. $\dfrac{3d - b}{b - 2c}$

47. $-3d \div \left| \dfrac{ab - 4c}{2b + c} \right|$

48. $-2bc + \left| \dfrac{bc + d}{ab - c} \right|$

49. $2(d - b) \div (3a - c)$

50. $(d - 4a)^2 \div c^3$

51. $-d^2 - c^3 a$

52. $a^2 c - d^3$

53. $-d^3 + 4ac$

54. b^a

55. $4^{(a^2)}$

56. a^b

Objective C **To simplify a variable expression**

For Exercises 57 to 90, simplify.

57. $5x + 7x$

58. $3x + 10x$

59. $-8ab - 5ab$

60. $-2x + 5x - 7x$

61. $3x - 5x + 9x$

62. $-2a + 7b + 9a$

63. $5b - 8a - 12b$

64. $12\left(\dfrac{1}{12} x \right)$

65. $\dfrac{1}{3}(3y)$

66. $-3(x - 2)$

67. $-5(x - 9)$

68. $(x + 2)5$

69. $-(x + y)$

70. $-(-x - y)$

71. $3(a - 5)$

72. $3(x - 2y) - 5$

73. $4x - 3(2y - 5)$

74. $-2a - 3(3a - 7)$

75. $3x - 2(5x - 7)$

76. $2x - 3(x - 2y)$

77. $3[a - 5(5 - 3a)]$

78. $5[-2 - 6(a - 5)]$

79. $3[x - 2(x + 2y)]$

80. $5[y - 3(y - 2x)]$

81. $-2(x - 3y) + 2(3y - 5x)$

82. $4(-a - 2b) - 2(3a - 5b)$

83. $5(3a - 2b) - 3(-6a + 5b)$

84. $-7(2a - b) + 2(-3b + a)$

85. $3x - 2[y - 2(x + 3[2x + 3y])]$

86. $2x - 4[x - 4(y - 2[5y + 3])]$

87. $4 - 2(7x - 2y) - 3(-2x + 3y)$

88. $3x + 8(x - 4) - 3(2x - y)$

89. $\dfrac{1}{3}[8x - 2(x - 12) + 3]$

90. $\dfrac{1}{4}[14x - 3(x - 8) - 7x]$

APPLYING THE CONCEPTS

In Exercises 91 to 98, it is possible that at least one of the properties of real numbers has been applied incorrectly. If the statement is incorrect, state the incorrect application of the properties of real numbers and correct the answer. If the statement is correct, state the property of real numbers that is being used.

91. $4(3y + 1) = 12y + 4$

92. $-4(5x - y) = -20x + 4y$

93. $2 + 3x = (2 + 3)x = 5x$

94. $6 - 6x = 0x = 0$

95. $2(3y) = (2 \cdot 3)(2y) = 12y$

96. $3a - 4b = 4b - 3a$

97. $-x^2 + y^2 = y^2 - x^2$

98. $x^4 \cdot \dfrac{1}{x^4} = 1, x \neq 0$

1.4 Verbal Expressions and Variable Expressions

Objective A **To translate a verbal expression into a variable expression**

One of the major skills required in applied mathematics is translating a verbal expression into a mathematical expression. Doing so requires recognizing the verbal phrases that translate into mathematical operations. Following is a partial list of the verbal phrases used to indicate the different mathematical operations.

Addition	more than	8 more than w	$w + 8$
	added to	x added to 9	$9 + x$
	the sum of	the sum of z and 9	$z + 9$
	the total of	the total of r and s	$r + s$
	increased by	x increased by 7	$x + 7$
Subtraction	less than	12 less than b	$b - 12$
	the difference between	the difference between x and 1	$x - 1$
	minus	z minus 7	$z - 7$
	decreased by	17 decreased by a	$17 - a$
Multiplication	times	negative 2 times c	$-2c$
	the product of	the product of x and y	xy
	multiplied by	3 multiplied by n	$3n$
	of	three-fourths of m	$\dfrac{3}{4}m$
	twice	twice d	$2d$
Division	divided by	v divided by 15	$\dfrac{v}{15}$
	the quotient of	the quotient of y and 3	$\dfrac{y}{3}$
	the ratio of	the ratio of x to 7	$\dfrac{x}{7}$
Power	the square of or the second power of	the square of x	x^2
	the cube of or the third power of	the cube of r	r^3
	the fifth power of	the fifth power of a	a^5

Translating a phrase that contains the word *sum, difference, product,* or *quotient* can sometimes cause a problem. In the examples at the right, note where the operation symbol is placed.

the *sum* of *x and y*	$x + y$
the *difference* between *x and y*	$x - y$
the *product* of *x and y*	$x \cdot y$
the *quotient* of *x and y*	$\dfrac{x}{y}$

HOW TO Translate "three times the sum of c and five" into a variable expression.

Identify words that indicate the mathematical operations.

3 <u>times</u> the <u>sum of</u> c and 5

Use the identified words to write the variable expression. Note that the phrase <u>times the sum of</u> requires parentheses.

$3(c + 5)$

HOW TO The sum of two numbers is thirty-seven. If x represents the smaller number, translate "twice the larger number" into a variable expression.

Write an expression for the larger number by subtracting the smaller number, x, from 37.

larger number: $37 - x$

Identify the words that indicate the mathematical operations.

<u>twice</u> the larger number

Use the identified words to write a variable expression.

$2(37 - x)$

HOW TO Translate "five less than twice the difference between a number and seven" into a variable expression. Then simplify.

Identify words that indicate the mathematical operations.

the unknown number: x

5 <u>less than</u> twice the <u>difference between</u> x and 7

Use the identified words to write the variable expression.

$2(x - 7) - 5$

Simplify the expression.

$2x - 14 - 5 = 2x - 19$

Example 1

Translate "the quotient of r and the sum of r and four" into a variable expression.

Solution

the quotient of r and the sum of r and four

$$\dfrac{r}{r+4}$$

Example 2

Translate "the sum of the square of y and six" into a variable expression.

Solution

the sum of the square of y and six

$y^2 + 6$

Example 3

The sum of two numbers is twenty-eight. Using x to represent the smaller number, translate "the sum of three times the larger number and the smaller number" into a variable expression. Then simplify.

Solution

The smaller number is x.
The larger number is $28 - x$.
the sum of three times the larger number and the smaller number

$3(28 - x) + x$ • This is the variable expression.
$84 - 3x + x$ • Simplify.
$84 - 2x$

Example 4

Translate "eight more than the product of four and the total of a number and twelve" into a variable expression. Then simplify.

Solution

Let the unknown number be x.
8 more than the product of 4 and the total of x and 12

$4(x + 12) + 8$ • This is the variable expression.
$4x + 48 + 8$ • Simplify.
$4x + 56$

You Try It 1

Translate "twice x divided by the difference between x and seven" into a variable expression.

Your solution

You Try It 2

Translate "the product of negative three and the square of d" into a variable expression.

Your solution

You Try It 3

The sum of two numbers is sixteen. Using x to represent the smaller number, translate "the difference between twice the smaller number and the larger number" into a variable expression. Then simplify.

Your solution

You Try It 4

Translate "the difference between fourteen and the sum of a number and seven" into a variable expression. Then simplify.

Your solution

Solutions on p. S2

Objective B **To solve application problems**

Many of the applications of mathematics require that you identify the unknown quantity, assign a variable to that quantity, and then attempt to express other unknowns in terms of that quantity.

HOW TO Ten gallons of paint were poured into two containers of different sizes. Express the amount of paint poured into the smaller container in terms of the amount poured into the larger container.

Assign a variable to the amount of paint poured into the larger container.

the number of gallons of paint poured into the larger container: g

Express the amount of paint in the smaller container in terms of g.
(g gallons of paint were poured into the larger container.)

the number of gallons of paint poured into the smaller container: $10 - g$

Example 5

A cyclist is riding at a rate that is twice the speed of a runner. Express the speed of the cyclist in terms of the speed of the runner.

Solution
the speed of the runner: r
the speed of the cyclist is twice r: $2r$

You Try It 5

The length of the Carnival cruise ship *Destiny* is 56 ft more than the height of the Empire State Building. Express the length of *Destiny* in terms of the height of the Empire State Building.

Your solution

Example 6

The length of a rectangle is 2 ft more than 3 times the width. Express the length of the rectangle in terms of the width.

Solution
the width of the rectangle: w
the length is 2 more than 3 times w: $3w + 2$

You Try It 6

The depth of the deep end of a swimming pool is 2 ft more than twice the depth of the shallow end. Express the depth of the deep end in terms of the depth of the shallow end.

Your solution

Example 7

In a survey of listener preferences for AM or FM radio stations, one-third of the number of people surveyed preferred AM stations. Express the number of people who preferred AM stations in terms of the number of people surveyed.

Solution
the number of people surveyed: x
the number of people who preferred AM

stations: $\frac{1}{3}x$

You Try It 7

A customer's recent credit card bill showed that one-fourth of the total bill was for the cost of restaurant meals. Express the cost of restaurant meals in terms of the total credit card bill.

Your solution

Solutions on p. S3

1.4 Exercises

To translate a verbal expression into a variable expression

For Exercises 1 to 6, translate into a variable expression.

1. eight less than a number

2. the product of negative six and a number

3. four-fifths of a number

4. the difference between a number and twenty

5. the quotient of a number and fourteen

6. a number increased by two hundred

For Exercises 7 to 20, translate into a variable expression. Then simplify.

7. a number minus the sum of the number and two

8. a number decreased by the difference between five and the number

9. five times the product of eight and a number

10. a number increased by two-thirds of the number

11. the difference between seventeen times a number and twice the number

12. one-half of the total of six times a number and twenty-two

13. the difference between the square of a number and the total of twelve and the square of the number

14. eleven more than the square of a number added to the difference between the number and seventeen

15. the sum of five times a number and twelve added to the product of fifteen and the number

16. four less than twice the sum of a number and eleven

17. The sum of two numbers is fifteen. Using x to represent the smaller of the two numbers, translate "the sum of two more than the larger number and twice the smaller number" into a variable expression. Then simplify.

18. The sum of two numbers is twenty. Using x to represent the smaller of the two numbers, translate "the difference between two more than the larger number and twice the smaller number" into a variable expression. Then simplify.

19. The sum of two numbers is thirty-four. Using x to represent the larger of the two numbers, translate "the quotient of five times the smaller number and the difference between the larger number and three" into a variable expression.

20. The sum of two numbers is thirty-three. Using x to represent the larger of the two numbers, translate "the difference between six more than twice the larger number and the sum of the smaller number and three" into a variable expression. Then simplify.

Objective B To solve application problems

21. **Astronomy** The distance from Earth to the sun is approximately 390 times the distance from Earth to the moon. Express the distance from Earth to the sun in terms of the distance from Earth to the moon.

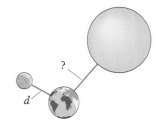

22. **Health** There are seven times as many deaths each year by heart disease as there are deaths by accidents. Express the number of deaths each year by heart disease in terms of the number of deaths by accidents. (*Source: Wall Street Journal Almanac*)

23. **Mixtures** A mixture of candy contains 3 lb more milk chocolate than caramel. Express the amount of milk chocolate in the mixture in terms of the amount of caramel in the mixture.

24. **Construction** The longest rail tunnel, from Hanshu to Hokkaido, Japan, is 23.36 mi longer than the longest road tunnel, from Goschenen to Airo, Switzerland. Express the length of the longest rail tunnel in terms of the longest road tunnel.

25. **Investments** A financial advisor has invested $10,000 in two accounts. If one account contains x dollars, express the amount in the second account in terms of x.

26. **Recreation** A fishing line 3 ft long is cut into two pieces, one shorter than the other. Express the length of the shorter piece in terms of the length of the longer piece.

27. **Geometry** The measure of angle A of a triangle is twice the measure of angle B. The measure of angle C is twice the measure of angle A. Write expressions for angle A and angle C in terms of angle B.

28. **Carpentry** A 12-foot board is cut into two pieces of different lengths. Express the length of the longer piece in terms of the length of the shorter piece.

APPLYING THE CONCEPTS

For each of the following, write a phrase that would translate into the given expression.

29. $2x + 3$ **30.** $5y - 4$ **31.** $2(x + 3)$ **32.** $5(y - 4)$

33. Translate each of the following into a variable expression. Each expression is part of a formula from the sciences, and its translation requires more than one variable.
 a. the product of one-half the acceleration due to gravity (g) and the time (t) squared
 b. the product of mass (m) and acceleration (a)
 c. the product of the area (A) and the square of the velocity (v)

Focus on Problem Solving

Polya's Four-Step Process

Your success in mathematics and your success in the workplace are heavily dependent on your ability to solve problems. One of the foremost mathematicians to study problem solving was George Polya (1887–1985). The basic structure that Polya advocated for problem solving has four steps, as outlined below.

1. Understand the Problem

You must have a clear understanding of the problem. To help you focus on understanding the problem, here are some questions to think about.

- Can you restate the problem in your own words?
- Can you determine what is known about this type of problem?
- Is there missing information that you need in order to solve the problem?
- Is there information given that is not needed?
- What is the goal?

2. Devise a Plan

Successful problem solvers use a variety of techniques when they attempt to solve a problem. Here are some frequently used strategies.

- Make a list of the known information.
- Make a list of information that is needed to solve the problem.
- Make a table or draw a diagram.
- Work backwards.
- Try to solve a similar but simpler problem.
- Research the problem to determine whether there are known techniques for solving problems of its kind.
- Try to determine whether some pattern exists.
- Write an equation.

3. Carry Out the Plan

Once you have devised a plan, you must carry it out.

- Work carefully.
- Keep an accurate and neat record of all your attempts.
- Realize that some of your initial plans will not work and that you may have to return to Step 2 and devise another plan or modify your existing plan.

4. Review Your Solution

Once you have found a solution, check the solution against the known facts.

- Make sure that the solution is consistent with the facts of the problem.
- Interpret the solution in the context of the problem.
- Ask yourself whether there are generalizations of the solution that could apply to other problems.
- Determine the strengths and weaknesses of your solution. For instance, is your solution only an approximation to the actual solution?
- Consider the possibility of alternative solutions.

We will use Polya's four-step process to solve the following problem.

1.5 in.

6 in.

1 in.

A large soft drink costs \$1.25 at a college cafeteria. The dimensions of the cup are shown at the left. Suppose you don't put any ice in the cup. Determine the cost per ounce for the soft drink.

1. *Understand the problem.* We must determine the cost per ounce for the soft drink. To do this, we need the dimensions of the cup (which are given), the cost of the drink (given), and a formula for the volume of the cup (unknown). Also, because the dimensions are given in inches, the volume will be in cubic inches. We need a conversion factor that will convert cubic inches to fluid ounces.

2. *Devise a plan.* Consult a resource book that gives an equation for the volume of the figure, which is called a **frustrum.** The formula for the volume is

$$V = \frac{\pi h}{3}(r^2 + rR + R^2)$$

where h is the height, r is the radius of the base, and R is the radius of the top. Also from a reference book, 1 in^3 ≈ 0.55 fl oz. The general plan is to calculate the volume, convert the answer to fluid ounces, and then divide the cost by the number of fluid ounces.

3. *Carry out the plan.* Using the information from the drawing, evaluate the formula for the volume.

$$V = \frac{6\pi}{3}[1^2 + 1(1.5) + 1.5^2] = 9.5\pi ≈ 29.8451 \text{ in}^3$$

$$V ≈ 29.8451(0.55) ≈ 16.4148 \text{ fl oz}$$ • **Convert to fluid ounces.**

$$\text{Cost per ounce} ≈ \frac{1.25}{16.4148} ≈ 0.07615$$ • **Divide the cost by the volume.**

The cost of the soft drink is approximately 7.62 cents per ounce.

4. *Review the solution.* The cost of a 12-ounce can of soda from a vending machine is generally about 75¢. Therefore, the cost of canned soda is 75¢ ÷ 12 = 6.25¢ per ounce. This is consistent with our solution. This does not mean our solution is correct, but it does indicate that it is at least reasonable. Why might soda from a cafeteria be more expensive per ounce than soda from a vending machine?

Is there an alternative way to obtain the solution? There are probably many, but one possibility is to get a measuring cup, pour the soft drink into it, and read the number of ounces. Name an advantage and a disadvantage of this method.

Use the four-step solution process to solve Exercises 1 and 2.

1. A cup dispenser next to a water cooler holds cups that have the shape of a right circular cone. The height of the cone is 4 in. and the radius of the circular top is 1.5 in. How many ounces of water can the cup hold?

2. Soft drink manufacturers research the preferences of consumers with regard to the look, feel, and size of a soft drink can. Suppose a manufacturer has determined that people want to have their hands reach around approximately 75% of the can. If this preference is to be achieved, how tall should the can be if it contains 12 oz of fluid? Assume the can is a right circular cylinder.

Projects and Group Activities

Water Displacement

When an object is placed in water, the object displaces an amount of water that is equal to the volume of the object.

> **HOW TO** A sphere with a diameter of 4 in. is placed in a rectangular tank of water that is 6 in. long and 5 in. wide. How much does the water level rise? Round to the nearest hundredth.
>
> $$V = \frac{4}{3}\pi r^3$$
> - Use the formula for the volume of a sphere.
>
> $$V = \frac{4}{3}\pi(2^3) = \frac{32}{3}\pi$$
> - $r = \frac{1}{2}d = \frac{1}{2}(4) = 2$

Let x represent the amount of the rise in water level. The volume of the sphere will equal the volume displaced by the water. As shown at the left, this volume is the rectangular solid with width 5 in., length 6 in., and height x in.

$$V = LWH$$
- Use the formula for the volume of a rectangular solid.

$$\frac{32}{3}\pi = (6)(5)x$$
- Substitute $\frac{32}{3}\pi$ for V, 5 for W, and 6 for L.

$$\frac{32}{90}\pi = x$$
- The exact height that the water will fill is $\frac{32}{90}\pi$.

$$1.12 \approx x$$
- Use a calculator to find an approximation.

The water will rise approximately 1.12 in.

| **Figure 1** | **Figure 2** | **Figure 3** |

1. A cylinder with a 2-centimeter radius and a height of 10 cm is submerged in a tank of water that is 20 cm wide and 30 cm long (see Figure 1). How much does the water level rise? Round to the nearest hundredth.

2. A sphere with a radius of 6 in. is placed in a rectangular tank of water that is 16 in. wide and 20 in. long (see Figure 2). The sphere displaces water until two-thirds of the sphere is submerged. How much does the water level rise? Round to the nearest hundredth.

3. A chemist wants to know the density of a statue that weighs 15 lb. The statue is placed in a rectangular tank of water that is 12 in. long and 12 in. wide (see Figure 3). The water level rises 0.42 in. Find the density of the statue. Round to the nearest hundredth. (*Hint:* Density = weight ÷ volume)

Chapter 1 Summary

Key Words	Examples
The *integers* are $\ldots, -4, -3, -2, -1, 0, 1, 2, 3, 4, \ldots$. The *negative integers* are the integers $\ldots, -4, -3, -2, -1$. The *positive integers*, or *natural numbers*, are the integers $1, 2, 3, 4, \ldots$. The positive integers and zero are called the *whole numbers*. [1.1A, p. 3]	$-58, -12, 0, 7$, and 46 are integers. -58 and -12 are negative integers. 7 and 46 are positive integers.
A *prime number* is a natural number greater than 1 that is divisible only by itself and 1. A natural number that is not a prime number is a *composite number*. [1.1A, p. 3]	2, 3, 5, 7, 11, and 13 are prime numbers. 4, 6, 8, 9, 10, and 12 are composite numbers.
A *rational number* can be written in the form $\frac{p}{q}$, where p and q are integers and $q \neq 0$. Every rational number can be written as either a *terminating decimal* or a *repeating decimal*. A number that cannot be written as a terminating or a repeating decimal is an *irrational number*. The rational numbers and the irrational numbers taken together are the *real numbers*. [1.1A, pp. 3–4]	$\frac{5}{6}, -\frac{3}{8}$, and 4 are rational numbers. $\frac{7}{\sqrt{2}}$ is not a rational number because $\sqrt{2}$ is not an integer. $\sqrt{2}$ is an irrational number. $-\frac{3}{8} = -0.375$, a terminating decimal. $\frac{5}{6} = 0.8\overline{3}$, a repeating decimal.
The *graph of a real number* is made by placing a heavy dot directly above the number on a number line. [1.1A, p. 4]	The graph of 3 is shown below.
Numbers that are the same distance from zero on the number line but are on opposite sides of zero are *additive inverses*, or *opposites*. [1.1A, p. 5]	8 and -8 are additive inverses.
The *absolute value* of a number is its distance from zero on the number line. [1.1A, p. 5]	The absolute value of 7 is 7. The absolute value of -7 is 7.
A *set* is a collection of objects. The objects are called the *elements* of the set. [1.1A, p. 3]	Natural numbers $= \{1, 2, 3, 4, 5, 6, \ldots\}$
The *roster method* of writing a set encloses the list of the elements of the set in braces. In an *infinite set*, the pattern of numbers continues without end. In a *finite set*, all the elements of the set can be listed. The set that contains no elements is the *empty set*, or *null set*, and is symbolized by \varnothing or { }. [1.1B, pp. 6–7]	$\{2, 4, 6, 8, \ldots\}$ is an infinite set. $\{2, 4, 6, 8\}$ is a finite set.
Another method of representing a set is *set-builder notation*, which makes use of a variable and a certain property that only elements of that set possess. [1.1B, p. 7]	$\{x \mid x < 7, x \in \text{integers}\}$ is read "the set of all x such that x is less than 7 and x is an element of the integers."

Sets can also be expressed using *interval notation*. A parenthesis is used to indicate that a number is not included in the set. A bracket is used to indicate that a number is included in the set. An interval is said to be *closed* if it includes both endpoints. It is *open* if it does not include either endpoint. An interval is *half-open* if one endpoint is included and the other is not. To indicate an interval that extends forever in one or both directions using interval notation, use the *infinity symbol* ∞ or the *negative infinity symbol* −∞. [1.1C, p. 10]

The interval notation $[−4, 2)$ indicates the interval of all real numbers greater than or equal to $−4$ and less than 2.
The interval $[−4, 2)$ has endpoints $−4$ and 2. It is an example of a half-open interval.
The interval notation $(−∞, 5]$ indicates the interval of all real numbers less than 5.

The *multiplicative inverse* or *reciprocal* of a nonzero real number a is $\frac{1}{a}$. [1.2A, p. 18]

The multiplicative inverse of $−6$ is $−\frac{1}{6}$.
The multiplicative inverse of $\frac{3}{8}$ is $\frac{8}{3}$.

The *least common multiple* (LCM) of two or more numbers is the smallest number that is a multiple of the numbers. The *greatest common factor* (GCF) of two or more numbers is the largest integer that divides evenly into all of the numbers. [1.2B, p. 20]

The LCM of 6 and 8 is 24.
The GCF of 6 and 8 is 2.

The expression a^n is in *exponential form*, where a is the base and n is the exponent. a^n is the *nth power of a* and represents the product of n factors of a. [1.2C, pp. 21–22]

In the exponential expression 5^3, 5 is the base and 3 is the exponent.
$5^3 = 5 \cdot 5 \cdot 5 = 125$

A *complex fraction* is a fraction whose numerator or denominator contains one or more fractions. [1.2D, p. 23]

$\dfrac{\frac{3}{5} + \frac{1}{2}}{-\frac{4}{7}}$ is a complex fraction.

A *variable* is a letter of the alphabet that is used to stand for a number. [1.1A, p. 4] An expression that contains one or more variables is a *variable expression*. The *terms* of a variable expression are the addends of the expression. A *variable term* is composed of a *numerical coefficient* and a *variable part*. A *constant term* has no variable part. [1.3B, p. 31]

The variable expression $−4x^2 + 3x − 5$ has three terms: $−4x^2$, $3x$, and $−5$.
$−4x^2$ and $3x$ are variable terms.
$−5$ is a constant term.
For the term $−4x^2$, the coefficient is $−4$ and the variable part is x^2.

Like terms of a variable expression have the same variable part. Constant terms are also like terms. Adding the coefficients of like terms is called *combining like terms*. [1.3C, p. 32]

$−6a^3b^2$ and $4a^3b^2$ are like terms.
$−6a^3b^2 + 4a^3b^2 = −2a^3b^2$

Replacing the variable or variables in a variable expression and then simplifying the resulting numerical expression is called *evaluating the variable expression*. [1.3B, p. 31]

Evaluate $5x^3 + |6 − 2y|$ when $x = −1$ and $y = 4$.
$5x^3 + |6 − 2y|$
$$5(−1)^3 + |6 − 2(4)| = 5(−1)^3 + |6 − 8|$$
$$= 5(−1)^3 + |−2|$$
$$= 5(−1)^3 + 2$$
$$= 5(−1) + 2$$
$$= −5 + 2 = −3$$

Essential Rules and Procedures	**Examples**

Definition of Inequality Symbols [1.1A, p. 5]

If a and b are two real numbers and a is to the left of b on the number line, then a is **less than** b. This is written $a < b$.

$-19 < 36$

If a and b are two real numbers and a is to the right of b on the number line, then a is **greater than** b. This is written $a > b$.

$-1 > -20$

Absolute Value [1.1A, p. 5]

The absolute value of a positive number is the number itself.

The absolute value of a negative number is the opposite of the negative number.

$|18| = 18$
$|-18| = 18$

The absolute value of zero is zero.

$|0| = 0$

Union of Two Sets [1.1C, p. 8]

The **union** of two sets, written $A \cup B$, is the set of all elements that belong to either set **A or** set B. In set-builder notation, this is written

$$A \cup B = \{x \mid x \in A \text{ or } x \in B\}$$

Given $A = \{0, 1, 2, 3, 4\}$ and $B = \{2, 4, 6, 8\}$,
$A \cup B = \{0, 1, 2, 3, 4, 6, 8\}$.

Intersection of Two Sets [1.1C, p. 8]

The **intersection** of two sets, written $A \cap B$, is the set of all elements that are common to both set **A and** set B. In set-builder notation, this is written

$$A \cap B = \{x \mid x \in A \text{ and } x \in B\}$$

Given $A = \{0, 1, 2, 3, 4\}$ and $B = \{2, 4, 6, 8\}$, $A \cap B = \{2, 4\}$.

Graphing Intervals on the Number Line [1.1C, p. 9]

A parenthesis on a graph indicates that the number is not included in a set. A bracket indicates that the number is included in the set.

The graph of $\{x \mid x \leq -2\}$ is shown below.

Rules for Addition of Real Numbers [1.2A, p. 17]

To add numbers with the same sign, add the absolute values of the numbers. Then attach the sign of the addends.

$-12 + (-18) = -30$

To add numbers with different signs, find the absolute value of each number. Subtract the lesser of the two numbers from the greater. Then attach the sign of the number with the greater absolute value.

$12 + (-18) = -6$

Rule for Subtraction of Real Numbers [1.2A, p. 17]

If a and b are real numbers, then $a - b = a + (-b)$.

$6 - 9 = 6 + (-9) = -3$

Sign Rules for Multiplication of Real Numbers [1.2A, p. 18]

The product of two numbers with the same sign is positive.

The product of two numbers with different signs is negative.

$-5(-9) = 45$
$-5(9) = -45$

Equivalent Fractions [1.2A, p. 18]

If a and b are real numbers and $b \neq 0$, then $\dfrac{-a}{b} = \dfrac{a}{-b} = -\dfrac{a}{b}$.

$\dfrac{-3}{4} = \dfrac{3}{-4} = -\dfrac{3}{4}$

Properties of Zero and One in Division [1.2A, pp. 18–19]

Zero divided by any number other than zero is zero.

$0 \div 4 = 0$

Division by zero is not defined.

$4 \div 0$ is undefined.

Any number other than zero divided by itself is 1.

$4 \div 4 = 1$

Any number divided by 1 is the number.

$4 \div 1 = 4$

Division of Fractions [1.2B, p. 21]

To divide two fractions, multiply by the reciprocal of the divisor.

$\dfrac{a}{b} \div \dfrac{c}{d} = \dfrac{a}{b} \cdot \dfrac{d}{c}$

$-\dfrac{3}{5} \div \dfrac{3}{10} = -\dfrac{3}{5} \cdot \dfrac{10}{3} = -2$

Order of Operations Agreement [1.2D, p. 22]

Step 1 Perform operations inside grouping symbols.

Step 2 Simplify exponential expressions.

Step 3 Do multiplication and division as they occur from left to right.

Step 4 Do addition and subtraction as they occur from left to right.

$$\begin{aligned}
-6^2 - 3(-2 + 4) &= -6^2 - 3(2) \\
&= -36 - 6 \\
&= -36 + (-6) \\
&= -42
\end{aligned}$$

Properties of Real Numbers [1.3A, pp. 29–30]

Commutative Property of Addition $a + b = b + a$

$3 + 8 = 8 + 3$

Commutative Property of Multiplication $a \cdot b = b \cdot a$

$4 \cdot 9 = 9 \cdot 4$

Associative Property of Addition $(a + b) + c = a + (b + c)$

$(-2 + 4) + 6 = -2 + (4 + 6)$

Associative Property of Multiplication $(a \cdot b) \cdot c = a \cdot (b \cdot c)$

$(-5 \cdot 3) \cdot 6 = -5 \cdot (3 \cdot 6)$

Addition Property of Zero $a + 0 = 0 + a = a$

$-9 + 0 = -9$

Multiplication Property of Zero $a \cdot 0 = 0 \cdot a = 0$

$-6(0) = 0$

Multiplication Property of One $a \cdot 1 = 1 \cdot a = a$

$-12(1) = -12$

Inverse Property of Addition $a + (-a) = (-a) + a = 0$

$-7 + 7 = 0$

Inverse Property of Multiplication $a \cdot \dfrac{1}{a} = \dfrac{1}{a} \cdot a = 1, a \neq 0$

$8 \cdot \dfrac{1}{8} = 1$

Distributive Property $a(b + c) = ab + ac$

$-2(4x + 5) = -8x - 10$

Chapter 1 Review Exercises

1. Use the roster method to write the set of integers between -3 and 4.

2. Find $A \cap B$ given $A = \{0, 1, 2, 3\}$ and $B = \{2, 3, 4, 5\}$.

3. Graph $(-2, 4]$.

4. Identify the property that justifies the statement.
$$2(3x) = (2 \cdot 3)x$$

5. Simplify: $-4.07 + 2.3 - 1.07$

6. Evaluate $(a - 2b^2) \div (ab)$ when $a = 4$ and $b = -3$.

7. Simplify: $-2 \cdot (4^2) \cdot (-3)^2$

8. Simplify: $4y - 3[x - 2(3 - 2x) - 4y]$

9. Find the additive inverse of $-\frac{3}{4}$.

10. Use set-builder notation to write the set of real numbers less than -3.

11. Graph $\{x \mid x < 1\}$.

12. Simplify: $-10 - (-3) - 8$

13. Simplify: $-\frac{2}{3} + \frac{3}{5} - \frac{1}{6}$

14. Use the Associative Property of Addition to complete the statement.
$$3 + (4 + y) = (3 + ?) + y$$

15. Simplify: $-\frac{3}{8} \div \frac{3}{5}$

16. Let $x \in \{-4, -2, 0, 2\}$. For what values of x is $x > -1$ true?

17. Evaluate $2a^2 - \frac{3b}{a}$ when $a = -3$ and $b = 2$.

18. Simplify: $18 - |-12 + 8|$

19. Simplify: $20 \div \dfrac{3^2 - 2^2}{3^2 + 2^2}$

20. Graph $[-3, \infty)$.

21. Find $A \cup B$ given $A = \{1, 3, 5, 7\}$ and $B = \{2, 4, 6, 8\}$.

22. Simplify: $-204 \div (-17)$

23. Write $[-2, 3]$ in set-builder notation.

24. Simplify: $\dfrac{3}{5}\left(-\dfrac{10}{21}\right)\left(-\dfrac{7}{15}\right)$

25. Use the Distributive Property to complete the statement.
$6x - 21y = ?(2x - 7y)$

26. Graph $\{x \,|\, x \leq -3\} \cup \{x \,|\, x > 0\}$.

27. Simplify: $-2(x - 3) + 4(2 - x)$

28. Let $p \in \{-4, 0, 7\}$. Evaluate $-|p|$ for each element of the set.

29. Identify the property that justifies the statement.
$-4 + 4 = 0$

30. Simplify: $-3.286 \div (-1.06)$

31. Find the additive inverse of -87.

32. Let $y \in \{-4, -1, 4\}$. For which values of y is $y > -2$ true?

33. Use the roster method to write the set of integers between -4 and 2.

34. Use set-builder notation to write the set of real numbers less than 7.

35. Given $A = \{-4, -2, 0, 2, 4\}$ and $B = \{0, 5, 10\}$, find $A \cup B$.

36. Given $A = \{-9, -6, -3\}$ and $B = \{3, 6, 9\}$, find $A \cap B$.

37. Graph $\{x \,|\, x \leq 3\} \cap \{x \,|\, x > -2\}$.

38. Graph $(-3, 4) \cup [-1, 5]$.

39. Simplify: $9 - (-3) - 7$

40. Simplify: $-\dfrac{2}{3} + \left(-\dfrac{1}{4}\right) - \left(-\dfrac{5}{12}\right)$

41. Simplify: $\left(\dfrac{2}{3}\right)^3 (-3)^4$

42. Simplify: $(-3)^3 - (2 - 6)^2 \cdot 5$

43. Evaluate $-8ac \div b^2$ when $a = -1$, $b = 2$, and $c = -3$.

44. Simplify: $-(3a + b) - 2(-4a - 5b)$

45. Translate "four times the sum of a number and four" into a variable expression. Then simplify.

46. **Travel** The total flying time for a round trip between New York and San Diego is 13 h. Because of the jet stream, the time going is not equal to the time returning. Express the flying time between New York and San Diego in terms of the flying time between San Diego and New York.

47. **Calories** For a 140-pound person, the number of calories burned by cross-country skiing for 1 h is 396 more than the number of calories burned by walking at 4 mph for 1 h. (*Source:* **Healthstatus.com**) Express the number of calories burned by cross-country skiing for 1 h in terms of the number of calories burned by walking at 4 mph for 1 h.

48. Translate "eight more than twice the difference between a number and two" into a variable expression. Then simplify.

49. A second integer is 5 more than four times the first integer. Express the second integer in terms of the first integer.

50. Translate "twelve minus the quotient of three more than a number and four" into a variable expression. Then simplify.

51. The sum of two numbers is forty. Using x to represent the smaller of the two numbers, translate "the sum of twice the smaller number and five more than the larger number" into a variable expression. Then simplify.

52. **Geometry** The length of a rectangle is three feet less than three times the width. Express the length of the rectangle in terms of the width.

Chapter 1 Test

1. Simplify: $(-2)(-3)(-5)$

2. Find $A \cap B$ given $A = \{1, 3, 5, 7\}$ and $B = \{5, 7, 9, 11\}$.

3. Simplify: $(-2)^3(-3)^2$

4. Graph $(-\infty, 1]$.

5. Find $A \cap B$ given $A = \{-3, -2, -1, 0, 1, 2, 3\}$ and $B = \{-1, 0, 1\}$.

6. Evaluate $(a - b)^2 \div (2b + 1)$ when $a = 2$ and $b = -3$.

7. Simplify: $\left|-3 - (-5)\right|$

8. Simplify: $2x - 4[2 - 3(x + 4y) - 2]$

9. Find the additive inverse of -12.

10. Simplify: $-5^2 \cdot 4$

11. Graph $\{x \mid x < 3\} \cap \{x \mid x > -2\}$.

12. Simplify: $2 - (-12) + 3 - 5$

13. Simplify: $\dfrac{2}{3} - \dfrac{5}{12} + \dfrac{4}{9}$

14. Use the Commutative Property of Addition to complete the statement.
$(3 + 4) + 2 = (? + 3) + 2$

15. Simplify: $-\dfrac{2}{3}\left(\dfrac{9}{15}\right)\left(\dfrac{10}{27}\right)$

16. Let $x \in \{-5, 3, 7\}$. For what values of x is $x < -1$ true?

17. Evaluate $\dfrac{b^2 - c^2}{a - 2c}$ when $a = 2$, $b = 3$, and $c = -1$.

18. Simplify: $-180 \div 12$

19. Simplify: $12 - 4\left(\dfrac{5^2 - 1}{3}\right) \div 16$

20. Graph $(3, \infty)$.

<!-- number line from -5 to 5 -->
$-5\ -4\ -3\ -2\ -1\ \ 0\ \ 1\ \ 2\ \ 3\ \ 4\ \ 5$

21. Find $A \cup B$ given $A = \{1, 3, 5, 7\}$ and $B = \{2, 3, 4, 5\}$.

22. Simplify: $3x - 2(x - y) - 3(y - 4x)$

23. Simplify: $8 - 4(2 - 3)^2 \div 2$

24. Simplify: $\dfrac{3}{5}\left(-\dfrac{10}{21}\right)\left(-\dfrac{7}{15}\right)$

25. Identify the property that justifies the statement.
$-2(x + y) = -2x - 2y$

26. Graph $\{x \,|\, x \le 3\} \cup \{x \,|\, x < -2\}$.

<!-- number line from -5 to 5 -->
$-5\ -4\ -3\ -2\ -1\ \ 0\ \ 1\ \ 2\ \ 3\ \ 4\ \ 5$

27. Simplify: $4.27 - 6.98 + 1.3$

28. Find $A \cup B$ given $A = \{-2, -1, 0, 1, 2, 3\}$ and $B = \{-1, 0, 1\}$.

29. The sum of two numbers is nine. Using x to represent the larger of the two numbers, translate "the difference between one more than the larger number and twice the smaller number" into a variable expression. Then simplify.

30. **Cocoa Production** The two countries with the highest cocoa production are the Ivory Coast and Ghana. The Ivory Coast produces three times the amount of cocoa produced in Ghana. (*Source:* International Cocoa Organization) Express the amount of cocoa produced in the Ivory Coast in terms of the amount of cocoa produced in Ghana.

chapter

2 First-Degree Equations and Inequalities

Hourly wage, salary, and commissions are three ways to receive payment for doing work. Commissions are usually paid to salespersons and are calculated as a percent of total sales. The salesperson in this photo receives a combination of an hourly wage and commissions. The sales personnel in **Exercises 94 and 95 on page 93** receive a combination of salary and commissions. In these exercises, you will be using first-degree inequalities to determine the amount of sales needed to reach target goals in income.

OBJECTIVES

Section 2.1

A To solve an equation using the Addition or the Multiplication Property of Equations

B To solve an equation using both the Addition and the Multiplication Properties of Equations

C To solve an equation containing parentheses

D To solve a literal equation for one of the variables

Section 2.2

A To solve integer problems

B To solve coin and stamp problems

Section 2.3

A To solve value mixture problems

B To solve percent mixture problems

C To solve uniform motion problems

Section 2.4

A To solve an inequality in one variable

B To solve a compound inequality

C To solve application problems

Section 2.5

A To solve an absolute value equation

B To solve an absolute value inequality

C To solve application problems

Need help? For online student resources, such as section quizzes, visit this textbook's website at **math.college.hmco.com/students**.

PREP TEST ● ● ●

Do these exercises to prepare for Chapter 2.

For Exercises 1 to 5, add, subtract, multiply, or divide.

1. $8 - 12$

2. $-9 + 3$

3. $\dfrac{-18}{-6}$

4. $-\dfrac{3}{4}\left(-\dfrac{4}{3}\right)$

5. $-\dfrac{5}{8}\left(\dfrac{4}{5}\right)$

For Exercises 6 to 9, simplify.

6. $3x - 5 + 7x$

7. $6(x - 2) + 3$

8. $n + (n + 2) + (n + 4)$

9. $0.08x + 0.05(400 - x)$

10. A twenty-ounce snack mixture contains nuts and pretzels. Let n represent the number of ounces of nuts in the mixture. Express the number of ounces of pretzels in the mixture in terms of n.

GO FIGURE ● ● ●

In a school election, one candidate for class president received more than 94%, but less than 100%, of the votes cast. What is the least possible number of votes cast?

2.1 Solving First-Degree Equations

Objective A **To solve an equation using the Addition or the Multiplication Property of Equations**

An **equation** expresses the equality of two mathematical expressions. The expressions can be either numerical or variable expressions.

$$2 + 8 = 10$$
$$x + 8 = 11$$
$$x^2 + 2y = 7$$
Equations

The equation at the right is a **conditional equation.** The equation is true if the variable is replaced by 3. The equation is false if the variable is replaced by 4. A conditional equation is true for at least one value of the variable.

$$x + 2 = 5$$ A conditional equation
$$3 + 2 = 5$$ A true equation
$$4 + 2 = 5$$ A false equation

The replacement value(s) of the variable that will make an equation true is (are) called the **root(s) of the equation** or the **solution(s) of the equation.** The solution of the equation $x + 2 = 5$ is 3 because $3 + 2 = 5$ is a true equation.

The equation at the right is an **identity.** Any replacement for x will result in a true equation.

$$x + 2 = x + 2$$

The equation at the right has no solution because there is no number that equals itself plus one. Any replacement value for x will result in a false equation. This equation is a **contradiction.**

$$x = x + 1$$

Each of the equations at the right is a **first-degree equation in one variable.** All variables have an exponent of 1.

$$x + 2 = 12$$
$$3y - 2 = 5y$$
$$3(a + 2) = 14a$$

Solving an equation means finding a root or solution of the equation. The simplest equation to solve is an equation of the form *variable = constant*, because the constant is the solution. If $x = 3$, then 3 is the solution of the equation, because $3 = 3$ is a true equation.

Equivalent equations are equations that have the same solution. For instance, $x + 4 = 6$ and $x = 2$ are equivalent equations because the solution of each equation is 2. In solving an equation, the goal is to produce *simpler* but equivalent equations until you reach the goal of *variable = constant*. The **Addition Property of Equations** can be used to rewrite an equation in this form.

> **The Addition Property of Equations**
>
> If a, b, and c are algebraic expressions, then the equation $a = b$ has the same solutions as the equation $a + c = b + c$.

The Addition Property of Equations states that the same quantity can be added to each side of an equation without changing the solution of the equation. This property is used to remove a term from one side of the equation by adding the opposite of that term to each side of the equation.

TAKE NOTE

The model of an equation as a balance scale applies.

Adding a weight to one side of the equation requires adding the same weight to the other side of the equation so that the pans remain in balance.

HOW TO Solve: $x - 3 = 7$

$$x - 3 = 7$$
$$x - 3 + 3 = 7 + 3$$
$$x + 0 = 10$$
$$x = 10$$

- Add 3 to each side of the equation.
- Simplify.
- The equation is in the form variable = constant.

Check:
$$\frac{x - 3 = 7}{10 - 3 \mid 7}$$
$$7 = 7$$

- Check the solution. Replace x with **10**.
- When simplified, the left side of the equation equals the right side. Therefore, **10** is the correct solution of the equation.

The solution is 10.

Because subtraction is defined in terms of addition, the Addition Property of Equations enables us to subtract the same number from each side of an equation.

HOW TO Solve: $x + \dfrac{7}{12} = \dfrac{1}{2}$

TAKE NOTE

Remember to check the solution.

$$\frac{x + \dfrac{7}{12} = \dfrac{1}{2}}{-\dfrac{1}{12} + \dfrac{7}{12} \mid \dfrac{1}{2}}$$
$$\frac{6}{12} \mid \dfrac{1}{2}$$
$$\dfrac{1}{2} = \dfrac{1}{2}$$

$$x + \frac{7}{12} = \frac{1}{2}$$
$$x + \frac{7}{12} - \frac{7}{12} = \frac{1}{2} - \frac{7}{12}$$
$$x + 0 = \frac{6}{12} - \frac{7}{12}$$
$$x = -\frac{1}{12}$$

- Subtract $\dfrac{7}{12}$ from each side of the equation.
- Simplify.

The solution is $-\dfrac{1}{12}$.

The **Multiplication Property of Equations** is also used to produce equivalent equations.

The Multiplication Property of Equations

If a, b, and c are algebraic expressions and $c \neq 0$, then the equation $a = b$ has the same solution as the equation $ac = bc$.

This property states that each side of an equation can be multiplied by the same nonzero number without changing the solution of the equation.

Recall that the goal of solving an equation is to rewrite the equation in the form *variable = constant*. The Multiplication Property of Equations is used to rewrite an equation in this form by multiplying each side by the reciprocal of the coefficient.

HOW TO Solve: $-\dfrac{3}{4}x = 12$

TAKE NOTE

Remember to check the solution.

$$\frac{-\dfrac{3}{4}x = 12}{-\dfrac{3}{4}(-16) \mid 12}$$
$$12 = 12$$

$$-\frac{3}{4}x = 12$$
$$\left(-\frac{4}{3}\right)\left(-\frac{3}{4}\right)x = \left(-\frac{4}{3}\right)12$$
$$1x = -16$$
$$x = -16$$

- Multiply each side of the equation by $-\dfrac{4}{3}$, the reciprocal of $-\dfrac{3}{4}$.
- Simplify.

The solution is -16.

Because division is defined in terms of multiplication, the Multiplication Property of Equations enables us to divide each side of an equation by the same nonzero quantity.

HOW TO Solve: $-5x = 9$

Multiplying each side of the equation by the reciprocal of -5 is equivalent to dividing each side of the equation by -5.

$$-5x = 9$$

$$\frac{-5x}{-5} = \frac{9}{-5}$$ • Divide each side of the equation by -5.

$$1x = -\frac{9}{5}$$ • Simplify.

$$x = -\frac{9}{5}$$

The solution is $-\frac{9}{5}$. You should check the solution.

When using the Multiplication Property of Equations, it is usually easier to multiply each side of the equation by the reciprocal of the coefficient when the coefficient is a fraction. Divide each side of the equation by the coefficient when the coefficient is an integer or a decimal.

Example 1

Solve: $x - 7 = -12$

Solution

$$x - 7 = -12$$
$$x - 7 + 7 = -12 + 7$$ • Add 7 to each side.
$$x = -5$$

The solution is -5.

You Try It 1

Solve: $x + 4 = -3$

Your solution

Example 2

Solve: $-\dfrac{2x}{7} = \dfrac{4}{21}$

Solution

$$-\frac{2x}{7} = \frac{4}{21}$$

$$\left(-\frac{7}{2}\right)\left(-\frac{2}{7}x\right) = \left(-\frac{7}{2}\right)\left(\frac{4}{21}\right)$$ • Multiply each side by $-\frac{7}{2}$.

$$x = -\frac{2}{3}$$

The solution is $-\dfrac{2}{3}$.

You Try It 2

Solve: $-3x = 18$

Your solution

Solutions on p. S3

Objective B **To solve an equation using both the Addition and the Multiplication Properties of Equations**

In solving an equation, it is often necessary to apply both the Addition and the Multiplication Properties of Equations.

HOW TO Solve: $4x - 3 + 2x = 8 + 9x - 12$

$$4x - 3 + 2x = 8 + 9x - 12$$
$$6x - 3 = -4 + 9x$$

• Simplify each side of the equation by combining like terms.

$$6x - 3 - 9x = -4 + 9x - 9x$$
$$-3x - 3 = -4$$

• Subtract 9x from each side of the equation. Then simplify.

$$-3x - 3 + 3 = -4 + 3$$
$$-3x = -1$$

• Add 3 to each side of the equation. Then simplify.

$$\frac{-3x}{-3} = \frac{-1}{-3}$$
$$x = \frac{1}{3}$$

• Divide each side of the equation by the coefficient −3. Then simplify.

Check:

$$
\begin{array}{c|c}
\multicolumn{2}{c}{4x - 3 + 2x = 8 + 9x - 12} \\
\hline
4\left(\dfrac{1}{3}\right) - 3 + 2\left(\dfrac{1}{3}\right) & 8 + 9\left(\dfrac{1}{3}\right) - 12 \\
\dfrac{4}{3} - 3 + \dfrac{2}{3} & 8 + 3 - 12 \\
-\dfrac{5}{3} + \dfrac{2}{3} & 11 - 12 \\
\end{array}
$$

$$-1 = -1$$

The solution is $\frac{1}{3}$.

TAKE NOTE

You should always check your work by substituting your answer into the original equation and simplifying the resulting numerical expressions. If the left and right sides are equal, your answer is the solution of the equation.

Example 3

Solve: $3x - 5 = x + 2 - 7x$

Solution

$$3x - 5 = x + 2 - 7x$$
$$3x - 5 = -6x + 2$$ • Combine like terms.
$$3x - 5 + 6x = -6x + 2 + 6x$$ • Add 6x to each side.
$$9x - 5 = 2$$
$$9x - 5 + 5 = 2 + 5$$ • Add 5 to each side.
$$9x = 7$$
$$\frac{9x}{9} = \frac{7}{9}$$ • Divide each side by 9.
$$x = \frac{7}{9}$$

The solution is $\frac{7}{9}$.

You Try It 3

Solve: $6x - 5 - 3x = 14 - 5x$

Your solution

Solution on p. S3

Objective C **To solve an equation containing parentheses**

When an equation contains parentheses, one of the steps required to solve the equation involves using the Distributive Property.

Integrating Technology

You can check the solution to this equation using a calculator. Evaluate the left side of the equation after substituting 3 for x. Press

The display reads 6. Evaluate the right side of the equation after substituting 3 for x.

2 **(** 6 **−** 3 **)**

ENTER

Again the display reads 6. The solution checks.

HOW TO Solve: $3(x - 2) + 3 = 2(6 - x)$

$$3(x - 2) + 3 = 2(6 - x)$$
$$3x - 6 + 3 = 12 - 2x$$
$$3x - 3 = 12 - 2x$$

- Use the Distributive Property to remove parentheses. Then simplify.

$$3x - 3 + 2x = 12 - 2x + 2x$$
$$5x - 3 = 12$$

- Add $2x$ to each side of the equation.

$$5x - 3 + 3 = 12 + 3$$
$$5x = 15$$

- Add 3 to each side of the equation.

$$\frac{5x}{5} = \frac{15}{5}$$
$$x = 3$$

- Divide each side of the equation by the coefficient 5.

The solution is 3.

To solve an equation containing fractions, first clear the denominators by multiplying each side of the equation by the least common multiple (LCM) of the denominators.

HOW TO Solve: $\dfrac{x}{2} - \dfrac{7}{9} = \dfrac{x}{6} + \dfrac{2}{3}$

Find the LCM of the denominators. The LCM of 2, 9, 6, and 3 is 18.

$$\frac{x}{2} - \frac{7}{9} = \frac{x}{6} + \frac{2}{3}$$

$$18\left(\frac{x}{2} - \frac{7}{9}\right) = 18\left(\frac{x}{6} + \frac{2}{3}\right)$$

- Multiply each side of the equation by the LCM of the denominators.

$$\frac{18x}{2} - \frac{18 \cdot 7}{9} = \frac{18x}{6} + \frac{18 \cdot 2}{3}$$

- Use the Distributive Property to remove parentheses. Then simplify.

$$9x - 14 = 3x + 12$$

$$6x - 14 = 12$$

- Subtract $3x$ from each side of the equation. Then simplify.

$$6x = 26$$

- Add 14 to each side of the equation. Then simplify.

$$\frac{6x}{6} = \frac{26}{6}$$

- Divide each side of the equation by the coefficient of x. Then simplify.

$$x = \frac{13}{3}$$

The solution is $\dfrac{13}{3}$.

Example 4 Solve:
$$5(2x - 7) + 2 = 3(4 - x) - 12$$

You Try It 4 Solve:
$$6(5 - x) - 12 = 2x - 3(4 + x)$$

Solution
$$5(2x - 7) + 2 = 3(4 - x) - 12$$
$$10x - 35 + 2 = 12 - 3x - 12$$
$$10x - 33 = -3x$$
$$-33 = -13x$$
$$\frac{-33}{-13} = \frac{-13x}{-13}$$
$$\frac{33}{13} = x$$

The solution is $\frac{33}{13}$.

Your solution

Solution on p. S3

Objective D **To solve a literal equation for one of the variables**

A **literal equation** is an equation that contains more than one variable. Some examples are shown at the right.

$$3x - 2y = 4$$
$$v^2 = v_0^2 + 2as$$

Formulas are used to express a relationship among physical quantities. A **formula** is a literal equation that states rules about measurement. Examples are shown at the right.

$$s = vt - 16t^2 \quad \text{(Physics)}$$
$$c^2 = a^2 + b^2 \quad \text{(Geometry)}$$
$$I = P(1 + r)^n \quad \text{(Business)}$$

The Addition and Multiplication Properties of Equations can be used to solve a literal equation for one of the variables. The goal is to rewrite the equation so that the variable being solved for is alone on one side of the equation and all the other numbers and variables are on the other side.

HOW TO Solve $A = P + Prt$ for t.

$$A = P + Prt$$
$$A - P = Prt$$
$$\frac{A - P}{Pr} = \frac{Prt}{Pr}$$
$$\frac{A - P}{Pr} = t$$

- Subtract *P* from each side of the equation.
- Divide each side of the equation by *Pr*.

Example 5

Solve $C = \frac{5}{9}(F - 32)$ for F.

You Try It 5

Solve $S = C - rC$ for r.

Solution

$$C = \frac{5}{9}(F - 32)$$
$$\frac{9}{5}C = \frac{9}{5} \cdot \frac{5}{9}(F - 32)$$
$$\frac{9}{5}C = F - 32$$
$$\frac{9}{5}C + 32 = F$$

- Multiply each side by $\frac{9}{5}$.

- Add 32 to each side.

Your solution

Solution on p. S3

2.1 Exercises

Objective A **To solve an equation using the Addition or the Multiplication Property of Equations**

1. How does an equation differ from an expression?

2. What is the solution of an equation?

3. What is the Addition Property of Equations and how is it used?

4. What is the Multiplication Property of Equations and how is it used?

5. Is 1 a solution of $7 - 3m = 4$?

6. Is 5 a solution of $4y - 5 = 3y$?

7. Is -2 a solution of $6x - 1 = 7x + 1$?

8. Is 3 a solution of $x^2 = 4x - 5$?

For Exercises 9 to 43, solve and check.

9. $x - 2 = 7$

10. $x - 8 = 4$

11. $a + 3 = -7$

12. $a + 5 = -2$

13. $b - 3 = -5$

14. $10 + m = 2$

15. $-7 = x + 8$

16. $-12 = x - 3$

17. $3x = 12$

18. $8x = 4$

19. $-3x = 2$

20. $-5a = 7$

21. $-\dfrac{3}{2} + x = \dfrac{4}{3}$

22. $\dfrac{2}{7} + x = \dfrac{17}{21}$

23. $x + \dfrac{2}{3} = \dfrac{5}{6}$

24. $\dfrac{5}{8} - y = \dfrac{3}{4}$

25. $\dfrac{2}{3}y = 5$

26. $\dfrac{3}{5}y = 12$

27. $-\dfrac{5}{8}x = \dfrac{4}{5}$

28. $-\dfrac{5}{12}x = \dfrac{7}{16}$

29. $\dfrac{4x}{7} = -12$

30. $\dfrac{3c}{10} = -9$

31. $-\dfrac{5y}{7} = \dfrac{10}{21}$

32. $-\dfrac{2d}{9} = \dfrac{4}{3}$

33. $-\dfrac{3b}{5} = -\dfrac{3}{5}$

34. $-\dfrac{7b}{12} = -\dfrac{7}{8}$

35. $-\dfrac{2}{3}x = -\dfrac{5}{8}$

36. $-\dfrac{3}{4}x = -\dfrac{4}{7}$

37. $-\dfrac{5}{8}x = 40$

38. $\dfrac{2}{7}y = -8$

39. $-\dfrac{5}{6}y = -\dfrac{25}{36}$

40. $-\dfrac{15}{24}x = -\dfrac{10}{27}$

41. $3x + 5x = 12$

42. $2x - 7x = 15$

43. $3y - 5y = 0$

Objective B **To solve an equation using both the Addition and the Multiplication Properties of Equations**

For Exercises 44 to 64, solve and check.

44. $5x + 9 = 6$

45. $2x - 4 = 12$

46. $2y - 9 = -9$

47. $4x - 6 = 3x$

48. $2a - 7 = 5a$

49. $7x + 12 = 9x$

50. $3x - 12 = 5x$

51. $4x + 2 = 4x$

52. $3m - 7 = 3m$

53. $2x + 2 = 3x + 5$

54. $7x - 9 = 3 - 4x$

55. $2 - 3t = 3t - 4$

56. $7 - 5t = 2t - 9$

57. $3b - 2b = 4 - 2b$

58. $3x - 5 = 6 - 9x$

59. $3x + 7 = 3 + 7x$

60. $\frac{5}{8}b - 3 = 12$

61. $\frac{1}{3} - 2b = 3$

62. $3.24a + 7.14 = 5.34a$

63. $5.3y + 0.35 = 5.02y$

64. $-1.27 - 4.6d = 7.93$

65. $\frac{2x}{3} - \frac{1}{2} = \frac{5}{6}$

66. $\frac{5x}{4} + \frac{3}{8} = -\frac{1}{4}$

67. $\frac{3}{2} - \frac{10x}{9} = -\frac{1}{6}$

68. If $3x + 1 = 2x - 3$, evaluate $5x - 8$.

69. If $2y - 6 = 3y + 2$, evaluate $7y + 1$.

Objective C **To solve an equation containing parentheses**

For Exercises 70 to 95, solve and check.

70. $2x + 2(x + 1) = 10$

71. $2x + 3(x - 5) = 15$

72. $2(a - 3) = 2(4 - 2a)$

73. $5(2 - b) = -3(b - 3)$

74. $3 - 2(y - 3) = 4y - 7$

75. $3(y - 5) - 5y = 2y + 9$

76. $2(3x - 2) - 5x = 3 - 2x$

77. $4 - 3x = 7x - 2(3 - x)$

78. $8 - 5(4 - 3x) = 2(4 - x) - 8x$

79. $-3x - 2(4 + 5x) = 14 - 3(2x - 3)$

80. $3[2 - 3(y - 2)] = 12$

81. $3y = 2[5 - 3(2 - y)]$

82. $4[3 + 5(3 - x) + 2x] = 6 - 2x$

83. $2[4 + 2(5 - x) - 2x] = 4x - 7$

84. $3[4 - 2(a - 2)] = 3(2 - 4a)$

85. $2[3 - 2(z + 4)] = 3(4 - z)$

86. $-3(x - 2) = 2[x - 4(x - 2) + x]$

87. $3[x - (2 - x) - 2x] = 3(4 - x)$

88. $\dfrac{5 - 2x}{5} + \dfrac{x - 4}{10} = \dfrac{3}{10}$

89. $\dfrac{2x - 5}{12} - \dfrac{3 - x}{6} = \dfrac{11}{12}$

90. $\dfrac{x - 2}{4} - \dfrac{x + 5}{6} = \dfrac{5x - 2}{9}$

91. $\dfrac{2x - 1}{4} + \dfrac{3x + 4}{8} = \dfrac{1 - 4x}{12}$

92. $-4.2(p + 3.4) = 11.13$

93. $-1.6(b - 2.35) = -11.28$

94. $0.08x + 0.06(200 - x) = 30$

95. $0.05(300 - x) + 0.07x = 45$

96. If $2x - 5(x + 1) = 7$, evaluate $x^2 - 1$.

97. If $3(2x + 1) = 5 - 2(x - 2)$, evaluate $2x^2 + 1$.

98. If $4 - 3(2x + 3) = 5 - 4x$, evaluate $x^2 - 2x$.

99. If $5 - 2(4x - 1) = 3x + 7$, evaluate $x^4 - x^2$.

Objective D **To solve a literal equation for one of the variables**

For Exercises 100 to 115, solve the formula for the given variable.

100. $I = Prt$; r (Business)

101. $C = 2\pi r$; r (Geometry)

102. $PV = nRT$; R (Chemistry)

103. $A = \dfrac{1}{2}bh$; h (Geometry)

104. $V = \dfrac{1}{3}\pi r^2 h;\ h$ (Geometry)

105. $I = \dfrac{100M}{C};\ M$ (Intelligence Quotient)

106. $P = 2L + 2W;\ W$ (Geometry)

107. $A = P + Prt;\ r$ (Business)

108. $s = V_0 t - 16t^2;\ V_0$ (Physics)

109. $s = \dfrac{1}{2}(a + b + c);\ c$ (Geometry)

110. $F = \dfrac{9}{5}C + 32;\ C$ (Temperature Conversion)

111. $S = 2\pi r^2 + 2\pi rh;\ h$ (Geometry)

112. $A = \dfrac{1}{2}h(b_1 + b_2);\ b_2$ (Geometry)

113. $P = \dfrac{R - C}{n};\ R$ (Business)

114. $a_n = a_1 + (n - 1)d;\ d$ (Mathematics)

115. $S = 2WH + 2WL + 2LH;\ H$ (Geometry)

APPLYING THE CONCEPTS

116. The following is offered as the solution of
$5x + 15 = 2x + 3(2x + 5)$.

$$5x + 15 = 2x + 3(2x + 5)$$
$$5x + 15 = 2x + 6x + 15$$
$$5x + 15 = 8x + 15$$
$$5x + 15 - 15 = 8x + 15 - 15$$

- Use the Distributive Property.
- Combine like terms.
- Subtract 15 from each side of the equation.

$$5x = 8x$$
$$\frac{5x}{x} = \frac{8x}{x}$$

- Divide each side of the equation by x.

$$5 = 8$$

Because $5 = 8$ is not a true equation, the equation has no solution.

If this is correct, so state. If not, explain why it is not correct and supply the correct answer.

117. Why, when the Multiplication Property of Equations is used, must the quantity that multiplies each side of the equation not be zero?

2.2 Applications: Puzzle Problems

Objective A To solve integer problems

An equation states that two mathematical expressions are equal. Therefore, to translate a sentence into an equation requires recognition of the words or phrases that mean "equals." A partial list of these phrases includes "is," "is equal to," "amounts to," and "represents."

Once the sentence is translated into an equation, the equation can be solved by rewriting it in the form *variable = constant*.

Recall that an **even integer** is an integer that is divisible by 2. An **odd integer** is an integer that is not divisible by 2.

Consecutive integers are integers that follow one another in order. Examples of consecutive integers are shown at the right. (Assume that the variable n represents an integer.)

$$8, 9, 10$$
$$-3, -2, -1$$
$$n, n + 1, n + 2$$

Examples of **consecutive even integers** are shown at the right. (Assume that the variable n represents an even integer.)

$$16, 18, 20$$
$$-6, -4, -2$$
$$n, n + 2, n + 4$$

Examples of **consecutive odd integers** are shown at the right. (Assume that the variable n represents an odd integer.)

$$11, 13, 15$$
$$-23, -21, -19$$
$$n, n + 2, n + 4$$

HOW TO The sum of three consecutive even integers is seventy-eight. Find the integers.

> **Strategy for Solving an Integer Problem**
>
> 1. Let a variable represent one of the integers. Express each of the other integers in terms of that variable. Remember that for consecutive integer problems, consecutive integers differ by 1. Consecutive even or consecutive odd integers differ by 2.

First even integer: n • Represent three consecutive even integers.
Second even integer: $n + 2$
Third even integer: $n + 4$

2. Determine the relationship among the integers.

$$n + (n + 2) + (n + 4) = 78$$ • The sum of the three even integers is 78.
$$3n + 6 = 78$$
$$3n = 72$$
$$n = 24$$ • The first integer is 24.

$$n + 2 = 24 + 2 = 26$$ • Find the second and third integers.
$$n + 4 = 24 + 4 = 28$$

The three consecutive even integers are 24, 26, and 28.

Example 1

One number is four more than another number. The sum of the two numbers is sixty-six. Find the two numbers.

Strategy

• The smaller number: n
 The larger number: $n + 4$
• The sum of the numbers is 66.

$n + (n + 4) = 66$

Solution

$n + (n + 4) = 66$
$2n + 4 = 66$ • Combine like terms.
$2n = 62$ • Subtract 4 from each side.
$n = 31$ • Divide each side by 2.

$n + 4 = 31 + 4 = 35$

The numbers are 31 and 35.

You Try It 1

The sum of three numbers is eighty-one. The second number is twice the first number, and the third number is three less than four times the first number. Find the numbers.

Your strategy

Your solution

Example 2

Five times the first of three consecutive even integers is five more than the product of four and the third integer. Find the integers.

Strategy

• First even integer: n
 Second even integer: $n + 2$
 Third even integer: $n + 4$
• Five times the first integer equals five more than the product of four and the third integer.

$5n = 4(n + 4) + 5$

Solution

$5n = 4(n + 4) + 5$
$5n = 4n + 16 + 5$ • Distributive Property
$5n = 4n + 21$ • Combine like terms.
$n = 21$ • Subtract $4n$ from each side.

Because 21 is not an even integer, there is no solution.

You Try It 2

Find three consecutive odd integers such that three times the sum of the first two integers is ten more than the product of the third integer and four.

Your strategy

Your solution

Solution on p. S3

Objective B **To solve coin and stamp problems**

In solving problems that deal with coins or stamps of different values, it is necessary to represent the value of the coins or stamps in the same unit of money. The unit of money is frequently cents. For example,

The value of five 8¢ stamps is 5 · 8, or 40 cents.
The value of four 20¢ stamps is 4 · 20, or 80 cents.
The value of n 10¢ stamps is n · 10, or 10n cents.

HOW TO A collection of stamps consists of 5¢, 13¢, and 18¢ stamps. The number of 13¢ stamps is two more than three times the number of 5¢ stamps. The number of 18¢ stamps is five less than the number of 13¢ stamps. The total value of all the stamps is $1.68. Find the number of 18¢ stamps.

> **Strategy for Solving a Stamp Problem**
>
> **1.** For each denomination of stamp, write a numerical or variable expression for the number of stamps, the value of the stamp, and the total value of the stamps in cents. The results can be recorded in a table.

The number of 5¢ stamps: x
The number of 13¢ stamps: $3x + 2$
The number of 18¢ stamps: $(3x + 2) - 5 = 3x - 3$

Stamp	Number of Stamps	·	Value of Stamp in Cents	=	Total Value in Cents
5¢	x	·	5	=	$5x$
13¢	$3x + 2$	·	13	=	$13(3x + 2)$
18¢	$3x - 3$	·	18	=	$18(3x - 3)$

> **2.** Determine the relationship between the total values of the stamps. Use the fact that the sum of the total values of each denomination of stamp is equal to the total value of all the stamps.

The sum of the total values of each denomination of stamp is equal to the total value of all the stamps (168 cents).

$$5x + 13(3x + 2) + 18(3x - 3) = 168$$
$$5x + 39x + 26 + 54x - 54 = 168$$
$$98x - 28 = 168$$
$$98x = 196$$
$$x = 2$$

● The sum of the total values equals 168.

The number of 18¢ stamps is $3x - 3$. Replace x by 2 and evaluate.

$$3x - 3 = 3(2) - 3 = 3$$

There are three 18¢ stamps in the collection.

Some of the problems in Section 4 of the chapter "Review of Real Numbers" involved using one variable to describe two numbers whose sum was known. For example, given that the sum of two numbers is 12, we let one of the two numbers be x. Then the other number is $12 - x$. Note that the sum of these two numbers, $x + 12 - x$, equals 12.

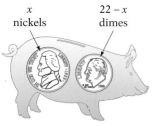

x
nickels

$22 - x$
dimes

$x + (22 - x) = 22$

In Example 3 below, we are told that there are only nickels and dimes in a coin bank, and that there is a total of twenty-two coins. This means that the sum of the number of nickels and the number of dimes is 22. Let the number of nickels be x. Then the number of dimes is $22 - x$. (If you let the number of dimes be x and the number of nickels be $22 - x$, the solution to the problem will be the same.)

Example 3

A coin bank contains $1.80 in nickels and dimes; in all, there are twenty-two coins in the bank. Find the number of nickels and the number of dimes in the bank.

Strategy

• Number of nickels: x
 Number of dimes: $22 - x$

Coin	Number	Value	Total Value
Nickel	x	5	$5x$
Dime	$22 - x$	10	$10(22 - x)$

• The sum of the total values of each denomination of coin equals the total value of all the coins (180 cents).

$5x + 10(22 - x) = 180$

Solution

$5x + 10(22 - x) = 180$
$5x + 220 - 10x = 180$ • Distributive Property
$-5x + 220 = 180$ • Combine like terms.
$-5x = -40$ • Subtract 220 from each side.
$x = 8$ • Divide each side by −5.

$22 - x = 22 - 8 = 14$

The bank contains 8 nickels and 14 dimes.

You Try It 3

A collection of stamps contains 3¢, 10¢, and 15¢ stamps. The number of 10¢ stamps is two more than twice the number of 3¢ stamps. There are three times as many 15¢ stamps as there are 3¢ stamps. The total value of the stamps is $1.56. Find the number of 15¢ stamps.

Your strategy

Your solution

Solution on p. S4

2.2 Exercises

Objective A **To solve integer problems**

1. What number must be added to the numerator of $\frac{3}{10}$ to produce the fraction $\frac{4}{5}$?

2. What number must be added to the numerator of $\frac{5}{12}$ to produce the fraction $\frac{2}{3}$?

3. The sum of two integers is ten. Three times the larger integer is three less than eight times the smaller integer. Find the integers.

4. The sum of two integers is thirty. Eight times the smaller integer is six more than five times the larger integer. Find the integers.

5. One integer is eight less than another integer. The sum of the two integers is fifty. Find the integers.

6. One integer is four more than another integer. The sum of the integers is twenty-six. Find the integers.

7. The sum of three numbers is one hundred twenty-three. The second number is two more than twice the first number. The third number is five less than the product of three and the first number. Find the three numbers.

8. The sum of three numbers is forty-two. The second number is twice the first number, and the third number is three less than the second number. Find the three numbers.

9. The sum of three consecutive integers is negative fifty-seven. Find the integers.

10. The sum of three consecutive integers is one hundred twenty-nine. Find the integers.

11. Five times the smallest of three consecutive odd integers is ten more than twice the largest. Find the integers.

12. Find three consecutive even integers such that twice the sum of the first and third integers is twenty-one more than the second integer.

13. Find three consecutive odd integers such that three times the middle integer is seven more than the sum of the first and third integers.

14. Find three consecutive even integers such that four times the sum of the first and third integers is twenty less than six times the middle integer.

Objective B **To solve coin and stamp problems**

15. A collection of fifty-three coins has a value of $3.70. The collection contains only nickels and dimes. Find the number of dimes in the collection.

16. A collection of twenty-two coins has a value of $4.75. The collection contains dimes and quarters. Find the number of quarters in the collection.

17. A coin bank contains twenty-two coins in nickels, dimes, and quarters. There are four times as many dimes as quarters. The value of the coins is $2.30. How many dimes are in the bank?

18. A coin collection contains nickels, dimes, and quarters. There are twice as many dimes as quarters and seven more nickels than dimes. The total value of all the coins is $2.00. How many quarters are in the collection?

19. A stamp collector has some 15¢ stamps and some 20¢ stamps. The number of 15¢ stamps is eight less than three times the number of 20¢ stamps. The total value is $4. Find the number of each type of stamp in the collection.

Frank Lloyd Wright 1867-1959 Fallingwater Mill Run PA
Architecture USA 20c

20. An office has some 20¢ stamps and some 28¢ stamps. All together the office has 140 stamps for a total value of $31.20. How many of each type of stamp does the office have?

21. A stamp collection consists of 3¢, 8¢, and 13¢ stamps. The number of 8¢ stamps is three less than twice the number of 3¢ stamps. The number of 13¢ stamps is twice the number of 8¢ stamps. The total value of all the stamps is $2.53. Find the number of 3¢ stamps in the collection.

22. An account executive bought 330 stamps for $79.50. The purchase included 15¢ stamps, 20¢ stamps, and 40¢ stamps. The number of 20¢ stamps is four times the number of 15¢ stamps. How many 40¢ stamps were purchased?

23. A stamp collector has 8¢, 13¢, and 18¢ stamps. The collector has twice as many 8¢ stamps as 18¢ stamps. There are three more 13¢ than 18¢ stamps. The total value of the stamps in the collection is $3.68. Find the number of 18¢ stamps in the collection.

24. A stamp collection consists of 3¢, 12¢, and 15¢ stamps. The number of 3¢ stamps is five times the number of 12¢ stamps. The number of 15¢ stamps is four less than the number of 12¢ stamps. The total value of the stamps in the collection is $3.18. Find the number of 15¢ stamps in the collection.

APPLYING THE CONCEPTS

25. **Integers** Find three consecutive odd integers such that the product of the second and third minus the product of the first and second is 42.

26. **Integers** The sum of the digits of a three-digit number is six. The tens digit is one less than the units digit, and the number is twelve more than one hundred times the hundreds digit. Find the number.

2.3 Applications: Mixture and Uniform Motion Problems

Objective A | To solve value mixture problems

> **TAKE NOTE**
>
> The equation $AC = V$ is used to find the value of an ingredient. For example, the value of 12 lb of coffee costing $5.25 per pound is
> $$AC = V$$
> $$12(\$5.25) = V$$
> $$\$63 = V$$

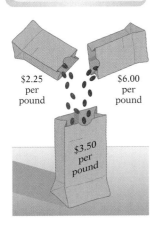

$2.25 per pound

$6.00 per pound

$3.50 per pound

A **value mixture problem** involves combining two ingredients that have different prices into a single blend. For example, a coffee manufacturer may blend two types of coffee into a single blend.

The solution of a value mixture problem is based on the equation $AC = V$, where A is the amount of the ingredient, C is the cost per unit of the ingredient, and V is the value of the ingredient.

HOW TO How many pounds of peanuts that cost $2.25 per pound must be mixed with 40 lb of cashews that cost $6.00 per pound to make a mixture that costs $3.50 per pound?

> **Strategy for Solving a Value Mixture Problem**
>
> **1.** For each ingredient in the mixture, write a numerical or variable expression for the amount of the ingredient used, the unit cost of the ingredient, and the value of the amount used. For the mixture, write a numerical or variable expression for the amount, the unit cost of the mixture, and the value of the amount. The results can be recorded in a table.

Pounds of peanuts: x
Pounds of cashews: 40
Pounds of mixture: $x + 40$

	Amount (A)	·	Unit Cost (C)	=	Value (V)
Peanuts	x	·	2.25	=	$2.25x$
Cashews	40	·	6.00	=	$6.00(40)$
Mixture	$x + 40$	·	3.50	=	$3.50(x + 40)$

> **2.** Determine how the values of the ingredients are related. Use the fact that the sum of the values of all the ingredients taken separately is equal to the value of the mixture.

The sum of the values of the peanuts and the cashews is equal to the value of the mixture.

$$2.25x + 6.00(40) = 3.50(x + 40)$$ • Value of peanuts plus value of cashews equals value of mixture.
$$2.25x + 240 = 3.50x + 140$$
$$-1.25x + 240 = 140$$
$$-1.25x = -100$$
$$x = 80$$

The mixture must contain 80 lb of peanuts.

Example 1

How many ounces of a gold alloy that costs $320 per ounce must be mixed with 100 oz of an alloy that costs $100 per ounce to make a mixture that costs $160 per ounce?

Strategy

- Ounces of $320 gold alloy: x
 Ounces of $100 gold alloy: 100
 Ounces of $160 mixture: $x + 100$

	Amount	*Cost*	*Value*
$320 alloy	x	320	$320x$
$100 alloy	100	100	$100(100)$
Mixture	$x + 100$	160	$160(x + 100)$

- The sum of the values before mixing equals the value after mixing.

 $320x + 100(100) = 160(x + 100)$

Solution

$$320x + 100(100) = 160(x + 100)$$
$$320x + 10{,}000 = 160x + 16{,}000$$
$$160x + 10{,}000 = 16{,}000$$
$$160x = 6000$$
$$x = 37.5$$

The mixture must contain 37.5 oz of the $320 gold alloy.

You Try It 1

A butcher combined hamburger that costs $3.00 per pound with hamburger that costs $1.80 per pound. How many pounds of each were used to make a 75-pound mixture costing $2.20 per pound?

Your strategy

Your solution

Solution on p. S4

Objective B To solve percent mixture problems

TAKE NOTE

The equation $Ar = Q$ is used to find the amount of a substance in a mixture. For example, the number of grams of silver in 50 g of a 40% alloy is:
$$Ar = Q$$
$$(50 \text{ g})(0.40) = Q$$
$$20 \text{ g} = Q$$

The amount of a substance in a solution or alloy can be given as a percent of the total solution or alloy. For example, in a 10% hydrogen peroxide solution, 10% of the total solution is hydrogen peroxide. The remaining 90% is water.

The solution of a **percent mixture problem** is based on the equation $Ar = Q$, where A is the amount of solution or alloy, r is the percent of concentration, and Q is the quantity of a substance in the solution or alloy.

HOW TO A chemist mixes an 11% acid solution with a 4% acid solution. How many milliliters of each solution should the chemist use to make a 700-milliliter solution that is 6% acid?

> **Strategy for Solving a Percent Mixture Problem**
>
> 1. For each solution, use the equation $Ar = Q$. Write a numerical or variable expression for the amount of solution, the percent of concentration, and the quantity of the substance in the solution. The results can be recorded in a table.

Amount of 11% solution: x
Amount of 4% solution: $700 - x$
Amount of 6% mixture: 700

	Amount of Solution (A)	·	Percent of Concentration (r)	=	Quantity of Substance (Q)
11% solution	x	·	0.11	=	$0.11x$
4% solution	$700 - x$	·	0.04	=	$0.04(700 - x)$
6% solution	700	·	0.06	=	$0.06(700)$

> 2. Determine how the quantities of the substance in each solution are related. Use the fact that the sum of the quantities of the substances being mixed is equal to the quantity of the substance after mixing.

The sum of the quantities of the substance in the 11% solution and the 4% solution is equal to the quantity of the substance in the 6% solution.

$0.11x + 0.04(700 - x) = 0.06(700)$ • Quantity in 11% solution plus quantity in 4% solution equals quantity in 6% solution.

$$0.11x + 28 - 0.04x = 42$$
$$0.07x + 28 = 42$$
$$0.07x = 14$$
$$x = 200$$

The amount of 4% solution is $700 - x$. Replace x by 200 and evaluate.

$700 - x = 700 - 200 = 500$ • $x = 200$

The chemist should use 200 ml of the 11% solution and 500 ml of the 4% solution.

Example 2

How many milliliters of pure acid must be added to 60 ml of an 8% acid solution to make a 20% acid solution?

Strategy

- Milliliters of pure acid: x

60 ml of + x ml of = (60 + x) ml
8% acid 100% acid of 20% acid

	Amount	Percent	Quantity
Pure Acid (100%)	x	1.00	x
8%	60	0.08	0.08(60)
20%	60 + x	0.20	0.20(60 + x)

- The sum of the quantities before mixing equals the quantity after mixing.

$$x + 0.08(60) = 0.20(60 + x)$$

Solution

$$x + 0.08(60) = 0.20(60 + x)$$
$$x + 4.8 = 12 + 0.20x$$
$$0.8x + 4.8 = 12$$

- Subtract 0.20x from each side.

$$0.8x = 7.2$$

- Subtract 4.8 from each side.

$$x = 9$$

- Divide each side by 0.8.

To make the 20% acid solution, 9 ml of pure acid must be used.

You Try It 2

A butcher has some hamburger that is 22% fat and some that is 12% fat. How many pounds of each should be mixed to make 80 lb of hamburger that is 18% fat?

Your strategy

Your solution

Solution on p. S4

Objective C ### To solve uniform motion problems

A car that travels constantly in a straight line at 55 mph is in uniform motion. **Uniform motion** means that the speed of an object does not change.

The solution of a **uniform motion problem** is based on the equation $rt = d$, where r is the rate of travel, t is the time spent traveling, and d is the distance traveled.

HOW TO An executive has an appointment 785 mi from the office. The executive takes a helicopter from the office to the airport and a plane from the airport to the business appointment. The helicopter averages 70 mph and the plane averages 500 mph. The total time spent traveling is 2 h. Find the distance from the executive's office to the airport.

> **Strategy for Solving a Uniform Motion Problem**
>
> **1.** For each object, write a numerical or variable expression for the distance, rate, and time. The results can be recorded in a table. It may also help to draw a diagram.

├─ 70 t ─┼─ 500 (2 − t) ─┤
Office Airport Appointment
├──────── 785 ────────┤

Unknown time in the helicopter: t
Time in the plane: $2 - t$

	Rate (r)	·	Time (t)	=	Distance (d)
Helicopter	70	·	t	=	$70t$
Plane	500	·	$2 - t$	=	$500(2 - t)$

> **2.** Determine how the distances traveled by each object are related. For example, the total distance traveled by both objects may be known, or it may be known that the two objects traveled the same distance.

The total distance traveled is 785 mi.

$70t + 500(2 - t) = 785$ • Distance by helicopter plus distance by plane equals 785.

$$70t + 1000 - 500t = 785$$
$$-430t + 1000 = 785$$
$$-430t = -215$$
$$t = 0.5$$

The time spent traveling from the office to the airport in the helicopter is 0.5 h. To find the distance between these two points, substitute the values of r and t into the equation $rt = d$.

$$rt = d$$
$$70 \cdot 0.5 = d \quad \text{• } r = 70; \; t = 0.5$$
$$35 = d$$

The distance from the office to the airport is 35 mi.

Example 3

A long-distance runner started a course running at an average speed of 6 mph. One and one-half hours later, a cyclist traveled the same course at an average speed of 12 mph. How long after the runner started did the cyclist overtake the runner?

Strategy

- Unknown time for the cyclist: t
 Time for the runner: $t + 1.5$

	Rate	Time	Distance
Runner	6	$t + 1.5$	$6(t + 1.5)$
Cyclist	12	t	$12t$

- The runner and the cyclist travel the same distance. Therefore the distances are equal.

 $6(t + 1.5) = 12t$

Solution

$6(t + 1.5) = 12t$

$6t + 9 = 12t$ • Distributive Property

$9 = 6t$ • Subtract $6t$ from each side.

$\dfrac{3}{2} = t$ • Divide each side by 6.

The cyclist traveled for 1.5 h.

$t + 1.5 = 1.5 + 1.5 = 3$

The cyclist overtook the runner 3 h after the runner started.

You Try It 3

Two small planes start from the same point and fly in opposite directions. The first plane is flying 30 mph faster than the second plane. In 4 h the planes are 1160 mi apart. Find the rate of each plane.

Your strategy

Your solution

Solution on p. S4

2.3 Exercises

Objective A **To solve value mixture problems**

1. Forty pounds of cashews costing $5.60 per pound were mixed with 100 lb of peanuts costing $1.89 per pound. Find the cost of the resulting mixture.

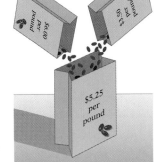

2. A coffee merchant combines coffee costing $6 per pound with coffee costing $3.50 per pound. How many pounds of each should be used to make 25 lb of a blend costing $5.25 per pound?

3. Adult tickets for a play cost $10.00 and children's tickets cost $4.00. For one performance, 460 tickets were sold. Receipts for the performance were $3760. Find the number of adult tickets sold.

4. Tickets for a school play sold for $7.50 for each adult and $3.00 for each child. The total receipts for 113 tickets sold were $663. Find the number of adult tickets sold.

5. A restaurant manager mixes 5 L of pure maple syrup that costs $9.50 per liter with imitation maple syrup that costs $4.00 per liter. How much imitation maple syrup is needed to make a mixture that costs $5.00 per liter?

6. To make a flour mixture, a miller combined soybeans that cost $8.50 per bushel with wheat that cost $4.50 per bushel. How many bushels of each were used to make a mixture of 1000 bushels costing $5.50 per bushel?

7. A goldsmith combined pure gold that cost $400 per ounce with an alloy of gold that cost $150 per ounce. How many ounces of each were used to make 50 oz of gold alloy costing $250 per ounce?

8. A silversmith combined pure silver that cost $5.20 per ounce with 50 oz of a silver alloy that cost $2.80 per ounce. How many ounces of the pure silver were used to make an alloy of silver costing $4.40 per ounce?

9. A tea mixture was made from 40 lb of tea costing $5.40 per pound and 60 lb of tea costing $3.25 per pound. Find the cost of the tea mixture.

10. Find the cost per ounce of a sunscreen made from 100 oz of lotion that cost $3.46 per ounce and 60 oz of lotion that cost $12.50 per ounce.

11. The owner of a fruit stand combined cranberry juice that cost $5.60 per gallon with 50 gal of apple juice that cost $4.24 per gallon. How much cranberry juice was used to make the cranapple juice if the mixture cost $5.00 per gallon?

12. Walnuts that cost $4.05 per kilogram were mixed with cashews that cost $7.25 per kilogram. How many kilograms of each were used to make a 50-kilogram mixture costing $6.25 per kilogram? Round to the nearest tenth.

Objective B **To solve percent mixture problems**

13. How many pounds of a 15% aluminum alloy must be mixed with 500 lb of a 22% aluminum alloy to make a 20% aluminum alloy?

14. A hospital staff mixed a 75% disinfectant solution with a 25% disinfectant solution. How many liters of each were used to make 20 L of a 40% disinfectant solution?

15. Rubbing alcohol is typically diluted with water to 70% strength. If you need 3.5 oz of 45% rubbing alcohol, how many ounces of 70% rubbing alcohol and how much water should you combine?

16. A silversmith mixed 25 g of a 70% silver alloy with 50 g of a 15% silver alloy. What is the percent concentration of the resulting alloy?

17. How many ounces of pure water must be added to 75 oz of an 8% salt solution to make a 5% salt solution?

18. How many quarts of water must be added to 5 qt of an 80% antifreeze solution to make a 50% antifreeze solution?

19. How many milliliters of alcohol must be added to 200 ml of a 25% iodine solution to make a 10% iodine solution?

20. A butcher has some hamburger that is 21% fat and some that is 15% fat. How many pounds of each should be mixed to make 84 lb of hamburger that is 17% fat?

21. Many fruit drinks are actually only 5% real fruit juice. If you let 2 oz of water evaporate from 12 oz of a drink that is 5% fruit juice, what is the percent concentration of the result?

22. How much water must be evaporated from 6 qt of a 50% antifreeze solution to produce a 75% solution?

23. A car radiator contains 12 qt of a 40% antifreeze solution. How many quarts will have to be replaced with pure antifreeze if the resulting solution is to be 60% antifreeze?

Objective C **To solve uniform motion problems**

24. A car traveling at 56 mph overtakes a cyclist who, traveling at 14 mph, had a 1.5-hour head start. How far from the starting point does the car overtake the cyclist?

25. A helicopter traveling 130 mph overtakes a speeding car traveling 80 mph. The car had a 0.5-hour head start. How far from the starting point does the helicopter overtake the car?

26. Two planes are 1620 mi apart and are traveling toward each other. One plane is traveling 120 mph faster than the other plane. The planes meet in 1.5 h. Find the speed of each plane.

27. Two cars are 310 mi apart and are traveling toward each other. One car travels 8 mph faster than the other car. The cars meet in 2.5 h. Find the speed of each car.

28. A ferry leaves a harbor and travels to a resort island at an average speed of 20 mph. On the return trip, the ferry travels at an average speed of 12 mph because of fog. The total time for the trip is 5 h. How far is the island from the harbor?

29. A commuter plane provides transportation from an international airport to the surrounding cities. One commuter plane averaged 250 mph flying to a city and 150 mph returning to the international airport. The total flying time was 4 h. Find the distance between the two airports.

30. A student rode a bicycle to the repair shop and then walked home. The student averaged 14 mph riding to the shop and 3.5 mph walking home. The round trip took 1 h. How far is it from the student's home to the repair shop?

31. A passenger train leaves a depot 1.5 h after a freight train leaves the same depot. The passenger train is traveling 18 mph faster than the freight train. Find the rate of each train if the passenger train overtakes the freight train in 2.5 h.

32. A plane leaves an airport at 3 P.M. At 4 P.M. another plane leaves the same airport traveling in the same direction at a speed 150 mph faster than that of the first plane. Four hours after the first plane takes off, the second plane is 250 mi ahead of the first plane. How far does the second plane travel?

33. A jogger and a cyclist set out at 9 A.M. from the same point headed in the same direction. The average speed of the cyclist is four times the average speed of the jogger. In 2 h, the cyclist is 33 mi ahead of the jogger. How far did the cyclist ride?

APPLYING THE CONCEPTS

34. Uniform Motion
 a. If a parade 2 mi long is proceeding at 3 mph, how long will it take a runner jogging at 6 mph to travel from the front of the parade to the end of the parade?
 b. If a parade 2 mi long is proceeding at 3 mph, how long will it take a runner jogging at 6 mph to travel from the end of the parade to the start of the parade?

35. Mixtures The concentration of gold in an alloy is measured in karats, which indicate how many parts out of 24 are pure gold. For example, 1 karat is $\frac{1}{24}$ pure gold. What amount of 12-karat gold should be mixed with 3 oz of 24-karat gold to create 14-karat gold, the most commonly used alloy?

36. Uniform Motion A student jogs 1 mi at a rate of 8 mph and jogs back at a rate of 6 mph. Does it seem reasonable that the average rate is 7 mph? Why or why not? Support your answer.

37. Uniform Motion Two cars are headed directly toward each other at rates of 40 mph and 60 mph. How many miles apart are they 2 min before impact?

38. Mixtures
 a. A radiator contains 6 qt of a 25% antifreeze solution. How much should be removed and replaced with pure antifreeze to yield a 33% solution?
 b. A radiator contains 6 qt of a 25% antifreeze solution. How much should be removed and replaced with pure antifreeze to yield a 60% solution?

39. Uniform Motion Two birds start flying, at the same time and at the same rate, from the tops of two towers that are 50 ft apart. One tower is 30 ft high, and the other tower is 40 ft high. At exactly the same time, the two birds reach a grass seed on the ground. How far is the grass seed from the base of the 40-foot tower? (This problem appeared in a math text written around A.D. 1200.)

2.4 First-Degree Inequalities

Objective A **To solve an inequality in one variable**

The **solution set of an inequality** is a set of numbers, each element of which, when substituted for the variable, results in a true inequality.

The inequality at the right is true if the variable is replaced by (for instance) 3, -1.98, or $\frac{2}{3}$.

$$x - 1 < 4$$
$$3 - 1 < 4$$
$$-1.98 - 1 < 4$$
$$\frac{2}{3} - 1 < 4$$

There are many values of the variable x that will make the inequality $x - 1 < 4$ true. The solution set of the inequality is any number less than 5. The solution set can be written in set-builder notation as $\{x \mid x < 5\}$.

Integrating Technology

See the Keystroke Guide: *Test* for instructions on using a graphing calculator to graph the solution set of an inequality.

The graph of the solution set of $x - 1 < 4$ is shown at the right.

When solving an inequality, we use the **Addition and Multiplication Properties of Inequalities** to rewrite the inequality in the form *variable < constant* or in the form *variable > constant*.

> **The Addition Property of Inequalities**
>
> If $a > b$, then $a + c > b + c$.
> If $a < b$, then $a + c < b + c$.

The Addition Property of Inequalities states that the same number can be added to each side of an inequality without changing the solution set of the inequality. This property is also true for an inequality that contains the symbol \leq or \geq.

The Addition Property of Inequalities is used to remove a term from one side of an inequality by adding the additive inverse of that term to each side of the inequality. Because subtraction is defined in terms of addition, the same number can be subtracted from each side of an inequality without changing the solution set of the inequality.

TAKE NOTE

The solution set of an inequality can be written in set-builder notation or in interval notation.

HOW TO Solve and graph the solution set: $x + 2 \geq 4$

$$x + 2 \geq 4$$
$$x + 2 - 2 \geq 4 - 2 \qquad \bullet \text{ Subtract 2 from each side of the inequality.}$$
$$x \geq 2 \qquad\qquad\quad \bullet \text{ Simplify.}$$

The solution set is $\{x \mid x \geq 2\}$ or $[2, \infty)$.

HOW TO Solve: $3x - 4 < 2x - 1$

Write the solution set in set-builder notation.

$$3x - 4 < 2x - 1$$

$$3x - 4 - 2x < 2x - 1 - 2x$$
- Subtract 2*x* from each side of the inequality.

$$x - 4 < -1$$

$$x - 4 + 4 < -1 + 4$$
- Add 4 to each side of the inequality.

$$x < 3$$

The solution set is $\{x \mid x < 3\}$.

The Multiplication Property of Inequalities is used to remove a coefficient from one side of an inequality by multiplying each side of the inequality by the reciprocal of the coefficient.

TAKE NOTE

$c > 0$ means c is a positive number. Note that the inequality symbols do not change.

$c < 0$ means c is a negative number. Note that the inequality symbols are reversed.

The Multiplication Property of Inequalities

Rule 1 If $a > b$ and $c > 0$, then $ac > bc$.
 If $a < b$ and $c > 0$, then $ac < bc$.

Rule 2 If $a > b$ and $c < 0$, then $ac < bc$.
 If $a < b$ and $c < 0$, then $ac > bc$.

Here are some examples of this property.

Rule 1		**Rule 2**	
$3 > 2$	$2 < 5$	$3 > 2$	$2 < 5$
$3(4) > 2(4)$	$2(4) < 5(4)$	$3(-4) < 2(-4)$	$2(-4) > 5(-4)$
$12 > 8$	$8 < 20$	$-12 < -8$	$-8 > -20$

Rule 1 states that when each side of an inequality is multiplied by a positive number, the inequality symbol remains the same. However, Rule 2 states that when each side of an inequality is multiplied by a negative number, the inequality symbol must be reversed. Because division is defined in terms of multiplication, when each side of an inequality is divided by a *positive* number, the inequality symbol remains the same. But when each side of an inequality is divided by a *negative* number, the inequality symbol must be reversed.

The Multiplication Property of Inequalities is also true for the symbols \leq and \geq.

HOW TO Solve: $-3x > 9$

Write the solution set in interval notation.

$$-3x > 9$$

$$\frac{-3x}{-3} < \frac{9}{-3}$$
- Divide each side of the inequality by the coefficient -3. Because -3 is a negative number, the inequality symbol must be reversed.

$$x < -3$$

The solution set is $(-\infty, -3)$.

TAKE NOTE

Each side of the inequality is divided *by* a negative number; the inequality symbol must be reversed.

HOW TO Solve: $3x + 2 < -4$
Write the solution set in set-builder notation.

$$3x + 2 < -4$$
$$3x < -6$$
$$\frac{3x}{3} < \frac{-6}{3}$$
$$x < -2$$

- Subtract 2 from each side of the inequality.
- Divide each side of the inequality by the coefficient 3.

The solution set is $\{x \mid x < -2\}$.

HOW TO Solve: $2x - 9 > 4x + 5$
Write the solution set in set-builder notation.

$$2x - 9 > 4x + 5$$
$$-2x - 9 > 5$$
$$-2x > 14$$
$$\frac{-2x}{-2} < \frac{14}{-2}$$
$$x < -7$$

- Subtract $4x$ from each side of the inequality.
- Add 9 to each side of the inequality.
- Divide each side of the inequality by the coefficient -2. Reverse the inequality symbol.

The solution set is $\{x \mid x < -7\}$.

HOW TO Solve: $5(x - 2) \geq 9x - 3(2x - 4)$
Write the solution set in interval notation.

$$5(x - 2) \geq 9x - 3(2x - 4)$$
$$5x - 10 \geq 9x - 6x + 12$$
$$5x - 10 \geq 3x + 12$$
$$2x - 10 \geq 12$$
$$2x \geq 22$$
$$\frac{2x}{2} \geq \frac{22}{2}$$
$$x \geq 11$$

- Use the Distributive Property to remove parentheses.
- Subtract $3x$ from each side of the inequality.
- Add 10 to each side of the inequality.
- Divide each side of the inequality by the coefficient 2.

The solution set is $[11, \infty)$.

Example 1

Solve and graph the solution set:
$x + 3 > 4x + 6$
Write the solution set in set-builder notation.

Solution

$$x + 3 > 4x + 6$$
$$-3x + 3 > 6$$
$$-3x > 3$$
$$\frac{-3x}{-3} < \frac{3}{-3}$$
$$x < -1$$

- Subtract $4x$ from each side.
- Subtract 3 from each side.
- Divide each side by -3.

The solution set is $\{x \mid x < -1\}$.

You Try It 1

Solve and graph the solution set:
$2x - 1 < 6x + 7$
Write the solution set in set-builder notation.

Your solution

Solution on p. S5

Example 2

Solve: $3x - 5 \le 3 - 2(3x + 1)$
Write the solution set in interval notation.

Solution

$3x - 5 \le 3 - 2(3x + 1)$
$3x - 5 \le 3 - 6x - 2$
$3x - 5 \le 1 - 6x$
$9x - 5 \le 1$
$\qquad 9x \le 6$
$\qquad \dfrac{9x}{9} \le \dfrac{6}{9}$
$\qquad\; x \le \dfrac{2}{3}$

$\left(-\infty, \dfrac{2}{3}\right]$

You Try It 2

Solve: $5x - 2 \le 4 - 3(x - 2)$
Write the solution set in interval notation.

Your solution

Solution on p. S5

Objective B **To solve a compound inequality**

A **compound inequality** is formed by joining two inequalities with a connective word such as *and* or *or*. The inequalities at the right are compound inequalities.

$2x < 4$ and $3x - 2 > -8$

$2x + 3 > 5$ or $x + 2 < 5$

The solution set of a compound inequality with the connective word *and* is the set of all elements that appear in the solution sets of both inequalities. Therefore, it is the intersection of the solution sets of the two inequalities.

HOW TO Solve: $2x < 6$ and $3x + 2 > -4$

$\quad 2x < 6 \quad$ and $\quad 3x + 2 > -4$
$\qquad x < 3 \qquad\qquad\quad 3x > -6$ • Solve each inequality.
$\{x \,|\, x < 3\} \qquad\qquad\quad x > -2$
$\qquad\qquad\qquad\qquad \{x \,|\, x > -2\}$

The solution set of a compound inequality with *and* is the intersection of the solution sets of the two inequalities.

$\{x \,|\, x < 3\} \cap \{x \,|\, x > -2\} = \{x \,|\, -2 < x < 3\}$ or $(-2, 3)$.

HOW TO Solve: $-3 < 2x + 1 < 5$

This inequality is equivalent to the compound inequality $-3 < 2x + 1$ and $2x + 1 < 5$.

$-3 < 2x + 1 \quad$ and $\quad 2x + 1 < 5$
$-4 < 2x \qquad\qquad\qquad 2x < 4$ • Solve each inequality.
$-2 < x \qquad\qquad\qquad\;\; x < 2$
$\{x \,|\, x > -2\} \qquad\qquad \{x \,|\, x < 2\}$

The solution set of a compound inequality with *and* is the intersection of the solution sets of the two inequalities.

$\{x \,|\, x > -2\} \cap \{x \,|\, x < 2\} = \{x \,|\, -2 < x < 2\}$ or $(-2, 2)$.

There is an alternative method for solving the inequality in the last example.

HOW TO Solve: $-3 < 2x + 1 < 5$

$$-3 < 2x + 1 < 5$$
$$-3 - 1 < 2x + 1 - 1 < 5 - 1$$
$$-4 < 2x < 4$$
$$\frac{-4}{2} < \frac{2x}{2} < \frac{4}{2}$$
$$-2 < x < 2$$

- Subtract 1 from each of the three parts of the inequality.

- Divide each of the three parts of the inequality by the coefficient 2.

The solution set is $\{x \mid -2 < x < 2\}$ or $(-2, 2)$.

The solution set of a compound inequality with the connective word *or* is the union of the solution sets of the two inequalities.

HOW TO Solve: $2x + 3 > 7$ or $4x - 1 < 3$

$$2x + 3 > 7 \quad \text{or} \quad 4x - 1 < 3$$
$$2x > 4 \qquad\qquad 4x < 4$$
$$x > 2 \qquad\qquad x < 1$$
$$\{x \mid x > 2\} \qquad \{x \mid x < 1\}$$

- Solve each inequality.

Find the union of the solution sets.

$$\{x \mid x > 2\} \cup \{x \mid x < 1\} = \{x \mid x > 2 \text{ or } x < 1\} \text{ or } (-\infty, 1) \cup (2, \infty).$$

Example 3

Solve: $1 < 3x - 5 < 4$
Write the solution set in interval notation.

Solution

$$1 < 3x - 5 < 4$$
$$1 + 5 < 3x - 5 + 5 < 4 + 5$$
$$6 < 3x < 9$$
$$\frac{6}{3} < \frac{3x}{3} < \frac{9}{3}$$
$$2 < x < 3$$
$$(2, 3)$$

- Add 5 to each of the three parts.

- Divide each of the three parts by 3.

You Try It 3

Solve: $-2 \le 5x + 3 \le 13$
Write the solution set in interval notation.

Your solution

Example 4

Solve: $11 - 2x > -3$ and $7 - 3x < 4$
Write the solution set in set-builder notation.

Solution

$$11 - 2x > -3 \quad \text{and} \quad 7 - 3x < 4$$
$$-2x > -14 \qquad\qquad -3x < -3$$
$$x < 7 \qquad\qquad\quad x > 1$$
$$\{x \mid x < 7\} \qquad \{x \mid x > 1\}$$
$$\{x \mid x < 7\} \cap \{x \mid x > 1\} = \{x \mid 1 < x < 7\}$$

You Try It 4

Solve: $2 - 3x > 11$ or $5 + 2x > 7$
Write the solution set in set-builder notation.

Your solution

Solutions on p. S5

Copyright © Houghton Mifflin Company. All rights reserved.

> Objective C **To solve application problems**

Example 5

A cellular phone company advertises two pricing plans. The first is $19.95 per month with 20 free minutes and $.39 per minute thereafter. The second is $23.95 per month with 20 free minutes and $.30 per minute thereafter. How many minutes can you talk per month for the first plan to cost less than the second?

Strategy

To find the number of minutes, write and solve an inequality using N to represent the number of minutes. Then $N - 20$ is the number of minutes for which you are charged after the first free 20 min.

Solution

Cost of first plan $<$ cost of second plan
$$19.95 + 0.39(N - 20) < 23.95 + 0.30(N - 20)$$
$$19.95 + 0.39N - 7.8 < 23.95 + 0.30N - 6$$
$$12.15 + 0.39N < 17.95 + 0.30N$$
$$12.15 + 0.09N < 17.95$$
$$0.09N < 5.8$$
$$N < 64.\overline{4}$$

The first plan costs less if you talk less than 65 min.

You Try It 5

The base of a triangle is 12 in. and the height is $(x + 2)$ in. Express as an integer the maximum height of the triangle when the area is less than 50 in².

Your strategy

Your solution

Example 6

Find three consecutive positive odd integers whose sum is between 27 and 51.

Strategy

To find the three integers, write and solve a compound inequality using n to represent the first odd integer.

Solution

Lower limit $<$ sum $<$ upper limit
of the sum of the sum
$$27 < n + (n + 2) + (n + 4) < 51$$
$$27 < 3n + 6 < 51$$
$$27 - 6 < 3n + 6 - 6 < 51 - 6$$
$$21 < 3n < 45$$
$$\frac{21}{3} < \frac{3n}{3} < \frac{45}{3}$$
$$7 < n < 15$$

The three odd integers are 9, 11, and 13; or 11, 13, and 15; or 13, 15, and 17.

You Try It 6

An average score of 80 to 89 in a history course receives a B. Luisa Montez has grades of 72, 94, 83, and 70 on four exams. Find the range of scores on the fifth exam that will give Luisa a B for the course.

Your strategy

Your solution

Solutions on p. S5

2.4 Exercises

Objective A **To solve an inequality in one variable**

1. State the Addition Property of Inequalities and give numerical examples of its use.

2. State the Multiplication Property of Inequalities and give numerical examples of its use.

3. Which numbers are solutions of the inequality $x + 7 \leq -3$?
 a. -17 **b.** 8 **c.** -10 **d.** 0

4. Which numbers are solutions of the inequality $2x - 1 > 5$?
 a. 6 **b.** -4 **c.** 3 **d.** 5

For Exercises 5 to 31, solve. Write the solution in set-builder notation. For Exercises 5 to 10, graph the solution set.

5. $x - 3 < 2$

6. $x + 4 \geq 2$

7. $4x \leq 8$

8. $6x > 12$

9. $-2x > 8$

10. $-3x \leq -9$

11. $3x - 1 > 2x + 2$

12. $5x + 2 \geq 4x - 1$

13. $2x - 1 > 7$

14. $3x + 2 < 8$

15. $5x - 2 \leq 8$

16. $4x + 3 \leq -1$

17. $6x + 3 > 4x - 1$

18. $7x + 4 < 2x - 6$

19. $8x + 1 \geq 2x + 13$

20. $5x - 4 < 2x + 5$

21. $4 - 3x < 10$

22. $2 - 5x > 7$

23. $7 - 2x \geq 1$

24. $3 - 5x \leq 18$

25. $-3 - 4x > -11$

26. $-2 - x < 7$

27. $4x - 2 < x - 11$

28. $6x + 5 \leq x - 10$

29. $x + 7 \geq 4x - 8$

30. $3x + 1 \leq 7x - 15$

31. $3x + 2 \leq 7x + 4$

For Exercises 32 to 47, solve. Write the solution in interval notation.

32. $3x - 5 \geq -2x + 5$

33. $\dfrac{3}{5}x - 2 < \dfrac{3}{10} - x$

34. $\dfrac{5}{6}x - \dfrac{1}{6} < x - 4$

35. $\dfrac{2}{3}x - \dfrac{3}{2} < \dfrac{7}{6} - \dfrac{1}{3}x$

36. $\dfrac{7}{12}x - \dfrac{3}{2} < \dfrac{2}{3}x + \dfrac{5}{6}$

37. $\dfrac{1}{2}x - \dfrac{3}{4} < \dfrac{7}{4}x - 2$

38. $6 - 2(x - 4) \leq 2x + 10$

39. $4(2x - 1) > 3x - 2(3x - 5)$

40. $2(1 - 3x) - 4 > 10 + 3(1 - x)$

41. $2 - 5(x + 1) \geq 3(x - 1) - 8$

42. $2 - 2(7 - 2x) < 3(3 - x)$

43. $3 + 2(x + 5) \geq x + 5(x + 1) + 1$

44. $10 - 13(2 - x) < 5(3x - 2)$

45. $3 - 4(x + 2) \leq 6 + 4(2x + 1)$

46. $3x - 2(3x - 5) \leq 2 - 5(x - 4)$

47. $12 - 2(3x - 2) \geq 5x - 2(5 - x)$

Objective B **To solve a compound inequality**

48. **a.** Which set operation is used when a compound inequality is combined with *or*?
 b. Which set operation is used when a compound inequality is combined with *and*?

49. Explain why writing $-3 > x > 4$ does not make sense.

For Exercises 50 to 63, solve. Write the solution set in interval notation.

50. $3x < 6$ and $x + 2 > 1$

51. $x - 3 \leq 1$ and $2x \geq -4$

52. $x + 2 \geq 5$ or $3x \leq 3$

53. $2x < 6$ or $x - 4 > 1$

54. $-2x > -8$ and $-3x < 6$

55. $\dfrac{1}{2}x > -2$ and $5x < 10$

56. $\dfrac{1}{3}x < -1$ or $2x > 0$

57. $\dfrac{2}{3}x > 4$ or $2x < -8$

58. $x + 4 \geq 5$ and $2x \geq 6$

59. $3x < -9$ and $x - 2 < 2$

60. $-5x > 10$ and $x + 1 > 6$

61. $2x - 3 > 1$ and $3x - 1 < 2$

62. $7x < 14$ and $1 - x < 4$

63. $4x + 1 < 5$ and $4x + 7 > -1$

For Exercises 64 to 83, solve. Write the solution set in set-builder notation.

64. $3x + 7 < 10$ or $2x - 1 > 5$

65. $6x - 2 < -14$ or $5x + 1 > 11$

66. $-5 < 3x + 4 < 16$

67. $5 < 4x - 3 < 21$

68. $0 < 2x - 6 < 4$

69. $-2 < 3x + 7 < 1$

70. $4x - 1 > 11$ or $4x - 1 \leq -11$

71. $3x - 5 > 10$ or $3x - 5 < -10$

72. $9x - 2 < 7$ and $3x - 5 > 10$

73. $8x + 2 \leq -14$ and $4x - 2 > 10$

74. $3x - 11 < 4$ or $4x + 9 \geq 1$

75. $5x + 12 \geq 2$ or $7x - 1 \leq 13$

76. $-6 \le 5x + 14 \le 24$

77. $3 \le 7x - 14 \le 31$

78. $3 - 2x > 7$ and $5x + 2 > -18$

79. $1 - 3x < 16$ and $1 - 3x > -16$

80. $5 - 4x > 21$ or $7x - 2 > 19$

81. $6x + 5 < -1$ or $1 - 2x < 7$

82. $3 - 7x \le 31$ and $5 - 4x > 1$

83. $9 - x \ge 7$ and $9 - 2x < 3$

Objective C **To solve application problems**

84. **Integers** Five times the difference between a number and two is greater than the quotient of two times the number and three. Find the smallest integer that will satisfy the inequality.

85. **Integers** Two times the difference between a number and eight is less than or equal to five times the sum of the number and four. Find the smallest number that will satisfy the inequality.

86. **Geometry** The length of a rectangle is two feet more than four times the width. Express as an integer the maximum width of the rectangle when the perimeter is less than thirty-four feet.

87. **Geometry** The length of a rectangle is five centimeters less than twice the width. Express as an integer the maximum width of the rectangle when the perimeter is less than sixty centimeters.

88. **Telecommunications** In 2003, the computer service America Online offered its customers a rate of $23.90 per month for unlimited use or $4.95 per month with 3 free hours plus $2.50 for each hour thereafter. Express as an integer the maximum number of hours you can use this service per month if the second plan is to cost you less than the first.

89. **Telecommunications** TopPage advertises local paging service for $6.95 per month for up to 400 pages, and $.10 per page thereafter. A competitor advertises service for $3.95 per month for up to 400 pages and $.15 per page thereafter. For what number of pages per month is the TopPage plan less expensive?

90. **Consumerism** Suppose PayRite Rental Cars rents compact cars for $32 per day with unlimited mileage, and Otto Rentals offers compact cars for $19.99 per day but charges $.19 for each mile beyond 100 mi driven per day. You want to rent a car for one week. How many miles can you drive during the week if Otto Rentals is to be less expensive than PayRite?

91. **Consumerism** During a weekday, to call a city 40 mi away from a certain pay phone costs $.70 for the first 3 min and $.15 for each additional minute. If you use a calling card, there is a $.35 fee and then the rates are $.196 for the first minute and $.126 for each additional minute. How long must a call be if it is to be cheaper to pay with coins rather than a calling card?

92. **Temperature** The temperature range for a week was between 14°F and 77°F. Find the temperature range in Celsius degrees. $F = \frac{9}{5}C + 32$

93. **Temperature** The temperature range for a week in a mountain town was between 0°C and 30°C. Find the temperature range in Fahrenheit degrees. $C = \frac{5(F - 32)}{9}$

94. **Compensation** You are a sales account executive earning $1200 per month plus 6% commission on the amount of sales. Your goal is to earn a minimum of $6000 per month. What amount of sales will enable you to earn $6000 or more per month?

95. **Compensation** George Stoia earns $1000 per month plus 5% commission on the amount of sales. George's goal is to earn a minimum of $3200 per month. What amount of sales will enable George to earn $3200 or more per month?

96. **Banking** Heritage National Bank offers two different checking accounts. The first charges $3 per month, and $.50 per check after the first 10 checks. The second account charges $8 per month with unlimited check writing. How many checks can be written per month if the first account is to be less expensive than the second account?

97. **Banking** Glendale Federal Bank offers a checking account to small businesses. The charge is $8 per month plus $.12 per check after the first 100 checks. A competitor is offering an account for $5 per month plus $.15 per check after the first 100 checks. If a business chooses the first account, how many checks does the business write monthly if it is assumed that the first account will cost less than the competitor's account?

98. **Education** An average score of 90 or above in a history class receives an A grade. You have scores of 95, 89, and 81 on three exams. Find the range of scores on the fourth exam that will give you an A grade for the course.

99. **Education** An average of 70 to 79 in a mathematics class receives a C grade. A student has scores of 56, 91, 83, and 62 on four tests. Find the range of scores on the fifth test that will give the student a C for the course.

100. **Integers** Find four consecutive integers whose sum is between 62 and 78.

101. **Integers** Find three consecutive even integers whose sum is between 30 and 51.

APPLYING THE CONCEPTS

102. Let $-2 \leq x \leq 3$ and $a \leq 2x + 1 \leq b$.
 a. Find the largest possible value of a.
 b. Find the smallest possible value of b.

103. Determine whether the following statements are always true, sometimes true, or never true.
 a. If $a > b$, then $-a < -b$.
 b. If $a < b$ and $a \neq 0, b \neq 0$, then $\frac{1}{a} < \frac{1}{b}$.
 c. When dividing both sides of an inequality by an integer, we must reverse the inequality symbol.
 d. If $a < 1$, then $a^2 < a$.
 e. If $a < b < 0$ and $c < d < 0$, then $ac > bd$.

104. The following is offered as the solution of $2 + 3(2x - 4) < 6x + 5$.

$$2 + 3(2x - 4) < 6x + 5$$

$2 + 6x - 12 < 6x + 5$	• **Use the Distributive Property.**
$6x - 10 < 6x + 5$	• **Simplify.**
$6x - 6x - 10 < 6x - 6x + 5$	• **Subtract 6x from each side.**
$-10 < 5$	

Because $-10 < 5$ is a true inequality, the solution set is all real numbers.

If this is correct, so state. If it is not correct, explain the incorrect step and supply the correct answer.

2.5 Absolute Value Equations and Inequalities

Objective A **To solve an absolute value equation**

Study Tip

Before the class meeting in which your professor begins a new section, you should read each objective statement for that section. Next, browse through the objective material. The purpose of browsing through the material is to set the stage for your brain to accept and organize new information when it is presented to you. See *AIM for Success,* page xxix.

The *absolute value* of a number is its distance from zero on the number line. Distance is always a positive number or zero. Therefore, the absolute value of a number is always a positive number or zero.

The distance from 0 to 3 or from 0 to -3 is 3 units.

$$|3| = 3 \qquad |-3| = 3$$

Absolute value can be used to represent the distance between any two points on the number line. The **distance between two points** on the number line is the absolute value of the difference between the coordinates of the two points.

The distance between point a and point b is given by $|b - a|$.

The distance between 4 and -3 on the number line is 7 units. Note that the order in which the coordinates are subtracted does not affect the distance.

$$\text{Distance} = |-3 - 4| \qquad \text{Distance} = |4 - (-3)|$$
$$= |-7| \qquad\qquad = |7|$$
$$= 7 \qquad\qquad = 7$$

For any two numbers a and b, $|b - a| = |a - b|$.

An equation containing an absolute value symbol is called an **absolute value equation.** Here are three examples.

$$|x| = 3 \qquad |x + 2| = 8 \qquad |3x - 4| = 5x - 9$$

Solutions of an Absolute Value Equation

If $a \geq 0$ and $|x| = a$, then $x = a$ or $x = -a$.

For instance, given $|x| = 3$, then $x = 3$ or $x = -3$ because $|3| = 3$ and $|-3| = 3$. We can solve this equation as follows:

$$|x| = 3$$
$$x = 3 \qquad x = -3$$

• Remove the absolute value sign from $|x|$ and let x equal 3 and the opposite of 3.

Check:

$$\begin{array}{c|c} |x| = 3 & |x| = 3 \\ \hline |3| \,|\, 3 & |-3| \,|\, 3 \\ 3 = 3 & 3 = 3 \end{array}$$

The solutions are 3 and -3.

HOW TO Solve: $|x + 2| = 8$

$$|x + 2| = 8$$

$x + 2 = 8$	$x + 2 = -8$
$x = 6$	$x = -10$

- Remove the absolute value sign and rewrite as two equations.
- Solve each equation.

Check:

| $|x + 2| = 8$ | | $|x + 2| = 8$ | |
|:---:|:---:|:---:|:---:|
| $|6 + 2|$ | 8 | $|-10 + 2|$ | 8 |
| $|8|$ | 8 | $|-8|$ | 8 |
| $8 = 8$ | | $8 = 8$ | |

The solutions are 6 and -10.

HOW TO Solve: $|5 - 3x| - 8 = -4$

$$|5 - 3x| - 8 = -4$$
$$|5 - 3x| = 4$$

$5 - 3x = 4$	$5 - 3x = -4$
$-3x = -1$	$-3x = -9$
$x = \dfrac{1}{3}$	$x = 3$

- Solve for the absolute value.
- Remove the absolute value sign and rewrite as two equations.
- Solve each equation.

Check:

| $|5 - 3x| - 8 = -4$ | | $|5 - 3x| - 8 = -4$ | |
|:---:|:---:|:---:|:---:|
| $\left|5 - 3\left(\dfrac{1}{3}\right)\right| - 8$ | -4 | $|5 - 3(3)| - 8$ | -4 |
| $|5 - 1| - 8$ | -4 | $|5 - 9| - 8$ | -4 |
| $4 - 8$ | -4 | $4 - 8$ | -4 |
| $-4 = -4$ | | $-4 = -4$ | |

The solutions are $\dfrac{1}{3}$ and 3.

Example 1

Solve: $|2 - x| = 12$

Solution

$$|2 - x| = 12$$

$2 - x = 12$	$2 - x = -12$	
$-x = 10$	$-x = -14$	• Subtract 2.
$x = -10$	$x = 14$	• Multiply by -1.

The solutions are -10 and 14.

Example 2

Solve: $|2x| = -4$

Solution

$$|2x| = -4$$

There is no solution to this equation because the absolute value of a number must be nonnegative.

You Try It 1

Solve: $|2x - 3| = 5$

Your solution

You Try It 2

Solve: $|x - 3| = -2$

Your solution

Solutions on p. S5

Example 3

Solve: $3 - |2x - 4| = -5$

Solution

$$3 - |2x - 4| = -5$$
$$-|2x - 4| = -8 \qquad \bullet \text{ Subtract 3.}$$
$$|2x - 4| = 8 \qquad \bullet \text{ Multiply by } -1.$$

$$\begin{array}{ll} 2x - 4 = 8 & 2x - 4 = -8 \\ 2x = 12 & 2x = -4 \\ x = 6 & x = -2 \end{array}$$

The solutions are 6 and -2.

You Try It 3

Solve: $5 - |3x + 5| = 3$

Your solution

Solution on p. S5

Objective B **To solve an absolute value inequality**

Recall that absolute value represents the distance between two points. For example, the solutions of the absolute value equation $|x - 1| = 3$ are the numbers whose distance from 1 is 3. Therefore, the solutions are -2 and 4. An **absolute value inequality** is an inequality that contains an absolute value symbol.

The solutions of the absolute value inequality $|x - 1| < 3$ are the numbers whose distance from 1 is less than 3. Therefore, the solutions are the numbers greater than -2 and less than 4. The solution set is $\{x | -2 < x < 4\}$.

To solve an absolute value inequality of the form $|ax + b| < c$, solve the equivalent compound inequality $-c < ax + b < c$.

HOW TO Solve: $|3x - 1| < 5$

$$|3x - 1| < 5$$
$$-5 < 3x - 1 < 5 \qquad \bullet \text{ Solve the equivalent compound inequality.}$$
$$-5 + 1 < 3x - 1 + 1 < 5 + 1$$
$$-4 < 3x < 6$$
$$\frac{-4}{3} < \frac{3x}{3} < \frac{6}{3}$$
$$-\frac{4}{3} < x < 2$$

The solution set is $\left\{x \left| -\frac{4}{3} < x < 2 \right.\right\}$.

TAKE NOTE

In this objective, we will write all solution sets in set-builder notation.

The solutions of the absolute value inequality $|x + 1| > 2$ are the numbers whose distance from -1 is greater than 2. Therefore, the solutions are the numbers that are less than -3 or greater than 1. The solution set of $|x + 1| > 2$ is $\{x | x < -3 \text{ or } x > 1\}$.

To solve an absolute value inequality of the form $|ax + b| > c$, solve the equivalent compound inequality $ax + b < -c$ or $ax + b > c$.

HOW TO Solve: $|3 - 2x| > 1$

$$
\begin{array}{lll}
3 - 2x < -1 & \text{or} & 3 - 2x > 1 \\
\quad -2x < -4 & & \quad -2x > -2 \\
\quad\quad x > 2 & & \quad\quad x < 1 \\
\{x \mid x > 2\} & & \{x \mid x < 1\}
\end{array}
$$

• **Solve each inequality.**

The solution of a compound inequality with *or* is the union of the solution sets of the two inequalities.

$$\{x \mid x > 2\} \cup \{x \mid x < 1\} = \{x \mid x > 2 \text{ or } x < 1\}$$

The rules for solving these absolute value inequalities are summarized below.

Solutions of Absolute Value Inequalities

To solve an absolute value inequality of the form $|ax + b| < c$, $c > 0$, solve the equivalent compound inequality $-c < ax + b < c$.

To solve an absolute value inequality of the form $|ax + b| > c$, solve the equivalent compound inequality $ax + b < -c$ or $ax + b > c$.

Example 4 Solve: $|4x - 3| < 5$

Solution Solve the equivalent compound inequality.

$$
\begin{aligned}
-5 &< 4x - 3 < 5 \\
-5 + 3 &< 4x - 3 + 3 < 5 + 3 \\
-2 &< 4x < 8 \\
\frac{-2}{4} &< \frac{4x}{4} < \frac{8}{4} \\
-\frac{1}{2} &< x < 2
\end{aligned}
$$

$$\left\{ x \,\middle|\, -\frac{1}{2} < x < 2 \right\}$$

You Try It 4 Solve: $|3x + 2| < 8$

Your solution

Example 5 Solve: $|x - 3| < 0$

Solution The absolute value of a number is greater than or equal to zero, since it measures the number's distance from zero on the number line. Therefore, the solution set of $|x - 3| < 0$ is the empty set.

You Try It 5 Solve: $|3x - 7| < 0$

Your solution

Solutions on p. S6

Example 6 Solve: $|x + 4| > -2$

Solution The absolute value of a number is greater than or equal to zero. Therefore, the solution set of $|x + 4| > -2$ is the set of real numbers.

You Try It 6 Solve: $|2x + 7| \geq -1$

Your solution

Example 7 Solve: $|2x - 1| > 7$

Solution Solve the equivalent compound inequality.

$$2x - 1 < -7 \quad \text{or} \quad 2x - 1 > 7$$
$$2x < -6 \qquad\qquad 2x > 8$$
$$x < -3 \qquad\qquad x > 4$$
$$\{x \mid x < -3\} \qquad \{x \mid x > 4\}$$
$$\{x \mid x < -3\} \cup \{x \mid x > 4\}$$
$$= \{x \mid x < -3 \text{ or } x > 4\}$$

You Try It 7 Solve: $|5x + 3| > 8$

Your solution

Solutions on p. S6

Objective C **To solve application problems**

piston

The **tolerance of a component,** or part, is the amount by which it is acceptable for the component to vary from a given measurement. For example, the diameter of a piston may vary from the given measurement of 9 cm by 0.001 cm. This is written 9 cm \pm 0.001 cm and is read "9 centimeters plus or minus 0.001 centimeter." The maximum diameter, or **upper limit,** of the piston is 9 cm + 0.001 cm = 9.001 cm. The minimum diameter, or **lower limit,** is 9 cm − 0.001 cm = 8.999 cm.

The lower and upper limits of the diameter of the piston could also be found by solving the absolute value inequality $|d - 9| \leq 0.001$, where d is the diameter of the piston.

$$|d - 9| \leq 0.001$$
$$-0.001 \leq d - 9 \leq 0.001$$
$$-0.001 + 9 \leq d - 9 + 9 \leq 0.001 + 9$$
$$8.999 \leq d \leq 9.001$$

The lower and upper limits of the diameter of the piston are 8.999 cm and 9.001 cm.

Example 8

The diameter of a piston for an automobile is $3\frac{5}{16}$ in. with a tolerance of $\frac{1}{64}$ in. Find the lower and upper limits of the diameter of the piston.

Strategy

To find the lower and upper limits of the diameter of the piston, let d represent the diameter of the piston, T the tolerance, and L the lower and upper limits of the diameter. Solve the absolute value inequality $|L - d| \le T$ for L.

Solution

$$|L - d| \le T$$

$$\left|L - 3\frac{5}{16}\right| \le \frac{1}{64}$$

$$-\frac{1}{64} \le L - 3\frac{5}{16} \le \frac{1}{64}$$

$$-\frac{1}{64} + 3\frac{5}{16} \le L - 3\frac{5}{16} + 3\frac{5}{16} \le \frac{1}{64} + 3\frac{5}{16}$$

$$3\frac{19}{64} \le L \le 3\frac{21}{64}$$

The lower and upper limits of the diameter of the piston are $3\frac{19}{64}$ in. and $3\frac{21}{64}$ in.

You Try It 8

A machinist must make a bushing that has a tolerance of 0.003 in. The diameter of the bushing is 2.55 in. Find the lower and upper limits of the diameter of the bushing.

Your strategy

Your solution

Solution on p. S6

2.5 Exercises

Objective A **To solve an absolute value equation**

1. Is 2 a solution of $|x - 8| = 6$?

2. Is -2 a solution of $|2x - 5| = 9$?

3. Is -1 a solution of $|3x - 4| = 7$?

4. Is 1 a solution of $|6x - 1| = -5$?

For Exercises 5 to 64, solve.

5. $|x| = 7$

6. $|a| = 2$

7. $|b| = 4$

8. $|c| = 12$

9. $|-y| = 6$

10. $|-t| = 3$

11. $|-a| = 7$

12. $|-x| = 3$

13. $|x| = -4$

14. $|y| = -3$

15. $|-t| = -3$

16. $|-y| = -2$

17. $|x + 2| = 3$

18. $|x + 5| = 2$

19. $|y - 5| = 3$

20. $|y - 8| = 4$

21. $|a - 2| = 0$

22. $|a + 7| = 0$

23. $|x - 2| = -4$

24. $|x + 8| = -2$

25. $|3 - 4x| = 9$

26. $|2 - 5x| = 3$

27. $|2x - 3| = 0$

28. $|5x + 5| = 0$

29. $|3x - 2| = -4$

30. $|2x + 5| = -2$

31. $|x - 2| - 2 = 3$

32. $|x - 9| - 3 = 2$

33. $|3a + 2| - 4 = 4$

34. $|2a + 9| + 4 = 5$

35. $|2 - y| + 3 = 4$

36. $|8 - y| - 3 = 1$

37. $|2x - 3| + 3 = 3$

38. $|4x - 7| - 5 = -5$

39. $|2x - 3| + 4 = -4$

40. $|3x - 2| + 1 = -1$

41. $|6x - 5| - 2 = 4$

42. $|4b + 3| - 2 = 7$

43. $|3t + 2| + 3 = 4$

44. $|5x - 2| + 5 = 7$

45. $3 - |x - 4| = 5$

46. $2 - |x - 5| = 4$

47. $8 - |2x - 3| = 5$

48. $8 - |3x + 2| = 3$

49. $|2 - 3x| + 7 = 2$

50. $|1 - 5a| + 2 = 3$

51. $|8 - 3x| - 3 = 2$

52. $|6 - 5b| - 4 = 3$

53. $|2x - 8| + 12 = 2$

54. $|3x - 4| + 8 = 3$

55. $2 + |3x - 4| = 5$

56. $5 + |2x + 1| = 8$

57. $5 - |2x + 1| = 5$

58. $3 - |5x + 3| = 3$

59. $6 - |2x + 4| = 3$

60. $8 - |3x - 2| = 5$

61. $8 - |1 - 3x| = -1$

62. $3 - |3 - 5x| = -2$

63. $5 + |2 - x| = 3$

64. $6 + |3 - 2x| = 2$

Objective B **To solve an absolute value inequality**

For Exercises 65 to 94, solve.

65. $|x| > 3$

66. $|x| < 5$

67. $|x + 1| > 2$

68. $|x - 2| > 1$

69. $|x - 5| \leq 1$

70. $|x - 4| \leq 3$

71. $|2 - x| \geq 3$

72. $|3 - x| \geq 2$

73. $|2x + 1| < 5$

74. $|3x - 2| < 4$ **75.** $|5x + 2| > 12$ **76.** $|7x - 1| > 13$

77. $|4x - 3| \leq -2$ **78.** $|5x + 1| \leq -4$ **79.** $|2x + 7| > -5$

80. $|3x - 1| > -4$ **81.** $|4 - 3x| \geq 5$ **82.** $|7 - 2x| > 9$

83. $|5 - 4x| \leq 13$ **84.** $|3 - 7x| < 17$ **85.** $|6 - 3x| \leq 0$

86. $|10 - 5x| \geq 0$ **87.** $|2 - 9x| > 20$ **88.** $|5x - 1| < 16$

89. $|2x - 3| + 2 < 8$ **90.** $|3x - 5| + 1 < 7$ **91.** $|2 - 5x| - 4 > -2$

92. $|4 - 2x| - 9 > -3$ **93.** $8 - |2x - 5| < 3$ **94.** $12 - |3x - 4| > 7$

Objective C **To solve application problems**

95. **Mechanics** The diameter of a bushing is 1.75 in. The bushing has a tolerance of 0.008 in. Find the lower and upper limits of the diameter of the bushing.

← 1.75 in. →

96. **Mechanics** A machinist must make a bushing that has a tolerance of 0.004 in. The diameter of the bushing is 3.48 in. Find the lower and upper limits of the diameter of the bushing.

97. **Appliances** An electric motor is designed to run on 220 volts plus or minus 25 volts. Find the lower and upper limits of voltage on which the motor will run.

98. Computers A power strip is utilized on a computer to prevent the loss of programming by electrical surges. The power strip is designed to allow 110 volts plus or minus 16.5 volts. Find the lower and upper limits of voltage to the computer.

99. Automobiles A piston rod for an automobile is $9\frac{5}{8}$ in. long with a tolerance of $\frac{1}{32}$ in. Find the lower and upper limits of the length of the piston rod.

100. Automobiles A piston rod for an automobile is $9\frac{3}{8}$ in. long with a tolerance of $\frac{1}{64}$ in. Find the lower and upper limits of the length of the piston rod.

Electronics The tolerance of the resistors used in electronics is given as a percent. Use your calculator for Exercises 101 to 104.

101. Find the lower and upper limits of a 29,000-ohm resistor with a 2% tolerance.

102. Find the lower and upper limits of a 15,000-ohm resistor with a 10% tolerance.

103. Find the lower and upper limits of a 25,000-ohm resistor with a 5% tolerance.

104. Find the lower and upper limits of a 56-ohm resistor with a 5% tolerance.

APPLYING THE CONCEPTS

105. For what values of the variable is the equation true? Write the solution set in set-builder notation.
 a. $|x + 3| = x + 3$ **b.** $|a - 4| = 4 - a$

106. Write an absolute value inequality to represent all real numbers within 5 units of 2.

107. Replace the question mark with \leq, \geq, or $=$.
 a. $|x + y|\ ?\ |x| + |y|$ **b.** $|x - y|\ ?\ |x| - |y|$
 c. $||x| - |y||\ ?\ |x| - |y|$ **d.** $\left|\dfrac{x}{y}\right|\ ?\ \dfrac{|x|}{|y|}, y \neq 0$
 e. $|xy|\ ?\ |x||y|$

108. Let $|x| \leq 2$ and $|3x - 2| \leq a$. Find the smallest possible value of a.

Focus on Problem Solving

Understand the Problem

The first of the four steps that Polya advocated to solve problems is to *understand the problem*. This aspect of problem solving is frequently not given enough attention. There are various exercises that you can try to achieve a good understanding of a problem. Some of these are stated in the Focus on Problem Solving in the chapter entitled "Review of Real Numbers" and are reviewed here.

- Try to restate the problem in your own words.
- Determine what is known about this type of problem.
- Determine what information is given.
- Determine what information is unknown.
- Determine whether any of the information given is unnecessary.
- Determine the goal.

To illustrate this aspect of problem solving, consider the following famous ancient limerick.

> As I was going to St. Ives,
> I met a man with seven wives;
> Each wife had seven sacks,
> Each sack had seven cats,
> Each cat had seven kits:
> Kits, cats, sacks, and wives,
> How many were going to St. Ives?

To answer the question in the limerick, we will ask and answer some of the questions listed above.

1. What is the goal?

The goal is to determine how many were going to St. Ives. (We know this from reading the last line of the limerick.)

2. What information is necessary and what information is unnecessary?

The first line indicates that the poet was going to St. Ives. The next five lines describe a man the poet *met* on the way. This information is irrelevant.

The answer to the question, then, is 1. Only the poet was going to St. Ives.

Point of Interest

Try this brain teaser: You have two U.S. coins that add up to $.55. One is not a nickel. What are the two coins?

There are many other examples of the importance, in problem solving, of recognizing *irrelevant information*. One that frequently makes the college circuit is posed in the form of a test. The first line of a 100-question test states, "Read the entire test before you begin." The last line of the test reads, "Choose any *one* question to answer." Many people ignore the information given in the first line and just begin the test, only to find out much later that they did a lot more work than was necessary.

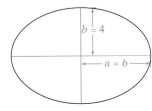

To illustrate another aspect of Polya's first step in the problem-solving process, consider the problem of finding the area of the oval-shaped region (called an *ellipse*) shown in the diagram at the left. This problem can be solved by doing some research to determine *what information is known about this type of problem*. Mathematicians have found a formula for the area of an ellipse. That formula is $A = \pi ab$, where a and b are as shown in the diagram. Therefore, $A = \pi(6)(4) = 24\pi \approx 75.40$ square units. Without the formula, this problem is difficult to solve. With the formula, it is fairly easy.

For Exercises 1 to 5, examine the problem in terms of the first step in Polya's problem-solving method. *Do not solve the problem.*

1. Johanna spent one-third of her allowance on a book. She then spent $5 for a sandwich and iced tea. The cost of the iced tea was one-fifth the cost of the sandwich. Find the cost of the iced tea.

2. A flight from Los Angeles to Boston took 6 h. What was the average speed of the plane?

3. A major league baseball is approximately 5 in. in diameter and is covered with cowhide. Approximately how much cowhide is used to cover 10 baseballs?

4. How many donuts are in seven baker's dozen?

5. The smallest prime number is 2. Twice the difference between the eighth and the seventh prime numbers is two more than the smallest prime number. How large is the smallest prime number?

Projects and Group Activities

Electricity

Since the Industrial Revolution at the turn of the century, technology has been a far-reaching and ever-changing phenomenon. Most of the technological advances that have been made, however, could not have been accomplished without electricity, and mathematics plays an integral part in the science of electricity.

Point of Interest

Ohm's law is named after Georg Simon Ohm (1789–1854), a German physicist whose work contributed to mathematics, acoustics, and the measurement of electrical resistance.

Central to the study of electricity is **Ohm's law,** one part of which states that $V = IR$, where V is voltage, I is current, and R is resistance.

The word *electricity* comes from the same root word as the word *electron.* Electrons are tiny particles in atoms. Each electron has an electric charge, and this is the fundamental cause of electricity. In order to move, electrons need a source of energy—for example, light, heat, pressure, or a chemical reaction. This is where voltage comes in. Basically, **voltage** is a measure of the amount of energy in a flow of electricity. Voltage is measured in units called **volts.**

Current is a measure of how many electrons pass a given point in a fixed amount of time in a flow of electricity. This means that current is a measure of the strength of the flow of electrons through a wire—that is, their speed. Picture a faucet: The more you turn the handle, the more water you get. Similarly, the more current in a wire, the stronger the flow of electricity. Current is measured in **amperes,** often simply called **amps.** A current of 1 ampere would be sufficient to light the bulb in a flashlight.

Point of Interest

The ampere is named after André Marie Ampère (1775–1836), a French physicist and mathematician who formulated Ampère's law, a mathematical description of the magnetic field produced by a current-carrying conductor.

Resistance is a measure of the amount of resistance, or opposition, to the flow of electricity. You might think of resistance as friction. The more friction, the slower the speed. Resistance is measured in **ohms.**

Watts measure the **power** in an electrical system, measuring both the strength and the speed of the flow of electrons. The power in an electrical system is equal to the voltage times the current, written $P = VI$, where P is the power measured in watts, V is voltage measured in volts, and I is current measured in amperes.

In an electrical circuit, a source of electricity, such as a battery or generator, drives electrons through a wire to the part of the machine that produces the output. Because we cannot see electrons, it may help to think of an electrical system as similar to water flowing through a pipe.

The water is like the electrons.

The drawbridge is like the resistance.

The source of energy that drives the water through the pipe is like the voltage.

The amount of water is like the current.

The power of the water as it falls is like the watts.

Here are the formulas introduced above, along with the meaning of each variable and the unit in which it is measured:

> **Formulas for Voltage and Power**
>
> $V = IR$
>
> Voltage $=$ Current \times Resistance
> (volts) (amperes) (ohms)
>
> $P = VI$
>
> Power $=$ Voltage \times Current
> (watts) (volts) (amperes)

HOW TO How many amperes of current pass through a wire in which the voltage is 90 volts and the resistance is 5 ohms?

$V = IR$
$90 = I \cdot 5$

- *V* is the voltage, measured in volts.
 R is the resistance, measured in ohms.

$\dfrac{90}{5} = \dfrac{I \cdot 5}{5}$
$18 = I$

- Solve for *I*, the current, measured in amperes.

18 amperes of current pass through the wire.

HOW TO How much voltage does a 120-watt light bulb with a current of 20 amperes require?

$P = VI$
$120 = V \cdot 20$

- *P* is the power, measured in watts.
 I is the current, measured in amperes.

$\dfrac{120}{20} = \dfrac{V \cdot 20}{20}$
$6 = V$

- Solve for *V*, the voltage, measured in volts.

The light bulb requires 6 volts.

HOW TO Find the power in an electrical system that has a voltage of 120 volts and resistance of 150 ohms.

First solve the formula $V = IR$ for I. This will give us the amperes.
Then solve the formula $P = VI$ for P. This will give us the watts.

$$V = IR \qquad\qquad\qquad P = VI$$
$$120 = I \cdot 150 \qquad\qquad P = 120(0.8)$$
$$\frac{120}{150} = \frac{I \cdot 150}{150} \qquad\qquad P = 96$$
$$0.8 = I$$

The power in the electrical system is 96 watts.

For Exercises 1 to 14, solve.

1. How many volts pass through a wire in which the current is 20 amperes and the resistance is 100 ohms?

2. Find the voltage in a system in which the current is 4.5 amperes and the resistance is 150 ohms.

3. How many amperes of current pass through a wire in which the voltage is 100 volts and the resistance is 10 ohms?

4. A lamp has a resistance of 70 ohms. How much current flows through the lamp when it is connected to a 115-volt circuit? Round to the nearest hundredth.

5. What is the resistance of a semiconductor that passes 0.12 ampere of current when 0.48 volt is applied to it?

6. Find the resistance when the current is 120 amperes and the voltage is 1.5 volts.

7. Determine the power in a light bulb when the current is 10 amperes and the voltage is 12 volts.

8. Find the power in a hand-held dryer that operates from a voltage of 115 volts and draws 2.175 amperes of current.

9. How much current flows through a 110-volt, 850-watt lamp? Round to the nearest hundredth.

10. A heating element in a clothes dryer is rated at 4500 watts. How much current is used by the dryer when the voltage is 240 volts?

Study Tip

Three important features of this text that can be used to prepare for a test are the following:

- Chapter Summary
- Chapter Review Exercises
- Chapter Test

See *AIM for Success,* page xxxii.

11. A miniature lamp pulls 0.08 ampere of current while lighting a 0.5-watt bulb. What voltage battery is needed for the lamp?

12. The power of a car sound system that pulls 15 amperes is 180 watts. Find the voltage of the system.

13. Find the power of a lamp that has a voltage of 160 volts and resistance of 80 ohms.

14. Find the power in an electrical system that has a voltage of 105 volts and resistance of 70 ohms.

Chapter 2 Summary

Key Words	Examples				
An *equation* expresses the equality of two mathematical expressions. [2.1A, p. 57]	$3 + 2 = 5$ $2x - 5 = 4$				
A *conditional equation* is one that is true for at least one value of the variable but not for all values of the variable. An *identity* is an equation that is true for all values of the variable. A *contradiction* is an equation for which no value of the variable produces a true equation. [2.1A, p. 57]	$x + 3 = 7$ is a conditional equation. $x + 4 = x + 4$ is an identity. $x = x + 2$ is a contradiction.				
An equation of the form $ax + b = c$, $a \neq 0$, is called a *first-degree equation* because all variables have an exponent of 1. [2.1A, p. 57]	$6x + 5 = -7$ is a first-degree equation. $a = 6$, $b = 5$, and $c = -7$.				
The *solution*, or *root, of an equation*, is a replacement value for the variable that will make the equation true. [2.1A, p. 57]	The solution, or root, of the equation $x + 3 = 7$ is 4 because $4 + 3 = 7$.				
To solve an equation means to find its solutions. The goal is to rewrite the equation in the form *variable = constant* because the constant is the solution. [2.1A, p. 57]	The equation $x = 12$ is in the form *variable = constant*. The constant 12 is the solution of the equation.				
Equivalent equations are equations that have the same solution. [2.1A, p. 57]	$x + 3 = 7$ and $x = 4$ are equivalent equations because the solution of each equation is 4.				
A *literal equation* is an equation that contains more than one variable. A *formula* is a literal equation that states a rule about measurement. [2.1D, p. 62]	$4x - 5y = 20$ is a literal equation. $A = \pi r^2$ is the formula for the area of a circle. It is also a literal equation.				
The *solution set of an inequality* is a set of numbers, each element of which, when substituted in the inequality, results in a true inequality. [2.4A, p. 83]	Any number greater than 4 is a solution of the inequality $x > 4$.				
A *compound inequality* is formed by joining two inequalities with a connective word such as *and* or *or*. [2.4B, p. 86]	$3x > 6$ and $2x + 5 < 7$ $2x + 1 < 3$ or $x + 2 > 4$				
An *absolute value equation* is an equation that contains an absolute value symbol. [2.5A, p. 95]	$	x - 2	= 3$		
An *absolute value inequality* is an inequality that contains an absolute value symbol. [2.5B, p. 97]	$	x - 4	< 5$ $	2x - 3	> 6$

The *tolerance* of a component or part is the amount by which it is acceptable for the component to vary from a given measurement. The maximum measurement is the *upper limit*. The minimum measurement is the *lower limit*. [2.5C, p. 99]

The diameter of a bushing is 1.5 in. with a tolerance of 0.005 in. The lower and upper limits of the diameter of the bushing are 1.5 in. ± 0.005 in.

Essential Rules and Procedures

Examples

Addition Property of Equations [2.1A, p. 57]
If $a = b$, then $a + c = b + c$.

$$x + 5 = -3$$
$$x + 5 - 5 = -3 - 5$$
$$x = -8$$

Multiplication Property of Equations [2.1A, p. 58]
If $a = b$ and $c \neq 0$, then $ac = bc$.

$$\frac{2}{3}x = 4$$
$$\left(\frac{3}{2}\right)\left(\frac{2}{3}x\right) = \left(\frac{3}{2}\right)4$$
$$x = 6$$

Consecutive Integers [2.2A, p. 67]
$n, n + 1, n + 2, \ldots$

The sum of three consecutive integers is 57.

$n + (n + 1) + (n + 2) = 57$

Consecutive Even or Consecutive Odd Integers [2.2A, p. 67]
$n, n + 2, n + 4, \ldots$

The sum of three consecutive even integers is 132.

$n + (n + 2) + (n + 4) = 132$

Coin and Stamp Equation [2,2B, p. 69]
$$\frac{\text{Number}}{\text{of items}} \cdot \frac{\text{Value of}}{\text{each item}} = \frac{\text{Total value}}{\text{of the items}}$$

A collection of stamps consists of 17¢ and 27¢ stamps. In all there are 15 stamps with a value of $3.55. How many 17¢ stamps are in the collection?

$17n + 27(15 - n) = 355$

Value Mixture Equation [2.3A, p. 73]
Amount · Unit Cost = Value
$$AC = V$$

A merchant combines coffee that costs $6 per pound with coffee that costs $3.20 per pound. How many pounds of each should be used to make 60 lb of a blend that costs $4.50 per pound?

$6x + 3.20(60 - x) = 4.50(60)$

Percent Mixture Problems [2.3B, p. 75]

$$\begin{array}{c} \text{Amount of} \\ \text{solution} \end{array} \cdot \begin{array}{c} \text{percent of} \\ \text{concentration} \end{array} = \begin{array}{c} \text{quantity of} \\ \text{substance} \end{array}$$

$$Ar = Q$$

A silversmith mixed 120 oz of an 80% silver alloy with 240 oz of a 30% silver alloy. Find the percent concentration of the resulting silver alloy.

$$0.80(120) + 0.30(240) = x(360)$$

Uniform Motion Equation [2.3C, p. 77]

Rate · Time = Distance

$$rt = d$$

Two planes are 1640 mi apart and are traveling toward each other. One plane is traveling 60 mph faster than the other plane. The planes meet in 2 h. Find the speed of each plane.

$$2r + 2(r + 60) = 1640$$

Addition Property of Inequalities [2.4A, p. 83]

If $a > b$, then $a + c > b + c$.
If $a < b$, then $a + c < b + c$.

$$x + 3 > -2$$
$$x + 3 - 3 > -2 - 3$$
$$x > -5$$

Multiplication Property of Inequalities [2.4A, p. 84]

Rule 1 If $a > b$ and $c > 0$, then $ac > bc$.
 If $a < b$ and $c > 0$, then $ac < bc$.

$$3x > 12$$
$$\left(\frac{1}{3}\right)(3x) > \left(\frac{1}{3}\right)12$$
$$x > 4$$

Rule 2 If $a > b$ and $c < 0$, then $ac < bc$.
 If $a < b$ and $c < 0$, then $ac > bc$.

$$-2x < 8$$
$$\frac{-2x}{-2} > \frac{8}{-2}$$
$$x > -4$$

Solutions of an Absolute Value Equation [2.5A, p. 95]

If $a \geq 0$ and $|x| = a$, then $x = a$ or $x = -a$.

$$|x - 3| = 7$$
$$x - 3 = 7 \qquad\qquad x - 3 = -7$$
$$x = 10 \qquad\qquad\quad x = -4$$

Solutions of Absolute Value Inequalities [2.5B, p. 98]

To solve an absolute value inequality of the form $|ax + b| < c$, $c > 0$, solve the equivalent compound inequality $-c < ax + b < c$.

$$|x - 5| < 9$$
$$-9 < x - 5 < 9$$
$$-9 + 5 < x - 5 + 5 < 9 + 5$$
$$-4 < x < 14$$

To solve an absolute value inequality of the form $|ax + b| > c$, solve the equivalent compound inequality $ax + b < -c$ or $ax + b > c$.

$$|x - 5| > 9$$
$$x - 5 < -9 \quad \text{or} \quad x - 5 > 9$$
$$x < -4 \quad \text{or} \qquad\quad x > 14$$

Chapter 2 Review Exercises

1. Solve: $3t - 3 + 2t = 7t - 15$

2. Solve: $3x - 7 > -2$
Write the solution set in interval notation.

3. Solve $P = 2L + 2W$ for L.

4. Solve: $x + 4 = -5$

5. Solve: $3x < 4$ and $x + 2 > -1$
Write the solution set in set-builder notation.

6. Solve: $\dfrac{3}{5}x - 3 = 2x + 5$

7. Solve: $-\dfrac{2}{3}x = \dfrac{4}{9}$

8. Solve: $|x - 4| - 8 = -3$

9. Solve: $|2x - 5| < 3$

10. Solve: $\dfrac{2x - 3}{3} + 2 = \dfrac{2 - 3x}{5}$

11. Solve: $2(a - 3) = 5(4 - 3a)$

12. Solve: $5x - 2 > 8$ or $3x + 2 < -4$
Write the solution set in set-builder notation.

13. Solve: $|4x - 5| \geq 3$

14. Solve $P = \dfrac{R - C}{n}$ for C.

15. Solve: $\dfrac{1}{2}x - \dfrac{5}{8} = \dfrac{3}{4}x + \dfrac{3}{2}$

16. Solve: $6 + |3x - 3| = 2$

17. Solve: $3x - 2 > x - 4$ or $7x - 5 < 3x + 3$
Write the solution set in interval notation.

18. Solve: $2x - (3 - 2x) = 4 - 3(4 - 2x)$

19. Solve: $x + 9 = -6$

20. Solve: $\dfrac{2}{3} = x + \dfrac{3}{4}$

21. Solve: $-3x = -21$

22. Solve: $\dfrac{2}{3}a = \dfrac{4}{9}$

23. Solve: $3y - 5 = 3 - 2y$

24. Solve: $4x - 5 + x = 6x - 8$

25. Solve: $3(x - 4) = -5(6 - x)$

26. Solve: $\dfrac{3x - 2}{4} + 1 = \dfrac{2x - 3}{2}$

27. Solve: $5x - 8 < -3$
Write the solution set in interval notation.

28. Solve: $2x - 9 \le 8x + 15$
Write the solution set in interval notation.

29. Solve: $\dfrac{2}{3}x - \dfrac{5}{8} \ge \dfrac{3}{4}x + 1$
Write the solution set in set-builder notation.

30. Solve: $2 - 3(2x - 4) \le 4x - 2(1 - 3x)$
Write the solution set in set-builder notation.

31. Solve: $-5 < 4x - 1 < 7$
Write the solution set in interval notation.

32. Solve: $|2x - 3| = 8$

33. Solve: $|5x + 8| = 0$

34. Solve: $|5x - 4| < -2$

35. **Uniform Motion** A ferry leaves a dock and travels to an island at an average speed of 16 mph. On the return trip, the ferry travels at an average speed of 12 mph. The total time for the trip is $2\dfrac{1}{3}$ h. How far is the island from the dock?

36. Mixtures A grocer mixed apple juice that costs $4.20 per gallon with 40 gal of cranberry juice that costs $6.50 per gallon. How much apple juice was used to make cranapple juice costing $5.20 per gallon?

37. Compensation A sales executive earns $800 per month plus 4% commission on the amount of sales. The executive's goal is to earn $3000 per month. What amount of sales will enable the executive to earn $3000 or more per month?

38. Coins A coin collection contains thirty coins in nickels, dimes, and quarters. There are three more dimes than nickels. The value of the coins is $3.55. Find the number of quarters in the collection.

39. Mechanics The diameter of a bushing is 2.75 in. The bushing has a tolerance of 0.003 in. Find the lower and upper limits of the diameter of the bushing.

40. Integers The sum of two integers is twenty. Five times the smaller integer is two more than twice the larger integer. Find the two integers.

41. Education An average score of 80 to 90 in a psychology class receives a B grade. A student has scores of 92, 66, 72, and 88 on four tests. Find the range of scores on the fifth test that will give the student a B for the course.

42. Uniform Motion Two planes are 1680 mi apart and are traveling toward each other. One plane is traveling 80 mph faster than the other plane. The planes meet in 1.75 h. Find the speed of each plane.

$d = 1680$ mi

43. Mixtures An alloy containing 30% tin is mixed with an alloy containing 70% tin. How many pounds of each were used to make 500 lb of an alloy containing 40% tin?

44. Automobiles A piston rod for an automobile is $10\frac{3}{8}$ in. long with a tolerance of $\frac{1}{32}$ in. Find the lower and upper limits of the length of the piston rod.

Chapter 2 Test

1. Solve: $x - 2 = -4$

2. Solve: $b + \dfrac{3}{4} = \dfrac{5}{8}$

3. Solve: $-\dfrac{3}{4}y = -\dfrac{5}{8}$

4. Solve: $3x - 5 = 7$

5. Solve: $\dfrac{3}{4}y - 2 = 6$

6. Solve: $2x - 3 - 5x = 8 + 2x - 10$

7. Solve: $2[a - (2 - 3a) - 4] = a - 5$

8. Solve $E = IR + Ir$ for R.

9. Solve: $\dfrac{2x + 1}{3} - \dfrac{3x + 4}{6} = \dfrac{5x - 9}{9}$

10. Solve: $3x - 2 \geq 6x + 7$
Write the solution set in set-builder notation.

11. Solve: $4 - 3(x + 2) < 2(2x + 3) - 1$
Write the solution set in interval notation.

12. Solve: $4x - 1 > 5$ or $2 - 3x < 8$
Write the solution set in set-builder notation.

13. Solve: $4 - 3x \geq 7$ and $2x + 3 \geq 7$
Write the solution set in set-builder notation.

14. Solve: $\left|3 - 5x\right| = 12$

15. Solve: $2 - \left|2x - 5\right| = -7$

16. Solve: $\left|3x - 5\right| \leq 4$

17. Solve: $\left|4x - 3\right| > 5$

18. **Consumerism** Gambelli Agency rents cars for $12 per day plus 10¢ for every mile driven. McDougal Rental rents cars for $24 per day with unlimited mileage. How many miles a day can you drive a Gambelli Agency car if it is to cost you less than a McDougal Rental car?

19. **Mechanics** A machinist must make a bushing that has a tolerance of 0.002 in. The diameter of the bushing is 2.65 in. Find the lower and upper limits of the diameter of the bushing.

20. **Integers** The sum of two integers is fifteen. Eight times the smaller integer is one less than three times the larger integer. Find the integers.

21. **Stamps** A stamp collection contains 11¢, 15¢, and 24¢ stamps. There are twice as many 11¢ stamps as 15¢ stamps. There are thirty stamps in all, with a value of $4.40. How many 24¢ stamps are in the collection?

22. **Mixtures** A butcher combines 100 lb of hamburger that costs $2.10 per pound with 60 lb of hamburger that costs $3.70 per pound. Find the cost of the hamburger mixture.

23. **Uniform Motion** A jogger runs a distance at a speed of 8 mph and returns the same distance running at a speed of 6 mph. Find the total distance that the jogger ran if the total time running was one hour and forty-five minutes.

24. **Uniform Motion** Two trains are 250 mi apart and are traveling toward each other. One train is traveling 5 mph faster than the other train. The trains pass each other in 2 h. Find the speed of each train.

25. **Mixtures** How many ounces of pure water must be added to 60 oz of an 8% salt solution to make a 3% salt solution?

Cumulative Review Exercises

1. Simplify: $-4 - (-3) - 8 + (-2)$

2. Simplify: $-2^2 \cdot 3^3$

3. Simplify: $4 - (2 - 5)^2 \div 3 + 2$

4. Simplify: $4 \div \dfrac{\frac{3}{8} - 1}{5} \cdot 2$

5. Evaluate $2a^2 - (b - c)^2$ when $a = 2$, $b = 3$, and $c = -1$.

6. Evaluate $\dfrac{a - b^2}{b - c}$ when $a = 2$, $b = -3$, and $c = 4$.

7. Identify the property that justifies the statement.
$(2x + 3y) + 2 = (3y + 2x) + 2$

8. Translate and simplify "the sum of three times a number and six added to the product of three and the number."

9. Solve $F = \dfrac{evB}{c}$ for B.

10. Simplify: $5[y - 2(3 - 2y) + 6]$

11. Find $A \cap B$, given $A = \{-4, -2, 0, 2\}$ and $B = \{-4, 0, 4, 8\}$.

12. Graph the solution set of $\{x \mid x \le 3\} \cap \{x \mid x > -1\}$.

$$\overset{\textstyle\leftarrow\!\!-\!\!+\!\!-\!\!+\!\!-\!\!+\!\!-\!\!+\!\!-\!\!+\!\!-\!\!+\!\!-\!\!+\!\!-\!\!+\!\!-\!\!+\!\!-\!\!+\!\!\rightarrow}{-5\ -4\ -3\ -2\ -1\ \ 0\ \ 1\ \ 2\ \ 3\ \ 4\ \ 5}$$

13. Solve $Ax + By + C = 0$ for y.

14. Solve: $-\dfrac{5}{6}b = -\dfrac{5}{12}$

15. Solve: $2x + 5 = 5x + 2$

16. Solve: $\dfrac{5}{12}x - 3 = 7$

17. Solve: $2[3 - 2(3 - 2x)] = 2(3 + x)$

18. Solve: $3[2x - 3(4 - x)] = 2(1 - 2x)$

19. Solve: $\dfrac{1}{2}y - \dfrac{2}{3}y + \dfrac{5}{12} = \dfrac{3}{4}y - \dfrac{1}{2}$

20. Solve: $\dfrac{3x - 1}{4} - \dfrac{4x - 1}{12} = \dfrac{3 + 5x}{8}$

21. Solve: $3 - 2(2x - 1) \geq 3(2x - 2) + 1$
Write the solution set in interval notation.

22. Solve: $3x + 2 \leq 5$ and $x + 5 > 1$
Write the solution set in set-builder notation.

23. Solve: $|3 - 2x| = 5$

24. Solve: $3 - |2x - 3| = -8$

25. Solve: $|3x - 1| > 5$

26. Solve: $|2x - 4| < 8$

27. Banking A bank offers two types of checking accounts. One account has a charge of $5 per month plus 4¢ for each check. The second account has a charge of $2 per month plus 10¢ for each check. How many checks can a customer who has the second type of account write if it is to cost the customer less than the first type of account?

28. Integers Four times the sum of the first and third of three consecutive odd integers is one less than seven times the middle integer. Find the first integer.

29. Coins A coin purse contains dimes and quarters. The number of dimes is five less than twice the number of quarters. The total value of the coins is $4.00. Find the number of dimes in the coin purse.

30. Mixtures A silversmith combined pure silver that costs $8.50 per ounce with 100 oz of a silver alloy that costs $4.00 per ounce. How many ounces of pure silver were used to make an alloy of silver costing $6.00 per ounce?

31. Uniform Motion Two planes are 1400 mi apart and are traveling toward each other. One plane is traveling 120 mph faster than the other plane. The planes meet in 2.5 h. Find the speed of the slower plane.

32. Mechanics The diameter of a bushing is 2.45 in. The bushing has a tolerance of 0.001 in. Find the lower and upper limits of the diameter of the bushing.

33. Mixtures How many liters of a 12% acid solution must be mixed with 4 L of a 5% acid solution to make an 8% acid solution?

3

Linear Functions and Inequalities in Two Variables

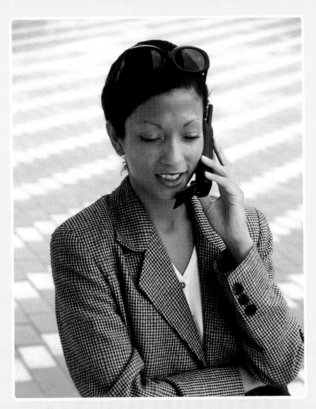

Do you have a cellular phone? Do you pay a monthly fee plus a charge for each minute you use? In this chapter you will learn how to write a linear function that models the monthly cost of a cell phone in terms of the number of minutes you use it. **Exercise 78 on page 173** asks you to write linear functions for certain cell phone options. Being able to use a linear function to model a relationship between two variables is an important skill in many fields, such as busness, economics, and nutrition.

OBJECTIVES

Section 3.1

A To graph points in a rectangular coordinate system
B To find the length and midpoint of a line segment
C To graph a scatter diagram

Section 3.2

A To evaluate a function

Section 3.3

A To graph a linear function
B To graph an equation of the form $Ax + By = C$
C To find the x- and the y-intercepts of a straight line
D To solve application problems

Section 3.4

A To find the slope of a line given two points
B To graph a line given a point and the slope

Section 3.5

A To find the equation of a line given a point and the slope
B To find the equation of a line given two points
C To solve application problems

Section 3.6

A To find parallel and perpendicular lines

Section 3.7

A To graph the solution set of an inequality in two variables

Need help? For online student resources, such as section quizzes, visit this textbook's Website at **math.college.hmco.com/students.**

PREP TEST • • •

Do these exercises to prepare for Chapter 3.

For Exercises 1 to 3, simplify.

1. $-4(x - 3)$

2. $\sqrt{(-6)^2 + (-8)^2}$

3. $\dfrac{3 - (-5)}{2 - 6}$

4. Evaluate $-2x + 5$ for $x = -3$.

5. Evaluate $\dfrac{2r}{r - 1}$ for $r = 5$.

6. Evaluate $2p^3 - 3p + 4$ for $p = -1$.

7. Evaluate $\dfrac{x_1 + x_2}{2}$ for $x_1 = 7$ and $x_2 = -5$.

8. Given $3x - 4y = 12$, find the value of x when $y = 0$.

9. Solve $2x - y = 7$ for y.

GO FIGURE • • •

If $\boxed{5} = 4$ and $\textcircled{5} = 6$ and $y = x - 1$, which of the following has the largest value?

 \boxed{x} \textcircled{x} \boxed{y} \textcircled{y}

3.1 The Rectangular Coordinate System

Objective A To graph points in a rectangular coordinate system

Point of Interest

A rectangular coordinate system is also called a **Cartesian coordinate system,** in honor of Descartes.

Before the 15th century, geometry and algebra were considered separate branches of mathematics. That all changed when René Descartes, a French mathematician who lived from 1596 to 1650, founded analytic geometry. In this geometry, a *coordinate system* is used to study relationships between variables.

A **rectangular coordinate system** is formed by two number lines, one horizontal and one vertical, that intersect at the zero point of each line. The point of intersection is called the **origin.** The two lines are called **coordinate axes,** or simply **axes.**

The axes determine a **plane,** which can be thought of as a large, flat sheet of paper. The two axes divide the plane into four regions called **quadrants.** The quadrants are numbered counterclockwise from I to IV.

Point of Interest

Gottfried Leibnitz introduced the words *abscissa* and *ordinate. Abscissa* is from Latin, meaning "to cut off." Originally, Leibnitz used the phrase *abscissa linea,* "cut off a line" (axis). The root of *ordinate* is also a Latin word used to suggest a sense of order.

Each point in the plane can be identified by a pair of numbers called an **ordered pair.** The first number of the pair measures a horizontal distance and is called the **abscissa.** The second number of the pair measures a vertical distance and is called the **ordinate.** The **coordinates of a point** are the numbers in the ordered pair associated with the point. The abscissa is also called the **first coordinate** of the ordered pair, and the ordinate is also called the **second coordinate** of the ordered pair.

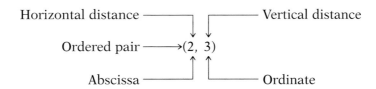

Graphing, or plotting, an ordered pair in the plane means placing a dot at the location given by the ordered pair. The **graph of an ordered pair** is the dot drawn at the coordinates of the point in the plane. The points whose coordinates are $(3, 4)$ and $(-2.5, -3)$ are graphed in the figure at the right.

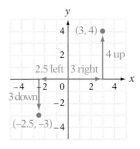

T A K E N O T E

The *ordered pair* is an important concept. Remember: There are two numbers (a *pair*), and the *order* in which they are given is important.

The points whose coordinates are $(3, -1)$ and $(-1, 3)$ are graphed at the right. Note that the graphs are in different locations. The *order* of the coordinates of an ordered pair is important.

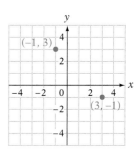

When drawing a rectangular coordinate system, we often label the horizontal axis x and the vertical axis y. In this case, the coordinate system is called an **xy-coordinate system.** The coordinates of the points are given by ordered pairs (x, y), where the abscissa is called the **x-coordinate** and the ordinate is called the **y-coordinate.**

The xy-coordinate system is used to graph **equations in two variables.** Examples of equations in two variables are shown at the right. A **solution of an equation in two variables** is an ordered pair (x, y) whose coordinates make the equation a true statement.

$$y = 3x + 7$$
$$y = x^2 - 4x + 3$$
$$x^2 + y^2 = 25$$
$$x = \frac{y}{y^2 + 4}$$

HOW TO Is the ordered pair $(-3, 7)$ a solution of the equation $y = -2x + 1$?

$$y = -2x + 1$$

7	$-2(-3) + 1$
7	$6 + 1$

$$7 = 7$$

Yes, the ordered pair $(-3, 7)$ is a solution of the equation.

- **Replace *x* by -3 and *y* by 7.**
- **Simplify.**
- **Compare the results. If the resulting equation is a true statement, the ordered pair is a solution of the equation. If it is not a true statement, the ordered pair is not a solution of the equation.**

Besides the ordered pair $(-3, 7)$, there are many other ordered-pair solutions of the equation $y = -2x + 1$. For example, $(-5, 11)$, $(0, 1)$, $\left(-\frac{3}{2}, 4\right)$, and $(4, -7)$ are also solutions of the equation.

In general, an equation in two variables has an infinite number of solutions. By choosing any value of x and substituting that value into the equation, we can calculate a corresponding value for y. The resulting ordered-pair solution (x, y) of the equation can be graphed in a rectangular coordinate system.

HOW TO Graph the solutions (x, y) of $y = x^2 - 1$ when x equals $-2, -1, 0, 1$, and 2.

Substitute each value of x into the equation and solve for y. It is convenient to record the ordered-pair solutions in a table similar to the one shown below. Then graph the ordered pairs, as shown at the left.

x	$y = x^2 - 1$	y	(x, y)
-2	$y = (-2)^2 - 1$	3	$(-2, 3)$
-1	$y = (-1)^2 - 1$	0	$(-1, 0)$
0	$y = 0^2 - 1$	-1	$(0, -1)$
1	$y = 1^2 - 1$	0	$(1, 0)$
2	$y = 2^2 - 1$	3	$(2, 3)$

Example 1

Determine the ordered-pair solution of

$y = \dfrac{x}{x-2}$ corresponding to $x = 4$.

Solution

$y = \dfrac{x}{x-2} = \dfrac{4}{4-2} = \dfrac{4}{2} = 2$ • Replace x by 4 and
 solve for y.

The ordered-pair solution is (4, 2).

You Try It 1

Determine the ordered-pair solution of

$y = \dfrac{3x}{x+1}$ corresponding to $x = -2$.

Your solution

Example 2

Graph the ordered-pair solutions of
$y = x^2 - x$ when $x = -1, 0, 1,$ and 2.

Solution

x	y
-1	2
0	0
1	0
2	2

You Try It 2

Graph the ordered-pair solutions of
$y = |x + 1|$ when $x = -3, -2, -1, 0,$ and 1.

Your solution

Solutions on p. S6

Objective B **To find the length and midpoint of a line segment**

The distance between two points in an xy-coordinate system can be calculated by using the Pythagorean Theorem.

> **TAKE NOTE**
>
> A **right triangle** contains one 90° angle. The side opposite the 90° angle is the **hypotenuse.** The other two sides are called **legs.**

> **Pythagorean Theorem**
>
> If a and b are the lengths of the legs of a right triangle and c is the length of the hypotenuse, then $a^2 + b^2 = c^2$.
>
>

Consider the two points and the right triangle shown at the right. The vertical distance between $P_1(x_1, y_1)$ and $P_2(x_2, y_2)$ is $|y_2 - y_1|$.

The horizontal distance between the points $P_1(x_1, y_1)$ and $P_2(x_2, y_2)$ is $|x_2 - x_1|$.

The quantity d^2 is calculated by applying the Pythagorean Theorem to the right triangle.

The distance d is the square root of d^2.

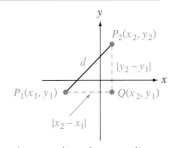

$d^2 = |x_2 - x_1|^2 + |y_2 - y_1|^2$

$d^2 = (x_2 - x_1)^2 + (y_2 - y_1)^2$

$d = \sqrt{(x_2 - x_1)^2 + (y_2 - y_1)^2}$

Because $(x_2 - x_1)^2 = (x_1 - x_2)^2$ and $(y_2 - y_1)^2 = (y_1 - y_2)^2$, the distance formula is usually written in the following form.

The Distance Formula

If $P_1(x_1, y_1)$ and $P_2(x_2, y_2)$ are two points in the plane, then the distance d between the two points is given by

$$d = \sqrt{(x_1 - x_2)^2 + (y_1 - y_2)^2}$$

HOW TO Find the distance between the points $(-6, 1)$ and $(-2, 4)$.

$d = \sqrt{(x_1 - x_2)^2 + (y_1 - y_2)^2}$ • Use the distance formula.

$\quad = \sqrt{[-6 - (-2)]^2 + (1 - 4)^2}$ • Let $(x_1, y_1) = (-6, 1)$ and $(x_2, y_2) = (-2, 4)$.

$\quad = \sqrt{(-4)^2 + (-3)^2} = \sqrt{16 + 9}$

$\quad = \sqrt{25} = 5$

The distance between the points is 5 units.

The **midpoint of a line segment** is equidistant from its endpoints. The coordinates of the midpoint of the line segment P_1P_2 are (x_m, y_m). The intersection of the horizontal line segment through P_1 and the vertical line segment through P_2 is Q, with coordinates (x_2, y_1).

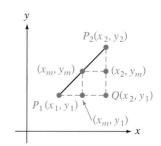

The x-coordinate x_m of the midpoint of the line segment P_1P_2 is the same as the x-coordinate of the midpoint of the line segment P_1Q. It is the average of the x-coordinates of the points P_1 and P_2.

$$x_m = \frac{x_1 + x_2}{2}$$

Similarly, the y-coordinate y_m of the midpoint of the line segment P_1P_2 is the same as the y-coordinate of the midpoint of the line segment P_2Q. It is the average of the y-coordinates of the points P_1 and P_2.

$$y_m = \frac{y_1 + y_2}{2}$$

The Midpoint Formula

If $P_1(x_1, y_1)$ and $P_2(x_2, y_2)$ are the endpoints of a line segment, then the coordinates of the midpoint (x_m, y_m) of the line segment are given by

$$x_m = \frac{x_1 + x_2}{2} \quad \text{and} \quad y_m = \frac{y_1 + y_2}{2}$$

HOW TO Find the coordinates of the midpoint of the line segment with endpoints $(-3, 5)$ and $(1, -7)$.

$$x_m = \frac{x_1 + x_2}{2} \qquad y_m = \frac{y_1 + y_2}{2}$$

• Use the midpoint formula.

$$= \frac{-3 + 1}{2} \qquad = \frac{5 + (-7)}{2}$$

• Let $(x_1, y_1) = (-3, 5)$ and $(x_2, y_2) = (1, -7)$.

$$= -1 \qquad = -1$$

The coordinates of the midpoint are $(-1, -1)$.

Example 3

Find the distance, to the nearest hundredth, between the points whose coordinates are $(-3, 2)$ and $(4, -1)$.

Solution
$(x_1, y_1) = (-3, 2); (x_2, y_2) = (4, -1)$
$$\begin{aligned} d &= \sqrt{(x_1 - x_2)^2 + (y_1 - y_2)^2} \\ &= \sqrt{(-3 - 4)^2 + [2 - (-1)]^2} \\ &= \sqrt{(-7)^2 + 3^2} = \sqrt{49 + 9} \\ &= \sqrt{58} \approx 7.62 \end{aligned}$$

You Try It 3

Find the distance, to the nearest hundredth, between the points whose coordinates are $(5, -2)$ and $(-4, 3)$.

Your solution

Example 4

Find the coordinates of the midpoint of the line segment with endpoints $(-5, 4)$ and $(-3, 7)$.

Solution
$(x_1, y_1) = (-5, 4); (x_2, y_2) = (-3, 7)$
$$x_m = \frac{x_1 + x_2}{2} \qquad y_m = \frac{y_1 + y_2}{2}$$
$$= \frac{-5 + (-3)}{2} \qquad = \frac{4 + 7}{2}$$
$$= -4 \qquad = \frac{11}{2}$$

The midpoint is $\left(-4, \frac{11}{2}\right)$.

You Try It 4

Find the coordinates of the midpoint of the line segment with endpoints $(-3, -5)$ and $(-2, 3)$.

Your solution

Solutions on p. S6

Objective C **To graph a scatter diagram**

Discovering a relationship between two variables is an important task in the study of mathematics. These relationships occur in many forms and in a wide variety of applications. Here are some examples.

• A botanist wants to know the relationship between the number of bushels of wheat yielded per acre and the amount of watering per acre.
• An environmental scientist wants to know the relationship between the incidence of skin cancer and the amount of ozone in the atmosphere.
• A business analyst wants to know the relationship between the price of a product and the number of products that are sold at that price.

A researcher may investigate the relationship between two variables by means of *regression analysis,* which is a branch of statistics. The study of the relationship between the two variables may begin with a **scatter diagram,** which is a graph of the ordered pairs of the known data.

The following table shows randomly selected data from the participants 40 years old and older and their times (in minutes) for a recent Boston Marathon.

Age (x)	55	46	53	40	40	44	54	44	41	50
Time (y)	254	204	243	194	281	197	238	300	232	216

TAKE NOTE

The jagged portion of the horizontal axis in the figure at the right indicates that the numbers between 0 and 40 are missing.

The scatter diagram for these data is shown at the right. Each ordered pair represents the age and time for a participant. For instance, the ordered pair $(53, 243)$ indicates that a 53-year-old participant ran the marathon in 243 min.

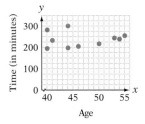

Example 5

The grams of sugar and the grams of fiber in a 1-ounce serving of six breakfast cereals are shown in the table below. Draw a scatter diagram of these data.

	Sugar (x)	Fiber (y)
Wheaties	4	3
Rice Krispies	3	0
Total	5	3
Life	6	2
Kix	3	1
Grape-Nuts	7	5

Strategy

To draw a scatter diagram:

• Draw a coordinate grid with the horizontal axis representing the grams of sugar and the vertical axis the grams of fiber.
• Graph the ordered pairs (4, 3), (3, 0), (5, 3), (6, 2), (3, 1), and (7, 5).

Solution

You Try It 5

According to the National Interagency Fire Center, the number of deaths in U.S. wildland fires is as shown in the table below. Draw a scatter diagram of these data.

Year	Number of Deaths
1998	14
1999	28
2000	17
2001	18
2002	23

Your strategy

Your solution

Solution on p. S7

3.1 Exercises

To graph points in a rectangular coordinate system

1. Graph the ordered pairs $(1, -1)$, $(2, 0)$, $(3, 2)$, and $(-1, 4)$.

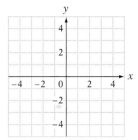

2. Graph the ordered pairs $(-1, -3)$, $(0, -4)$, $(0, 4)$, and $(3, -2)$.

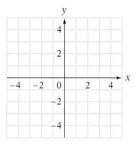

3. Find the coordinates of each of the points.

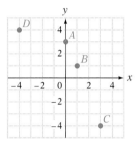

4. Find the coordinates of each of the points.

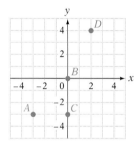

5. Draw a line through all points with an abscissa of 2.

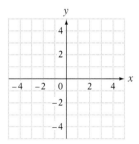

6. Draw a line through all points with an abscissa of -3.

7. Draw a line through all points with an ordinate of -3.

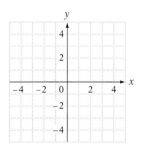

8. Draw a line through all points with an ordinate of 4.

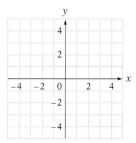

9. Graph the ordered-pair solutions of $y = x^2$ when $x = -2, -1, 0, 1,$ and 2.

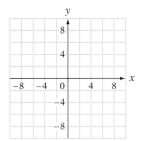

10. Graph the ordered-pair solutions of $y = -x^2 + 1$ when $x = -2, -1, 0, 1,$ and 2.

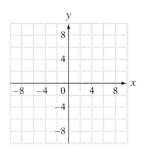

11. Graph the ordered-pair solutions of $y = |x + 1|$ when $x = -5, -3, 0, 3,$ and 5.

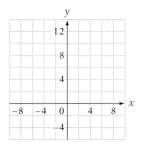

12. Graph the ordered-pair solutions of $y = -2|x|$ when $x = -3, -1, 0, 1,$ and 3.

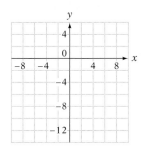

13. Graph the ordered-pair solutions of $y = -x^2 + 2$ when $x = -2, -1, 0, 1,$ and 2.

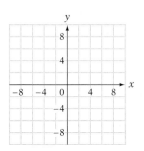

14. Graph the ordered-pair solutions of $y = -x^2 + 4$ when $x = -3, -1, 0, 1,$ and 3.

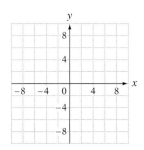

15. Graph the ordered-pair solutions of $y = x^3 - 2$ when $x = -1, 0, 1,$ and 2.

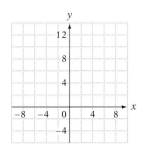

16. Graph the ordered-pair solutions of $y = -x^3 + 1$ when $x = -1, 0, 1,$ and $\frac{3}{2}$.

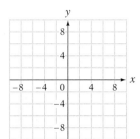

Objective B **To find the length and midpoint of a line segment**

For Exercises 17 to 28, find the distance to the nearest hundredth between the given points. Then find the coordinates of the midpoint of the line segment connecting the points.

17. $P_1(3, 5)$ and $P_2(5, 1)$

18. $P_1(-2, 3)$ and $P_2(4, -1)$

19. $P_1(0, 3)$ and $P_2(-2, 4)$

20. $P_1(6, -1)$ and $P_2(-3, -2)$

21. $P_1(-3, -5)$ and $P_2(2, -4)$

22. $P_1(-7, -5)$ and $P_2(-2, -1)$

23. $P_1(5, -2)$ and $P_2(-2, 5)$

24. $P_1(3, -6)$ and $P_2(6, 0)$

25. $P_1(5, -5)$ and $P_2(2, -5)$

26. $P_1(-2, -3)$ and $P_2(-2, 5)$

27. $P_1\left(\dfrac{3}{2}, -\dfrac{4}{3}\right)$ and $P_2\left(-\dfrac{1}{2}, \dfrac{7}{3}\right)$

28. $P_1(4.5, -6.3)$ and $P_2(-1.7, -4.5)$

Objective C **To graph a scatter diagram**

29. **Chemistry** The temperature of a chemical reaction is measured at intervals of 10 min and recorded in the scatter diagram at the right.
 a. Find the temperature of the reaction after 20 min.
 b. After how many minutes is the temperature 160°F?

The jagged line means that the numbers between 0 and 140 are missing.

30. **Chemistry** The amount of a substance that can be dissolved in a fixed amount of water usually increases as the temperature of the water increases. Cerium selenate, however, does not behave in this manner. The graph at the right shows the number of grams of cerium selenate that will dissolve in 100 mg of water for various temperatures, in degrees Celsius.
 a. Determine the temperature at which 25 g of cerium selenate will dissolve.
 b. Determine the number of grams of cerium selenate that will dissolve when the temperature is 80°C.

31. **Business** Past experience of executives of a car company shows that the profit of a dealership will depend on the total income of all the residents of the town in which the dealership is located. The table below shows the profits of several dealerships and the total incomes of the towns. Draw a scatter diagram for these data.

Profit (in thousands of dollars)	65	85	81	77	89	69
Total Income (in billions of dollars)	2.2	2.6	2.5	2.4	2.7	2.3

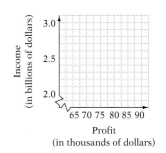

32. **Utilities** A power company suggests that a larger power plant can produce energy more efficiently and therefore at lower cost to consumers. The table below shows the output and average cost for power plants of various sizes. Draw a scatter diagram for these data.

Output (in millions of watts)	0.7	2.2	2.6	3.2	2.8	3.5
Average Cost (in dollars)	6.9	6.5	6.3	6.4	6.5	6.1

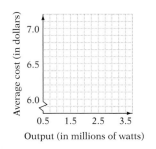

APPLYING THE CONCEPTS

33. Graph the ordered pairs (x, x^2), where $x \in \{-2, -1, 0, 1, 2\}$.

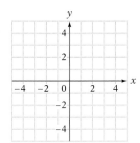

34. Graph the ordered pairs $\left(x, \frac{1}{x}\right)$, where $x \in \left\{-2, -1, -\frac{1}{2}, -\frac{1}{3}, \frac{1}{3}, \frac{1}{2}, 1, 2\right\}$.

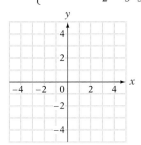

35. Describe the graph of all the ordered pairs (x, y) that are 5 units from the origin.

36. Consider two distinct fixed points in the plane. Describe the graph of all the points (x, y) that are equidistant from these fixed points.

37. Draw a line passing through every point whose abscissa equals its ordinate.

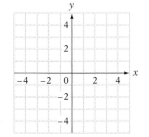

38. Draw a line passing through every point whose ordinate is the additive inverse of its abscissa.

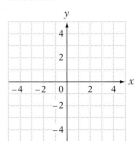

3.2 Introduction to Functions

Objective A **To evaluate a function**

In mathematics and its applications, there are many times when it is necessary to investigate a relationship between two quantities. Here is a financial application: Consider a person who is planning to finance the purchase of a car. If the current interest rate for a 5-year loan is 5%, the equation that describes the relationship between the amount that is borrowed B and the monthly payment P is $P = 0.018871B$.

For each amount the purchaser may borrow (B), there is a certain monthly payment (P). The relationship between the amount borrowed and the payment can be recorded as ordered pairs, where the first coordinate is the amount borrowed and the second coordinate is the monthly payment. Some of these ordered pairs are shown at the right.

$$0.018871B = P$$

(6000, 113.23)
(7000, 132.10)
(8000, 150.97)
(9000, 169.84)

A relationship between two quantities is not always given by an equation. The table at the right describes a grading scale that defines a relationship between a score on a test and a letter grade. For each score, the table assigns only one letter grade. The ordered pair (84, B) indicates that a score of 84 receives a letter grade of B.

Score	Grade
90–100	A
80–89	B
70–79	C
60–69	D
0–59	F

The graph at the right also shows a relationship between two quantities. It is a graph of the viscosity V of SAE 40 motor oil at various temperatures T. Ordered pairs can be approximated from the graph. The ordered pair (120, 250) indicates that the viscosity of the oil at 120°F is 250 units.

In each of these examples, there is a rule (an equation, a table, or a graph) that determines a certain set of ordered pairs.

> **Definition of Relation**
>
> A **relation** is a set of ordered pairs.

Here are some of the ordered pairs for the relations given above.

Relation	Some of the Ordered Pairs of the Relation
Car Payment	(7500, 141.53), (8750, 165.12), (9390, 177.20)
Grading Scale	(78, C), (98, A), (70, C), (81, B), (94, A)
Oil Viscosity	(100, 500), (120, 250), (130, 200), (150, 180)

Each of these three relations is actually a special type of relation called a *function*. Functions play an important role in mathematics and its applications.

> **Definition of Function**
>
> A **function** is a relation in which no two ordered pairs have the same first coordinate and different second coordinates.

The **domain** of a function is the set of the first coordinates of all the ordered pairs of the function. The **range** is the set of the second coordinates of all the ordered pairs of the function.

For the function defined by the ordered pairs

$$\{(2, 3), (4, 5), (6, 7), (8, 9)\}$$

the domain is $\{2, 4, 6, 8\}$ and the range is $\{3, 5, 7, 9\}$.

HOW TO Find the domain and range of the function $\{(2, 3), (4, 6), (6, 8), (10, 6)\}$.

The domain is $\{2, 4, 6, 10\}$.

- The domain of the function is the set of the first coordinates of the ordered pairs.

The range is $\{3, 6, 8\}$.

- The range of the function is the set of the second coordinates of the ordered pairs.

For each element of the domain of a function there is a corresponding element in the range of the function. A possible diagram for the function above is

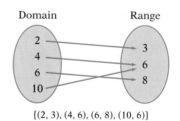

$$\{(2, 3), (4, 6), (6, 8), (10, 6)\}$$

Functions defined by tables or graphs, such as those described at the beginning of this section, have important applications. However, a major focus of this text is functions defined by equations in two variables.

The **square function,** which pairs each real number with its square, can be defined by the equation

$$y = x^2$$

This equation states that for a given value of x in the domain, the corresponding value of y in the range is the square of x. For instance, if $x = 6$, then $y = 36$ and if $x = -7$, then $y = 49$. Because the value of y *depends* on the value of x, y is called the **dependent variable** and x is called the **independent variable.**

TAKE NOTE

A pictorial representation of the square function is shown at the right. The function acts as a machine that changes a number from the domain into the square of the number.

A function can be thought of as a rule that pairs one number with another number. For instance, the *square function* pairs a number with its square. The ordered pairs for the values shown at the right are $(-5, 25)$, $\left(\frac{3}{5}, \frac{9}{25}\right)$, $(0, 0)$, and $(3, 9)$. For this function, the second coordinate is the square of the first coordinate. If we let x represent the first coordinate, then the second coordinate is x^2 and we have the ordered pair (x, x^2).

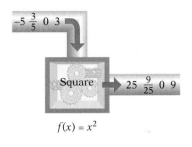

$f(x) = x^2$

A function cannot have two ordered pairs with *different* second coordinates and the same first coordinate. However, a function may contain ordered pairs with the *same* second coordinate. For instance, the square function has the ordered pairs $(-3, 9)$ and $(3, 9)$; the second coordinates are the same but the first coordinates are different.

The **double function** pairs a number with twice that number. The ordered pairs for the values shown at the right are $(-5, -10)$, $\left(\frac{3}{5}, \frac{6}{5}\right)$, $(0, 0)$, and $(3, 6)$. For this function, the second coordinate is twice the first coordinate. If we let x represent the first coordinate, then the second coordinate is $2x$ and we have the ordered pair $(x, 2x)$.

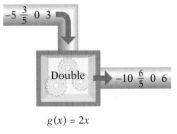

$g(x) = 2x$

Not every equation in two variables defines a function. For instance, consider the equation

$$y^2 = x^2 + 9$$

Because

$$5^2 = 4^2 + 9 \qquad \text{and} \qquad (-5)^2 = 4^2 + 9$$

the ordered pairs $(4, 5)$ and $(4, -5)$ are both solutions of the equation. Consequently, there are two ordered pairs that have the same first coordinate (4) but *different* second coordinates (5 and -5). Therefore, the equation does not define a function. Other ordered pairs for this equation are $(0, -3)$, $(0, 3)$, $\left(\sqrt{7}, -4\right)$, and $\left(\sqrt{7}, 4\right)$. A graphical representation of these ordered pairs is shown below.

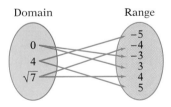

Note from this graphical representation that each element from the domain has two arrows pointing to two different elements in the range. Any time this occurs, the situation does not represent a function. However, this diagram *does* represent a relation. The relation for the values shown is $\left\{(0, -3), (0, 3), (4, -5), (4, 5), \left(\sqrt{7}, -4\right), \left(\sqrt{7}, 4\right)\right\}$.

The phrase "*y* is a function of *x*," or the same phrase with different variables, is used to describe an equation in two variables that defines a function. To emphasize that the equation represents a function, **functional notation** is used.

Just as the variable x is commonly used to represent a number, the letter f is commonly used to name a function. The square function is written in functional notation as follows:

This is the value of the function.
It is the number that is paired with x.

$$f(x) = x^2$$

The name of the function is f. This is an algebraic expression that defines the relationship between the dependent and independent variables.

The symbol $f(x)$ is read "the value of f at x" or "f of x."

It is important to note that $f(x)$ does *not* mean f times x. The symbol $f(x)$ is the **value of the function** and represents the value of the dependent variable for a given value of the independent variable. We often write $y = f(x)$ to emphasize the relationship between the independent variable x and the dependent variable y. Remember that y and $f(x)$ are different symbols for the same number.

The letters used to represent a function are somewhat arbitrary. All of the following equations represent the same function.

$$\left. \begin{array}{l} f(x) = x^2 \\ s(t) = t^2 \\ P(v) = v^2 \end{array} \right\} \text{Each equation represents the square function.}$$

The process of determining $f(x)$ for a given value of x is called **evaluating a function**. For instance, to evaluate $f(x) = x^2$ when $x = 4$, replace x by 4 and simplify.

$$f(x) = x^2$$
$$f(4) = 4^2 = 16$$

The *value* of the function is 16 when $x = 4$. An ordered pair of the function is $(4, 16)$.

HOW TO Evaluate $g(t) = 3t^2 - 5t + 1$ when $t = -2$.

$$g(t) = 3t^2 - 5t + 1$$
$$g(-2) = 3(-2)^2 - 5(-2) + 1 \quad \bullet \text{ Replace } t \text{ by } -2 \text{ and then simplify.}$$
$$= 3(4) - 5(-2) + 1$$
$$= 12 + 10 + 1 = 23$$

When t is -2, the value of the function is 23. Therefore, an ordered pair of the function is $(-2, 23)$.

It is possible to evaluate a function for a variable expression.

HOW TO Evaluate $P(z) = 3z - 7$ when $z = 3 + h$.

$$P(z) = 3z - 7$$
$$P(3 + h) = 3(3 + h) - 7 \quad \bullet \text{ Replace } z \text{ by } 3 + h \text{ and then simplify.}$$
$$= 9 + 3h - 7$$
$$= 3h + 2$$

When z is $3 + h$, the value of the function is $3h + 2$. Therefore, an ordered pair of the function is $(3 + h, 3h + 2)$.

Recall that the range of a function is found by applying the function to each element of the domain. If the domain contains an infinite number of elements, it may be difficult to find the range. However, if the domain has a finite number of elements, then the range can be found by evaluating the function for each element in the domain.

HOW TO Find the range of $f(x) = x^3 + x$ if the domain is $\{-2, -1, 0, 1, 2\}$.

$f(x) = x^3 + x$
$f(-2) = (-2)^3 + (-2) = -10$ • Replace x by each member of the
$f(-1) = (-1)^3 + (-1) = -2$ domain. The range includes the
$f(0) = 0^3 + 0 = 0$ values of $f(-2)$, $f(-1)$, $f(0)$, $f(1)$,
$f(1) = 1^3 + 1 = 2$ and $f(2)$.
$f(2) = 2^3 + 2 = 10$

The range is $\{-10, -2, 0, 2, 10\}$.

When a function is represented by an equation, the domain of the function is all real numbers for which the value of the function is a real number. For instance:

• The domain of $f(x) = x^2$ is all real numbers, because the square of every real number is a real number.

• The domain of $g(x) = \dfrac{1}{x - 2}$ is all real numbers except 2, because when $x = 2$,

$g(2) = \dfrac{1}{2 - 2} = \dfrac{1}{0}$, which is not a real number.

The domain of the grading-scale function is the set of whole numbers from 0 to 100. In set-builder notation, this is written $\{x | 0 \le x \le 100, x \in \text{whole numbers}\}$. The range is $\{A, B, C, D, F\}$.

Score	Grade
90–100	A
80–89	B
70–79	C
60–69	D
0–59	F

HOW TO What values, if any, are excluded from the domain of $f(x) = 2x^2 - 7x + 1$?

Because the value of $2x^2 - 7x + 1$ is a real number for any value of x, the domain of the function is all real numbers. No values are excluded from the domain of $f(x) = 2x^2 - 7x + 1$.

Example 1

Find the domain and range of the function $\{(5, 3), (9, 7), (13, 7), (17, 3)\}$.

Solution

Domain: $\{5, 9, 13, 17\}$; • The domain is the set
Range: $\{3, 7\}$ of first coordinates.

You Try It 1

Find the domain and range of the function $\{(-1, 5), (3, 5), (4, 5), (6, 5)\}$.

Your solution

Solution on p. S7

Example 2

Given $p(r) = 5r^3 - 6r - 2$, find $p(-3)$.

Solution

$$p(r) = 5r^3 - 6r - 2$$
$$p(-3) = 5(-3)^3 - 6(-3) - 2$$
$$= 5(-27) + 18 - 2$$
$$= -135 + 18 - 2 = -119$$

You Try It 2

Evaluate $G(x) = \dfrac{3x}{x + 2}$ when $x = -4$.

Your solution

Example 3

Evaluate $Q(r) = 2r + 5$ when $r = h + 3$.

Solution

$$Q(r) = 2r + 5$$
$$Q(h + 3) = 2(h + 3) + 5$$
$$= 2h + 6 + 5$$
$$= 2h + 11$$

You Try It 3

Evaluate $f(x) = x^2 - 11$ when $x = 3h$.

Your solution

Example 4

Find the range of $f(x) = x^2 - 1$ if the domain is $\{-2, -1, 0, 1, 2\}$.

Solution

To find the range, evaluate the function at each element of the domain.

$$f(x) = x^2 - 1$$
$$f(-2) = (-2)^2 - 1 = 4 - 1 = 3$$
$$f(-1) = (-1)^2 - 1 = 1 - 1 = 0$$
$$f(0) = 0^2 - 1 = 0 - 1 = -1$$
$$f(1) = 1^2 - 1 = 1 - 1 = 0$$
$$f(2) = 2^2 - 1 = 4 - 1 = 3$$

The range is $\{-1, 0, 3\}$. Note that 0 and 3 are listed only once.

You Try It 4

Find the range of $h(z) = 3z + 1$ if the domain is $\left\{0, \dfrac{1}{3}, \dfrac{2}{3}, 1\right\}$.

Your solution

Example 5

What is the domain of $f(x) = 2x^2 - 7x + 1$?

Solution

Because $2x^2 - 7x + 1$ evaluates to a real number for any value of x, the domain of the function is all real numbers.

You Try It 5

What value is excluded from the domain of $f(x) = \dfrac{2}{x - 5}$?

Your solution

Solutions on p. S7

3.2 Exercises

Objective A　　**To evaluate a function**

1. In your own words, explain what a function is.

2. What is the domain of a function? What is the range of a function?

3. Does the diagram below represent a function? Explain your answer.

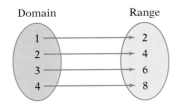

4. Does the diagram below represent a function? Explain your answer.

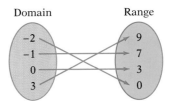

5. Does the diagram below represent a function? Explain your answer.

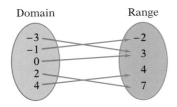

6. Does the diagram below represent a function? Explain your answer.

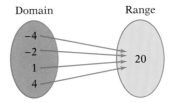

7. Does the diagram below represent a function? Explain your answer.

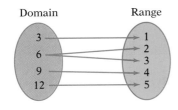

8. Does the diagram below represent a function? Explain your answer.

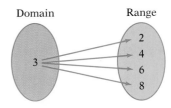

For Exercises 9 to 16, state whether the relation is a function.

9. $\{(0, 0), (2, 4), (3, 6), (4, 8), (5, 10)\}$

10. $\{(1, 3), (3, 5), (5, 7), (7, 9)\}$

11. $\{(-2, -1), (-4, -5), (0, -1), (3, 5)\}$

12. $\{(-3, -1), (-1, -1), (0, 1), (2, 6)\}$

13. $\{(-2, 3), (-1, 3), (0, -3), (1, 3), (2, 3)\}$

14. $\{(0, 0), (1, 0), (2, 0), (3, 0), (4, 0)\}$

15. $\{(1, 1), (4, 2), (9, 3), (1, -1), (4, -2)\}$

16. $\{(3, 1), (3, 2), (3, 3), (3, 4)\}$

17. 🔵 **Shipping** The table at the right shows the cost to send an overnight package using United Parcel Service.
 a. Does this table define a function?
 b. Given $x = 2.75$ lb, find y.

Weight in pounds (x)	Cost (y)
$0 < x \leq 1$	$28.25
$1 < x \leq 2$	$31.25
$2 < x \leq 3$	$34.75
$3 < x \leq 4$	$37.75
$4 < x \leq 5$	$40.75

18. 🔵 **Shipping** The table at the right shows the cost to send an "Express Mail" package using the U.S. Postal Service.
 a. Does this table define a function?
 b. Given $x = 0.5$ lb, find y.

Weight in pounds (x)	Cost (y)
$0 < x \leq 0.5$	$13.65
$0.5 < x \leq 2$	$17.85
$2 < x \leq 3$	$21.05
$3 < x \leq 4$	$24.20
$4 < x \leq 5$	$27.30

For Exercises 19 to 22, given $f(x) = 5x - 4$, evaluate:

19. $f(3)$ **20.** $f(-2)$ **21.** $f(0)$ **22.** $f(-1)$

For Exercises 23 to 26, given $G(t) = 4 - 3t$, evaluate:

23. $G(0)$ **24.** $G(-3)$ **25.** $G(-2)$ **26.** $G(4)$

For Exercises 27 to 30, given $q(r) = r^2 - 4$, evaluate:

27. $q(3)$ **28.** $q(4)$ **29.** $q(-2)$ **30.** $q(-5)$

For Exercises 31 to 34, given $F(x) = x^2 + 3x - 4$, evaluate:

31. $F(4)$ **32.** $F(-4)$ **33.** $F(-3)$ **34.** $F(-6)$

For Exercises 35 to 38, given $H(p) = \dfrac{3p}{p + 2}$, evaluate:

35. $H(1)$ **36.** $H(-3)$ **37.** $H(t)$ **38.** $H(v)$

For Exercises 39 to 42, given $s(t) = t^3 - 3t + 4$, evaluate:

39. $s(-1)$ **40.** $s(2)$ **41.** $s(a)$ **42.** $s(w)$

43. Given $P(x) = 4x + 7$, write $P(-2 + h) - P(-2)$ in simplest form.

44. Given $G(t) = 9 - 2t$, write $G(-3 + h) - G(-3)$ in simplest form.

45. Business Game Engineering has just completed the programming and testing for a new computer game. The cost to manufacture and package the game depends on the number of units Game Engineering plans to sell. The table at the right shows the cost per game for packaging various quantities.
 a. Evaluate this function when $x = 7000$.
 b. Evaluate this function when $x = 20,000$.

Number of Games Manufactured	Cost to Manufacture One Game
$0 < x \leq 2500$	$6.00
$2500 < x \leq 5000$	$5.50
$5000 < x \leq 10,000$	$4.75
$10,000 < x \leq 20,000$	$4.00
$20,000 < x \leq 40,000$	$3.00

46. Airports Airport administrators have a tendency to price airport parking at a rate that discourages people from using the parking lot for long periods of time. The rate structure for an airport is given in the table at the right.
 a. Evaluate this function when $t = 2.5$ h.
 b. Evaluate this function when $t = 7$ h.

Hours Parked	Cost
$0 < t \leq 1$	$1.00
$1 < t \leq 2$	$3.00
$2 < t \leq 4$	$6.50
$4 < t \leq 7$	$10.00
$7 < t \leq 12$	$14.00

47. Real Estate A real estate appraiser charges a fee that depends on the estimated value, V, of the property. A table giving the fees charged for various estimated values of the real estate appears at the right.
 a. Evaluate this function when $V = \$5,000,000$.
 b. Evaluate this function when $V = \$767,000$.

Value of Property	Appraisal Fee
$V < 100,000$	$350
$100,000 \leq V < 500,000$	$525
$500,000 \leq V < 1,000,000$	$950
$1,000,000 \leq V < 5,000,000$	$2500
$5,000,000 \leq V < 10,000,000$	$3000

48. Shipping The cost to mail a priority overnight package by Federal Express depends on the weight, w, of the package. A table of the costs for selected weights is given at the right.
 a. Evaluate this function when $w = 2$ lb 3 oz.
 b. Evaluate this function when $w = 1.9$ lb.

Weight (lb)	Cost
$0 < w \leq 1$	$27.00
$1 < w \leq 2$	$29.75
$2 < w \leq 3$	$32.75
$3 < w \leq 4$	$35.75
$4 < w \leq 5$	$39.00

For Exercises 49 to 58, find the domain and range of the function.

49. $\{(1, 1), (2, 4), (3, 7), (4, 10), (5, 13)\}$

50. $\{(2, 6), (4, 18), (6, 38), (8, 66), (10, 102)\}$

51. $\{(0, 1), (2, 2), (4, 3), (6, 4)\}$

52. $\{(0, 1), (1, 2), (4, 3), (9, 4)\}$

53. $\{(1, 0), (3, 0), (5, 0), (7, 0), (9, 0)\}$

54. $\{(-2, -4), (2, 4), (-1, 1), (1, 1), (-3, 9), (3, 9)\}$

55. $\{(0, 0), (1, 1), (-1, 1), (2, 2), (-2, 2)\}$

56. $\{(0, -5), (5, 0), (10, 5), (15, 10)\}$

57. $\{(-2, -3), (-1, 6), (0, 7), (2, 3), (1, 9)\}$

58. $\{(-8, 0), (-4, 2), (-2, 4), (0, -4), (4, 4)\}$

For Exercises 59 to 76, what values are excluded from the domain of the function?

59. $f(x) = \dfrac{1}{x - 1}$

60. $g(x) = \dfrac{1}{x + 4}$

61. $h(x) = \dfrac{x + 3}{x + 8}$

62. $F(x) = \dfrac{2x - 5}{x - 4}$

63. $f(x) = 3x + 2$

64. $g(x) = 4 - 2x$

65. $G(x) = x^2 + 1$

66. $H(x) = \dfrac{1}{2}x^2$

67. $f(x) = \dfrac{x - 1}{x}$

68. $g(x) = \dfrac{2x + 5}{7}$

69. $H(x) = x^2 - x + 1$

70. $f(x) = 3x^2 + x + 4$

71. $f(x) = \dfrac{2x - 5}{3}$

72. $g(x) = \dfrac{3 - 5x}{5}$

73. $H(x) = \dfrac{x - 2}{x + 2}$

74. $h(x) = \dfrac{3 - x}{6 - x}$

75. $f(x) = \dfrac{x - 2}{2}$

76. $G(x) = \dfrac{2}{x - 2}$

For Exercises 77 to 90, find the range of the function defined by the equation and the given domain.

77. $f(x) = 4x - 3$; domain = $\{0, 1, 2, 3\}$

78. $G(x) = 3 - 5x$; domain = $\{-2, -1, 0, 1, 2\}$

79. $g(x) = 5x - 8$; domain = $\{-3, -1, 0, 1, 3\}$

80. $h(x) = 3x - 7$; domain = $\{-4, -2, 0, 2, 4\}$

81. $h(x) = x^2$; domain = $\{-2, -1, 0, 1, 2\}$

82. $H(x) = 1 - x^2$; domain = $\{-2, -1, 0, 1, 2\}$

83. $f(x) = 2x^2 - 2x + 2$;
 domain = $\{-4, -2, 0, 4\}$

84. $G(x) = -2x^2 + 5x - 2$;
 domain = $\{-3, -1, 0, 1, 3\}$

85. $H(x) = \dfrac{5}{1-x}$; domain = $\{-2, 0, 2\}$

86. $g(x) = \dfrac{4}{4-x}$; domain = $\{-5, 0, 3\}$

87. $f(x) = \dfrac{2}{x-4}$; domain = $\{-2, 0, 2, 6\}$

88. $g(x) = \dfrac{x}{3-x}$; domain = $\{-2, -1, 0, 1, 2\}$

89. $H(x) = 2 - 3x - x^2$; domain = $\{-5, 0, 5\}$

90. $G(x) = 4 - 3x - x^3$; domain = $\{-3, 0, 3\}$

APPLYING THE CONCEPTS

91. Explain the words *relation* and *function*. Include in your explanation how the meanings of the two words differ.

92. Give a real-world example of a relation that is not a function. Is it possible to give an example of a function that is not a relation? If so, give one. If not, explain why it is not possible.

93. **a.** Find the set of ordered pairs (x, y) determined by the equation $y = x^3$, where $x \in \{-2, -1, 0, 1, 2\}$.
 b. Does the set of ordered pairs define a function? Why or why not?

94. **a.** Find the set of ordered pairs (x, y) determined by the equation $|y| = x$, where $x \in \{0, 1, 2, 3\}$.
 b. Does the set of ordered pairs define a function? Why or why not?

95. **Energy** The power a windmill can generate is a function of the velocity of the wind. The function can be approximated by $P = f(v) = 0.015v^3$, where P is the power in watts and v is the velocity of the wind in meters per second. How much power will be produced by a windmill when the velocity of the wind is 15 m/s?

96. **Automotive Technology** The distance, *s* (in feet), a car will skid on a certain road surface after the brakes are applied is a function of the car's velocity, *v* (in miles per hour). The function can be approximated by $s = f(v) = 0.017v^2$. How far will a car skid after its brakes are applied if it is traveling 60 mph?

For Exercises 97 to 100, each graph defines a function. Evaluate the function by estimating the ordinate (which is the value of the function) for the given value of *t*.

97. **Parachuting** The graph at the right shows the speed, *v*, in feet per second that a parachutist is falling during the first 20 s after jumping out of a plane.
 a. Estimate the speed at which the parachutist is falling when *t* = 5 s.
 b. Estimate the speed at which the parachutist is falling when *t* = 15 s.

98. **Psychology** The graph at the right shows what an industrial psychologist has determined to be the average percent score, *P*, for an employee taking a performance test *t* weeks after training begins.
 a. Estimate the score an employee would receive on this test when *t* = 4 weeks.
 b. Estimate the score an employee would receive on this test when *t* = 10 weeks.

99. **Temperature** The graph at the right shows the temperature, *T*, in degrees Fahrenheit, of a can of cola *t* hours after it is placed in a refrigerator.
 a. Use the graph to estimate the temperature of the cola when *t* = 5 h.
 b. Use the graph to estimate the temperature of the cola when *t* = 15 h.

100. **Athletics** The graph at the right shows the decrease in the heart rate, *r*, of a runner (in beats per minute) *t* minutes after the completion of a race.
 a. Use the graph to estimate the heart rate of a runner when *t* = 5 min.
 b. Use the graph to estimate the heart rate of a runner when *t* = 20 min.

3.3 Linear Functions

Objective A **To graph a linear function**

Recall that the ordered pairs of a function can be written as $(x, f(x))$ or (x, y). The **graph of a function** is a graph of the ordered pairs (x, y) that belong to the function. Certain functions have characteristic graphs. A function that can be written in the form $f(x) = mx + b$ (or $y = mx + b$) is called a **linear function** because its graph is a straight line.

Examples of linear functions are shown at the right. Note that the exponent on each variable is 1.

$$f(x) = 2x + 5 \qquad (m = 2, b = 5)$$
$$P(t) = 3t - 2 \qquad (m = 3, b = -2)$$
$$y = -2x \qquad (m = -2, b = 0)$$
$$y = -\frac{2}{3}x + 1 \qquad \left(m = -\frac{2}{3}, b = 1\right)$$
$$g(z) = z - 2 \qquad (m = 1, b = -2)$$

The equation $y = x^2 + 4x + 3$ is not a linear function because it includes a term with a variable squared. The equation $f(x) = \dfrac{3}{x - 2}$ is not a linear function because a variable occurs in the denominator. Another example of an equation that is not a linear function is $y = \sqrt{x} + 4$; this equation contains a variable within a radical and so is not a linear function.

Consider $f(x) = 2x + 1$. Evaluating the linear function when $x = -2$, $-1, 0, 1,$ and 2 produces some of the ordered pairs of the function. It is convenient to record the results in a table similar to the one at the right. The graph of the ordered pairs is shown in the leftmost figure below.

x	$f(x) = 2x + 1$	y	(x, y)
-2	$2(-2) + 1$	-3	$(-2, -3)$
-1	$2(-1) + 1$	-1	$(-1, -1)$
0	$2(0) + 1$	1	$(0, 1)$
1	$2(1) + 1$	3	$(1, 3)$
2	$2(2) + 1$	5	$(2, 5)$

Evaluating the function when x is not an integer produces more ordered pairs to graph, such as $\left(-\dfrac{5}{2}, -4\right)$ and $\left(\dfrac{3}{2}, 4\right)$, as shown in the middle figure below. Evaluating the function for still other values of x would result in more and more ordered pairs being graphed. The result would be so many dots that the graph would look like the straight line shown in the rightmost figure, which is the graph of $f(x) = 2x + 1$.

 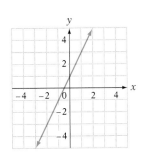

No matter what value of x is chosen, $2x + 1$ is a real number. This means the domain of $f(x) = 2x + 1$ is all real numbers. Therefore, we can use any real number when evaluating the function. Normally, however, values such as π or $\sqrt{5}$ are not used because it is difficult to graph the resulting ordered pairs.

Note from the graph of $f(x) = 2x + 1$ shown at the right that $(-1.5, -2)$ and $(3, 7)$ are the coordinates of points on the graph and that $f(-1.5) = -2$ and $f(3) = 7$. Note also that the point whose coordinates are $(2, 1)$ is not a point on the graph and that $f(2) \neq 1$. Every point on the graph is an ordered pair that belongs to the function, and every ordered pair that belongs to the function corresponds to a point on the graph.

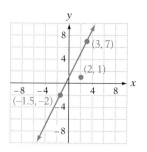

Integrating Technology

See the Projects and Group Activities at the end of this chapter for instructions on using a graphing calculator to graph a linear function. Instructions are also provided in the Keystroke Guide: *Graph*.

Whether an equation is written as $f(x) = mx + b$ or as $y = mx + b$, the equation represents a linear function, and the graph of the equation is a straight line.

Because the graph of a linear function is a straight line, and a straight line is determined by two points, the graph of a linear function can be drawn by finding only two of the ordered pairs of the function. However, it is recommended that you find at least *three* ordered pairs to ensure accuracy.

HOW TO Graph: $f(x) = -\dfrac{1}{2}x + 3$

x	$y = f(x)$
-4	5
0	3
2	2

- Find at least three ordered pairs. When the coefficient of x is a fraction, choose values of x that will simplify the calculations. The ordered pairs can be displayed in a table.

- Graph the ordered pairs and draw a line through the points.

TAKE NOTE

When the coefficient of x is a fraction, choose values of x that are multiples of the denominator of the fraction. This will result in coordinates that are integers.

Example 1 Graph: $f(x) = -\dfrac{3}{2}x - 3$

Solution

x	$y = f(x)$
0	-3
-2	0
-4	3

You Try It 1 Graph: $f(x) = \dfrac{3}{5}x - 4$

Your solution

Solution on p. S7

Example 2 Graph: $y = \dfrac{2}{3}x$

Solution

x	y
0	0
3	2
-3	-2

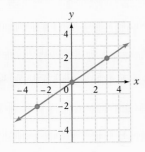

You Try It 2 Graph: $y = -\dfrac{3}{4}x$

Your solution

Solution on p. S7

Objective B **To graph an equation of the form $Ax + By = C$**

An equation of the form $Ax + By = C$, where A and B are coefficients and C is a constant, is also a *linear equation in two variables*. This equation can be written in the form $y = mx + b$.

> **HOW TO** Write $4x - 3y = 6$ in the form $y = mx + b$.
>
> $$4x - 3y = 6$$
> $$-3y = -4x + 6$$ • Subtract $4x$ from each side of the equation.
> $$y = \frac{4}{3}x - 2$$ • Divide each side of the equation by -3. This is the form
> $$y = mx + b.\ m = \frac{4}{3},\ b = -2$$

To graph an equation of the form $Ax + By = C$, first solve the equation for y. Then follow the same procedure used for graphing an equation of the form $y = mx + b$.

> **HOW TO** Graph: $3x + 2y = 6$
>
> $$3x + 2y = 6$$
> $$2y = -3x + 6$$ • Solve the equation for y.
> $$y = -\frac{3}{2}x + 3$$
>
x	y
> | 0 | 3 |
> | 2 | 0 |
> | 4 | -3 |
>
> • Find at least three solutions.
>
>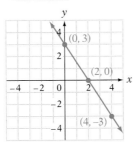
>
> • Graph the ordered pairs in a rectangular coordinate system. Draw a straight line through the points.

An equation in which one of the variables is missing has a graph that is either a horizontal or a vertical line.

The equation $y = -2$ can be written

$$0 \cdot x + y = -2$$

Because $0 \cdot x = 0$ for all values of x, y is always -2 for every value of x.

Some of the possible ordered pairs are given in the table below. The graph is shown at the right.

x	y
-2	-2
0	-2
3	-2

Graph of $y = b$

The **graph of $y = b$** is a horizontal line passing through the point $(0, b)$.

The equation $y = -2$ represents a function. Some of the ordered pairs of this function are $(-2, -2)$, $(0, -2)$, and $(3, -2)$. In functional notation we would write $f(x) = -2$. This function is an example of a constant function. No matter what value of x is selected, $f(x) = -2$.

Graph of a Constant Function

A function given by $f(x) = b$, where b is a constant, is a **constant function**. The graph of the constant function is a horizontal line passing through $(0, b)$.

For each value in the domain of a constant function, the value of the function is the same (constant). For instance, if $f(x) = 4$, then $f(2) = 4$, $f(3) = 4$, $f(\sqrt{3}) = 4$, $f(\pi) = 4$, and so on. The value of $f(x)$ is 4 for all values of x.

HOW TO Graph: $y + 4 = 0$

Solve for y.
$y + 4 = 0$
$\quad y = -4$

The graph of $y = -4$ is a horizontal line passing through $(0, -4)$.

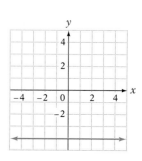

HOW TO Evaluate $P(t) = -7$ when $t = 6$.

$P(t) = -7$ • The value of the constant function is the
$P(6) = -7$ same for all values of the variable.

For the equation $y = -2$, the coefficient of x was zero. For the equation $x = 2$, the coefficient of y is zero. For instance, the equation $x = 2$ can be written

$$x + 0 \cdot y = 2$$

No matter what value of y is chosen, $0 \cdot y = 0$ and therefore x is always 2.

Some of the possible ordered pairs are given in the table below. The graph is shown at the right.

x	y
2	6
2	1
2	−4

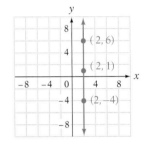

Graph of $x = a$

The **graph of $x = a$** is a vertical line passing through the point $(a, 0)$.

TAKE NOTE

The equation $y = b$ represents a function. The equation $x = a$ does *not* represent a function. Remember, not all equations represent functions.

Recall that a function is a set of ordered pairs in which no two ordered pairs have the same first coordinate and different second coordinates. Because $(2, 6)$, $(2, 1)$, and $(2, -4)$ are ordered pairs belonging to the equation $x = 2$, this equation does not represent a function, and the graph is not the graph of a function.

Example 3 Graph: $2x + 3y = 9$

Solution Solve the equation for y.

$2x + 3y = 9$
$\quad 3y = -2x + 9$
$\quad\ \ y = -\dfrac{2}{3}x + 3$

x	y
−3	5
0	

You Try It 3 Graph: $-3x + 2y = 4$

Your solution

Solution on p. S7

Example 4 Graph: $x = -4$

Solution

The graph of an equation of the form $x = a$ is a vertical line passing through the point whose coordinates are $(a, 0)$.

- The graph of $x = -4$ goes through the point $(-4, 0)$.

You Try It 4 Graph: $y - 3 = 0$

Your solution

Solution on p. S7

Objective C **To find the *x*- and the *y*-intercepts of a straight line**

The graph of the equation $x - 2y = 4$ is shown at the right. The graph crosses the x-axis at the point $(4, 0)$. This point is called the **x-intercept.** The graph also crosses the y-axis at the point $(0, -2)$. This point is called the **y-intercept.**

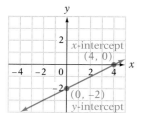

TAKE NOTE

The x-intercept occurs when $y = 0$.

The y-intercept occurs when $x = 0$.

HOW TO Find the x- and y-intercepts of the graph of the equation $3x + 4y = -12$.

To find the x-intercept, let $y = 0$. (Any point on the x-axis has y-coordinate 0.)

$$3x + 4y = -12$$
$$3x + 4(0) = -12$$
$$3x = -12$$
$$x = -4$$

The x-intercept is $(-4, 0)$.

To find the y-intercept, let $x = 0$. (Any point on the y-axis has x-coordinate 0.)

$$3x + 4y = -12$$
$$3(0) + 4y = -12$$
$$4y = -12$$
$$y = -3$$

The y-intercept is $(0, -3)$.

TAKE NOTE

Note that the y-coordinate of the y-intercept $(0, 7)$ has the same value as the constant term of $y = \frac{2}{3}x + 7$.

As shown below, this is always true:

$$y = mx + b$$
$$y = m(0) + b$$
$$y = b$$

Thus, the y-intercept is $(0, b)$.

HOW TO Find the y-intercept of $y = \frac{2}{3}x + 7$.

$$y = \frac{2}{3}x + 7$$

$$y = \frac{2}{3}(0) + 7$$

- To find the y-intercept, let $x = 0$.

$$y = 7$$

The y-intercept is $(0, 7)$.

For any equation of the form $y = mx + b$, the y-intercept is $(0, b)$.

A linear equation can be graphed by finding the *x*- and *y*-intercepts and then drawing a line through the two points.

HOW TO Graph $3x - 2y = 6$ by using the *x*- and *y*-intercepts.

$$3x - 2y = 6$$
$$3x - 2(0) = 6$$
$$3x = 6$$
$$x = 2$$

• To find the *x*-intercept, let $y = 0$. Then solve for *x*.

$$3x - 2y = 6$$
$$3(0) - 2y = 6$$
$$-2y = 6$$
$$y = -3$$

• To find the *y*-intercept, let $x = 0$. Then solve for *y*.

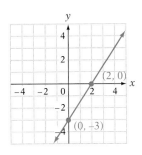

The *x*-intercept is $(2, 0)$. The *y*-intercept is $(0, -3)$.

Example 5

Graph $4x - y = 4$ by using the *x*- and *y*-intercepts.

Solution

x-intercept:		*y*-intercept:

$$4x - y = 4$$
$$4x - 0 = 4 \quad \bullet \text{ Let}$$
$$4x = 4 \qquad\quad y = 0.$$
$$x = 1$$
$$(1, 0)$$

$$4x - y = 4$$
$$4(0) - y = 4 \quad \bullet \text{ Let}$$
$$-y = 4 \qquad\quad x = 0.$$
$$y = -4$$
$$(0, -4)$$

You Try It 5

Graph $3x - y = 2$ by using the *x*- and *y*-intercepts.

Your solution

Example 6

Graph $y = \dfrac{2}{3}x - 2$ by using the *x*- and *y*-intercepts.

Solution

x-intercept:

$$y = \frac{2}{3}x - 2$$
$$0 = \frac{2}{3}x - 2$$
$$-\frac{2}{3}x = -2$$
$$x = 3$$
$$(3, 0)$$

y-intercept:

$(0, b)$
$b = -2$
$(0, -2)$

You Try It 6

Graph $y = \dfrac{1}{4}x + 1$ by using the *x*- and *y*-intercepts.

Your solution

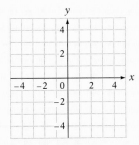

Solutions on pp. S7–S8

Objective D **To solve application problems**

There are a variety of applications of linear functions.

HOW TO The heart rate, R, after t minutes for a person taking a brisk walk can be approximated by the equation $R = 2t + 72$.

a. Graph this equation for $0 \le t \le 10$.
b. The point whose coordinates are (5, 82) is on the graph. Write a sentence that describes the meaning of this ordered pair.

a.

b. The ordered pair (5, 82) means that after 5 min the person's heart rate is 82 beats per minute.

Example 7

An electronics technician charges $45 plus $1 per minute to repair defective wiring in a home or apartment. The equation that describes the total cost, C, to have defective wiring repaired is given by $C = t + 45$, where t is the number of minutes the technician works. Graph this equation for $0 \le t \le 60$. The point whose coordinates are (15, 60) is on this graph. Write a sentence that describes the meaning of this ordered pair.

Solution

• Graph $C = t + 45$.
When $t = 0$, $C = 45$.
When $t = 50$, $C = 95$.

The ordered pair (15, 60) means that it costs $60 for the technician to work 15 min.

You Try It 7

The height h (in inches) of a person and the length L (in inches) of that person's stride while walking are related. The equation $h = \frac{3}{4}L + 50$ approximates this relationship. Graph this equation for $15 \le L \le 40$. The point whose coordinates are (32, 74) is on this graph. Write a sentence that describes the meaning of this ordered pair.

Your solution

Solution on p. S8

3.3 Exercises

Objective A **To graph a linear function**

For Exercises 1 to 9, graph.

1. $y = 3x - 4$

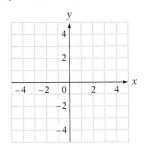

2. $y = -2x + 3$

3. $y = -\dfrac{2}{3}x$

4. $y = \dfrac{3}{2}x$

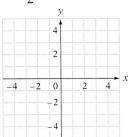

5. $y = \dfrac{2}{3}x - 4$

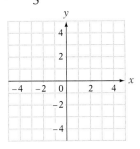

6. $y = \dfrac{3}{4}x + 2$

7. $y = -\dfrac{1}{3}x + 2$

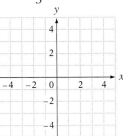

8. $y = -\dfrac{3}{2}x - 3$

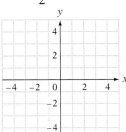

9. $y = \dfrac{3}{5}x - 1$

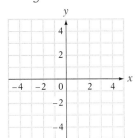

Objective B **To graph an equation of the form *Ax + By = C***

For Exercises 10 to 18, graph.

10. $2x - y = 3$

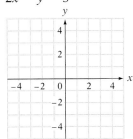

11. $2x + y = -3$

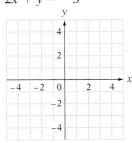

12. $2x + 5y = 10$

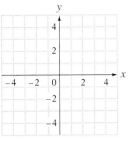

13. $x - 4y = 8$

14. $y = -2$

15. $y = \dfrac{1}{3}x$

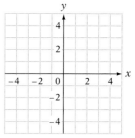

16. $2x - 3y = 12$

17. $3x - y = -2$

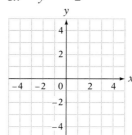

18. $3x - 2y = 8$

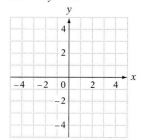

Objective C **To find the *x*- and the *y*-intercepts of a straight line**

For Exercises 19 to 27, find the *x*- and *y*-intercepts and graph.

19. $x - 2y = -4$

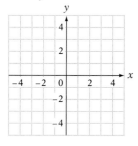

20. $3x + y = 3$

21. $2x - 3y = 9$

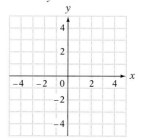

22. $2x - y = 4$

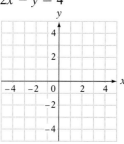

23. $2x + y = 3$

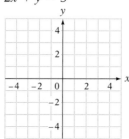

24. $4x - 3y = 8$

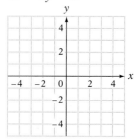

25. $3x + 2y = 4$

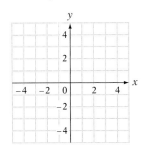

26. $2x - 3y = 4$

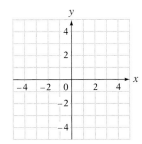

27. $4x - 3y = 6$

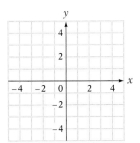

Objective D **To solve application problems**

28. **Animal Science** A bee beats its wings approximately 100 times per second. The equation that describes the total number of times a bee beats its wings is given by $B = 100t$. Graph this equation for $0 \le t \le 60$. The point (35, 3500) is on this graph. Write a sentence that describes the meaning of this ordered pair.

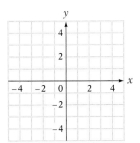

29. **Compensation** Marlys receives $11 per hour as a mathematics department tutor. The equation that describes Marlys's wages is $W = 11t$, where t is the number of hours she spends tutoring. Graph this equation for $0 \le t \le 20$. The ordered pair (15, 165) is on the graph. Write a sentence that describes the meaning of this ordered pair.

30. **Roller Coasters** The Superman Krypton Coaster at Fiesta Texas® in San Antonio has a maximum speed of approximately 105 ft/s. The equation that describes the total number of feet traveled by the roller coaster in t seconds at this speed is given by $D = 105t$. Graph this equation for $0 \le t \le 10$. The point (5, 525) is on this graph. Write a sentence that describes the meaning of this ordered pair.

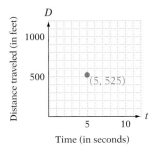

31. **Telecommunication** The monthly cost for receiving messages from a telephone answering service is $8.00 plus $.20 a message. The equation that describes the cost is $C = 0.20n + 8.00$, where n is the number of messages received. Graph this equation for $0 \le n \le 40$. The point (32, 14.40) is on the graph. Write a sentence that describes the meaning of this ordered pair.

32. **Manufacturing** The cost of manufacturing skis is $5000 for startup and $80 per pair of skis manufactured. The equation that describes the cost of manufacturing n pairs of skis is $C = 80n + 5000$. Graph this equation for $0 \le n \le 2000$. The point (50, 9000) is on the graph. Write a sentence that describes the meaning of this ordered pair.

33. 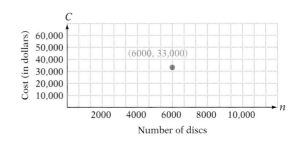 **Manufacturing** The cost of manufacturing compact discs is $3000 for startup and $5 per disc. The equation that describes the cost of manufacturing n compact discs is $C = 5n + 3000$. Graph this equation for $0 \le n \le 10,000$. The point (6000, 33,000) is on the graph. Write a sentence that describes the meaning of this ordered pair.

APPLYING THE CONCEPTS

34. Explain what the graph of an equation represents.

35. Explain how to graph the equation of a straight line by plotting points.

36. Explain how to graph the equation of a straight line by using its x- and y-intercepts.

37. Explain why you cannot graph the equation $4x + 3y = 0$ by using just its intercepts.

An equation of the form $\frac{x}{a} + \frac{y}{b} = 1$, where $a \ne 0$ and $b \ne 0$, is called the *intercept form of a straight line* because $(a, 0)$ and $(0, b)$ are the x- and y-intercepts of the graph of the equation. Graph the equations in Exercises 38 to 40.

38. $\dfrac{x}{3} + \dfrac{y}{5} = 1$

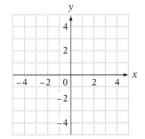

39. $\dfrac{x}{2} + \dfrac{y}{3} = 1$

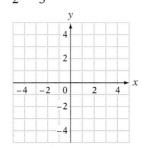

40. $\dfrac{x}{3} - \dfrac{y}{2} = 1$

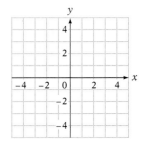

3.4 Slope of a Straight Line

Objective A **To find the slope of a line given two points**

The graphs of $y = 3x + 2$ and $y = \frac{2}{3}x + 2$ are shown at the left. Each graph crosses the y-axis at the point $(0, 2)$, but the graphs have different slants. The **slope** of a line is a measure of the slant of the line. The symbol for slope is m.

The slope of a line containing two points is the ratio of the change in the y values between the two points to the change in the x values. The line containing the points whose coordinates are $(-1, -3)$ and $(5, 2)$ is shown below.

The change in the y values is the difference between the y-coordinates of the two points.

$$\text{Change in } y = 2 - (-3) = 5$$

The change in the x values is the difference between the x-coordinates of the two points.

$$\text{Change in } x = 5 - (-1) = 6$$

The slope of the line between the two points is the ratio of the change in y to the change in x.

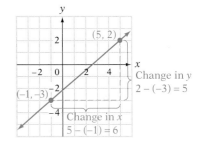

$$\text{Slope} = m = \frac{\text{change in } y}{\text{change in } x} = \frac{5}{6} \qquad m = \frac{2 - (-3)}{5 - (-1)} = \frac{5}{6}$$

In general, if $P_1(x_1, y_1)$ and $P_2(x_2, y_2)$ are two points on a line, then

$$\text{Change in } y = y_2 - y_1 \qquad \text{Change in } x = x_2 - x_1$$

Using these ideas, we can state a formula for slope.

Slope Formula

The slope of the line containing the two points $P_1(x_1, y_1)$ and $P_2(x_2, y_2)$ is given by

$$m = \frac{y_2 - y_1}{x_2 - x_1}, \ x_1 \neq x_2$$

Frequently, the Greek letter Δ is used to designate the change in a variable. Using this notation, we can write the equations for the change in y and the change in x as follows:

$$\text{Change in } y = \Delta y = y_2 - y_1 \qquad \text{Change in } x = \Delta x = x_2 - x_1$$

Using this notation, the slope formula is written $m = \dfrac{\Delta y}{\Delta x}$.

Study Tip

To learn mathematics, you must be an active participant. Listening and watching your professor do mathematics is not enough. Take notes in class, mentally think through every question your instructor asks, and try to answer it even if you are not called on to answer it verbally. Ask questions when you have them. See *AIM for Success*, pages xxix and xxx, for other ways to be an active learner.

HOW TO Find the slope of the line containing the points whose coordinates are $(-2, 0)$ and $(4, 5)$.

Let $P_1 = (-2, 0)$ and $P_2 = (4, 5)$. (It does not matter which point is named P_1 or P_2; the slope will be the same.)

$$m = \frac{y_2 - y_1}{x_2 - x_1} = \frac{5 - 0}{4 - (-2)} = \frac{5}{6}$$

A line that slants upward to the right always has a **positive slope.**

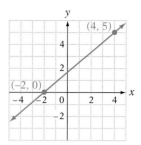

Positive slope

HOW TO Find the slope of the line containing the points whose coordinates are $(-3, 4)$ and $(4, 2)$.

Let $P_1 = (-3, 4)$ and $P_2 = (4, 2)$.

$$m = \frac{y_2 - y_1}{x_2 - x_1} = \frac{2 - 4}{4 - (-3)} = \frac{-2}{7} = -\frac{2}{7}$$

A line that slants downward to the right always has a **negative slope.**

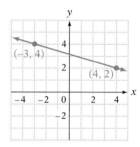

Negative slope

HOW TO Find the slope of the line containing the points whose coordinates are $(-2, 2)$ and $(4, 2)$.

Let $P_1 = (-2, 2)$ and $P_2 = (4, 2)$.

$$m = \frac{y_2 - y_1}{x_2 - x_1} = \frac{2 - 2}{4 - (-2)} = \frac{0}{6} = 0$$

A horizontal line has **zero slope.**

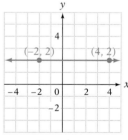

Zero slope

HOW TO Find the slope of the line containing the points whose coordinates are $(1, -2)$ and $(1, 3)$.

Let $P_1 = (1, -2)$ and $P_2 = (1, 3)$.

$$m = \frac{y_2 - y_1}{x_2 - x_1} = \frac{3 - (-2)}{1 - 1} = \frac{5}{0} \qquad \text{Not a real number}$$

The slope of a vertical line has **undefined slope.**

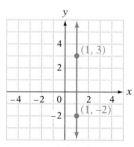

Undefined

Point of Interest

One of the motivations for the discovery of calculus was to solve a more complicated version of the distance-rate problem at the right.

You may be familiar with twirling a ball on the end of a string. If you release the string, the ball flies off in a path as shown below.

The question that mathematicians tried to answer was essentially, "What is the slope of the line represented by the arrow?"

Answering questions similar to this led to the development of one aspect of calculus.

There are many applications of slope. Here are two examples.

The first record for the one-mile run was recorded in 1865 in England. Richard Webster ran the mile in 4 min 36.5 s. His average speed was approximately 19 feet per second.

The graph at the right shows the distance Webster ran during that run. From the graph, note that after 60 s (1 min) he had traveled 1140 ft and that after 180 s (3 min) he had traveled 3420 ft.

Let the point (60, 1140) be (x_1, y_1) and the point (180, 3420) be (x_2, y_2). The slope of the line between these two points is

$$m = \frac{y_2 - y_1}{x_2 - x_1}$$

$$= \frac{3420 - 1140}{180 - 60} = \frac{2280}{120} = 19$$

Note that the slope of the line is the same as Webster's average speed, 19 feet per second.

Average speed is related to slope.

The following example is related to agriculture.

The number of farm workers in the United States declined from 10 million in 1950 to 3 million in 2000. (*Source*: U.S. Department of Agriculture)

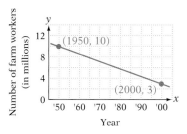

Let the point (1950, 10) be (x_1, y_1) and the point (2000, 3) be (x_2, y_2). The slope of the line between these two points is

$$m = \frac{y_2 - y_1}{x_2 - y_1}$$

$$= \frac{3 - 10}{2000 - 1950} = \frac{-7}{50} = -0.14$$

Note that if we interpret negative slope as decreasing population, then the slope of the line represents the annual rate at which the number of farm workers was decreasing, 0.14 million or 140,000 workers per year.

In general, any quantity that is expressed by using the word *per* is represented mathematically as slope. In the first example, the slope represented the average speed, 19 feet per second. In the second example, the slope represented the rate at which the number of farm workers was decreasing, -0.14 million workers per year.

Example 1

Find the slope of the line containing the points $(2, -5)$ and $(-4, 2)$.

Solution

Let $P_1 = (2, -5)$ and $P_2 = (-4, 2)$.

$$m = \frac{y_2 - y_1}{x_2 - x_1} = \frac{2 - (-5)}{-4 - 2} = \frac{7}{-6}$$

The slope is $-\dfrac{7}{6}$.

You Try It 1

Find the slope of the line containing the points $(4, -3)$ and $(2, 7)$.

Your solution

Example 2

Find the slope of the line containing the points $(-3, 4)$ and $(5, 4)$.

Solution

Let $P_1 = (-3, 4)$ and $P_2 = (5, 4)$.

$$m = \frac{y_2 - y_1}{x_2 - x_1}$$

$$= \frac{4 - 4}{5 - (-3)}$$

$$= \frac{0}{8} = 0$$

The slope of the line is zero.

You Try It 2

Find the slope of the line containing the points $(6, -1)$ and $(6, 7)$.

Your solution

Example 3

The graph below shows the relationship between the cost of an item and the sales tax. Find the slope of the line between the two points shown on the graph. Write a sentence that states the meaning of the slope.

Solution

$$m = \frac{5.25 - 3.50}{75 - 50}$$

$$= \frac{1.75}{25}$$

$$= 0.07$$

• $(x_1, y_1) = (50, 3.50)$,
 $(x_2, y_2) = (75, 5.25)$

A slope of 0.07 means that the sales tax is $.07 per dollar.

You Try It 3

The graph below shows the decrease in the value of a recycling truck for a period of 6 years. Find the slope of the line between the two points shown on the graph. Write a sentence that states the meaning of the slope.

Your solution

Solutions on p. S8

Objective B **To graph a line given a point and the slope**

The graph of the equation $y = -\frac{3}{4}x + 4$ is shown at the right. The points $(-4, 7)$ and $(4, 1)$ are on the graph. The slope of the line is

$$m = \frac{7 - 1}{-4 - 4} = \frac{6}{-8} = -\frac{3}{4}$$

Note that the slope of the line has the same value as the coefficient of x. The y-intercept is $(0, 4)$.

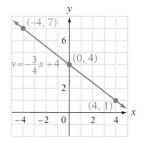

Slope-Intercept Form of a Straight Line

The equation $y = mx + b$ is called the **slope-intercept form of a straight line.** The slope of the line is m, the coefficient of x. The y-intercept is $(0, b)$.

When the equation of a straight line is in the form $y = mx + b$, the graph can be drawn by using the slope and y-intercept. First locate the y-intercept. Use the slope to find a second point on the line. Then draw a line through the two points.

HOW TO Graph $y = \frac{5}{3}x - 4$ by using the slope and y-intercept.

The slope is the coefficient of x:

$$m = \frac{5}{3} = \frac{\text{change in } y}{\text{change in } x}$$

The y-intercept is $(0, -4)$.

Beginning at the y-intercept $(0, -4)$, move right 3 units (change in x) and then up 5 units (change in y).

The point whose coordinates are $(3, 1)$ is a second point on the graph. Draw a line through the points $(0, -4)$ and $(3, 1)$.

TAKE NOTE

When graphing a line by using its slope and y-intercept, *always* start at the y-intercept.

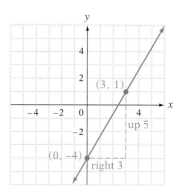

HOW TO Graph $x + 2y = 4$ by using the slope and y-intercept.

Solve the equation for y.

$x + 2y = 4$

$2y = -x + 4$

$y = -\frac{1}{2}x + 2$ • $m = -\frac{1}{2} = \frac{-1}{2}$,

$y\text{-intercept} = (0, 2)$

Beginning at the y-intercept $(0, 2)$, move right 2 units (change in x) and then down 1 unit (change in y).

The point whose coordinates are $(2, 1)$ is a second point on the graph. Draw a line through the points $(0, 2)$ and $(2, 1)$.

The graph of a line can be drawn when any point on the line and the slope of the line are given.

HOW TO Graph the line that passes through the point $(-4, -4)$ and has slope 2.

When the slope is an integer, write it as a fraction with denominator 1.

$$m = 2 = \frac{2}{1} = \frac{\text{change in } y}{\text{change in } x}$$

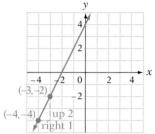

Locate $(-4, -4)$ in the coordinate plane. Beginning at that point, move right 1 unit (change in x) and then up 2 units (change in y).

The point whose coordinates are $(-3, -2)$ is a second point on the graph. Draw a line through the points $(-4, -4)$ and $(-3, -2)$.

TAKE NOTE

This example differs from the preceding two in that a point other than the y-intercept is used. In this case, start at the given point.

Example 4 Graph $y = -\frac{3}{2}x + 4$ by using the slope and y-intercept.

Solution $m = -\frac{3}{2} = \frac{-3}{2}$

y-intercept $= (0, 4)$

You Try It 4 Graph $2x + 3y = 6$ by using the slope and y-intercept.

Your solution

Example 5 Graph the line that passes through the point $(-2, 3)$ and has slope $-\frac{4}{3}$.

Solution $(x_1, y_1) = (-2, 3)$

$m = -\frac{4}{3} = \frac{-4}{3}$

You Try It 5 Graph the line that passes through the point $(-3, -2)$ and has slope 3.

Your solution

Solutions on p. S8

3.4 Exercises

Objective A **To find the slope of a line given two points**

For Exercises 1 to 18, find the slope of the line containing the points.

1. $P_1(1, 3), P_2(3, 1)$

2. $P_1(2, 3), P_2(5, 1)$

3. $P_1(-1, 4), P_2(2, 5)$

4. $P_1(3, -2), P_2(1, 4)$

5. $P_1(-1, 3), P_2(-4, 5)$

6. $P_1(-1, -2), P_2(-3, 2)$

7. $P_1(0, 3), P_2(4, 0)$

8. $P_1(-2, 0), P_2(0, 3)$

9. $P_1(2, 4), P_2(2, -2)$

10. $P_1(4, 1), P_2(4, -3)$

11. $P_1(2, 5), P_2(-3, -2)$

12. $P_1(4, 1), P_2(-1, -2)$

13. $P_1(2, 3), P_2(-1, 3)$

14. $P_1(3, 4), P_2(0, 4)$

15. $P_1(0, 4), P_2(-2, 5)$

16. $P_1(-2, 3), P_2(-2, 5)$

17. $P_1(-3, -1), P_2(-3, 4)$

18. $P_1(-2, -5), P_2(-4, -1)$

19. **Travel** The graph below shows the relationship between the distance traveled by a motorist and the time of travel. Find the slope of the line between the two points shown on the graph. Write a sentence that states the meaning of the slope.

20. **Media** The graph below shows the number of people subscribing to a sports magazine of increasing popularity. Find the slope of the line between the two points shown on the graph. Write a sentence that states the meaning of the slope.

21. **Temperature** The graph below shows the relationship between the temperature inside an oven and the time since the oven was turned off. Find the slope of the line. Write a sentence that states the meaning of the slope.

22. **Home Maintenance** The graph below shows the number of gallons of water remaining in a pool x minutes after a valve is opened to drain the pool. Find the slope of the line. Write a sentence that states the meaning of the slope.

23. **Fuel Consumption** The graph below shows how the amount of gas in the tank of a car decreases as the car is driven. Find the slope of the line. Write a sentence that states the meaning of the slope.

24. 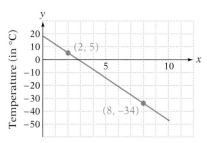 **Meteorology** The troposphere extends from Earth's surface to an elevation of about 11 km. The graph below shows the decrease in the temperature of the troposphere as altitude increases. Find the slope of the line. Write a sentence that states the meaning of the slope.

25. **Sports** The graph below shows the relationship between distance and time for the 5000-meter run for the world record by Deena Drossin in 2002. Find the slope of the line between the two points shown on the graph. Round to the nearest tenth. Write a sentence that states the meaning of the slope.

26. **Sports** The graph below shows the relationship between distance and time for the 10,000-meter run for the world record by Sammy Kipketer in 2002. Find the slope of the line between the two points shown on the graph. Round to the nearest tenth. Write a sentence that states the meaning of the slope.

Construction For Exercises 27 and 28, use the fact that the American National Standards Institute (ANSI) states that the slope for a wheelchair ramp must not exceed $\frac{1}{12}$.

27. Does a ramp that is 6 in. high and 5 ft long meet the requirements of ANSI?

28. Does a ramp that is 12 in. high and 170 in. long meet the requirements of ANSI?

Objective B **To graph a line given a point and the slope**

For Exercises 29 to 32, complete the table.

	Equation	Value of m	Value of b	Slope	y-intercept
29.	$y = -3x + 5$				
30.	$y = \dfrac{2}{5}x - 8$				
31.	$y = 4x$				
32.	$y = -7$				

For Exercises 33 to 44, graph by using the slope and the y-intercept.

33. $y = \dfrac{1}{2}x + 2$

34. $y = \dfrac{2}{3}x - 3$

35. $y = -\dfrac{3}{2}x$

36. $y = \dfrac{3}{4}x$

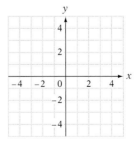

37. $y = -\dfrac{1}{2}x + 2$

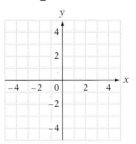

38. $y = \dfrac{2}{3}x - 1$

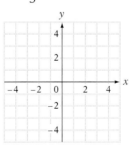

39. $y = 2x - 4$

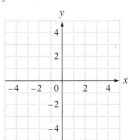

40. $y = -3x + 1$

41. $4x - y = 1$

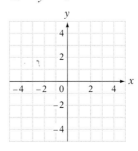

42. $4x + y = 2$

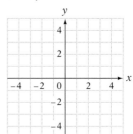

43. $x - 3y = 3$

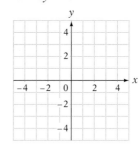

44. $3x + 2y = 8$

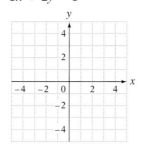

45. Graph the line that passes through the point $(-1, -3)$ and has slope $\frac{4}{3}$.

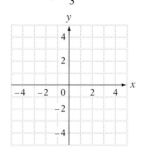

46. Graph the line that passes through the point $(-2, -3)$ and has slope $\frac{5}{4}$.

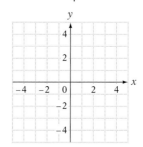

47. Graph the line that passes through the point $(-3, 0)$ and has slope -3.

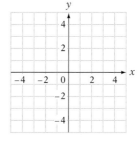

48. Graph the line that passes through the point $(2, -4)$ and has slope $-\frac{1}{2}$.

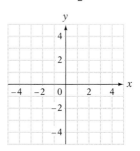

49. Graph the line that passes through the point $(-4, 1)$ and has slope $\frac{2}{3}$.

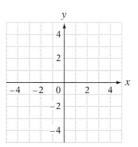

50. Graph the line that passes through the point $(1, 5)$ and has slope -4.

APPLYING THE CONCEPTS

Complete the following sentences.

51. If a line has a slope of 2, then the value of y <u>increases/decreases</u> by _____ as the value of x increases by 1.

52. If a line has a slope of -3, then the value of y <u>increases/decreases</u> by _____ as the value of x increases by 1.

53. If a line has a slope of -2, then the value of y <u>increases/decreases</u> by _____ as the value of x decreases by 1.

54. If a line has a slope of 3, then the value of y <u>increases/decreases</u> by _____ as the value of x decreases by 1.

55. If a line has a slope of $-\frac{2}{3}$, then the value of y increases/decreases by _____ as the value of x increases by 1.

56. If a line has a slope of $\frac{1}{2}$, then the value of y increases/decreases by _____ as the value of x increases by 1.

57. Match each equation with its graph.

　　i. $y = -2x + 4$

　　ii. $y = 2x - 4$

　　iii. $y = 2$

　　iv. $2x + 4y = 0$

　　v. $y = \frac{1}{2}x + 4$

　　vi. $y = -\frac{1}{4}x - 2$

A. 　　**B.** 　　**C.**

D. 　　**E.** 　　**F.**

58. Match each equation with its graph.

　　i. $y = 3x - 2$

　　ii. $y = -3x + 2$

　　iii. $y = -3x$

　　iv. $y = -3$

　　v. $3x - 2y = 0$

　　vi. $y = -\frac{1}{3}x + 2$

A. 　　**B.** 　　**C.**

D. 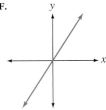　　**E.**　　**F.**

59. Explain how you can use the slope of a line to determine whether three given points lie on the same line. Then use your procedure to determine whether each of the following sets of points lie on the same line.
a. $(2, 5), (-1, -1), (3, 7)$　　　　**b.** $(-1, 5), (0, 3), (-3, 4)$

For Exercises 60 to 63, determine the value of k such that the points whose coordinates are given lie on the same line.

60. $(3, 2), (4, 6), (5, k)$

61. $(-2, 3), (1, 0), (k, 2)$

62. $(k, 1), (0, -1), (2, -2)$

63. $(4, -1), (3, -4), (k, k)$

3.5 Finding Equations of Lines

Objective A **To find the equation of a line given a point and the slope**

When the slope of a line and a point on the line are known, the equation of the line can be determined. If the particular point is the y-intercept, use the slope-intercept form, $y = mx + b$, to find the equation.

> **HOW TO** Find the equation of the line that contains the point $(0, 3)$ and has slope $\frac{1}{2}$.
>
> The known point is the y-intercept, $(0, 3)$.
>
> $y = mx + b$ • Use the slope-intercept form.
>
> $y = \dfrac{1}{2}x + 3$ • Replace m with $\dfrac{1}{2}$, the given slope.
>
> Replace b with 3, the y-coordinate of the y-intercept.
>
> The equation of the line is $y = \dfrac{1}{2}x + 3$.

One method of finding the equation of a line when the slope and *any* point on the line are known involves using the *point-slope* formula. This formula is derived from the formula for the slope of a line as follows.

Let (x_1, y_1) be the given point on the line, and let (x, y) be any other point on the line. See the graph at the left.

$\dfrac{y - y_1}{x - x_1} = m$ • Use the formula for the slope of a line.

$\dfrac{y - y_1}{x - x_1}(x - x_1) = m(x - x_1)$ • Multiply each side by $(x - x_1)$.

$y - y_1 = m(x - x_1)$ • Simplify.

> **Point-Slope Formula**
>
> Let m be the slope of a line, and let (x_1, y_1) be the coordinates of a point on the line. The equation of the line can be found from the **point-slope formula:**
>
> $$y - y_1 = m(x - x_1)$$

> **HOW TO** Find the equation of the line that contains the point whose coordinates are $(4, -1)$ and has slope $-\dfrac{3}{4}$.
>
> $y - y_1 = m(x - x_1)$ • Use the point-slope formula.
>
> $y - (-1) = \left(-\dfrac{3}{4}\right)(x - 4)$ • $m = -\dfrac{3}{4}$, $(x_1, y_1) = (4, -1)$
>
> $y + 1 = -\dfrac{3}{4}x + 3$ • Simplify.
>
> $y = -\dfrac{3}{4}x + 2$ • Write the equation in the form $y = mx + b$.
>
> The equation of the line is $y = -\dfrac{3}{4}x + 2$.

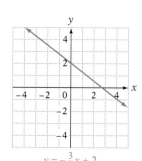

$y = -\dfrac{3}{4}x + 2$

HOW TO Find the equation of the line that passes through the point whose coordinates are (4, 3) and whose slope is undefined.

Because the slope is undefined, the point-slope formula cannot be used to find the equation. Instead, recall that when the slope is undefined, the line is vertical and that the equation of a vertical line is $x = a$, where a is the x-coordinate of the x-intercept. Because the line is vertical and passes through (4, 3), the x-intercept is (4, 0).
The equation of the line is $x = 4$.

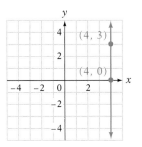

Example 1

Find the equation of the line that contains the point (3, 0) and has slope -4.

Solution
$m = -4$ $\qquad (x_1, y_1) = (3, 0)$

$y - y_1 = m(x - x_1)$
$y - 0 = -4(x - 3)$
$\qquad y = -4x + 12$

The equation of the line is $y = -4x + 12$.

You Try It 1

Find the equation of the line that contains the point $(-3, -2)$ and has slope $-\dfrac{1}{3}$.

Your solution

Example 2

Find the equation of the line that contains the point $(-2, 4)$ and has slope 2.

Solution
$m = 2$ $\qquad (x_1, y_1) = (-2, 4)$

$y - y_1 = m(x - x_1)$
$y - 4 = 2[x - (-2)]$
$y - 4 = 2(x + 2)$
$y - 4 = 2x + 4$
$\qquad y = 2x + 8$

The equation of the line is $y = 2x + 8$.

You Try It 2

Find the equation of the line that contains the point $(4, -3)$ and has slope -3.

Your solution

Solutions on p. S8

Objective B **To find the equation of a line given two points**

The point-slope formula and the formula for slope are used to find the equation of a line when two points are known.

HOW TO Find the equation of the line containing the points $(3, 2)$ and $(-5, 6)$.

To use the point-slope formula, we must know the slope. Use the formula for slope to determine the slope of the line between the two given points. Let $(x_1, y_1) = (3, 2)$ and $(x_2, y_2) = (-5, 6)$.

$$m = \frac{y_2 - y_1}{x_2 - x_1} = \frac{6 - 2}{-5 - 3} = \frac{4}{-8} = -\frac{1}{2}$$

Now use the point-slope formula with $m = -\frac{1}{2}$ and $(x_1, y_1) = (3, 2)$.

$y - y_1 = m(x - x_1)$ • Use the point-slope formula.

$y - 2 = \left(-\frac{1}{2}\right)(x - 3)$ • $m = -\frac{1}{2}, (x_1, y_1) = (3, 2)$

$y - 2 = -\frac{1}{2}x + \frac{3}{2}$ • Simplify.

$y = -\frac{1}{2}x + \frac{7}{2}$

The equation of the line is $y = -\frac{1}{2}x + \frac{7}{2}$.

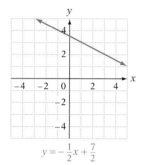

$y = -\frac{1}{2}x + \frac{7}{2}$

Example 3

Find the equation of the line containing the points $(2, 3)$ and $(4, 1)$.

Solution
Let $(x_1, y_1) = (2, 3)$ and $(x_2, y_2) = (4, 1)$.

$$m = \frac{y_2 - y_1}{x_2 - x_1} = \frac{1 - 3}{4 - 2} = \frac{-2}{2} = -1$$

$y - y_1 = m(x - x_1)$
$y - 3 = -1(x - 2)$
$y - 3 = -x + 2$
$\quad\ y = -x + 5$

The equation of the line is $y = -x + 5$.

You Try It 3

Find the equation of the line containing the points $(2, 0)$ and $(5, 3)$.

Your solution

Example 4

Find the equation of the line containing the points $(2, -3)$ and $(2, 5)$.

Solution
Let $(x_1, y_1) = (2, -3)$ and $(x_2, y_2) = (2, 5)$.

$$m = \frac{y_2 - y_1}{x_2 - x_1} = \frac{5 - (-3)}{2 - 2} = \frac{8}{0}$$

The slope is undefined, so the graph of the line is vertical.

The equation of the line is $x = 2$.

You Try It 4

Find the equation of the line containing the points $(2, 3)$ and $(-5, 3)$.

Your solution

Solutions on p. S9

Objective C **To solve application problems**

Linear functions can be used to model a variety of applications in science and business. For each application, data are collected and the independent and dependent variables are selected. Then a linear function is determined that models the data.

Example 5

Suppose a manufacturer has determined that at a price of $115, consumers will purchase 1 million portable CD players and that at a price of $90, consumers will purchase 1.25 million portable CD players. Describe this situation with a linear function. Use this function to predict how many portable CD players consumers will purchase if the price is $80.

Strategy

- Select the independent and dependent variables. Because you are trying to determine the number of CD players, that quantity is the *dependent* variable, y. The price of the CD players is the *independent* variable, x.
- From the given data, two ordered pairs are (115, 1) and (90, 1.25). (The ordinates are in millions of units.) Use these ordered pairs to determine the linear function.
- Evaluate this function when $x = 80$ to predict how many CD players consumers will purchase if the price is $80.

Solution

Let $(x_1, y_1) = (115, 1)$ and $(x_2, y_2) = (90, 1.25)$.

$$m = \frac{y_2 - y_1}{x_2 - x_1} = \frac{1.25 - 1}{90 - 115} = -\frac{0.25}{25} = -0.01$$

$$y - y_1 = m(x - x_1)$$

$$y - 1 = -0.01(x - 115)$$

$$y = -0.01x + 2.15$$

The linear function is $f(x) = -0.01x + 2.15$.

$$f(80) = -0.01(80) + 2.15 = 1.35$$

Consumers will purchase 1.35 million CD players at a price of $80.

You Try It 5

Gabriel Daniel Fahrenheit invented the mercury thermometer in 1717. In terms of readings on this thermometer, water freezes at 32°F and boils at 212°F. In 1742, Anders Celsius invented the Celsius temperature scale. On this scale, water freezes at 0°C and boils at 100°C. Determine a linear function that can be used to predict the Celsius temperature when the Fahrenheit temperature is known.

Your strategy

Your solution

Solutions on p. S9

3.5 Exercises

Objective A **To find the equation of a line given a point and the slope**

1. Explain how to find the equation of a line given its slope and its y-intercept.

2. What is the point-slope formula and how is it used?

For Exercises 3 to 38, find the equation of the line that contains the given point and has the given slope.

3. Point $(0, 5)$, $m = 2$

4. Point $(0, 3)$, $m = 1$

5. Point $(2, 3)$, $m = \dfrac{1}{2}$

6. Point $(5, 1)$, $m = \dfrac{2}{3}$

7. Point $(-1, 4)$, $m = \dfrac{5}{4}$

8. Point $(-2, 1)$, $m = \dfrac{3}{2}$

9. Point $(3, 0)$, $m = -\dfrac{5}{3}$

10. Point $(-2, 0)$, $m = \dfrac{3}{2}$

11. Point $(2, 3)$, $m = -3$

12. Point $(1, 5)$, $m = -\dfrac{4}{5}$

13. Point $(-1, 7)$, $m = -3$

14. Point $(-2, 4)$, $m = -4$

15. Point $(-1, -3)$, $m = \dfrac{2}{3}$

16. Point $(-2, -4)$, $m = \dfrac{1}{4}$

17. Point $(0, 0)$, $m = \dfrac{1}{2}$

18. Point $(0, 0)$, $m = \dfrac{3}{4}$

19. Point $(2, -3)$, $m = 3$

20. Point $(4, -5)$, $m = 2$

21. Point $(3, 5)$, $m = -\dfrac{2}{3}$

22. Point $(5, 1)$, $m = -\dfrac{4}{5}$

23. Point $(0, -3)$, $m = -1$

24. Point $(2, 0)$, $m = \dfrac{5}{6}$

25. Point $(1, -4)$, $m = \dfrac{7}{5}$

26. Point $(3, 5)$, $m = -\dfrac{3}{7}$

27. Point $(4, -1)$, $m = -\dfrac{2}{5}$

28. Point $(-3, 5)$, $m = -\dfrac{1}{4}$

29. Point $(3, -4)$, slope is undefined

30. Point $(-2, 5)$, slope is undefined

31. Point $(-2, -5)$, $m = -\dfrac{5}{4}$

32. Point $(-3, -2)$, $m = -\dfrac{2}{3}$

33. Point $(-2, -3)$, $m = 0$

34. Point $(-3, -2)$, $m = 0$

35. Point $(4, -5)$, $m = -2$

36. Point $(-3, 5)$, $m = 3$

37. Point $(-5, -1)$, slope is undefined

38. Point $(0, 4)$, slope is undefined

Objective B **To find the equation of a line given two points**

For Exercises 39 to 74, find the equation of the line that contains the given points.

39. $P_1(0, 2)$, $P_2(3, 5)$

40. $P_1(0, 4)$, $P_2(1, 5)$

41. $P_1(0, -3)$, $P_2(-4, 5)$

42. $P_1(0, -2)$, $P_2(-3, 4)$

43. $P_1(2, 3)$, $P_2(5, 5)$

44. $P_1(4, 1)$, $P_2(6, 3)$

45. $P_1(-1, 3)$, $P_2(2, 4)$

46. $P_1(-1, 1)$, $P_2(4, 4)$

47. $P_1(-1, -2)$, $P_2(3, 4)$

48. $P_1(-3, -1)$, $P_2(2, 4)$

49. $P_1(0, 3)$, $P_2(2, 0)$

50. $P_1(0, 4)$, $P_2(2, 0)$

51. $P_1(-3, -1)$, $P_2(2, -1)$

52. $P_1(-3, -5)$, $P_2(4, -5)$

53. $P_1(-2, -3)$, $P_2(-1, -2)$

54. $P_1(4, 1), P_2(3, -2)$ **55.** $P_1(-2, 3), P_2(2, -1)$ **56.** $P_1(3, 1), P_2(-3, -2)$

57. $P_1(2, 3), P_2(5, -5)$ **58.** $P_1(7, 2), P_2(4, 4)$ **59.** $P_1(2, 0), P_2(0, -1)$

60. $P_1(0, 4), P_2(-2, 0)$ **61.** $P_1(3, -4), P_2(-2, -4)$ **62.** $P_1(-3, 3), P_2(-2, 3)$

63. $P_1(0, 0), P_2(4, 3)$ **64.** $P_1(2, -5), P_2(0, 0)$ **65.** $P_1(2, -1), P_2(-1, 3)$

66. $P_1(3, -5), P_2(-2, 1)$ **67.** $P_1(-2, 5), P_2(-2, -5)$ **68.** $P_1(3, 2), P_2(3, -4)$

69. $P_1(2, 1), P_2(-2, -3)$ **70.** $P_1(-3, -2), P_2(1, -4)$ **71.** $P_1(-4, -3), P_2(2, 5)$

72. $P_1(4, 5), P_2(-4, 3)$ **73.** $P_1(0, 3), P_2(3, 0)$ **74.** $P_1(1, -3), P_2(-2, 4)$

Objective C **To solve application problems**

75. Aviation The pilot of a Boeing 747 jet takes off from Boston's Logan Airport, which is at sea level, and climbs to a cruising altitude of 32,000 ft at a constant rate of 1200 ft/min.
 a. Write a linear equation for the height of the plane in terms of the time after takeoff.
 b. Use your equation to find the height of the plane 11 min after takeoff.

76. Calories A jogger running at 9 mph burns approximately 14 Calories per minute.
 a. Write a linear equation for the number of Calories burned by the jogger in terms of the number of minutes run.
 b. Use your equation to find the number of Calories that the jogger has burned after jogging for 32 min.

77. Construction A building contractor estimates that the cost to build a home is $30,000 plus $85 for each square foot of floor space in the house.
 a. Determine a linear function that will give the cost of building a house that contains a given number of square feet.
 b. Use this model to determine the cost to build a house that contains 1800 square feet.

78. Telecommunications A cellular phone company offers several different options for using a cellular telephone. One option, for people who plan on using the phone only in emergencies, costs the user $4.95 per month plus $.59 per minute for each minute the phone is used.
 a. Write a linear function for the monthly cost of the phone in terms of the number of minutes the phone is used.
 b. Use your equation to find the cost of using the cellular phone for 13 minutes in one month.

79. Fuel Consumption The gas tank of a certain car contains 16 gal when the driver of the car begins a trip. Each mile driven by the driver decreases the amount of gas in the tank by 0.032 gal.
 a. Write a linear function for the number of gallons of gas in the tank in terms of the number of miles driven.
 b. Use your equation to find the number of gallons in the tank after 150 mi are driven.

80. Boiling Points At sea level, the boiling point of water is 100°C. At an altitude of 2 kilometers, the boiling point of water is 93°C.
 a. Write a linear function for the boiling point of water in terms of the altitude above sea level.
 b. Use your equation to predict the boiling point of water on top of Mount Everest, which is approximately 8.85 kilometers above sea level. Round to the nearest degree.

Mt. Everest

81. Business A manufacturer of economy cars has determined that 50,000 cars per month can be sold at a price of $9000. At a price of $8750, the number of cars sold per month would increase to 55,000.
 a. Determine a linear function that will predict the number of cars that would be sold each month at a given price.
 b. Use this model to predict the number of cars that would be sold at a price of $8500.

82. Business A manufacturer of graphing calculators has determined that 10,000 calculators per week will be sold at a price of $95. At a price of $90, it is estimated that 12,000 calculators would be sold.
 a. Determine a linear function that will predict the number of calculators that would be sold at a given price.
 b. Use this model to predict the number of calculators that would be sold each week at a price of $75.

83. Calories There are approximately 126 Calories in a 2-ounce serving of lean hamburger and approximately 189 Calories in a 3-ounce serving.
 a. Determine a linear function for the number of Calories in lean hamburger in terms of the size of the serving.
 b. Use your equation to estimate the number of Calories in a 5-ounce serving of lean hamburger.

84. Compensation An account executive receives a base salary plus a commission. On $20,000 in monthly sales, the account executive receives $1800. On $50,000 in monthly sales, the account executive receives $3000.
 a. Determine a linear function that will yield the compensation of the sales executive for a given amount of monthly sales.
 b. Use this model to determine the account executive's compensation for $85,000 in monthly sales.

85. Let f be a linear function. If $f(2) = 5$ and $f(0) = 3$, find $f(x)$.

86. Let f be a linear function. If $f(-3) = 4$ and $f(1) = -8$, find $f(x)$.

87. Given that f is a linear function for which $f(1) = 3$ and $f(-1) = 5$, determine $f(4)$.

88. Given that f is a linear function for which $f(-3) = 2$ and $f(2) = 7$, determine $f(0)$.

89. A line with slope $\frac{4}{3}$ passes through the point $(3, 2)$.
 a. What is y when $x = -6$?
 b. What is x when $y = 6$?

90. A line with slope $-\frac{3}{4}$ passes through the point $(8, -2)$.
 a. What is y when $x = -4$?
 b. What is x when $y = 1$?

APPLYING THE CONCEPTS

91. Explain the similarities and differences between the point-slope formula and the slope-intercept form of a straight line.

92. Explain why the point-slope formula cannot be used to find the equation of a line that is parallel to the y-axis.

93. Refer to Example 5 in this section for each of the following.
 a. Explain the meaning of the slope of the graph of the linear function given in the example.
 b. Explain the meaning of the y-intercept.
 c. Explain the meaning of the x-intercept.

94. For an equation of the form $y = mx + b$, how does the graph of this equation change if the value of b changes and the value of m remains constant?

95. A line contains the points $(-3, 6)$ and $(6, 0)$. Find the coordinates of three other points that are on this line.

96. A line contains the points $(4, -1)$ and $(2, 1)$. Find the coordinates of three other points that are on this line.

97. Find the equation of the line that passes through the midpoint of the line segment between $P_1(2, 5)$ and $P_2(-4, 1)$ and has slope -2.

3.6 Parallel and Perpendicular Lines

Objective A **To find parallel and perpendicular lines**

Two lines that have the same slope do not intersect and are called **parallel lines.**

The slope of each of the lines at the right is $\frac{2}{3}$.

The lines are parallel.

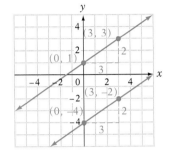

Slopes of Parallel Lines

Two nonvertical lines with slopes of m_1 and m_2 are parallel if and only if $m_1 = m_2$. Any two vertical lines are parallel.

HOW TO Is the line containing the points $(-2, 1)$ and $(-5, -1)$ parallel to the line that contains the points $(1, 0)$ and $(4, 2)$?

$$m_1 = \frac{-1 - 1}{-5 - (-2)} = \frac{-2}{-3} = \frac{2}{3}$$

- Find the slope of the line through $(-2, 1)$ and $(-5, -1)$.

$$m_2 = \frac{2 - 0}{4 - 1} = \frac{2}{3}$$

- Find the slope of the line through $(1, 0)$ and $(4, 2)$.

Because $m_1 = m_2$, the lines are parallel.

HOW TO Find the equation of the line that contains the point $(2, 3)$ and is parallel to the line $y = \frac{1}{2}x - 4$.

The slope of the given line is $\frac{1}{2}$. Because parallel lines have the same slope, the slope of the unknown line is also $\frac{1}{2}$.

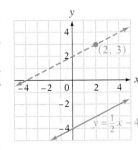

$$y - y_1 = m(x - x_1)$$
- Use the point-slope formula.

$$y - 3 = \frac{1}{2}(x - 2)$$
- $m = \frac{1}{2}$, $(x_1, y_1) = (2, 3)$

$$y - 3 = \frac{1}{2}x - 1$$
- Simplify.

$$y = \frac{1}{2}x + 2$$
- Write the equation in the form $y = mx + b$.

The equation of the line is $y = \frac{1}{2}x + 2$.

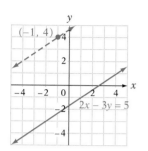

HOW TO Find the equation of the line that contains the point $(-1, 4)$ and is parallel to the line $2x - 3y = 5$.

Because the lines are parallel, the slope of the unknown line is the same as the slope of the given line. Solve $2x - 3y = 5$ for y and determine its slope.

$$2x - 3y = 5$$
$$-3y = -2x + 5$$
$$y = \frac{2}{3}x - \frac{5}{3}$$

The slope of the given line is $\frac{2}{3}$. Because the lines are parallel, this is the slope of the unknown line. Use the point-slope formula to determine the equation.

$y - y_1 = m(x - x_1)$ • Use the point-slope formula.

$y - 4 = \frac{2}{3}[x - (-1)]$ • $m = \frac{2}{3}, (x_1, y_1) = (-1, 4)$

$y - 4 = \frac{2}{3}x + \frac{2}{3}$ • Simplify.

$y = \frac{2}{3}x + \frac{14}{3}$ • Write the equation in the form $y = mx + b$.

The equation of the line is $y = \frac{2}{3}x + \frac{14}{3}$.

Two lines that intersect at right angles are **perpendicular lines.**

Any horizontal line is perpendicular to any vertical line. For example, $x = 3$ is perpendicular to $y = -2$.

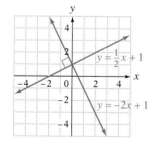

Slopes of Perpendicular Lines

If m_1 and m_2 are the slopes of two lines, neither of which is vertical, then the lines are perpendicular if and only if $m_1 \cdot m_2 = -1$.

A vertical line is perpendicular to a horizontal line.

Solving $m_1 \cdot m_2 = -1$ for m_1 gives $m_1 = -\frac{1}{m_2}$. This last equation states that the slopes of perpendicular lines are **negative reciprocals** of each other.

HOW TO Is the line that contains the points $(4, 2)$ and $(-2, 5)$ perpendicular to the line that contains the points $(-4, 3)$ and $(-3, 5)$?

$m_1 = \dfrac{5 - 2}{-2 - 4} = \dfrac{3}{-6} = -\dfrac{1}{2}$ • Find the slope of the line through $(4, 2)$ and $(-2, 5)$.

$m_2 = \dfrac{5 - 3}{-3 - (-4)} = \dfrac{2}{1} = 2$ • Find the slope of the line through $(-4, 3)$ and $(-3, 5)$.

$m_1 \cdot m_2 = -\dfrac{1}{2}(2) = -1$ • Find the product of the two slopes.

Because $m_1 \cdot m_2 = -1$, the lines are perpendicular.

HOW TO Are the graphs of the lines whose equations are $3x + 4y = 8$ and $8x + 6y = 5$ perpendicular?

To determine whether the lines are perpendicular, solve each equation for y and find the slope of each line. Then use the equation $m_1 \cdot m_2 = -1$.

$$3x + 4y = 8 \qquad\qquad\qquad 8x + 6y = 5$$
$$4y = -3x + 8 \qquad\qquad\qquad 6y = -8x + 5$$
$$y = -\frac{3}{4}x + 2 \qquad\qquad\qquad y = -\frac{4}{3}x + \frac{5}{6}$$
$$m_1 = -\frac{3}{4} \qquad\qquad\qquad\qquad m_2 = -\frac{4}{3}$$

$$m_1 \cdot m_2 = \left(-\frac{3}{4}\right)\left(-\frac{4}{3}\right) = 1$$

Because $m_1 \cdot m_2 = 1 \neq -1$, the lines are not perpendicular.

HOW TO Find the equation of the line that contains the point $(-2, 1)$ and is perpendicular to the line $y = -\frac{2}{3}x + 2$.

The slope of the given line is $-\frac{2}{3}$. The slope of the line perpendicular to the given line is the negative reciprocal of $-\frac{2}{3}$, which is $\frac{3}{2}$. Substitute this slope and the coordinates of the given point, $(-2, 1)$, into the point-slope formula.

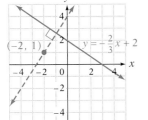

$$y - y_1 = m(x - x_1)$$

- The point-slope formula

$$y - 1 = \frac{3}{2}[x - (-2)]$$

- $m = \frac{3}{2}$, $(x_1, y_1) = (-2, 1)$

$$y - 1 = \frac{3}{2}x + 3$$

- Simplify.

$$y = \frac{3}{2}x + 4$$

- Write the equation in the form $y = mx + b$.

The equation of the perpendicular line is $y = \frac{3}{2}x + 4$.

HOW TO Find the equation of the line that contains the point $(3, -4)$ and is perpendicular to the line $2x - y = -3$.

$$2x - y = -3$$
$$-y = -2x - 3$$
$$y = 2x + 3$$

- Determine the slope of the given line by solving the equation for y.
- The slope is 2.

The slope of the line perpendicular to the given line is $-\frac{1}{2}$, the negative reciprocal of 2. Now use the point-slope formula to find the equation of the line.

$$y - y_1 = m(x - x_1)$$

- The point-slope formula

$$y - (-4) = -\frac{1}{2}(x - 3)$$

- $m = -\frac{1}{2}$, $(x_1, y_1) = (3, -4)$

$$y + 4 = -\frac{1}{2}x + \frac{3}{2}$$

- Simplify.

$$y = -\frac{1}{2}x - \frac{5}{2}$$

- Write the equation in the form $y = mx + b$.

The equation of the perpendicular line is $y = -\frac{1}{2}x - \frac{5}{2}$.

Example 1

Is the line that contains the points $(-4, 2)$ and $(1, 6)$ parallel to the line that contains the points $(2, -4)$ and $(7, 0)$?

Solution

$m_1 = \dfrac{6 - 2}{1 - (-4)} = \dfrac{4}{5}$ • $(x_1, y_1) = (-4, 2)$, $(x_2, y_2) = (1, 6)$

$m_2 = \dfrac{0 - (-4)}{7 - 2} = \dfrac{4}{5}$ • $(x_1, y_1) = (2, -4)$, $(x_2, y_2) = (7, 0)$

$m_1 = m_2 = \dfrac{4}{5}$

The lines are parallel.

You Try It 1

Is the line that contains the points $(-2, -3)$ and $(7, 1)$ perpendicular to the line that contains the points $(4, 1)$ and $(6, -5)$?

Your solution

Example 2

Are the lines $4x - y = -2$ and $x + 4y = -12$ perpendicular?

Solution

$4x - y = -2$ $x + 4y = -12$
$-y = -4x - 2$ $4y = -x - 12$
$y = 4x + 2$ $y = -\dfrac{1}{4}x - 3$
$m_1 = 4$

$m_2 = -\dfrac{1}{4}$

$m_1 \cdot m_2 = 4\left(-\dfrac{1}{4}\right) = -1$

The lines are perpendicular.

You Try It 2

Are the lines $5x + 2y = 2$ and $5x + 2y = -6$ parallel?

Your solution

Example 3

Find the equation of the line that contains the point $(3, -1)$ and is parallel to the line $3x - 2y = 4$.

Solution

$3x - 2y = 4$
$-2y = -3x + 4$
$y = \dfrac{3}{2}x - 2$ • $m = \dfrac{3}{2}$

$y - y_1 = m(x - x_1)$
$y - (-1) = \dfrac{3}{2}(x - 3)$ • $(x_1, y_1) = (3, -1)$

$y + 1 = \dfrac{3}{2}x - \dfrac{9}{2}$

$y = \dfrac{3}{2}x - \dfrac{11}{2}$

The equation of the line is $y = \dfrac{3}{2}x - \dfrac{11}{2}$.

You Try It 3

Find the equation of the line that contains the point $(-2, 2)$ and is perpendicular to the line $x - 4y = 3$.

Your solution

Solutions on p. S9

3.6 Exercises

Objective A **To find parallel and perpendicular lines**

1. Explain how to determine whether the graphs of two lines are parallel.

2. Explain how to determine whether the graphs of two lines are perpendicular.

3. The slope of a line is -5. What is the slope of any line parallel to this line?

4. The slope of a line is $\frac{3}{2}$. What is the slope of any line parallel to this line?

5. The slope of a line is 4. What is the slope of any line perpendicular to this line?

6. The slope of a line is $-\frac{4}{5}$. What is the slope of any line perpendicular to this line?

7. Is the line $x = -2$ perpendicular to the line $y = 3$?

8. Is the line $y = \frac{1}{2}$ perpendicular to the line $y = -4$?

9. Is the line $x = -3$ parallel to the line $y = \frac{1}{3}$?

10. Is the line $x = 4$ parallel to the line $x = -4$?

11. Is the line $y = \frac{2}{3}x - 4$ parallel to the line $y = -\frac{3}{2}x - 4$?

12. Is the line $y = -2x + \frac{2}{3}$ parallel to the line $y = -2x + 3$?

13. Is the line $y = \frac{4}{3}x - 2$ perpendicular to the line $y = -\frac{3}{4}x + 2$?

14. Is the line $y = \frac{1}{2}x + \frac{3}{2}$ perpendicular to the line $y = -\frac{1}{2}x + \frac{3}{2}$?

15. Are the lines $2x + 3y = 2$ and $2x + 3y = -4$ parallel?

16. Are the lines $2x - 4y = 3$ and $2x + 4y = -3$ parallel?

17. Are the lines $x - 4y = 2$ and $4x + y = 8$ perpendicular?

18. Are the lines $4x - 3y = 2$ and $4x + 3y = -7$ perpendicular?

19. Is the line that contains the points $(3, 2)$ and $(1, 6)$ parallel to the line that contains the points $(-1, 3)$ and $(-1, -1)$?

20. Is the line that contains the points $(4, -3)$ and $(2, 5)$ parallel to the line that contains the points $(-2, -3)$ and $(-4, 1)$?

21. Is the line that contains the points $(-3, 2)$ and $(4, -1)$ perpendicular to the line that contains the points $(1, 3)$ and $(-2, -4)$?

22. Is the line that contains the points $(-1, 2)$ and $(3, 4)$ perpendicular to the line that contains the points $(-1, 3)$ and $(-4, 1)$?

23. Find the equation of the line containing the point $(-2, -4)$ and parallel to the line $2x - 3y = 2$.

24. Find the equation of the line containing the point $(3, 2)$ and parallel to the line $3x + y = -3$.

25. Find the equation of the line containing the point $(4, 1)$ and perpendicular to the line $y = -3x + 4$.

26. Find the equation of the line containing the point $(2, -5)$ and perpendicular to the line $y = \frac{5}{2}x - 4$.

27. Find the equation of the line containing the point $(-1, -3)$ and perpendicular to the line $3x - 5y = 2$.

28. Find the equation of the line containing the point $(-1, 3)$ and perpendicular to the line $2x + 4y = -1$.

APPLYING THE CONCEPTS

Physics For Exercises 29 and 30, suppose a ball is being twirled at the end of a string and the center of rotation is the origin of a coordinate system. If the string breaks, the initial path of the ball is on a line that is perpendicular to the radius of the circle.

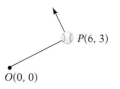

29. Suppose the string breaks when the ball is at the point whose coordinates are $P(6, 3)$. Find the equation of the line on which the initial path lies.

30. Suppose the string breaks when the ball is at the point whose coordinates are $P(2, 8)$. Find the equation of the line on which the initial path lies.

31. If the graphs of $A_1x + B_1y = C_1$ and $A_2x + B_2y = C_2$ are parallel, express $\frac{A_1}{B_1}$ in terms of A_2 and B_2.

32. If the graphs of $A_1x + B_1y = C_1$ and $A_2x + B_2y = C_2$ are perpendicular, express $\frac{A_1}{B_1}$ in terms of A_2 and B_2.

33. The graphs of $y = -\frac{1}{2}x + 2$ and $y = \frac{2}{3}x - 5$ intersect at the point whose coordinates are $(6, -1)$. Find the equation of a line whose graph intersects the graphs of the given lines to form a right triangle. (*Hint:* There is more than one answer to this question.)

34. Geometry A theorem from geometry states that a line passing through the center of a circle and through a point P on the circle is perpendicular to the tangent line at P. (See the figure at the right.)
 a. If the coordinates of P are $(5, 4)$ and the coordinates of C are $(3, 2)$, what is the equation of the tangent line?
 b. What is the x-intercept of the tangent line?
 c. What is the y-intercept of the tangent line?

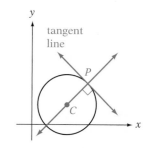

3.7 Inequalities in Two Variables

Objective A To graph the solution set of an inequality in two variables

The graph of the linear equation $y = x - 1$ separates the plane into three sets: the set of points on the line, the set of points above the line, and the set of points below the line.

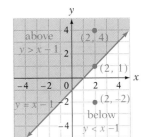

The point whose coordinates are $(2, 1)$ is a solution of $y = x - 1$ and is a point on the line.

The point whose coordinates are $(2, 4)$ is a solution of $y > x - 1$ and is a point above the line.

The point whose coordinates are $(2, -2)$ is a solution of $y < x - 1$ and is a point below the line.

The set of points on the line are the solutions of the equation $y = x - 1$. The set of points above the line are the solutions of the inequality $y > x - 1$. These points form a **half-plane.** The set of points below the line are solutions of the inequality $y < x - 1$. These points also form a half-plane.

An inequality of the form $y > mx + b$ or $Ax + By > C$ is a **linear inequality in two variables.** (The inequality symbol > could be replaced by ≥, <, or ≤.) The solution set of a linear inequality in two variables is a half-plane.

The following illustrates the procedure for graphing the solution set of a linear inequality in two variables.

> **TAKE NOTE**
>
> When solving the inequality at the right for y, both sides of the inequality are divided by -4, so the inequality symbol must be reversed.

HOW TO Graph the solution set of $3x - 4y < 12$.

$$3x - 4y < 12$$
$$-4y < -3x + 12 \qquad \bullet \text{ Solve the inequality for } y.$$
$$y > \frac{3}{4}x - 3$$

Change the inequality $y > \frac{3}{4}x - 3$ to the equality $y = \frac{3}{4}x - 3$, and graph the line.

> **TAKE NOTE**
>
> As shown below, $(0, 0)$ is a solution of the inequality in the example at the right.
>
> $$y > \frac{3}{4}x - 3$$
> $$0 > \frac{3}{4}(0) - 3$$
> $$0 > 0 - 3$$
> $$0 > -3$$
>
> Because $(0, 0)$ is a solution of the inequality, $(0, 0)$ should be in the shaded region. The solution set as graphed is correct.

If the inequality contains ≤ or ≥, the line belongs to the solution set and is shown by a *solid line.* If the inequality contains < or >, the line is not part of the solution set and is shown by a *dotted line.*

If the inequality contains > or ≥, shade the upper half-plane. If the inequality contains < or ≤, shade the lower half-plane.

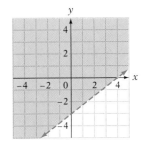

As a check, use the ordered pair $(0, 0)$ to determine whether the correct region of the plane has been shaded. If $(0, 0)$ is a solution of the inequality, then $(0, 0)$ should be in the shaded region. If $(0, 0)$ is not a solution of the inequality, then $(0, 0)$ should not be in the shaded region.

Integrating Technology

See the Keystroke Guide: *Graphing Inequalities* for instructions on using a graphing calculator to graph the solution set of an inequality in two variables.

If the line passes through the point $(0, 0)$, another point, such as $(0, 1)$, must be used as a check.

From the graph of $y > \frac{3}{4}x - 3$, note that for a given value of x, more than one value of y can be paired with that value of x. For instance, $(4, 1)$, $(4, 3)$, $(5, 1)$, and $\left(5, \frac{9}{4}\right)$ are all ordered pairs that belong to the graph. Because there are ordered pairs with the same first coordinate and different second coordinates, the inequality does not represent a function. The inequality is a relation but not a function.

Example 1 Graph the solution set of $x + 2y \le 4$.

Solution
$$x + 2y \le 4$$
$$2y \le -x + 4$$
$$y \le -\frac{1}{2}x + 2$$

Graph $y = -\frac{1}{2}x + 2$ as a solid line.

Shade the lower half-plane.

Check:
$$y \le -\frac{1}{2}x + 2$$
$$0 \le -\frac{1}{2}(0) + 2$$
$$0 \le 0 + 2$$
$$0 \le 2$$
The point $(0, 0)$ should be in the shaded region.

You Try It 1 Graph the solution set of $x + 3y > 6$.

Your solution

Example 2 Graph the solution set of $x \ge -1$.

Solution Graph $x = -1$ as a solid line.

Shade the half-plane to the right of the line.

You Try It 2 Graph the solution set of $y < 2$.

Your solution

Solutions on p. S9

3.7 Exercises

Objective A **To graph the solution set of an inequality in two variables**

1. ✎ What is a half-plane?

2. ✎ Explain a method you can use to check that the graph of a linear inequality in two variables has been shaded correctly.

3. Is (0, 0) a solution of $y > 2x - 7$?

4. Is (0, 0) a solution of $y < 5x + 3$?

5. Is (0, 0) a solution of $y \leq -\dfrac{2}{3}x - 8$?

6. Is (0, 0) a solution of $y \geq -\dfrac{3}{4}x + 9$?

For Exercises 7 to 24, graph the solution set.

7. $3x - 2y \geq 6$

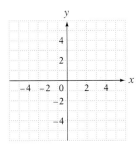

8. $4x - 3y \leq 12$

9. $x + 3y < 4$

10. $3x - 5y > 15$

11. $4x - 5y > 10$

12. $4x + 3y < 9$

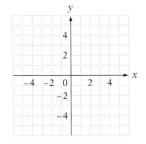

13. $x + 3y < 6$

14. $2x - 5y \leq 10$

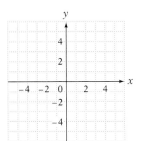

15. $2x + 3y \geq 6$

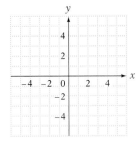

16. $3x + 2y < 4$

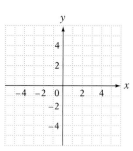

17. $-x + 2y > -8$

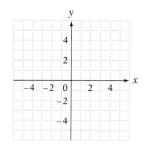

18. $-3x + 2y > 2$

19. $y - 4 < 0$

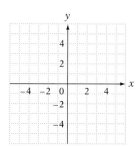

20. $x + 2 \geq 0$

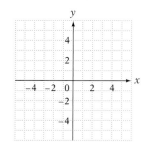

21. $6x + 5y < 15$

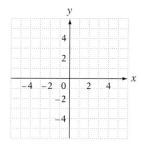

22. $3x - 5y < 10$

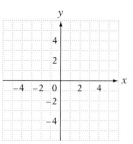

23. $-5x + 3y \geq -12$

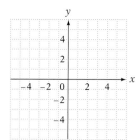

24. $3x + 4y \geq 12$

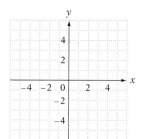

APPLYING THE CONCEPTS

25. Does the inequality $y < 3x - 1$ represent a function? Explain your answer.

26. Are there any points whose coordinates satisfy both $y \leq x + 3$ and $y \geq -\frac{1}{2}x + 1$? If so, give the coordinates of three such points. If not, explain why not.

27. Are there any points whose coordinates satisfy both $y \leq x - 1$ and $y \geq x + 2$? If so, give the coordinates of three such points. If not, explain why not.

Focus on Problem Solving

Find a Pattern Polya's four recommended problem-solving steps are stated below.

1. Understand the problem. **3.** Carry out the plan.

2. Devise a plan. **4.** Review your solution.

Karl Friedrich Gauss

Karl Friedrich Gauss (1777–1855) has been called the "Prince of Mathematicians" by some historians. He applied his genius to many areas of mathematics and science. A unit of magnetism, the *gauss*, is named in his honor. Some electronic equipment (televisions, for instance) contains a *degausser* that controls magnetic fields.

One of the several ways of devising a plan is first to try to find a pattern. Karl Friedrich Gauss supposedly used this method to solve a problem that was given to his math class when he was in elementary school. As the story goes, his teacher wanted to grade some papers while the class worked on a math problem. The problem given to the class was to find the sum

$$1 + 2 + 3 + 4 + \cdots + 100$$

Gauss quickly solved the problem by seeing a pattern. Here is what he saw.

Note that

$1 + 100 = 101$
$2 + 99 = 101$
$3 + 98 = 101$
$4 + 97 = 101$

Gauss noted that there were 50 sums of 101. Therefore, the sum of the first 100 natural numbers is

$$1 + 2 + 3 + 4 + \cdots + 97 + 98 + 99 + 100 = 50(101) = 5050$$

Try to solve Exercises 1 to 6 by finding a pattern.

1. Find the sum $2 + 4 + 6 + \cdots + 96 + 98 + 100$.

2. Find the sum $1 + 3 + 5 + \cdots + 97 + 99 + 101$.

3. Find another method of finding the sum $1 + 3 + 5 + \cdots + 97 + 99 + 101$ given in the preceding exercise.

4. Find the sum $\dfrac{1}{1 \cdot 2} + \dfrac{1}{2 \cdot 3} + \dfrac{1}{3 \cdot 4} + \cdots + \dfrac{1}{49 \cdot 50}$.

Hint: $\dfrac{1}{1 \cdot 2} = \dfrac{1}{2}, \dfrac{1}{1 \cdot 2} + \dfrac{1}{2 \cdot 3} = \dfrac{2}{3}, \dfrac{1}{1 \cdot 2} + \dfrac{1}{2 \cdot 3} + \dfrac{1}{3 \cdot 4} = \dfrac{3}{4}$

5. A *polynomial number* is a number that can be represented by arranging that number of dots in rows to form a geometric figure such as a triangle, square, pentagon, or hexagon. For instance, the first four *triangular* numbers, 3, 6, 10, and 15, are shown below. What are the next two triangular numbers?

3 6 10 15

2 points, 2 regions 3 points, 4 regions

4 points, 8 regions 5 points, ? regions

6. The following problem shows that checking a few cases does not always result in a conjecture that is true for *all* cases. Select any two points on a circle (see the drawing in the left margin) and draw a *chord*, a line connecting the points. The chord divides the circle into 2 regions. Now select 3 different points and draw chords connecting each of the three points with every other point. The chords divide the circle into 4 regions. Now select 4 points and connect each of the points with every other point. Make a conjecture as to the relationship between the number of regions and the number of points on the circle. Does your conjecture work for 5 points? 6 points?

Projects and Group Activities

Evaluating a Function with a Graphing Calculator

You can use a graphing calculator to evaluate some functions. Shown below are the keystrokes needed to evaluate $f(x) = 3x^2 + 2x - 1$ for $x = -2$ on a TI-83. (There are other methods of evaluating functions. This is just one of them.) Try these keystrokes. The calculator should display 7 as the value of the function.

Introduction to Graphing Calculators

There are a variety of computer programs and calculators that can graph an equation. A computer or graphing calculator screen is divided into pixels. Depending on the computer or calculator, there are approximately 6000 to 790,000 pixels available on the screen. The greater the number of pixels, the smoother the graph will appear. A portion of a screen is shown at the left. Each little rectangle represents one pixel.

Calculator Screen

A graphing calculator draws a graph in a manner similar to the method we have used in this chapter. Values of x are chosen and ordered pairs calculated. Then a graph is drawn through those points by illuminating pixels (an abbreviation for "picture element") on the screen.

Graphing utilities can display only a portion of the xy-plane, called a window. The window [Xmin, Xmax] by [Ymin, Ymax] consists of those points (x, y) that satisfy both of the following inequalities:

Xmin $\leq x \leq$ Xmax and Ymin $\leq y \leq$ Ymax

The user sets these values before a graph is drawn.

The numbers Xscl and Yscl are the distances between the tick marks that are drawn on the x- and y-axes. If you do not want tick marks on the axes, set Xscl = 0 and Yscl = 0.

The graph at the right is a portion of the graph of $y = \frac{1}{2}x + 1$, as it was drawn with a graphing calculator. The window is

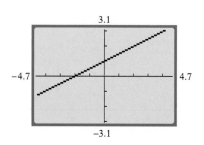

Xmin = −4.7, Xmax = 4.7
Ymin = −3.1, Ymax = 3.1
Xscl = 1 and Yscl = 1

Using interval notation, this is written $[-4.7, 4.7]$ by $[-3.1, 3.1]$. The window $[-4.7, 4.7]$ by $[-3.1, 3.1]$ gives "nice" coordinates in the sense that each time the ◉ or the ◉ is pressed, the change in x is 0.1. The reason for this is that the horizontal distance from the middle of the first pixel to the middle of the last pixel is 94 units. By using Xmin $= -4.7$ and Xmax $= 4.7$, we have[1]

$$\text{Change in } x = \frac{\text{Xmax} - \text{Xmin}}{94} = \frac{4.7 - (-4.7)}{94} = \frac{9.4}{94} = 0.1$$

Similarly, the vertical distance from the middle of the first pixel to the middle of the last pixel is 62 units. Therefore, using Ymin $= -3.1$ and Ymax $= 3.1$ will give nice coordinates in the vertical direction.

Graph the equations in Exercises 1 to 6 by using a graphing calculator.

Integrating Technology

See the Keystroke Guide: *Windows* for instructions on changing the viewing window.

1. $y = 2x + 1$ — For $2x$, you may enter $2 \times x$ or just $2x$. The times sign \times is not necessary on many graphing calculators.

2. $y = -x + 2$ — Many calculators use the $(-)$ key to enter a negative sign.

3. $3x + 2y = 6$ — Solve for y. Then enter the equation.

4. $y = 50x$ — You must adjust the viewing window. Try the window $[-4.7, 4.7]$ by $[-250, 250]$ with Yscl $= 50$.

5. $y = \frac{2}{3}x - 3$ — You may enter $\frac{2}{3}x$ as $2x/3$ or $(2/3)x$. Although entering $2/3x$ works on some calculators, it is not recommended.

6. $4x + 3y = 75$ — You must adjust the viewing window.

Wind-Chill Index

The wind-chill index is the temperature of still air that would have the same effect on exposed human skin as a given combination of wind speed and air temperature. For example, given a wind speed of 10 mph and a temperature reading of 20°F, the wind-chill index is 9°F.

In the fall of 2001, the U.S. National Weather Service began using a new wind-chill formula. The following ordered pairs are derived from the new formula. In each ordered pair, the abscissa is the air temperature in degrees Fahrenheit, and the ordinate is the wind-chill index when the wind speed is 10 mph.

$\{(35, 27), (30, 21), (25, 15), (20, 9), (15, 3), (10, -4), (5, -10), (0, -16), (-5, -22),$
$(-10, -28), (-15, -35), (-20, -41), (-25, -47), (-30, -53), (-35, -59)\}$

1. Use the coordinate axes at the left to graph the ordered pairs.

2. List the domain of the relation. List the range of the relation.

3. Is the set of ordered pairs a function?

4. The equation of the line that approximately models the graph of the ordered pairs above is $y = 1.2357x - 16$, where x is the air temperature in degrees Fahrenheit and y is the wind-chill index. Evaluate the function $f(x) = 1.2357x - 16$ to determine whether any of the ordered pairs listed above do not satisfy this function. Round to the nearest integer.

5. What does the model predict the wind-chill index is when the air temperature is -40°F and the wind speed is 10 mph?

Wind-chill factor (in °F)

Temperature (in °F)

[1]Some calculators have screen widths of 126 pixels. For those calculators, use Xmin $= -6.3$ and Xmax $= 6.3$ to obtain nice coordinates.

6. a. What is the x-intercept of the graph of $f(x) = 1.2357x - 16$? Round to the nearest integer. What does the x-intercept represent?

 b. What is the y-intercept of the graph of $f(x) = 1.2357x - 16$? What does the y-intercept represent?

In the set of ordered pairs that follows, the abscissa is the air temperature in degrees Fahrenheit, and the ordinate is the wind-chill index when the wind speed is 20 mph. Note that the abscissa in each case is the same as in the ordered pairs above. However, the wind speed has increased to 20 mph.

$$\{(35, 24), (30, 17), (25, 11), (20, 4), (15, -2), (10, -9), (5, -15), (0, -22), (-5, -29),$$
$$(-10, -35), (-15, -42), (-20, -48), (-25, -55), (-30, -61), (-35, -68)\}$$

7. Use the same coordinate axes to graph these ordered pairs.

8. Where do the points representing this relation lie in relation to the points drawn for Exercise 1? What does this mean with respect to the temperature remaining constant and the wind speed increasing?

9. The equation of the line that approximately models the graph of the ordered pairs in this second relation is $y = 1.3114x - 22$, where x is the air temperature in degrees Fahrenheit and y is the wind-chill index. Compare this with the model for the first relation. Are the lines parallel? Explain why that might be the case.

10. Using the old wind-chill formula and a wind speed of 10 mph gave a wind-chill index of $-10°F$ when the air temperature was $20°F$ and a wind-chill index of $-67°F$ when the air temperature was $-20°F$.

 a. Find the linear equation that models these data. [*Hint:* Use the ordered pairs $(20, -10)$ and $(-20, -67)$.]

 b. Compare the y-intercept of your equation to the y-intercept of the equation $y = 1.2357x - 16$, which models the new wind-chill index for a wind speed of 10 mph. Use the difference to determine whether the new formula yields a wind-chill index that is warmer or colder, given the same air temperature and wind speed.

Chapter 3 Summary

Key Words

Examples

A *rectangular coordinate system* is formed by two numbers lines, one horizontal and one vertical, that intersect at the zero point of each line. The point of intersection is called the *origin*. The number lines that make up a rectangular coordinate system are called *coordinate axes*. A rectangular coordinate system divides the plane into four regions called *quadrants*. [3.1A, p. 121]

An *ordered pair* (x, y) is used to locate a point in a rectangular coordinate system. The first number of the pair measures a horizontal distance and is called the *abscissa* or *x-coordinate*. The second number of the pair measures a vertical distance and is called the *ordinate* or *y-coordinate*. The *coordinates* of the point are the numbers in the ordered pair associated with the point. To *graph*, or *plot*, a point in the plane, place a dot at the location given by the ordered pair. The *graph of an ordered pair* is the dot drawn at the coordinates of the point in the plane. [3.1A, pp. 121–122]

(3, 4) is an ordered pair.
3 is the abscissa.
4 is the ordinate.
The graph of (3, 4) is shown below.

An equation of the form $y = mx + b$, where m and b are constants, is a *linear equation in two variables*. A solution of a linear equation in two variables is an ordered pair (x, y) whose coordinates make the equation a true statement. The graph of a linear equation in two variables is a straight line. [3.1A, p. 122]

$y = 3x + 2$ is a linear equation in two variables; $m = 3$ and $b = 2$. Ordered-pair solutions of $y = 3x + 2$ are shown below, along with the graph of the equation.

x	y
1	5
0	2
−1	−1

A *scatter diagram* is a graph of ordered-pair data. [3.1C, p. 126]

A *relation* is any set of ordered pairs. [3.2A, p. 131]

{(2, 3), (2, 4), (3, 4), (5, 7)}

A *function* is a relation in which no two ordered pairs have the same first coordinate and different second coordinates. The *domain* of a function is the set of the first coordinates of all the ordered pairs of the function. The *range* is the set of the second coordinates of all the ordered pairs of the function. [3.2A, p. 132]

{(2, 3), (3, 5), (5, 7), (6, 9)}
The domain is {2, 3, 5, 6}.
The range is {3, 5, 7, 9}.

Functional notation is used for those equations that represent functions. For the equation at the right, x is the *independent variable* and y is the *dependent variable*. The symbol $f(x)$ is the *value of the function* and represents the value of the dependent variable for a given value of the independent variable. [3.2A, pp. 132–134]

In functional notation, $y = 3x + 7$ is written as $f(x) = 3x + 7$.

The process of determining $f(x)$ for a given value of x is called *evaluating the function*. [3.2A, p. 134]

Evaluate $f(x) = 2x - 3$ when $x = 4$.

$f(x) = 2x - 3$
$f(4) = 2(4) - 3$
$f(4) = 5$

The *graph of a function* is a graph of the ordered pairs (x, y) that belong to the function. A function that can be written in the form $f(x) = mx + b$ (or $y = mx + b$) is a *linear function* because its graph is a straight line. [3.3A, p. 143]

$f(x) = -\frac{2}{3}x + 3$ is an example of a linear function. In this case, $m = -\frac{2}{3}$ and $b = 3$.

The point at which a graph crosses the x-axis is called the *x-intercept*, and the point at which a graph crosses the y-axis is called the *y-intercept*. [3.3C, p. 148]

The x-intercept of $x + y = 4$ is $(4, 0)$. The y-intercept of $x + y = 4$ is $(0, 4)$.

The *slope* of a line is a measure of the slant, or tilt, of the line. The symbol for slope is m. A line that slants upward to the right has a *positive slope*, and a line that slants downward to the right has a *negative slope*. A horizontal line has *zero slope*. The slope of a vertical line is *undefined*. [3.4A, pp. 155–156]

The line $y = 2x - 3$ has a slope of 2 and slants upward to the right. The line $y = -5x + 2$ has a slope of -5 and slants downward to the right. The line $y = 4$ has a slope of 0.

An inequality of the form $y > mx + b$ or of the form $Ax + By > C$ is a *linear inequality in two variables*. (The symbol > can be replaced by ≥, <, or ≤.) The solution set of an inequality in two variables is a *half-plane*. [3.7A, p. 181]

$4x - 3y < 12$ and $y \geq 2x + 6$ are linear inequalities in two variables.

Essential Rules and Procedures

Examples

Pythagorean Theorem [3.1B, p. 123]
If a and b are the lengths of the legs of a right triangle and c is the length of the hypotenuse, then $a^2 + b^2 = c^2$.

A triangle with legs that measure 3 in. and 4 in. and a hypotenuse that measures 5 in. is a right triangle because 3, 4, and 5 satisfy the Pythagorean Theorem.

$$a^2 + b^2 = c^2$$
$$3^2 + 4^2 = 5^2$$
$$9 + 16 = 25$$
$$25 = 25$$

Distance Formula [3.1B, p. 124]
If $P_1(x_1, y_1)$ and $P_2(x_2, y_2)$ are two points in the plane, then the distance between the two points is given by $d = \sqrt{(x_1 - x_2)^2 + (y_1 - y_2)^2}$.

$(x_1, y_1) = (-3, 5)$, $(x_2, y_2) = (2, 4)$
$d = \sqrt{(x_1 - x_2)^2 + (y_1 - y_2)^2}$
$\quad = \sqrt{(-3 - 2)^2 + (5 - 4)^2}$
$\quad = \sqrt{25 + 1}$
$\quad = \sqrt{26}$

Midpoint Formula [3.1B, p. 124]
If $P_1(x_1, y_1)$ and $P_2(x_2, y_2)$ are the endpoints of a line segment, then the coordinates of the midpoint of the line segment $P_m(x_m, y_m)$ are given by $x_m = \frac{x_1 + x_2}{2}$ and $y_m = \frac{y_1 + y_2}{2}$.

$(x_1, y_1) = (2, 3)$, $(x_2, y_2) = (-6, -1)$
$x_m = \frac{x_1 + x_2}{2} = \frac{2 + (-6)}{2} = -2$

$y_m = \frac{y_1 + y_2}{2} = \frac{3 + (-1)}{2} = 1$
$(x_m, y_m) = (-2, 1)$

Graph of $y = b$ [3.3B, p. 146]
The graph of $y = b$ is a horizontal line passing through the point $(0, b)$.

The graph of $y = -5$ is a horizontal line passing through the point $(0, -5)$.

Graph of a Constant Function [3.3B, p. 146]
A function given by $f(x) = b$, where b is a constant, is a *constant function*. The graph of the constant function is a horizontal line passing through $(0, b)$.

The graph of $f(x) = -5$ is a horizontal line passing through the point $(0, -5)$. Note that this is the same as the graph of $y = -5$.

Graph of $x = a$ [3.3B, p. 147]
The graph of $x = a$ is a vertical line passing through the point $(a, 0)$.

The graph of $x = 4$ is a vertical line passing through the point $(4, 0)$.

Finding Intercepts of Graphs of Linear Equations
[3.3C, p. 148]
To find the x-intercept, let $y = 0$.
To find the y-intercept, let $x = 0$.
For any equation of the form $y = mx + b$, the y-intercept is $(0, b)$.

$3x + 4y = 12$

Let $y = 0$:
$$3x + 4(0) = 12$$
$$3x = 12$$
$$x = 4$$

Let $x = 0$:
$$3(0) + 4y = 12$$
$$4y = 12$$
$$y = 3$$

The x-intercept is $(4, 0)$.

The y-intercept is $(0, 3)$.

Slope Formula [3.4A, p. 155]
The slope of the line containing the two points $P_1(x_1, y_1)$ and $P_2(x_2, y_2)$ is given by $m = \frac{y_2 - y_1}{x_2 - x_1}$, $x_1 \neq x_2$.

$(x_1, y_1) = (-3, 2)$, $(x_2, y_2) = (1, 4)$
$$m = \frac{y_2 - y_1}{x_2 - x_1} = \frac{4 - 2}{1 - (-3)} = \frac{1}{2}$$
The slope of the line through the points $(-3, 2)$ and $(1, 4)$ is $\frac{1}{2}$.

Slope-Intercept Form of a Straight Line [3.4B, p. 159]
The equation $y = mx + b$ is called the *slope-intercept form* of a straight line. The slope of the line is m, the coefficient of x. The y-intercept is $(0, b)$.

For the equation $y = -3x + 2$, the slope is -3 and the y-intercept is $(0, 2)$.

Point-Slope Formula [3.5A, p. 166]
Let m be the slope of a line, and let (x_1, y_1) be the coordinates of a point on the line. The equation of the line can be found from the point-slope formula: $y - y_1 = m(x - x_1)$.

The equation of the line that passes through the point $(4, 2)$ and has slope -3 is
$$y - y_1 = m(x - x_1)$$
$$y - 2 = -3(x - 4)$$
$$y - 2 = -3x + 12$$
$$y = -3x + 14$$

Slopes of Parallel Lines [3.6A, p. 175]
Two nonvertical lines with slopes of m_1 and m_2 are parallel if and only if $m_1 = m_2$. Any two vertical lines are parallel.

$y = 3x - 4$, $m_1 = 3$
$y = 3x + 2$, $m_2 = 3$
Because $m_1 = m_2$, the lines are parallel.

Slopes of Perpendicular Lines [3.6A, p. 176]
If m_1 and m_2 are the slopes of two lines, neither of which is vertical, then the lines are perpendicular if and only if $m_1 \cdot m_2 = -1$. This states that the slopes of perpendicular lines are *negative reciprocals* of each other. A vertical line is perpendicular to a horizontal line.

$y = \frac{1}{2}x - 1$, $m_1 = \frac{1}{2}$
$y = -2x + 2$, $m_2 = -2$
Because $m_1 \cdot m_2 = -1$, the lines are perpendicular.

Chapter 3 Review Exercises

1. Determine the ordered-pair solution of $y = \dfrac{x}{x-2}$ that corresponds to $x = 4$.

2. Given $P(x) = 3x + 4$, evaluate $P(-2)$ and $P(a)$.

3. Graph the ordered-pair solutions of $y = 2x^2 - 5$ when $x = -2, -1, 0, 1,$ and 2.

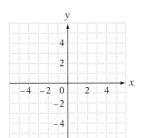

4. Draw a line through all the points with an abscissa of -3.

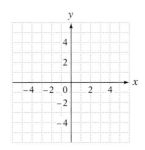

5. Find the range of $f(x) = x^2 + x - 1$ if the domain is $\{-2, -1, 0, 1, 2\}$.

6. Find the domain and range of the function $\{(-1, 0), (0, 2), (1, 2), (5, 4)\}$.

7. Find the midpoint and the length (to the nearest hundredth) of the line segment with endpoints $(-2, 4)$ and $(3, 5)$.

8. What value of x is excluded from the domain of $f(x) = \dfrac{x}{x+4}$?

9. Find the x- and y-intercepts and graph $y = -\dfrac{2}{3}x - 2$.

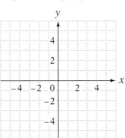

10. Graph $3x + 2y = -6$ by using the x- and y-intercepts.

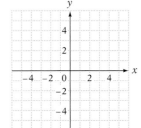

11. Graph: $y = -2x + 2$

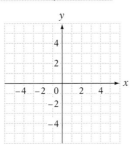

12. Graph: $4x - 3y = 12$

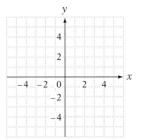

13. Find the slope of the line that contains the ordered pairs $(3, -2)$ and $(-1, 2)$.

14. Find the equation of the line that contains the ordered pair $(-3, 4)$ and has slope $\dfrac{5}{2}$.

15. Draw a line through all points with an ordinate of −2.

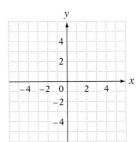

16. Graph the line that passes through the point (−2, 3) and has slope $-\frac{1}{4}$.

17. Find the range of $f(x) = x^2 - 2$ if the domain is $\{-2, -1, 0, 1, 2\}$.

18. **The Hospitality Industry** The manager of a hotel determines that 200 rooms will be occupied if the rate is $95 per night. For each $10 increase in the rate, 10 fewer rooms will be occupied.
 a. Determine a linear function that will predict the number of rooms that will be occupied at a given rate.
 b. Use the model to predict occupancy when the rate is $120.

19. Find the equation of the line that contains the point (−2, 3) and is parallel to the line $y = -4x + 3$.

20. Find the equation of the line that contains the point (−2, 3) and is perpendicular to the line $y = -\frac{2}{5}x - 3$.

21. Graph $y = 1$.

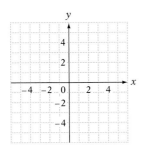

22. Graph $x = -1$.

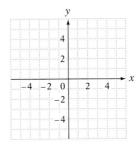

23. Find the equation of the line that contains the ordered pair (−3, 3) and has slope $-\frac{2}{3}$.

24. Find the equation of the line that contains the ordered pairs (−8, 2) and (4, 5).

25. Find the distance between the points whose coordinates are (4, −5) and (−2, 3).

26. Find the coordinates of the midpoint of the line segment with endpoints (−3, 8) and (5, −2).

27. **Life Expectancy** The table at the right shows the average number of years that women and men were estimated to live beyond age 65 in 1950, 1960, 1970, 1980, 1990, and 2000. (*Source:* 2000 Social Security Trustees Report) Figures are rounded to the nearest whole number. Graph the scatter diagram for these data.

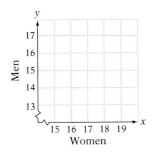

Women, x	15	16	17	18	19	19
Men, y	13	13	13	14	15	16

28. Graph the solution set of $y \geq 2x - 3$.

29. Graph the solution set of $3x - 2y < 6$.

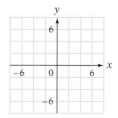

30. Find the equation of the line that contains the ordered pairs $(-2, 4)$ and $(4, -3)$.

31. Find the equation of the line that contains the ordered pair $(-2, -4)$ and is parallel to the graph of $4x - 2y = 7$.

32. Find the equation of the line that contains the ordered pair $(3, -2)$ and is parallel to the graph of $y = -3x + 4$.

33. Find the equation of the line that contains the ordered pair $(2, 5)$ and is perpendicular to the graph of the line $y = -\frac{2}{3}x + 6$.

34. Graph the line that passes through the point whose coordinates are $(-1, 4)$ and has slope $-\frac{1}{3}$.

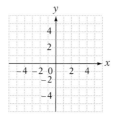

35. **Travel** A car is traveling at 55 mph. The equation that describes the distance traveled is $d = 55t$. Graph this equation for $0 \leq t \leq 6$. The point whose coordinates are $(4, 220)$ is on the graph. Write a sentence that explains the meaning of this ordered pair.

36. **Manufacturing** The graph at the right shows the relationship between the cost of manufacturing calculators and the number of calculators manufactured. Find the slope of the line between the two points shown on the graph. Write a sentence that states the meaning of the slope.

37. **Construction** A building contractor estimates that the cost to build a new home is $25,000 plus $80 for each square foot of floor space.
a. Determine a linear function that will give the cost to build a house that contains a given number of square feet.
b. Use the model to determine the cost to build a house that contains 2000 ft².

Chapter 3 Test

1. Graph the ordered-pair solutions of $P(x) = 2 - x^2$ when $x = -2, -1, 0, 1,$ and 2.

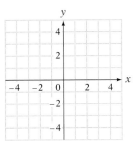

2. Find the ordered-pair solution of $y = 2x + 6$ that corresponds to $x = -3$.

3. Graph: $y = \dfrac{2}{3}x - 4$

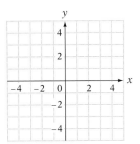

4. Graph: $2x + 3y = -3$

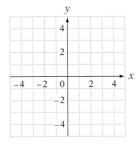

5. Find the equation of the vertical line that contains the point $(-2, 3)$.

6. Find the length, to the nearest hundredth, and the midpoint of the line segment with endpoints $(4, 2)$ and $(-5, 8)$.

7. Find the slope of the line that contains the points $(-2, 3)$ and $(4, 2)$.

8. Given $P(x) = 3x^2 - 2x + 1$, evaluate $P(2)$.

9. Graph $2x - 3y = 6$ by using the x- and the y-intercepts.

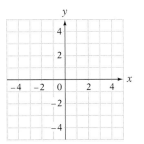

10. Graph the line that passes through the point $(-2, 3)$ and has slope $-\dfrac{3}{2}$.

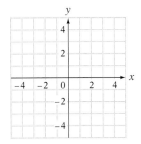

11. Find the equation of the line that contains the point $(-5, 2)$ and has slope $\frac{2}{5}$.

12. What value of x is excluded from the domain of $f(x) = \frac{2x + 1}{x}$?

13. Find the equation of the line that contains the points $(3, -4)$ and $(-2, 3)$.

14. Find the equation of the horizontal line that contains the point $(4, -3)$.

15. Find the domain and range of the function $\{(-4, 2), (-2, 2), (0, 0), (3, 5)\}$.

16. Find the equation of the line that contains the point $(1, 2)$ and is parallel to the line $y = -\frac{3}{2}x - 6$.

17. Find the equation of the line that contains the point $(-2, -3)$ and is perpendicular to the line $y = -\frac{1}{2}x - 3$.

18. Graph the solution set of $3x - 4y > 8$.

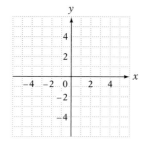

19. Depreciation The graph below shows the relationship between the cost of a rental house and the depreciation allowed for income tax purposes. Find the slope between the two points shown on the graph. Write a sentence that states the meaning of the slope.

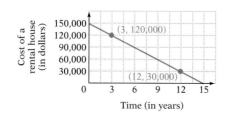

20. Summer Camp The director of a baseball camp estimates that 100 students will enroll if the tuition is $250. For each $20 increase in tuition, six fewer students will enroll.
a. Determine a linear function that will predict the number of students who will enroll at a given tuition.
b. Use this model to predict enrollment when the tuition is $300.

Cumulative Review Exercises

1. Identify the property that justifies the statement $(x + y) \cdot 2 = 2 \cdot (x + y)$.

2. Solve: $3 - \dfrac{x}{2} = \dfrac{3}{4}$

3. Solve:
$2[y - 2(3 - y) + 4] = 4 - 3y$

4. Solve:
$\dfrac{1 - 3x}{2} + \dfrac{7x - 2}{6} = \dfrac{4x + 2}{9}$

5. Solve:
$x - 3 < -4 \quad \text{or} \quad 2x + 2 > 3$
Write the solution set in set-builder notation.

6. Solve:
$8 - |2x - 1| = 4$

7. Solve: $|3x - 5| < 5$

8. Simplify: $4 - 2(4 - 5)^3 + 2$

9. Evaluate $(a - b)^2 \div (ab)$ when $a = 4$ and $b = -2$.

10. Graph: $\{x \mid x < -2\} \cup \{x \mid x > 0\}$

11. Solve $P = \dfrac{R - C}{n}$ for C.

12. Solve $2x + 3y = 6$ for x.

13. Solve: $3x - 1 < 4$ and $x - 2 > 2$

14. Given $P(x) = x^2 + 5$, evaluate $P(-3)$.

15. Find the ordered-pair solution of $y = -\dfrac{5}{4}x + 3$ that corresponds to $x = -8$.

16. Find the slope of the line that contains the points $(-1, 3)$ and $(3, -4)$.

17. Find the equation of the line that contains the point $(-1, 5)$ and has slope $\dfrac{3}{2}$.

18. Find the equation of the line that contains the points $(4, -2)$ and $(0, 3)$.

19. Find the equation of the line that contains the point (2, 4) and is parallel to the line $y = -\frac{3}{2}x + 2$.

20. Find the equation of the line that contains the point (4, 0) and is perpendicular to the line $3x - 2y = 5$.

21. **Coins** A coin purse contains 17 coins with a value of $1.60. The purse contains nickels, dimes, and quarters. There are four times as many nickels as quarters. Find the number of dimes in the purse.

22. **Uniform Motion** Two planes are 1800 mi apart and are traveling toward each other. One plane is traveling twice as fast as the other plane. The planes meet in 3 h. Find the speed of each plane.

23. **Mixtures** A grocer combines coffee costing $9 per pound with coffee costing $6 per pound. How many pounds of each should be used to make 60 pounds of a blend costing $8 per pound?

24. Graph $3x - 5y = 15$ by using the x- and y-intercepts.

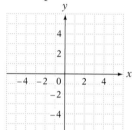

25. Graph the line that passes through the point $(-3, 1)$ and has slope $-\frac{3}{2}$.

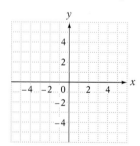

26. Graph the inequality $3x - 2y \geq 6$.

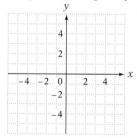

27. **Depreciation** The relationship between the cost of a truck and the depreciation allowed for income tax purposes is shown in the graph at the right.
 a. Write the equation for the line that represents the depreciated value of the truck.
 b. Write a sentence that states the meaning of the slope.

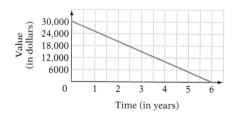

4 Systems of Linear Equations and Inequalities

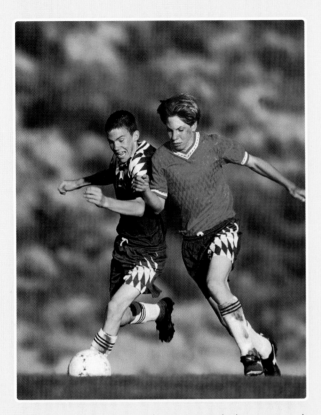

Soccer is one of the most popular team sports in the country, and is played by people of all ages and at every skill level. Children all over the country play in community leagues like AYSO, students play for their high schools and colleges, and the U.S. has an Olympic team. Suppose you volunteer to be the manager of your community's soccer league, which consists of 12 teams. You would need to create a schedule in which each team plays every other team once. How many games must be scheduled? The **Focus on Problem Solving on page 245** discusses methods of solving difficult problems, such as trying an easier version of the problem first, drawing diagrams, and identifying patterns.

OBJECTIVES

Section 4.1

A To solve a system of linear equations by graphing

B To solve a system of linear equations by the substitution method

C To solve investment problems

Section 4.2

A To solve a system of two linear equations in two variables by the addition method

B To solve a system of three linear equations in three variables by the addition method

Section 4.3

A To evaluate a determinant

B To solve a system of equations by using Cramer's Rule

Section 4.4

A To solve rate-of-wind or rate-of-current problems

B To solve application problems

Section 4.5

A To graph the solution set of a system of linear inequalities

Need help? For online student resources, such as section quizzes, visit this textbook's website at **math.college.hmco.com/students.**

Do these exercises to prepare for Chapter 4.

1. Simplify: $10\left(\dfrac{3}{5}x + \dfrac{1}{2}y\right)$

2. Evaluate $3x + 2y - z$ for $x = -1$, $y = 4$, and $z = -2$.

3. Given $3x - 2z = 4$, find the value of x when $z = -2$.

4. Solve: $3x + 4(-2x - 5) = -5$

5. Solve: $0.45x + 0.06(-x + 4000) = 630$

6. Graph: $y = \dfrac{1}{2}x - 4$

7. Graph: $3x - 2y = 6$

8. Graph: $y > -\dfrac{3}{5}x + 1$

GO FIGURE • • • •

Chris can beat Pat by 100 m in a 1000-meter race. Pat can beat Leslie by 10 m in a 100-meter race. If both run at this given rate, by how many meters will Chris beat Leslie in a 1000-meter race?

4.1 Solving Systems of Linear Equations by Graphing and by the Substitution Method

Objective A To solve a system of linear equations by graphing

A **system of equations** is two or more equations considered together. The system at the right is a system of two linear equations in two variables. The graphs of the equations are straight lines.

$$3x + 4y = 7$$
$$2x - 3y = 6$$

A **solution of a system of equations in two variables** is an ordered pair that is a solution of each equation of the system.

> **HOW TO** Is $(3, -2)$ a solution of the system
> $$2x - 3y = 12$$
> $$5x + 2y = 11?$$
>
$2x - 3y = 12$
> | $2(3) - 3(-2)$ \vert 12 |
> | $6 - (-6)$ \vert 12 |
> | $12 = 12$ |
>
$5x + 2y = 11$
> | $5(3) + 2(-2)$ \vert 11 |
> | $15 + (-4)$ \vert 11 |
> | $11 = 11$ |
>
> • Replace x by 3 and y by -2.
>
> Yes, because $(3, -2)$ is a solution of each equation, it is a solution of the system of equations.

A solution of a system of linear equations can be found by graphing the lines of the system on the same set of coordinate axes. Three examples of linear equations in two variables are shown below, along with the graphs of the equations of the systems.

System I
$$x + 2y = 4$$
$$2x + y = -1$$

System II
$$2x + 3y = 6$$
$$4x + 6y = -12$$

System III
$$x - 2y = 4$$
$$2x - 4y = 8$$

 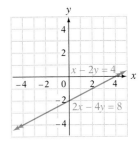

Check:

$x + 2y = 4$
$-2 + 2(3)$ \vert 4
$-2 + 6$ \vert 4
$4 = 4$

$2x + y = -1$
$2(-2) + 3$ \vert -1
$-4 + 3$ \vert -1
$-1 = -1$

In System I, the two lines intersect at a single point whose coordinates are $(-2, 3)$. Because this point lies on both lines, it is a solution of each equation of the system of equations. We can check this by replacing x by -2 and y by 3. The check is shown at the left. The ordered pair $(-2, 3)$ is a solution of System I.

When the graphs of a system of equations intersect at only one point, the system is called an **independent system of equations.** System I is an independent system of equations.

System II from the preceding page and the graph of the equations of that system are shown again at the right. Note in this case that the graphs of the lines are parallel and do not intersect. Since the graphs do not intersect, there is no point that is on both lines. Therefore, the system of equations has no solution.

When a system of equations has no solution, it is called an **inconsistent system of equations.** System II is an inconsistent system of equations.

$$2x + 3y = 6$$
$$4x + 6y = -12$$

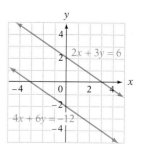

System III from the preceding page and the graph of the equations of that system are shown again at the right. Note that the graph of $x - 2y = 4$ lies directly on top of the graph of $2x - 4y = 8$. Thus the two lines intersect at an infinite number of points. Because the graphs intersect at an infinite number of points, there is an infinite number of solutions of this system of equations. Since each equation represents the same set of points, the solutions of the system of equations can be stated by using the ordered pairs of either one of the equations. Therefore, we can say, "The solutions are the ordered pairs that satisfy $x - 2y = 4$," or we can solve the equation for y and say, "The solutions are the ordered pairs that satisfy $y = \frac{1}{2}x - 2$."

We normally state this solution using ordered pairs—

"The solutions are the ordered pairs $\left(x, \frac{1}{2}x - 2\right)$."

$$x - 2y = 4$$
$$2x - 4y = 8$$

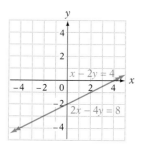

When the two equations in a system of equations represent the same line, the system is called a **dependent system of equations.** System III is a dependent system of equations.

Summary of the Three Possibilities for a System of Linear Equations in Two Variables

1. The graphs intersect at one point.
 The solution of the system of equations is the ordered pair (x, y) whose coordinates name the point of intersection. The system of equations is independent.
2. The lines are parallel and never intersect.
 There is no solution of the system of equations. The system of equations is inconsistent.
3. The graphs are the same line, and they intersect at infinitely many points.
 There are infinitely many solutions of the system of equations. The system of equations is dependent.

Integrating Technology

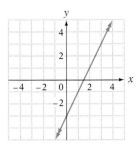

See the Projects and Group Activities at the end of this chapter for instructions on using a graphing calculator to solve a system of equations. Instructions are also provided in the Keystroke Guide: *Intersect.*

HOW TO Solve by graphing: $2x - y = 3$
$\qquad\qquad\qquad\qquad\qquad\qquad\quad 4x - 2y = 6$

Graph each line.
The system of equations is dependent.
Solve one of the equations for y.

$$2x - y = 3$$
$$-y = -2x + 3$$
$$y = 2x - 3$$

The solutions are the ordered pairs $(x, 2x - 3)$.

Example 1 Solve by graphing:
$2x - y = 3$
$3x + y = 2$

Solution

• Find the point of intersection of the graphs of the equations.

The solution is $(1, -1)$.

You Try It 1 Solve by graphing:
$\quad x + y = 1$
$\quad 2x + y = 0$

Your solution

Example 2 Solve by graphing:
$2x + 3y = 6$
$\qquad\qquad y = -\dfrac{2}{3}x + 1$

Solution

• Graph the two equations.

The lines are parallel and therefore do not intersect. The system of equations has no solution. The system of equations is inconsistent.

You Try It 2 Solve by graphing:
$2x + 5y = 10$
$\qquad\qquad y = -\dfrac{2}{5}x - 2$

Your solution

Solutions on p. S10

Example 3 Solve by graphing:
$$x - 2y = 6$$
$$y = \frac{1}{2}x - 3$$

Solution

• Graph the two equations.

The system of equations is dependent. The solutions are the ordered pairs $\left(x, \frac{1}{2}x - 3\right)$.

You Try It 3 Solve by graphing:
$$3x - 4y = 12$$
$$y = \frac{3}{4}x - 3$$

Your solution

Solution on p. S10

Objective B To solve a system of linear equations by the substitution method

The graphical solution of a system of equations is based on approximating the coordinates of a point of intersection. An algebraic method called the **substitution method** can be used to find an exact solution of a system of equations. To use the substitution method, we must write one of the equations of the system in terms of x or in terms of y.

HOW TO Solve by the substitution method: (1) $3x + y = 5$
(2) $4x + 5y = 3$

$$3x + y = 5$$
(3) $$y = -3x + 5$$

• Solve Equation (1) for y. This is Equation (3).

(2) $$4x + 5y = 3$$
$$4x + 5(-3x + 5) = 3$$

• This is Equation (2).
• Equation (3) states that $y = -3x + 5$. Substitute $-3x + 5$ for y in Equation (2).

$$4x - 15x + 25 = 3$$
$$-11x + 25 = 3$$
$$-11x = -22$$
$$x = 2$$

• Solve for x.

(3) $$y = -3x + 5$$
$$y = -3(2) + 5$$
$$y = -6 + 5$$
$$y = -1$$

• Substitute the value of x into Equation (3) and solve for y.

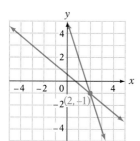

The solution is the ordered pair $(2, -1)$.

The graph of the system of equations is shown at the left. Note that the graphs intersect at the point whose coordinates are $(2, -1)$, the solution of the system of equations.

HOW TO Solve by the substitution method: (1) $6x + 2y = 8$
(2) $3x + y = 2$

(3) $\quad\quad \begin{aligned} 3x + y &= 2 \\ y &= -3x + 2 \end{aligned}$

- We will solve Equation (2) for y.
- This is Equation (3).

(1) $\quad\quad \begin{aligned} 6x + 2y &= 8 \\ 6x + 2(-3x + 2) &= 8 \end{aligned}$

- This is Equation (1).
- Equation (3) states that $y = -3x + 2$. Substitute $-3x + 2$ for y in Equation (1).
- Solve for x.

$$6x - 6x + 4 = 8$$
$$0x + 4 = 8$$
$$4 = 8$$

This is not a true equation.
The system of equations has no solution.
The system of equations is inconsistent.

The graph of the system of equations is shown at the left. Note that the lines are parallel.

Example 4 Solve by substitution:
(1) $3x - 2y = 4$
(2) $-x + 4y = -3$

You Try It 4 Solve by substitution:
$$3x - y = 3$$
$$6x + 3y = -4$$

Solution Solve Equation (2) for x.
$$\begin{aligned} -x + 4y &= -3 \\ -x &= -4y - 3 \\ x &= 4y + 3 \end{aligned}$$ • Equation (3)

Your solution

Substitute $4y + 3$ for x in
Equation (1).
$$\begin{aligned} 3x - 2y &= 4 \\ 3(4y + 3) - 2y &= 4 \\ 12y + 9 - 2y &= 4 \\ 10y + 9 &= 4 \\ 10y &= -5 \\ y &= -\frac{5}{10} = -\frac{1}{2} \end{aligned}$$
• Equation (1)
• $x = 4y + 3$

Substitute the value of y into
Equation (3).
$$\begin{aligned} x &= 4y + 3 \\ &= 4\left(-\frac{1}{2}\right) + 3 \\ &= -2 + 3 = 1 \end{aligned}$$
• Equation (3)
• $y = -\frac{1}{2}$

The solution is $\left(1, -\frac{1}{2}\right)$.

Solution on p. S10

Example 5 Solve by substitution and graph:
$$3x - 3y = 2$$
$$y = x + 2$$

You Try It 5 Solve by substitution and graph:
$$y = 2x - 3$$
$$3x - 2y = 6$$

Solution
$$3x - 3y = 2$$
$$3x - 3(x + 2) = 2$$
$$3x - 3x - 6 = 2$$
$$-6 = 2$$

Your solution

• Graph the two equations.

This is not a true equation. The lines are parallel, so the system is inconsistent. The system does not have a solution.

Example 6 Solve by substitution and graph:
$$9x + 3y = 12$$
$$y = -3x + 4$$

You Try It 6 Solve by substitution and graph:
$$6x - 3y = 6$$
$$2x - y = 2$$

Solution
$$9x + 3y = 12$$
$$9x + 3(-3x + 4) = 12$$
$$9x - 9x + 12 = 12$$
$$12 = 12$$
This is a true equation. The system is dependent.

Your solution

• Graph the two equations.

The solutions are the ordered pairs $(x, -3x + 4)$.

Solutions on pp. S10–S11

Objective C **To solve investment problems**

The annual simple interest that an investment earns is given by the equation **Pr = I**, where *P* is the principal, or the amount invested, *r* is the simple interest rate, and *I* is the simple interest.

For instance, if you invest $500 at a simple interest rate of 5%, then the interest earned after one year is calculated as follows:

$$Pr = I$$
$$500(0.05) = I$$ • Replace *P* by 500 and *r* by 0.05 (5%).
$$25 = I$$ • Simplify.

The amount of interest earned is $25.

HOW TO You have a total of $5000 to invest in two simple interest accounts. On one account, the money market fund, the annual simple interest rate is 3.5%. On the second account, the bond fund, the annual simple interest rate is 7.5%. If you earn $245 per year from these two investments, how much do you have invested in each account?

Strategy for Solving Simple-Interest Investment Problems

1. For each amount invested, use the equation $Pr = I$. Write a numerical or variable expression for the principal, the interest rate, and the interest earned.

Amount invested at 3.5%: x
Amount invested at 7.5%: y

	Principal, P	·	Interest rate, r	=	Interest earned, I
Amount at 3.5%	x	·	0.035	=	$0.035x$
Amount at 7.5%	y	·	0.075	=	$0.075y$

2. Write a system of equations. One equation will express the relationship between the amounts invested. The second equation will express the relationship between the amounts of interest earned by the investments.

The total amount invested is $5000: $x + y = 5000$
The total annual interest earned is $245: $0.035x + 0.075y = 245$

Solve the system of equations. (1) $x + y = 5000$
 (2) $0.035x + 0.075y = 245$

Solve Equation (1) for y: (3) $y = -x + 5000$
Substitute into Equation (2): (2) $0.035x + 0.075(-x + 5000) = 245$
 $0.035x - 0.075x + 375 = 245$
 $-0.04x = -130$
 $x = 3250$

Substitute the value of x into Equation (3) and solve for y.

$$y = -x + 5000$$
$$y = -3250 + 5000 = 1750$$

The amount invested at 3.5% is $3250.
The amount invested at 7.5% is $1750.

Example 7

An investment of $4000 is made at an annual simple interest rate of 4.9%. How much additional money must be invested at an annual simple interest rate of 7.4% so that the total interest earned is 6.4% of the total investment?

Strategy

- Amount invested at 4.9%: 4000
 Amount invested at 7.4%: x
 Amount invested at 6.4%: y

	Principal	*Rate*	*Interest*
Amount at 4.9%	4000	0.049	0.049(4000)
Amount at 7.4%	x	0.074	0.074x
Amount at 6.4%	y	0.064	0.064y

- The amount invested at 6.4% (y) is $4000 more than the amount invested at 7.4% (x):
 $y = x + 4000$
- The sum of the interest earned at 4.9% and the interest earned at 7.4% equals the interest earned at 6.4%:
 $0.049(4000) + 0.074x = 0.064y$

Solution

$$y = x + 4000 \quad (1)$$
$$0.049(4000) + 0.074x = 0.064y \quad (2)$$

Replace y in Equation (2) by $x + 4000$ from Equation (1). Then solve for x.

$$0.049(4000) + 0.074x = 0.064(x + 4000)$$
$$196 + 0.074x = 0.064x + 256$$
$$0.01x = 60$$
$$x = 6000$$

$6000 must be invested at an annual simple interest rate of 7.4%.

You Try It 7

An investment club invested $13,600 in two simple interest accounts. On one account, the annual simple interest rate is 4.2%. On the other, the annual simple interest rate is 6%. How much should be invested in each account so that both accounts earn the same annual interest?

Your strategy

Your solution

Solution on p. S11

4.1 Exercises

Objective A To solve a system of linear equations by graphing

For Exercises 1 to 4, determine whether the ordered pair is a solution of the system of equations.

1. $(0, -1)$
$3x - 2y = 2$
$x + 2y = 6$

2. $(2, 1)$
$x + y = 3$
$2x - 3y = 1$

3. $(-3, -5)$
$x + y = -8$
$2x + 5y = -31$

4. $(1, -1)$
$3x - y = 4$
$7x + 2y = -5$

For Exercises 5 to 8, state whether the system of equations is independent, inconsistent, or dependent.

5.

6.

7.

8.

For Exercises 9 to 26, solve by graphing.

9. $x + y = 2$
$x - y = 4$

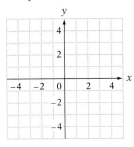

10. $x + y = 1$
$3x - y = -5$

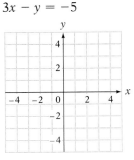

11. $x - y = -2$
$x + 2y = 10$

12. $2x - y = 5$
$3x + y = 5$

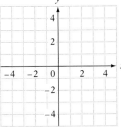

13. $3x - 2y = 6$
$y = 3$

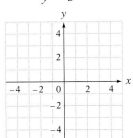

14. $x = 4$
$3x - 2y = 4$

15. $x = 4$
$y = -1$

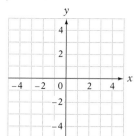

16. $x + 2 = 0$
$y - 1 = 0$

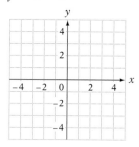

17. $2x + y = 3$
$x - 2 = 0$

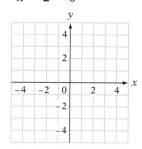

18. $x - 3y = 6$
$y + 3 = 0$

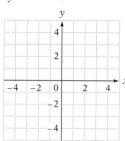

19. $x - y = 6$
$x + y = 2$

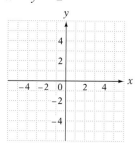

20. $2x + y = 2$
$-x + y = 5$

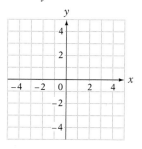

21. $y = x - 5$
$2x + y = 4$

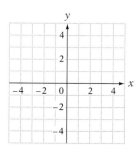

22. $2x - 5y = 4$
$y = x + 1$

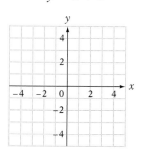

23. $y = \frac{1}{2}x - 2$
$x - 2y = 8$

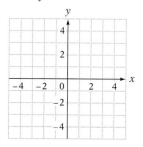

24. $2x + 3y = 6$
$y = -\frac{2}{3}x + 1$

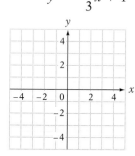

25. $2x - 5y = 10$
$y = \frac{2}{5}x - 2$

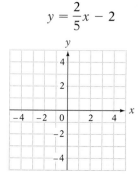

26. $3x - 2y = 6$
$y = \frac{3}{2}x - 3$

Objective B **To solve a system of linear equations by the substitution method**

For Exercises 27 to 56, solve by the substitution method.

27. $y = -x + 1$
$2x - y = 5$

28. $x = 3y + 1$
$x - 2y = 6$

29. $x = 2y - 3$
$3x + y = 5$

30. $4x - 3y = 5$
$y = 2x - 3$

31. $3x + 5y = -1$
$y = 2x - 8$

32. $5x - 2y = 9$
$y = 3x - 4$

33. $4x - 3y = 2$
$y = 2x + 1$

34. $x = 2y + 4$
$4x + 3y = -17$

35. $3x - 2y = -11$
$x = 2y - 9$

36. $5x + 4y = -1$
$y = 2 - 2x$

37. $3x + 2y = 4$
$y = 1 - 2x$

38. $2x - 5y = -9$
$y = 9 - 2x$

39. $5x + 2y = 15$
$x = 6 - y$

40. $7x - 3y = 3$
$x = 2y + 2$

41. $3x - 4y = 6$
$x = 3y + 2$

42. $2x + 2y = 7$
$y = 4x + 1$

43. $3x + 7y = -5$
$y = 6x - 5$

44. $3x + y = 5$
$2x + 3y = 8$

45. $3x - y = 10$
$6x - 2y = 5$

46. $6x - 4y = 3$
$3x - 2y = 9$

47. $3x + 4y = 14$
$2x + y = 1$

48. $5x + 3y = 8$
$3x + y = 8$

49. $3x + 5y = 0$
$x - 4y = 0$

50. $2x - 7y = 0$
$3x + y = 0$

51. $2x - 4y = 16$
$-x + 2y = -8$

52. $3x - 12y = -24$
$-x + 4y = 8$

53. $y = 3x + 2$
$y = 2x + 3$

54. $y = 3x - 7$
$y = 2x - 5$

55. $y = 3x + 1$
$y = 6x - 1$

56. $y = 2x - 3$
$y = 4x - 4$

Objective C **To solve investment problems**

57. The Community Relief Charity Group is earning 3.5% simple interest on the $2800 it invested in a savings account. It also earns 4.2% simple interest on an insured bond fund. The annual interest earned from both accounts is $329. How much is invested in the insured bond fund?

58. Two investments earn an annual income of $575. One investment earns an annual simple interest rate of 8.5%, and the other investment earns an annual simple interest rate of 6.4%. The total amount invested is $8000. How much is invested in each account?

59. An investment club invested $6000 at an annual simple interest rate of 4.0%. How much additional money must be invested at an annual simple interest rate of 6.5% so that the total annual interest earned will be 5% of the total investment?

60. A small company invested $30,000 by putting part of it into a savings account that earned 3.2% annual simple interest and the remainder in a risky stock fund that earned 12.6% annual simple interest. If the company earned $1665 annually from the investments, how much was in each account?

61. An account executive divided $42,000 between two simple interest accounts. On the tax-free account the annual simple interest rate is 3.5%, and on the money market fund the annual simple interest rate is 4.5%. How much should be invested in each account so that both accounts earn the same annual interest?

62. The Ridge Investment Club placed $33,000 into two simple interest accounts. On one account, the annual simple interest rate is 6.5%. On the other, the annual simple interest rate is 4.5%. How much should be invested in each account so that both accounts earn the same annual interest?

63. The Cross Creek Investment Club has $20,000 to invest. The members of the club decided to invest $16,000 of their money in two bond funds. The first, a mutual bond fund, earns annual simple interest of 4.5%. The second account, a corporate bond fund, earns 8% annual simple interest. If the members earned $1070 from these two accounts, how much was invested in the mutual bond fund?

64. Cabin Financial Service Group recommends that a client purchase for $10,000 a corporate bond that earns 5% annual simple interest. How much additional money must be placed in U.S. government securities that earn a simple interest rate of 3.5% so that the total annual interest earned from the two investments is 4% of the total investment?

APPLYING THE CONCEPTS

 For Exercises 65 to 67, use a graphing calculator to estimate the solution to the system of equations. Round coordinates to the nearest hundredth. See the Projects and Group Activities at the end of this chapter for assistance.

65. $y = -\dfrac{1}{2}x + 2$

$y = 2x - 1$

66. $y = \sqrt{2}x - 1$

$y = -\sqrt{3}x + 1$

67. $y = \pi x - \dfrac{2}{3}$

$y = -x + \dfrac{\pi}{2}$

4.2 Solving Systems of Linear Equations by the Addition Method

Objective A **To solve a system of two linear equations in two variables by the addition method**

The **addition method** is an alternative method for solving a system of equations. This method is based on the Addition Property of Equations. Use the addition method when it is not convenient to solve one equation for one variable in terms of another variable.

Note for the system of equations at the right the effect of adding Equation (2) to Equation (1). Because $-3y$ and $3y$ are additive inverses, adding the equations results in an equation with only one variable.

$$(1)\ 5x - 3y = 14$$
$$(2)\ 2x + 3y = -7$$
$$7x + 0y = 7$$
$$7x = 7$$

The solution of the resulting equation is the first coordinate of the ordered-pair solution of the system.

$$7x = 7$$
$$x = 1$$

The second coordinate is found by substituting the value of x into Equation (1) or (2) and then solving for y. Equation (1) is used here.

$$(1)\quad 5x - 3y = 14$$
$$5(1) - 3y = 14$$
$$5 - 3y = 14$$
$$-3y = 9$$
$$y = -3$$

The solution is $(1, -3)$.

Point of Interest

There are records of Babylonian mathematicians solving systems of equations 3600 years ago. Here is a system of equations from that time (in our modern notation):

$$\frac{2}{3}x = \frac{1}{2}y + 500$$
$$x + y = 1800$$

We say *modern notation* for many reasons. Foremost is the fact that using variables did not become widespread until the 17th century. There are many other reasons, however. The equals sign had not been invented, 2 and 3 did not look like they do today, and zero had not even been considered as a possible number.

Sometimes each equation of the system of equations must be multiplied by a constant so that the coefficients of one of the variable terms are opposites.

HOW TO Solve by the addition method: $(1)\ 3x + 4y = 2$
$(2)\ 2x + 5y = -1$

To eliminate x, multiply Equation (1) by 2 and Equation (2) by -3. Note at the right how the constants are chosen.

$$2(3x + 4y) = 2 \cdot 2$$
$$-3(2x + 5y) = -3(-1)$$

• The negative is used so that the coefficients will be opposites.

Study Tip

Always check the proposed solution of a system of equations. For the equation at the right:

$$3x + 4y = 2$$
$$\begin{array}{c|c} 3(2) + 4(-1) & 2 \\ 6 - 4 & 2 \\ & 2 = 2 \end{array}$$

$$2x + 5y = -1$$
$$\begin{array}{c|c} 2(2) + 5(-1) & -1 \\ 4 - 5 & -1 \\ & -1 = -1 \end{array}$$

The solution checks.

$$\begin{array}{r} 6x + 8y = 4 \\ -6x - 15y = 3 \\ \hline -7y = 7 \\ y = -1 \end{array}$$

• 2 times Equation (1).
• -3 times Equation (2).
• Add the equations.
• Solve for y.

Substitute the value of y into Equation (1) or Equation (2) and solve for x. Equation (1) will be used here.

$$(1)\quad 3x + 4y = 2$$
$$3x + 4(-1) = 2$$
$$3x - 4 = 2$$
$$3x = 6$$
$$x = 2$$

• Substitute -1 for y.
• Solve for x.

The solution is $(2, -1)$.

HOW TO Solve by the addition method: (1) $\dfrac{2}{3}x + \dfrac{1}{2}y = 4$

(2) $\dfrac{1}{4}x - \dfrac{3}{8}y = -\dfrac{3}{4}$

Clear fractions. Multiply each equation by the LCM of the denominators.

$6\left(\dfrac{2}{3}x + \dfrac{1}{2}y\right) = 6(4)$

$8\left(\dfrac{1}{4}x - \dfrac{3}{8}y\right) = 8\left(-\dfrac{3}{4}\right)$

$\begin{aligned} 4x + 3y &= 24 \\ \underline{2x - 3y} &= \underline{-6} \\ 6x &= 18 \\ x &= 3 \end{aligned}$ • **Eliminate *y*. Add the equations. Then solve for *x*.**

$\dfrac{2}{3}x + \dfrac{1}{2}y = 4$ • **This is Equation (1).**

$\dfrac{2}{3}(3) + \dfrac{1}{2}y = 4$ • **Substitute *x* = 3 into Equation (1) and solve for *y*.**

$2 + \dfrac{1}{2}y = 4$

$\dfrac{1}{2}y = 2$

$y = 4$

The solution is (3, 4).

HOW TO Solve by the addition method: (1) $2x - y = 3$
(2) $4x - 2y = 6$

Eliminate *y*. Multiply Equation (1) by -2.

(1) $-2(2x - y) = -2(3)$ • **-2 times Equation (1).**
(3) $-4x + 2y = -6$ • **This is Equation (3).**

Add Equation (3) to Equation (2).

(2) $4x - 2y = 6$
(3) $\underline{-4x + 2y = -6}$
 $0 = 0$

TAKE NOTE

The result of adding
Equations (3) and (2) is
$0 = 0$. It is not $x = 0$ and
it is not $y = 0$. There is
no variable in the equa-
tion $0 = 0$. This result
does not indicate that the
solution is (0, 0); rather,
it indicates a dependent
system of equations.

The equation $0 = 0$ indicates that the system of
equations is dependent. This means that the
graphs of the two lines are the same. Therefore,
the solutions of the system of equations are the
ordered-pair solutions of the equation of the line.
Solve Equation (1) for *y*.

$2x - y = 3$
$-y = -2x + 3$
$y = 2x - 3$

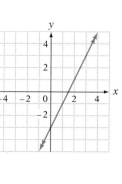

The ordered-pair solutions are (x, y), where
$y = 2x - 3$. These ordered pairs are usually
written $(x, 2x - 3)$, where $2x - 3$ is substituted
for *y*.

Example 1 Solve by the addition method:
(1) $3x - 2y = 2x + 5$
(2) $2x + 3y = -4$

You Try It 1 Solve by the addition method:
$2x + 5y = 6$
$3x - 2y = 6x + 2$

Solution

Write Equation (1) in the form $Ax + By = C$.

$3x - 2y = 2x + 5$
$x - 2y = 5$

Solve the system:

$x - 2y = 5$
$2x + 3y = -4$

Eliminate x.

$-2(x - 2y) = -2(5)$
$2x + 3y = -4$

$-2x + 4y = -10$
$\underline{2x + 3y = -4}$
$7y = -14$ • **Add the equations.**
$y = -2$ • **Solve for y.**

Replace y in Equation (2).

$2x + 3y = -4$
$2x + 3(-2) = -4$
$2x - 6 = -4$
$2x = 2$
$x = 1$

The solution is $(1, -2)$.

Your solution

Example 2 Solve by the addition method:
(1) $4x - 8y = 36$
(2) $3x - 6y = 27$

You Try It 2 Solve by the addition method:
$2x + y = 5$
$4x + 2y = 6$

Solution

Eliminate x.

$3(4x - 8y) = 3(36)$
$-4(3x - 6y) = -4(27)$

$12x - 24y = 108$
$\underline{-12x + 24y = -108}$
$0 = 0$ • **Add the equations.**

The system of equations is dependent. The solutions are the ordered pairs $\left(x, \frac{1}{2}x - \frac{9}{2}\right)$.

Your solution

Solutions on p. S11

Objective B **To solve a system of three linear equations in three variables by the addition method**

An equation of the form $Ax + By + Cz = D$, where A, B, and C are the coefficients of the variables and D is a constant, is a **linear equation in three variables.** Examples of this type of equation are shown at the right.

$$2x + 4y - 3z = 7$$
$$x - 6y + z = -3$$

Graphing an equation in three variables requires a third coordinate axis perpendicular to the xy-plane. The third axis is commonly called the z-axis. The result is a three-dimensional coordinate system called the **xyz-coordinate system.** To help visualize a three-dimensional coordinate system, think of a corner of a room: The floor is the xy-plane, one wall is the yz-plane, and the other wall is the xz-plane. A three-dimensional coordinate system is shown at the right.

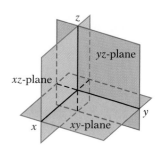

The graph of a point in an xyz-coordinate system is an **ordered triple** (x, y, z). Graphing an ordered triple requires three moves, the first in the direction of the x-axis, the second in the direction of the y-axis, and the third in the direction of the z-axis. The graph of the points $(-4, 2, 3)$ and $(3, 4, -2)$ is shown at the right.

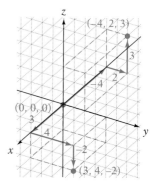

The graph of a linear equation in three variables is a plane. That is, if all the solutions of a linear equation in three variables were plotted in an xyz-coordinate system, the graph would look like a large piece of paper extending infinitely. The graph of $x + y + z = 3$ is shown at the right.

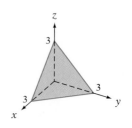

There are different ways in which three planes can be oriented in an *xyz*-coordinate system. The systems of equations represented by the planes below are inconsistent.

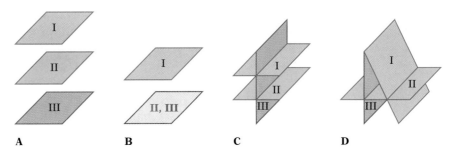

Graphs of Inconsistent Systems of Equations

For a system of three equations in three variables to have a solution, the graphs of the planes must intersect at a single point, they must intersect along a common line, or all equations must have a graph that is the same plane. These situations are shown in the figures below.

The three planes shown in Figure E intersect at a point. A system of equations represented by planes that intersect at a point is independent.

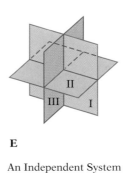

E

An Independent System
of Equations

The planes shown in Figures F and G intersect along a common line. The system of equations represented by the planes in Figure H has a graph that is the same plane. The systems of equations represented by the graphs below are dependent.

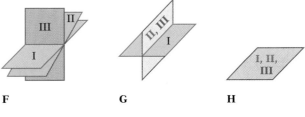

Dependent Systems of Equations

Point of Interest

In the early 1980s, Stephen Hoppe became interested in winning Monopoly strategies. Finding these strategies required solving a system that contained 123 equations in 123 variables!

Just as a solution of an equation in two variables is an ordered pair (x, y), a **solution of an equation in three variables** is an ordered triple (x, y, z). For example, $(2, 1, -3)$ is a solution of the equation $2x - y - 2z = 9$. The ordered triple $(1, 3, 2)$ is not a solution.

A **system of linear equations in three variables** is shown at the right. A **solution of a system of equations in three variables** is an ordered triple that is a solution of each equation of the system.

$$x - 2y + z = 6$$
$$3x + y - 2z = 2$$
$$2x - 3y + 5z = 1$$

A system of linear equations in three variables can be solved by using the addition method. First, eliminate one variable from any two of the given equations. Then eliminate the same variable from any other two equations. The result will be a system of two equations in two variables. Solve this system by the addition method.

HOW TO Solve: (1) $x + 4y - z = 10$
(2) $3x + 2y + z = 4$
(3) $2x - 3y + 2z = -7$

Eliminate z from Equations (1) and (2) by adding the two equations.

$$x + 4y - z = 10$$
$$3x + 2y + z = 4$$
(4) $4x + 6y = 14$ • Add the equations.

Eliminate z from Equations (1) and (3). Multiply Equation (1) by 2 and add to Equation (3).

$$2x + 8y - 2z = 20$$ • 2 times Equation (1).
$$2x - 3y + 2z = -7$$ • This is Equation (3).
(5) $4x + 5y = 13$ • Add the equations.

Using Equations (4) and (5), solve the system of two equations in two variables.

(4) $4x + 6y = 14$
(5) $4x + 5y = 13$

Eliminate x. Multiply Equation (5) by -1 and add to Equation (4).

$$4x + 6y = 14$$ • This is Equation (4).
$$-4x - 5y = -13$$ • -1 times Equation (5).
$$y = 1$$ • Add the equations.

Substitute the value of y into Equation (4) or Equation (5) and solve for x. Equation (4) is used here.

$$4x + 6y = 14$$ • This is Equation (4).
$$4x + 6(1) = 14$$ • $y = 1$
$$4x = 8$$ • Solve for x.
$$x = 2$$

Substitute the value of y and the value of x into one of the equations in the original system. Equation (2) is used here.

$$3x + 2y + z = 4$$
$$3(2) + 2(1) + z = 4$$ • $x = 2, y = 1$
$$6 + 2 + z = 4$$
$$8 + z = 4$$
$$z = -4$$

The solution is $(2, 1, -4)$.

Study Tip

Always check the proposed solution of a system of equtions. For the equation at the right:

$$\frac{x + 4y - z = 10}{2 + 4(1) - (-4) \mid 10}$$
$$10 = 10$$

$$\frac{3x + 2y + z = 4}{3(2) + 2(1) + (-4) \mid 4}$$
$$4 = 4$$

$$\frac{2x - 3y + 2z = -7}{2(2) - 3(1) + 2(-4) \mid -7}$$
$$-7 = -7$$

The solution checks.

HOW TO Solve: (1) $2x - 3y - z = 1$
(2) $x + 4y + 3z = 2$
(3) $4x - 6y - 2z = 5$

Eliminate x from Equations (1) and (2).

$$\begin{array}{l} 2x - 3y - z = 1 \\ -2x - 8y - 6z = -4 \\ \hline -11y - 7z = -3 \end{array}$$

- This is Equation (1).
- -2 times Equation (2).
- Add the equations.

Eliminate x from Equations (1) and (3).

$$\begin{array}{l} -4x + 6y + 2z = -2 \\ 4x - 6y - 2z = 5 \\ \hline 0 = 3 \end{array}$$

- -2 times Equation (1).
- This is Equation (3).
- Add the equations.

The equation $0 = 3$ is not a true equation. The system of equations is inconsistent and therefore has no solution.

HOW TO Solve: (1) $3x - z = -1$
(2) $2y - 3z = 10$
(3) $x + 3y - z = 7$

Eliminate x from Equations (1) and (3). Multiply Equation (3) by -3 and add to Equation (1).

$$\begin{array}{l} 3x - z = -1 \\ -3x - 9y + 3z = -21 \\ \hline -9y + 2z = -22 \end{array}$$

(4)

- This is Equation (1).
- -3 times Equation (3).
- Add the equations.

Use Equations (2) and (4) to form a system of equations in two variables.

(2) $2y - 3z = 10$
(4) $-9y + 2z = -22$

Eliminate z. Multiply Equation (2) by 2 and Equation (4) by 3.

$$\begin{array}{l} 4y - 6z = 20 \\ -27y + 6z = -66 \\ \hline -23y = -46 \\ y = 2 \end{array}$$

- 2 times Equation (2).
- 3 times Equation (4).
- Add the equations.
- Solve for y.

Substitute the value of y into Equation (2) or Equation (4) and solve for z. Equation (2) is used here.

(2) $\begin{aligned} 2y - 3z &= 10 \\ 2(2) - 3z &= 10 \\ 4 - 3z &= 10 \\ -3z &= 6 \\ z &= -2 \end{aligned}$

- This is Equation (2).
- $y = 2$
- Solve for z.

Substitute the value of z into Equation (1) and solve for x.

(1) $\begin{aligned} 3x - z &= -1 \\ 3x - (-2) &= -1 \\ 3x + 2 &= -1 \\ 3x &= -3 \\ x &= -1 \end{aligned}$

- This is Equation (1).
- $z = -2$
- Solve for x.

The solution is $(-1, 2, -2)$.

Example 3 Solve: (1) $3x - y + 2z = 1$
(2) $2x + 3y + 3z = 4$
(3) $x + y - 4z = -9$

You Try It 3 Solve: $x - y + z = 6$
$2x + 3y - z = 1$
$x + 2y + 2z = 5$

Solution Eliminate y. Add
Equations (1) and (3).

$3x - y + 2z = 1$
$x + y - 4z = -9$
$4x - 2z = -8$

Multiply each side of the
equation by $\frac{1}{2}$.

$2x - z = -4$ • **Equation (4)**

Multiply Equation (1) by 3 and
add to Equation (2).

$9x - 3y + 6z = 3$
$2x + 3y + 3z = 4$
$11x + 9z = 7$ • **Equation (5)**

Solve the system of two
equations.

(4) $2x - z = -4$
(5) $11x + 9z = 7$

Multiply Equation (4) by 9 and
add to Equation (5).

$18x - 9z = -36$
$11x + 9z = 7$
$29x = -29$
$x = -1$

Replace x by -1 in
Equation (4).

$2x - z = -4$
$2(-1) - z = -4$
$-2 - z = -4$
$-z = -2$
$z = 2$

Replace x by -1 and z by 2 in
Equation (3).

$x + y - 4z = -9$
$-1 + y - 4(2) = -9$
$-9 + y = -9$
$y = 0$

The solution is $(-1, 0, 2)$.

Your solution

Solution on pp. S11–S12

4.2 Exercises

Objective A **To solve a system of two linear equations in two variables by the addition method**

For Exercises 1 to 42, solve by the addition method.

1. $x - y = 5$
$x + y = 7$

2. $x + y = 1$
$2x - y = 5$

3. $3x + y = 4$
$x + y = 2$

4. $x - 3y = 4$
$x + 5y = -4$

5. $3x + y = 7$
$x + 2y = 4$

6. $x - 2y = 7$
$3x - 2y = 9$

7. $2x + 3y = -1$
$x + 5y = 3$

8. $x + 5y = 7$
$2x + 7y = 8$

9. $3x - y = 4$
$6x - 2y = 8$

10. $x - 2y = -3$
$-2x + 4y = 6$

11. $2x + 5y = 9$
$4x - 7y = -16$

12. $8x - 3y = 21$
$4x + 5y = -9$

13. $4x - 6y = 5$
$2x - 3y = 7$

14. $3x + 6y = 7$
$2x + 4y = 5$

15. $3x - 5y = 7$
$x - 2y = 3$

16. $3x + 4y = 25$
$2x + y = 10$

17. $x + 3y = 7$
$-2x + 3y = 22$

18. $2x - 3y = 14$
$5x - 6y = 32$

19. $3x + 2y = 16$
$2x - 3y = -11$

20. $2x - 5y = 13$
$5x + 3y = 17$

21. $4x + 4y = 5$
$2x - 8y = -5$

22. $3x + 7y = 16$
$4x - 3y = 9$

23. $5x + 4y = 0$
$3x + 7y = 0$

24. $3x - 4y = 0$
$4x - 7y = 0$

25. $5x + 2y = 1$
$2x + 3y = 7$

26. $3x + 5y = 16$
$5x - 7y = -4$

27. $3x - 6y = 6$
$9x - 3y = 8$

28. $\dfrac{2}{3}x - \dfrac{1}{2}y = 3$
$\dfrac{1}{3}x - \dfrac{1}{4}y = \dfrac{3}{2}$

29. $\dfrac{3}{4}x + \dfrac{1}{3}y = -\dfrac{1}{2}$
$\dfrac{1}{2}x - \dfrac{5}{6}y = -\dfrac{7}{2}$

30. $\dfrac{2}{5}x - \dfrac{1}{3}y = 1$
$\dfrac{3}{5}x + \dfrac{2}{3}y = 5$

31. $\dfrac{5x}{6} + \dfrac{y}{3} = \dfrac{4}{3}$
$\dfrac{2x}{3} - \dfrac{y}{2} = \dfrac{11}{6}$

32. $\dfrac{3x}{4} + \dfrac{2y}{5} = -\dfrac{3}{20}$
$\dfrac{3x}{2} - \dfrac{y}{4} = \dfrac{3}{4}$

33. $\dfrac{2x}{5} - \dfrac{y}{2} = \dfrac{13}{2}$
$\dfrac{3x}{4} - \dfrac{y}{5} = \dfrac{17}{2}$

34. $\dfrac{x}{2} + \dfrac{y}{3} = \dfrac{5}{12}$
$\dfrac{x}{2} - \dfrac{y}{3} = \dfrac{1}{12}$

35. $\dfrac{3x}{2} - \dfrac{y}{4} = -\dfrac{11}{12}$
$\dfrac{x}{3} - y = -\dfrac{5}{6}$

36. $\dfrac{3x}{4} - \dfrac{2y}{3} = 0$
$\dfrac{5x}{4} - \dfrac{y}{3} = \dfrac{7}{12}$

37. $4x - 5y = 3y + 4$
$2x + 3y = 2x + 1$

38. $5x - 2y = 8x - 1$
$2x + 7y = 4y + 9$

39. $2x + 5y = 5x + 1$
$3x - 2y = 3y + 3$

40. $4x - 8y = 5$
$8x + 2y = 1$

41. $5x + 2y = 2x + 1$
$2x - 3y = 3x + 2$

42. $3x + 3y = y + 1$
$x + 3y = 9 - x$

Objective B **To solve a system of three linear equations in three variables by the addition method**

For Exercises 43 to 66, solve by the addition method.

43. $x + 2y - z = 1$
$2x - y + z = 6$
$x + 3y - z = 2$

44. $x + 3y + z = 6$
$3x + y - z = -2$
$2x + 2y - z = 1$

45. $2x - y + 2z = 7$
$x + y + z = 2$
$3x - y + z = 6$

46. $x - 2y + z = 6$
$x + 3y + z = 16$
$3x - y - z = 12$

47. $3x + y = 5$
$3y - z = 2$
$x + z = 5$

48. $2y + z = 7$
$2x - z = 3$
$x - y = 3$

49. $x - y + z = 1$
$2x + 3y - z = 3$
$-x + 2y - 4z = 4$

50. $2x + y - 3z = 7$
$x - 2y + 3z = 1$
$3x + 4y - 3z = 13$

51. $2x + 3z = 5$
$3y + 2z = 3$
$3x + 4y = -10$

52. $3x + 4z = 5$
$2y + 3z = 2$
$2x - 5y = 8$

53. $2x + 4y - 2z = 3$
$x + 3y + 4z = 1$
$x + 2y - z = 4$

54. $x - 3y + 2z = 1$
$x - 2y + 3z = 5$
$2x - 6y + 4z = 3$

55. $2x + y - z = 5$
$x + 3y + z = 14$
$3x - y + 2z = 1$

56. $3x - y - 2z = 11$
$2x + y - 2z = 11$
$x + 3y - z = 8$

57. $3x + y - 2z = 2$
$x + 2y + 3z = 13$
$2x - 2y + 5z = 6$

58. $4x + 5y + z = 6$
$2x - y + 2z = 11$
$x + 2y + 2z = 6$

59. $2x - y + z = 6$
$3x + 2y + z = 4$
$x - 2y + 3z = 12$

60. $3x + 2y - 3z = 8$
$2x + 3y + 2z = 10$
$x + y - z = 2$

61. $3x - 2y + 3z = -4$
$2x + y - 3z = 2$
$3x + 4y + 5z = 8$

62. $3x - 3y + 4z = 6$
$4x - 5y + 2z = 10$
$x - 2y + 3z = 4$

63. $3x - y + 2z = 2$
$4x + 2y - 7z = 0$
$2x + 3y - 5z = 7$

64. $2x + 2y + 3z = 13$
$-3x + 4y - z = 5$
$5x - 3y + z = 2$

65. $2x - 3y + 7z = 0$
$x + 4y - 4z = -2$
$3x + 2y + 5z = 1$

66. $5x + 3y - z = 5$
$3x - 2y + 4z = 13$
$4x + 3y + 5z = 22$

APPLYING THE CONCEPTS

67. Explain, graphically, the following situations when they are related to a system of three linear equations in three variables.
a. The system of equations has no solution.
b. The system of equations has exactly one solution.
c. The system of equations has infinitely many solutions.

68. Describe the graph of each of the following equations in an xyz-coordinate system.
a. $x = 3$ **b.** $y = 4$ **c.** $z = 2$ **d.** $y = x$

In Exercises 69 to 72, the systems are not linear systems of equations. However, they can be solved by using a modification of the addition method. Solve each system of equations.

69. $\dfrac{1}{x} - \dfrac{2}{y} = 3$
$\dfrac{2}{x} + \dfrac{3}{y} = -1$

70. $\dfrac{1}{x} + \dfrac{2}{y} = 3$
$\dfrac{1}{x} - \dfrac{3}{y} = -2$

71. $\dfrac{3}{x} + \dfrac{2}{y} = 1$
$\dfrac{2}{x} + \dfrac{4}{y} = -2$

72. $\dfrac{3}{x} - \dfrac{5}{y} = -\dfrac{3}{2}$
$\dfrac{1}{x} - \dfrac{2}{y} = -\dfrac{2}{3}$

73. For Exercise 69, solve each equation of the system for y. Then use a graphing calculator to verify the solution you determined algebraically. Note that the graphs are not straight lines.

74. For Exercise 70, solve each equation of the system for y. Then use a graphing calculator to verify the solution you determined algebraically. Note that the graphs are not straight lines.

4.3 Solving Systems of Equations by Using Determinants

Objective A To evaluate a determinant

Point of Interest

The word *matrix* was first used in a mathematical context in 1850. The root of this word is the Latin *mater*, meaning *mother*. A matrix was thought of as an object from which something else originates. The idea was that a determinant, discussed below, originated (was born) from a matrix. Today, matrices are one of the most widely used tools of applied mathematics.

A **matrix** is a rectangular array of numbers. Each number in a matrix is called an **element of a matrix.** The matrix at the right, with three rows and four columns, is called a 3×4 (read "3 by 4") matrix.

$$A = \begin{bmatrix} 1 & -3 & 2 & 4 \\ 0 & 4 & -3 & 2 \\ 6 & -5 & 4 & -1 \end{bmatrix}$$

A matrix of m rows and n columns is said to be of **order $m \times n$.** The matrix above has order 3×4. The notation a_{ij} refers to the element of a matrix in the ith row and the jth column. For matrix A, $a_{23} = -3$, $a_{31} = 6$, and $a_{13} = 2$.

A **square matrix** is one that has the same number of rows as columns. A 2×2 matrix and a 3×3 matrix are shown at the right.

$$\begin{bmatrix} -1 & 3 \\ 5 & 2 \end{bmatrix} \qquad \begin{bmatrix} 4 & 0 & 1 \\ 5 & -3 & 7 \\ 2 & 1 & 4 \end{bmatrix}$$

Associated with every square matrix is a number called its **determinant.**

Determinant of a 2 × 2 Matrix

The **determinant** of a 2×2 matrix $\begin{bmatrix} a_{11} & a_{12} \\ a_{21} & a_{22} \end{bmatrix}$ is written $\begin{vmatrix} a_{11} & a_{12} \\ a_{21} & a_{22} \end{vmatrix}$. The value of this determinant is given by the formula

$$\begin{vmatrix} a_{11} & a_{12} \\ a_{21} & a_{22} \end{vmatrix} = a_{11} a_{22} - a_{12} a_{21}$$

TAKE NOTE

Note that vertical bars are used to represent the determinant and that brackets are used to represent the matrix.

HOW TO Find the value of the determinant $\begin{vmatrix} 3 & 4 \\ -1 & 2 \end{vmatrix}$.

$$\begin{vmatrix} 3 & 4 \\ -1 & 2 \end{vmatrix} = 3 \cdot 2 - 4(-1) = 6 - (-4) = 10$$

The value of the determinant is 10.

For a square matrix whose order is 3×3 or greater, the value of the determinant is found by using 2×2 determinants.

The **minor of an element** in a 3×3 determinant is the 2×2 determinant that is obtained by eliminating the row and column that contain that element.

HOW TO Find the minor of -3 for the determinant $\begin{vmatrix} 2 & -3 & 4 \\ 0 & 4 & 8 \\ -1 & 3 & 6 \end{vmatrix}$.

The minor of -3 is the 2×2 determinant created by eliminating the row and column that contain -3.

Eliminate the row and column as shown: $\begin{vmatrix} 2 & -3 & 4 \\ 0 & 4 & 8 \\ -1 & 3 & 6 \end{vmatrix}$

The minor of -3 is $\begin{vmatrix} 0 & 8 \\ -1 & 6 \end{vmatrix}$.

Cofactor of an Element of a Matrix

The **cofactor of an element of a matrix** is $(-1)^{i+j}$ times the minor of that element, where i is the row number of the element and j is the column number of the element.

HOW TO For the determinant $\begin{vmatrix} 3 & -2 & 1 \\ 2 & -5 & -4 \\ 0 & 3 & 1 \end{vmatrix}$, find the cofactor of -2 and of -5.

Because -2 is in the first row and the second column, $i = 1$ and $j = 2$. Thus $(-1)^{i+j} = (-1)^{1+2} = (-1)^3 = -1$. The cofactor of -2 is $(-1)\begin{vmatrix} 2 & -4 \\ 0 & 1 \end{vmatrix}$.

Because -5 is in the second row and the second column, $i = 2$ and $j = 2$.

Thus $(-1)^{i+j} = (-1)^{2+2} = (-1)^4 = 1$. The cofactor of -5 is $1 \cdot \begin{vmatrix} 3 & 1 \\ 0 & 1 \end{vmatrix}$.

Note from this example that the cofactor of an element is -1 times the minor of that element or 1 times the minor of that element, depending on whether the sum $i + j$ is an odd or an even integer.

The value of a 3×3 or larger determinant can be found by **expanding by cofactors** of *any* row or *any* column. The result of expanding by cofactors using the first row of a 3×3 matrix is shown below.

$$\begin{vmatrix} a_{11} & a_{12} & a_{13} \\ a_{21} & a_{22} & a_{23} \\ a_{31} & a_{32} & a_{33} \end{vmatrix} = a_{11}(-1)^{1+1}\begin{vmatrix} a_{22} & a_{23} \\ a_{32} & a_{33} \end{vmatrix} + a_{12}(-1)^{1+2}\begin{vmatrix} a_{21} & a_{23} \\ a_{31} & a_{33} \end{vmatrix} + a_{13}(-1)^{1+3}\begin{vmatrix} a_{21} & a_{22} \\ a_{31} & a_{32} \end{vmatrix}$$

$$= a_{11}\begin{vmatrix} a_{22} & a_{23} \\ a_{32} & a_{33} \end{vmatrix} - a_{12}\begin{vmatrix} a_{21} & a_{23} \\ a_{31} & a_{33} \end{vmatrix} + a_{13}\begin{vmatrix} a_{21} & a_{22} \\ a_{31} & a_{32} \end{vmatrix}$$

HOW TO Find the value of the determinant $\begin{vmatrix} 2 & -3 & 2 \\ 1 & 3 & -1 \\ 0 & -2 & 2 \end{vmatrix}$.

Expand by cofactors of the first row.

$$\begin{vmatrix} 2 & -3 & 2 \\ 1 & 3 & -1 \\ 0 & -2 & 2 \end{vmatrix} = 2\begin{vmatrix} 3 & -1 \\ -2 & 2 \end{vmatrix} - (-3)\begin{vmatrix} 1 & -1 \\ 0 & 2 \end{vmatrix} + 2\begin{vmatrix} 1 & 3 \\ 0 & -2 \end{vmatrix}$$

$$= 2(6 - 2) - (-3)(2 - 0) + 2(-2 - 0)$$
$$= 2(4) - (-3)(2) + 2(-2) = 8 - (-6) + (-4) = 10$$

To illustrate a statement made earlier, the value of this determinant will now be found by expanding by cofactors using the second column.

$$\begin{vmatrix} 2 & -3 & 2 \\ 1 & 3 & -1 \\ 0 & -2 & 2 \end{vmatrix} = -3(-1)^{1+2}\begin{vmatrix} 1 & -1 \\ 0 & 2 \end{vmatrix} + 3(-1)^{2+2}\begin{vmatrix} 2 & 2 \\ 0 & 2 \end{vmatrix} + (-2)(-1)^{3+2}\begin{vmatrix} 2 & 2 \\ 1 & -1 \end{vmatrix}$$

$$= -3(-1)\begin{vmatrix} 1 & -1 \\ 0 & 2 \end{vmatrix} + 3(1)\begin{vmatrix} 2 & 2 \\ 0 & 2 \end{vmatrix} + (-2)(-1)\begin{vmatrix} 2 & 2 \\ 1 & -1 \end{vmatrix}$$

$$= 3(2 - 0) + 3(4 - 0) + 2(-2 - 2)$$

$$= 3(2) + 3(4) + 2(-4) = 6 + 12 + (-8) = 10$$

Note that the value of the determinant is the same whether the first row or the second column is used to expand by cofactors. *Any row or column* can be used to evaluate a determinant by expanding by cofactors.

Example 1

Find the value of $\begin{vmatrix} 3 & -2 \\ 6 & -4 \end{vmatrix}$.

Solution

$\begin{vmatrix} 3 & -2 \\ 6 & -4 \end{vmatrix} = 3(-4) - (-2)(6) = -12 + 12 = 0$

The value of the determinant is 0.

You Try It 1

Find the value of $\begin{vmatrix} -1 & -4 \\ 3 & -5 \end{vmatrix}$.

Your solution

Example 2

Find the value of $\begin{vmatrix} -2 & 3 & 1 \\ 4 & -2 & 0 \\ 1 & -2 & 3 \end{vmatrix}$.

Solution
Expand by cofactors of the first row.

$\begin{vmatrix} -2 & 3 & 1 \\ 4 & -2 & 0 \\ 1 & -2 & 3 \end{vmatrix}$

$= -2 \begin{vmatrix} -2 & 0 \\ -2 & 3 \end{vmatrix} - 3 \begin{vmatrix} 4 & 0 \\ 1 & 3 \end{vmatrix} + 1 \begin{vmatrix} 4 & -2 \\ 1 & -2 \end{vmatrix}$

$= -2(-6 - 0) - 3(12 - 0) + 1(-8 + 2)$

$= -2(-6) - 3(12) + 1(-6)$

$= 12 - 36 - 6$

$= -30$

The value of the determinant is -30.

You Try It 2

Find the value of $\begin{vmatrix} 1 & 4 & -2 \\ 3 & 1 & 1 \\ 0 & -2 & 2 \end{vmatrix}$.

Your solution

Example 3

Find the value of $\begin{vmatrix} 0 & -2 & 1 \\ 1 & 4 & 1 \\ 2 & -3 & 4 \end{vmatrix}$.

Solution

$\begin{vmatrix} 0 & -2 & 1 \\ 1 & 4 & 1 \\ 2 & -3 & 4 \end{vmatrix}$

$= 0 \begin{vmatrix} 4 & 1 \\ -3 & 4 \end{vmatrix} - (-2) \begin{vmatrix} 1 & 1 \\ 2 & 4 \end{vmatrix} + 1 \begin{vmatrix} 1 & 4 \\ 2 & -3 \end{vmatrix}$

$= 0 - (-2)(4 - 2) + 1(-3 - 8)$

$= 2(2) + 1(-11)$

$= 4 - 11$

$= -7$

The value of the determinant is -7.

You Try It 3

Find the value of $\begin{vmatrix} 3 & -2 & 0 \\ 1 & 4 & 2 \\ -2 & 1 & 3 \end{vmatrix}$.

Your solution

Solutions on p. S12

Objective B **To solve a system of equations by using Cramer's Rule**

The connection between determinants and systems of equations can be understood by solving a general system of linear equations.

Solve: $(1)\ a_1x + b_1y = c_1$
$\quad\quad\ (2)\ a_2x + b_2y = c_2$

Eliminate y. Multiply Equation (1) by b_2 and Equation (2) by $-b_1$.

$$a_1b_2x + b_1b_2y = c_1b_2$$
$$-a_2b_1x - b_1b_2y = -c_2b_1$$

• b_2 times Equation (1).
• $-b_1$ times Equation (2).

Add the equations.

$$a_1b_2x - a_2b_1x = c_1b_2 - c_2b_1$$

$$(a_1b_2 - a_2b_1)x = c_1b_2 - c_2b_1$$

• Solve for x, assuming $a_1b_2 - a_2b_1 \neq 0$.

$$x = \frac{c_1b_2 - c_2b_1}{a_1b_2 - a_2b_1}$$

The denominator $a_1b_2 - a_2b_1$ is the determinant of the coefficients of x and y. This is called the **coefficient determinant.**

$$a_1b_2 - a_2b_1 = \begin{vmatrix} a_1 & b_1 \\ a_2 & b_2 \end{vmatrix}$$

coefficients of x ⟶
coefficients of y ⟶

The numerator $c_1b_2 - c_2b_1$ is the determinant obtained by replacing the first column in the coefficient determinant by the constants c_1 and c_2. This is called a **numerator determinant.**

$$c_1b_2 - c_2b_1 = \begin{vmatrix} c_1 & b_1 \\ c_2 & b_2 \end{vmatrix}$$

constants of ⟶
the equations

Following a similar procedure and eliminating x, it is also possible to express the y-component of the solution in determinant form. These results are summarized in Cramer's Rule.

Point of Interest

Cramer's Rule is named after Gabriel Cramer, who used it in a book he published in 1750. However, this rule was also published in 1683 by the Japanese mathematician Seki Kown. That publication occurred seven years before Cramer's birth.

> **Cramer's Rule**
>
> The solution of the system of equations $\begin{aligned} a_1x + b_1y &= c_1 \\ a_2x + b_2y &= c_2 \end{aligned}$ is given by $x = \dfrac{D_x}{D}$ and $y = \dfrac{D_y}{D}$,
>
> where $D = \begin{vmatrix} a_1 & b_1 \\ a_2 & b_2 \end{vmatrix}$, $D_x = \begin{vmatrix} c_1 & b_1 \\ c_2 & b_2 \end{vmatrix}$, $D_y = \begin{vmatrix} a_1 & c_1 \\ a_2 & c_2 \end{vmatrix}$, and $D \neq 0$.

HOW TO Solve by using Cramer's Rule: $3x - 2y = 1$
$\quad\quad\quad\quad\quad\quad\quad\quad\quad\quad\quad\quad\quad\quad\quad 2x + 5y = 3$

$$D = \begin{vmatrix} 3 & -2 \\ 2 & 5 \end{vmatrix} = 19$$

• Find the value of the coefficient determinant.

$$D_x = \begin{vmatrix} 1 & -2 \\ 3 & 5 \end{vmatrix} = 11, \ D_y = \begin{vmatrix} 3 & 1 \\ 2 & 3 \end{vmatrix} = 7$$

• Find the value of each of the numerator determinants.

$$x = \frac{D_x}{D} = \frac{11}{19}, \ y = \frac{D_y}{D} = \frac{7}{19}$$

• Use Cramer's Rule to write the solution.

The solution is $\left(\dfrac{11}{19}, \dfrac{7}{19}\right)$.

A procedure similar to that followed for two equations in two variables can be used to extend Cramer's Rule to three equations in three variables.

Cramer's Rule for a System of Three Equations in Three Variables

The solution of the system of equations

$$a_1x + b_1y + c_1z = d_1$$
$$a_2x + b_2y + c_2z = d_2$$
$$a_3x + b_3y + c_3z = d_3$$

is given by $x = \dfrac{D_x}{D}$, $y = \dfrac{D_y}{D}$, and $z = \dfrac{D_z}{D}$, where

$$D = \begin{vmatrix} a_1 & b_1 & c_1 \\ a_2 & b_2 & c_2 \\ a_3 & b_3 & c_3 \end{vmatrix}, \; D_x = \begin{vmatrix} d_1 & b_1 & c_1 \\ d_2 & b_2 & c_2 \\ d_3 & b_3 & c_3 \end{vmatrix}, \; D_y = \begin{vmatrix} a_1 & d_1 & c_1 \\ a_2 & d_2 & c_2 \\ a_3 & d_3 & c_3 \end{vmatrix}, \; D_z = \begin{vmatrix} a_1 & b_1 & d_1 \\ a_2 & b_2 & d_2 \\ a_3 & b_3 & d_3 \end{vmatrix}, \text{ and } D \neq 0.$$

HOW TO Solve by using Cramer's Rule: $2x - y + z = 1$
$x + 3y - 2z = -2$
$3x + y + 3z = 4$

Find the value of the coefficient determinant.

$$D = \begin{vmatrix} 2 & -1 & 1 \\ 1 & 3 & -2 \\ 3 & 1 & 3 \end{vmatrix} = 2\begin{vmatrix} 3 & -2 \\ 1 & 3 \end{vmatrix} - (-1)\begin{vmatrix} 1 & -2 \\ 3 & 3 \end{vmatrix} + 1\begin{vmatrix} 1 & 3 \\ 3 & 1 \end{vmatrix}$$
$$= 2(11) + 1(9) + 1(-8)$$
$$= 23$$

Find the value of each of the numerator determinants.

$$D_x = \begin{vmatrix} 1 & -1 & 1 \\ -2 & 3 & -2 \\ 4 & 1 & 3 \end{vmatrix} = 1\begin{vmatrix} 3 & -2 \\ 1 & 3 \end{vmatrix} - (-1)\begin{vmatrix} -2 & -2 \\ 4 & 3 \end{vmatrix} + 1\begin{vmatrix} -2 & 3 \\ 4 & 1 \end{vmatrix}$$
$$= 1(11) + 1(2) + 1(-14)$$
$$= -1$$

$$D_y = \begin{vmatrix} 2 & 1 & 1 \\ 1 & -2 & -2 \\ 3 & 4 & 3 \end{vmatrix} = 2\begin{vmatrix} -2 & -2 \\ 4 & 3 \end{vmatrix} - 1\begin{vmatrix} 1 & -2 \\ 3 & 3 \end{vmatrix} + 1\begin{vmatrix} 1 & -2 \\ 3 & 4 \end{vmatrix}$$
$$= 2(2) - 1(9) + 1(10)$$
$$= 5$$

$$D_z = \begin{vmatrix} 2 & -1 & 1 \\ 1 & 3 & -2 \\ 3 & 1 & 4 \end{vmatrix} = 2\begin{vmatrix} 3 & -2 \\ 1 & 4 \end{vmatrix} - (-1)\begin{vmatrix} 1 & -2 \\ 3 & 4 \end{vmatrix} + 1\begin{vmatrix} 1 & 3 \\ 3 & 1 \end{vmatrix}$$
$$= 2(14) + 1(10) + 1(-8)$$
$$= 30$$

Use Cramer's Rule to write the solution.

$$x = \frac{D_x}{D} = \frac{-1}{23}, \qquad y = \frac{D_y}{D} = \frac{5}{23}, \qquad z = \frac{D_z}{D} = \frac{30}{23}$$

The solution is $\left(-\dfrac{1}{23}, \dfrac{5}{23}, \dfrac{30}{23}\right)$.

Example 4

Solve by using Cramer's Rule.
$6x - 9y = 5$
$4x - 6y = 4$

Solution

$$D = \begin{vmatrix} 6 & -9 \\ 4 & -6 \end{vmatrix} = 0$$

Because $D = 0$, $\frac{D_x}{D}$ is undefined. Therefore, the system is dependent or inconsistent.

You Try It 4

Solve by using Cramer's Rule.
$3x - y = 4$
$6x - 2y = 5$

Your solution

Example 5

Solve by using Cramer's Rule.
$3x - y + z = 5$
$x + 2y - 2z = -3$
$2x + 3y + z = 4$

Solution

$$D = \begin{vmatrix} 3 & -1 & 1 \\ 1 & 2 & -2 \\ 2 & 3 & 1 \end{vmatrix} = 28,$$

$$D_x = \begin{vmatrix} 5 & -1 & 1 \\ -3 & 2 & -2 \\ 4 & 3 & 1 \end{vmatrix} = 28,$$

$$D_y = \begin{vmatrix} 3 & 5 & 1 \\ 1 & -3 & -2 \\ 2 & 4 & 1 \end{vmatrix} = 0,$$

$$D_z = \begin{vmatrix} 3 & -1 & 5 \\ 1 & 2 & -3 \\ 2 & 3 & 4 \end{vmatrix} = 56$$

$$x = \frac{D_x}{D} = \frac{28}{28} = 1$$

$$y = \frac{D_y}{D} = \frac{0}{28} = 0$$

$$z = \frac{D_z}{D} = \frac{56}{28} = 2$$

The solution is (1, 0, 2).

You Try It 5

Solve by using Cramer's Rule.
$2x - y + z = -1$
$3x + 2y - z = 3$
$x + 3y + z = -2$

Your solution

Solutions on p. S12

4.3 Exercises

Objective A **To evaluate a determinant**

1. How do you find the value of the determinant associated with a 2 × 2 matrix?

2. What is the cofactor of a given element in a matrix?

For Exercises 3 to 14, evaluate the determinant.

3. $\begin{vmatrix} 2 & -1 \\ 3 & 4 \end{vmatrix}$

4. $\begin{vmatrix} 5 & 1 \\ -1 & 2 \end{vmatrix}$

5. $\begin{vmatrix} 6 & -2 \\ -3 & 4 \end{vmatrix}$

6. $\begin{vmatrix} -3 & 5 \\ 1 & 7 \end{vmatrix}$

7. $\begin{vmatrix} 3 & 6 \\ 2 & 4 \end{vmatrix}$

8. $\begin{vmatrix} 5 & -10 \\ 1 & -2 \end{vmatrix}$

9. $\begin{vmatrix} 1 & -1 & 2 \\ 3 & 2 & 1 \\ 1 & 0 & 4 \end{vmatrix}$

10. $\begin{vmatrix} 4 & 1 & 3 \\ 2 & -2 & 1 \\ 3 & 1 & 2 \end{vmatrix}$

11. $\begin{vmatrix} 3 & -1 & 2 \\ 0 & 1 & 2 \\ 3 & 2 & -2 \end{vmatrix}$

12. $\begin{vmatrix} 4 & 5 & -2 \\ 3 & -1 & 5 \\ 2 & 1 & 4 \end{vmatrix}$

13. $\begin{vmatrix} 4 & 2 & 6 \\ -2 & 1 & 1 \\ 2 & 1 & 3 \end{vmatrix}$

14. $\begin{vmatrix} 3 & 6 & -3 \\ 4 & -1 & 6 \\ -1 & -2 & 3 \end{vmatrix}$

Objective B **To solve a system of equations by using Cramer's Rule**

For Exercises 15 to 32, solve by using Cramer's Rule.

15. $2x - 5y = 26$
$5x + 3y = 3$

16. $3x + 7y = 15$
$2x + 5y = 11$

17. $x - 4y = 8$
$3x + 7y = 5$

18. $5x + 2y = -5$
$3x + 4y = 11$

19. $2x + 3y = 4$
$6x - 12y = -5$

20. $5x + 4y = 3$
$15x - 8y = -21$

21. $2x + 5y = 6$
$6x - 2y = 1$

22. $7x + 3y = 4$
$5x - 4y = 9$

23. $-2x + 3y = 7$
$4x - 6y = 9$

24. $9x + 6y = 7$
$3x + 2y = 4$

25. $2x - 5y = -2$
$3x - 7y = -3$

26. $8x + 7y = -3$
$2x + 2y = 5$

27. $2x - y + 3z = 9$
$x + 4y + 4z = 5$
$3x + 2y + 2z = 5$

28. $3x - 2y + z = 2$
$2x + 3y + 2z = -6$
$3x - y + z = 0$

29. $3x - y + z = 11$
$x + 4y - 2z = -12$
$2x + 2y - z = -3$

30. $x + 2y + 3z = 8$
$2x - 3y + z = 5$
$3x - 4y + 2z = 9$

31. $4x - 2y + 6z = 1$
$3x + 4y + 2z = 1$
$2x - y + 3z = 2$

32. $x - 3y + 2z = 1$
$2x + y - 2z = 3$
$3x - 9y + 6z = -3$

APPLYING THE CONCEPTS

33. Determine whether the following statements are always true, sometimes true, or never true.
 a. The determinant of a matrix is a positive number.
 b. A determinant can be evaluated by expanding about any row or column of the matrix.
 c. Cramer's Rule can be used to solve a system of linear equations in three variables.

34. Show that $\begin{vmatrix} a & b \\ c & d \end{vmatrix} = -\begin{vmatrix} c & d \\ a & b \end{vmatrix}$.

35. **Surveying** Surveyors use a formula to find the area of a plot of land. The *surveyor's area formula* states that if the vertices (x_1, y_1), $(x_2, y_2), \ldots, (x_n, y_n)$ of a simple polygon are listed counterclockwise around the perimeter, then the area of the polygon is

$$A = \frac{1}{2}\left\{ \begin{vmatrix} x_1 & x_2 \\ y_1 & y_2 \end{vmatrix} + \begin{vmatrix} x_2 & x_3 \\ y_2 & y_3 \end{vmatrix} + \begin{vmatrix} x_3 & x_4 \\ y_3 & y_4 \end{vmatrix} + \cdots + \begin{vmatrix} x_n & x_1 \\ y_n & y_1 \end{vmatrix} \right\}$$

Use the surveyor's area formula to find the area of the polygon with vertices $(9, -3)$, $(26, 6)$, $(18, 21)$, $(16, 10)$, and $(1, 11)$. Measurements are given in feet.

36. **Television** The graph at the right shows the number of hours annually that the average American watched free broadcast television and the number of hours annually that the average American watched cable or satellite dish television. (*Sources:* Veronis Shuler Stevenson *Communication Industry Forecast*; U.S. Census Bureau) Assume that the data model linear functions.
 a. Predict the year in which the number of hours spent watching free broadcast TV will be the same as the number of hours spent watching cable or satellite dish TV. Round to the nearest year.
 b. What is the number of hours spent watching free broadcast TV during that year? (The same number of hours will be spent watching cable or satellite dish TV.) Round to the nearest whole number.

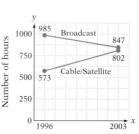

Average number of hours per year spent watching TV

4.4 Application Problems

Objective A To solve rate-of-wind or rate-of-current problems

Motion problems that involve an object moving with or against a wind or current normally require two variables to solve.

A motorboat traveling with the current can go 24 mi in 2 h. Against the current it takes 3 h to go the same distance. Find the rate of the motorboat in calm water and the rate of the current.

> **Strategy for Solving Rate-of-Wind or Rate-of-Current Problems**
>
> **1.** Choose one variable to represent the rate of the object in calm conditions and a second variable to represent the rate of the wind or current. Using these variables, express the rate of the object with and against the wind or current. Use the equation $rt = d$ to write expressions for the distance traveled by the object. The results can be recorded in a table.

Rate of the boat in calm water: x
Rate of the current: y

	Rate	·	Time	=	Distance
With the current	$x + y$	·	2	=	$2(x + y)$
Against the current	$x - y$	·	3	=	$3(x - y)$

> **2.** Determine how the expressions for distance are related.

The distance traveled with the current is 24 mi: $2(x + y) = 24$
The distance traveled against the current is 24 mi: $3(x - y) = 24$

Solve the system of equations.

$$2(x + y) = 24 \xrightarrow{\text{Multiply by } \frac{1}{2}} \frac{1}{2} \cdot 2(x + y) = \frac{1}{2} \cdot 24 \longrightarrow x + y = 12$$

$$3(x - y) = 24 \xrightarrow{\text{Multiply by } \frac{1}{3}} \frac{1}{3} \cdot 3(x - y) = \frac{1}{3} \cdot 24 \longrightarrow \underline{x - y = 8}$$

$$2x = 20 \quad \bullet \text{ Add the}$$
$$x = 10 \qquad \text{equations.}$$

Replace x by 10 in the equation $x + y = 12$. $x + y = 12$
Solve for y. $10 + y = 12$
 $y = 2$

The rate of the boat in calm water is 10 mph.
The rate of the current is 2 mph.

Example 1

Flying with the wind, a plane flew 1000 mi in 5 h. Flying against the wind, the plane could fly only 500 mi in the same amount of time. Find the rate of the plane in calm air and the rate of the wind.

Strategy

• Rate of the plane in still air: p
 Rate of the wind: w

	Rate	*Time*	*Distance*
With wind	$p + w$	5	$5(p + w)$
Against wind	$p - w$	5	$5(p - w)$

• The distance traveled with the wind is 1000 mi.
 The distance traveled against the wind is 500 mi.

$$5(p + w) = 1000$$
$$5(p - w) = 500$$

Solution

$5(p + w) = 1000$ \qquad $\dfrac{1}{5} \cdot 5(p + w) = \dfrac{1}{5} \cdot 1000$

$5(p - w) = 500$ \qquad $\dfrac{1}{5} \cdot 5(p - w) = \dfrac{1}{5} \cdot 500$

$$p + w = 200$$
$$\underline{p - w = 100}$$
$$2p = 300$$
$$p = 150$$

$p + w = 200$
$150 + w = 200$ \qquad • Substitute 150 for p.
$w = 50$

The rate of the plane in calm air is 150 mph.
The rate of the wind is 50 mph.

You Try It 1

A rowing team rowing with the current traveled 18 mi in 2 h. Against the current, the team rowed 10 mi in 2 h. Find the rate of the rowing team in calm water and the rate of the current.

Your strategy

Your solution

Solution on p. S12

Objective B To solve application problems

The application problems in this section are varieties of problems solved earlier in the text. Each of the strategies for the problems in this section will result in a system of equations.

A store owner purchased twenty 60-watt light bulbs and 30 fluorescent bulbs for a total cost of $80. A second purchase, at the same prices, included thirty 60-watt light bulbs and 10 fluorescent bulbs for a total cost of $50. Find the cost of a 60-watt bulb and that of a fluorescent bulb.

> **Strategy for Solving Application Problems**
>
> **1.** Choose a variable to represent each of the unknown quantities. Write numerical or variable expressions for all the remaining quantities. These results may be recorded in tables, one for each condition.

Cost of 60-watt bulb: b
Cost of fluorescent bulb: f

First Purchase

	Amount	·	*Unit Cost*	=	*Value*
60-watt	20	·	b	=	$20b$
Fluorescent	30	·	f	=	$30f$

Second Purchase

	Amount	·	*Unit Cost*	=	*Value*
60-watt	30	·	b	=	$30b$
Fluorescent	10	·	f	=	$10f$

> **2.** Determine a system of equations. The strategies presented in the chapter on First-Degree Equations and Inequalities can be used to determine the relationships among the expressions in the tables. Each table will give one equation of the system of equations.

The total of the first purchase was $80: $20b + 30f = 80$
The total of the second purchase was $50: $30b + 10f = 50$

Solve the system of equations: (1) $20b + 30f = 80$
 (2) $30b + 10f = 50$

$$
\begin{array}{l}
60b + 90f = 240 \\
\underline{-60b - 20f = -100} \\
70f = 140 \\
f = 2
\end{array}
$$

- **3 times Equation (1).**
- **−2 times Equation (2).**

Replace f by 2 in Equation (1) and solve for b.

$$
\begin{array}{r}
20b + 30f = 80 \\
20b + 30(2) = 80 \\
20b + 60 = 80 \\
20b = 20 \\
b = 1
\end{array}
$$

The cost of a 60-watt bulb was $1.00.
The cost of a fluorescent bulb was $2.00.

Some application problems may require more than two variables, as shown in Example 2 and You Try It 2 on the next page.

Example 2

An investor has a total of $20,000 deposited in three different accounts, which earn annual interest rates of 9%, 7%, and 5%. The amount deposited in the 9% account is twice the amount in the 7% account. If the total annual interest earned for the three accounts is $1300, how much is invested in each account?

Strategy

- Amount invested at 9%: x
 Amount invested at 7%: y
 Amount invested at 5%: z

	Principal	*Rate*	*Interest*
Amount at 9%	x	0.09	$0.09x$
Amount at 7%	y	0.07	$0.07y$
Amount at 5%	z	0.05	$0.05z$

- The amount invested at 9% (x) is twice the amount invested at 7% (y): $x = 2y$
 The sum of the interest earned for all three accounts is $1300:
 $0.09x + 0.07y + 0.05z = 1300$
 The total amount invested is $20,000:
 $x + y + z = 20{,}000$

Solution

(1) $x = 2y$
(2) $0.09x + 0.07y + 0.05z = 1300$
(3) $x + y + z = 20{,}000$

Solve the system of equations. Substitute $2y$ for x in Equation (2) and Equation (3).

$0.09(2y) + 0.07y + 0.05z = 1300$
 $2y + y + z = 20{,}000$

(4) $0.25y + 0.05z = 1300$ • $0.09(2y) + 0.07y = 0.25y$
(5) $3y + z = 20{,}000$ • $2y + y = 3y$

Solve the system of equations in two variables by multiplying Equation (5) by -0.05 and adding to Equation (4).

$0.25y + 0.05z = 1300$
$\underline{-0.15y - 0.05z = -1000}$

$0.10y = 300$
$y = 3000$

Substituting the value of y into Equation (1), $x = 6000$.

Substituting the values of x and y into Equation (3), $z = 11{,}000$.

The investor placed $6000 in the 9% account, $3000 in the 7% account, and $11,000 in the 5% account.

You Try It 2

A coin bank contains only nickels, dimes, and quarters. The value of the 19 coins in the bank is $2. If there are twice as many nickels as dimes, find the number of each type of coin in the bank.

Your strategy

Your solution

Solution on p. S13

4.4 Exercises

Objective A **To solve rate-of-wind or rate-of-current problems**

1. A motorboat traveling with the current went 36 mi in 2 h. Against the current, it took 3 h to travel the same distance. Find the rate of the boat in calm water and the rate of the current.

2. A cabin cruiser traveling with the current went 45 mi in 3 h. Against the current, it took 5 h to travel the same distance. Find the rate of the cabin cruiser in calm water and the rate of the current.

3. A jet plane flying with the wind went 2200 mi in 4 h. Against the wind, the plane could fly only 1820 mi in the same amount of time. Find the rate of the plane in calm air and the rate of the wind.

4. Flying with the wind, a small plane flew 300 mi in 2 h. Against the wind, the plane could fly only 270 mi in the same amount of time. Find the rate of the plane in calm air and the rate of the wind.

5. A rowing team rowing with the current traveled 20 km in 2 h. Rowing against the current, the team rowed 12 km in the same amount of time. Find the rate of the team in calm water and the rate of the current.

6. A motorboat traveling with the current went 72 km in 3 h. Against the current, the boat could go only 48 km in the same amount of time. Find the rate of the boat in calm water and the rate of the current.

7. A turboprop plane flying with the wind flew 800 mi in 4 h. Flying against the wind, the plane required 5 h to travel the same distance. Find the rate of the wind and the rate of the plane in calm air.

8. Flying with the wind, a pilot flew 600 mi between two cities in 4 h. The return trip against the wind took 5 h. Find the rate of the plane in calm air and the rate of the wind.

9. A plane flying with a tailwind flew 600 mi in 5 h. Against the wind, the plane required 6 h to fly the same distance. Find the rate of the plane in calm air and the rate of the wind.

10. Flying with the wind, a plane flew 720 mi in 3 h. Against the wind, the plane required 4 h to fly the same distance. Find the rate of the plane in calm air and the rate of the wind.

Objective B **To solve application problems**

11. Coins A coin bank contains only nickels and dimes. The total value of the coins in the bank is $2.50. If the nickels were dimes and the dimes were nickels, the total value of the coins would be $3.50. Find the number of nickels in the bank.

12. Business A merchant mixed 10 lb of cinnamon tea with 5 lb of spice tea. The 15-pound mixture cost $40. A second mixture included 12 lb of the cinnamon tea and 8 lb of the spice tea. The 20-pound mixture cost $54. Find the cost per pound of the cinnamon tea and of the spice tea.

13. Purchasing A carpenter purchased 60 ft of redwood and 80 ft of pine for a total cost of $286. A second purchase, at the same prices, included 100 ft of redwood and 60 ft of pine for a total cost of $396. Find the cost per foot of redwood and of pine.

14. Coins The total value of the quarters and dimes in a coin bank is $5.75. If the quarters were dimes and the dimes were quarters, the total value of the coins would be $6.50. Find the number of quarters in the bank.

15. Purchasing A contractor buys 16 yd of nylon carpet and 20 yd of wool carpet for $1840. A second purchase, at the same prices, includes 18 yd of nylon carpet and 25 yd of wool carpet for $2200. Find the cost per yard of the wool carpet.

16. Finances During one month, a homeowner used 500 units of electricity and 100 units of gas for a total cost of $352. The next month, 400 units of electricity and 150 units of gas were used for a total cost of $304. Find the cost per unit of gas.

17. Manufacturing A company manufactures both mountain bikes and trail bikes. The cost of materials for a mountain bike is $70, and the cost of materials for a trail bike is $50. The cost of labor to manufacture a mountain bike is $80, and the cost of labor to manufacture a trail bike is $40. During a week in which the company has budgeted $2500 for materials and $2600 for labor, how many mountain bikes does the company plan to manufacture?

18. **Manufacturing** A company manufactures both liquid crystal display (LCD) and cathode ray tube (CRT) color monitors. The cost of materials for a CRT monitor is $50, whereas the cost of materials for an LCD monitor is $150. The cost of labor to manufacture a CRT monitor is $80, whereas the cost of labor to manufacture an LCD monitor is $130. During a week when the company has budgeted $9600 for materials and $8760 for labor, how many LCD monitors does the company plan to manufacture?

19. **Chemistry** A chemist has two alloys, one of which is 10% gold and 15% lead, and the other of which is 30% gold and 40% lead. How many grams of each of the two alloys should be used to make an alloy that contains 60 g of gold and 88 g of lead?

20. **Health Science** A pharmacist has two vitamin-supplement powders. The first powder is 20% vitamin B1 and 10% vitamin B2. The second is 15% vitamin B1 and 20% vitamin B2. How many milligrams of each of the two powders should the pharmacist use to make a mixture that contains 130 mg of vitamin B1 and 80 mg of vitamin B2?

21. **Business** On Monday, a computer manufacturing company sent out three shipments. The first order, which contained a bill for $114,000, was for 4 Model II, 6 Model VI, and 10 Model IX computers. The second shipment, which contained a bill for $72,000, was for 8 Model II, 3 Model VI, and 5 Model IX computers. The third shipment, which contained a bill for $81,000, was for 2 Model II, 9 Model VI, and 5 Model IX computers. What does the manufacturer charge for each Model VI computer?

22. **Purchasing** A relief organization supplies blankets, cots, and lanterns to victims of fires, floods, and other natural disasters. One week the organization purchased 15 blankets, 5 cots, and 10 lanterns for a total cost of $1250. The next week, at the same prices, the organization purchased 20 blankets, 10 cots, and 15 lanterns for a total cost of $2000. The next week, at the same prices, the organization purchased 10 blankets, 15 cots, and 5 lanterns for a total cost of $1625. Find the cost of one blanket, the cost of one cot, and the cost of one lantern.

23. **Ticket Sales** A science museum charges $10 for a regular admission ticket, but members receive a discount of $3 and students are admitted for $5. Last Saturday, 750 tickets were sold for a total of $5400. If 20 more student tickets than regular tickets were sold, how many of each type of ticket were sold?

24. **Coins** The total value of the nickels, dimes, and quarters in a coin bank is $4. If the nickels were dimes and the dimes were nickels, the total value of the coins would be $3.75. If the quarters were dimes and the dimes were quarters, the total value would be $6.25. Find the number of nickels, dimes, and quarters in the bank.

25. Investments An investor has a total of $25,000 deposited in three different accounts, which earn annual interest rates of 8%, 6%, and 4%. The amount deposited in the 8% account is twice the amount in the 6% account. If the three accounts earn total annual interest of $1520, how much money is deposited in each account?

APPLYING THE CONCEPTS

26. Geometry Two angles are complementary. The larger angle is 9° more than eight times the measure of the smaller angle. Find the measures of the two angles. (Complementary angles are two angles whose sum is 90°.)

27. Geometry Two angles are supplementary. The larger angle is 40° more than three times the measure of the smaller angle. Find the measures of the two angles. (Supplementary angles are two angles whose sum is 180°.)

$x + y = 180°$

28. Coins The sum of the ages of a gold coin and a silver coin is 75 years. The age of the gold coin 10 years from now will be 5 years less than the age of the silver coin 10 years ago. Find the present ages of the two coins.

29. Art The difference between the ages of an oil painting and a watercolor is 35 years. The age of the oil painting 5 years from now will be twice the age that the watercolor was five years ago. Find the present ages of each.

30. Coins A coin bank contains only dimes and quarters. The total value of all the coins is $1. How many of each type of coin are in the bank? (*Hint:* There is more than one answer to this problem.)

31. Investments An investor has a total of $25,000 deposited in three different accounts, which earn annual interest rates of 8%, 6%, and 4%. The amount deposited in the 8% account is $1000 more than the amount in the 4% account. If the investor wishes to earn $1520 from the three accounts, show that there is more than one way the investor can allocate money to each of the accounts to reach the goal. Explain why this happens here, whereas in Exercise 25 there is only one way to allocate the $25,000. *Suggestion:* Suppose the amount invested at 4% is $8000 and then find the other amounts. Now suppose $10,000 is invested at 4% and then find the other amounts.

4.5 Solving Systems of Linear Inequalities

Objective A

To graph the solution set of a system of linear inequalities

Point of Interest

Large systems of linear inequalities containing over 100 inequalities have been used to solve application problems in such diverse areas as providing health care and hardening a nuclear missile silo.

Two or more inequalities considered together are called a **system of inequalities**. The **solution set of a system of inequalities** is the intersection of the solution sets of the individual inequalities. To graph the solution set of a system of inequalities, first graph the solution set of each inequality. The solution set of the system of inequalities is the region of the plane represented by the intersection of the shaded areas.

HOW TO Graph the solution set: $2x - y \leq 3$
$3x + 2y > 8$

Solve each inequality for y.

$$2x - y \leq 3$$
$$-y \leq -2x + 3$$
$$y \geq 2x - 3$$

$$3x + 2y > 8$$
$$2y > -3x + 8$$
$$y > -\frac{3}{2}x + 4$$

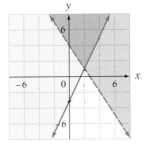

Graph $y = 2x - 3$ as a solid line. Because the inequality is \geq, shade above the line.

Graph $y = -\frac{3}{2}x + 4$ as a dotted line. Because the inequality is $>$, shade above the line.

TAKE NOTE

You can use a test point to check that the correct region has been denoted as the solution set. We can see from the graph that the point $(2, 4)$ is in the solution set and, as shown below, it is a solution of each inequality in the system. This indicates that the solution set as graphed is correct.

$$2x - y \leq 3$$
$$2(2) - (4) \leq 3$$
$$0 \leq 3 \quad \text{True}$$

$$3x + 2y > 8$$
$$3(2) + 2(4) > 8$$
$$14 > 8 \quad \text{True}$$

The solution set of the system is the region of the plane represented by the intersection of the solution sets of the individual inequalities.

HOW TO Graph the solution set: $-x + 2y \geq 4$
$x - 2y \geq 6$

Solve each inequality for y.

$$-x + 2y \geq 4$$
$$2y \geq x + 4$$
$$y \geq \frac{1}{2}x + 2$$

$$x - 2y \geq 6$$
$$-2y \geq -x + 6$$
$$y \leq \frac{1}{2}x - 3$$

Shade above the solid line $y = \frac{1}{2}x + 2$.

Shade below the solid line $y = \frac{1}{2}x - 3$.

Because the solution sets of the two inequalities do not intersect, the solution of the system is the empty set.

Integrating Technology

See the Keystroke Guide: *Graphing Inequalities* for instructions on using a graphing calculator to graph the solution set of a system of inequalities.

Example 1

Graph the solution set: $y \geq x - 1$
 $y < -2x$

Solution

Shade above the solid line $y = x - 1$.
Shade below the dotted line $y = -2x$.

The solution of the system is the intersection of the solution sets of the individual inequalities.

You Try It 1

Graph the solution set: $y \geq 2x - 3$
 $y > -3x$

Your solution

Example 2

Graph the solution set: $2x + 3y > 9$
 $y < -\dfrac{2}{3}x + 1$

Solution

$2x + 3y > 9$
$\quad 3y > -2x + 9$
$\quad\ y > -\dfrac{2}{3}x + 3$

Graph above the dotted line $y = -\dfrac{2}{3}x + 3$.

Graph below the dotted line $y = -\dfrac{2}{3}x + 1$.

The intersection of the system is the empty set, because the solution sets of the two inequalities do not intersect.

You Try It 2

Graph the solution set: $3x + 4y > 12$
 $y < \dfrac{3}{4}x - 1$

Your solution

Solutions on p. S13

4.5 Exercises

Objective A **To graph the solution set of a system of linear inequalities**

For Exercises 1 to 18, graph the solution set.

1. $x - y \geq 3$
 $x + y \leq 5$

2. $2x - y < 4$
 $x + y < 5$

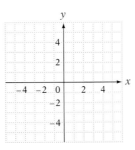

3. $3x - y < 3$
 $2x + y \geq 2$

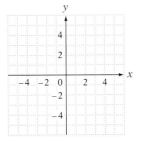

4. $x + 2y \leq 6$
 $x - y \leq 3$

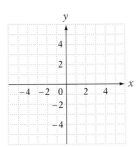

5. $2x + y \geq -2$
 $6x + 3y \leq 6$

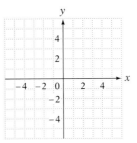

6. $x + y \geq 5$
 $3x + 3y \leq 6$

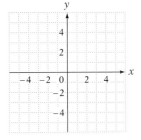

7. $3x - 2y < 6$
 $y \leq 3$

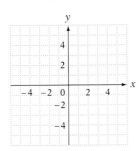

8. $x \leq 2$
 $3x + 2y > 4$

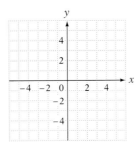

9. $y > 2x - 6$
 $x + y < 0$

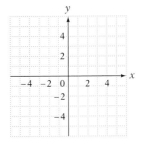

10. $x < 3$
 $y < -2$

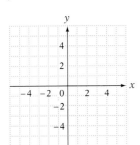

11. $x + 1 \geq 0$
 $y - 3 \leq 0$

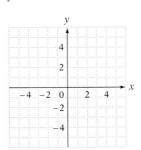

12. $5x - 2y \geq 10$
 $3x + 2y \geq 6$

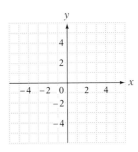

13. $2x + y \geq 4$
$3x - 2y < 6$

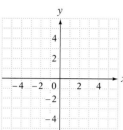

14. $3x - 4y < 12$
$x + 2y < 6$

15. $x - 2y \leq 6$
$2x + 3y \leq 6$

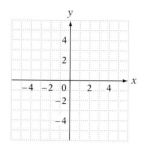

16. $x - 3y > 6$
$2x + y > 5$

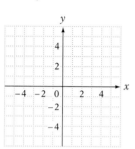

17. $x - 2y \leq 4$
$3x + 2y \leq 8$

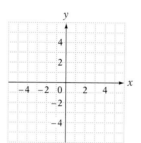

18. $3x - 2y < 0$
$5x + 3y > 9$

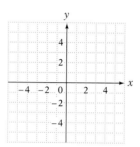

APPLYING THE CONCEPTS

For Exercises 19 to 24, graph the solution set.

19. $2x + 3y \leq 15$
$3x - y \leq 6$
$y \geq 0$

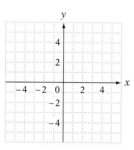

20. $x + y \leq 6$
$x - y \leq 2$
$x \geq 0$

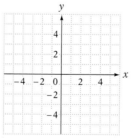

21. $x - y \leq 5$
$2x - y \geq 6$
$y \geq 0$

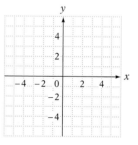

22. $x - 3y \leq 6$
$5x - 2y \geq 4$
$y \geq 0$

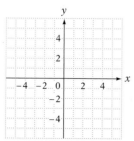

23. $2x - y \leq 4$
$3x + y < 1$
$y \leq 0$

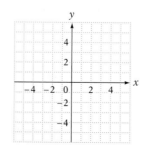

24. $x - y \leq 4$
$2x + 3y > 6$
$x \geq 0$

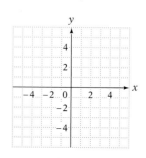

Focus on Problem Solving

Solve an Easier Problem

We begin this problem-solving feature with a restatement of the four steps of Polya's recommended problem-solving model.

1. Understand the problem.
2. Devise a plan.
3. Carry out the plan.
4. Review your solution.

One of the several methods of devising a plan is to try to solve an easier problem. Suppose you are in charge of your softball league, which consists of 15 teams. You must devise a schedule in which each team plays every other team once. How many games must be scheduled?

To solve this problem, we will attempt an easier problem first. Suppose that your league contains only a small number of teams. For instance, if there were only 1 team, you would schedule 0 games. If there were 2 teams, you would schedule 1 game. If there were 3 teams, you would schedule 3 games. The diagram at the left shows that 6 games must be scheduled when there are 4 teams in the league.

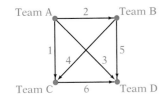

Here is a table of our results so far. (Remember that making a table is another strategy to be used in problem solving.)

Number of Terms	Number of Games	Possible Pattern
1	0	0
2	1	1
3	3	1 + 2
4	6	1 + 2 + 3

1. Draw a diagram with 5 dots to represent the teams. Draw lines from each dot to a second dot, and determine the number of games required.
2. What is the apparent pattern for the number of games required?
3. Assuming that the pattern continues, how many games must be scheduled for the 15 teams of the original problem?

After solving a problem, good problem solvers ask whether it is possible to solve the problem in a different manner. Here is a possible alternative method of solving the scheduling problem.

Begin with one of the 15 teams (say team A) and ask, "How many games must this team play?" Because there are 14 teams left to play, you must schedule 14 games. Now move to team B. It is already scheduled to play team A, and it does not play itself, so there are 13 teams left for it to play. Consequently, you must schedule 14 + 13 games.

4. Continue this reasoning for the remaining teams and determine the number of games that must be scheduled. Does this answer correspond to the answer you obtained using the first method?
5. Making connections to other problems you have solved is an important step toward becoming an excellent problem solver. How does the answer to the scheduling of teams relate to the triangular numbers discussed in the Focus on Problem Solving in the chapter titled Linear Functions and Inequalities in Two Variables?

Projects and Group Activities

Using a Graphing Calculator to Solve a System of Equations

A graphing calculator can be used to solve a system of equations. For this procedure to work on most calculators, it is necessary that the point of intersection be on the screen. This means that you may have to experiment with Xmin, Xmax, Ymin, and Ymax values until the graphs intersect on the screen.

To solve a system of equations graphically, solve each equation for y. Then graph the equations of the system. Their point of intersection is the solution.

For instance, to solve the system of equations

$$4x - 3y = 7$$
$$5x + 4y = 2$$

first solve each equation for y.

$$4x - 3y = 7 \Rightarrow y = \frac{4}{3}x - \frac{7}{3}$$

$$5x + 4y = 2 \Rightarrow y = -\frac{5}{4}x + \frac{1}{2}$$

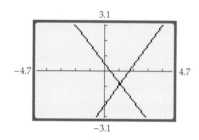

The keystrokes needed to solve this system using a TI-83 are given below. We are using a viewing window of $[-4.7, 4.7]$ by $[-3.1, 3.1]$. The approximate solution is $(1.096774, -0.870968)$.

Once the calculator has drawn the graphs, use the TRACE feature and move the cursor to the approximate point of intersection. This will give you an approximate solution of the system of equations. A more accurate solution can be found by using the following keystrokes.

Some of the exercises in the first section of this chapter asked you to solve a system of equations by graphing. Try those exercises again, this time using your graphing calculator.

Here is an example of using a graphing calculator to solve an investment problem.

HOW TO A marketing manager deposited $8000 in two simple interest accounts, one with an interest rate of 3.5% and the other with an interest rate of 4.5%. How much is deposited in each account if both accounts earn the same interest?

Amounted invested at 3.5%: x
Amount invested at 4.5%: $8000 - x$

Interest earned on the 3.5% account: $0.035x$
Interest earned on the 4.5% account: $0.045(8000 - x)$

Enter $0.035x$ into Y1. Enter
$0.045(8000 - x)$ into Y2. Graph the
equations. (We used a window of
Xmin = 0, Xmax = 8000,
Ymin = 0, Ymax = 400.) Use the
intersect feature to find the
point of intersection.

At the point of intersection,
$x = 4500$. This is the amount in
the 3.5% account.
The amount in the 4.5% account is
$8000 - x = 8000 - 4500 = 3500$.

$4500 is invested at 3.5%. $3500 is invested at 4.5%.

As shown on the graphing calculator screen in the example above, at the point of
intersection, $y = 157.5$. This is the interest earned, $157.50, on $4500 invested at
3.5%. It is also the interest earned on $3500 invested at 4.5%. In other words, it
is the interest earned on each account when the interest earned on each account
is the same.

For Exercises 1 to 4, solve using a graphing calculator.

1. **Finances** Suppose that a breadmaker costs $190, and that the ingredients
 and electricity to make one loaf of bread cost $.95. If a comparable loaf of
 bread at a grocery store costs $1.98, how many loaves of bread must you
 make before the breadmaker pays for itself?

2. **Finances** Suppose a natural gas clothes washer costs $295 and uses $.25 of
 gas and water to wash a large load of clothes. If a laundromat charges $1.75
 to do a load of laundry, how many loads of clothes must you wash before hav-
 ing purchased the washing machine becomes more economical?

3. **Finances** Suppose a natural gas clothes dryer costs $260 and uses $.40 of
 gas to dry a load of clothes for 1 h. The laundromat charges $1.75 to use a
 dryer for 1 h.
 a. How many loads of clothes must you dry before having purchased the gas
 dryer becomes more economical?
 b. What is the y-coordinate at the point of intersection? What does this rep-
 resent in the context of the problem?

4. **Investments** When Mitch Deerfield changed jobs, he rolled over the $7500
 in his retirement account into two simple interest accounts. On one account,
 the annual simple interest rate is 6.25%; on the second account, the annual
 simple interest rate is 5.75%.
 a. How much is invested in each account if the accounts earn the same
 amount of annual interest?
 b. What is the y-coordinate at the point of intersection? What does this rep-
 resent in the context of the problem?

Chapter 4 Summary

Key Words	Examples
A *system of equations* is two or more equations considered together. A *solution of a system of equations in two variables* is an ordered pair that is a solution of each equation in the system. [4.1A, p. 201]	The solution of the system $$x + y = 2$$ $$x - y = 4$$ is the ordered pair $(3, -1)$. $(3, -1)$ is the only ordered pair that is a solution of both equations.
When the graphs of a system of equations intersect at only one point, the system is called an *independent system of equations*. [4.1A, p. 201]	
When the graphs of a system of equations do not intersect, the system has no solution and is called an *inconsistent system of equations*. [4.1A, p. 202]	
When the graphs of a system of equations coincide, the system is called a *dependent system of equations*. [4.1A, p. 202]	
An equation of the form $Ax + By + Cz = D$, where A, B, and C are coefficients of the variables and D is a constant, is a *linear equation in three variables*. A *solution of an equation in three variables* is an **ordered triple** (x, y, z). [4.2B, p. 216]	$3x + 2y - 5z = 12$ is a linear equation in three variables. One solution of this equation is the ordered triple $(0, 1, -2)$.
A *solution of a system of equations in three variables* is an ordered triple that is a solution of each equation of the system. [4.2B, p. 218]	The solution of the system $$3x + y - 3z = 2$$ $$-x + 2y + 3z = 6$$ $$2x + 2y - 2z = 4$$ is the ordered triple $(1, 2, 1)$. $(1, 2, 1)$ is the only ordered triple that is a solution of all three equations.
A *matrix* is a rectangular array of numbers. Each number in the matrix is called an *element* of the matrix. A matrix of m rows and n columns is said to be of *order* $m \times n$. [4.3A, p. 225]	$\begin{bmatrix} 2 & 3 & 6 \\ -1 & 2 & 4 \end{bmatrix}$ is a 2×3 matrx. The elements in the first row are 2, 3, and 6.
A *square matrix* has the same number of rows as columns. [4.3A, p. 225]	$\begin{bmatrix} 2 & 3 \\ 4 & -1 \end{bmatrix}$ is a square matrix.

The *minor of an element* of a 3×3 determinant is the 2×2 determinant obtained by eliminating the row and column that contain the element. [4.3A, p. 225]

$\begin{vmatrix} 2 & 1 & 3 \\ 4 & 6 & 2 \\ 1 & 8 & 5 \end{vmatrix}$ The minor of 4 is $\begin{vmatrix} 1 & 3 \\ 8 & 5 \end{vmatrix}$.

The *cofactor of an element of a determinant* is $(-1)^{i+j}$ times the minor of the element, where i is the row number of the element and j is its column number. [4.3A, p. 226]

In the determinant above, 4 is from the second row and the first column. The cofactor of 4 is

$$(-1)^{2+1} \begin{vmatrix} 1 & 3 \\ 8 & 5 \end{vmatrix} = - \begin{vmatrix} 1 & 3 \\ 8 & 5 \end{vmatrix}$$

Two or more inequalities considered together are called a *system of inequalities*. The *solution set of a system of inequalities* is the intersection of the solution sets of the individual inequalities. [4.5A, p. 241]

$x + y > 3$
$x - y > -2$

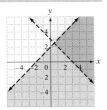

Essential Rules and Procedures

Examples

Solving Systems of Equations
A system of equations can be solved by:
a. *Graphing* [4.1A, p. 201]

$y = \dfrac{1}{2}x + 2$

$y = \dfrac{5}{2}x - 2$

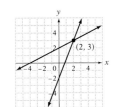

b. *The substitution method* [4.1B, p. 204]

(1) $2x - 3y = 4$
(2) $y = -x + 2$

Substitute the value of y into equation (1).

$2x - 3(-x + 2) = 4$

c. *The addition method* [4.2A, p. 213]

$\begin{array}{r} -2x + 3y = 7 \\ 2x - 5y = 2 \\ \hline -2y = 9 \end{array}$ • **Add the equations.**

Annual Simple Interest Equation [4.1C, p. 207]
Principal · Simple interest rate = Simple interest
$$Pr = I$$

You have a total of $10,000 to invest in two simple interest accounts, one earning 4.5% annual simple interest and the other earning 5% annual simple interest. If you earn $485 per year in interest from these two investments, how much do you have invested in each account?

$x + y = 10,000$
$0.045x + 0.05y = 485$

Determinant of a 2 × 2 Matrix [4.3A, p. 225]

The *determinant* of a 2 × 2 matrix $\begin{bmatrix} a_{11} & a_{12} \\ a_{21} & a_{22} \end{bmatrix}$ is written

$\begin{vmatrix} a_{11} & a_{12} \\ a_{21} & a_{22} \end{vmatrix}$.

The value of this determinant is given by the formula

$\begin{vmatrix} a_{11} & a_{12} \\ a_{21} & a_{22} \end{vmatrix} = a_{11}a_{22} - a_{12}a_{21}$.

$$\begin{vmatrix} 2 & -3 \\ -4 & 5 \end{vmatrix} = 2(5) - (-3)(-4)$$
$$= 10 - 12$$
$$= -2$$

Determinant of a 3 × 3 Matrix [4.3A, p. 226]

The evaluation of a 3 × 3 matrix is accomplished by *expanding by cofactors*.

Expand by cofactors of the first column.

$$\begin{vmatrix} 2 & 1 & 3 \\ 4 & 6 & 2 \\ 1 & 8 & 5 \end{vmatrix} = 2\begin{vmatrix} 6 & 2 \\ 8 & 5 \end{vmatrix} - 4\begin{vmatrix} 1 & 3 \\ 8 & 5 \end{vmatrix}$$
$$+ 1\begin{vmatrix} 1 & 3 \\ 6 & 2 \end{vmatrix}$$

Cramer's Rule [4.3B, pp. 228, 229]

For a system of two equations in two variables:
The solution of the system of equations

$$a_1x + b_1y = c_1$$
$$a_2x + b_2y = c_2$$

is given by $x = \dfrac{D_x}{D}$ and $y = \dfrac{D_y}{D}$, where $D = \begin{vmatrix} a_1 & b_1 \\ a_2 & b_2 \end{vmatrix}$,

$D_x = \begin{vmatrix} c_1 & b_1 \\ c_2 & b_2 \end{vmatrix}$, $D_y = \begin{vmatrix} a_1 & c_1 \\ a_2 & c_2 \end{vmatrix}$, and $D \neq 0$.

$$2x - y = 6$$
$$x + 3y = 4$$

$$D = \begin{vmatrix} 2 & -1 \\ 1 & 3 \end{vmatrix} = 7,$$

$$D_x = \begin{vmatrix} 6 & -1 \\ 4 & 3 \end{vmatrix} = 22,$$

$$D_y = \begin{vmatrix} 2 & 6 \\ 1 & 4 \end{vmatrix} = 2$$

$$x = \frac{D_x}{D} = \frac{22}{7}, \ y = \frac{D_y}{D} = \frac{2}{7}$$

For a system of three equations in three variables:
The solution of the system of equations

$$a_1x + b_1y + c_1z = d_1$$
$$a_2x + b_2y + c_2z = d_2$$
$$a_3x + b_3y + c_3z = d_3$$

is given by $x = \dfrac{D_x}{D}$, $y = \dfrac{D_y}{D}$, and $z = \dfrac{D_z}{D}$, where $D = \begin{vmatrix} a_1 & b_1 & c_1 \\ a_2 & b_2 & c_2 \\ a_3 & b_3 & c_3 \end{vmatrix}$,

$D_x = \begin{vmatrix} d_1 & b_1 & c_1 \\ d_2 & b_2 & c_2 \\ d_3 & b_3 & c_3 \end{vmatrix}$, $D_y = \begin{vmatrix} a_1 & d_1 & c_1 \\ a_2 & d_2 & c_2 \\ a_3 & d_3 & c_3 \end{vmatrix}$, $D_z = \begin{vmatrix} a_1 & b_1 & d_1 \\ a_2 & b_2 & d_2 \\ a_3 & b_3 & d_3 \end{vmatrix}$, and

$D \neq 0$.

$$x - y + z = 2$$
$$2x + y - 2z = -2$$
$$x - 2y + 3z = 6$$

$$D = \begin{vmatrix} 1 & -1 & 1 \\ 2 & 1 & -2 \\ 1 & -2 & 3 \end{vmatrix} = 2,$$

$$D_x = \begin{vmatrix} 2 & -1 & 1 \\ -2 & 1 & -2 \\ 6 & -2 & 3 \end{vmatrix} = 2,$$

$$D_y = \begin{vmatrix} 1 & 2 & 1 \\ 2 & -2 & -2 \\ 1 & 6 & 3 \end{vmatrix} = 4,$$

$$D_z = \begin{vmatrix} 1 & -1 & 2 \\ 2 & 1 & -2 \\ 1 & -2 & 6 \end{vmatrix} = 6$$

$$x = \frac{D_x}{D} = \frac{2}{2} = 1, \ y = \frac{D_y}{D} = \frac{4}{2} = 2,$$

$$z = \frac{D_z}{D} = \frac{6}{2} = 3$$

Chapter 4 Review Exercises

1. Solve by substitution:
$$2x - 6y = 15$$
$$x = 4y + 8$$

2. Solve by the addition method:
$$3x + 2y = 2$$
$$x + y = 3$$

3. Solve by graphing:
$$x + y = 3$$
$$3x - 2y = -6$$

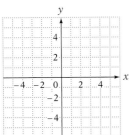

4. Solve by graphing:
$$2x - y = 4$$
$$y = 2x - 4$$

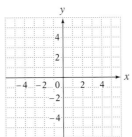

5. Solve by substitution:
$$3x + 12y = 18$$
$$x + 4y = 6$$

6. Solve by the addition method:
$$5x - 15y = 30$$
$$x - 3y = 6$$

7. Solve by the addition method:
$$3x - 4y - 2z = 17$$
$$4x - 3y + 5z = 5$$
$$5x - 5y + 3z = 14$$

8. Solve by the addition method:
$$3x + y = 13$$
$$2y + 3z = 5$$
$$x + 2z = 11$$

9. Evaluate the determinant:
$$\begin{vmatrix} 6 & 1 \\ 2 & 5 \end{vmatrix}$$

10. Evaluate the determinant:
$$\begin{vmatrix} 1 & 5 & -2 \\ -2 & 1 & 4 \\ 4 & 3 & -8 \end{vmatrix}$$

11. Solve by using Cramer's Rule:
$$2x - y = 7$$
$$3x + 2y = 7$$

12. Solve by using Cramer's Rule:
$$3x - 4y = 10$$
$$2x + 5y = 15$$

13. Solve by using Cramer's Rule:
$$x + y + z = 0$$
$$x + 2y + 3z = 5$$
$$2x + y + 2z = 3$$

14. Solve by using Cramer's Rule:
$$x + 3y + z = 6$$
$$2x + y - z = 12$$
$$x + 2y - z = 13$$

15. Graph the solution set:
$$x + 3y \leq 6$$
$$2x - y \geq 4$$

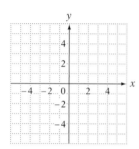

16. Graph the solution set:
$$2x + 4y \geq 8$$
$$x + y \leq 3$$

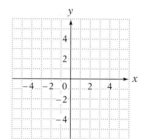

17. **Boating** A cabin cruiser traveling with the current went 60 mi in 3 h. Against the current, it took 5 h to travel the same distance. Find the rate of the cabin cruiser in calm water and the rate of the current.

18. **Aeronautics** A pilot flying with the wind flew 600 mi in 3 h. Flying against the wind, the pilot required 4 h to travel the same distance. Find the rate of the plane in calm air and the rate of the wind.

19. **Ticket Sales** At a movie theater, admission tickets are $5 for children and $8 for adults. The receipts for one Friday evening were $2500. The next day there were three times as many children as the preceding evening and only half the number of adults as the night before, yet the receipts were still $2500. Find the number of children who attended on Friday evening.

20. **Investments** A trust administrator divides $20,000 between two accounts. One account earns an annual simple interest rate of 3%, and a second account earns an annual simple interest rate of 7%. The total annual income from the two accounts is $1200. How much is invested in each account?

Chapter 4 Test

1. Solve by graphing: $2x - 3y = -6$
$2x - y = 2$

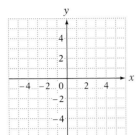

2. Solve by graphing: $x - 2y = -5$
$3x + 4y = -15$

3. Graph the solution set: $2x - y < 3$
$4x + 3y < 11$

4. Graph the solution set: $x + y > 2$
$2x - y < -1$

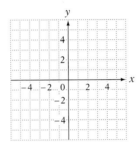

5. Solve by substitution: $3x + 2y = 4$
$x = 2y - 1$

6. Solve by substitution: $5x + 2y = -23$
$2x + y = -10$

7. Solve by substitution: $y = 3x - 7$
$y = -2x + 3$

8. Solve by the addition method:
$3x + 4y = -2$
$2x + 5y = 1$

9. Solve by the addition method:
$4x - 6y = 5$
$6x - 9y = 4$

10. Solve by the addition method:
$3x - y = 2x + y - 1$
$5x + 2y = y + 6$

11. Solve by the addition method:
$2x + 4y - z = 3$
$x + 2y + z = 5$
$4x + 8y - 2z = 7$

12. Solve by the addition method:
$x - y - z = 5$
$2x + z = 2$
$3y - 2z = 1$

13. Evaluate the determinant: $\begin{vmatrix} 3 & -1 \\ -2 & 4 \end{vmatrix}$

14. Evaluate the determinant: $\begin{vmatrix} 1 & -2 & 3 \\ 3 & 1 & 1 \\ 2 & -1 & -2 \end{vmatrix}$

15. Solve by using Cramer's Rule:
$$x - y = 3$$
$$2x + y = -4$$

16. Solve by using Cramer's Rule:
$$5x + 2y = 9$$
$$3x + 5y = -7$$

17. Solve by using Cramer's Rule:
$$x - y + z = 2$$
$$2x - y - z = 1$$
$$x + 2y - 3z = -4$$

18. **Aeronautics** A plane flying with the wind went 350 mi in 2 h. The return trip, flying against the wind, took 2.8 h. Find the rate of the plane in calm air and the rate of the wind.

19. **Purchasing** A clothing manufacturer purchased 60 yd of cotton and 90 yd of wool for a total cost of $1800. Another purchase, at the same prices, included 80 yd of cotton and 20 yd of wool for a total cost of $1000. Find the cost per yard of the cotton and of the wool.

20. **Investments** The annual interest earned on two investments is $549. One investment is in a 2.7% tax-free annual simple interest account, and the other investment is in a 5.1% annual simple interest CD. The total amount invested is $15,000. How much is invested in each account?

Cumulative Review Exercises

1. Solve: $\dfrac{3}{2}x - \dfrac{3}{8} + \dfrac{1}{4}x = \dfrac{7}{12}x - \dfrac{5}{6}$

2. Find the equation of the line that contains the points $(2, -1)$ and $(3, 4)$.

3. Simplify: $3[x - 2(5 - 2x) - 4x] + 6$

4. Evaluate $a + bc \div 2$ when $a = 4$, $b = 8$, and $c = -2$.

5. Solve: $2x - 3 < 9$ or $5x - 1 < 4$
 Write the solution set in interval notation.

6. Solve: $|x - 2| - 4 < 2$

7. Solve: $|2x - 3| > 5$

8. Given $f(x) = 3x^3 - 2x^2 + 1$, evaluate $f(-3)$.

9. Find the range of $f(x) = 3x^2 - 2x$ if the domain is $\{-2, -1, 0, 1, 2\}$.

10. Given $F(x) = x^2 - 3$, find $F(2)$.

11. Given $f(x) = 3x - 4$, write $f(2 + h) - f(2)$ in simplest form.

12. Graph the solution set of $\{x \mid x \le 2\} \cap \{x \mid x > -3\}$.

13. Find the equation of the line that contains the point $(-2, 3)$ and has slope $-\dfrac{2}{3}$.

14. Find the equation of the line that contains the point $(-1, 2)$ and is perpendicular to the line $2x - 3y = 7$.

15. Find the distance, to the nearest hundredth, between the points $(-4, 2)$ and $(2, 0)$.

16. Find the midpoint of the line segment connecting the points $(-4, 3)$ and $(3, 5)$.

17. Graph $2x - 5y = 10$ by using the slope and y-intercept.

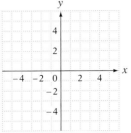

18. Graph the solution set of the inequality $3x - 4y \ge 8$.

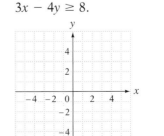

19. Solve by graphing.
$5x - 2y = 10$
$3x + 2y = 6$

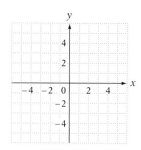

20. Solve by graphing.
$3x - 2y \geq 4$
$x + y < 3$

21. Solve by the addition method:
$3x + 2z = 1$
$2y - z = 1$
$x + 2y = 1$

22. Evaluate the determinant:
$$\begin{vmatrix} 2 & -5 & 1 \\ 3 & 1 & 2 \\ 6 & -1 & 4 \end{vmatrix}$$

23. Solve by using Cramer's Rule:
$4x - 3y = 17$
$3x - 2y = 12$

24. Solve by substitution:
$3x - 2y = 7$
$y = 2x - 1$

25. Coins A coin purse contains 40 coins in nickels, dimes, and quarters. There are three times as many dimes as quarters. The total value of the coins is $4.10. How many nickels are in the coin purse?

26. Mixtures How many milliliters of pure water must be added to 100 ml of a 4% salt solution to make a 2.5% salt solution?

27. Aeronautics Flying with the wind, a small plane required 2 h to fly 150 mi. Against the wind, it took 3 h to fly the same distance. Find the rate of the wind.

28. Purchasing A restaurant manager buys 100 lb of hamburger and 50 lb of steak for a total cost of $540. A second purchase, at the same prices, includes 150 lb of hamburger and 100 lb of steak. The total cost is $960. Find the price of 1 lb of steak.

29. Electronics Find the lower and upper limits of a 12,000-ohm resistor with a 15% tolerance.

30. Compensation The graph shows the relationship between the monthly income and the sales of an account executive. Find the slope of the line between the two points shown on the graph. Write a sentence that states the meaning of the slope.

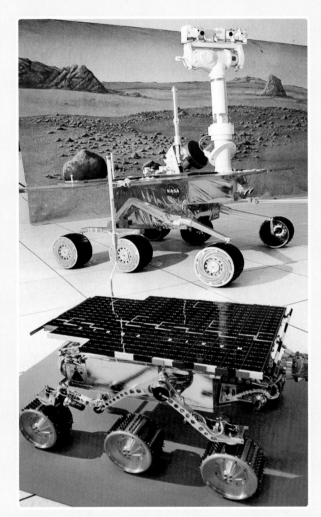

The twin robots Spirit and Opportunity are part of the Mars Exploration Rover mission. They were launched toward Mars in 2003 to see if the Red Planet once had water. Since their landing in 2004, the NASA Deep Space Network, which is an international network of antennas, has provided the communication links between the scientists on Earth and the rovers on Mars. The communications transmitted travel a distance of over 100 million miles. Distances this great, as well as distances very small, are generally expressed in scientific notation. **Exercises 115 to 125 on pages 269 and 270** ask you to express distances as well as other measurements in scientific notation.

Need help? For online student resources, such as section quizzes, visit this textbook's website at **math.college.hmco.com/students.**

OBJECTIVES

Section 5.1
A To multiply monomials
B To divide monomials and simplify expressions with negative exponents
C To write a number using scientific notation
D To solve application problems

Section 5.2
A To evaluate polynomial functions
B To add or subtract polynomials

Section 5.3
A To multiply a polynomial by a monomial
B To multiply two polynomials
C To multiply polynomials that have special products
D To solve application problems

Section 5.4
A To divide a polynomial by a monomial
B To divide polynomials
C To divide polynomials by using synthetic division
D To evaluate a polynomial function using synthetic division

Section 5.5
A To factor a monomial from a polynomial
B To factor by grouping
C To factor a trinomial of the form $x^2 + bx + c$
D To factor $ax^2 + bx + c$

Section 5.6
A To factor the difference of two perfect squares or a perfect-square trinomial
B To factor the sum or the difference of two perfect cubes
C To factor a trinomial that is quadratic in form
D To factor completely

Section 5.7
A To solve an equation by factoring
B To solve application problems

Do these exercises to prepare for Chapter 5.

For Exercises 1 to 5, simplify.

1. $-4(3y)$

2. $(-2)^3$

3. $-4a - 8b + 7a$

4. $3x - 2[y - 4(x + 1) + 5]$

5. $-(x - y)$

6. Write 40 as a product of prime numbers.

7. Find the GCF of 16, 20, and 24.

8. Evaluate $x^3 - 2x^2 + x + 5$ for $x = -2$.

9. Solve: $3x + 1 = 0$

If I travel at 57 mph, I will arrive 20 min late. If I travel at 64 mph, I will arrive 15 min early. At what average speed should I travel to arrive exactly on time?

5.1 Exponential Expressions

Objective A

To multiply monomials

A **monomial** is a number, a variable, or a product of a number and variables.

The examples at the right are monomials. The **degree of a monomial** is the sum of the exponents of the variables.

x degree 1 ($x = x^1$)
$3x^2$ degree 2
$4x^2y$ degree 3
$6x^3y^4z^2$ degree 9

In this chapter, the variable n is considered a positive integer when used as an exponent.

x^n degree n

The degree of a nonzero constant term is zero.

6 degree 0

The expression $5\sqrt{x}$ is not a monomial because \sqrt{x} cannot be written as a product of variables. The expression $\dfrac{x}{y}$ is not a monomial because it is a quotient of variables.

The expression x^4 is an exponential expression. The exponent, 4, indicates the number of times the base, x, occurs as a factor.

The product of exponential expressions with the *same* base can be simplified by writing each expression in factored form and writing the result with an exponent.

$$x^3 \cdot x^4 = \overbrace{(x \cdot x \cdot x)}^{3 \text{ factors}} \cdot \overbrace{(x \cdot x \cdot x \cdot x)}^{4 \text{ factors}}$$
$$\underbrace{\qquad\qquad\qquad}_{7 \text{ factors}}$$
$$= x^7$$

Note that adding the exponents results in the same product.

$$x^3 \cdot x^4 = x^{3+4} = x^7$$

Rule for Multiplying Exponential Expressions

If m and n are positive integers, then $x^m \cdot x^n = x^{m+n}$.

HOW TO Simplify: $(-4x^5y^3)(3xy^2)$

$(-4x^5y^3)(3xy^2) = (-4 \cdot 3)(x^5 \cdot x)(y^3 \cdot y^2)$

- Use the Commutative and Associative Properties of Multiplication to rearrange and group factors.

$= -12(x^{5+1})(y^{3+2})$

- Multiply variables with the same base by adding their exponents.

$= -12x^6y^5$

- Simplify.

As shown below, the power of a monomial can be simplified by writing the power in factored form and then using the Rule for Multiplying Exponential Expressions. It can also be simplified by multiplying each exponent inside the parentheses by the exponent outside the parentheses.

$$(a^2)^3 = a^2 \cdot a^2 \cdot a^2$$
$$= a^{2+2+2}$$
$$= a^6$$

$$(x^3y^4)^2 = (x^3y^4)(x^3y^4)$$
$$= x^{3+3}y^{4+4}$$
$$= x^6y^8$$

- Write in factored form. Then use the Rule for Multiplying Exponential Expressions.

$$(a^2)^3 = a^{2 \cdot 3} = a^6$$

$$(x^3y^4)^2 = x^{3 \cdot 2}y^{4 \cdot 2} = x^6y^8$$

- Multiply each exponent inside the parentheses by the exponent outside the parentheses.

Rule for Simplifying the Power of an Exponential Expression

If m and n are positive integers, then $(x^m)^n = x^{mn}$.

Rule for Simplifying Powers of Products

If m, n, and p are positive integers, then $(x^m y^n)^p = x^{mp} y^{np}$.

HOW TO Simplify: $(x^4)^5$

$$(x^4)^5 = x^{4 \cdot 5}$$
$$= x^{20}$$

- Use the Rule for Simplifying Powers of Exponential Expressions to multiply the exponents.

HOW TO Simplify: $(2a^3b^4)^3$

$$(2a^3b^4)^3 = 2^{1 \cdot 3}a^{3 \cdot 3}b^{4 \cdot 3}$$
$$= 2^3 a^9 b^{12}$$
$$= 8a^9 b^{12}$$

- Use the Rule for Simplifying Powers of Products to multiply each exponent inside the parentheses by the exponent outside the parentheses.

Example 1 Simplify: $(2xy^2)(-3xy^4)^3$

Solution

$$(2xy^2)(-3xy^4)^3 = (2xy^2)[(-3)^3 x^3 y^{12}]$$
$$= (2xy^2)(-27x^3 y^{12})$$
$$= -54x^4 y^{14}$$

You Try It 1 Simplify: $(-3a^2b^4)(-2ab^3)^4$

Your solution

Example 2 Simplify: $(x^{n+2})^5$

Solution

$$(x^{n+2})^5 = x^{5n+10}$$ • Multiply the exponents.

You Try It 2 Simplify: $(y^{n-3})^2$

Your solution

Example 3 Simplify: $[(2xy^2)^2]^3$

Solution

$$[(2xy^2)^2]^3 = [2^2 x^2 y^4]^3 = [4x^2 y^4]^3$$
$$= 4^3 x^6 y^{12} = 64x^6 y^{12}$$

You Try It 3 Simplify: $[(ab^3)^3]^4$

Your solution

Solutions on p. S13

Objective B **To divide monomials and simplify expressions with negative exponents**

The quotient of two exponential expressions with the same base can be simplified by writing each expression in factored form, dividing by the common factors, and then writing the result with an exponent.

$$\frac{x^5}{x^2} = \frac{\overset{1}{\cancel{x}} \cdot \overset{1}{\cancel{x}} \cdot x \cdot x \cdot x}{\underset{1}{\cancel{x}} \cdot \underset{1}{\cancel{x}}} = x^3$$

Note that subtracting the exponents gives the same result.

$$\frac{x^5}{x^2} = x^{5-2} = x^3$$

To divide two monomials with the same base, subtract the exponents of the like bases.

HOW TO Simplify: $\dfrac{z^8}{z^2}$

$\dfrac{z^8}{z^2} = z^{8-2}$ • The bases are the same. Subtract the exponents.

$\quad = z^6$

HOW TO Simplify: $\dfrac{a^5 b^9}{a^4 b}$

$\dfrac{a^5 b^9}{a^4 b} = a^{5-4} b^{9-1}$ • Subtract the exponents of the like bases.

$\quad = a b^8$

Consider the expression $\dfrac{x^4}{x^4}$, $x \neq 0$. This expression can be simplified, as shown below, by subtracting exponents or dividing by common factors.

$$\frac{x^4}{x^4} = x^{4-4} = x^0 \qquad \frac{x^4}{x^4} = \frac{\overset{1}{\cancel{x}} \cdot \overset{1}{\cancel{x}} \cdot \overset{1}{\cancel{x}} \cdot \overset{1}{\cancel{x}}}{\underset{1}{\cancel{x}} \cdot \underset{1}{\cancel{x}} \cdot \underset{1}{\cancel{x}} \cdot \underset{1}{\cancel{x}}} = 1$$

The equations $\dfrac{x^4}{x^4} = x^0$ and $\dfrac{x^4}{x^4} = 1$ suggest the following definition of x^0.

TAKE NOTE

In the example at the right, we indicated that $z \neq 0$. If we try to evaluate $(16z^5)^0$ when $z = 0$, we have $[16(0)^5]^0 = [16(0)]^0 = 0^0$. However, 0^0 is not defined, so we must assume that $z \neq 0$. To avoid stating this for every example or exercise, we will assume that variables cannot take on values that result in the expression 0^0.

> **Definition of Zero as an Exponent**
>
> If $x \neq 0$, then $x^0 = 1$. The expression 0^0 is not defined.

HOW TO Simplify: $(16z^5)^0$, $z \neq 0$

$(16z^5)^0 = 1$ • Any nonzero expression to the zero power is 1.

HOW TO Simplify: $-(7x^4 y^3)^0$

$-(7x^4 y^3)^0 = -(1) = -1$ • The negative outside the parentheses is not affected by the exponent.

Point of Interest

In the 15th century, the expression $12^{\overline{2m}}$ was used to mean $12x^{-2}$. The use of \overline{m} reflects an Italian influence, where m was used for minus and p was used for plus. It was understood that $2\overline{m}$ referred to an unnamed variable. Isaac Newton, in the 17th century, advocated the use of a negative exponent, the symbol we use today.

Consider the expression $\dfrac{x^4}{x^6}$, $x \neq 0$. This expression can be simplified, as shown below, by subtracting exponents or by dividing by common factors.

$$\frac{x^4}{x^6} = x^{4-6} = x^{-2} \qquad \frac{x^4}{x^6} = \frac{\overset{1}{\cancel{x}} \cdot \overset{1}{\cancel{x}} \cdot \overset{1}{\cancel{x}} \cdot \overset{1}{\cancel{x}}}{\underset{1}{\cancel{x}} \cdot \underset{1}{\cancel{x}} \cdot \underset{1}{\cancel{x}} \cdot \underset{1}{\cancel{x}} \cdot x \cdot x} = \frac{1}{x^2}$$

The equations $\dfrac{x^4}{x^6} = x^{-2}$ and $\dfrac{x^4}{x^6} = \dfrac{1}{x^2}$ suggest that $x^{-2} = \dfrac{1}{x^2}$.

Definition of a Negative Exponent

If $x \neq 0$ and n is a positive integer, then

$$x^{-n} = \frac{1}{x^n} \quad \text{and} \quad \frac{1}{x^{-n}} = x^n$$

TAKE NOTE

Note from the example at the right that 2^{-4} is a *positive* number. A negative exponent does not change the sign of a number.

HOW TO Evaluate: 2^{-4}

$$2^{-4} = \frac{1}{2^4}$$

- Use the Definition of a Negative Exponent.

$$= \frac{1}{16}$$

- Evaluate the expression.

The expression $\left(\dfrac{x^3}{y^4}\right)^2$, $y \neq 0$, can be simplified by squaring $\dfrac{x^3}{y^4}$ or by multiplying each exponent in the quotient by the exponent outside the parentheses.

$$\left(\frac{x^3}{y^4}\right)^2 = \left(\frac{x^3}{y^4}\right)\left(\frac{x^3}{y^4}\right) = \frac{x^3 \cdot x^3}{y^4 \cdot y^4} = \frac{x^{3+3}}{y^{4+4}} = \frac{x^6}{y^8} \qquad \left(\frac{x^3}{y^4}\right)^2 = \frac{x^{3\cdot2}}{y^{4\cdot2}} = \frac{x^6}{y^8}$$

Rule for Simplifying Powers of Quotients

If m, n, and p are integers and $y \neq 0$, then $\left(\dfrac{x^m}{y^n}\right)^p = \dfrac{x^{mp}}{y^{np}}$.

HOW TO Simplify: $\left(\dfrac{a^2}{b^3}\right)^{-2}$

$$\left(\frac{a^2}{b^3}\right)^{-2} = \frac{a^{2(-2)}}{b^{3(-2)}}$$

- Use the Rule for Simplifying Powers of Quotients.

$$= \frac{a^{-4}}{b^{-6}} = \frac{b^6}{a^4}$$

- Use the Definition of a Negative Exponent to write the expression with positive exponents.

The preceding example suggests the following rule.

> ### Rule for Negative Exponents on Fractional Expressions
>
> If $a \neq 0$, $b \neq 0$, and n is a positive integer, then $\left(\dfrac{a}{b}\right)^{-n} = \left(\dfrac{b}{a}\right)^{n}$.

An exponential expression is in simplest form when it is written with only positive exponents.

HOW TO Simplify: $7d^{-5}$, $d \neq 0$

$$7d^{-5} = 7 \cdot \frac{1}{d^5} = \frac{7}{d^5}$$

- Use the Definition of a Negative Exponent to rewrite the expression with a positive exponent.

HOW TO Simplify: $\dfrac{2}{5a^{-4}}$

$$\frac{2}{5a^{-4}} = \frac{2}{5} \cdot \frac{1}{a^{-4}} = \frac{2}{5} \cdot a^4 = \frac{2a^4}{5}$$

- Use the Definition of a Negative Exponent to rewrite the expression with a positive exponent.

Now that zero as an exponent and negative exponents have been defined, a rule for dividing exponential expressions can be stated.

> ### Rule for Dividing Exponential Expressions
>
> If m and n are integers and $x \neq 0$, then $\dfrac{x^m}{x^n} = x^{m-n}$.

HOW TO Simplify: $\dfrac{x^4}{x^9}$

$$\frac{x^4}{x^9} = x^{4-9}$$
$$= x^{-5}$$
$$= \frac{1}{x^5}$$

- Use the Rule for Dividing Exponential Expressions.
- Subtract the exponents.
- Use the Definition of a Negative Exponent to rewrite the expression with a positive exponent.

The rules for simplifying exponential expressions and powers of exponential expressions are true for all integers. These rules are restated here for convenience.

> ### Rules of Exponents
>
> If m, n, and p are integers, then
>
> $x^m \cdot x^n = x^{m+n}$ $(x^m)^n = x^{mn}$ $(x^m y^n)^p = x^{mp} y^{np}$
>
> $\dfrac{x^m}{x^n} = x^{m-n}$, $x \neq 0$ $\left(\dfrac{x^m}{y^n}\right)^p = \dfrac{x^{mp}}{y^{np}}$, $y \neq 0$ $x^{-n} = \dfrac{1}{x^n}$, $x \neq 0$
>
> $x^0 = 1$, $x \neq 0$

HOW TO Simplify: $(3ab^{-4})(-2a^{-3}b^7)$

$$(3ab^{-4})(-2a^{-3}b^7) = [3 \cdot (-2)](a^{1+(-3)}b^{-4+7})$$
$$= -6a^{-2}b^3$$
$$= -\frac{6b^3}{a^2}$$

- When multiplying expressions, add the exponents on like bases.

HOW TO Simplify: $\dfrac{4a^{-2}b^5}{6a^5b^2}$

$$\frac{4a^{-2}b^5}{6a^5b^2} = \frac{\cancel{2} \cdot 2a^{-2}b^5}{\cancel{2} \cdot 3a^5b^2} = \frac{2a^{-2}b^5}{3a^5b^2}$$

$$= \frac{2a^{-2-5}b^{5-2}}{3}$$

$$= \frac{2a^{-7}b^3}{3} = \frac{2b^3}{3a^7}$$

- Divide the coefficients by their common factor.
- Use the Rule for Dividing Exponential Expressions.
- Use the Definition of a Negative Exponent to rewrite the expression with a positive exponent.

HOW TO Simplify: $\left[\dfrac{6m^2n^3}{8m^7n^2}\right]^{-3}$

$$\left[\frac{6m^2n^3}{8m^7n^2}\right]^{-3} = \left[\frac{3m^{2-7}n^{3-2}}{4}\right]^{-3}$$

$$= \left[\frac{3m^{-5}n}{4}\right]^{-3}$$

$$= \frac{3^{-3}m^{15}n^{-3}}{4^{-3}}$$

$$= \frac{4^3m^{15}}{3^3n^3} = \frac{64m^{15}}{27n^3}$$

- Simplify inside the brackets.
- Subtract the exponents.
- Use the Rule for Simplifying Powers of Quotients.
- Use the Definition of a Negative Exponent to rewrite the expression with positive exponents. Then simplify.

Example 4

Simplify: $\dfrac{-28x^6z^{-3}}{42x^{-1}z^4}$

Solution

$$\frac{-28x^6z^{-3}}{42x^{-1}z^4} = -\frac{14 \cdot 2x^{6-(-1)}z^{-3-4}}{14 \cdot 3}$$

$$= -\frac{2x^7z^{-7}}{3} = -\frac{2x^7}{3z^7}$$

You Try It 4

Simplify: $\dfrac{20r^{-2}t^{-5}}{-16r^{-3}s^{-2}}$

Your solution

Example 5

Simplify: $\dfrac{(3a^{-1}b^4)^{-3}}{(6^{-1}a^{-3}b^{-4})^3}$

Solution

$$\frac{(3a^{-1}b^4)^{-3}}{(6^{-1}a^{-3}b^{-4})^3} = \frac{3^{-3}a^3b^{-12}}{6^{-3}a^{-9}b^{-12}} = 3^{-3} \cdot 6^3a^{12}b^0$$

$$= \frac{6^3a^{12}}{3^3} = \frac{216a^{12}}{27} = 8a^{12}$$

You Try It 5

Simplify: $\dfrac{(9u^{-6}v^4)^{-1}}{(6u^{-3}v^{-2})^{-2}}$

Your solution

Solutions on p. S13

Example 6

Simplify: $\dfrac{x^{4n-2}}{x^{2n-5}}$

Solution

$\dfrac{x^{4n-2}}{x^{2n-5}} = x^{4n-2-(2n-5)}$ • Subtract the exponents.

$= x^{4n-2-2n+5}$

$= x^{2n+3}$

You Try It 6

Simplify: $\dfrac{a^{2n+1}}{a^{n+3}}$

Your solution

Solution on p. S14

Objective C **To write a number using scientific notation**

Integer exponents are used to represent the very large and very small numbers encountered in the fields of science and engineering. For example, the mass of the electron is 0.0000000000000000000000000009 g. Numbers such as this are difficult to read and write, so a more convenient system for writing such numbers has been developed. It is called **scientific notation.**

To express a number in scientific notation, write the number as the product of a number between 1 and 10 and a power of 10. The form for scientific notation is $a \times 10^n$, where $1 \le a < 10$.

Point of Interest

Astronomers measure the distance of some stars by using the parsec. One parsec is approximately 1.91×10^{13} mi.

For numbers greater than 10, move the decimal point to the right of the first digit. The exponent n is positive and equal to the number of places the decimal point has been moved.

965,000	$= 9.65 \times 10^5$
3,600,000	$= 3.6 \times 10^6$
92,000,000,000	$= 9.2 \times 10^{10}$

TAKE NOTE

There are two steps involved in writing a number in scientific notation: (1) determine the number between 1 and 10, and (2) determine the exponent on 10.

For numbers less than 1, move the decimal point to the right of the first nonzero digit. The exponent n is negative. The absolute value of the exponent is equal to the number of places the decimal point has been moved.

0.0002	$= 2 \times 10^{-4}$
0.0000000974	$= 9.74 \times 10^{-8}$
0.000000000086	$= 8.6 \times 10^{-11}$

Converting a number written in scientific notation to decimal notation requires moving the decimal point.

When the exponent is positive, move the decimal point to the right the same number of places as the exponent.

$1.32 \times 10^4 = 13,200$

$1.4 \times 10^8 = 140,000,000$

When the exponent is negative, move the decimal point to the left the same number of places as the absolute value of the exponent.

$1.32 \times 10^{-2} = 0.0132$

$1.4 \times 10^{-4} = 0.00014$

Numerical calculations involving numbers that have more digits than a handheld calculator is able to handle can be performed using scientific notation.

Integrating Technology

See the Keystroke Guide: *Scientific Notation* for instructions on entering a number that is in scientific notation into a graphing calculator.

HOW TO Simplify: $\dfrac{220,000 \times 0.000000092}{0.0000011}$

$\dfrac{220,000 \times 0.000000092}{0.0000011} = \dfrac{2.2 \times 10^5 \times 9.2 \times 10^{-8}}{1.1 \times 10^{-6}}$ • Write the numbers in scientific notation.

$= \dfrac{(2.2)(9.2) \times 10^{5+(-8)-(-6)}}{1.1}$ • Simplify.

$= 18.4 \times 10^3 = 18,400$

Example 7 Write 0.000041 in scientific notation.

Solution $0.000041 = 4.1 \times 10^{-5}$

You Try It 7 Write 942,000,000 in scientific notation.

Your solution

Example 8 Write 3.3×10^7 in decimal notation.

Solution $3.3 \times 10^7 = 33{,}000{,}000$

You Try It 8 Write 2.7×10^{-5} in decimal notation.

Your solution

Example 9 Simplify:
$$\frac{2{,}400{,}000{,}000 \times 0.0000063}{0.00009 \times 480}$$

Solution
$$\frac{2{,}400{,}000{,}000 \times 0.0000063}{0.00009 \times 480}$$
$$= \frac{2.4 \times 10^9 \times 6.3 \times 10^{-6}}{9 \times 10^{-5} \times 4.8 \times 10^2}$$
$$= \frac{(2.4)(6.3) \times 10^{9+(-6)-(-5)-2}}{(9)(4.8)}$$
$$= 0.35 \times 10^6 = 350{,}000$$

You Try It 9 Simplify:
$$\frac{5{,}600{,}000 \times 0.000000081}{900 \times 0.000000028}$$

Your solution

Solutions on p. S14

Objective D **To solve application problems**

Example 10

How many miles does light travel in one day? The speed of light is 186,000 mi/s. Write the answer in scientific notation.

Strategy
To find the distance traveled:
• Write the speed of light in scientific notation.
• Write the number of seconds in one day in scientific notation.
• Use the equation $d = rt$, where r is the speed of light and t is the number of seconds in one day.

Solution
$r = 186{,}000 = 1.86 \times 10^5$
$t = 24 \cdot 60 \cdot 60 = 86{,}400 = 8.64 \times 10^4$
$d = rt$
$d = (1.86 \times 10^5)(8.64 \times 10^4)$
 $= 1.86 \times 8.64 \times 10^9$
 $= 16.0704 \times 10^9$
 $= 1.60704 \times 10^{10}$

Light travels 1.60704×10^{10} mi in one day.

You Try It 10

A computer can do an arithmetic operation in 1×10^{-7} s. In scientific notation, how many arithmetic operations can the computer perform in 1 min?

Your strategy

Your solution

Solution on p. S14

5.1 Exercises

To multiply monomials

For Exercises 1 to 37, simplify.

1. $(ab^3)(a^3b)$

2. $(-2ab^4)(-3a^2b^4)$

3. $(9xy^2)(-2x^2y^2)$

4. $(x^2y)^2$

5. $(x^2y^4)^4$

6. $(-2ab^2)^3$

7. $(-3x^2y^3)^4$

8. $(2^2a^2b^3)^3$

9. $(3^3a^5b^3)^2$

10. $(xy)(x^2y)^4$

11. $(x^2y^2)(xy^3)^3$

12. $[(2x)^4]^2$

13. $[(3x)^3]^2$

14. $[(x^2y)^4]^5$

15. $[(ab)^3]^6$

16. $[(2ab)^3]^2$

17. $[(2xy)^3]^4$

18. $[(3x^2y^3)^2]^2$

19. $[(2a^4b^3)^3]^2$

20. $y^n \cdot y^{2n}$

21. $x^n \cdot x^{n+1}$

22. $y^{2n} \cdot y^{4n+1}$

23. $y^{3n} \cdot y^{3n-2}$

24. $(a^n)^{2n}$

25. $(a^{n-3})^{2n}$

26. $(y^{2n-1})^3$

27. $(x^{3n+2})^5$

28. $(b^{2n-1})^n$

29. $(2xy)(-3x^2yz)(x^2y^3z^3)$

30. $(x^2z^4)(2xyz^4)(-3x^3y^2)$

31. $(3b^5)(2ab^2)(-2ab^2c^2)$

32. $(-c^3)(-2a^2bc)(3a^2b)$

33. $(-2x^2y^3z)(3x^2yz^4)$

34. $(2a^2b)^3(-3ab^4)^2$

35. $(-3ab^3)^3(-2^2a^2b)^2$

36. $(4ab)^2(-2ab^2c^3)^3$

37. $(-2ab^2)(-3a^4b^5)^3$

To divide monomials and simplify expressions with negative exponents

For Exercises 38 to 88, simplify.

38. 2^{-3}

39. $\dfrac{1}{3^{-5}}$

40. $\dfrac{1}{x^{-4}}$

41. $\dfrac{1}{y^{-3}}$

42. $\dfrac{2x^{-2}}{y^4}$

43. $\dfrac{a^3}{4b^{-2}}$

44. $x^{-3}y$

45. xy^{-4}

46. $-5x^0$

47. $\dfrac{1}{2x^0}$

48. $\dfrac{(2x)^0}{-2^3}$

49. $\dfrac{-3^{-2}}{(2y)^0}$

50. $\dfrac{y^{-7}}{y^{-8}}$

51. $\dfrac{y^{-2}}{y^6}$

52. $(x^2y^{-4})^2$

53. $(x^3y^5)^{-2}$

54. $\dfrac{x^{-2}y^{-11}}{xy^{-2}}$

55. $\dfrac{x^4y^3}{x^{-1}y^{-2}}$

56. $\dfrac{a^{-1}b^{-3}}{a^4b^{-5}}$

57. $\dfrac{a^6b^{-4}}{a^{-2}b^5}$

58. $(2a^{-1})^{-2}(2a^{-1})^4$

59. $(3a)^{-3}(9a^{-1})^{-2}$

60. $(x^{-2}y)^2(xy)^{-2}$

61. $(x^{-1}y^2)^{-3}(x^2y^{-4})^{-3}$

62. $\dfrac{50b^{10}}{70b^5}$

63. $\dfrac{x^3y^6}{x^6y^2}$

64. $\dfrac{x^{17}y^5}{-x^7y^{10}}$

65. $\dfrac{-6x^2y}{12x^4y}$

66. $\dfrac{2x^2y^4}{(3xy^2)^3}$

67. $\dfrac{-3ab^2}{(9a^2b^4)^3}$

68. $\left(\dfrac{-12a^2b^3}{9a^5b^9}\right)^3$

69. $\left(\dfrac{12x^3y^2z}{18xy^3z^4}\right)^4$

70. $\dfrac{(4x^2y)^2}{(2xy^3)^3}$

71. $\dfrac{(3a^2b)^3}{(-6ab^3)^2}$

72. $\dfrac{(-4x^2y^3)^2}{(2xy^2)^3}$

73. $\dfrac{(-3a^2b^3)^2}{(-2ab^4)^3}$

74. $\dfrac{(-4xy^3)^3}{(-2x^7y)^4}$

75. $\dfrac{(-8x^2y^2)^4}{(16x^3y^7)^2}$

76. $\dfrac{a^{5n}}{a^{3n}}$

77. $\dfrac{b^{6n}}{b^{10n}}$

78. $\dfrac{-x^{5n}}{x^{2n}}$

79. $\dfrac{y^{2n}}{-y^{8n}}$

80. $\dfrac{x^{2n-1}}{x^{n-3}}$

81. $\dfrac{y^{3n+2}}{y^{2n+4}}$

82. $\dfrac{a^{3n}b^n}{a^nb^{2n}}$

83. $\dfrac{x^ny^{3n}}{x^ny^{5n}}$

84. $\dfrac{a^{3n-2}b^{n+1}}{a^{2n+1}b^{2n+2}}$

85. $\dfrac{x^{2n-1}y^{n-3}}{x^{n+4}y^{n+3}}$

86. $\left(\dfrac{4^{-2}xy^{-3}}{x^{-3}y}\right)^3\left(\dfrac{8^{-1}x^{-2}y}{x^4y^{-1}}\right)^{-2}$

87. $\left(\dfrac{9ab^{-2}}{8a^{-2}b}\right)^{-2}\left(\dfrac{3a^{-2}b}{2a^2b^{-2}}\right)^3$

88. $\left(\dfrac{2ab^{-1}}{ab}\right)^{-1}\left(\dfrac{3a^{-2}b}{a^2b^2}\right)^{-2}$

Objective C **To write a number using scientific notation**

For Exercises 89 to 94, write in scientific notation.

89. 0.00000467

90. 0.00000005

91. 0.00000000017

92. 4,300,000

93. 200,000,000,000

94. 9,800,000,000

For Exercises 95 to 100, write in decimal notation.

95. 1.23×10^{-7}

96. 6.2×10^{-12}

97. 8.2×10^{15}

98. 6.34×10^5

99. 3.9×10^{-2}

100. 4.35×10^9

For Exercises 101 to 114, simplify. Write the answer in decimal notation.

101. $(3 \times 10^{-12})(5 \times 10^{16})$

102. $(8.9 \times 10^{-5})(3.2 \times 10^{-6})$

103. $(0.0000065)(3,200,000,000,000)$

104. $(480,000)(0.0000000096)$

105. $\dfrac{9 \times 10^{-3}}{6 \times 10^5}$

106. $\dfrac{2.7 \times 10^4}{3 \times 10^{-6}}$

107. $\dfrac{0.0089}{500,000,000}$

108. $\dfrac{4800}{0.00000024}$

109. $\dfrac{0.00056}{0.000000000004}$

110. $\dfrac{0.000000346}{0.0000005}$

111. $\dfrac{(3.2 \times 10^{-11})(2.9 \times 10^{15})}{8.1 \times 10^{-3}}$

112. $\dfrac{(6.9 \times 10^{27})(8.2 \times 10^{-13})}{4.1 \times 10^{15}}$

113. $\dfrac{(0.00000004)(84,000)}{(0.0003)(1,400,000)}$

114. $\dfrac{(720)(0.0000000039)}{(26,000,000,000)(0.018)}$

Objective D **To solve application problems**

For Exercises 115 to 125, solve. Write the answer in scientific notation.

115. **Astronomy** Our galaxy is estimated to be 5.6×10^{19} mi across. How long (in hours) would it take a space ship to cross the galaxy traveling at 25,000 mph?

116. **Astronomy** How long does it take light to travel to Earth from the sun? The sun is 9.3×10^7 mi from Earth, and light travels 1.86×10^5 mi/s

The Milky Way

117. **Astronomy** The distance to Saturn is 8.86×10^8 mi. A satellite leaves Earth traveling at a constant rate of 1×10^5 mph. How long does it take for the satellite to reach Saturn?

118. **The Federal Government** In 2004, the gross national debt was approximately 7×10^{12} dollars. How much would each American have to pay in order to pay off the debt? Use 3×10^8 as the number of citizens.

119. **Physics** The mass of an electron is 9.109×10^{-31} kg. The mass of a proton is 1.673×10^{-27} kg. How many times heavier is a proton than an electron?

120. **Geology** The mass of Earth is 5.9×10^{24} kg. The mass of the sun is 2×10^{30} kg. How many times heavier is the sun than Earth?

121. **Physics** How many meters does light travel in 8 h? The speed of light is 3×10^8 m/s.

122. **Physics** How many meters does light travel in 1 day? The speed of light is 3×10^8 m/s.

123. **Astronomy** It took 11 min for the commands from a computer on Earth to travel to the rover Sojourner on Mars, a distance of 119 million miles. How fast did the signals from Earth to Mars travel?

124. **Measurement** The weight of 31 million orchid seeds is 1 oz. Find the weight of one orchid seed.

Sojourner

125. Physics A high-speed centrifuge makes 4×10^8 revolutions each minute. Find the time in seconds for the centrifuge to make one revolution.

Centrifuge

APPLYING THE CONCEPTS

126. Correct the error in the following expressions. Explain which rule or property was used incorrectly.
a. $x^0 = 0$
b. $(x^4)^5 = x^9$
c. $x^2 \cdot x^3 = x^6$

127. Simplify. **a.** $1 + [1 + (1 + 2^{-1})^{-1}]^{-1}$
b. $2 - [2 - (2 - 2^{-1})^{-1}]^{-1}$

5.2 Introduction to Polynomial Functions

Objective A **To evaluate polynomial functions**

A **polynomial** is a variable expression in which the terms are monomials.

A polynomial of one term is a **monomial.** $5x$

A polynomial of two terms is a **binomial.** $5x^2y + 6x$

A polynomial of three terms is a **trinomial.** $3x^2 + 9xy - 5y$

Polynomials with more than three terms do not have special names.

The **degree of a polynomial** is the greatest of the degrees of any of its terms.

$$3x + 2 \qquad \text{degree } 1$$
$$3x^2 + 2x - 4 \qquad \text{degree } 2$$
$$4x^3y^2 + 6x^4 \qquad \text{degree } 5$$
$$3x^{2n} - 5x^n - 2 \qquad \text{degree } 2n$$

The terms of a polynomial in one variable are usually arranged so that the exponents of the variable decrease from left to right. This is called **descending order.**

$$2x^2 - x + 8$$

$$3y^3 - 3y^2 + y - 12$$

For a polynomial in more than one variable, descending order may refer to any one of the variables.

The polynomial at the right is shown first in descending order of the x variable and then in descending order of the y variable.

$$2x^2 + 3xy + 5y^2$$

$$5y^2 + 3xy + 2x^2$$

Polynomial functions have many applications in mathematics. In general, a **polynomial function** is an expression whose terms are monomials. The **linear function** given by $f(x) = mx + b$ is an example of a polynomial function. It is a polynomial function of degree 1. A second-degree polynomial function, called a **quadratic function,** is given by the equation $f(x) = ax^2 + bx + c$, $a \neq 0$. A third-degree polynomial function is called a **cubic function.**

To **evaluate a polynomial function,** replace the variable by its value and simplify.

HOW TO Given $P(x) = x^3 - 3x^2 + 4$, evaluate $P(-3)$.

$$P(x) = x^3 - 3x^2 + 4$$
$$P(-3) = (-3)^3 - 3(-3)^2 + 4 \qquad \bullet \text{ Substitute } -3 \text{ for } x \text{ and simplify.}$$
$$= -27 - 27 + 4$$
$$= -50$$

The **leading coefficient** of a polynomial function is the coefficient of the variable with the largest exponent. The **constant term** is the term without a variable.

HOW TO Find the leading coefficient, the constant term, and the degree of the polynomial function $P(x) = 7x^4 - 3x^2 + 2x - 4$.

The leading coefficient is 7, the constant term is -4, and the degree is 4.

The three equations below do not represent polynomial functions.

$f(x) = 3x^2 + 2x^{-1}$ A polynomial function does not have a variable raised to a negative power.

$g(x) = 2\sqrt{x} - 3$ A polynomial function does not have a variable expression within a radical.

$h(x) = \dfrac{x}{x-1}$ A polynomial function does not have a variable in the denominator of a fraction.

The graph of a linear function is a straight line and can be found by plotting just two points. The graph of a polynomial function of degree greater than 1 is a curve. Consequently, many points may have to be found before an accurate graph can be drawn.

Evaluating the quadratic function given by the equation $f(x) = x^2 - x - 6$ when $x = -3, -2, -1, 0, 1, 2, 3$, and 4 gives the points shown in Figure 1 below. For instance, $f(-3) = 6$, so $(-3, 6)$ is graphed; $f(2) = -4$, so $(2, -4)$ is graphed; and $f(4) = 6$, so $(4, 6)$ is graphed. Evaluating the function when x is not an integer, such as when $x = -\dfrac{3}{2}$ and $x = \dfrac{5}{2}$, produces more points to graph, as shown in Figure 2. Connecting the points with a smooth curve results in Figure 3, which is the graph of f.

Figure 1

Figure 2

Figure 3

Here is an example of graphing a cubic function, $P(x) = x^3 - 2x^2 - 5x + 6$. Evaluating the function when $x = -2, -1, 0, 1, 2, 3$, and 4 gives the graph in Figure 4 below. Evaluating for some noninteger values gives the graph in Figure 5. Finally, connecting the dots with a smooth curve gives the graph in Figure 6.

Figure 4

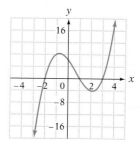

Figure 5

Figure 6

Example 1

Given $P(x) = x^3 + 3x^2 - 2x + 8$, evaluate $P(-2)$.

Solution

$$P(x) = x^3 + 3x^2 - 2x + 8$$
$$P(-2) = (-2)^3 + 3(-2)^2 - 2(-2) + 8 \quad \bullet \; \text{Replace } x$$
$$= (-8) + 3(4) + 4 + 8 \qquad\qquad \text{by } -2.$$
$$= -8 + 12 + 4 + 8 \qquad\qquad\; \text{Simplify.}$$
$$= 16$$

You Try It 1

Given $R(x) = -2x^4 - 5x^3 + 2x - 8$, evaluate $R(2)$.

Your solution

Example 2

Find the leading coefficient, the constant term, and the degree of the polynomial.
$P(x) = 5x^6 - 4x^5 - 3x^2 + 7$

Solution

The leading coefficient is 5, the constant term is 7, and the degree is 6.

You Try It 2

Find the leading coefficient, the constant term, and the degree of the polynomial.
$r(x) = -3x^4 + 3x^3 + 3x^2 - 2x - 12$

Your solution

Example 3

Which of the following is a polynomial function?
a. $P(x) = 3x^{\frac{1}{2}} + 2x^2 - 3$
b. $T(x) = 3\sqrt{x} - 2x^2 - 3x + 2$
c. $R(x) = 14x^3 - \pi x^2 + 3x + 2$

Solution

a. This is not a polynomial function. A polynomial function does not have a variable raised to a fractional power.
b. This is not a polynomial function. A polynomial function does not have a variable expression within a radical.
c. This is a polynomial function.

You Try It 3

Which of the following is a polynomial function?
a. $R(x) = 5x^{14} - 5$
b. $V(x) = -x^{-1} + 2x - 7$
c. $P(x) = 2x^4 - 3\sqrt{x} - 3$

Your solution

Example 4

Graph $f(x) = x^2 - 2$.

Solution

x	$y = f(x)$
-3	7
-2	2
-1	-1
0	-2
1	-1
2	2
3	7

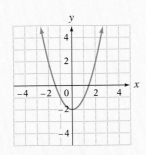

You Try It 4

Graph $f(x) = x^2 + 2x - 3$.

Your solution

Solutions on p. S14

Example 5

Graph $f(x) = x^3 - 1$.

Solution

x	$y = f(x)$
-2	-9
-1	-2
0	-1
1	0
2	7

You Try It 5

Graph $f(x) = -x^3 + 1$.

Your solution

Solution on p. S14

Objective B **To add or subtract polynomials**

Polynomials can be added by combining like terms. Either a vertical or a horizontal format can be used.

HOW TO Simplify $(3x^2 + 2x - 7) + (7x^3 - 3 + 4x^2)$. Use a horizontal format.

Use the Commutative and Associative Properties of Addition to rearrange and group like terms.

$(3x^2 + 2x - 7) + (7x^3 - 3 + 4x^2)$
$= 7x^3 + (3x^2 + 4x^2) + 2x + (-7 - 3)$
$= 7x^3 + 7x^2 + 2x - 10$ • Combine like terms.

HOW TO Simplify $(4x^2 + 5x - 3) + (7x^3 - 7x + 1) + (2x - 3x^2 + 4x^3 + 1)$. Use a vertical format.

Arrange the terms of each polynomial in descending order with like terms in the same column.

$$
\begin{array}{r}
4x^2 + 5x - 3 \\
7x^3 - 7x + 1 \\
4x^3 - 3x^2 + 2x + 1 \\
\hline
11x^3 + x^2 - 1
\end{array}
$$
 • Add the terms in each column.

TAKE NOTE

The additive inverse of a polynomial is that polynomial with the sign of every term changed.

The **additive inverse of the polynomial** $x^2 + 5x - 4$ is $-(x^2 + 5x - 4)$.

To simplify the additive inverse of a polynomial, change the sign of every term inside the parentheses.

$-(x^2 + 5x - 4) = -x^2 - 5x + 4$

TAKE NOTE

This is the same definition used for subtraction of integers: subtraction is addition of the opposite.

To subtract two polynomials, add the additive inverse of the second polynomial to the first.

> **HOW TO** Simplify $(3x^2 - 7xy + y^2) - (-4x^2 + 7xy - 3y^2)$. Use a horizontal format.
>
> Rewrite the subtraction as the addition of the additive inverse.
>
> $(3x^2 - 7xy + y^2) - (-4x^4 + 7xy - 3y^2)$
> $= (3x^2 - 7xy + y^2) + (4x^2 - 7xy + 3y^2)$
> $= 7x^2 - 14xy + 4y^2$ • Combine like terms.

> **HOW TO** Simplify $(6x^3 - 3x + 7) - (3x^2 - 5x + 12)$. Use a vertical format.
>
> Rewrite subtraction as the addition of the additive inverse.
>
> $(6x^3 - 3x + 7) - (3x^2 - 5x + 12) = (6x^3 - 3x + 7) + (-3x^2 + 5x - 12)$
>
> Arrange the terms of each polynomial in descending order with like terms in the same column.
>
> $$\begin{array}{r} 6x^3 \qquad - 3x + 7 \\ -3x^2 + 5x - 12 \\ \hline 6x^3 - 3x^2 + 2x - 5 \end{array}$$ • Combine the terms in each column.

Function notation can be used when adding or subtracting polynomials.

> **HOW TO** Given $P(x) = 3x^2 - 2x + 4$ and $R(x) = -5x^3 + 4x + 7$, find $P(x) + R(x)$.
>
> $P(x) + R(x) = (3x^2 - 2x + 4) + (-5x^3 + 4x + 7)$
> $= -5x^3 + 3x^2 + 2x + 11$

> **HOW TO** Given $P(x) = -5x^2 + 8x - 4$ and $R(x) = -3x^2 - 5x + 9$, find $P(x) - R(x)$.
>
> $P(x) - R(x) = (-5x^2 + 8x - 4) - (-3x^2 - 5x + 9)$
> $= (-5x^2 + 8x - 4) + (3x^2 + 5x - 9)$
> $= -2x^2 + 13x - 13$

> **HOW TO** Given $P(x) = 3x^2 - 5x + 6$ and $R(x) = 2x^2 - 5x - 7$, find $S(x)$, the sum of the two polynomials.
>
> $S(x) = P(x) + R(x) = (3x^2 - 5x + 6) + (2x^2 - 5x - 7)$
> $= 5x^2 - 10x - 1$

Note from the preceding example that evaluating $P(x) = 3x^2 - 5x + 6$ and $R(x) = 2x^2 - 5x - 7$ at, for example, $x = 3$ and then adding the values is the same as evaluating $S(x) = 5x^2 - 10x - 1$ at 3.

$P(3) = 3(3)^2 - 5(3) + 6 = 27 - 15 + 6 = 18$
$R(3) = 2(3)^2 - 5(3) - 7 = 18 - 15 - 7 = -4$

$P(3) + R(3) = 18 + (-4) = 14$

$S(3) = 5(3)^2 - 10(3) - 1 = 45 - 30 - 1 = 14$

Example 6

Simplify:
$(4x^2 - 3xy + 7y^2) + (-3x^2 + 7xy + y^2)$
Use a vertical format.

Solution

$$\begin{array}{r} 4x^2 - 3xy + 7y^2 \\ -3x^2 + 7xy + y^2 \\ \hline x^2 + 4xy + 8y^2 \end{array}$$

You Try It 6

Simplify:
$(-3x^2 - 4x + 9) + (-5x^2 - 7x + 1)$
Use a vertical format.

Your solution

Example 7

Simplify:
$(3x^2 - 2x + 4) - (7x^2 + 3x - 12)$
Use a vertical format.

Solution

Add the additive inverse of $7x^2 + 3x - 12$ to $3x^2 - 2x + 4$.

$$\begin{array}{r} 3x^2 - 2x + 4 \\ -7x^2 - 3x + 12 \\ \hline -4x^2 - 5x + 16 \end{array}$$

You Try It 7

Simplify:
$(-5x^2 + 2x - 3) - (6x^2 + 3x - 7)$
Use a vertical format.

Your solution

Example 8

Given $P(x) = -3x^2 + 2x - 6$ and
$R(x) = 4x^3 - 3x + 4$, find $S(x) = P(x) + R(x)$.
Evaluate $S(-2)$.

Solution

$S(x) = (-3x^2 + 2x - 6) + (4x^3 - 3x + 4)$
$ = 4x^3 - 3x^2 - x - 2$

$S(-2) = 4(-2)^3 - 3(-2)^2 - (-2) - 2$
$ = 4(-8) - 3(4) + 2 - 2$
$ = -44$

You Try It 8

Given $P(x) = 4x^3 - 3x^2 + 2$ and
$R(x) = -2x^2 + 2x - 3$, find
$S(x) = P(x) + R(x)$. Evaluate $S(-1)$.

Your solution

Example 9

Given $P(x) = (2x^{2n} - 3x^n + 7)$ and
$R(x) = (3x^{2n} + 3x^n + 5)$, find
$D(x) = P(x) - R(x)$.

Solution

$D(x) = P(x) - R(x)$
$D(x) = (2x^{2n} - 3x^n + 7) - (3x^{2n} + 3x^n + 5)$
$ = (2x^{2n} - 3x^n + 7) + (-3x^{2n} - 3x^n - 5)$
$ = -x^{2n} - 6x^n + 2$

You Try It 9

Given $P(x) = (5x^{2n} - 3x^n - 7)$ and
$R(x) = (-2x^{2n} - 5x^n + 8)$, find
$D(x) = P(x) - R(x)$.

Your solution

Solutions on p. S14

5.2 Exercises

Objective A **To evaluate polynomial functions**

1. Given $P(x) = 3x^2 - 2x - 8$, evaluate $P(3)$.

2. Given $P(x) = -3x^2 - 5x + 8$, evaluate $P(-5)$.

3. Given $R(x) = 2x^3 - 3x^2 + 4x - 2$, evaluate $R(2)$.

4. Given $R(x) = -x^3 + 2x^2 - 3x + 4$, evaluate $R(-1)$.

5. Given $f(x) = x^4 - 2x^2 - 10$, evaluate $f(-1)$.

6. Given $f(x) = x^5 - 2x^3 + 4x$, evaluate $f(2)$.

In Exercises 7 to 18, indicate which define a polynomial function. For those that are polynomial functions: **a.** Identify the leading coefficient. **b.** Identify the constant term. **c.** State the degree.

7. $P(x) = -x^2 + 3x + 8$

8. $P(x) = 3x^4 - 3x - 7$

9. $R(x) = \dfrac{x}{x + 1}$

10. $R(x) = \dfrac{3x^2 - 2x + 1}{x}$

11. $f(x) = \sqrt{x} - x^2 + 2$

12. $f(x) = x^2 - \sqrt{x + 2} - 8$

13. $g(x) = 3x^5 - 2x^2 + \pi$

14. $g(x) = -4x^5 - 3x^2 + x - \sqrt{7}$

15. $P(x) = 3x^2 - 5x^3 + 2$

16. $P(x) = x^2 - 5x^4 - x^6$

17. $R(x) = 14$

18. $R(x) = \dfrac{1}{x} + 2$

For Exercises 19 to 24, graph.

19. $P(x) = x^2 - 1$

20. $P(x) = 2x^2 + 3$

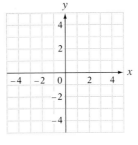

21. $R(x) = x^3 + 2$

22. $R(x) = x^4 + 1$

23. $f(x) = x^3 - 2x$

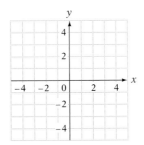

24. $f(x) = x^2 - x - 2$

Objective B **To add or subtract polynomials**

Simplify. Use a vertical format for Exercises 25 to 28.

25. $(5x^2 + 2x - 7) + (x^2 - 8x + 12)$

26. $(3x^2 - 2x + 7) + (-3x^2 + 2x - 12)$

27. $(x^2 - 3x + 8) - (2x^2 - 3x + 7)$

28. $(2x^2 + 3x - 7) - (5x^2 - 8x - 1)$

Simplify. Use a horizontal format for Exercises 29 to 32.

29. $(3y^2 - 7y) + (2y^2 - 8y + 2)$

30. $(-2y^2 - 4y - 12) + (5y^2 - 5y)$

31. $(2a^2 - 3a - 7) - (-5a^2 - 2a - 9)$

32. $(3a^2 - 9a) - (-5a^2 + 7a - 6)$

33. Given $P(x) = x^2 - 3xy + y^2$ and $R(x) = 2x^2 - 3y^2$, find $P(x) + R(x)$.

34. Given $P(x) = x^{2n} + 7x^n - 3$ and $R(x) = -x^{2n} + 2x^n + 8$, find $P(x) + R(x)$.

35. Given $P(x) = 3x^2 + 2y^2$ and $R(x) = -5x^2 + 2xy - 3y^2$, find $P(x) - R(x)$.

36. Given $P(x) = 2x^{2n} - x^n - 1$ and $R(x) = 5x^{2n} + 7x^n + 1$, find $P(x) - R(x)$.

37. Given $P(x) = 3x^4 - 3x^3 - x^2$ and $R(x) = 3x^3 - 7x^2 + 2x$, find $S(x) = P(x) + R(x)$. Evaluate $S(2)$.

38. Given $P(x) = 3x^4 - 2x + 1$ and $R(x) = 3x^5 - 5x - 8$, find $S(x) = P(x) + R(x)$. Evaluate $S(-1)$.

APPLYING THE CONCEPTS

39. For what value of k is the given equation an identity?
a. $(2x^3 + 3x^2 + kx + 5) - (x^3 + 2x^2 + 3x + 7) = x^3 + x^2 + 5x - 2$
b. $(6x^3 + kx^2 - 2x - 1) - (4x^3 - 3x^2 + 1) = 2x^3 - x^2 - 2x - 2$

40. If $P(x)$ is a third-degree polynomial and $Q(x)$ is a fourth-degree polynomial, what can be said about the degree of $P(x) + Q(x)$? Give some examples of polynomials that support your answer.

41. If $P(x)$ is a fifth-degree polynomial and $Q(x)$ is a fourth-degree polynomial, what can be said about the degree of $P(x) - Q(x)$? Give some examples of polynomials that support your answer.

42. Sports The deflection D (in inches) of a beam that is uniformly loaded is given by the polynomial function $D(x) = 0.005x^4 - 0.1x^3 + 0.5x^2$, where x is the distance from one end of the beam. See the figure at the right. The maximum deflection occurs when x is the midpoint of the beam. Determine the maximum deflection for the beam in the diagram.

5.3 Multiplication of Polynomials

Objective A **To multiply a polynomial by a monomial**

To multiply a polynomial by a monomial, use the Distributive Property and the Rule for Multiplying Exponential Expressions.

HOW TO Multiply: $-3x^2(2x^2 - 5x + 3)$

$-3x^2(2x^2 - 5x + 3)$

$= -3x^2(2x^2) - (-3x^2)(5x) + (-3x^2)(3)$ • Use the Distributive Property.

$= -6x^4 + 15x^3 - 9x^2$ • Use the Rule for Multiplying Exponential Expressions.

HOW TO Simplify: $2x^2 - 3x[2 - x(4x + 1) + 2]$

$2x^2 - 3x[2 - x(4x + 1) + 2]$

$= 2x^2 - 3x[2 - 4x^2 - x + 2]$ • Use the Distributive Property to remove the parentheses.

$= 2x^2 - 3x[-4x^2 - x + 4]$ • Simplify.

$= 2x^2 + 12x^3 + 3x^2 - 12x$ • Use the Distributive Property to remove the brackets.

$= 12x^3 + 5x^2 - 12x$ • Simplify.

HOW TO Multiply: $x^n(x^n - x^2 + 3)$

$x^n(x^n - x^2 + 3)$

$= x^n(x^n) - x^n(x^2) + x^n(3)$ • Use the Distributive Property.

$= x^{2n} - x^{n+2} + 3x^n$

Example 1

Multiply: $(3a^2 - 2a + 4)(-3a)$

Solution
$(3a^2 - 2a + 4)(-3a)$

$= 3a^2(-3a) - 2a(-3a) + 4(-3a)$ • Use the Distributive Property.

$= -9a^3 + 6a^2 - 12a$

You Try It 1

Multiply: $(2b^2 - 7b - 8)(-5b)$

Your solution

Solution on p. S14

Example 2

Simplify: $y - 3y[y - 2(3y - 6) + 2]$

Solution

$y - 3y[y - 2(3y - 6) + 2]$
$= y - 3y[y - 6y + 12 + 2]$
$= y - 3y[-5y + 14]$
$= y + 15y^2 - 42y$
$= 15y^2 - 41y$

You Try It 2

Simplify: $x^2 - 2x[x - x(4x - 5) + x^2]$

Your solution

Example 3

Multiply: $x^{n+2}(x^{n-1} + 2x - 1)$

Solution

$x^{n+2}(x^{n-1} + 2x - 1)$
$= x^{n+2}(x^{n-1}) + (x^{n+2})(2x) - (x^{n+2})(1)$
$= x^{n+2+(n-1)} + 2x^{n+2+1} - x^{n+2}$
$= x^{2n+1} + 2x^{n+3} - x^{n+2}$

• Use the Distributive Property.

You Try It 3

Multiply: $y^{n+3}(y^{n-2} - 3y^2 + 2)$

Your solution

Solutions on p. S15

Objective B To multiply two polynomials

The product of two polynomials is the polynomial obtained by multiplying each term of one polynomial by each term of the other polynomial and then combining like terms.

HOW TO Multiply: $(2x^2 - 2x + 1)(3x + 2)$

Use the Distributive Property to multiply the trinomial by each term of the binomial.

$(2x^2 - 2x + 1)(3x + 2) = (2x^2 - 2x + 1)(3x) + (2x^2 - 2x + 1)(2)$

$= (6x^3 - 6x^2 + 3x) + (4x^2 - 4x + 2)$

$= 6x^3 - 2x^2 - x + 2$

A convenient method of multiplying two polynomials is to use a vertical format similar to that used for multiplication of whole numbers.

HOW TO Multiply: $(3x^2 - 4x + 8)(2x - 7)$

$$
\begin{array}{r}
3x^2 - 4x + 8 \\
2x - 7 \\
\hline
-21x^2 + 28x - 56 \\
6x^3 - 8x^2 + 16x \\
\hline
6x^3 - 29x^2 + 44x - 56
\end{array}
$$

$= -7(3x^2 - 4x + 8)$

$= 2x(3x^2 - 4x + 8)$

• Like terms are in the same column.

• Combine like terms.

It is frequently necessary to find the product of two binomials. The product can be found by using a method called **FOIL,** which is based on the Distributive Property. The letters of FOIL stand for **F**irst, **O**uter, **I**nner, and **L**ast.

Multiply: $(3x - 2)(2x + 5)$

Multiply the First terms.	$(3x - 2)(2x + 5)$	$3x \cdot 2x = 6x^2$
Multiply the Outer terms.	$(3x - 2)(2x + 5)$	$3x \cdot 5 = 15x$
Multiply the Inner terms.	$(3x - 2)(2x + 5)$	$-2 \cdot 2x = -4x$
Multiply the Last terms.	$(3x - 2)(2x + 5)$	$-2 \cdot 5 = -10$

$$\qquad\qquad\qquad\qquad\qquad\quad \text{F} \quad\quad \text{O} \quad\quad \text{I} \quad\quad \text{L}$$

Add the products. $\qquad\qquad (3x - 2)(2x + 5) \quad = \quad 6x^2 + 15x - 4x - 10$

Combine like terms. $\qquad\qquad\qquad\qquad\qquad\qquad\quad = \quad 6x^2 + 11x - 10$

HOW TO Multiply: $(6x - 5)(3x - 4)$

$$(6x - 5)(3x - 4) = 6x(3x) + 6x(-4) + (-5)(3x) + (-5)(-4)$$
$$= 18x^2 - 24x - 15x + 20$$
$$= 18x^2 - 39x + 20$$

Example 4

Multiply: $(4a^3 - 3a + 7)(a - 5)$

Solution

$$
\begin{array}{r}
4a^3 \qquad - 3a + 7 \\
a - 5 \\
\hline
-20a^3 \qquad + 15a - 35 \\
4a^4 \qquad\quad - 3a^2 + 7a \qquad\qquad \\
\hline
4a^4 - 20a^3 - 3a^2 + 22a - 35
\end{array}
$$

• $-5(4a^3 - 3a + 7)$
• $a(4a^3 - 3a + 7)$

You Try It 4

Multiply: $(-2b^2 + 5b - 4)(-3b + 2)$

Your solution

Example 5

Multiply: $(5a - 3b)(2a + 7b)$

Solution

$(5a - 3b)(2a + 7b)$
$= 10a^2 + 35ab - 6ab - 21b^2$ • FOIL
$= 10a^2 + 29ab - 21b^2$

You Try It 5

Multiply: $(3x - 4)(2x - 3)$

Your solution

Example 6

Multiply: $(a^n - 2b^n)(3a^n - b^n)$

Solution

$(a^n - 2b^n)(3a^n - b^n)$
$= 3a^{2n} - a^n b^n - 6a^n b^n + 2b^{2n}$ • FOIL
$= 3a^{2n} - 7a^n b^n + 2b^{2n}$

You Try It 6

Multiply: $(2x^n + y^n)(x^n - 4y^n)$

Your solution

Solutions on p. S15

Objective C **To multiply polynomials that have special products**

Using FOIL, a pattern can be found for the **product of the sum and difference of two terms** [that is, a polynomial that can be expressed in the form $(a + b)(a - b)$] and for the **square of a binomial** [that is, a polynomial that can be expressed in the form $(a + b)^2$].

The Sum and Difference of Two Terms

$$(a + b)(a - b) = a^2 - ab + ab - b^2$$
$$= a^2 - b^2$$

Square of the first term ⟶

Square of the second term ⟶

The Square of a Binomial

$$(a + b)^2 = (a + b)(a + b) = a^2 + ab + ab + b^2$$
$$= a^2 + 2ab + b^2$$

Square of the first term ⟶

Twice the product of the two terms ⟶

Square of the second term ⟶

HOW TO Multiply: $(4x + 3)(4x - 3)$

$(4x + 3)(4x - 3)$ is the sum and difference of the same two terms. The product is the difference of the squares of the terms.

$$(4x + 3)(4x - 3) = (4x)^2 - 3^2$$
$$= 16x^2 - 9$$

HOW TO Expand: $(2x - 3y)^2$

$(2x - 3y)^2$ is the square of a binomial.

$$(2x - 3y)^2 = (2x)^2 + 2(2x)(-3y) + (-3y)^2$$
$$= 4x^2 - 12xy + 9y^2$$

Example 7

Multiply: $(2a - 3)(2a + 3)$

Solution

$(2a - 3)(2a + 3) = 4a^2 - 9$ • The sum and difference of two terms

You Try It 7

Multiply: $(3x - 7)(3x + 7)$

Your solution

Example 8

Multiply: $(x^n + 5)(x^n - 5)$

Solution

$(x^n + 5)(x^n - 5) = x^{2n} - 25$ • The sum and difference of two terms

You Try It 8

Multiply: $(2x^n + 3)(2x^n - 3)$

Your solution

Solutions on p. S15

Example 9

Expand: $(2x + 7y)^2$

Solution

$(2x + 7y)^2 = 4x^2 + 28xy + 49y^2$ • The square of
a binomial

You Try It 9

Expand: $(3x - 4y)^2$

Your solution

Example 10

Expand: $(x^{2n} - 2)^2$

Solution

$(x^{2n} - 2)^2 = x^{4n} - 4x^{2n} + 4$ • The square of
a binomial

You Try It 10

Expand: $(2x^n - 8)^2$

Your solution

Solutions on p. S15

Objective D **To solve application problems**

Example 11

The length of a rectangle is $(2x + 3)$ ft. The width is $(x - 5)$ ft. Find the area of the rectangle in terms of the variable x.

$x - 5$

$2x + 3$

Strategy

To find the area, replace the variables L and W in the equation $A = L \cdot W$ by the given values and solve for A.

Solution

$A = L \cdot W$
$A = (2x + 3)(x - 5)$
$\quad = 2x^2 - 10x + 3x - 15$ • FOIL
$\quad = 2x^2 - 7x - 15$

The area is $(2x^2 - 7x - 15)$ ft^2.

You Try It 11

The base of a triangle is $(2x + 6)$ ft. The height is $(x - 4)$ ft. Find the area of the triangle in terms of the variable x.

Your strategy

Your solution

Solution on p. S15

Example 12

The corners are cut from a rectangular piece of cardboard measuring 8 in. by 12 in. The sides are folded up to make a box. Find the volume of the box in terms of the variable x, where x is the length of the side of the square cut from each corner of the rectangle.

8 in.

12 in.

Strategy

Length of the box: $12 - 2x$
Width of the box: $8 - 2x$
Height of the box: x
To find the volume, replace the variables L, W, and H in the equation $V = L \cdot W \cdot H$ and solve for V.

Solution

$V = L \cdot W \cdot H$
$V = (12 - 2x)(8 - 2x)x$
$\quad = (96 - 24x - 16x + 4x^2)x \quad$ • FOIL
$\quad = (96 - 40x + 4x^2)x$
$\quad = 96x - 40x^2 + 4x^3$
$\quad = 4x^3 - 40x^2 + 96x$

The volume is $(4x^3 - 40x^2 + 96x)$ in^3.

You Try It 12

Find the volume of the rectangular solid shown in the diagram below. All dimensions are in feet.

$5x - 4$

$2x$

x

$7x + 2$

$12x$

Your strategy

Your solution

Example 13

The radius of a circle is $(3x - 2)$ cm. Find the area of the circle in terms of the variable x. Use 3.14 for π.

Strategy

To find the area, replace the variable r in the equation $A = \pi r^2$ by the given value and solve for A.

Solution

$A = \pi r^2$
$A \approx 3.14(3x - 2)^2$
$\quad = 3.14(9x^2 - 12x + 4)$
$\quad = 28.26x^2 - 37.68x + 12.56$

The area is $(28.26x^2 - 37.68x + 12.56)$ cm^2.

You Try It 13

The radius of a circle is $(2x + 3)$ cm. Find the area of the circle in terms of the variable x. Use 3.14 for π.

Your strategy

Your solution

Solutions on p. S15

5.3 Exercises

Objective A To multiply a polynomial by a monomial

For Exercises 1 to 28, simplify.

1. $2x(x - 3)$

2. $2a(2a + 4)$

3. $3x^2(2x^2 - x)$

4. $-4y^2(4y - 6y^2)$

5. $3xy(2x - 3y)$

6. $-4ab(5a - 3b)$

7. $x^n(x + 1)$

8. $y^n(y^{2n} - 3)$

9. $x^n(x^n + y^n)$

10. $x - 2x(x - 2)$

11. $2b + 4b(2 - b)$

12. $-2y(3 - y) + 2y^2$

13. $-2a^2(3a^2 - 2a + 3)$

14. $4b(3b^3 - 12b^2 - 6)$

15. $(-3y^2 - 4y + 2)(y^2)$

16. $(6b^4 - 5b^2 - 3)(-2b^3)$

17. $-5x^2(4 - 3x + 3x^2 + 4x^3)$

18. $-2y^2(3 - 2y - 3y^2 + 2y^3)$

19. $-2x^2y(x^2 - 3xy + 2y^2)$

20. $3ab^2(3a^2 - 2ab + 4b^2)$

21. $x^n(x^{2n} + x^n + x)$

22. $x^{2n}(x^{2n-2} + x^{2n} + x)$

23. $a^{n+1}(a^n - 3a + 2)$

24. $a^{n+4}(a^{n-2} + 5a^2 - 3)$

25. $2y^2 - y[3 - 2(y - 4) - y]$

26. $3x^2 - x[x - 2(3x - 4)]$

27. $2y - 3[y - 2y(y - 3) + 4y]$

28. $4a^2 - 2a[3 - a(2 - a + a^2)]$

29. Given $P(b) = 3b$ and $Q(b) = 3b^4 - 3b^2 + 8$, find $P(b) \cdot Q(b)$.

30. Given $P(x) = -2x^2$ and $Q(x) = 2x^2 - 3x - 7$, find $P(x) \cdot Q(x)$.

Objective B **To multiply two polynomials**

For Exercises 31 to 62, multiply.

31. $(x - 2)(x + 7)$

32. $(y + 8)(y + 3)$

33. $(2y - 3)(4y + 7)$

34. $(5x - 7)(3x - 8)$

35. $2(2x - 3y)(2x + 5y)$

36. $-3(7x - 3y)(2x - 9y)$

37. $(xy + 4)(xy - 3)$

38. $(xy - 5)(2xy + 7)$

39. $(2x^2 - 5)(x^2 - 5)$

40. $(x^2 - 4)(x^2 - 6)$

41. $(5x^2 - 5y)(2x^2 - y)$

42. $(x^2 - 2y^2)(x^2 + 4y^2)$

43. $(x^n + 2)(x^n - 3)$

44. $(x^n - 4)(x^n - 5)$

45. $(2a^n - 3)(3a^n + 5)$

46. $(5b^n - 1)(2b^n + 4)$

47. $(2a^n - b^n)(3a^n + 2b^n)$

48. $(3x^n + b^n)(x^n + 2b^n)$

49. $(x + 5)(x^2 - 3x + 4)$

50. $(a + 2)(a^2 - 3a + 7)$

51. $(2a - 3b)(5a^2 - 6ab + 4b^2)$

52. $(3a + b)(2a^2 - 5ab - 3b^2)$

53. $(2x - 5)(2x^4 - 3x^3 - 2x + 9)$

54. $(2a - 5)(3a^4 - 3a^2 + 2a - 5)$

55. $(x^2 + 2x - 3)(x^2 - 5x + 7)$

56. $(x^2 - 3x + 1)(x^2 - 2x + 7)$

57. $(a - 2)(2a - 3)(a + 7)$

58. $(b - 3)(3b - 2)(b - 1)$

59. $(x^n + 1)(x^{2n} + x^n + 1)$

60. $(a^{2n} - 3)(a^{5n} - a^{2n} + a^n)$

61. $(x^n + y^n)(x^n - 2x^ny^n + 3y^n)$

62. $(x^n - y^n)(x^{2n} - 3x^ny^n - y^{2n})$

63. Given $P(y) = 2y^2 - 1$ and $Q(y) = y^3 - 5y^2 - 3$, find $P(y) \cdot Q(y)$.

64. Given $P(b) = 2b^2 - 3$ and $Q(b) = 3b^2 - 3b + 6$, find $P(b) \cdot Q(b)$.

Objective C **To multiply polynomials that have special products**

For Exercises 65 to 88, simplify or expand.

65. $(3x - 2)(3x + 2)$

66. $(4y + 1)(4y - 1)$

67. $(6 - x)(6 + x)$

68. $(10 + b)(10 - b)$

69. $(2a - 3b)(2a + 3b)$

70. $(5x - 7y)(5x + 7y)$

71. $(x^2 + 1)(x^2 - 1)$

72. $(x^2 + y^2)(x^2 - y^2)$

73. $(x^n + 3)(x^n - 3)$

74. $(x^n + y^n)(x^n - y^n)$

75. $(x - 5)^2$

76. $(y + 2)^2$

77. $(3a + 5b)^2$

78. $(5x - 4y)^2$

79. $(x^2 - 3)^2$

80. $(x^2 + y^2)^2$

81. $(2x^2 - 3y^2)^2$

82. $(x^n - 1)^2$

83. $(a^n - b^n)^2$

84. $(2x^n + 5y^n)^2$

85. $y^2 - (x - y)^2$

86. $a^2 + (a + b)^2$

87. $(x - y)^2 - (x + y)^2$

88. $(a + b)^2 + (a - b)^2$

Objective D **To solve application problems**

89. **Geometry** The length of a rectangle is $(3x - 2)$ ft. The width is $(x + 4)$ ft. Find the area of the rectangle in terms of the variable x.

90. **Geometry** The base of a triangle is $(x - 4)$ ft. The height is $(3x + 2)$ ft. Find the area of the triangle in terms of the variable x.

91. **Geometry** Find the area of the figure shown below. All dimensions given are in meters.

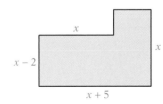

92. **Geometry** Find the area of the figure shown below. All dimensions given are in feet.

93. Geometry The length of the side of a cube is $(x + 3)$ cm. Find the volume of the cube in terms of the variable x.

94. Geometry The length of a box is $(3x + 2)$ cm, the width is $(x - 4)$ cm, and the height is x cm. Find the volume of the box in terms of the variable x.

95. Geometry Find the volume of the figure shown below. All dimensions given are in inches.

96. Geometry Find the volume of the figure shown below. All dimensions given are in centimeters.

97. Geometry The radius of a circle is $(5x + 4)$ in. Find the area of the circle in terms of the variable x. Use 3.14 for π.

98. Geometry The radius of a circle is $(x - 2)$ in. Find the area of the circle in terms of the variable x. Use 3.14 for π.

APPLYING THE CONCEPTS

99. Find the product.
 a. $(a - b)(a^2 + ab + b^2)$
 b. $(x + y)(x^2 - xy + y^2)$

100. Correct the error in each of the following.
 a. $(x + 3)^2 = x^2 + 9$
 b. $(a - b)^2 = a^2 - b^2$

101. For what value of k is the given equation an identity?
 a. $(3x - k)(2x + k) = 6x^2 + 5x - k^2$
 b. $(4x + k)^2 = 16x^2 + 8x + k^2$

102. Complete.
 a. If $m = n + 1$, then $\dfrac{a^m}{a^n} = $ _____ .
 b. If $m = n + 2$, then $\dfrac{a^m}{a^n} = $ _____ .

103. Subtract the product of $4a + b$ and $2a - b$ from $9a^2 - 2ab$.

104. Subtract the product of $5x - y$ and $x + 3y$ from $6x^2 + 12xy - 2y^2$.

5.4 Division of Polynomials

Objective A **To divide a polynomial by a monomial**

As shown below, $\dfrac{6 + 4}{2}$ can be simplified by first adding the terms in the numerator and then dividing the result by the denominator. It can also be simplified by first dividing each term in the numerator by the denominator and then adding the results.

$$\frac{6 + 4}{2} = \frac{10}{2} = 5 \qquad\qquad \frac{6 + 4}{2} = \frac{6}{2} + \frac{4}{2} = 3 + 2 = 5$$

It is this second method that is used to divide a polynomial by a monomial: Divide each term in the numerator by the denominator, and then write the sum of the quotients.

To divide $\dfrac{6x^2 + 4x}{2x}$, divide each term of the polynomial $6x^2 + 4x$ by the monomial $2x$. Then simplify each quotient.

> **TAKE NOTE**
> Recall that the fraction bar can be read "divided by."

$$\frac{6x^2 + 4x}{2x} = \frac{6x^2}{2x} + \frac{4x}{2x}$$ • Divide each term in the numerator by the denominator.

$$= 3x + 2$$ • Simplify each quotient.

We can check this quotient by multiplying it by the divisor.

$$2x(3x + 2) = 6x^2 + 4x$$ • The product is the dividend. The quotient checks.

HOW TO Divide and check: $\dfrac{16x^5 - 8x^3 + 4x}{2x}$

$$\frac{16x^5 - 8x^3 + 4x}{2x} = \frac{16x^5}{2x} - \frac{8x^3}{2x} + \frac{4x}{2x}$$ • Divide each term in the numerator by the denominator.

$$= 8x^4 - 4x^2 + 2$$ • Simplify each quotient.

Check:

$$2x(8x^4 - 4x^2 + 2) = 16x^5 - 8x^3 + 4x$$ • The quotient checks.

Example 1

Divide and check: $\dfrac{6x^3 - 3x^2 + 9x}{3x}$

Solution

$$\frac{6x^3 - 3x^2 + 9x}{3x}$$

$$= \frac{6x^3}{3x} - \frac{3x^2}{3x} + \frac{9x}{3x}$$ • Divide each term in the numerator by the denominator.

$$= 2x^2 - x + 3$$ • Simplify each quotient.

Check: $3x(2x^2 - x + 3) = 6x^3 - 3x^2 + 9x$

You Try It 1

Divide and check: $\dfrac{4x^3y + 8x^2y^2 - 4xy^3}{2xy}$

Your solution

Solution on p. S15

Objective B **To divide polynomials**

The division method illustrated in Objective A is appropriate only when the divisor is a monomial. To divide two polynomials in which the divisor is not a monomial, use a method similar to that used for division of whole numbers.

To check division of polynomials, use

$$\textbf{Dividend = (quotient} \times \textbf{divisor) + remainder}$$

HOW TO Divide: $(x^2 + 5x - 7) \div (x + 3)$

Step 1

$$\begin{array}{r} x \phantom{{}+5x-7} \\ x + 3 \overline{)x^2 + 5x - 7} \\ \underline{x^2 + 3x} \downarrow \\ 2x - 7 \end{array}$$

Think: $x\overline{)x^2} = \dfrac{x^2}{x} = x$

Multiply: $x(x + 3) = x^2 + 3x$

Subtract: $(x^2 + 5x) - (x^2 + 3x) = 2x$

Step 2

$$\begin{array}{r} x + 2 \\ x + 3 \overline{)x^2 + 5x - 7} \\ \underline{x^2 + 3x} \\ 2x - 7 \\ \underline{2x + 6} \\ -13 \end{array}$$

Think: $x\overline{)2x} = \dfrac{2x}{x} = 2$

Multiply: $2(x + 3) = 2x + 6$

Subtract: $(2x - 7) - (2x + 6) = -13$

The remainder is -13.

Check: $(x + 2)(x + 3) + (-13) = x^2 + 3x + 2x + 6 - 13 = x^2 + 5x - 7$

$(x^2 + 5x - 7) \div (x + 3) = x + 2 - \dfrac{13}{x + 3}$

HOW TO Divide: $\dfrac{6 - 6x^2 + 4x^3}{2x + 3}$

Arrange the terms in descending order. Note that there is no term of x in $4x^3 - 6x^2 + 6$. Insert a zero for the missing term so that like terms will be in the same columns.

$$\begin{array}{r} 2x^2 - 6x + 9 \\ 2x + 3 \overline{)4x^3 - 6x^2 + 0x + 6} \\ \underline{4x^3 + 6x^2} \\ -12x^2 + 0x \\ \underline{-12x^2 - 18x} \\ 18x + 6 \\ \underline{18x + 27} \\ -21 \end{array}$$

$\dfrac{4x^3 - 6x^2 + 6}{2x + 3} = 2x^2 - 6x + 9 - \dfrac{21}{2x + 3}$

Example 2

Divide: $\dfrac{12x^2 - 11x + 10}{4x - 5}$

Solution

$$
\begin{array}{r}
3x + 1 \\
4x - 5 \overline{)12x^2 - 11x + 10} \\
\underline{12x^2 - 15x} \\
4x + 10 \\
\underline{4x - 5} \\
15
\end{array}
$$

$$\frac{12x^2 - 11x + 10}{4x - 5} = 3x + 1 + \frac{15}{4x - 5}$$

You Try It 2

Divide: $\dfrac{15x^2 + 17x - 20}{3x + 4}$

Your solution

Example 3

Divide: $\dfrac{x^3 + 1}{x + 1}$

Solution

$$
\begin{array}{r}
x^2 - x + 1 \\
x + 1 \overline{)x^3 + 0x^2 + 0x + 1} \\
\underline{x^3 + x^2} \\
-x^2 + 0x \\
\underline{-x^2 - x} \\
x + 1 \\
\underline{x + 1} \\
0
\end{array}
$$

• **Insert zeros for the missing terms.**

$$\frac{x^3 + 1}{x + 1} = x^2 - x + 1$$

You Try It 3

Divide: $\dfrac{3x^3 + 8x^2 - 6x + 2}{3x - 1}$

Your solution

Example 4

Divide:
$(2x^4 - 7x^3 + 3x^2 + 4x - 5) \div (x^2 - 2x - 2)$

Solution

$$
\begin{array}{r}
2x^2 - 3x + 1 \\
x^2 - 2x - 2 \overline{)2x^4 - 7x^3 + 3x^2 + 4x - 5} \\
\underline{2x^4 - 4x^3 - 4x^2} \\
-3x^3 + 7x^2 + 4x \\
\underline{-3x^3 + 6x^2 + 6x} \\
x^2 - 2x - 5 \\
\underline{x^2 - 2x - 2} \\
-3
\end{array}
$$

$(2x^4 - 7x^3 + 3x^2 + 4x - 5) \div (x^2 - 2x - 2)$

$$= 2x^2 - 3x + 1 - \frac{3}{x^2 - 2x - 2}$$

You Try It 4

Divide:
$(3x^4 - 11x^3 + 16x^2 - 16x + 8) \div (x^2 - 3x + 2)$

Your solution

Solutions on pp. S15–16

Objective C **To divide polynomials by using synthetic division**

Synthetic division is a shorter method of dividing a polynomial by a binomial of the form $x - a$.

Divide $(3x^2 - 4x + 6) \div (x - 2)$ by using long division.

$$
\begin{array}{r}
3x + 2 \\
x - 2 \overline{)3x^2 - 4x + 6} \\
\underline{3x^2 - 6x} \\
2x + 6 \\
\underline{2x - 4} \\
10
\end{array}
$$

$(3x^2 - 4x + 6) \div (x - 2) = 3x + 2 + \dfrac{10}{x - 2}$

The variables can be omitted because the position of a term indicates the power of the term.

$$
\begin{array}{r}
3 \quad\; 2 \\
-2 \overline{)3 \; -4 \quad 6} \\
\underline{3 \; -6} \\
2 \quad\; 6 \\
\underline{2 \; -4} \\
10
\end{array}
$$

Each number shown in color is exactly the same as the number above it. Removing the colored numbers condenses the vertical spacing.

$$
\begin{array}{r}
3 \quad\; 2 \\
-2 \overline{)3 \; -4 \quad 6} \\
\underline{-6 \; -4} \\
2 \quad 10
\end{array}
$$

The number in color in the top row is the same as the one in the bottom row. Writing the 3 from the top row in the bottom row allows the spacing to be condensed even further.

$$
\begin{array}{c|ccc}
-2 & 3 & -4 & 6 \\
 & & -6 & -4 \\
\hline
 & 3 & 2 & 10
\end{array}
$$

$\underbrace{\qquad\qquad}_{\text{Terms of the quotient}} \qquad \underbrace{\quad}_{\text{Remainder}}$

Because the degree of the dividend $(3x^2 - 4x + 6)$ is 2 and the degree of the divisor $(x - 2)$ is 1, the degree of the quotient is $2 - 1 = 1$. This means that, using the terms of the quotient given above, that quotient is $3x + 2$. The remainder is 10.

In general, the degree of the quotient of two polynomials is the difference between the degree of the dividend and the degree of the divisor.

By replacing the constant term in the divisor by its additive inverse, we may add rather than subtract terms. This is illustrated in the following example.

HOW TO Divide: $(3x^3 + 6x^2 - x - 2) \div (x + 3)$

The additive inverse of the binomial constant

Coefficients of the polynomial

-3	3	6	-1	-2
	↓			
	3			

• Bring down the 3.

-3	3	6	-1	-2
		-9		
	3	-3		

• Multiply $-3(3)$ and add the product to 6.

-3	3	6	-1	-2
		-9	9	
	3	-3	8	

• Multiply $-3(-3)$ and add the product to -1.

-3	3	6	-1	-2
		-9	9	-24
	3	-3	8	-26

• Multiply $-3(8)$ and add the product to -2.

Terms of the quotient Remainder

The degree of the dividend is 3 and the degree of the divisor is 1. Therefore, the degree of the quotient is $3 - 1 = 2$.

$$(3x^3 + 6x^2 - x - 2) \div (x + 3) = 3x^2 - 3x + 8 - \frac{26}{x + 3}$$

HOW TO Divide: $(2x^3 - x + 2) \div (x - 2)$

The additive inverse of the binomial constant

Coefficients of the polynomial

2	2	0	-1	2
	↓			
	2			

• Insert a 0 for the missing term and bring down the 2.

2	2	0	-1	2
		4		
	2	4		

• Multiply $2(2)$ and add the product to 0.

2	2	0	-1	2
		4	8	
	2	4	7	

• Multiply $2(4)$ and add the product to -1.

2	2	0	-1	2
		4	8	14
	2	4	7	16

• Multiply $2(7)$ and add the product to 2.

Terms of the quotient Remainder

$$(2x^3 - x + 2) \div (x - 2) = 2x^2 + 4x + 7 + \frac{16}{x - 2}$$

Example 5

Divide: $(7 - 3x + 5x^2) \div (x - 1)$

Solution

Arrange the coefficients in decreasing powers of x.

$$
\begin{array}{r|rrr}
1 & 5 & -3 & 7 \\
 & & 5 & 2 \\
\hline
 & 5 & 2 & 9
\end{array}
$$

$(5x^2 - 3x + 7) \div (x - 1) = 5x + 2 + \dfrac{9}{x - 1}$

You Try It 5

Divide: $(6x^2 + 8x - 5) \div (x + 2)$

Your solution

Example 6

Divide: $(2x^3 + 4x^2 - 3x + 12) \div (x + 4)$

Solution

$$
\begin{array}{r|rrrr}
-4 & 2 & 4 & -3 & 12 \\
 & & -8 & 16 & -52 \\
\hline
 & 2 & -4 & 13 & -40
\end{array}
$$

$(2x^3 + 4x^2 - 3x + 12) \div (x + 4)$

$\quad = 2x^2 - 4x + 13 - \dfrac{40}{x + 4}$

You Try It 6

Divide: $(5x^3 - 12x^2 - 8x + 16) \div (x - 2)$

Your solution

Example 7

Divide: $(3x^4 - 8x^2 + 2x + 1) \div (x + 2)$

Solution

Insert a zero for the missing term.

$$
\begin{array}{r|rrrrr}
-2 & 3 & 0 & -8 & 2 & 1 \\
 & & -6 & 12 & -8 & 12 \\
\hline
 & 3 & -6 & 4 & -6 & 13
\end{array}
$$

$(3x^4 - 8x^2 + 2x + 1) \div (x + 2)$

$\quad = 3x^3 - 6x^2 + 4x - 6 + \dfrac{13}{x + 2}$

You Try It 7

Divide: $(2x^4 - 3x^3 - 8x^2 - 2) \div (x - 3)$

Your solution

Solutions on p. S16

Objective D **To evaluate a polynomial function using synthetic division**

A polynomial can be evaluated by using synthetic division. Consider the polynomial $P(x) = 2x^4 - 3x^3 + 4x^2 - 5x + 1$. One way to evaluate the polynomial when $x = 2$ is to replace x by 2 and then simplify the numerical expression.

$P(x) = 2x^4 - 3x^3 + 4x^2 - 5x + 1$

$P(2) = 2(2)^4 - 3(2)^3 + 4(2)^2 - 5(2) + 1$

$\quad = 2(16) - 3(8) + 4(4) - 5(2) + 1$

$\quad = 32 - 24 + 16 - 10 + 1$

$\quad = 15$

Now use synthetic division to divide $(2x^4 - 3x^3 + 4x^2 - 5x + 1) \div (x - 2)$.

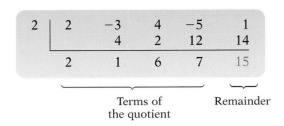

$$\begin{array}{c|ccccc} 2 & 2 & -3 & 4 & -5 & 1 \\ & & 4 & 2 & 12 & 14 \\ \hline & 2 & 1 & 6 & 7 & 15 \end{array}$$

Terms of Remainder
the quotient

Note that the remainder is 15, which is the same value as $P(2)$. This is not a coincidence. The following theorem states that this situation is always true.

> **Remainder Theorem**
>
> If the polynomial $P(x)$ is divided by $x - a$, the remainder is $P(a)$.

HOW TO Evaluate $P(x) = x^4 - 3x^2 + 4x - 5$ when $x = -2$ by using the Remainder Theorem.

The value at which the polynomial is evaluated

$$\begin{array}{c|ccccc} -2 & 1 & 0 & -3 & 4 & -5 \\ & & -2 & 4 & -2 & -4 \\ \hline & 1 & -2 & 1 & 2 & -9 \end{array}$$

• A 0 is inserted for the x^3 term.

←— The remainder

$P(-2) = -9$

Example 8

Use synthetic division to evaluate $P(x) = x^2 - 6x + 4$ when $x = 3$.

Solution

$$\begin{array}{c|ccc} 3 & 1 & -6 & 4 \\ & & 3 & -9 \\ \hline & 1 & -3 & -5 \end{array}$$

$P(3) = -5$

You Try It 8

Use synthetic division to evaluate $P(x) = 2x^2 - 3x - 5$ when $x = 2$.

Your solution

Example 9

Use synthetic division to evaluate $P(x) = -x^4 + 3x^3 + 2x^2 - x - 5$ when $x = -2$.

Solution

$$\begin{array}{c|ccccc} -2 & -1 & 3 & 2 & -1 & -5 \\ & & 2 & -10 & 16 & -30 \\ \hline & -1 & 5 & -8 & 15 & -35 \end{array}$$

$P(-2) = -35$

You Try It 9

Use synthetic division to evaluate $P(x) = 2x^3 - 5x^2 + 7$ when $x = -3$.

Your solution

Solutions on p. S16

5.4 Exercises

Objective A **To divide a polynomial by a monomial**

For Exercises 1 to 12, divide and check.

1. $\dfrac{3x^2 - 6x}{3x}$

2. $\dfrac{10y^2 - 6y}{2y}$

3. $\dfrac{5x^2 - 10x}{-5x}$

4. $\dfrac{3y^2 - 27y}{-3y}$

5. $\dfrac{5x^2y^2 + 10xy}{5xy}$

6. $\dfrac{8x^2y^2 - 24xy}{8xy}$

7. $\dfrac{x^3 + 3x^2 - 5x}{x}$

8. $\dfrac{a^3 - 5a^2 + 7a}{a}$

9. $\dfrac{9b^5 + 12b^4 + 6b^3}{3b^2}$

10. $\dfrac{a^8 - 5a^5 - 3a^3}{a^2}$

11. $\dfrac{a^5b - 6a^3b + ab}{ab}$

12. $\dfrac{5c^3d + 10c^2d^2 - 15cd^3}{5cd}$

Objective B **To divide polynomials**

For Exercises 13 to 34, divide by using long division.

13. $(x^2 + 3x - 40) \div (x - 5)$

14. $(x^2 - 14x + 24) \div (x - 2)$

15. $(x^3 - 3x + 2) \div (x - 3)$

16. $(x^3 + 4x^2 - 8) \div (x + 4)$

17. $(6x^2 + 13x + 8) \div (2x + 1)$

18. $(12x^2 + 13x - 14) \div (3x - 2)$

19. $(10x^2 + 9x - 5) \div (2x - 1)$

20. $(18x^2 - 3x + 2) \div (3x + 2)$

21. $(8x^3 - 9) \div (2x - 3)$

22. $(64x^3 + 4) \div (4x + 2)$

23. $(6x^4 - 13x^2 - 4) \div (2x^2 - 5)$

24. $(12x^4 - 11x^2 + 10) \div (3x^2 + 1)$

25. $\dfrac{-10 - 33x + 3x^3 - 8x^2}{3x + 1}$

26. $\dfrac{10 - 49x + 38x^2 - 8x^3}{1 - 4x}$

27. $\dfrac{x^3 - 5x^2 + 7x - 4}{x - 3}$

28. $\dfrac{2x^3 - 3x^2 + 6x + 4}{2x + 1}$

29. $\dfrac{16x^2 - 13x^3 + 2x^4 + 20 - 9x}{x - 5}$

30. $\dfrac{x - x^2 + 5x^3 + 3x^4 - 2}{x + 2}$

31. $\dfrac{2x^3 + 4x^2 - x + 2}{x^2 + 2x - 1}$

32. $\dfrac{3x^3 - 2x^2 + 5x - 4}{x^2 - x + 3}$

33. $\dfrac{x^4 + 2x^3 - 3x^2 - 6x + 2}{x^2 - 2x - 1}$

34. $\dfrac{x^4 - 3x^3 + 4x^2 - x + 1}{x^2 + x - 3}$

35. Given $Q(x) = 2x + 1$ and $P(x) = 2x^3 + x^2 + 8x + 7$, find $\dfrac{P(x)}{Q(x)}$.

36. Given $Q(x) = 3x - 2$ and $P(x) = 3x^3 - 2x^2 + 3x - 5$, find $\dfrac{P(x)}{Q(x)}$.

Objective C **To divide polynomials by using synthetic division**

For Exercises 37 to 52, divide by using synthetic division.

37. $(2x^2 - 6x - 8) \div (x + 1)$

38. $(3x^2 + 19x + 20) \div (x + 5)$

39. $(3x^2 - 14x + 16) \div (x - 2)$

40. $(4x^2 - 23x + 28) \div (x - 4)$

41. $(3x^2 - 4) \div (x - 1)$

42. $(4x^2 - 8) \div (x - 2)$

43. $(2x^3 - x^2 + 6x + 9) \div (x + 1)$

44. $(3x^3 + 10x^2 + 6x - 4) \div (x + 2)$

45. $(18 + x - 4x^3) \div (2 - x)$

46. $(12 - 3x^2 + x^3) \div (x + 3)$

47. $(2x^3 + 5x^2 - 5x + 20) \div (x + 4)$

48. $(5x^3 + 3x^2 - 17x + 6) \div (x + 2)$

49. $\dfrac{5 + 5x - 8x^2 + 4x^3 - 3x^4}{2 - x}$

50. $\dfrac{3 - 13x - 5x^2 + 9x^3 - 2x^4}{3 - x}$

51. $\dfrac{3x^4 + 3x^3 - x^2 + 3x + 2}{x + 1}$

52. $\dfrac{4x^4 + 12x^3 - x^2 - x + 2}{x + 3}$

53. Given $Q(x) = x - 2$ and $P(x) = 3x^2 - 5x + 6$, find $\dfrac{P(x)}{Q(x)}$.

54. Given $Q(x) = x + 5$ and $P(x) = 2x^2 + 7x - 12$, find $\dfrac{P(x)}{Q(x)}$.

Objective D **To evaluate a polynomial function using synthetic division**

For Exercises 55 to 72, use the Remainder Theorem to evaluate the polynomial function.

55. $P(x) = 2x^2 - 3x - 1$; $P(3)$

56. $Q(x) = 3x^2 - 5x - 1$; $Q(2)$

57. $R(x) = x^3 - 2x^2 + 3x - 1; R(4)$

58. $F(x) = x^3 + 4x^2 - 3x + 2; F(3)$

59. $P(z) = 2z^3 - 4z^2 + 3z - 1; P(-2)$

60. $R(t) = 3t^3 + t^2 - 4t + 2; R(-3)$

61. $Z(p) = 2p^3 - p^2 + 3; Z(-3)$

62. $P(y) = 3y^3 + 2y^2 - 5; P(-2)$

63. $Q(x) = x^4 + 3x^3 - 2x^2 + 4x - 9; Q(2)$

64. $Y(z) = z^4 - 2z^3 - 3z^2 - z + 7; Y(3)$

65. $F(x) = 2x^4 - x^3 + 2x - 5; F(-3)$

66. $Q(x) = x^4 - 2x^3 + 4x - 2; Q(-2)$

67. $P(x) = x^3 - 3; P(5)$

68. $S(t) = 4t^3 + 5; S(-4)$

69. $R(t) = 4t^4 - 3t^2 + 5; R(-3)$

70. $P(z) = 2z^4 + z^2 - 3; P(-4)$

71. $Q(x) = x^5 - 4x^3 - 2x^2 + 5x - 2; Q(2)$

72. $R(x) = 2x^5 - x^3 + 4x - 1; R(-2)$

APPLYING THE CONCEPTS

73. Divide by using long division.
 a. $\dfrac{a^3 + b^3}{a + b}$

 b. $\dfrac{x^5 + y^5}{x + y}$

 c. $\dfrac{x^6 - y^6}{x + y}$

74. For what value of k will the remainder be zero?
 a. $(x^3 - x^2 - 3x + k) \div (x + 3)$

 b. $(2x^3 - x + k) \div (x - 1)$

75. Show how synthetic division can be modified so that the divisor can be of the form $ax + b$.

 Factoring Polynomials

To factor a monomial from a polynomial

One number is a **factor** of another when it can be divided into that other number with a remainder of zero.

The greatest common factor (GCF) of two or more monomials is the product of the common factors with the smallest exponents.

$$16a^4b = 2^4a^4b$$
$$40a^2b^5 = 2^3 \cdot 5a^2b^5$$
$$GCF = 2^3a^2b = 8a^2b$$

Note that the exponent of each variable in the GCF is the same as the smallest exponent of the variable in either of the monomials.

Factoring a polynomial means writing the polynomial as a product of other polynomials.

In the example at the right, $3x$ is the GCF of the terms $3x^2$ and $6x$. $3x$ is a **common monomial factor** of the terms of the binomial. $x - 2$ is a **binomial factor** of $3x^2 - 6x$.

HOW TO Factor: $4x^3y^2 + 12x^3y + 20xy^3$

The GCF of $4x^3y^2$, $12x^3y$, and $20xy^3$ is $4xy$.

$4x^3y^2 + 12x^3y + 20xy^3$

$$= 4xy\left(\frac{4x^3y^2}{4xy} + \frac{12x^3y}{4xy} + \frac{20xy^3}{4xy}\right)$$

$$= 4xy(x^2y + 3x^2 + 5y^2)$$

- Find the GCF of the terms of the polynomial.

- Factor the GCF from each term of the polynomial. Think of this as dividing each term of the polynomial by the GCF.

Example 1

Factor: $4x^2y^2 - 6xy^2 + 12xy^3$

Solution
The GCF of $4x^2y^2$, $6xy^2$, and $12xy^3$ is $2xy^2$.

$4x^2y^2 - 6xy^2 + 12xy^3 = 2xy^2(2x - 3 + 6y)$

You Try It 1

Factor: $3x^3y - 6x^2y^2 - 3xy^3$

Your solution

Solution on p. S16

Example 2

Factor: $x^{2n} + x^{n+1} + x^n$

Solution

The GCF of x^{2n}, x^{n+1}, and x^n is x^n.

$x^{2n} + x^{n+1} + x^n = x^n(x^n + x + 1)$

You Try It 2

Factor: $6t^{2n} - 9t^n$

Your solution

Solution on p. S16

Objective B **To factor by grouping**

In the examples at the right, the binomials in parentheses are binomial factors.

$$4x^4(2x - 3)$$
$$-2r^2s(5r + 2s)$$

The Distributive Property is used to factor a common binomial factor from an expression.

> **HOW TO** Factor: $4a(2b + 3) - 5(2b + 3)$
>
> The common binomial factor is $(2b + 3)$. Use the Distributive Property to write the expression as a product of factors.
>
> $4a(2b + 3) - 5(2b + 3) = (2b + 3)(4a - 5)$

Consider the binomial $y - x$. Factoring -1 from this binomial gives

$$y - x = -(x - y)$$

This equation is used to factor a common binomial from an expression.

TAKE NOTE

For the simplification at the right,

$6r(r - s) - 7(s - r)$
$= 6r(r - s) - 7[(-1)(r - s)]$
$= 6r(r - s) + 7(r - s)$

> **HOW TO** Factor: $6r(r - s) - 7(s - r)$
>
> $6r(r - s) - 7(s - r) = 6r(r - s) + 7(r - s)$ • $s - r = -(r - s)$
> $\qquad\qquad\qquad\qquad = (r - s)(6r + 7)$

Some polynomials can be factored by grouping terms so that a common binomial factor is found. This is called **factoring by grouping.**

> **HOW TO** Factor: $3xz - 4yz - 3xa + 4ya$
>
> $3xz - 4yz - 3xa + 4ya$
> $\quad = (3xz - 4yz) - (3xa - 4ya)$
>
>
> $\quad = z(3x - 4y) - a(3x - 4y)$
> $\quad = (3x - 4y)(z - a)$

• Group the first two terms and the last two terms. Note that $-3xa + 4ya = -(3xa - 4ya)$.
• Factor the GCF from each group.
• Write the expression as the product of factors.

HOW TO Factor: $8y^2 + 4y - 6ay - 3a$

$8y^2 + 4y - 6ay - 3a$
$= (8y^2 + 4y) - (6ay + 3a)$

$= 4y(2y + 1) - 3a(2y + 1)$
$= (2y + 1)(4y - 3a)$

- Group the first two terms and the last two terms. Note that $-6ay - 3a = -(6ay + 3a)$.
- Factor the GCF from each group.
- Write the expression as the product of factors.

Example 3

Factor: $x^2(5y - 2) - 7(5y - 2)$

Solution
$x^2(5y - 2) - 7(5y - 2)$
$= (5y - 2)(x^2 - 7)$

- The common binomial factor is $(5y - 2)$.

You Try It 3

Factor: $3(6x - 7y) - 2x^2(6x - 7y)$

Your solution

Example 4

Factor: $15x^2 + 6x - 5xz - 2z$

Solution
$15x^2 + 6x - 5xz - 2z$
$= (15x^2 + 6x) - (5xz + 2z)$
$= 3x(5x + 2) - z(5x + 2)$
$= (5x + 2)(3x - z)$

You Try It 4

Factor: $4a^2 - 6a - 6ax + 9x$

Your solution

Solutions on p. S16

Objective C **To factor a trinomial of the form $x^2 + bx + c$**

A **quadratic trinomial** is a trinomial of the form $ax^2 + bx + c$, where a, b, and c are nonzero integers. The degree of a quadratic trinomial is 2. Here are examples of quadratic trinomials:

$$4x^2 - 3x - 7 \qquad z^2 + z + 10 \qquad 2y^2 + 4y - 9$$
$$(a = 4, b = -3, c = -7) \qquad (a = 1, b = 1, c = 10) \qquad (a = 2, b = 4, c = -9)$$

Factoring a quadratic trinomial means expressing the trinomial as the product of two binomials. For example,

Trinomial		Factored Form
$2x^2 - x - 1$	$=$	$(2x + 1)(x - 1)$
$y^2 - 3y + 2$	$=$	$(y - 1)(y - 2)$

In this objective, trinomials of the form $x^2 + bx + c$ ($a = 1$) will be factored. The next objective deals with trinomials where $a \neq 1$.

The method by which factors of a trinomial are found is based on FOIL. Consider the following binomial products, noting the relationship between the constant terms of the binomials and the terms of the trinomial.

> sum of binomial constants

> product of binomial constants

$$F \quad O \quad I \quad L$$

$$(x + 4)(x + 5) = x \cdot x + 5x + 4x + 4 \cdot 5 \quad = x^2 + 9x + 20$$
$$(x - 6)(x + 8) = x \cdot x + 8x - 6x + (-6) \cdot 8 = x^2 + 2x - 48$$
$$(x - 3)(x - 2) = x \cdot x - 2x - 3x + (-3)(-2) = x^2 - 5x + 6$$

Observe two important points from these examples.

1. The constant term of the trinomial is the *product* of the constant terms of the binomials. The coefficient of x in the trinomial is the *sum* of the constant terms of the binomials.

2. When the constant term of the trinomial is positive, the constant terms of the binomials have the *same* sign. When the constant term of the trinomial is negative, the constant terms of the binomials have *opposite* signs.

HOW TO Factor: $x^2 - 7x + 12$

The constant term is positive. The signs of the binomial constants will be the same.

Find two negative factors of 12 whose sum is -7.

Negative Factors of 12	Sum
$-1, -12$	-13
$-2, -6$	-8
$-3, -4$	-7

Write the trinomial in factored form. $\quad x^2 - 7x + 12 = (x - 3)(x - 4)$

Check: $(x - 3)(x - 4) = x^2 - 4x - 3x + 12 = x^2 - 7x + 12$

HOW TO Factor: $y^2 + 10y - 24$

The constant term is negative. The signs of the binomial constants will be opposite.

Find two factors of -24 that have opposite signs and whose sum is 10. All of the possible factors are shown at the right. In practice, once the correct pair is found, the remaining choices need not be checked.

Factors of -24	Sum
$-1, \quad 24$	23
$1, -24$	-23
$-2, \quad 12$	10
$2, -12$	-10
$-3, \quad 8$	5
$3, -8$	-5
$-4, \quad 6$	2
$4, -6$	-2

Write the trinomial in factored form. $\quad y^2 + 10y - 24 = (y - 2)(y + 12)$

Check: $(y - 2)(y + 12) = y^2 + 12y - 2y - 24 = y^2 + 10y - 24$

When only integers are used, some trinomials do not factor. For example, to factor $x^2 + 11x + 5$, it would be necessary to find two positive integers whose product is 5 and whose sum is 11. This is not possible, because the only positive factors of 5 are 1 and 5, and the sum of 1 and 5 is 6. The polynomial $x^2 + 11x + 5$ is a **prime polynomial.** Such a polynomial is said to be **nonfactorable over the integers.** Binomials of the form $x + a$ or $x - a$ are also prime polynomials.

Example 5

Factor: $x^2 + 2x - 4$

Solution

There are no factors of -4 whose signs are opposite and whose sum is 2.

The trinomial is nonfactorable over the integers.

You Try It 5

Factor: $x^2 - x - 20$

Your solution

Example 6

Factor: $10 - 3x - x^2$

Solution

When the coefficient of x^2 is -1, factoring -1 from the trinomial may make factoring the trinomial easier.

$$10 - 3x - x^2 = -x^2 - 3x + 10$$
$$= -(x^2 + 3x - 10)$$
$$= -(x - 2)(x + 5)$$

You Try It 6

Factor: $x^2 + 5xy + 6y^2$

Your solution

Solutions on p. S16

Objective D **To factor $ax^2 + bx + c$**

There are various methods of factoring trinomials of the form $ax^2 + bx + c$, where $a \neq 1$. Factoring by using trial factors and factoring by grouping will be discussed in this objective. Factoring by using trial factors is illustrated first.

To use the trial factor method, use the factors of a and the factors of c to write all of the possible binomial factors of the trinomial. Then use FOIL to determine the correct factorization. To reduce the number of trial factors that must be considered, remember the following.

1. Use the signs of the constant term and the coefficient of x in the trinomial to determine the signs of the binomial factors. If the constant term is positive, the signs of the binomial factors will be the same as the sign of the coefficient of x in the trinomial. If the sign of the constant term is negative, the constant terms in the binomials will have opposite signs.

2. If the terms of the trinomial do not have a common factor, then the terms in either one of the binomial factors will not have a common factor.

HOW TO Factor: $3x^2 - 8x + 4$

The terms have no common factor.
The constant term is positive.
The coefficient of x is negative.
The binomial constants will be negative.

Positive Factors of 3 (coefficient of x^2)	Negative Factors of 4 (constant term)
1, 3	$-1, -4$ $-2, -2$

Write trial factors. Use the *O*uter and *I*nner products of FOIL to determine the middle term of the trinomial.

Trial Factors	Middle Term
$(x - 1)(3x - 4)$	$-4x - 3x = -7x$
$(x - 4)(3x - 1)$	$-x - 12x = -13x$
$(x - 2)(3x - 2)$	$-2x - 6x = -8x$

Write the trinomial in factored form.

$$3x^2 - 8x + 4 = (x - 2)(3x - 2)$$

HOW TO Factor: $5 - 3x - 2x^2$

Factor -1 from the trinomial.
$5 - 3x - 2x^2 = -(2x^2 + 3x - 5)$
The constant term, -5, is negative; the signs of the binomial constants will be opposites.

Positive Factors of 2 (coefficient of x^2)	Factors of -5 (constant term)
1, 2	1, -5 $-1, 5$

Write trial factors. Use the *O*uter and *I*nner products of FOIL to determine the middle term of the trinomial.

Trial Factors	Middle Term
$(x + 1)(2x - 5)$	$-5x + 2x = -3x$
$(x - 5)(2x + 1)$	$x - 10x = -9x$
$(x - 1)(2x + 5)$	$5x - 2x = 3x$
$(x + 5)(2x - 1)$	$-x + 10x = 9x$

Write the trinomial in factored form.

$$5 - 3x - 2x^2 = -(x - 1)(2x + 5)$$

HOW TO Factor: $10y^3 + 44y^2 - 30y$

The GCF is $2y$. Factor the GCF from the terms.

$$10y^3 + 44y^2 - 30y = 2y(5y^2 + 22y - 15)$$

Factor the trinomial $5y^2 + 22y - 15$.
The constant term is negative.
The binomial constants will have opposite signs.

Positive Factors of 5 (coefficient of y^2)	Factors of -15 (constant term)
1, 5	$-1,\ \ 15$ $1, -15$ $-3,\ \ 5$ $3,\ -5$

Write trial factors. Use the *O*uter and *I*nner products of FOIL to determine the middle term of the trinomial.

It is not necessary to test trial factors that have a common factor.

Trial Factors	Middle Term
$(y - 1)(5y + 15)$	common factor
$(y + 15)(5y - 1)$	$-y + 75y = 74y$
$(y + 1)(5y - 15)$	common factor
$(y - 15)(5y + 1)$	$y - 75y = -74y$
$(y - 3)(5y + 5)$	common factor
$(y + 5)(5y - 3)$	$-3y + 25y = 22y$
$(y + 3)(5y - 5)$	common factor
$(y - 5)(5y + 3)$	$3y - 25y = -22y$

Write the trinomial in factored form.

$$10y^3 + 44y^2 - 30y = 2y(y + 5)(5y - 3)$$

For previous examples, all the trial factors were listed. Once the correct factors have been found, however, the remaining trial factors can be omitted.

Trinomials of the form $ax^2 + bx + c$ can also be factored by grouping. This method is an extension of the method discussed in the preceding objective.

To factor $ax^2 + bx + c$, first find two factors of $a \cdot c$ whose sum is b. Then use factoring by grouping to write the factorization of the trinomial.

For the trinomial $3x^2 + 11x + 8$, $a = 3$, $b = 11$, and $c = 8$. To find two factors of $a \cdot c$ whose sum is b, first find the product $a \cdot c$ ($a \cdot c = 3 \cdot 8 = 24$). Then find two factors of 24 whose sum is 11. (3 and 8 are two factors of 24 whose sum is 11.)

HOW TO Factor: $3x^2 + 11x + 8$

Find two positive factors of 24 ($ac = 3 \cdot 8$) whose sum is 11, the coefficient of x.

Positive Factors of 24	Sum
1, 24	25
2, 12	14
3, 8	11

The required sum has been found. The remaining factors need not be checked.

Use the factors of 24 whose sum is 11 to write $11x$ as $3x + 8x$.
Factor by grouping.

$$3x^2 + 11x + 8 = 3x^2 + 3x + 8x + 8$$
$$= (3x^2 + 3x) + (8x + 8)$$
$$= 3x(x + 1) + 8(x + 1)$$
$$= (x + 1)(3x + 8)$$

Check: $(x + 1)(3x + 8) = 3x^2 + 8x + 3x + 8 = 3x^2 + 11x + 8$

HOW TO Factor: $4z^2 - 17z - 21$

Find two factors of -84 [$ac = 4 \cdot (-21)$] whose sum is -17, the coefficient of z.

Once the required sum is found, the remaining factors need not be checked.

Factors of −84	Sum
1, −84	−83
−1, 84	83
2, −42	−40
−2, 42	40
3, −28	−25
−3, 28	25
4, −21	−17

Use the factors of -84 whose sum is -17 to write $-17z$ as $4z - 21z$.

$$4z^2 - 17z - 21 = 4z^2 + 4z - 21z - 21$$

Factor by grouping. Recall that $-21z - 21 = -(21z + 21)$.

$$= (4z^2 + 4z) - (21z + 21)$$
$$= 4z(z + 1) - 21(z + 1)$$
$$= (z + 1)(4z - 21)$$

Check: $(z + 1)(4z - 21) = 4z^2 - 21z + 4z - 21 = 4z^2 - 17z - 21$

HOW TO Factor: $3x^2 - 11x + 4$

Find two negative factors of 12 ($3 \cdot 4$) whose sum is -11.

Negative Factors of 12	Sum
$-1, -12$	-13
$-2, -6$	-8
$-3, -4$	-7

Because no integer factors of 12 have a sum of -11, $3x^2 - 11x + 4$ is nonfactorable over the integers. $3x^2 - 11x + 4$ is a prime polynomial over the integers.

Example 7

Factor: $6x^2 + 11x - 10$

Solution
$6x^2 + 11x - 10 = (2x + 5)(3x - 2)$

You Try It 7

Factor: $4x^2 + 15x - 4$

Your solution

Example 8

Factor: $12x^2 - 32x + 5$

Solution
$12x^2 - 32x + 5 = (6x - 1)(2x - 5)$

You Try It 8

Factor: $10x^2 + 39x + 14$

Your solution

Example 9

Factor: $30y + 2xy - 4x^2y$

Solution
The GCF of $30y$, $2xy$, and $4x^2y$ is $2y$.

$$30y + 2xy - 4x^2y = 2y(15 + x - 2x^2)$$
$$= -2y(2x^2 - x - 15)$$
$$= -2y(2x + 5)(x - 3)$$

You Try It 9

Factor: $3a^3b^3 + 3a^2b^2 - 60ab$

Your solution

Example 10

Find two linear functions, f and g, such that $f(x) \cdot g(x) = 2x^2 + 9x - 5$.

Solution
We are looking for two polynomials that, when multiplied, equal $2x^2 + 9x - 5$. We must factor $2x^2 + 9x - 5$.

$f(x) = 2x - 1$ and $g(x) = x + 5$

You Try It 10

Find two linear functions, f and g, such that $f(x) \cdot g(x) = 3x^2 - 17x - 6$.

Your solution

Solutions on p. S16

5.5 Exercises

Objective A To factor a monomial from a polynomial

For Exercises 1 to 24, factor.

1. $6a^2 - 15a$

2. $32b^2 + 12b$

3. $4x^3 - 3x^2$

4. $12a^5b^2 + 16a^4b$

5. $3a^2 - 10b^3$

6. $9x^2 + 14y^4$

7. $x^5 - x^3 - x$

8. $y^4 - 3y^2 - 2y$

9. $16x^2 - 12x + 24$

10. $2x^5 + 3x^4 - 4x^2$

11. $5b^2 - 10b^3 + 25b^4$

12. $x^2y^4 - x^2y - 4x^2$

13. $x^{2n} - x^n$

14. $2a^{5n} + a^{2n}$

15. $x^{3n} - x^{2n}$

16. $y^{4n} + y^{2n}$

17. $a^{2n+2} + a^2$

18. $b^{n+5} - b^5$

19. $12x^2y^2 - 18x^3y + 24x^2y$

20. $14a^4b^4 - 42a^3b^3 + 28a^3b^2$

21. $24a^3b^2 - 4a^2b^2 - 16a^2b^4$

22. $10x^2y + 20x^2y^2 + 30x^2y^3$

23. $y^{2n+2} + y^{n+2} - y^2$

24. $a^{2n+2} + a^{2n+1} + a^n$

Objective B To factor by grouping

For Exercises 25 to 45, factor.

25. $x(a + 2) - 2(a + 2)$

26. $3(x + y) + a(x + y)$

27. $a(x - 2) - b(2 - x)$

28. $3(a - 7) - b(7 - a)$

29. $x(a - 2b) + y(2b - a)$

30. $b(3 - 2c) - 5(2c - 3)$

31. $xy + 4y - 2x - 8$

32. $ab + 7b - 3a - 21$

33. $ax + bx - ay - by$

34. $2ax - 3ay - 2bx + 3by$

35. $x^2y - 3x^2 - 2y + 6$

36. $a^2b + 3a^2 + 2b + 6$

37. $6 + 2y + 3x^2 + x^2y$

38. $15 + 3b - 5a^2 - a^2b$

39. $2ax^2 + bx^2 - 4ay - 2by$

40. $4a^2x + 2a^2y - 6bx - 3by$

41. $6xb + 3ax - 4by - 2ay$

42. $a^2x - 3a^2y + 2x - 6y$

43. $x^ny - 5x^n + y - 5$

44. $a^nx^n + 2a^n + x^n + 2$

45. $2x^3 - x^2 + 4x - 2$

Objective C **To factor a trinomial of the form $x^2 + bx + c$**

For Exercises 46 to 72, factor.

46. $x^2 - 8x + 15$

47. $x^2 + 12x + 20$

48. $a^2 + 12a + 11$

49. $a^2 + a - 72$

50. $b^2 + 2b - 35$

51. $a^2 + 7a + 6$

52. $y^2 - 16y + 39$

53. $y^2 - 18y + 72$

54. $b^2 + 4b - 32$

55. $x^2 + x - 132$

56. $a^2 - 15a + 56$

57. $x^2 + 15x + 50$

58. $y^2 + 13y + 12$

59. $b^2 - 6b - 16$

60. $x^2 + 4x - 5$

61. $a^2 - 3ab + 2b^2$

62. $a^2 + 11ab + 30b^2$

63. $a^2 + 8ab - 33b^2$

64. $x^2 - 14xy + 24y^2$

65. $x^2 + 5xy + 6y^2$

66. $y^2 + 2xy - 63x^2$

67. $2 + x - x^2$

68. $21 - 4x - x^2$

69. $5 + 4x - x^2$

70. $50 + 5a - a^2$

71. $x^2 - 5x + 6$

72. $x^2 - 7x - 12$

Objective D To factor $ax^2 + bx + c$

For Exercises 73 to 104, factor.

73. $2x^2 + 7x + 3$

74. $2x^2 - 11x - 40$

75. $6y^2 + 5y - 6$

76. $4y^2 - 15y + 9$

77. $6b^2 - b - 35$

78. $2a^2 + 13a + 6$

79. $3y^2 - 22y + 39$

80. $12y^2 - 13y - 72$

81. $6a^2 - 26a + 15$

82. $5x^2 + 26x + 5$

83. $4a^2 - a - 5$

84. $11x^2 - 122x + 11$

85. $10x^2 - 29x + 10$

86. $2x^2 + 5x + 12$

87. $4x^2 - 6x + 1$

88. $6x^2 + 5xy - 21y^2$

89. $6x^2 + 41xy - 7y^2$

90. $4a^2 + 43ab + 63b^2$

91. $7a^2 + 46ab - 21b^2$

92. $10x^2 - 23xy + 12y^2$

93. $18x^2 + 27xy + 10y^2$

94. $24 + 13x - 2x^2$

95. $6 - 7x - 5x^2$

96. $8 - 13x + 6x^2$

97. $30 + 17a - 20a^2$

98. $15 - 14a - 8a^2$

99. $35 - 6b - 8b^2$

100. $12y^3 + 22y^2 - 70y$

101. $5y^4 - 29y^3 + 20y^2$

102. $30a^2 + 85ab + 60b^2$

103. $20x^2 - 38x^3 - 30x^4$

104. $12x + x^2 - 6x^3$

105. $3y - 16y^2 + 16y^3$

106. Find two linear functions, f and g, such that $f(x) \cdot g(x) = 2x^2 + 9x - 18$.

107. Find two linear functions, f and h, such that $f(x) \cdot h(x) = 2x^2 - 5x + 2$.

108. Find two linear functions, F and G, such that $F(x) \cdot G(x) = 2x^2 - 9x - 5$.

109. Find two linear functions, f and g, such that $f(a) \cdot g(a) = 3a^2 + 11a - 4$.

110. Find two linear functions, f and g, such that $f(b) \cdot g(b) = 4b^2 - 17b + 15$.

111. Find two linear functions, f and g, such that $f(x) \cdot g(x) = 2x^2 + 13x - 24$.

112. Find two linear functions, f and h, such that $f(x) \cdot h(x) = 4x^2 - 12x - 7$.

113. Find two linear functions, g and h, such that $g(x) \cdot h(x) = 6x^2 + 7x - 5$.

114. Find two linear functions, f and g, such that $f(x) \cdot g(x) = 4x^2 + 23x + 15$.

115. Find two linear functions, f and g, such that $f(t) \cdot g(t) = 6t^2 - 17t - 3$.

APPLYING THE CONCEPTS

116. Geometry Write the areas of the shaded regions in factored form.

a.

b.

c.

d.

117. Find all integers k such that the trinomial can be factored.
 a. $x^2 + kx + 8$
 b. $x^2 + kx - 6$
 c. $2x^2 + kx + 3$

 d. $2x^2 + kx - 5$
 e. $3x^2 + kx + 5$
 f. $2x^2 + kx - 3$

5.6 Special Factoring

Objective A

To factor the difference of two perfect squares or a perfect-square trinomial

The product of a term and itself is called a **perfect square.** The exponents on variables of perfect squares are always even numbers.

Term		Perfect Square
5	$5 \cdot 5 =$	25
x	$x \cdot x =$	x^2
$3y^4$	$3y^4 \cdot 3y^4 =$	$9y^8$
x^n	$x^n \cdot x^n =$	x^{2n}

The **square root of a perfect square** is one of the two equal factors of the perfect square. "$\sqrt{}$" is the symbol for square root. To find the exponent of the square root of a variable term, multiply the exponent by $\frac{1}{2}$.

$\sqrt{25} = 5$
$\sqrt{x^2} = x$
$\sqrt{9y^8} = 3y^4$
$\sqrt{x^{2n}} = x^n$

The **difference of two perfect squares** is the **product of the sum and difference of two terms.** The factors of the difference of two perfect squares are the sum and difference of the square roots of the perfect squares.

Factors of the Difference of Two Perfect Squares

$a^2 - b^2 = (a + b)(a - b)$

The **sum of two perfect squares,** $a^2 + b^2$, is nonfactorable over the integers.

HOW TO Factor: $4x^2 - 81y^2$

Write the binomial as the difference of two perfect squares.

$$4x^2 - 81y^2 = (2x)^2 - (9y)^2$$

The factors are the sum and difference of the square roots of the perfect squares.

$$= (2x + 9y)(2x - 9y)$$

A **perfect-square trinomial** is the square of a binomial.

Factors of a Perfect-Square Trinomial

$a^2 + 2ab + b^2 = (a + b)^2$
$a^2 - 2ab + b^2 = (a - b)^2$

In factoring a perfect-square trinomial, remember that the terms of the binomial are the square roots of the perfect squares of the trinomial. The sign in the binomial is the sign of the middle term of the trinomial.

HOW TO Factor: $4x^2 + 12x + 9$

Because $4x^2$ is a perfect square $[4x^2 = (2x)^2]$ and 9 is a perfect square ($9 = 3^2$), try factoring $4x^2 + 12x + 9$ as the square of a binomial.

$4x^2 + 12x + 9 \overset{?}{=} (2x + 3)^2$

Check: $(2x + 3)^2 = (2x + 3)(2x + 3) = 4x^2 + 6x + 6x + 9 = 4x^2 + 12x + 9$

The check verifies that $4x^2 + 12x + 9 = (2x + 3)^2$.

It is important to check a proposed factorization as we did above. The next example illustrates the importance of this check.

HOW TO Factor: $x^2 + 13x + 36$

Because x^2 is a perfect square and 36 is a perfect square, try factoring $x^2 + 13x + 36$ as the square of a binomial.

$x^2 + 13x + 36 \overset{?}{=} (x + 6)^2$

Check: $(x + 6)^2 = (x + 6)(x + 6) = x^2 + 6x + 6x + 36 = x^2 + 12x + 36$

In this case, the proposed factorization of $x^2 + 13x + 36$ does *not* check. Try another factorization. The numbers 4 and 9 are factors of 36 whose sum is 13.

$x^2 + 13x + 36 = (x + 4)(x + 9)$

Example 1

Factor: $25x^2 - 1$

Solution
$25x^2 - 1 = (5x)^2 - (1)^2$ • Difference of
$ = (5x + 1)(5x - 1)$ two squares

Example 2

Factor: $4x^2 - 20x + 25$

Solution
$4x^2 - 20x + 25 = (2x - 5)^2$ • Perfect square
$$ trinomial

Example 3

Factor: $(x + y)^2 - 4$

Solution
$(x + y)^2 - 4 = (x + y)^2 - (2)^2$ • Difference of
$ = (x + y)^2 - (2)^2$ two squares
$ = (x + y + 2)(x + y - 2)$

You Try It 1

Factor: $x^2 - 36y^4$

Your solution

You Try It 2

Factor: $9x^2 + 12x + 4$

Your solution

You Try It 3

Factor: $(a + b)^2 - (a - b)^2$

Your solution

Solutions on pp. S16–S17

Objective B **To factor the sum or the difference of two perfect cubes**

The product of the same three factors is called a **perfect cube.** The exponents on variables of perfect cubes are always divisible by 3.

Term		Perfect Cube
2	$2 \cdot 2 \cdot 2 = 2^3 =$	8
$3y$	$3y \cdot 3y \cdot 3y = (3y)^3 =$	$27y^3$
y^2	$y^2 \cdot y^2 \cdot y^2 = (y^2)^3 =$	y^6

The **cube root** of a perfect cube is one of the three equal factors of the perfect cube. "$\sqrt[3]{}$" is the symbol for cube root. To find the exponent of the cube root of a variable term, multiply the exponent by $\frac{1}{3}$.

$$\sqrt[3]{8} = 2$$
$$\sqrt[3]{27y^3} = 3y$$
$$\sqrt[3]{y^6} = y^2$$

The following rules are used to factor the sum or difference of two perfect cubes.

Factoring the Sum or Difference of Two Perfect Cubes

$a^3 + b^3 = (a + b)(a^2 - ab + b^2)$
$a^3 - b^3 = (a - b)(a^2 + ab + b^2)$

To factor $8x^3 - 27$:

Write the binomial as the difference of two perfect cubes.

$$8x^3 - 27 = (2x)^3 - 3^3$$

The terms of the binomial factor are the cube roots of the perfect cubes. The sign of the binomial factor is the same sign as in the given binomial. The trinomial factor is obtained from the binomial factor.

$$= (2x - 3)(4x^2 + 6x + 9)$$

Square of the first term ⟶

Opposite of the product of the two terms ⟶

Square of the last term ⟶

HOW TO Factor: $a^3 + 64y^3$

$a^3 + 64y^3 = a^3 + (4y)^3$
• Write the binomial as the sum of two perfect cubes.

$= (a + 4y)(a^2 - 4ay + 16y^2)$
• Factor.

HOW TO Factor: $64y^4 - 125y$

$64y^4 - 125y = y(64y^3 - 125)$
• Factor out y, the GCF.

$= y[(4y)^3 - 5^3]$
• Write the binomial as the difference of two perfect cubes.

$= y(4y - 5)(16y^2 + 20y + 25)$
• Factor.

Example 4

Factor: $x^3y^3 - 1$

Solution

$$x^3y^3 - 1 = (xy)^3 - 1^3 \qquad \bullet \text{ Difference of}$$
$$= (xy - 1)(x^2y^2 + xy + 1) \qquad \text{two cubes}$$

You Try It 4

Factor: $a^3b^3 - 27$

Your solution

Example 5

Factor: $64c^3 + 8d^3$

Solution

$$64c^3 + 8d^3$$
$$= 8(8c^3 + d^3) \qquad \bullet \text{ GCF}$$
$$= 8[(2c)^3 + d^3] \qquad \bullet \text{ Sum of}$$
$$= 8(2c + d)(4c^2 - 2cd + d^2) \qquad \text{two cubes}$$

You Try It 5

Factor: $8x^3 + y^3z^3$

Your solution

Example 6

Factor: $(x + y)^3 - x^3$

Solution

$$(x + y)^3 - x^3 \qquad \bullet \text{ Difference of two cubes}$$
$$= [(x + y) - x][(x + y)^2 + x(x + y) + x^2]$$
$$= y(x^2 + 2xy + y^2 + x^2 + xy + x^2)$$
$$= y(3x^2 + 3xy + y^2)$$

You Try It 6

Factor: $(x - y)^3 + (x + y)^3$

Your solution

Solutions on p. S17

Objective C **To factor a trinomial that is quadratic in form**

Certain trinomials that are not quadratic can be expressed as quadratic trinomials by making suitable variable substitutions. A trinomial is **quadratic in form** if it can be written as $au^2 + bu + c$.

As shown below, the trinomials $x^4 + 5x^2 + 6$ and $2x^2y^2 + 3xy - 9$ are quadratic in form.

$$x^4 + 5x^2 + 6 \qquad\qquad 2x^2y^2 + 3xy - 9$$

$$(x^2)^2 + 5(x^2) + 6 \qquad\qquad 2(xy)^2 + 3(xy) - 9$$

Let $u = x^2$.　　$u^2 + 5u + 6$　　　　Let $u = xy$.　　$2u^2 + 3u - 9$

When we use this method to factor a trinomial that is quadratic in form, the variable part of the first term in each binomial will be u.

HOW TO Factor: $x^4 + 5x^2 + 6$

$$x^4 + 5x^2 + 6 = u^2 + 5u + 6 \qquad \bullet \text{ Let } u = x^2.$$

$$= (u + 3)(u + 2) \qquad \bullet \text{ Factor.}$$

$$= (x^2 + 3)(x^2 + 2) \qquad \bullet \text{ Replace } u \text{ by } x^2.$$

Here is an example in which $u = \sqrt{x}$.

HOW TO Factor: $x - 2\sqrt{x} - 15$

$$x - 2\sqrt{x} - 15 = u^2 - 2u - 15 \qquad \bullet \text{ Let } u = \sqrt{x}. \text{ Then } u^2 = x.$$

$$= (u - 5)(u + 3) \qquad \bullet \text{ Factor.}$$

$$= (\sqrt{x} - 5)(\sqrt{x} + 3) \qquad \bullet \text{ Replace } u \text{ by } \sqrt{x}.$$

Example 7

Factor: $6x^2y^2 - xy - 12$

Solution
Let $u = xy$.

$$6x^2y^2 - xy - 12 = 6u^2 - u - 12$$
$$= (3u + 4)(2u - 3)$$
$$= (3xy + 4)(2xy - 3)$$

You Try It 7

Factor: $3x^4 + 4x^2 - 4$

Your solution

Solution on p. S17

Objective D To factor completely

General Factoring Strategy

1. Is there a common factor? If so, factor out the GCF.
2. If the polynomial is a binomial, is it the difference of two perfect squares, the sum of two perfect cubes, or the difference of two perfect cubes? If so, factor.
3. If the polynomial is a trinomial, is it a perfect-square trinomial or the product of two binomials? If so, factor.
4. Can the polynomial be factored by grouping? If so, factor.
5. Is each factor nonfactorable over the integers? If not, factor.

Example 8

Factor: $6a^3 + 15a^2 - 36a$

Solution

$6a^3 + 15a^2 - 36a = 3a(2a^2 + 5a - 12)$ • GCF
$\qquad\qquad\qquad\qquad = 3a(2a - 3)(a + 4)$

You Try It 8

Factor: $18x^3 - 6x^2 - 60x$

Your solution

Example 9

Factor: $x^2y + 2x^2 - y - 2$

Solution

$x^2y + 2x^2 - y - 2$
$\quad = (x^2y + 2x^2) - (y + 2)$ • Factor by
$\quad = x^2(y + 2) - (y + 2)$ grouping.
$\quad = (y + 2)(x^2 - 1)$
$\quad = (y + 2)(x + 1)(x - 1)$

You Try It 9

Factor: $4x - 4y - x^3 + x^2y$

Your solution

Example 10

Factor: $x^{4n} - y^{4n}$

Solution

$x^{4n} - y^{4n} = (x^{2n})^2 - (y^{2n})^2$ • Difference of
$\quad = (x^{2n} + y^{2n})(x^{2n} - y^{2n})$ two squares
$\quad = (x^{2n} + y^{2n})[(x^n)^2 - (y^n)^2]$
$\quad = (x^{2n} + y^{2n})(x^n + y^n)(x^n - y^n)$

You Try It 10

Factor: $x^{4n} - x^{2n}y^{2n}$

Your solution

Example 11

Factor: $x^{n+3} + x^n y^3$

Solution

$x^{n+3} + x^n y^3 = x^n(x^3 + y^3)$ • GCF
$\quad = x^n(x + y)(x^2 - xy + y^2)$ • Sum of
$\qquad\qquad\qquad\qquad\qquad\qquad$ two cubes

You Try It 11

Factor: $ax^5 - ax^2y^6$

Your solution

Solutions on p. S17

5.6 Exercises

Objective A **To factor the difference of two perfect squares or a perfect-square trinomial**

For Exercises 1 and 2, determine which expressions are perfect squares.

1. $4; 8; 25x^6; 12y^{10}; 100x^4y^4$

2. $9; 18; 15a^8; 49b^{12}; 64a^{16}b^2$

For Exercises 3 to 6, find the square root of the expression.

3. $16z^8$

4. $36d^{10}$

5. $81a^4b^6$

6. $25m^2n^{12}$

For Exercises 7 to 42, factor.

7. $x^2 - 16$

8. $y^2 - 49$

9. $4x^2 - 1$

10. $81x^2 - 4$

11. $16x^2 - 121$

12. $49y^2 - 36$

13. $1 - 9a^2$

14. $16 - 81y^2$

15. $x^2y^2 - 100$

16. $a^2b^2 - 25$

17. $x^2 + 4$

18. $a^2 + 16$

19. $25 - a^2b^2$

20. $64 - x^2y^2$

21. $a^{2n} - 1$

22. $b^{2n} - 16$

23. $x^2 - 12x + 36$

24. $y^2 - 6y + 9$

25. $b^2 - 2b + 1$

26. $a^2 + 14a + 49$

27. $16x^2 - 40x + 25$

28. $49x^2 + 28x + 4$

29. $4a^2 + 4a - 1$

30. $9x^2 + 12x - 4$

31. $b^2 + 7b + 14$

32. $y^2 - 5y + 25$

33. $x^2 + 6xy + 9y^2$

34. $4x^2y^2 + 12xy + 9$

35. $25a^2 - 40ab + 16b^2$

36. $4a^2 - 36ab + 81b^2$

37. $x^{2n} + 6x^n + 9$ **38.** $y^{2n} - 16y^n + 64$ **39.** $(x - 4)^2 - 9$

40. $16 - (a - 3)^2$ **41.** $(x - y)^2 - (a + b)^2$ **42.** $(x - 2y)^2 - (x + y)^2$

Objective B **To factor the sum or the difference of two perfect cubes**

For Exercises 43 and 44, determine which expressions are perfect cubes.

43. $4; 8; x^9; a^8b^8; 27c^{15}d^{18}$ **44.** $9; 27; y^{12}; m^3n^6; 64mn^9$

For Exercises 45 to 48, find the cube root of the expression.

45. $8x^9$ **46.** $27y^{15}$ **47.** $64a^6b^{18}$ **48.** $125c^{12}d^3$

For Exercises 49 to 72, factor.

49. $x^3 - 27$ **50.** $y^3 + 125$ **51.** $8x^3 - 1$

52. $64a^3 + 27$ **53.** $x^3 - y^3$ **54.** $x^3 - 8y^3$

55. $m^3 + n^3$ **56.** $27a^3 + b^3$ **57.** $64x^3 + 1$

58. $1 - 125b^3$ **59.** $27x^3 - 8y^3$ **60.** $64x^3 + 27y^3$

61. $x^3y^3 + 64$ **62.** $8x^3y^3 + 27$ **63.** $16x^3 - y^3$

64. $27x^3 - 8y^2$ **65.** $8x^3 - 9y^3$ **66.** $27a^3 - 16$

67. $(a - b)^3 - b^3$ **68.** $a^3 + (a + b)^3$ **69.** $x^{6n} + y^{3n}$

70. $x^{3n} + y^{3n}$ **71.** $x^{3n} + 8$ **72.** $a^{3n} + 64$

Objective C To factor a trinomial that is quadratic in form

For Exercises 73 to 93, factor.

73. $x^2y^2 - 8xy + 15$

74. $x^2y^2 - 8xy - 33$

75. $x^2y^2 - 17xy + 60$

76. $a^2b^2 + 10ab + 24$

77. $x^4 - 9x^2 + 18$

78. $y^4 - 6y^2 - 16$

79. $b^4 - 13b^2 - 90$

80. $a^4 + 14a^2 + 45$

81. $x^4y^4 - 8x^2y^2 + 12$

82. $a^4b^4 + 11a^2b^2 - 26$

83. $x^{2n} + 3x^n + 2$

84. $a^{2n} - a^n - 12$

85. $3x^2y^2 - 14xy + 15$

86. $5x^2y^2 - 59xy + 44$

87. $6a^2b^2 - 23ab + 21$

88. $10a^2b^2 + 3ab - 7$

89. $2x^4 - 13x^2 - 15$

90. $3x^4 + 20x^2 + 32$

91. $2x^{2n} - 7x^n + 3$

92. $4x^{2n} + 8x^n - 5$

93. $6a^{2n} + 19a^n + 10$

Objective D To factor completely

For Exercises 94 to 129, factor.

94. $5x^2 + 10x + 5$

95. $12x^2 - 36x + 27$

96. $3x^4 - 81x$

97. $27a^4 - a$

98. $7x^2 - 28$

99. $20x^2 - 5$

100. $y^4 - 10y^3 + 21y^2$

101. $y^5 + 6y^4 - 55y^3$

102. $x^4 - 16$

103. $16x^4 - 81$

104. $8x^5 - 98x^3$

105. $16a - 2a^4$

106. $x^3y^3 - x^3$

107. $a^3b^6 - b^3$

108. $x^6y^6 - x^3y^3$

109. $8x^4 - 40x^3 + 50x^2$

110. $6x^5 + 74x^4 + 24x^3$

111. $x^4 - y^4$

112. $16a^4 - b^4$

113. $x^6 + y^6$

114. $x^4 - 5x^2 - 4$

115. $a^4 - 25a^2 - 144$

116. $3b^5 - 24b^2$

117. $16a^4 - 2a$

118. $x^4y^2 - 5x^3y^3 + 6x^2y^4$

119. $a^4b^2 - 8a^3b^3 - 48a^2b^4$

120. $16x^3y + 4x^2y^2 - 42xy^3$

121. $24a^2b^2 - 14ab^3 - 90b^4$

122. $x^3 - 2x^2 - x + 2$

123. $x^3 - 2x^2 - 4x + 8$

124. $4x^2y^2 - 4x^2 - 9y^2 + 9$

125. $4x^4 - x^2 - 4x^2y^2 + y^2$

126. $a^{2n+2} - 6a^{n+2} + 9a^2$

127. $x^{2n+1} + 2x^{n+1} + x$

128. $2x^{n+2} - 7x^{n+1} + 3x^n$

129. $3b^{n+2} + 4b^{n+1} - 4b^n$

APPLYING THE CONCEPTS

130. Factor: $x^2(x - 3) - 3x(x - 3) + 2(x - 3)$

131. Given that $(x - 3)$ and $(x + 4)$ are factors of $x^3 + 6x^2 - 7x - 60$, explain how you can find a third *first-degree* factor of $x^3 + 6x^2 - 7x - 60$. Then find the factor.

5.7 Solving Equations by Factoring

Objective A To solve an equation by factoring

Consider the equation $ab = 0$. If a is not zero, then b must be zero. Conversely, if b is not zero, then a must be zero. This is summarized in the **Principle of Zero Products.**

> **Principle of Zero Products**
>
> If the product of two factors is zero, then at least one of the factors must be zero.
>
> If $ab = 0$, then $a = 0$ or $b = 0$.

The Principle of Zero Products is used to solve equations.

HOW TO Solve: $(x - 4)(x + 2) = 0$

By the Principle of Zero Products, if $(x - 4)(x + 2) = 0$, then $x - 4 = 0$ or $x + 2 = 0$.

$$\begin{aligned} x - 4 &= 0 & x + 2 &= 0 \\ x &= 4 & x &= -2 \end{aligned}$$

Check:

$$\frac{\begin{array}{c|c} (x - 4)(x + 2) = 0 & \\ (4 - 4)(4 + 2) & 0 \\ 0 \cdot 6 & 0 \\ & 0 = 0 \end{array}}{} \qquad \frac{\begin{array}{c|c} (x - 4)(x + 2) = 0 & \\ (-2 - 4)(-2 + 2) & 0 \\ -6 \cdot 0 & 0 \\ & 0 = 0 \end{array}}{}$$

-2 and 4 check as solutions. The solutions are -2 and 4.

An equation of the form $ax^2 + bx + c = 0$, $a \neq 0$, is a **quadratic equation.** In a **quadratic equation in standard form,** the polynomial is written in descending order and equal to zero.

Some quadratic equations can be solved by factoring and then using the Principle of Zero Products.

TAKE NOTE

Note the steps involved in solving a quadratic equation by factoring:
1. Write in standard form.
2. Factor.
3. Set each factor equal to 0.
4. Solve each equation.
5. Check the solutions.

HOW TO Solve: $2x^2 - x = 1$

$$\begin{aligned} 2x^2 - x &= 1 \\ 2x^2 - x - 1 &= 0 \\ (2x + 1)(x - 1) &= 0 \end{aligned}$$

- Write the equation in standard form.
- Factor.

$$\begin{aligned} 2x + 1 &= 0 & x - 1 &= 0 \\ 2x &= -1 & x &= 1 \\ x &= -\frac{1}{2} \end{aligned}$$

- Use the Principle of Zero Products.
- Solve each equation.

The solutions are $-\dfrac{1}{2}$ and 1. **You should check these solutions.**

TAKE NOTE

The Principle of Zero Products can be extended to more than two factors. For example, if $abc = 0$, then $a = 0$, $b = 0$, or $c = 0$.

HOW TO Solve: $t^4 - 2t^2 + 1 = 0$

$$t^4 - 2t^2 + 1 = 0$$
$$(t^2 - 1)(t^2 - 1) = 0$$
$$(t - 1)(t + 1)(t - 1)(t + 1) = 0$$

- Factor the trinomial.
- Factor completely.

$$t - 1 = 0, \quad t + 1 = 0, \quad t - 1 = 0, \quad t + 1 = 0$$
$$t = 1 \qquad t = -1 \qquad t = 1 \qquad t = -1$$

- Use the Principle of Zero Products.

The solutions are -1 and 1. These solutions are called **repeated solutions.**

Example 1

Solve: $x^2 + (x + 2)^2 = 100$

Solution
$$x^2 + (x + 2)^2 = 100$$
$$x^2 + x^2 + 4x + 4 = 100 \quad \bullet \text{ Square } (x + 2).$$
$$2x^2 + 4x - 96 = 0 \quad \bullet \text{ Write in standard form.}$$
$$2(x^2 + 2x - 48) = 0 \quad \bullet \text{ Factor the left side.}$$
$$2(x - 6)(x + 8) = 0$$

$$x - 6 = 0 \qquad x + 8 = 0 \quad \bullet \text{ Principle of}$$
$$x = 6 \qquad\qquad x = -8 \qquad \text{ Zero Products}$$

The solutions are -8 and 6.

You Try It 1

Solve: $(x + 4)(x - 1) = 14$

Your solution

Solution on p. S17

Objective B **To solve application problems**

Example 2

The length of a rectangle is 8 in. more than the width. The area of the rectangle is 240 in². Find the width of the rectangle.

Strategy
Draw a diagram. Then use the formula for the area of a rectangle.

x

$x + 8$

Solution
$$A = LW$$
$$240 = (x + 8)x$$
$$240 = x^2 + 8x$$
$$0 = x^2 + 8x - 240$$
$$0 = (x + 20)(x - 12)$$

$$x + 20 = 0 \qquad x - 12 = 0$$
$$x = -20 \qquad\quad x = 12$$

The width cannot be negative.
The width is 12 in.

You Try It 2

The height of a triangle is 3 cm more than the length of the base of the triangle. The area of the triangle is 54 cm². Find the height of the triangle and the length of the base.

Your strategy

Your solution

Solution on p. S17

5.7 Exercises

To solve an equation by factoring

For Exercises 1 to 35, solve.

1. $(x - 5)(x + 3) = 0$

2. $(x - 2)(x + 6) = 0$

3. $(x + 7)(x - 8) = 0$

4. $x(2x + 5)(x - 6) = 0$

5. $2x(3x - 2)(x + 4) = 0$

6. $6x(3x - 7)(x + 7) = 0$

7. $x^2 + 2x - 15 = 0$

8. $t^2 + 3t - 10 = 0$

9. $z^2 - 4z + 3 = 0$

10. $6x^2 - 9x = 0$

11. $r^2 - 10 = 3r$

12. $t^2 - 12 = 4t$

13. $4t^2 = 4t + 3$

14. $5y^2 + 11y = 12$

15. $4v^2 - 4v + 1 = 0$

16. $9s^2 - 6s + 1 = 0$

17. $x^2 - 9 = 0$

18. $t^2 - 16 = 0$

19. $4y^2 - 1 = 0$

20. $9z^2 - 4 = 0$

21. $x + 15 = x(x - 1)$

22. $x^2 - x - 2 = (2x - 1)(x - 3)$

23. $v^2 + v + 5 = (3v + 2)(v - 4)$

24. $z^2 + 5z - 4 = (2z + 1)(z - 4)$

25. $4x^2 + x - 10 = (x - 2)(x + 1)$

26. $x^3 + 2x^2 - 15x = 0$

27. $c^3 + 3c^2 - 10c = 0$

28. $z^4 - 18z^2 + 81 = 0$

29. $y^4 - 8y^2 + 16 = 0$

30. $x^3 + x^2 - 4x - 4 = 0$

31. $a^3 + a^2 - 9a - 9 = 0$

32. $2x^3 - x^2 - 2x + 1 = 0$

33. $3x^3 + 2x^2 - 12x - 8 = 0$

34. $2x^3 + 3x^2 - 18x - 27 = 0$

35. $5x^3 + 2x^2 - 20x - 8 = 0$

Objective B **To solve application problems**

36. **Mathematics** The sum of a number and its square is 72. Find the number.

37. **Mathematics** The sum of a number and its square is 210. Find the number.

38. **Geometry** The length of a rectangle is 2 ft more than twice the width. The area of the rectangle is 84 ft². Find the length and width of the rectangle.

w

$2w + 2$

39. **Geometry** The length of a rectangle is 8 cm more than three times the width. The area of the rectangle is 380 cm². Find the length and width of the rectangle.

40. **Geometry** The height of a triangle is 8 cm more than the length of the base. The area of the triangle is 64 cm². Find the base and height of the triangle.

$b + 8$

b

41. **Geometry** One leg of a right triangle is 2 ft more than twice the other leg. The hypotenuse is 1 ft more than the longer leg. Find the length of the hypotenuse of the right triangle.

42. **Physics** An object is thrown downward, with an initial speed of 16 ft/s, from the top of a building 480 ft high. How many seconds later will the object hit the ground? Use the equation $d = vt + 16t^2$, where d is the distance in feet, v is the initial speed, and t is the time in seconds.

43. **Physics** A penny is thrown into a wishing well with an initial speed of 8 ft/s. The well is 80 ft deep. How many seconds later will the penny hit the bottom of the well? Use the equation $d = vt + 16t^2$, where d is the distance in feet, v is the initial speed, and t is the time in seconds.

APPLYING THE CONCEPTS

44. Write an equation whose solutions are 3, 2, and -1.

45. **Construction** A rectangular piece of cardboard is 10 in. longer than it is wide. Squares 2 in. on a side are to be cut from each corner, and then the sides are to be folded up to make an open box with a volume of 112 in³. Find the length and width of the piece of cardboard.

46. The following seems to show that $1 = 2$. Explain the error.

$$a = b$$
$$a^2 = ab$$
$$a^2 - b^2 = ab - b^2$$
$$(a - b)(a + b) = b(a - b)$$
$$a + b = b$$
$$b + b = b$$
$$2b = b$$
$$2 = 1$$

- Multiply each side of the equation by *a*.
- Subtract *b*² from each side of the equation.
- Factor.
- Divide each side by *a* − *b*.
- Because *a* = *b*, substitute *b* for *a*.

- Divide both sides by *b*.

Focus on Problem Solving

Find a CounterExample

When you are faced with an assertion, it may be that the assertion is false. For instance, consider the statement "Every prime number is an odd number." This assertion is false because the prime number 2 is an even number.

Finding an example that illustrates that an assertion is false is called finding a *counterexample*. The number 2 is a counterexample to the assertion that every prime number is an odd number.

If you are given an unfamiliar problem, one strategy to consider as a means of solving the problem is to try to find a counterexample. For each of the following problems, answer *true* if the assertion is always true. If the assertion is not true, answer *false* and give a counterexample. If there are terms used that you do not understand, consult a reference to find the meaning of the term.

1. If x is a real number, then x^2 is always positive.

2. The product of an odd integer and an even integer is an even integer.

3. If m is a positive integer, then $2m + 1$ is always a positive odd integer.

4. If $x < y$, then $x^2 < y^2$.

5. Given any three positive numbers a, b, and c, it is possible to construct a triangle whose sides have lengths a, b, and c.

6. The product of two irrational numbers is an irrational number.

7. If n is a positive integer greater than 2, then $1 \cdot 2 \cdot 3 \cdot 4 \cdot \cdots \cdot n + 1$ is a prime number.

8. Draw a polygon with more than three sides. Select two different points inside the polygon and join the points with a line segment. The line segment always lies completely inside the polygon.

9. Let A, B, and C be three points in the plane that are not collinear. Let d_1 be the distance from A to B, and let d_2 be the distance from A to C. Then the distance between B and C is less than $d_1 + d_2$.

10. Consider the line segment AB shown at the left. Two points, C and D, are randomly selected on the line segment and three new segments are formed: AC, CD, and DB. The three new line segments can always be connected to form a triangle.

It may not be easy to establish that an assertion is true or to find a counterexample to the assertion. For instance, consider the assertion that every positive integer greater than 3 can be written as the sum of two primes. For example, $6 = 3 + 3$, $8 = 3 + 5$, $9 = 2 + 7$. Is this assertion always true? (*Note:* This assertion, called Goldbach's conjecture, has never been proved, nor has a counterexample been found!)

Projects and Group Activities

Astronomical Distances and Scientific Notation

Astronomers have units of measurement that are useful for measuring vast distances in space. Two of these units are the astronomical unit and the light-year. An **astronomical unit** is the average distance between Earth and the sun. A **light-year** is the distance a ray of light travels in 1 year.

1. Light travels at a speed of 1.86×10^5 mi/s. Find the measure of 1 light-year in miles. Use a 365-day year.

2. The distance between Earth and the star Alpha Centauri is approximately 25 trillion miles. Find the distance between Earth and Alpha Centauri in light-years. Round to the nearest hundredth.

3. The Coma cluster of galaxies is approximately 2.8×10^8 light-years from Earth. Find the distance, in miles, from the Coma Cluster to Earth. Write the answer in scientific notation.

Gemini

4. One astronomical unit (A.U.) is 9.3×10^7 mi. The star Pollux in the constellation Gemini is 1.8228×10^{12} mi from Earth. Find the distance from Pollux to Earth in astronomical units.

5. One light-year is equal to approximately how many astronomical units? Round to the nearest thousand.

Shown below are data on the planets in our solar system. The planets are listed in alphabetical order.

Point of Interest

In November 2001, the Hubble Space Telescope took photos of the atmosphere of a planet orbiting a star 150 light-years from Earth in the constellation Pegasus. The planet is about the size of Jupiter and orbits close to the star HD209458. It was the first discovery of an atmosphere around a planet outside our solar system.

Planet	Distance from the Sun (in kilometers)	Mass (in kilograms)
Earth	1.50×10^8	5.97×10^{24}
Jupiter	7.79×10^8	1.90×10^{27}
Mars	2.28×10^8	6.42×10^{23}
Mercury	5.79×10^7	3.30×10^{23}
Neptune	4.50×10^9	1.02×10^{26}
Pluto	5.87×10^9	1.25×10^{22}
Saturn	1.43×10^9	5.68×10^{26}
Uranus	2.87×10^9	8.68×10^{25}
Venus	1.08×10^8	4.87×10^{24}

Jupiter

6. Arrange the planets in order from closest to the sun to farthest from the sun.

7. Arrange the planets in order from the one with the greatest mass to the one with the least mass.

8. Write a rule for ordering numbers written in scientific notation.

Chapter 5 Summary

Key Words	**Examples**
A *monomial* is a number, a variable, or a product of numbers and variables. [5.1A, p. 259]	5 is a number, y is a variable. $8a^2b^2$ is a product of numbers and variables. 5, y, and $8a^2b^2$ are monomials.
The *degree of a monomial* is the sum of the exponents of the variables. [5.1A, p. 259]	The degree of $8x^4y^5z$ is 10.
A *polynomial* is a variable expression in which the terms are monomials. [5.2A, p. 271]	$x^4 - 2xy - 32x + 8$ is a polynomial. The terms are x^4, $-2xy$, $-32x$, and 8.
A polynomial of one term is a *monomial*, a polynomial of two terms is a *binomial*, and a polynomial of three terms is a *trinomial*. [5.2A, p. 271]	$5x^4$ is a monomial. $6y^3 - 2y$ is a binomial. $2x^2 - 5x + 3$ is a trinomial.
The *degree of a polynomial* is the greatest of the degrees of any of its terms. [5.2A, p. 271]	The degree of the polynomial $x^3 + 3x^2y^2 - 4xy - 3$ is 4.
The terms of a polynomial in one variable are usually arranged so that the exponents of the variable decrease from left to right. This is called *descending order*. [5.2A, p. 271]	The polynomial $4x^3 + 5x^2 - x + 7$ is written in descending order.
A *polynomial function* is an expression whose terms are monomials. Polynomial functions include the *linear function* given by $f(x) = mx + b$; the *quadratic function* given by $f(x) = ax^2 + bx + c$, $a \neq 0$; and the *cubic function*, which is a third-degree polynomial function. The *leading coefficient* of a polynomial function is the coefficient of the variable with the largest exponent. The *constant term* is the term without a variable. [5.2A, p. 271]	$f(x) = 5x - 4$ is a linear function. $f(x) = 3x^2 - 2x + 1$ is a quadratic function. 3 is the leading coefficient, and 1 is the constant term. $f(x) = x^3 - 1$ is a cubic function.
To *factor a polynomial* means to write the polynomial as the product of other polynomials. [5.5A, p. 300]	$x^2 + 5x + 6 = (x + 2)(x + 3)$
A *quadratic trinomial* is a polynomial of the form $ax^2 + bx + c$, where a and b are nonzero coefficients and c is a nonzero constant. To *factor a quadratic trinomial* means to express the trinomial as a product of two binomials. [5.5C, p. 302]	$3x^2 + 10x - 8$ is a quadratic trinomial in which $a = 3$, $b = 10$, and $c = -8$. $3x^2 + 10x - 8 = (3x - 2)(x + 4)$

A polynomial is *nonfactorable over the integers* if it does not factor using only integers. [5.5C, p. 304]

$x^2 + x + 1$ is nonfactorable over the integers.

The product of a term and itself is called a *perfect square*. The *square root* of a perfect square is one of the two equal factors of the perfect square. [5.6A, p. 312]

$(5x)(5x) = 25x^2$; $25x^2$ is a perfect square.
$\sqrt{25x^2} = 5x$

The product of the same three factors is called a *perfect cube*. The *cube root* of a perfect cube is one of the three equal factors of the perfect cube. [5.6B, p. 314]

$(2x)(2x)(2x) = 8x^3$; $8x^3$ is a perfect cube.
$\sqrt[3]{8x^3} = 2x$

A trinomial is *quadratic in form* if it can be written as $au^2 + bu + c$. [5.6C, p. 315]

$6x^4 - 5x^2 - 4$
$6(x^2)^2 - 5(x^2) - 4$
$6u^2 - 5u - 4$

An equation of the form $ax^2 + bx + c = 0$, $a \neq 0$, is a *quadratic equation*. A quadratic equation is in *standard form* when the polynomial is written in descending order and is equal to zero. [5.7A, p. 322]

$3x^2 + 3x + 8 = 0$ is a quadratic equation in standard form.

Essential Rules and Procedures

Examples

Rule for Multiplying Exponential Expressions [5.1A, p. 259]
$x^m \cdot x^n = x^{m+n}$

$b^5 \cdot b^4 = b^{5+4} = b^9$

Rule for Simplifying the Power of an Exponential Expression [5.1A, p. 260]
$(x^m)^n = x^{mn}$

$(y^3)^7 = y^{3(7)} = y^{21}$

Rule for Simplifying Powers of Products [5.1A, p. 260]
$(x^m y^n)^p = x^{mp} y^{np}$

$(x^6 y^4 z^5)^2 = x^{6(2)} y^{4(2)} z^{5(2)} = x^{12} y^8 z^{10}$

Definition of Zero as an Exponent [5.1B, p. 261]
For $x \neq 0$, $x^0 = 1$. The expression 0^0 is not defined.

$17^0 = 1$
$(5y)^0 = 1, y \neq 0$

Definition of Negative Exponents [5.1B, p. 262]
For $x \neq 0$, $x^{-n} = \dfrac{1}{x^n}$ and $\dfrac{1}{x^{-n}} = x^n$.

$x^{-6} = \dfrac{1}{x^6}$ and $\dfrac{1}{x^{-6}} = x^6$

Rule for Simplifying Powers of Quotients [5.1B, p. 262]

For $y \neq 0$, $\left(\dfrac{x^m}{y^n}\right)^p = \dfrac{x^{mp}}{y^{np}}$.

$\left(\dfrac{x^2}{y^4}\right)^5 = \dfrac{x^{2 \cdot 5}}{y^{4 \cdot 5}} = \dfrac{x^{10}}{y^{20}}$

Rule for Negative Exponents on Fractional Expressions [5.1B, p. 263]

For $a \neq 0$, $b \neq 0$, $\left(\dfrac{a}{b}\right)^{-n} = \left(\dfrac{b}{a}\right)^{n}$.

$\left(\dfrac{3}{8}\right)^{-4} = \left(\dfrac{8}{3}\right)^{4}$

Rule for Dividing Exponential Expressions [5.1B, p. 263]

For $x \neq 0$, $\dfrac{x^m}{x^n} = x^{m-n}$.

$\dfrac{y^8}{y^3} = y^{8-3} = y^5$

Scientific Notation [5.1C, p. 265]

To express a number in scientific notation, write it in the form $a \times 10^n$, where a is a number between 1 and 10 and n is an integer.
If the number is greater than 10, the exponent on 10 will be positive.
If the number is less than 1, the exponent on 10 will be negative.

$367{,}000{,}000 = 3.67 \times 10^8$
$0.0000059 = 5.9 \times 10^{-6}$

To change a number in scientific notation to decimal notation, move the decimal point to the right if the exponent on 10 is positive and to the left if the exponent on 10 is negative. Move the decimal point the same number of places as the absolute value of the exponent on 10.

$2.418 \times 10^7 = 24{,}180{,}000$
$9.06 \times 10^{-5} = 0.0000906$

To write the additive inverse of a polynomial, change the sign of every term of the polynomial. [5.2B, p. 274]

The additive inverse of $-y^2 + 4y - 5$ is $y^2 - 4y + 5$.

To add polynomials, combine like terms, which means to add the coefficients of the like terms. [5.2B, p. 274]

$(8x^2 + 2x - 9) + (-3x^2 + 5x - 7)$
$= (8x^2 - 3x^2) + (2x + 5x) + (-9 - 7)$
$= 5x^2 + 7x - 16$

To subtract two polynomials, add the additive inverse of the second polynomial to the first polynomial. [5.2B, p. 275]

$(3y^2 - 8y + 6) - (-y^2 + 4y - 5)$
$= (3y^2 - 8y + 6) + (y^2 - 4y + 5)$
$= 4y^2 - 12y + 11$

To multiply a polynomial by a monomial, use the Distributive Property and the Rule for Multiplying Exponential Expressions. [5.3A, p. 279]

$-2x^3(4x^2 + 5x - 1)$
$= -8x^5 - 10x^4 + 2x^3$

The FOIL Method [5.3B, p. 281]

The product of two binomials can be found by adding the products of the **F**irst terms, the **O**uter terms, the **I**nner terms, and the **L**ast terms.

$(4x + 3)(2x - 5)$
$= (4x)(2x) + (4x)(-5)$
$\quad + (3)(2x) + (3)(-5)$
$= 8x^2 - 20x + 6x - 15$
$= 8x^2 - 14x - 15$

To divide a polynomial by a monomial, divide each term of the polynomial by the monomial. [5.4A, p. 289]

$$\frac{12x^5 + 8x^3 - 6x}{4x^2} = 3x^3 + 2x - \frac{3}{2x}$$

Synthetic Division [5.4C, p. 292]
Synthetic division is a shorter method of dividing a polynomial by a binomial of the form $x - a$. This method uses only the coefficients of the variable terms.

$(3x^3 - 9x - 5) \div (x - 2)$

$$
\begin{array}{r|rrrr}
2 & 3 & 0 & -9 & -5 \\
 & & 6 & 12 & 6 \\
\hline
 & 3 & 6 & 3 & 1
\end{array}
$$

$(3x^3 - 9x - 5) \div (x - 2)$

$= 3x^2 + 6x + 3 + \dfrac{1}{x - 2}$

Remainder Theorem [5.4D, p. 295]
If the polynomial $P(x)$ is divided by $x - a$, the remainder is $P(a)$.

$P(x) = x^3 - x^2 + x - 1$

$$
\begin{array}{r|rrrr}
-2 & 1 & -1 & 1 & -1 \\
 & & -2 & 6 & -14 \\
\hline
 & 1 & -3 & 7 & -15
\end{array}
$$

$P(-2) = -15$

Factoring Patterns [5.6A, p. 312]
The difference of two perfect squares equals the sum and difference of two terms: $a^2 - b^2 = (a + b)(a - b)$.
A perfect-square trinomial equals the square of a binomial:
$a^2 + 2ab + b^2 = (a + b)^2$.
$a^2 - 2ab + b^2 = (a - b)^2$.

$4x^2 - 9 = (2x + 3)(2x - 3)$

$4x^2 + 12x + 9 = (2x + 3)^2$

Factoring the Sum or the Difference of Two Cubes
[5.6B, p. 314]
$a^3 + b^3 = (a + b)(a^2 - ab + b^2)$
$a^3 - b^3 = (a - b)(a^2 + ab + b^2)$

$x^3 + 64 = (x + 4)(x^2 - 4x + 16)$
$8b^3 - 1 = (2b - 1)(4b^2 + 2b + 1)$

To Factor Completely [5.6D, p. 316]
When factoring a polynomial completely, ask the following questions about the polynomial.
1. Is there a common factor? If so, factor out the GCF.
2. If the polynomial is a binomial, is it the difference of two perfect squares, the sum of two perfect cubes, or the difference of two perfect cubes? If so, factor.
3. If the polynomial is a trinomial, is it a perfect-square trinomial or the product of two binomials? If so, factor.
4. Can the polynomial be factored by grouping? If so, factor.
5. Is each factor nonfactorable over the integers? If not, factor.

$54x^3 - 6x = 6x(9x^2 - 1)$
$= 6x(3x + 1)(3x - 1)$

Principle of Zero Products [5.7A, p. 322]
If the product of two factors is zero, then at least one of the factors must be zero.
If $ab = 0$, then $a = 0$ or $b = 0$.

$(x - 4)(x + 2) = 0$
$x - 4 = 0 \qquad x + 2 = 0$

Chapter 5 Review Exercises

1. Factor: $18a^5b^2 - 12a^3b^3 + 30a^2b$

2. Simplify: $\dfrac{15x^2 + 2x - 2}{3x - 2}$

3. Simplify: $(2x^{-1}y^2z^5)^4(-3x^3yz^{-3})^2$

4. Factor: $2ax + 4bx - 3ay - 6by$

5. Factor: $12 + x - x^2$

6. Use the Remainder Theorem to evaluate $P(x) = x^3 - 2x^2 + 3x - 5$ when $x = 2$.

7. Simplify: $(5x^2 - 8xy + 2y^2) - (x^2 - 3y^2)$

8. Factor: $24x^2 + 38x + 15$

9. Factor: $4x^2 + 12xy + 9y^2$

10. Simplify: $(-2a^2b^4)(3ab^2)$

11. Factor: $64a^3 - 27b^3$

12. Simplify: $\dfrac{4x^3 + 27x^2 + 10x + 2}{x + 6}$

13. Given $P(x) = 2x^3 - x + 7$, evaluate $P(-2)$.

14. Factor: $x^2 - 3x - 40$

15. Factor: $x^2y^2 - 9$

16. Multiply: $4x^2y(3x^3y^2 + 2xy - 7y^3)$

17. Factor: $x^{2n} - 12x^n + 36$

18. Solve: $6x^2 + 60 = 39x$

19. Simplify: $5x^2 - 4x[x - 3(3x + 2) + x]$

20. Factor: $3a^6 - 15a^4 - 18a^2$

21. Expand: $(4x - 3y)^2$

22. Simplify: $\dfrac{x^4 - 4}{x - 4}$

23. Add: $(3x^2 - 2x - 6) + (-x^2 - 3x + 4)$

24. Multiply: $(5x^2yz^4)(2xy^3z^{-1})(7x^{-2}y^{-2}z^3)$

25. Simplify: $\dfrac{3x^4yz^{-1}}{-12xy^3z^2}$

26. Write 948,000,000 in scientific notation.

27. Simplify: $\dfrac{3 \times 10^{-3}}{15 \times 10^2}$

28. Use the Remainder Theorem to evaluate $P(x) = -2x^3 + 2x^2 - 4$ when $x = -3$.

29. Divide: $\dfrac{16x^5 - 8x^3 + 20x}{4x}$

30. Divide: $\dfrac{12x^2 - 16x - 7}{6x + 1}$

31. Multiply: $a^{2n+3}(a^n - 5a + 2)$

32. Multiply: $(x + 6)(x^3 - 3x^2 - 5x + 1)$

33. Factor: $10a^3b^3 - 20a^2b^4 + 35ab^2$

34. Factor: $5x^{n+5} + x^{n+3} + 4x^2$

35. Factor: $x(y - 3) + 4(3 - y)$

36. Factor: $x^2 - 16x + 63$

37. Factor: $24x^2 + 61x - 8$

38. Find two linear functions, f and g, such that $f(x) \cdot g(x) = 5x^2 + 3x - 2$.

39. Factor: $36 - a^{2n}$

40. Factor: $8 - y^{3n}$

41. Factor: $36x^8 - 36x^4 + 5$

42. Factor: $3a^4b - 3ab^4$

43. Factor: $x^{4n} - 8x^{2n} + 16$

44. Solve: $x^3 - x^2 - 6x = 0$

45. Solve: $x^3 - 16x = 0$

46. Solve: $y^3 + y^2 - 36y - 36 = 0$

47. Factor: $15x^4 + x^2 - 6$

48. Simplify: $\dfrac{(2a^4b^{-3}c^2)^3}{(2a^3b^2c^{-1})^4}$

49. Multiply: $(x - 4)(3x + 2)(2x - 3)$

50. Factor: $21x^4y^4 + 23x^2y^2 + 6$

51. Solve: $x^3 + 16 = x(x + 16)$

52. Multiply: $(5a + 2b)(5a - 2b)$

53. Write 2.54×10^{-3} in decimal notation.

54. Factor: $6x^2 - 31x + 18$

55. Graph $y = x^2 + 1$.

56. For the polynomial $P(x) = 3x^5 - 6x^2 + 7x + 8$:
a. Identify the leading coefficient.
b. Identify the constant term.
c. State the degree.

57. **Physics** The mass of the moon is 3.7×10^{-8} times the mass of the sun. The mass of the sun is 2.19×10^{27} tons. Find the mass of the moon. Write the answer in scientific notation.

58. **Mathematics** The sum of a number and its square is 56. Find the number.

The Moon

59. **Astronomy** The most distant object visible from Earth without the aid of a telescope is the Great Galaxy of Andromeda. It takes light from the Great Galaxy of Andromeda 2.2×10^6 years to travel to Earth. Light travels about 6.7×10^8 mph. How far from Earth is the Great Galaxy of Andromeda? Use a 365-day year.

60. **Geometry** The length of a rectangle is $(5x + 3)$ cm. The width is $(2x - 7)$ cm. Find the area of the rectangle in terms of the variable x.

Chapter 5 Test

1. Factor: $16t^2 + 24t + 9$

2. Multiply: $-6rs^2(3r - 2s - 3)$

3. Given $P(x) = 3x^2 - 8x + 1$, evaluate $P(2)$.

4. Factor: $27x^3 - 8$

5. Factor: $16x^2 - 25$

6. Multiply: $(3t^3 - 4t^2 + 1)(2t^2 - 5)$

7. Simplify: $-5x[3 - 2(2x - 4) - 3x]$

8. Factor: $12x^3 + 12x^2 - 45x$

9. Solve: $6x^3 + x^2 - 6x - 1 = 0$

10. Simplify:
$(6x^3 - 7x^2 + 6x - 7) - (4x^3 - 3x^2 + 7)$

11. Write the number 0.000000501 in scientific notation.

12. Simplify: $\dfrac{14x^2 + x + 1}{7x - 3}$

13. Multiply: $(7 - 5x)(7 + 5x)$

14. Factor: $6a^4 - 13a^2 - 5$

15. Multiply: $(3a + 4b)(2a - 7b)$

16. Factor: $3x^4 - 23x^2 - 36$

17. Simplify: $(-4a^2b)^3(-ab^4)$

18. Solve: $6x^2 = x + 1$

19. Use the Remainder Theorem to evaluate $P(x) = -x^3 + 4x - 8$ when $x = -2$.

20. Simplify: $\dfrac{(2a^{-4}b^2)^3}{4a^{-2}b^{-1}}$

21. Simplify: $\dfrac{x^3 - 2x^2 - 5x + 7}{x + 3}$

22. Factor: $12 - 17x + 6x^2$

23. Factor: $6x^2 - 4x - 3xa + 2a$

24. Write the number of seconds in one week in scientific notation.

25. **Sports** An arrow is shot into the air with an upward velocity of 48 ft/s from a hill 32 ft high. How many seconds later will the arrow be 64 ft above the ground? Use the equation $h = 32 + 48t - 16t^2$, where h is the height in feet and t is the time in seconds.

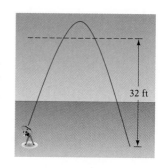

32 ft

26. **Geometry** The length of a rectangle is $(5x + 1)$ ft. The width is $(2x - 1)$ ft. Find the area of the rectangle in terms of the variable x.

Cumulative Review Exercises

1. Simplify: $8 - 2[-3 - (-1)]^2 + 4$

2. Evaluate $\dfrac{2a - b}{b - c}$ when $a = 4$, $b = -2$, and $c = 6$.

3. Identify the property that justifies the statement $2x + (-2x) = 0$.

4. Simplify: $2x - 4[x - 2(3 - 2x) + 4]$

5. Solve: $\dfrac{2}{3} - y = \dfrac{5}{6}$

6. Solve: $8x - 3 - x = -6 + 3x - 8$

7. Divide: $\dfrac{x^3 - 3}{x - 3}$

8. Solve: $3 - |2 - 3x| = -2$

9. Given $P(x) = 3x^2 - 2x + 2$, evaluate $P(-2)$.

10. What values of x are excluded from the domain of the function $f(x) = \dfrac{x + 1}{x + 2}$?

11. Find the range of the function given by $F(x) = 3x^2 - 4$ if the domain is $\{-2, -1, 0, 1, 2\}$.

12. Find the slope of the line containing the points $(-2, 3)$ and $(4, 2)$.

13. Find the equation of the line that contains the point $(-1, 2)$ and has slope $-\dfrac{3}{2}$.

14. Find the equation of the line that contains the point $(-2, 4)$ and is perpendicular to the line $3x + 2y = 4$.

15. Solve by using Cramer's Rule:
$$2x - 3y = 2$$
$$x + y = -3$$

16. Solve by the addition method:
$$x - y + z = 0$$
$$2x + y - 3z = -7$$
$$-x + 2y + 2z = 5$$

17. Graph $3x - 4y = 12$ by using the x- and y-intercepts.

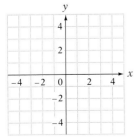

18. Graph the solution set: $-3x + 2y < 6$

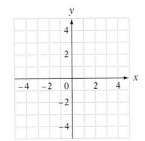

19. Solve by graphing:
$$x - 2y = 3$$
$$-2x + y = -3$$

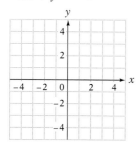

20. Graph the solution set:
$$2x + y < 3$$
$$-2x + y \geq 1$$

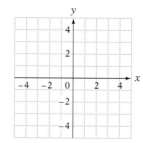

21. Simplify: $(4a^{-2}b^3)(2a^3b^{-1})^{-2}$

22. Simplify: $\dfrac{(5x^3y^{-3}z)^{-2}}{y^4z^{-2}}$

23. Simplify: $3 - (3 - 3^{-1})^{-1}$

24. Multiply: $(2x + 3)(2x^2 - 3x + 1)$

25. Factor: $-4x^3 + 14x^2 - 12x$

26. Factor: $a(x - y) - b(y - x)$

27. Factor: $x^4 - 16$

28. Factor: $2x^3 - 16$

29. Mathematics The sum of two integers is twenty-four. The difference between four times the smaller integer and nine is three less than twice the larger integer. Find the integers.

30. Mixtures How many ounces of pure gold that costs $360 per ounce must be mixed with 80 oz of an alloy that costs $120 per ounce to make a mixture that costs $200 per ounce?

31. Uniform Motion Two bicycles are 25 mi apart and are traveling toward each other. One cyclist is traveling at $\frac{2}{3}$ the rate of the other cyclist. They pass in 2 h. Find the rate of each cyclist.

32. Astronomy A space vehicle travels 2.4×10^5 mi from Earth to the moon at an average velocity of 2×10^4 mph. How long does it take the vehicle to reach the moon?

33. Uniform Motion The graph shows the relationship between the distance traveled and the time of travel. Find the slope of the line between the two points on the graph. Write a sentence that states the meaning of the slope.

<div style="margin-left: auto;">
chapter
</div>

6

Rational Expressions

This woman is a landscape architect. Her work involves drawing architectural plans for municipal parks and corporate landscapes. These architectural drawings are called scale drawings, and each dimension is carefully produced using ratios and proportions. She must be very accurate and consistent when creating her scale drawings, since contractors will depend on them as they create the real park or landscape. **Exercise 29 on page 366** asks you to determine the dimensions of a room based on a scale drawing in which one-quarter inch represents one foot.

OBJECTIVES

Section 6.1

A To find the domain of a rational function
B To simplify a rational function
C To multiply rational expressions
D To divide rational expressions

Section 6.2

A To rewrite rational expressions in terms of a common denominator
B To add or subtract rational expressions

Section 6.3

A To simplify a complex fraction

Section 6.4

A To solve a proportion
B To solve application problems

Section 6.5

A To solve a fractional equation
B To solve work problems
C To solve uniform motion problems

Section 6.6

A To solve variation problems

Need help? For online student resources, such as section quizzes, visit this textbook's website at **math.college.hmco.com/students.**

Do these exercises to prepare for Chapter 6.

1. Find the LCM of 10 and 25.

For Exercises 2 to 5, add, subtract, multiply, or divide.

2. $-\dfrac{3}{8} \cdot \dfrac{4}{9}$

3. $-\dfrac{4}{5} \div \dfrac{8}{15}$

4. $-\dfrac{5}{6} + \dfrac{7}{8}$

5. $-\dfrac{3}{8} - \left(-\dfrac{7}{12}\right)$

6. Simplify: $\dfrac{\dfrac{2}{3} - \dfrac{1}{4}}{\dfrac{1}{8} - 2}$

7. Evaluate $\dfrac{2x - 3}{x^2 - x + 1}$ for $x = 2$.

8. Solve: $4(2x + 1) = 3(x - 2)$

9. Solve: $10\left(\dfrac{t}{2} + \dfrac{t}{5}\right) = 10(1)$

10. Two planes start from the same point and fly in opposite directions. The first plane is flying 20 mph slower than the second plane. In 2 h, the planes are 480 mi apart. Find the rate of each plane.

GO FIGURE • • •

If 6 machines can fill 12 boxes of cereal in 7 minutes, how many boxes of cereal can be filled by 14 machines in 12 minutes?

6.1 Multiplication and Division of Rational Expressions

Objective A

To find the domain of a rational function

An expression in which the numerator and denominator are polynomials is called a **rational expression.** Examples of rational expressions are shown at the right.

$$\frac{9}{z} \qquad \frac{3x + 4}{2x^2 + 1} \qquad \frac{x^3 - x + 1}{x^2 - 3x - 5}$$

The expression $\dfrac{\sqrt{x} + 3}{x}$ is not a rational expression because $\sqrt{x} + 3$ is not a polynomial.

A function that is written in terms of a rational expression is a **rational function.** Each of the following equations represents a rational function.

$$f(x) = \frac{x^2 + 3}{2x - 1} \qquad g(t) = \frac{3}{t^2 - 4} \qquad R(z) = \frac{z^2 + 3z - 1}{z^2 + z - 12}$$

To evaluate a rational function, replace the variable by its value. Then simplify.

Integrating Technology

See the Keystroke Guide: *Evaluating Functions* for instructions on using a graphing calculator to evaluate a function.

> **HOW TO** Evaluate $f(-2)$ given $f(x) = \dfrac{x^2}{3x^2 - x - 9}$.
>
> $$f(x) = \frac{x^2}{3x^2 - x - 9}$$
> $$f(-2) = \frac{(-2)^2}{3(-2)^2 - (-2) - 9} = \frac{4}{12 + 2 - 9} = \frac{4}{5}$$
>
> • Replace x by -2. Then simplify.

Because division by zero is not defined, the domain of a rational function must exclude those numbers for which the value of the polynomial in the denominator is zero.

Integrating Technology

See the Keystroke Guide: *Graph* for instructions on using a graphing calculator to graph a function.

The graph of $f(x) = \dfrac{1}{x - 2}$ is shown at the right. Note that the graph never intersects the line $x = 2$ (shown as a dotted line). This value of x is excluded from the domain of $f(x) = \dfrac{1}{x - 2}$.

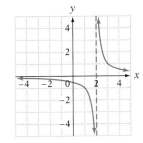

> **HOW TO** Determine the domain of $g(x) = \dfrac{x^2 + 4}{3x - 6}$.

The domain of g must exclude values of x for which the denominator is zero. To find these values, set the denominator equal to zero and solve for x.

$g(x)$

> $3x - 6 = 0$ • Set the denominator equal to zero.
> $3x = 6$ • Solve for x.
> $x = 2$ • This value must be *excluded* from the domain.

The domain of g is $\{x \,|\, x \neq 2\}$.

Example 1

Given $f(x) = \dfrac{3x - 4}{x^2 - 2x + 1}$, find $f(-2)$.

Solution

$$f(x) = \frac{3x - 4}{x^2 - 2x + 1}$$

$$f(-2) = \frac{3(-2) - 4}{(-2)^2 - 2(-2) + 1}$$

- Replace x by -2.

$$= \frac{-6 - 4}{4 + 4 + 1}$$

$$= \frac{-10}{9} = -\frac{10}{9}$$

You Try It 1

Given $f(x) = \dfrac{3 - 5x}{x^2 + 5x + 6}$, find $f(2)$.

Your solution

Example 2

Find the domain of $f(x) = \dfrac{x^2 - 1}{x^2 - 2x - 15}$.

Solution

Set the denominator equal to zero. Then solve for x.

$$x^2 - 2x - 15 = 0$$
$$(x - 5)(x + 3) = 0$$

- Solve the quadratic equation by factoring.

$$x - 5 = 0 \quad x + 3 = 0$$
$$x = 5 \qquad x = -3$$

The domain is $\{x \mid x \neq -3, 5\}$.

You Try It 2

Find the domain of $f(x) = \dfrac{2x + 1}{2x^2 - 7x + 3}$.

Your solution

Solutions on p. S17

Objective B **To simplify a rational function**

The Multiplication Property of One is used to write the **simplest form of a rational expression,** which means that the numerator and denominator of the rational expression have no common factors.

HOW TO Simplify: $\dfrac{x^2 - 25}{x^2 + 13x + 40}$

$$\frac{x^2 - 25}{x^2 + 13x + 40} = \frac{(x - 5)}{(x + 8)} \cdot \frac{(x + 5)}{(x + 5)} = \frac{x - 5}{x + 8} \cdot 1 = \frac{x - 5}{x + 8}, \qquad x \neq -8, -5$$

The requirement $x \neq -8, -5$ is necessary because division by 0 is not allowed.

The simplification above is usually shown with slashes to indicate that a common factor has been removed:

$$\frac{x^2 - 25}{x^2 + 13x + 40} = \frac{(x - 5)\cancel{(x + 5)}}{(x + 8)\cancel{(x + 5)}} = \frac{x - 5}{x + 8}, \qquad x \neq -8, -5$$

We will show a simplification with slashes. We will not show the restrictions that prevent division by zero. Nonetheless, those restrictions *always* are implied.

TAKE NOTE

Recall that
$b - a = -(a - b)$.

Therefore,
$4 - x = -(x - 4)$.

In general,
$$\dfrac{b - a}{a - b} = \dfrac{\overset{1}{-(a - b)}}{\underset{1}{a - b}} = \dfrac{-1}{1}$$
$$= -1.$$

HOW TO Simplify: $\dfrac{12 + 5x - 2x^2}{2x^2 - 3x - 20}$

$$\dfrac{12 + 5x - 2x^2}{2x^2 - 3x - 20} = \dfrac{(4 - x)(3 + 2x)}{(x - 4)(2x + 5)}$$

• Factor the numerator and denominator.

$$= \dfrac{\overset{-1}{\cancel{(4 - x)}}(3 + 2x)}{\underset{1}{\cancel{(x - 4)}}(2x + 5)}$$

• $\dfrac{4 - x}{x - 4} = \dfrac{-\overset{1}{\cancel{(x - 4)}}}{\underset{1}{\cancel{x - 4}}} = \dfrac{-1}{1} = -1$

$$= -\dfrac{2x + 3}{2x + 5}$$

• Write the answer in simplest form.

Example 3

Simplify: $\dfrac{6x^3 - 9x^2}{12x^2 - 18x}$

Solution

$$\dfrac{6x^3 - 9x^2}{12x^2 - 18x} = \dfrac{3x^2(2x - 3)}{6x(2x - 3)}$$

• Factor out the GCF.

$$= \dfrac{3x^2\cancel{(2x - 3)}}{6x\cancel{(2x - 3)}} = \dfrac{x}{2}$$

You Try It 3

Simplify: $\dfrac{6x^4 - 24x^3}{12x^3 - 48x^2}$

Your solution

Example 4

Simplify: $\dfrac{12x^3y^2 + 6x^3y^3}{6x^2y^2}$

Solution

$$\dfrac{12x^3y^2 + 6x^3y^3}{6x^2y^2} = \dfrac{6x^3y^2(2 + y)}{6x^2y^2}$$

• Factor out the GCF.

$$= x(2 + y)$$

You Try It 4

Simplify: $\dfrac{21a^3b - 14a^3b^2}{7a^2b}$

Your solution

Example 5

Simplify: $\dfrac{2x - 8x^2}{16x^3 - 28x^2 + 6x}$

Solution

$$\dfrac{2x - 8x^2}{16x^3 - 28x^2 + 6x} = \dfrac{2x(1 - 4x)}{2x(8x^2 - 14x + 3)}$$

$$= \dfrac{2x(1 - 4x)}{2x(4x - 1)(2x - 3)}$$

$$= \dfrac{2x\overset{-1}{\cancel{(1 - 4x)}}}{2x\underset{1}{\cancel{(4x - 1)}}(2x - 3)}$$

$$= -\dfrac{1}{2x - 3}$$

You Try It 5

Simplify: $\dfrac{20x - 15x^2}{15x^3 - 5x^2 - 20x}$

Your solution

Solutions on pp. S17–S18

Example 6

Simplify: $\dfrac{x^{2n} + x^n - 2}{x^{2n} - 1}$

Solution

$$\dfrac{x^{2n} + x^n - 2}{x^{2n} - 1} = \dfrac{(x^n - 1)(x^n + 2)}{(x^n - 1)(x^n + 1)}$$

$$= \dfrac{\cancel{(x^n - 1)}^{1}(x^n + 2)}{\cancel{(x^n - 1)}_{1}(x^n + 1)}$$

$$= \dfrac{x^n + 2}{x^n + 1}$$

You Try It 6

Simplify: $\dfrac{x^{2n} + x^n - 12}{x^{2n} - 3x^n}$

Your solution

Solution on p. S18

Objective C **To multiply rational expressions**

The product of two fractions is a fraction whose numerator is the product of the numerators of the two fractions and whose denominator is the product of the denominators of the two fractions.

$$\dfrac{a}{b} \cdot \dfrac{c}{d} = \dfrac{ac}{bd} \qquad \dfrac{5}{a + 2} \cdot \dfrac{b - 3}{3} = \dfrac{5(b - 3)}{(a + 2)3} = \dfrac{5b - 15}{3a + 6}$$

The product of two rational expressions can often be simplified by factoring the numerator and the denominator.

HOW TO Simplify: $\dfrac{x^2 - 2x}{2x^2 + x - 15} \cdot \dfrac{2x^2 - x - 10}{x^2 - 4}$

$$\dfrac{x^2 - 2x}{2x^2 + x - 15} \cdot \dfrac{2x^2 - x - 10}{x^2 - 4}$$

$$= \dfrac{x(x - 2)}{(x + 3)(2x - 5)} \cdot \dfrac{(x + 2)(2x - 5)}{(x + 2)(x - 2)}$$
- Factor the numerator and the denominator of each fraction.

$$= \dfrac{x(x - 2)(x + 2)(2x - 5)}{(x + 3)(2x - 5)(x + 2)(x - 2)}$$
- Multiply.

$$= \dfrac{x\cancel{(x - 2)}^{1}\cancel{(x + 2)}^{1}\cancel{(2x - 5)}^{1}}{(x + 3)\cancel{(2x - 5)}_{1}\cancel{(x + 2)}_{1}\cancel{(x - 2)}_{1}}$$
- Simplify.

$$= \dfrac{x}{x + 3}$$
- Write the answer in simplest form.

Example 7

Simplify: $\dfrac{2x^2 - 6x}{3x - 6} \cdot \dfrac{6x - 12}{8x^3 - 12x^2}$

Solution

$\dfrac{2x^2 - 6x}{3x - 6} \cdot \dfrac{6x - 12}{8x^3 - 12x^2}$

$= \dfrac{2x(x - 3)}{3(x - 2)} \cdot \dfrac{6(x - 2)}{4x^2(2x - 3)}$

$= \dfrac{2x(x - 3) \cdot 6(x - 2)}{3(x - 2) \cdot 4x^2(2x - 3)}$

$= \dfrac{2x(x - 3) \cdot 6\overset{1}{\cancel{(x - 2)}}}{3\underset{1}{\cancel{(x - 2)}} \cdot 4x^2(2x - 3)}$

$= \dfrac{x - 3}{x(2x - 3)}$

You Try It 7

Simplify: $\dfrac{12 + 5x - 3x^2}{x^2 + 2x - 15} \cdot \dfrac{2x^2 + x - 45}{3x^2 + 4x}$

Your solution

Example 8

Simplify: $\dfrac{6x^2 + x - 2}{6x^2 + 7x + 2} \cdot \dfrac{2x^2 + 9x + 4}{4 - 7x - 2x^2}$

Solution

$\dfrac{6x^2 + x - 2}{6x^2 + 7x + 2} \cdot \dfrac{2x^2 + 9x + 4}{4 - 7x - 2x^2}$

$= \dfrac{(2x - 1)(3x + 2)}{(3x + 2)(2x + 1)} \cdot \dfrac{(2x + 1)(x + 4)}{(1 - 2x)(4 + x)}$

$= \dfrac{(2x - 1)(3x + 2) \cdot (2x + 1)(x + 4)}{(3x + 2)(2x + 1) \cdot (1 - 2x)(4 + x)}$

$= \dfrac{\overset{-1}{\cancel{(2x - 1)}}\overset{1}{\cancel{(3x + 2)}}\overset{1}{\cancel{(2x + 1)}}\overset{1}{\cancel{(x + 4)}}}{\underset{1}{\cancel{(3x + 2)}}\underset{1}{\cancel{(2x + 1)}}\underset{1}{\cancel{(1 - 2x)}}\underset{1}{\cancel{(x + 4)}}}$

$= -1$

You Try It 8

Simplify: $\dfrac{2x^2 - 13x + 20}{x^2 - 16} \cdot \dfrac{2x^2 + 9x + 4}{6x^2 - 7x - 5}$

Your solution

Solutions on p. S18

Objective D **To divide rational expressions**

The **reciprocal of a rational expression** is the rational expression with the numerator and denominator interchanged.

$$\text{Rational Expression} \quad \left[\begin{array}{cc} \dfrac{a}{b} & \dfrac{b}{a} \\[2mm] \dfrac{a^2 - 2y}{4} & \dfrac{4}{a^2 - 2y} \end{array} \right\} \quad \text{Reciprocal}$$

To divide two rational expressions, multiply by the reciprocal of the divisor.

$$\dfrac{a}{b} \div \dfrac{c}{d} = \dfrac{a}{b} \cdot \dfrac{d}{c} = \dfrac{ad}{bc}$$

$$\dfrac{2}{a} \div \dfrac{5}{b} = \dfrac{2}{a} \cdot \dfrac{b}{5} = \dfrac{2b}{5a}$$

$$\dfrac{x + y}{2} \div \dfrac{x - y}{5} = \dfrac{x + y}{2} \cdot \dfrac{5}{x - y} = \dfrac{(x + y)5}{2(x - y)} = \dfrac{5x + 5y}{2x - 2y}$$

Example 9

Simplify: $\dfrac{12x^2y}{25a^2b^5} \div \dfrac{9x^3y^4}{10a^3b^4}$

Solution

$\dfrac{12x^2y}{25a^2b^5} \div \dfrac{9x^3y^4}{10a^3b^4} = \dfrac{12x^2y}{25a^2b^5} \cdot \dfrac{10a^3b^4}{9x^3y^4}$

$= \dfrac{12x^2y \cdot 10a^3b^4}{25a^2b^5 \cdot 9x^3y^4}$

$= \dfrac{8a}{15xy^3b}$

• Rewrite division as multiplication by the reciprocal.

You Try It 9

Simplify: $\dfrac{7a^3b^7}{15x^2y^3} \div \dfrac{21a^5b^2}{20x^4y}$

Your solution

Example 10

Simplify: $\dfrac{12x^2y^2 - 24xy^2}{5z^2} \div \dfrac{4x^3y - 8x^2y}{3z^4}$

Solution

$\dfrac{12x^2y^2 - 24xy^2}{5z^2} \div \dfrac{4x^3y - 8x^2y}{3z^4}$

$= \dfrac{12x^2y^2 - 24xy^2}{5z^2} \cdot \dfrac{3z^4}{4x^3y - 8x^2y}$

$= \dfrac{12xy^2(x - 2)}{5z^2} \cdot \dfrac{3z^4}{4x^2y(x - 2)}$

$= \dfrac{12xy^2(x - 2)3z^4}{5z^2 \cdot 4x^2y(x - 2)}$

$= \dfrac{12xy^2(x \overset{1}{\cancel{- 2}})3z^4}{5z^2 \cdot 4x^2y(x \underset{1}{\cancel{- 2}})}$

$= \dfrac{9yz^2}{5x}$

• Rewrite division as multiplication by the reciprocal.

You Try It 10

Simplify: $\dfrac{6x^2 - 3xy}{10ab^4} \div \dfrac{16x^2y^2 - 8xy^3}{15a^2b^2}$

Your solution

Example 11

Simplify: $\dfrac{3y^2 - 10y + 8}{3y^2 + 8y - 16} \div \dfrac{2y^2 - 7y + 6}{2y^2 + 5y - 12}$

Solution

$\dfrac{3y^2 - 10y + 8}{3y^2 + 8y - 16} \div \dfrac{2y^2 - 7y + 6}{2y^2 + 5y - 12}$

$= \dfrac{3y^2 - 10y + 8}{3y^2 + 8y - 16} \cdot \dfrac{2y^2 + 5y - 12}{2y^2 - 7y + 6}$

$= \dfrac{(y - 2)(3y - 4)}{(3y - 4)(y + 4)} \cdot \dfrac{(y + 4)(2y - 3)}{(y - 2)(2y - 3)}$

$= \dfrac{(y - 2)(3y - 4)(y + 4)(2y - 3)}{(3y - 4)(y + 4)(y - 2)(2y - 3)}$

$= \dfrac{\overset{1}{\cancel{(y - 2)}}\overset{1}{\cancel{(3y - 4)}}\overset{1}{\cancel{(y + 4)}}\overset{1}{\cancel{(2y - 3)}}}{\underset{1}{\cancel{(3y - 4)}}\underset{1}{\cancel{(y + 4)}}\underset{1}{\cancel{(y - 2)}}\underset{1}{\cancel{(2y - 3)}}}$

$= 1$

You Try It 11

Simplify: $\dfrac{6x^2 - 7x + 2}{3x^2 + x - 2} \div \dfrac{4x^2 - 8x + 3}{5x^2 + x - 4}$

Your solution

Solutions on p. S18

6.1 Exercises

Objective A **To find the domain of a rational function**

1. What is a rational function? Give an example of a rational function.

2. What values are excluded from the domain of a rational function?

3. Given $f(x) = \dfrac{2}{x-3}$, find $f(4)$.

4. Given $f(x) = \dfrac{-7}{5-x}$, find $f(-2)$.

5. Given $f(x) = \dfrac{x-2}{x+4}$, find $f(-2)$.

6. Given $f(x) = \dfrac{x-3}{2x-1}$, find $f(3)$.

7. Given $f(x) = \dfrac{1}{x^2 - 2x + 1}$, find $f(-2)$.

8. Given $f(x) = \dfrac{-3}{x^2 - 4x + 2}$, find $f(-1)$.

9. Given $f(x) = \dfrac{x-2}{2x^2 + 3x + 8}$, find $f(3)$.

10. Given $f(x) = \dfrac{x^2}{3x^2 - 3x + 5}$, find $f(4)$.

11. Given $f(x) = \dfrac{x^2 - 2x}{x^3 - x + 4}$, find $f(-1)$.

12. Given $f(x) = \dfrac{8 - x^2}{x^3 - x^2 + 4}$, find $f(-3)$.

For Exercises 13 to 24, find the domain of the function.

13. $f(x) = \dfrac{4}{x-3}$

14. $G(x) = \dfrac{-2}{x+2}$

15. $H(x) = \dfrac{x}{x+4}$

16. $F(x) = \dfrac{3x}{x-5}$

17. $h(x) = \dfrac{5x}{3x+9}$

18. $f(x) = \dfrac{-2x}{6-2x}$

19. $q(x) = \dfrac{4-x}{(x-4)(3x-2)}$

20. $p(x) = \dfrac{2x+1}{(2x+5)(3x-6)}$

21. $f(x) = \dfrac{2x-1}{x^2 + x - 6}$

22. $G(x) = \dfrac{3-4x}{x^2 + 4x - 5}$

23. $f(x) = \dfrac{x+1}{x^2 + 1}$

24. $g(x) = \dfrac{x^2 + 1}{x^2}$

Objective B **To simplify a rational function**

25. ✏️ When is a rational expression in simplest form?

26. ✏️ Are the rational expressions $\dfrac{x(x-2)}{2(x-2)}$ and $\dfrac{x}{2}$ equal for all values of x? Why or why not?

For Exercises 27 to 59, simplify.

27. $\dfrac{4-8x}{4}$

28. $\dfrac{8y+2}{2}$

29. $\dfrac{6x^2-2x}{2x}$

30. $\dfrac{3y-12y^2}{3y}$

31. $\dfrac{8x^2(x-3)}{4x(x-3)}$

32. $\dfrac{16y^4(y+8)}{12y^3(y+8)}$

33. $\dfrac{-36a^2-48a}{18a^3+24a^2}$

34. $\dfrac{a^2+4a}{4a-16}$

35. $\dfrac{3x-6}{x^2+2x}$

36. $\dfrac{16x^3-8x^2+12x}{4x}$

37. $\dfrac{3x^3y^3-12x^2y^2+15xy}{3xy}$

38. $\dfrac{-10a^4-20a^3+30a^2}{-10a^2}$

39. $\dfrac{x^{2n}+x^ny^n}{x^{2n}-y^{2n}}$

40. $\dfrac{a^{2n}-b^{2n}}{5a^{3n}+5a^{2n}b^n}$

41. $\dfrac{x^2-7x+12}{x^2-9x+20}$

42. $\dfrac{6-x-x^2}{3x^2-10x+8}$

43. $\dfrac{3x^2+10x-8}{8-14x+3x^2}$

44. $\dfrac{14-19x-3x^2}{3x^2-23x+14}$

45. $\dfrac{a^2-b^2}{a^3+b^3}$

46. $\dfrac{x^4-y^4}{x^2+y^2}$

47. $\dfrac{8x^3-y^3}{4x^2-y^2}$

48. $\dfrac{x^2-4}{a(x+2)-b(x+2)}$

49. $\dfrac{x^2(a-2)-a+2}{ax^2-ax}$

50. $\dfrac{x^4+3x^2+2}{x^4-1}$

51. $\dfrac{x^4-2x^2-3}{x^4+2x^2+1}$

52. $\dfrac{x^2y^2+4xy-21}{x^2y^2-10xy+21}$

53. $\dfrac{6x^2y^2+11xy+4}{9x^2y^2+9xy-4}$

54. $\dfrac{a^{2n}-a^n-2}{a^{2n}+3a^n+2}$

55. $\dfrac{a^{2n}+a^n-12}{a^{2n}-2a^n-3}$

56. $\dfrac{a^{2n}-1}{a^{2n}-2a^n+1}$

57. $\dfrac{a^{2n} + 2a^n b^n + b^{2n}}{a^{2n} - b^{2n}}$

58. $\dfrac{(x - 3) - b(x - 3)}{b(x + 3) - x - 3}$

59. $\dfrac{x^2(a + b) + a + b}{x^4 - 1}$

Objective C **To multiply rational expressions**

For Exercises 60 to 75, simplify.

60. $\dfrac{27a^2 b^5}{16xy^2} \cdot \dfrac{20x^2 y^3}{9a^2 b}$

61. $\dfrac{15x^2 y^4}{24ab^3} \cdot \dfrac{28a^2 b^4}{35xy^4}$

62. $\dfrac{3x - 15}{4x^2 - 2x} \cdot \dfrac{20x^2 - 10x}{15x - 75}$

63. $\dfrac{2x^2 + 4x}{8x^2 - 40x} \cdot \dfrac{6x^3 - 30x^2}{3x^2 + 6x}$

64. $\dfrac{x^2 y^3}{x^2 - 4x - 5} \cdot \dfrac{2x^2 - 13x + 15}{x^4 y^3}$

65. $\dfrac{2x^2 - 5x + 3}{x^6 y^3} \cdot \dfrac{x^4 y^4}{2x^2 - x - 3}$

66. $\dfrac{x^2 - 3x + 2}{x^2 - 8x + 15} \cdot \dfrac{x^2 + x - 12}{8 - 2x - x^2}$

67. $\dfrac{x^2 + x - 6}{12 + x - x^2} \cdot \dfrac{x^2 + x - 20}{x^2 - 4x + 4}$

68. $\dfrac{x^{n+1} + 2x^n}{4x^2 - 6x} \cdot \dfrac{8x^2 - 12x}{x^{n+1} - x^n}$

69. $\dfrac{x^{2n} + 2x^n}{x^{n+1} + 2x} \cdot \dfrac{x^2 - 3x}{x^{n+1} - 3x^n}$

70. $\dfrac{x^{2n} - x^n - 6}{x^{2n} + x^n - 2} \cdot \dfrac{x^{2n} - 5x^n - 6}{x^{2n} - 2x^n - 3}$

71. $\dfrac{x^{2n} + 3x^n + 2}{x^{2n} - x^n - 6} \cdot \dfrac{x^{2n} + x^n - 12}{x^{2n} - 1}$

72. $\dfrac{x^3 - y^3}{2x^2 + xy - 3y^2} \cdot \dfrac{2x^2 + 5xy + 3y^2}{x^2 + xy + y^2}$

73. $\dfrac{x^4 - 5x^2 + 4}{3x^2 - 4x - 4} \cdot \dfrac{3x^2 - 10x - 8}{x^2 - 4}$

74. $\dfrac{x^2 + x - 6}{3x^2 + 5x - 12} \cdot \dfrac{2x^2 - 14x}{16x^2 - 4x} \cdot \dfrac{12x^2 - 19x + 4}{x^2 + 5x - 14}$

75. $\dfrac{x^2 - y^2}{x^2 + xy + y^2} \cdot \dfrac{x^2 - xy}{3x^2 - 3xy} \cdot \dfrac{x^3 - y^3}{x^2 - 2xy + y^2}$

Objective D **To divide rational expressions**

For Exercises 76 to 93, simplify.

76. $\dfrac{6x^2y^4}{35a^2b^5} \div \dfrac{12x^3y^3}{7a^4b^5}$

77. $\dfrac{12a^4b^7}{13x^2y^2} \div \dfrac{18a^5b^6}{26xy^3}$

78. $\dfrac{2x - 6}{6x^2 - 15x} \div \dfrac{4x^2 - 12x}{18x^3 - 45x^2}$

79. $\dfrac{4x^2 - 4y^2}{6x^2y^2} \div \dfrac{3x^2 + 3xy}{2x^2y - 2xy^2}$

80. $\dfrac{2x^2 - 2y^2}{14x^2y^4} \div \dfrac{x^2 + 2xy + y^2}{35xy^3}$

81. $\dfrac{8x^3 + 12x^2y}{4x^2 - 9y^2} \div \dfrac{16x^2y^2}{4x^2 - 12xy + 9y^2}$

82. $\dfrac{x^2 - 8x + 15}{x^2 + 2x - 35} \div \dfrac{15 - 2x - x^2}{x^2 + 9x + 14}$

83. $\dfrac{2x^2 + 13x + 20}{8 - 10x - 3x^2} \div \dfrac{6x^2 - 13x - 5}{9x^2 - 3x - 2}$

84. $\dfrac{x^{2n} + x^n}{2x - 2} \div \dfrac{4x^n + 4}{x^{n+1} - x^n}$

85. $\dfrac{x^{2n} - 4}{4x^n + 8} \div \dfrac{x^{n+1} - 2x}{4x^3 - 12x^2}$

86. $\dfrac{14 + 17x - 6x^2}{3x^2 + 14x + 8} \div \dfrac{4x^2 - 49}{2x^2 + 15x + 28}$

87. $\dfrac{16x^2 - 9}{6 - 5x - 4x^2} \div \dfrac{16x^2 + 24x + 9}{4x^2 + 11x + 6}$

88. $\dfrac{2x^{2n} - x^n - 6}{x^{2n} - x^n - 2} \div \dfrac{2x^{2n} + x^n - 3}{x^{2n} - 1}$

89. $\dfrac{x^{4n} - 1}{x^{2n} + x^n - 2} \div \dfrac{x^{2n} + 1}{x^{2n} + 3x^n + 2}$

90. $\dfrac{6x^2 + 6x}{3x + 6x^2 + 3x^3} \div \dfrac{x^2 - 1}{1 - x^3}$

91. $\dfrac{x^3 + y^3}{2x^3 + 2x^2y} \div \dfrac{3x^3 - 3x^2y + 3xy^2}{6x^2 - 6y^2}$

92. $\dfrac{x^2 - 9}{x^2 - x - 6} \div \dfrac{2x^2 + x - 15}{x^2 + 7x + 10} \cdot \dfrac{2x^2 - x - 10}{x^2 - 25}$

93. $\dfrac{3x^2 + 10x - 8}{x^2 + 4x + 3} \cdot \dfrac{x^2 + 6x + 9}{3x^2 - 5x + 2} \div \dfrac{x^2 + x - 12}{x^2 - 1}$

APPLYING THE CONCEPTS

For Exercises 94 and 95, simplify.

94. $\dfrac{3x^2 + 6x}{4x^2 - 16} \cdot \dfrac{2x + 8}{x^2 + 2x} \div \dfrac{3x - 9}{5x - 20}$

95. $\dfrac{5y^2 - 20}{3y^2 - 12y} \cdot \dfrac{9y^3 + 6y}{2y^2 - 4y} \div \dfrac{y^3 + 2y^2}{2y^2 - 8y}$

6.2 Addition and Subtraction of Rational Expressions

Objective A

To rewrite rational expressions in terms of a common denominator

Study Tip

As you know, often in mathematics you learn one skill in order to perform another. This is true of this objective. You are learning to rewrite rational expressions in terms of a common denominator in order to add and subtract rational expressions in the next objective. To ensure success, be certain you understand this lesson before studying the next.

In adding or subtracting rational expressions, it is frequently necessary to express the rational expressions in terms of a common denominator. This common denominator is the least common multiple (LCM) of the denominators.

The **least common multiple (LCM) of two or more polynomials** is the simplest polynomial that contains the factors of each polynomial. To find the LCM, first factor each polynomial completely. The LCM is the product of each factor the greatest number of times it occurs in any one factorization.

HOW TO Find the LCM of $3x^2 + 15x$ and $6x^4 + 24x^3 - 30x^2$.

Factor each polynomial.

$$3x^2 + 15x = 3x(x + 5)$$
$$6x^4 + 24x^3 - 30x^2 = 6x^2(x^2 + 4x - 5)$$
$$= 6x^2(x - 1)(x + 5)$$

The LCM is the product of the LCM of the numerical coefficients and each variable factor the greatest number of times it occurs in any one factorization.

$$\text{LCM} = 6x^2(x - 1)(x + 5)$$

TAKE NOTE

This is the only section in this text in which the numerator in the final answer is in unfactored form. In building fractions, the fraction is not in simplest form, and leaving the fractions in unfactored form removes the temptation to cancel.

HOW TO Write the fractions $\frac{5}{6x^2y}$ and $\frac{a}{9xy^3}$ in terms of the LCM of the denominators.

The LCM is $18x^2y^3$.

$$\frac{5}{6x^2y} = \frac{5}{6x^2y} \cdot \frac{3y^2}{3y^2} = \frac{15y^2}{18x^2y^3}$$

$$\frac{a}{9xy^3} = \frac{a}{9xy^3} \cdot \frac{2x}{2x} = \frac{2ax}{18x^2y^3}$$

- Find the LCM of the denominators.
- For each fraction, multiply the numerator and denominator by the factor whose product with the denominator is the LCM.

HOW TO Write the fractions $\frac{x + 2}{x^2 - 2x}$ and $\frac{5x}{3x - 6}$ in terms of the LCM of the denominators.

$x(x - 2)$ $3(x - 2)$

The LCM is $3x(x - 2)$.

$$\frac{x + 2}{x^2 - 2x} = \frac{x + 2}{x(x - 2)} \cdot \frac{3}{3} = \frac{3x + 6}{3x(x - 2)}$$

$$\frac{5x}{3x - 6} = \frac{5x}{3(x - 2)} \cdot \frac{x}{x} = \frac{5x^2}{3x(x - 2)}$$

- Find the LCM of the denominators.
- For each fraction, multiply the numerator and denominator by the factor whose product with the denominator is the LCM.

TAKE NOTE

$x^2 - 2x = x(x - 2)$
$3x - 6 = 3(x - 2)$
The LCM of $x(x - 2)$ and $3(x - 2)$ is $3x(x - 2)$.

Example 1

Write the fractions $\dfrac{3x}{x-1}$ and $\dfrac{4}{2x+5}$ in terms of the LCM of the denominators.

Solution

The LCM of $x - 1$ and $2x + 5$ is $(x - 1)(2x + 5)$.

$$\dfrac{3x}{x-1} = \dfrac{3x}{x-1} \cdot \dfrac{2x+5}{2x+5} = \dfrac{6x^2 + 15x}{(x-1)(2x+5)}$$

$$\dfrac{4}{2x+5} = \dfrac{4}{2x+5} \cdot \dfrac{x-1}{x-1} = \dfrac{4x-4}{(x-1)(2x+5)}$$

You Try It 1

Write the fractions $\dfrac{2x}{2x-5}$ and $\dfrac{3}{x+4}$ in terms of the LCM of the denominators.

Your solution

Example 2

Write the fractions $\dfrac{2a-3}{a^2-2a}$ and $\dfrac{a+1}{2a^2-a-6}$ in terms of the LCM of the denominators.

Solution

$a^2 - 2a = a(a - 2)$;
$2a^2 - a - 6 = (2a + 3)(a - 2)$

The LCM is $a(a - 2)(2a + 3)$.

$$\dfrac{2a-3}{a^2-2a} = \dfrac{2a-3}{a(a-2)} \cdot \dfrac{2a+3}{2a+3}$$

$$= \dfrac{4a^2-9}{a(a-2)(2a+3)}$$

$$\dfrac{a+1}{2a^2-a-6} = \dfrac{a+1}{(a-2)(2a+3)} \cdot \dfrac{a}{a}$$

$$= \dfrac{a^2+a}{a(a-2)(2a+3)}$$

You Try It 2

Write the fractions $\dfrac{3x}{2x^2-11x+15}$ and $\dfrac{x-2}{x^2-3x}$ in terms of the LCM of the denominators.

Your solution

Example 3

Write the fractions $\dfrac{2x-3}{3x-x^2}$ and $\dfrac{3x}{x^2-4x+3}$ in terms of the LCM of the denominators.

Solution

$3x - x^2 = x(3 - x) = -x(x - 3)$;
$x^2 - 4x + 3 = (x - 3)(x - 1)$

The LCM is $x(x - 3)(x - 1)$.

$$\dfrac{2x-3}{3x-x^2} = -\dfrac{2x-3}{x(x-3)} \cdot \dfrac{x-1}{x-1}$$

$$= -\dfrac{2x^2-5x+3}{x(x-3)(x-1)}$$

$$\dfrac{3x}{x^2-4x+3} = \dfrac{3x}{(x-3)(x-1)} \cdot \dfrac{x}{x}$$

$$= \dfrac{3x^2}{x(x-3)(x-1)}$$

You Try It 3

Write the fractions $\dfrac{2x-7}{2x-x^2}$ and $\dfrac{3x-2}{3x^2-5x-2}$ in terms of the LCM of the denominators.

Your solution

Solutions on p. S18

Objective B **To add or subtract rational expressions**

$$\frac{a}{c} + \frac{b}{c} = \frac{a+b}{c}$$

When adding rational expressions in which the denominators are the same, add the numerators. The denominator of the sum is the common denominator.

$$\frac{4x}{15} + \frac{8x}{15} = \frac{4x+8x}{15} = \frac{12x}{15} = \frac{4x}{5}$$ • Note that the sum is written in simplest form.

$$\frac{a}{a^2 - b^2} + \frac{b}{a^2 - b^2} = \frac{a+b}{a^2 - b^2} = \frac{a+b}{(a-b)(a+b)} = \frac{\overset{1}{\cancel{(a+b)}}}{(a-b)\underset{1}{\cancel{(a+b)}}} = \frac{1}{a-b}$$

$$\frac{a}{c} - \frac{b}{c} = \frac{a-b}{c}$$

When subtracting rational expressions in which the denominators are the same, subtract the numerators. The denominator of the difference is the common denominator. Write the answer in simplest form.

$$\frac{7x-12}{2x^2 + 5x - 12} - \frac{3x-6}{2x^2 + 5x - 12} = \frac{(7x-12)-(3x-6)}{2x^2 + 5x - 12} = \frac{4x-6}{2x^2 + 5x - 12}$$

$$= \frac{2\overset{1}{\cancel{(2x-3)}}}{\underset{1}{\cancel{(2x-3)}}(x+4)} = \frac{2}{x+4}$$

Before two rational expressions with *different* denominators can be added or subtracted, each rational expression must be expressed in terms of a common denominator. This common denominator is the LCM of the denominators of the rational expressions.

TAKE NOTE

Note the steps involved in adding or subtracting rational expressions:

1. Find the LCM of the denominators.
2. Rewrite each fraction in terms of the common denominator.
3. Add or subtract the rational expressions.
4. Simplify the resulting sum or difference.

HOW TO Simplify: $\dfrac{x}{x-3} - \dfrac{x+1}{x-2}$

The LCM is $(x-3)(x-2)$. • Find the LCM of the denominators.

$$\frac{x}{x-3} - \frac{x+1}{x-2} = \frac{x}{x-3} \cdot \frac{x-2}{x-2} - \frac{x+1}{x-2} \cdot \frac{x-3}{x-3}$$ • Express each fraction in terms of the LCM.

$$= \frac{x(x-2) - (x+1)(x-3)}{(x-3)(x-2)}$$ • Subtract the fractions.

$$= \frac{(x^2 - 2x) - (x^2 - 2x - 3)}{(x-3)(x-2)}$$

$$= \frac{3}{(x-3)(x-2)}$$ • Simplify.

HOW TO Simplify: $\dfrac{3x}{2x-3} + \dfrac{3x+6}{2x^2 + x - 6}$

The LCM of $2x - 3$ and $2x^2 + x - 6$ is $(2x-3)(x+2)$. • Find the LCM of the denominators.

$$\frac{3x}{2x-3} + \frac{3x+6}{2x^2 + x - 6} = \frac{3x}{2x-3} \cdot \frac{x+2}{x+2} + \frac{3x+6}{(2x-3)(x+2)}$$ • Express each fraction in terms of the LCM.

$$= \frac{3x(x+2) + (3x+6)}{(2x-3)(x+2)}$$ • Add the fractions.

$$= \frac{(3x^2 + 6x) + (3x+6)}{(2x-3)(x+2)}$$

$$= \frac{3x^2 + 9x + 6}{(2x-3)(x+2)} = \frac{3\cancel{(x+2)}(x+1)}{(2x-3)\cancel{(x+2)}}$$ • Simplify.

$$= \frac{3(x+1)}{2x-3}$$

Example 4

Simplify: $\dfrac{2}{x} - \dfrac{3}{x^2} + \dfrac{1}{xy}$

Solution

The LCM is x^2y.

$\dfrac{2}{x} - \dfrac{3}{x^2} + \dfrac{1}{xy} = \dfrac{2}{x} \cdot \dfrac{xy}{xy} - \dfrac{3}{x^2} \cdot \dfrac{y}{y} + \dfrac{1}{xy} \cdot \dfrac{x}{x}$

$\qquad = \dfrac{2xy}{x^2y} - \dfrac{3y}{x^2y} + \dfrac{x}{x^2y}$

$\qquad = \dfrac{2xy - 3y + x}{x^2y}$

You Try It 4

Simplify: $\dfrac{2}{b} - \dfrac{1}{a} + \dfrac{4}{ab}$

Your solution

Example 5

Simplify: $\dfrac{x}{2x - 4} - \dfrac{4 - x}{x^2 - 2x}$

Solution

$2x - 4 = 2(x - 2); \; x^2 - 2x = x(x - 2)$

The LCM is $2x(x - 2)$.

$\dfrac{x}{2x - 4} - \dfrac{4 - x}{x^2 - 2x} = \dfrac{x}{2(x - 2)} \cdot \dfrac{x}{x} - \dfrac{4 - x}{x(x - 2)} \cdot \dfrac{2}{2}$

$\qquad = \dfrac{x^2 - (4 - x)2}{2x(x - 2)}$

$\qquad = \dfrac{x^2 - (8 - 2x)}{2x(x - 2)} = \dfrac{x^2 + 2x - 8}{2x(x - 2)}$

$\qquad = \dfrac{(x + 4)(x - 2)}{2x(x - 2)} = \dfrac{(x + 4)\overset{1}{\cancel{(x - 2)}}}{2x\underset{1}{\cancel{(x - 2)}}}$

$\qquad = \dfrac{x + 4}{2x}$

You Try It 5

Simplify: $\dfrac{a - 3}{a^2 - 5a} + \dfrac{a - 9}{a^2 - 25}$

Your solution

Example 6

Simplify: $\dfrac{x}{x + 1} - \dfrac{2}{x - 2} - \dfrac{3}{x^2 - x - 2}$

Solution

The LCM is $(x + 1)(x - 2)$.

$\dfrac{x}{x + 1} - \dfrac{2}{x - 2} - \dfrac{3}{x^2 - x - 2}$

$\qquad = \dfrac{x}{x + 1} \cdot \dfrac{x - 2}{x - 2} - \dfrac{2}{x - 2} \cdot \dfrac{x + 1}{x + 1}$

$\qquad\quad - \dfrac{3}{(x + 1)(x - 2)}$

$\qquad = \dfrac{x(x - 2) - 2(x + 1) - 3}{(x + 1)(x - 2)}$

$\qquad = \dfrac{x^2 - 2x - 2x - 2 - 3}{(x + 1)(x - 2)}$

$\qquad = \dfrac{x^2 - 4x - 5}{(x + 1)(x - 2)} = \dfrac{\overset{1}{\cancel{(x + 1)}}(x - 5)}{\underset{1}{\cancel{(x + 1)}}(x - 2)} = \dfrac{x - 5}{x - 2}$

You Try It 6

Simplify: $\dfrac{2x}{x - 4} - \dfrac{x - 1}{x + 1} + \dfrac{2}{x^2 - 3x - 4}$

Your solution

Solutions on p. S19

6.2 Exercises

Objective A **To rewrite rational expressions in terms of a common denominator**

For Exercises 1 to 25, write each fraction in terms of the LCM of the denominators.

1. $\dfrac{3}{4x^2y}, \dfrac{17}{12xy^4}$

2. $\dfrac{5}{16a^3b^3}, \dfrac{7}{30a^5b}$

3. $\dfrac{x-2}{3x(x-2)}, \dfrac{3}{6x^2}$

4. $\dfrac{5x-1}{4x(2x+1)}, \dfrac{2}{5x^3}$

5. $\dfrac{3x-1}{2x^2-10x}, -3x$

6. $\dfrac{4x-3}{3x(x-2)}, 2x$

7. $\dfrac{3x}{2x-3}, \dfrac{5x}{2x+3}$

8. $\dfrac{2}{7y-3}, \dfrac{-3}{7y+3}$

9. $\dfrac{2x}{x^2-9}, \dfrac{x+1}{x-3}$

10. $\dfrac{3x}{16-x^2}, \dfrac{2x}{16-4x}$

11. $\dfrac{3}{3x^2-12y^2}, \dfrac{5}{6x-12y}$

12. $\dfrac{2x}{x^2-36}, \dfrac{x-1}{6x-36}$

13. $\dfrac{3x}{x^2-1}, \dfrac{5x}{x^2-2x+1}$

14. $\dfrac{x^2+2}{x^3-1}, \dfrac{3}{x^2+x+1}$

15. $\dfrac{x-3}{8-x^3}, \dfrac{2}{4+2x+x^2}$

16. $\dfrac{2x}{x^2+x-6}, \dfrac{-4x}{x^2+5x+6}$

17. $\dfrac{2x}{x^2+2x-3}, \dfrac{-x}{x^2+6x+9}$

18. $\dfrac{3x}{2x^2-x-3}, \dfrac{-2x}{2x^2-11x+12}$

19. $\dfrac{-4x}{4x^2-16x+15}, \dfrac{3x}{6x^2-19x+10}$

20. $\dfrac{3}{2x^2+5x-12}, \dfrac{2x}{3-2x}, \dfrac{3x-1}{x+4}$

21. $\dfrac{5}{6x^2-17x+12}, \dfrac{2x}{4-3x}, \dfrac{x+1}{2x-3}$

22. $\dfrac{3x}{x-4}, \dfrac{4}{x+5}, \dfrac{x+2}{20-x-x^2}$

23. $\dfrac{2x}{x-3}, \dfrac{-2}{x+5}, \dfrac{x-1}{15-2x-x^2}$

24. $\dfrac{2}{x^{2n}-1}, \dfrac{5}{x^{2n}+2x^n+1}$

25. $\dfrac{x-5}{x^{2n}+3x^n+2}, \dfrac{2x}{x^n+2}$

Objective B **To add or subtract rational expressions**

For Exercises 26 to 81, simplify.

26. $\dfrac{3}{2xy} - \dfrac{7}{2xy} - \dfrac{9}{2xy}$

27. $-\dfrac{3}{4x^2} + \dfrac{8}{4x^2} - \dfrac{3}{4x^2}$

28. $\dfrac{x}{x^2-3x+2} - \dfrac{2}{x^2-3x+2}$

29. $\dfrac{3x}{3x^2+x-10} - \dfrac{5}{3x^2+x-10}$

30. $\dfrac{3}{2x^2y} - \dfrac{8}{5x} - \dfrac{9}{10xy}$

31. $\dfrac{2}{5ab} - \dfrac{3}{10a^2b} + \dfrac{4}{15ab^2}$

32. $\dfrac{2}{3x} - \dfrac{3}{2xy} + \dfrac{4}{5xy} - \dfrac{5}{6x}$

33. $\dfrac{3}{4ab} - \dfrac{2}{5a} + \dfrac{3}{10b} - \dfrac{5}{8ab}$

34. $\dfrac{2x-1}{12x} - \dfrac{3x+4}{9x}$

35. $\dfrac{3x-4}{6x} - \dfrac{2x-5}{4x}$

36. $\dfrac{3x+2}{4x^2y} - \dfrac{y-5}{6xy^2}$

37. $\dfrac{2y-4}{5xy^2} + \dfrac{3-2x}{10x^2y}$

38. $\dfrac{2x}{x-3} - \dfrac{3x}{x-5}$

39. $\dfrac{3a}{a-2} - \dfrac{5a}{a+1}$

40. $\dfrac{3}{2a-3} + \dfrac{2a}{3-2a}$

41. $\dfrac{x}{2x-5} - \dfrac{2}{5x-2}$

42. $\dfrac{1}{x+h} - \dfrac{1}{h}$

43. $\dfrac{1}{a-b} + \dfrac{1}{b}$

44. $\dfrac{2}{x} - 3 - \dfrac{10}{x-4}$

45. $\dfrac{6a}{a-3} - 5 + \dfrac{3}{a}$

46. $\dfrac{1}{2x-3} - \dfrac{5}{2x} + 1$

47. $\dfrac{5}{x} - \dfrac{5x}{5 - 6x} + 2$

48. $\dfrac{3}{x^2 - 1} + \dfrac{2x}{x^2 + 2x + 1}$

49. $\dfrac{1}{x^2 - 6x + 9} - \dfrac{1}{x^2 - 9}$

50. $\dfrac{x}{x + 3} - \dfrac{3 - x}{x^2 - 9}$

51. $\dfrac{1}{x + 2} - \dfrac{3x}{x^2 + 4x + 4}$

52. $\dfrac{2x - 3}{x + 5} - \dfrac{x^2 - 4x - 19}{x^2 + 8x + 15}$

53. $\dfrac{-3x^2 + 8x + 2}{x^2 + 2x - 8} - \dfrac{2x - 5}{x + 4}$

54. $\dfrac{x^n}{x^{2n} - 1} - \dfrac{2}{x^n + 1}$

55. $\dfrac{2}{x^n - 1} + \dfrac{x^n}{x^{2n} - 1}$

56. $\dfrac{2}{x^n - 1} - \dfrac{6}{x^{2n} + x^n - 2}$

57. $\dfrac{2x^n - 6}{x^{2n} - x^n - 6} + \dfrac{x^n}{x^n + 2}$

58. $\dfrac{2x - 2}{4x^2 - 9} - \dfrac{5}{3 - 2x}$

59. $\dfrac{x^2 + 4}{4x^2 - 36} - \dfrac{13}{x + 3}$

60. $\dfrac{x - 2}{x + 1} - \dfrac{3 - 12x}{2x^2 - x - 3}$

61. $\dfrac{3x - 4}{4x + 1} + \dfrac{3x + 6}{4x^2 + 9x + 2}$

62. $\dfrac{x + 1}{x^2 + x - 6} - \dfrac{x + 2}{x^2 + 4x + 3}$

63. $\dfrac{x + 1}{x^2 + x - 12} - \dfrac{x - 3}{x^2 + 7x + 12}$

64. $\dfrac{x^2 + 6x}{x^2 + 3x - 18} - \dfrac{2x - 1}{x + 6} + \dfrac{x - 2}{3 - x}$

65. $\dfrac{2x^2 - 2x}{x^2 - 2x - 15} - \dfrac{2}{x + 3} + \dfrac{x}{5 - x}$

66. $\dfrac{7 - 4x}{2x^2 - 9x + 10} + \dfrac{x - 3}{x - 2} - \dfrac{x + 1}{2x - 5}$

67. $\dfrac{x}{3x + 4} + \dfrac{3x + 2}{x - 5} - \dfrac{7x^2 + 24x + 28}{3x^2 - 11x - 20}$

68. $\dfrac{32x - 9}{2x^2 + 7x - 15} + \dfrac{x - 2}{3 - 2x} + \dfrac{3x + 2}{x + 5}$

69. $\dfrac{x + 1}{1 - 2x} - \dfrac{x + 3}{4x - 3} + \dfrac{10x^2 + 7x - 9}{8x^2 - 10x + 3}$

70. $\dfrac{x^2}{x^3 - 8} - \dfrac{x + 2}{x^2 + 2x + 4}$

71. $\dfrac{2x}{4x^2 + 2x + 1} + \dfrac{4x + 1}{8x^3 - 1}$

72. $\dfrac{2x^2}{x^4 - 1} - \dfrac{1}{x^2 - 1} + \dfrac{1}{x^2 + 1}$

73. $\dfrac{x^2 - 12}{x^4 - 16} + \dfrac{1}{x^2 - 4} - \dfrac{1}{x^2 + 4}$

74. $\left(\dfrac{x + 8}{4} + \dfrac{4}{x}\right) \div \dfrac{x + 4}{16x^2}$

75. $\left(\dfrac{a - 3}{a^2} - \dfrac{a - 3}{9}\right) \div \dfrac{a^2 - 9}{3a}$

76. $\dfrac{3}{x - 2} - \dfrac{x^2 + x}{2x^3 + 3x^2} \cdot \dfrac{2x^2 + x - 3}{x^2 + 3x + 2}$

77. $\dfrac{x^2 - 4x + 4}{2x + 1} \cdot \dfrac{2x^2 + x}{x^3 - 4x} - \dfrac{3x - 2}{x + 1}$

78. $\left(\dfrac{x - y}{x^2} - \dfrac{x - y}{y^2}\right) \div \dfrac{x^2 - y^2}{xy}$

79. $\left(\dfrac{a - 2b}{b} + \dfrac{b}{a}\right) \div \left(\dfrac{b + a}{a} - \dfrac{2a}{b}\right)$

80. $\dfrac{2}{x - 3} - \dfrac{x}{x^2 - x - 6} \cdot \dfrac{x^2 - 2x - 3}{x^2 - x}$

81. $\dfrac{2x}{x^2 - x - 6} - \dfrac{6x - 6}{2x^2 - 9x + 9} \div \dfrac{x^2 + x - 2}{2x - 3}$

APPLYING THE CONCEPTS

82. Correct the following expressions.

 a. $\dfrac{3}{4} + \dfrac{x}{5} = \dfrac{3 + x}{4 + 5}$
 b. $\dfrac{4x + 5}{4} = x + 5$
 c. $\dfrac{1}{x} + \dfrac{1}{y} = \dfrac{1}{x + y}$

83. Simplify.

 a. $\left(\dfrac{b}{6} - \dfrac{6}{b}\right) \div \left(\dfrac{6}{b} - 4 + \dfrac{b}{2}\right)$
 b. $\left(\dfrac{x + 1}{2x - 1} - \dfrac{x - 1}{2x + 1}\right) \cdot \left(\dfrac{2x - 1}{x} - \dfrac{2x - 1}{x^2}\right)$

84. Rewrite each fraction as the sum of two fractions in simplest form.

 a. $\dfrac{3x + 6y}{xy}$
 b. $\dfrac{4a^2 + 3ab}{a^2b^2}$
 c. $\dfrac{3m^2n + 2mn^2}{12m^3n^2}$

85. Let $f(x) = \dfrac{x}{x + 2}$, $g(x) = \dfrac{4}{x - 3}$, and $S(x) = \dfrac{x^2 + x + 8}{x^2 - x - 6}$. **a.** Evaluate $f(4)$, $g(4)$, and $S(4)$. Does $f(4) + g(4) = S(4)$? **b.** Let a be a real number $(a \neq -2, a \neq 3)$. Express $S(a)$ in terms of $f(a)$ and $g(a)$.

6.3 Complex Fractions

Objective A **To simplify a complex fraction**

A **complex fraction** is a fraction whose numerator or denominator contains one or more fractions. Examples of complex fractions are shown below.

$$\frac{5}{2+\dfrac{1}{2}}, \qquad \frac{5+\dfrac{1}{y}}{5-\dfrac{1}{y}}, \qquad \frac{x+4+\dfrac{1}{x+2}}{x-2+\dfrac{1}{x+2}}$$

HOW TO Simplify: $\dfrac{\dfrac{1}{x}+\dfrac{1}{y}}{\dfrac{1}{x}-\dfrac{1}{y}}$

The LCM of x and y is xy.

- Find the LCM of the denominators of the fractions in the numerator and denominator.
- Multiply the numerator and denominator of the complex fraction by the LCM.

$$\frac{\dfrac{1}{x}+\dfrac{1}{y}}{\dfrac{1}{x}-\dfrac{1}{y}} = \frac{\dfrac{1}{x}+\dfrac{1}{y}}{\dfrac{1}{x}-\dfrac{1}{y}} \cdot \frac{xy}{xy}$$

$$= \frac{\dfrac{1}{x}\cdot xy + \dfrac{1}{y}\cdot xy}{\dfrac{1}{x}\cdot xy - \dfrac{1}{y}\cdot xy} = \frac{y+x}{y-x}$$

The method shown above of simplifying a complex fraction by multiplying the numerator and denominator by the LCM of the denominators is used in Examples 1 and 2 on the next page. However, a different approach is to rewrite the numerator and denominator of the complex fraction as single fractions and then divide the numerator by the denominator. The example shown above is simplified below by using this alternative method.

$$\frac{\dfrac{1}{x}+\dfrac{1}{y}}{\dfrac{1}{x}-\dfrac{1}{y}} = \frac{\dfrac{1}{x}\cdot\dfrac{y}{y}+\dfrac{1}{y}\cdot\dfrac{x}{x}}{\dfrac{1}{x}\cdot\dfrac{y}{y}-\dfrac{1}{y}\cdot\dfrac{x}{x}} = \frac{\dfrac{y}{xy}+\dfrac{x}{xy}}{\dfrac{y}{xy}-\dfrac{x}{xy}} = \frac{\dfrac{y+x}{xy}}{\dfrac{y-x}{xy}}$$

- Rewrite the numerator and denominator of the complex fraction as single fractions.

$$= \frac{y+x}{xy} \div \frac{y-x}{xy} = \frac{y+x}{xy} \cdot \frac{xy}{y-x}$$

- Divide the numerator of the complex fraction by the denominator.

$$= \frac{(y+x)xy}{xy(y-x)} = \frac{y+x}{y-x}$$

- Multiply the fractions. Simplify.

Note that this is the same result as shown above.

Example 1

Simplify: $\dfrac{2x - 1 + \dfrac{7}{x + 4}}{3x - 8 + \dfrac{17}{x + 4}}$

Solution

The LCM of $x + 4$ and $x + 4$ is $x + 4$.

$$\dfrac{2x - 1 + \dfrac{7}{x + 4}}{3x - 8 + \dfrac{17}{x + 4}} = \dfrac{2x - 1 + \dfrac{7}{x + 4}}{3x - 8 + \dfrac{17}{x + 4}} \cdot \dfrac{x + 4}{x + 4}$$

$$= \dfrac{(2x - 1)(x + 4) + \dfrac{7}{x + 4}(x + 4)}{(3x - 8)(x + 4) + \dfrac{17}{x + 4}(x + 4)}$$

$$= \dfrac{2x^2 + 7x - 4 + 7}{3x^2 + 4x - 32 + 17}$$

$$= \dfrac{2x^2 + 7x + 3}{3x^2 + 4x - 15}$$

$$= \dfrac{(2x + 1)(x + 3)}{(3x - 5)(x + 3)}$$

$$= \dfrac{(2x + 1)\cancel{(x + 3)}}{(3x - 5)\cancel{(x + 3)}} = \dfrac{2x + 1}{3x - 5}$$

You Try It 1

Simplify: $\dfrac{2x + 5 + \dfrac{14}{x - 3}}{4x + 16 + \dfrac{49}{x - 3}}$

Your solution

Example 2

Simplify: $1 + \dfrac{a}{2 + \dfrac{1}{a}}$

Solution

The LCM of the denominators is a.

$$1 + \dfrac{a}{2 + \dfrac{1}{a}} = 1 + \dfrac{a}{2 + \dfrac{1}{a}} \cdot \dfrac{a}{a}$$

$$= 1 + \dfrac{a \cdot a}{2 \cdot a + \dfrac{1}{a} \cdot a} = 1 + \dfrac{a^2}{2a + 1}$$

The LCM of 1 and $2a + 1$ is $2a + 1$.

$$1 + \dfrac{a^2}{2a + 1} = 1 \cdot \dfrac{2a + 1}{2a + 1} + \dfrac{a^2}{2a + 1}$$

$$= \dfrac{2a + 1}{2a + 1} + \dfrac{a^2}{2a + 1} = \dfrac{2a + 1 + a^2}{2a + 1}$$

$$= \dfrac{a^2 + 2a + 1}{2a + 1} = \dfrac{(a + 1)^2}{2a + 1}$$

You Try It 2

Simplify: $2 - \dfrac{1}{2 - \dfrac{1}{x}}$

Your solution

Solutions on p. S19

6.3 Exercises

Objective A **To simplify a complex fraction**

1. What is a complex fraction?

2. What is the general goal of simplifying a complex fraction?

For Exercises 3 to 46, simplify.

3. $\dfrac{2 - \dfrac{1}{3}}{4 + \dfrac{11}{3}}$

4. $\dfrac{3 + \dfrac{5}{2}}{8 - \dfrac{3}{2}}$

5. $\dfrac{3 - \dfrac{2}{3}}{5 + \dfrac{5}{6}}$

6. $\dfrac{5 - \dfrac{3}{4}}{2 + \dfrac{1}{2}}$

7. $\dfrac{1 + \dfrac{1}{x}}{1 - \dfrac{1}{x^2}}$

8. $\dfrac{\dfrac{1}{y^2} - 1}{1 + \dfrac{1}{y}}$

9. $\dfrac{a - 2}{\dfrac{4}{a} - a}$

10. $\dfrac{\dfrac{25}{a} - a}{5 + a}$

11. $\dfrac{\dfrac{1}{a^2} - \dfrac{1}{a}}{\dfrac{1}{a^2} + \dfrac{1}{a}}$

12. $\dfrac{\dfrac{1}{b} + \dfrac{1}{2}}{\dfrac{4}{b^2} - 1}$

13. $\dfrac{2 - \dfrac{4}{x + 2}}{5 - \dfrac{10}{x + 2}}$

14. $\dfrac{4 + \dfrac{12}{2x - 3}}{5 + \dfrac{15}{2x - 3}}$

15. $\dfrac{\dfrac{3}{2a - 3} + 2}{\dfrac{-6}{2a - 3} - 4}$

16. $\dfrac{\dfrac{-5}{b - 5} - 3}{\dfrac{10}{b - 5} + 6}$

17. $\dfrac{\dfrac{x}{x + 1} - \dfrac{1}{x}}{\dfrac{x}{x + 1} + \dfrac{1}{x}}$

18. $\dfrac{\dfrac{2a}{a - 1} - \dfrac{3}{a}}{\dfrac{1}{a - 1} + \dfrac{2}{a}}$

19. $\dfrac{1 - \dfrac{1}{x} - \dfrac{6}{x^2}}{1 - \dfrac{4}{x} + \dfrac{3}{x^2}}$

20. $\dfrac{1 - \dfrac{3}{x} - \dfrac{10}{x^2}}{1 + \dfrac{11}{x} + \dfrac{18}{x^2}}$

21. $\dfrac{1 + \dfrac{1}{x} - \dfrac{12}{x^2}}{\dfrac{9}{x^2} + \dfrac{3}{x} - 2}$

22. $\dfrac{\dfrac{15}{x^2} - \dfrac{2}{x} - 1}{\dfrac{4}{x^2} - \dfrac{5}{x} + 4}$

23. $a + \dfrac{a}{a + \dfrac{1}{a}}$

24. $4 - \dfrac{2}{2 - \dfrac{3}{x}}$

25. $\dfrac{1 - \dfrac{1}{x - 4}}{1 - \dfrac{6}{x + 1}}$

26. $\dfrac{1 + \dfrac{3}{x + 2}}{1 + \dfrac{6}{x - 1}}$

27. $\dfrac{x - \dfrac{1}{x}}{x + \dfrac{1}{x}}$

28. $\dfrac{a - \dfrac{1}{a}}{\dfrac{1}{a} + a}$

29. $\dfrac{\dfrac{1}{x + h} - \dfrac{1}{x}}{h}$

30. $\dfrac{\dfrac{1}{(x + h)^2} - \dfrac{1}{x^2}}{h}$

31. $\dfrac{1 - \dfrac{2}{x - 3}}{1 + \dfrac{3}{2 - x}}$

32. $\dfrac{1 + \dfrac{x}{x + 1}}{1 + \dfrac{x - 1}{x + 2}}$

33. $\dfrac{x - 4 + \dfrac{9}{2x + 3}}{x + 3 - \dfrac{5}{2x + 3}}$

34. $\dfrac{2x - 3 - \dfrac{10}{4x - 5}}{3x + 2 + \dfrac{11}{4x - 5}}$

35. $\dfrac{3x - 2 - \dfrac{5}{2x - 1}}{x - 6 + \dfrac{9}{2x - 1}}$

36. $\dfrac{x + 4 - \dfrac{7}{2x - 5}}{2x + 7 - \dfrac{28}{2x - 5}}$

37. $\dfrac{\dfrac{1}{a} - \dfrac{3}{a - 2}}{\dfrac{2}{a} + \dfrac{5}{a - 2}}$

38. $\dfrac{\dfrac{2}{b} - \dfrac{5}{b + 3}}{\dfrac{3}{b} + \dfrac{3}{b + 3}}$

39. $\dfrac{\dfrac{1}{y^2} - \dfrac{1}{xy} - \dfrac{2}{x^2}}{\dfrac{1}{y^2} - \dfrac{3}{xy} + \dfrac{2}{x^2}}$

40. $\dfrac{\dfrac{2}{b^2} - \dfrac{5}{ab} - \dfrac{3}{a^2}}{\dfrac{2}{b^2} + \dfrac{7}{ab} + \dfrac{3}{a^2}}$

41. $\dfrac{\dfrac{x - 1}{x + 1} - \dfrac{x + 1}{x - 1}}{\dfrac{x - 1}{x + 1} + \dfrac{x + 1}{x - 1}}$

42. $\dfrac{\dfrac{y}{y + 2} - \dfrac{y}{y - 2}}{\dfrac{y}{y + 2} + \dfrac{y}{y - 2}}$

43. $a - \dfrac{a}{1 - \dfrac{a}{1 - a}}$

44. $3 - \dfrac{3}{3 - \dfrac{3}{3 - x}}$

45. $3 - \dfrac{2}{1 - \dfrac{2}{3 - \dfrac{2}{x}}}$

46. $a + \dfrac{a}{2 + \dfrac{1}{1 - \dfrac{2}{a}}}$

APPLYING THE CONCEPTS

47. Integers Find the sum of the reciprocals of three consecutive even integers.

48. If $a = \dfrac{b^2 + 4b + 4}{b^2 - 4}$ and $b = \dfrac{1}{c}$, express a in terms of c.

49. Simplify.

a. $\dfrac{x^{-1}}{x^{-1} + 2^{-1}}$

b. $\dfrac{-x^{-1} + x}{x^{-1} - x}$

c. $\dfrac{x^{-1} - x^{-2} - 6x^{-3}}{x^{-1} - 4x^{-3}}$

6.4 Ratio and Proportion

Objective A ### To solve a proportion

Quantities such as 3 feet, 5 liters, and 2 miles are number quantities written with units. In these examples, the units are feet, liters, and miles.

A **ratio** is the quotient of two quantities that have the same unit.

The weekly wages of a painter are $800. The painter spends $150 a week for food. The ratio of the wages spent for food to the total weekly wages is written

$$\frac{\$150}{\$800} = \frac{150}{800} = \frac{3}{16}$$ A ratio is in simplest form when the two numbers do not have a common factor. Note that the units are not written.

A **rate** is the quotient of two quantities that have different units.

A car travels 180 mi on 3 gal of gas. The miles-to-gallon rate is

$$\frac{180 \text{ mi}}{3 \text{ gal}} = \frac{60 \text{ mi}}{1 \text{ gal}}$$ A rate is in simplest form when the two numbers do not have a common factor. The units are written as part of the rate.

A **proportion** is an equation that states the equality of two ratios or rates. For example, $\frac{90 \text{ km}}{4 \text{ L}} = \frac{45 \text{ km}}{2 \text{ L}}$ and $\frac{3}{4} = \frac{x + 2}{16}$ are proportions.

Study Tip

Always check the proposed solution of an equation. For the equation at the right:

$$\frac{2}{7} = \frac{x}{5}$$

$$\frac{2}{7} \quad\bigg|\quad \frac{\frac{10}{7}}{5}$$

$$\frac{2}{7} \quad\bigg|\quad \frac{10}{7} \cdot \frac{1}{5}$$

$$\frac{2}{7} = \frac{2}{7}$$

The solution checks.

HOW TO Solve: $\dfrac{2}{7} = \dfrac{x}{5}$

$$\frac{2}{7} = \frac{x}{5}$$

$$35 \cdot \frac{2}{7} = 35 \cdot \frac{x}{5}$$

$$10 = 7x$$

$$\frac{10}{7} = x$$

• Multiply each side of the proportion by the LCM of the denominators.
• Solve the equation.

The solution is $\dfrac{10}{7}$.

Example 1

Solve: $\dfrac{3}{12} = \dfrac{5}{x + 5}$

Solution

$$\frac{3}{12} = \frac{5}{x + 5}$$

$$12(x + 5) \cdot \frac{3}{12} = 12(x + 5) \cdot \frac{5}{x + 5}$$

$$3(x + 5) = 5 \cdot 12$$

$$3x + 15 = 60$$

$$3x = 45$$

$$x = 15$$

• Multiply each side by the LCM of 12 and x + 5.

The solution is 15.

You Try It 1

Solve: $\dfrac{5}{x - 2} = \dfrac{3}{4}$

Your solution

Solution on p. S19

Example 2

Solve: $\dfrac{3}{x - 2} = \dfrac{4}{2x + 1}$

Solution

$$\frac{3}{x - 2} = \frac{4}{2x + 1}$$

$$\frac{3}{x - 2}(2x + 1)(x - 2) = \frac{4}{2x + 1}(2x + 1)(x - 2)$$

$$3(2x + 1) = 4(x - 2)$$

$$6x + 3 = 4x - 8$$

$$2x + 3 = -8$$

$$2x = -11$$

$$x = -\frac{11}{2}$$

The solution is $-\dfrac{11}{2}$.

You Try It 2

Solve: $\dfrac{5}{2x - 3} = \dfrac{-2}{x + 1}$

Your solution

Solution on p. S19

Objective B **To solve application problems**

Example 3

A stock investment of 50 shares pays a dividend of $106. At this rate, how many additional shares are required to earn a dividend of $424?

Strategy

To find the additional number of shares that are required, write and solve a proportion using x to represent the additional number of shares. Then $50 + x$ is the total number of shares of stock.

Solution

$$\frac{106}{50} = \frac{424}{50 + x}$$

$$\frac{53}{25} = \frac{424}{50 + x} \qquad \bullet \text{ Simplify } \frac{106}{50}.$$

$$\frac{53}{25}(25)(50 + x) = \frac{424}{50 + x}(25)(50 + x)$$

$$53(50 + x) = 424(25)$$

$$2650 + 53x = 10{,}600$$

$$53x = 7950$$

$$x = 150$$

An additional 150 shares of stock are required.

You Try It 3

Two pounds of cashews cost $5.80. At this rate, how much would 15 lb of cashews cost?

Your strategy

Your solution

Solution on p. S20

6.4 Exercises

Objective A **To solve a proportion**

1. How does a ratio differ from a rate?

2. What is a proportion?

For Exercises 3 to 22, solve.

3. $\dfrac{x}{30} = \dfrac{3}{10}$

4. $\dfrac{5}{15} = \dfrac{x}{75}$

5. $\dfrac{2}{x} = \dfrac{8}{30}$

6. $\dfrac{80}{16} = \dfrac{15}{x}$

7. $\dfrac{x+1}{10} = \dfrac{2}{5}$

8. $\dfrac{5-x}{10} = \dfrac{3}{2}$

9. $\dfrac{4}{x+2} = \dfrac{3}{4}$

10. $\dfrac{8}{3} = \dfrac{24}{x+3}$

11. $\dfrac{x}{4} = \dfrac{x-2}{8}$

12. $\dfrac{8}{x-5} = \dfrac{3}{x}$

13. $\dfrac{16}{2-x} = \dfrac{4}{x}$

14. $\dfrac{6}{x-5} = \dfrac{1}{x}$

15. $\dfrac{8}{x-2} = \dfrac{4}{x+1}$

16. $\dfrac{4}{x-4} = \dfrac{2}{x-2}$

17. $\dfrac{x}{3} = \dfrac{x+1}{7}$

18. $\dfrac{x-3}{2} = \dfrac{3+x}{5}$

19. $\dfrac{8}{3x-2} = \dfrac{2}{2x+1}$

20. $\dfrac{3}{2x-4} = \dfrac{-5}{x+2}$

21. $\dfrac{3x+1}{3x-4} = \dfrac{x}{x-2}$

22. $\dfrac{x-2}{x-5} = \dfrac{2x}{2x+5}$

Objective B **To solve application problems**

23. Nutrition If a 56-gram serving of pasta contains 7 g of protein, how many grams of protein are in a 454-gram box of the pasta?

24. Parking A movie theatre that can seat 1500 people has a parking lot with 600 spaces. At the same rate, how many parking spaces should a new theatre with 2200 seats provide?

25. Conservation A team of biologists captured and tagged approximately 2400 northern squawfish at the Bonneville Dam on the Colorado River. Later, 225 squawfish were recaptured and 9 of them had tags. How many squawfish would you estimate are in the area?

26. **USPS** The United States Postal Service delivers 200 billion pieces of mail each year. How many pieces of mail does the Postal Service deliver per week? Assume the Postal Service delivers mail 300 days during the year and 6 days per week.

27. **Computers** Of 250 people who purchased home computers from a national company, 45 received machines with defective CD-ROM drives. At this rate, how many of the 2800 computers the company sold nationwide would you expect to have defective CD-ROM drives?

28. **Crime** During a recent year, 3293 cars were stolen in Montgomery County, Maryland. Police were able to recover 2305 of these. At this rate, if 19,000 cars were stolen throughout the state of Maryland, how many of these would you expect to be recovered?

29. **Architecture** On an architectural drawing, $\frac{1}{4}$ in. represents 1 ft. Using this scale, find the dimensions of a room that measures $4\frac{1}{2}$ in. by 6 in. on the drawing.

30. **Interior Decorating** One hundred forty-four ceramic tiles are required to tile a 25-square-foot area. At this rate, how many tiles are required to tile 275 ft²?

31. **Medicine** One and one-half ounces of a medication are required for a 140-pound adult. At the same rate, how many additional ounces of medication are required for a 210-pound adult?

32. **Employment Benefits** After working for a company for 9 months, an employee accrued 6 vacation days. At this rate, how much longer will it be before she can take a 3-week vacation (15 work days)?

33. **Exercise** Walking 4 mi in 2 h will use up 650 calories. Walking at the same rate, how many miles would a person need to walk to lose 1 lb? (Burning 3500 calories is equivalent to losing 1 pound.) Round to the nearest hundredth.

34. **Fundraising** To raise money, a school is sponsoring a magazine subscription drive. By visiting 90 homes in the neighborhood, one student was able to sell $375 worth of subscriptions. At this rate, how many additional homes will the student have to visit in order to meet a $1000 goal?

APPLYING THE CONCEPTS

35. If $\frac{a}{b} = \frac{c}{d}$, show that $\frac{a+b}{b} = \frac{c+d}{d}$.

36. **Lotteries** Three people put their money together to buy lottery tickets. The first person put in $20, the second person put in $25, and the third person put in $30. One of the tickets was a winning ticket. If they won $4.5 million, what was the first person's share of the winnings?

6.5 Rational Equations

Objective A ## To solve a fractional equation

To solve an equation containing fractions, the first step is **clearing denominators**—that is, removing them by multiplying each side of the equation by the LCM of the denominators. Then solve for the variable. The solutions to the resulting equation must be checked, because multiplying each side of an equation by a variable expression may produce an equation that has a solution that is not a solution of the original equation.

After each side of a fractional equation has been multiplied by the LCM of the denominators, the resulting equation is sometimes a quadratic equation.

HOW TO Solve: $\dfrac{3x}{x-5} = x + \dfrac{12}{x-5}$

$$\dfrac{3x}{x-5} = x + \dfrac{12}{x-5}$$

$$(x-5)\left(\dfrac{3x}{x-5}\right) = (x-5)\left(x + \dfrac{12}{x-5}\right)$$

- **Multiply each side of the equation by the LCM of the denominators.**

$$3x = (x-5)x + (x-5)\left(\dfrac{12}{x-5}\right)$$

$$3x = x^2 - 5x + 12$$

$$0 = x^2 - 8x + 12$$

- **Solve the quadratic equation by factoring.**

$$0 = (x-2)(x-6)$$

$$x = 2 \qquad x = 6$$

2 and 6 check as solutions.

The solutions are 2 and 6.

HOW TO Solve the literal equation $\dfrac{1}{R_1} + \dfrac{1}{R_2} = \dfrac{1}{R}$ for R.

$$\dfrac{1}{R_1} + \dfrac{1}{R_2} = \dfrac{1}{R}$$

$$RR_1R_2\left(\dfrac{1}{R_1} + \dfrac{1}{R_2}\right) = RR_1R_2\left(\dfrac{1}{R}\right)$$

- **Multiply each side of the equation by the LCM of the denominators.**

$$\dfrac{RR_1R_2}{R_1} + \dfrac{RR_1R_2}{R_2} = \dfrac{RR_1R_2}{R}$$

$$RR_2 + RR_1 = R_1R_2$$

$$R(R_2 + R_1) = R_1R_2$$

- **Factor.**

$$\dfrac{R(R_1 + R_2)}{R_1 + R_2} = \dfrac{R_1R_2}{R_1 + R_2}$$

- **Divide each side by $R_1 + R_2$.**

$$R = \dfrac{R_1R_2}{R_1 + R_2}$$

This literal equation is true for all real numbers, $R_1 + R_2 \neq 0$.

Example 1

Solve: $\dfrac{3x}{x-3} = 2 + \dfrac{9}{x-3}$

Solution

$\dfrac{3x}{x-3} = 2 + \dfrac{9}{x-3}$

$(x-3)\left(\dfrac{3x}{x-3}\right) = (x-3)\left(2 + \dfrac{9}{x-3}\right)$

$3x = (x-3)2 + (x-3)\left(\dfrac{9}{x-3}\right)$

$3x = 2x - 6 + 9$

$3x = 2x + 3$

$x = 3$

3 does not check as a solution.
The equation has no solution.

You Try It 1

Solve: $\dfrac{x}{x-2} + x = \dfrac{6}{x-2}$

Your solution

Example 2

Solve: $\dfrac{1}{a} + \dfrac{1}{a+1} = \dfrac{3}{2}$

Solution

$\dfrac{1}{a} + \dfrac{1}{a+1} = \dfrac{3}{2}$

$2a(a+1)\left(\dfrac{1}{a} + \dfrac{1}{a+1}\right) = 2a(a+1)\dfrac{3}{2}$

$2(a+1) + 2a = a(a+1)\cdot 3$

$2a + 2 + 2a = 3a^2 + 3a$

$0 = 3a^2 - a - 2$

$0 = (3a+2)(a-1)$

$3a + 2 = 0 \qquad a - 1 = 0$

$3a = -2 \qquad\quad a = 1$

$a = -\dfrac{2}{3}$

$-\dfrac{2}{3}$ and 1 check as solutions.

The solutions are $-\dfrac{2}{3}$ and 1.

You Try It 2

Solve: $\dfrac{5}{x-2} = 2x + \dfrac{3x-1}{x-2}$

Your solution

Example 3

Solve for f: $\dfrac{1}{f} = \dfrac{1}{a} + \dfrac{1}{b}$

Solution

$\dfrac{1}{f} = \dfrac{1}{a} + \dfrac{1}{b}$

$fab\left(\dfrac{1}{f}\right) = fab\left(\dfrac{1}{a} + \dfrac{1}{b}\right)$ • Multiply by the LCM of f, a, and b.

$\dfrac{fab}{f} = \dfrac{fab}{a} + \dfrac{fab}{b}$

$ab = fb + fa$

$ab = f(b+a)$

$\dfrac{ab}{b+a} = \dfrac{f(b+a)}{b+a}$

$\dfrac{ab}{b+a} = f$

You Try It 3

Solve for T_1: $\dfrac{P_1 V_1}{T_1} = \dfrac{P_2 V_2}{T_2}$

Your solution

Objective B **To solve work problems**

Point of Interest

The following problem was recorded in the *Jiuzhang*, a Chinese text that dates to the Han dynasty (about 200 B.C. to A.D. 200). "A reservoir has five channels bringing water to it. The first can fill the reservoir in $\frac{1}{3}$ day, the second in 1 day, the third in $2\frac{1}{2}$ days, the fourth in 3 days, and the fifth in 5 days. If all channels are open, how long does it take to fill the reservoir?" This problem is the earliest known work problem.

Rate of work is that part of a task that is completed in one unit of time. If a mason can build a retaining wall in 12 h, then in 1 h the mason can build $\frac{1}{12}$ of the wall. The mason's rate of work is $\frac{1}{12}$ of the wall each hour. If an apprentice can build the wall in x hours, the rate of work for the apprentice is $\frac{1}{x}$ of the wall each hour.

In solving a work problem, the goal is to determine the time it takes to complete a task. The basic equation that is used to solve work problems is

Rate of work × time worked = part of task completed

For example, if a pipe can fill a tank in 5 h, then in 2 h the pipe will fill $\frac{1}{5} \times 2 = \frac{2}{5}$ of the tank. In t hours, the pipe will fill $\frac{1}{5} \times t = \frac{t}{5}$ of the tank.

HOW TO A mason can build a wall in 10 h. An apprentice can build a wall in 15 h. How long will it take them to build the wall when they work together?

> **Strategy for Solving a Work Problem**
>
> 1. For each person or machine, write a numerical or variable expression for the rate of work, the time worked, and the part of the task completed. The results can be recorded in a table.

Unknown time to build the wall working together: t

	Rate of Work	·	*Time Worked*	=	*Part of Task Completed*
Mason	$\dfrac{1}{10}$	·	t	=	$\dfrac{t}{10}$
Apprentice	$\dfrac{1}{15}$	·	t	=	$\dfrac{t}{15}$

Study Tip

> 2. Determine how the parts of the task completed are related. Use the fact that the sum of the parts of the task completed must equal 1, the complete task.

The sum of the part of the task completed by the mason and the part of the task completed by the apprentice is 1.

$$\frac{t}{10} + \frac{t}{15} = 1$$
$$30\left(\frac{t}{10} + \frac{t}{15}\right) = 30(1)$$
$$3t + 2t = 30$$
$$5t = 30$$
$$t = 6$$

Working together, they will build the wall in 6 h.

Example 4

An electrician requires 12 h to wire a house. The electrician's apprentice can wire a house in 16 h. After working alone on a job for 4 h, the electrician quits, and the apprentice completes the task. How long does it take the apprentice to finish wiring the house?

Strategy

- Time required for the apprentice to finish wiring the house: t

	Rate	Time	Part
Electrician	$\dfrac{1}{12}$	4	$\dfrac{4}{12}$
Apprentice	$\dfrac{1}{16}$	t	$\dfrac{t}{16}$

- The sum of the part of the task completed by the electrician and the part of the task completed by the apprentice is 1.

Solution

$$\frac{4}{12} + \frac{t}{16} = 1$$

$$\frac{1}{3} + \frac{t}{16} = 1 \qquad \bullet \text{ Simplify } \frac{4}{12}.$$

$$48\left(\frac{1}{3} + \frac{t}{16}\right) = 48(1) \qquad \bullet \text{ Multiply by the LCM of 3 and 16.}$$

$$16 + 3t = 48$$

$$3t = 32$$

$$t = \frac{32}{3}$$

It will take the apprentice $10\frac{2}{3}$ h to finish wiring the house.

You Try It 4

Two water pipes can fill a tank with water in 6 h. The larger pipe working alone can fill the tank in 9 h. How long will it take the smaller pipe working alone to fill the tank?

Your strategy

Your solution

Solution on p. S20

Objective C **To solve uniform motion problems**

A car that travels constantly in a straight line at 55 mph is in uniform motion. **Uniform motion** means that the speed of an object does not change.

The basic equation used to solve uniform motion problems is

Distance = rate × time

An alternative form of this equation can be written by solving the equation for time.

$$\frac{\textbf{Distance}}{\textbf{Rate}} = \textbf{time}$$

This form of the equation is useful when the total time of travel for two objects or the time of travel between two points is known.

50 mi
r

150 mi
$3r$

HOW TO A motorist drove 150 mi on country roads before driving 50 mi on mountain roads. The rate of speed on the country roads was three times the rate on the mountain roads. The time spent traveling the 200 mi was 5 h. Find the rate of the motorist on the country roads.

> **Strategy for Solving a Uniform Motion Problem**
>
> **1.** For each object, write a numerical or variable expression for the distance, rate, and time. The results can be recorded in a table.

Unknown rate of speed on the mountain roads: r
Rate of speed on the country roads: $3r$

	Distance	÷	*Rate*	=	*Time*
Country roads	150	÷	$3r$	=	$\frac{150}{3r}$
Mountain roads	50	÷	r	=	$\frac{50}{r}$

> **2.** Determine how the times traveled by the different objects are related. For example, it may be known that the times are equal, or the total time may be known.

The total time of the trip is 5 h.

$$\frac{150}{3r} + \frac{50}{r} = 5$$

$$3r\left(\frac{150}{3r} + \frac{50}{r}\right) = 3r(5)$$
$$150 + 150 = 15r$$
$$300 = 15r$$
$$20 = r$$

The rate of speed on the country roads was $3r$.
Replace r with 20 and evaluate.

$$3r = 3(20) = 60$$

The rate of speed on the country roads was 60 mph.

Example 5

A marketing executive traveled 810 mi on a corporate jet in the same amount of time as it took to travel an additional 162 mi by helicopter. The rate of the jet was 360 mph greater than the rate of the helicopter. Find the rate of the jet.

Strategy

• Rate of the helicopter: r
 Rate of the jet: $r + 360$

	Distance	Rate	Time
Jet	810	$r + 360$	$\dfrac{810}{r + 360}$
Helicopter	162	r	$\dfrac{162}{r}$

• The time traveled by jet is equal to the time traveled by helicopter.

Solution

$$\frac{810}{r + 360} = \frac{162}{r}$$

$$r(r + 360)\left(\frac{810}{r + 360}\right) = r(r + 360)\left(\frac{162}{r}\right)$$

• Multiply by $r(r + 360)$.

$$810r = (r + 360)162$$
$$810r = 162r + 58{,}320$$
$$648r = 58{,}320$$
$$r = 90$$
$$r + 360 = 90 + 360 = 450$$

The rate of the jet was 450 mph.

You Try It 5

A plane can fly at a rate of 150 mph in calm air. Traveling with the wind, the plane flew 700 mi in the same amount of time that it flew 500 mi against the wind. Find the rate of the wind.

Your strategy

Your solution

Solution on pp. S20–S21

6.5 Exercises

Objective A **To solve a fractional equation**

For Exercises 1 to 21, solve.

1. $\dfrac{x}{2} + \dfrac{5}{6} = \dfrac{x}{3}$

2. $\dfrac{x}{5} - \dfrac{2}{9} = \dfrac{x}{15}$

3. $\dfrac{8}{2x - 1} = 2$

4. $3 = \dfrac{18}{3x - 4}$

5. $1 - \dfrac{3}{y} = 4$

6. $7 + \dfrac{6}{y} = 5$

7. $\dfrac{3}{x - 2} = \dfrac{4}{x}$

8. $\dfrac{5}{x} = \dfrac{2}{x + 3}$

9. $\dfrac{6}{2y + 3} = \dfrac{6}{y}$

10. $\dfrac{3}{x - 4} + 2 = \dfrac{5}{x - 4}$

11. $\dfrac{5}{y + 3} - 2 = \dfrac{7}{y + 3}$

12. $5 + \dfrac{8}{a - 2} = \dfrac{4a}{a - 2}$

13. $\dfrac{-4}{a - 4} = 3 - \dfrac{a}{a - 4}$

14. $\dfrac{x}{x + 1} - x = \dfrac{-4}{x + 1}$

15. $\dfrac{2x}{x + 2} + 3x = \dfrac{-5}{x + 2}$

16. $\dfrac{x}{x - 3} + x = \dfrac{3x - 4}{x - 3}$

17. $\dfrac{x}{2x - 9} - 3x = \dfrac{10}{9 - 2x}$

18. $\dfrac{2}{4y^2 - 9} + \dfrac{1}{2y - 3} = \dfrac{3}{2y + 3}$

19. $\dfrac{5}{x - 2} - \dfrac{2}{x + 2} = \dfrac{3}{x^2 - 4}$

20. $\dfrac{5}{x^2 - 7x + 12} = \dfrac{2}{x - 3} + \dfrac{5}{x - 4}$

21. $\dfrac{9}{x^2 + 7x + 10} = \dfrac{5}{x + 2} - \dfrac{3}{x + 5}$

For Exercises 22 to 25, solve the formula for the given variable.

22. $F = \dfrac{Gm_1 m_2}{r^2}; \; m_2$ (Physics)

23. $\dfrac{P_1 V_1}{T_1} = \dfrac{P_2 V_2}{T_2}; \; P_2$ (Chemistry)

24. $\dfrac{1}{R} = \dfrac{1}{R_1} + \dfrac{1}{R_2}; \; R_2$ (Physics)

25. $\dfrac{1}{f} = \dfrac{1}{a} + \dfrac{1}{b}; \; b$ (Physics)

To solve work problems

26. A ski resort can manufacture enough machine-made snow to open its steepest run in 12 h, whereas naturally falling snow would take 36 h to provide enough snow. If the resort makes snow at the same time that it is snowing naturally, how long will it take until the run can open?

27. An experienced bricklayer can work twice as fast as an apprentice bricklayer. After they work together on a job for 6 h, the experienced bricklayer quits. The apprentice requires 10 more hours to finish the job. How long would it take the experienced bricklayer working alone to do the job?

28. A roofer requires 8 h to shingle a roof. After the roofer and an apprentice work on a roof for 2 h, the roofer moves on to another job. The apprentice requires 10 more hours to finish the job. How long would it take the apprentice working alone to do the job?

29. An employee is faxing a long document that will take 40 min to send on her fastest machine. After 15 min, she realizes the fax won't make her deadline and she starts sending some of the pages with a second, slower fax machine. Twenty minutes later the document finishes faxing. How long would it have taken the slower machine alone to send the fax?

30. The larger of two printers being used to print the payroll for a major corporation requires 30 min to print the payroll. After both printers have been operating for 10 min, the larger printer malfunctions. The smaller printer requires 40 more minutes to complete the payroll. How long would it take the smaller printer working alone to print the payroll?

31. Three machines are filling water bottles. The machines can fill the daily quota of water bottles in 10 h, 12 h, and 15 h, respectively. How long would it take to fill the daily quota of water bottles with all three machines working?

32. A goat can eat all the grass in a farmer's field in 12 days, whereas a cow can finish it in 15 days and a horse in 20 days. How long will it be before all the grass is eaten if all three animals graze in the field?

33. The inlet pipe can fill a water tank in 45 min. The outlet pipe can empty the tank in 30 min. How long would it take to empty a full tank with both pipes open?

34. Three computers can print out a task in 20 min, 30 min, and 60 min, respectively. How long would it take to complete the task with all three computers working?

35. Two circus clowns are blowing up balloons, but some of the balloons are popping before they can be sold. The first clown can blow up a balloon every 2 min, the second clown requires 3 min for each balloon, and every 5 min one balloon pops. How long will it take the clowns, working together, to have 76 balloons?

36. An oil tank has two inlet pipes and one outlet pipe. One inlet pipe can fill the tank in 12 h, and the other inlet pipe can fill the tank in 20 h. The outlet pipe can empty the tank in 10 h. How long would it take to fill the tank with all three pipes open?

37. Two clerks are addressing advertising envelopes for a company. One clerk can address one envelope every 30 s, whereas it takes 40 s for the second clerk to address one envelope. How long will it take them, working together, to address 140 envelopes?

38. A single-engine airplane carries enough fuel for an 8-hour flight. After the airplane has been flying for 1 h, the fuel tank begins to leak at a rate that would empty the tank in 12 h. How long after the leak begins does the plane have until it will run out of fuel?

Objective C **To solve uniform motion problems**

39. The rate of a bicyclist is 7 mph faster than the rate of a long-distance runner. The bicyclist travels 30 mi in the same amount of time as it takes the runner to travel 16 mi. Find the rate of the runner.

40. A passenger train travels 240 mi in the same amount of time as it takes a freight train to travel 168 mi. The rate of the passenger train is 18 mph greater than the rate of the freight train. Find the rate of each train.

41. A tortoise and a hare have joined a 360-foot race. Since the hare can run 180 times faster than the tortoise, it reaches the finish line 14 min and 55 s before the tortoise. How fast was each animal running?

42. A cabin cruiser travels 20 mi in the same amount of time it takes a power boat to travel 45 mi. The rate of the cabin cruiser is 10 mph less than the rate of the power boat. Find the rate of the cabin cruiser.

43. A cyclist and a jogger start from a town at the same time and head for a destination 30 mi away. The rate of the cyclist is twice the rate of the jogger. The cyclist arrives 3 h ahead of the jogger. Find the rate of the cyclist.

44. A Porsche 911 Turbo has a top speed that is 20 mph faster than a Dodge Viper's top speed. At top speed, the Porsche can travel 630 mi in the same amount of time that the Viper can travel 560 mi. What is the top speed of each car?

45. An insurance representative traveled 735 mi by commercial jet and then an additional 105 mi by helicopter. The rate of the jet was four times the rate of the helicopter. The entire trip took 2.2 h. Find the rate of the jet.

46. A canoe can travel 8 mph in still water. Rowing with the current of a river, the canoe can travel 15 mi in the same amount of time as it takes to travel 9 mi against the current. Find the rate of the current.

47. A tour boat used for river excursions can travel 7 mph in calm water. The amount of time it takes to travel 20 mi with the current is the same as the amount of time it takes to travel 8 mi against the current. Find the rate of the current.

48. Some military fighter planes are capable of flying at a rate of 950 mph. One of these planes flew 6150 mi with the wind in the same amount of time it took to fly 5250 mi against the wind. Find the rate of the wind.

49. A jet can travel 550 mph in calm air. Flying with the wind, the jet can travel 3059 mi in the same amount of time that it takes to fly 2450 mi against the wind. Find the rate of the wind. Round to the nearest hundredth.

50. A jet-ski can comfortably travel across calm water at 35 mph. If a rider traveled 8 mi down a river in the same amount of time it took to travel 6 mi back up the river, find the rate of the river's current.

51. A river excursion motorboat can travel 6 mph in calm water. A recent trip, going 16 mi down a river and then returning, took 6 h. Find the rate of the river's current.

52. A pilot can fly a plane at 125 mph in calm air. A recent trip of 300 mi flying with the wind and 300 mi returning against the wind took 5 h. Find the rate of the wind.

APPLYING THE CONCEPTS

53. Solve for x.

a. $\dfrac{x-2}{y} = \dfrac{x+2}{5y}$

b. $\dfrac{x}{x+y} = \dfrac{2x}{4y}$

c. $\dfrac{x-y}{x} = \dfrac{2x}{9y}$

54. **Mathematics** The denominator of a fraction is 4 more than the numerator. If the numerator and denominator of the fraction are increased by 3, the new fraction is $\dfrac{5}{6}$. Find the original fraction.

55. **Uniform Motion** Because of bad weather conditions, a bus driver reduced the usual speed along a 165-mile bus route by 5 mph. The bus arrived only 15 min later than its usual arrival time. How fast does the bus usually travel?

56. Explain the difference between dealing with the denominator in simplifying a rational expression and dealing with the denominator in solving a rational equation.

6.6 Variation

Objective A **To solve variation problems**

Direct variation is a special function that can be expressed as the equation $y = kx$, where k is a constant. The equation $y = kx$ is read "y varies directly as x" or "y is directly proportional to x." The constant k is called the **constant of variation** or the **constant of proportionality.**

The circumference (C) of a circle varies directly as the diameter (d). The direct variation equation is written $C = \pi d$. The constant of variation is π.

A nurse earns \$30 per hour. The total wage (w) of the nurse is directly proportional to the number of hours (h) worked. The equation of variation is $w = 30h$. The constant of proportionality is 30.

A direct variation equation can be written in the form $y = kx^n$, where n is a positive number. For example, the equation $y = kx^2$ is read "y varies directly as the square of x."

The area (A) of a circle varies directly as the square of the radius (r) of the circle. The direct variation equation is $A = \pi r^2$. The constant of variation is π.

HOW TO Given that V varies directly as r and that $V = 20$ when $r = 4$, find the constant of variation and the variation equation.

$V = kr$ • Write the basic direct variation equation.
$20 = k \cdot 4$ • Replace V and r by the given values. Then solve for k.
$5 = k$ • This is the constant of variation.
$V = 5r$ • Write the direct variation equation by substituting the value of k into the basic direct variation equation.

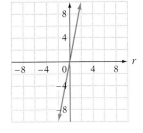
The graph of $V = 5r$

HOW TO The tension (T) in a spring varies directly as the distance (x) it is stretched. If $T = 8$ lb when $x = 2$ in., find T when $x = 4$ in.

$T = kx$ • Write the basic direct variation equation.
$8 = k \cdot 2$ • Replace T and x by the given values.
$4 = k$ • Solve for the constant of variation.
$T = 4x$ • Write the direct variation equation.

To find T when $x = 4$ in., substitute 4 for x in the equation and solve for T.

$T = 4x$
$T = 4 \cdot 4 = 16$

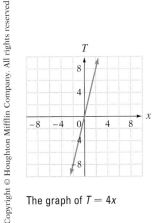
The graph of $T = 4x$

The tension is 16 lb.

Inverse variation is a function that can be expressed as the equation $y = \frac{k}{x}$, where k is a constant. The equation $y = \frac{k}{x}$ is read "y varies inversely as x" or "y is inversely proportional to x." In general, an inverse variation equation can be written $y = \frac{k}{x^n}$, where n is a positive number. For example, the equation $y = \frac{k}{x^2}$ is read "y varies inversely as the square of x."

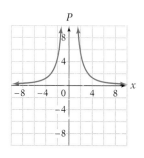

The graph of $P = \frac{20}{x^2}$

HOW TO Given that P varies inversely as the square of x and that $P = 5$ when $x = 2$, find the constant of variation and the variation equation.

$P = \dfrac{k}{x^2}$ • Write the basic inverse variation equation.

$5 = \dfrac{k}{2^2}$ • Replace P and x by the given values. Then solve for k.

$20 = k$ • This is the constant of variation.

$P = \dfrac{20}{x^2}$ • Write the inverse variation equation by substituting the value of k into the basic inverse variation equation.

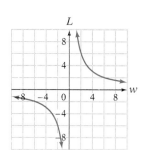

The graph of $L = \frac{12}{w}$

HOW TO The length (L) of a rectangle with fixed area is inversely proportional to the width (w). If $L = 6$ ft when $w = 2$ ft, find L when $w = 3$ ft.

$L = \dfrac{k}{w}$ • Write the basic inverse variation equation.

$6 = \dfrac{k}{2}$ • Replace L and w by the given values.

$12 = k$ • Solve for the constant of variation.

$L = \dfrac{12}{w}$ • Write the inverse variation equation.

To find L when $w = 3$ ft, substitute 3 for w in the equation and solve for L.

$L = \dfrac{12}{w}$

$L = \dfrac{12}{3} = 4$

The length is 4 ft.

Joint variation is a variation in which a variable varies directly as the product of two or more other variables. A joint variation can be expressed as the equation $z = kxy$, where k is a constant. The equation $z = kxy$ is read "z varies jointly as x and y."

The area (A) of a triangle varies jointly as the base (b) and the height (h). The joint variation equation is written $A = \frac{1}{2}bh$. The constant of variation is $\frac{1}{2}$.

A **combined variation** is a variation in which two or more types of variation occur at the same time. For example, in physics, the volume (V) of a gas varies directly as the temperature (T) and inversely as the pressure (P). This combined variation is written $V = \dfrac{kT}{P}$.

HOW TO A ball is being twirled on the end of a string. The tension (T) in the string is directly proportional to the square of the speed (v) of the ball and inversely proportional to the length (r) of the string. If the tension is 96 lb when the length of the string is 0.5 ft and the speed is 4 ft/s, find the tension when the length of the string is 1 ft and the speed is 5 ft/s.

$$T = \frac{kv^2}{r}$$ • Write the basic combined variation equation.

$$96 = \frac{k \cdot 4^2}{0.5}$$ • Replace *T*, *v*, and *r* by the given values.

$$96 = \frac{k \cdot 16}{0.5}$$

$$96 = k \cdot 32$$ • Solve for the constant of variation.

$$3 = k$$

$$T = \frac{3v^2}{r}$$ • Write the combined variation equation.

To find T when $r = 1$ ft and $v = 5$ ft/s, substitute 1 for r and 5 for v and solve for T.

$$T = \frac{3v^2}{r} = \frac{3 \cdot 5^2}{1} = 3 \cdot 25 = 75$$

The tension is 75 lb.

Example 1

The amount (A) of medication prescribed for a person is directly related to the person's weight (W). For a 50-kilogram person, 2 ml of medication is prescribed. How many milliliters of medication are required for a person who weighs 75 kg?

Strategy

To find the required amount of medication:
• Write the basic direct variation equation, replace the variables by the given values, and solve for k.
• Write the direct variation equation, replacing k by its value. Substitute 75 for W and solve for A.

Solution

$$A = kW$$ • Direct variation equation

$$2 = k \cdot 50$$ • Replace *A* by 2 and
 W by 50.

$$\frac{1}{25} = k$$

$$A = \frac{1}{25}W = \frac{1}{25} \cdot 75 = 3$$ • $k = \frac{1}{25}$, *W* = 75

The required amount of medication is 3 ml.

You Try It 1

The distance (s) a body falls from rest varies directly as the square of the time (t) of the fall. An object falls 64 ft in 2 s. How far will it fall in 5 s?

Your strategy

Your solution

Solution on p. S21

Example 2

A company that produces personal computers has determined that the number of computers it can sell (s) is inversely proportional to the price (P) of the computer. Two thousand computers can be sold when the price is $2500. How many computers can be sold when the price of a computer is $2000?

Strategy
To find the number of computers:
• Write the basic inverse variation equation, replace the variables by the given values, and solve for k.
• Write the inverse variation equation, replacing k by its value. Substitute 2000 for P and solve for s.

Solution

$$s = \frac{k}{P}$$ • Inverse variation equation

$$2000 = \frac{k}{2500}$$ • Replace s by 2000 and P by 2500.

$$5{,}000{,}000 = k$$

$$s = \frac{5{,}000{,}000}{P} = \frac{5{,}000{,}000}{2000} = 2500$$ • $k = 5{,}000{,}000$, $P = 2000$

At $2000 each, 2500 computers can be sold.

Example 3

The pressure (P) of a gas varies directly as the temperature (T) and inversely as the volume (V). When $T = 50°$ and $V = 275$ in³, $P = 20$ lb/in². Find the pressure of a gas when $T = 60°$ and $V = 250$ in³.

Strategy
To find the pressure:
• Write the basic combined variation equation, replace the variables by the given values, and solve for k.
• Write the combined variation equation, replacing k by its value. Substitute 60 for T and 250 for V, and solve for P.

Solution

$$P = \frac{kT}{V}$$ • Combined variation equation

$$20 = \frac{k \cdot 50}{275}$$ • Replace P by 20, T by 50, and V by 275.

$$110 = k$$

$$P = \frac{110T}{V} = \frac{110 \cdot 60}{250} = 26.4$$ • $k = 110$, $T = 60$ $V = 250$

The pressure is 26.4 lb/in².

You Try It 2

The resistance (R) to the flow of electric current in a wire of fixed length is inversely proportional to the square of the diameter (d) of the wire. If a wire of diameter 0.01 cm has a resistance of 0.5 ohm, what is the resistance in a wire that is 0.02 cm in diameter?

Your strategy

Your solution

You Try It 3

The strength (s) of a rectangular beam varies jointly as its width (W) and as the square of its depth (d) and inversely as its length (L). If the strength of a beam 2 in. wide, 12 in. deep, and 12 ft long is 1200 lb, find the strength of a beam 4 in. wide, 8 in. deep, and 16 ft long.

Your strategy

Your solution

Solutions on p. S21

6.6 Exercises

Objective A **To solve variation problems**

1. **Business** The profit (P) realized by a company varies directly as the number of products it sells (s). If a company makes a profit of $4000 on the sale of 250 products, what is the profit when the company sells 5000 products?

2. **Compensation** The income (I) of a computer analyst varies directly as the number of hours (h) worked. If the analyst earns $336 for working 8 h, how much will the analyst earn for working 36 h?

3. **Recreation** The pressure (p) on a diver in the water varies directly as the depth (d). If the pressure is 4.5 lb/in² when the depth is 10 ft, what is the pressure when the depth is 15 ft?

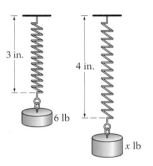

4. **Physics** The distance (d) that a spring will stretch varies directly as the force (f) applied to the spring. If a force of 6 lb is required to stretch a spring 3 in., what force is required to stretch the spring 4 in.?

5. **Physics** The distance (d) an object will fall is directly proportional to the square of the time (t) of the fall. If an object falls 144 ft in 3 s, how far will the object fall in 10 s?

6. **Physics** The period (p) of a pendulum, or the time it takes the pendulum to make one complete swing, varies directly as the square root of the length (L) of the pendulum. If the period of a pendulum is 1.5 s when the length is 2 ft, find the period when the length is 5 ft. Round to the nearest hundredth.

7. **Physics** The distance (s) a ball will roll down an inclined plane is directly proportional to the square of the time (t). If the ball rolls 6 ft in 1 s, how far will it roll in 3 s?

8. **Safety** The stopping distance (s) of a car varies directly as the square of its speed (v). If a car traveling 30 mph requires 63 ft to stop, find the stopping distance for a car traveling 55 mph.

9. **Geometry** The length (L) of a rectangle of fixed area varies inversely as the width (w). If the length of a rectangle is 8 ft when the width is 5 ft, find the length of the rectangle when the width is 4 ft.

10. **Travel** The time (t) for a car to travel between two cities is inversely proportional to the rate (r) of travel. If it takes 5 h to travel between the cities at a rate of 55 mph, find the time to travel between the two cities at a rate of 65 mph. Round to the nearest tenth.

11. **Mechanics** The speed (v) of a gear varies inversely as the number of teeth (t). If a gear that has 45 teeth makes 24 revolutions per minute, how many revolutions per minute will a gear that has 36 teeth make?

12. **Physics** The pressure (p) of a liquid varies directly as the product of the depth (d) and the density (D) of the liquid. If the pressure is 150 lb/in² when the depth is 100 in. and the density is 1.2, find the pressure when the density remains the same and the depth is 75 in.

13. **Electronics** The current (I) in a wire varies directly as the voltage (v) and inversely as the resistance (r). If the current is 10 amps when the voltage is 110 volts and the resistance is 11 ohms, find the current when the voltage is 180 volts and the resistance is 24 ohms.

14. **Magnetism** The repulsive force (f) between the north poles of two magnets is inversely proportional to the square of the distance (d) between them. If the repulsive force is 20 lb when the distance is 4 in., find the repulsive force when the distance is 2 in.

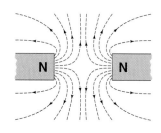

15. **Light** The intensity (I) of a light source is inversely proportional to the square of the distance (d) from the source. If the intensity is 12 foot-candles at a distance of 10 ft, what is the intensity when the distance is 5 ft?

APPLYING THE CONCEPTS

16. In the inverse variation equation $y = \dfrac{k}{x}$, what is the effect on x when y doubles?

17. In the direct variation equation $y = kx$, what is the effect on y when x doubles?

For Exercises 18 to 21, complete using the word *directly* or *inversely*.

18. If a varies directly as b and inversely as c, then c varies _____ as b and _____ as a.

19. If a varies _____ as b and c, then abc is constant.

20. If the length of a rectangle is held constant, the area of the rectangle varies _____ as the width.

21. If the area of a rectangle is held constant, the length of the rectangle varies _____ as the width.

Focus on Problem Solving

Implication Sentences that are constructed using "If..., then..." occur frequently in problem-solving situations. These sentences are called **implications.** The sentence "If it rains, then I will stay home" is an implication. The phrase *it rains* is called the **antecedent** of the implication. The phrase *I will stay home* is the **consequent** of the implication.

The sentence "If $x = 4$, then $x^2 = 16$" is a true sentence. The **contrapositive** of an implication is formed by switching the antecedent and the consequent and then negating each one. The contrapositive of "If $x = 4$, then $x^2 = 16$" is "If $x^2 \neq 16$, then $x \neq 4$." This sentence is also true. It is a principle of logic that an implication and its contrapositive are either both true or both false.

The **converse** of an implication is formed by switching the antecedent and the consequent. The converse of "If $x = 4$, then $x^2 = 16$" is "If $x^2 = 16$, then $x = 4$." Note that the converse is not a true statement, because if $x^2 = 16$, then x could equal -4. The converse of an implication may or may not be true.

Those statements for which the implication and the converse are both true are very important. They can be stated in the form "x if and only if y." For instance, a number is divisible by 5 if and only if the last digit of the number is 0 or 5. The implication "If a number is divisible by 5, then it ends in 0 or 5" and the converse "If a number ends in 0 or 5, then it is divisible by 5" are both true.

For Exercises 1 to 14, state the contrapositive and the converse of the implication. If the converse and the implication are both true statements, write a sentence using the phrase *if and only if.*

1. If I live in Chicago, then I live in Illinois.

2. If today is June 1, then yesterday was May 31.

3. If today is not Thursday, then tomorrow is not Friday.

4. If a number is divisible by 8, then it is divisible by 4.

5. If a number is an even number, then it is divisible by 2.

6. If a number is a multiple of 6, then it is a multiple of 3.

7. If $4z = 20$, then $z = 5$.

8. If an angle measures 90°, then it is a right angle.

9. If p is a prime number greater than 2, then p is an odd number.

10. If the equation of a graph is $y = mx + b$, then the graph of the equation is a straight line.

11. If $a = 0$ or $b = 0$, then $ab = 0$.

12. If the coordinates of a point are $(5, 0)$, then the point is on the x-axis.

13. If a quadrilateral is a square, then the quadrilateral has four sides of equal length.

14. If $x = y$, then $x^2 = y^2$.

Projects and Group Activities

Graphing Variation Equations

1. Graph $y = kx$ when $k = 2$.
 What kind of function does the graph represent?

2. Graph $y = kx$ when $k = \frac{1}{2}$.
 What kind of function does the graph represent?

3. Graph $y = \frac{k}{x}$ when $k = 2$ and $x > 0$.
 Is this the graph of a function?

Transformers

Transformers are devices used to increase or decrease the voltage in an electrical system. For example, if the voltage in a wire is 120 volts, it can pass through a transformer and "come out the other side" with a voltage of 60 volts. Transformers have extensive applications; for instance, they are commonly used to decrease the voltage from a mainframe to individual houses (look for the big gray boxes on telephone poles). Transformers are also used in wall adapters to decrease wall voltage—for example, for an answering machine. A transformer must have voltage and current entering and leaving. The voltage and current of both sides can be determined by an equation.

$$V_1 = \frac{V_2 I_2}{I_1}$$

where V_1 and I_1 are the voltage and current on one side of a transformer, and V_2 and I_2 are the voltage and current on the other side of the transformer.

HOW TO A transformer takes 1000 volts and reduces it to 100 volts. If the current is 30 amperes when the voltage is 1000 volts, find the current after the voltage is reduced.

$$V_1 = \frac{V_2 I_2}{I_1}$$
$$1000 = \frac{100 I_2}{30}$$
$$30{,}000 = 100 I_2$$
$$300 = I_2$$

• The volts are 1000 when the amperes are 30, so $V_1 = 1000$ and $I_1 = 30$. $V_2 = 100$. Find I_2.

After the voltage is reduced, the current is 300 amperes.

HOW TO On one side of a transformer, the current is 40 amperes and the voltage is 500 volts. If the current on the other side is 400 amperes, what is the voltage on that side?

$$V_1 = \frac{V_2 I_2}{I_1}$$

$$500 = \frac{V_2 \cdot 400}{40}$$

$$20{,}000 = 400V_2$$

$$50 = V_2$$

• The volts are 500 when the amperes are 40, so $V_1 = 500$ and $I_1 = 40$. $I_2 = 400$. Find V_2.

After the current is increased, the voltage is 50 volts.

Solve Exercises 1 to 6.

1. A transformer takes 2000 volts and reduces it to 1000 volts. If the current is 40 amperes when the voltage is 2000 volts, find the current after the voltage is reduced.

2. A wall adapter for a cassette tape player reduces 180 volts to 9 volts. If the current was 15 amperes at 180 volts, find the current at 9 volts.

3. A wall adapter for a portable CD player reduces 120 volts to 9 volts. If the current was 12 amperes at 120 volts, find the current at 9 volts.

4. On one side of a transformer, the current is 12 amperes and the voltage is 120 volts. If the current on the other side is 40 amperes, what is the voltage on that side?

5. On one side of a transformer, the voltage is 12 volts and the current is 4.5 amperes. Find the voltage on the other side if the current on that side is 60 amperes.

6. On one side of a transformer, the current is 500 amperes and the voltage is 8 volts. If the current on the other side is 400 amperes, what is the voltage on that side?

Chapter 6 Summary

Key Words	Examples
An expression in which the numerator and denominator are polynomials is called a *rational expression*. A function that is written in terms of a rational expression is a *rational function*. [6.1A, p. 341]	$\frac{x^2 + 2x + 1}{x - 3}$ is a rational expression.
A rational expression is in *simplest form* when the numerator and denominator have no common factors. [6.1B, p. 342]	$\dfrac{x^2 + 3x + 2}{(x + 1)^2} = \dfrac{(x + 1)(x + 2)}{(x + 1)(x + 1)}$ $= \dfrac{x + 2}{x + 1}$

The *reciprocal of a rational expression* is the rational expression with the numerator and denominator interchanged. [6.1D, p. 345]

The reciprocal of $\frac{a^2 - b^2}{x + y}$ is $\frac{x + y}{a^2 - b^2}$.

The *least common multiple (LCM) of two or more polynomials* is the simplest polynomial that contains the factors of each polynomial. [6.2A, p. 351]

$4x^2 - 12x = 4x(x - 3)$
$3x^3 - 21x^2 + 36x = 3x(x - 3)(x - 4)$
The LCM $= 12x(x - 3)(x - 4)$

A *complex fraction* is a fraction whose numerator or denominator contains one or more fractions. [6.3A, p. 359]

$\dfrac{\dfrac{1}{x} + \dfrac{1}{y}}{\dfrac{1}{x}}$ is a complex fraction.

A *ratio* is the quotient of two quantities that have the same unit. When a ratio is in simplest form, the units are not written. [6.4A, p. 363]

$\dfrac{\$35}{\$100}$ written as a ratio in simplest form is $\dfrac{7}{20}$.

A *rate* is the quotient of two quantities that have different units. The units are written as part of the rate. [6.4A, p. 363]

$\dfrac{65 \text{ mi}}{2 \text{ gal}}$ is a rate.

A *proportion* is an equation that states the equality of two ratios or rates. [6.4A, p. 363]

$\dfrac{75}{3} = \dfrac{x}{42}$ is a proportion.

Direct variation is a special function that can be expressed as the equation $y = kx^n$, where k is a constant called the *constant of variation* or the *constant of proportionality*. [6.6A, p. 377]

$E = mc^2$ is Einstein's formula relating energy and mass. c^2 is the constant of proportionality.

Inverse variation is a function that can be expressed as the equation $y = \dfrac{k}{x^n}$, where k is a constant. [6.6A, p. 378]

$I = \dfrac{k}{d^2}$ gives the intensity of a light source at a distance d from the source.

A *joint variation* is a variation in which a variable varies directly as the product of two or more variables. A joint variation can be expressed as the equation $z = kxy$, where k is a constant. [6.6A, p. 378]

$C = kAT$ is a formula for the cost of insulation, where A is the area to be insulated and T is the thickness of the insulation.

A *combined variation* is a variation in which two or more types of variation occur at the same time. [6.6A, p. 378]

$V = \dfrac{kT}{P}$ is a formula that states that the volume of a gas is directly proportional to the temperature and inversely proportional to the pressure.

Essential Rules and Procedures	Examples

To Multiply Rational Expressions [6.1C, p. 344]

$$\frac{a}{b} \cdot \frac{c}{d} = \frac{ac}{bd}$$

$$\frac{x^2 + x}{3} \cdot \frac{2}{5x} = \frac{(x^2 + x) \cdot 2}{3(5x)}$$
$$= \frac{2x(x + 1)}{15x} = \frac{2(x + 1)}{15}$$

To Divide Rational Expressions [6.1D, p. 345]

$$\frac{a}{b} \div \frac{c}{d} = \frac{a}{b} \cdot \frac{d}{c} = \frac{ad}{bc}$$

$$\frac{3x}{x - 5} \div \frac{3}{x + 4} = \frac{3x}{x - 5} \cdot \frac{x + 4}{3}$$
$$= \frac{3x(x + 4)}{3(x - 5)} = \frac{x(x + 4)}{x - 5}$$

To Add Rational Expressions [6.2B, p. 353]

$$\frac{a}{c} + \frac{b}{c} = \frac{a + b}{c}$$

$$\frac{2x - 7}{x^2 + 4} + \frac{x + 2}{x^2 + 4} = \frac{(2x - 7) + (x + 2)}{x^2 + 4}$$
$$= \frac{3x - 5}{x^2 + 4}$$

To Subtract Rational Expressions [6.2B, p. 353]

$$\frac{a}{c} - \frac{b}{c} = \frac{a - b}{c}$$

$$\frac{5x - 6}{x + 2} - \frac{2x + 4}{x + 2} = \frac{(5x - 6) - (2x + 4)}{x + 2}$$
$$= \frac{3x - 10}{x + 2}$$

To Solve an Equation Containing Fractions [6.5A, p. 367]
Clear denominators by multiplying each side of the equation by the LCM of the denominators.

$$\frac{3}{x} + 5 = \frac{7}{2x}$$
$$2x\left(\frac{3}{x} + 5\right) = 2x\left(\frac{7}{2x}\right)$$
$$6 + 10x = 7$$
$$10x = 1$$
$$x = \frac{1}{10}$$

Equation for Work Problems [6.5B, p. 369]
Rate of work × time worked = part of task completed

A roofer requires 24 h to shingle a roof. An apprentice can shingle the roof in 36 h. How long would it take to shingle the roof if both roofers worked together?

$$\frac{t}{24} + \frac{t}{36} = 1$$

Equation for Uniform Motion Problems [6.5C, p. 371]

$$\text{Distance} = \text{rate} \times \text{time} \quad \text{or} \quad \frac{\text{Distance}}{\text{Rate}} = \text{time}$$

A motorcycle travels 195 mi in the same amount of time it takes a car to travel 159 mi. The rate of the motorcycle is 12 mph greater than the rate of the car. Find the rate of the car.

$$\frac{195}{r + 12} = \frac{159}{r}$$

Chapter 6 Review Exercises

1. Simplify: $\dfrac{a^6 b^4 + a^4 b^6}{a^5 b^4 - a^4 b^4} \cdot \dfrac{a^2 - b^2}{a^4 - b^4}$

2. Simplify: $\dfrac{x}{x-3} - 4 - \dfrac{2x-5}{x+2}$

3. Given $P(x) = \dfrac{x}{x-3}$, find $P(4)$.

4. Solve: $\dfrac{3x-2}{x+6} = \dfrac{3x+1}{x+9}$

5. Simplify: $\dfrac{\dfrac{3x+4}{3x-4} + \dfrac{3x-4}{3x+4}}{\dfrac{3x-4}{3x+4} - \dfrac{3x+4}{3x-4}}$

6. Write each fraction in terms of the LCM of the denominators.
$$\dfrac{4x}{4x-1}, \ \dfrac{3x-1}{4x+1}$$

7. Solve $S = \dfrac{a}{1-r}$ for r.

8. Given $P(x) = \dfrac{x^2 - 2}{3x^2 - 2x + 5}$, find $P(-2)$.

9. Solve: $\dfrac{10}{5x+3} = \dfrac{2}{10x-3}$

10. Determine the domain of
$$f(x) = \dfrac{2x-7}{3x^2 + 3x - 18}.$$

11. Simplify: $\dfrac{3x^4 + 11x^2 - 4}{3x^4 + 13x^2 + 4}$

12. Determine the domain of $g(x) = \dfrac{2x}{x-3}$.

13. Simplify: $\dfrac{x^3 - 8}{x^3 + 2x^2 + 4x} \cdot \dfrac{x^3 + 2x^2}{x^2 - 4}$

14. Simplify: $\dfrac{3x^2 + 2}{x^2 - 4} - \dfrac{9x - x^2}{x^2 - 4}$

15. Solve $Q = \dfrac{N - S}{N}$ for N.

16. Solve: $\dfrac{30}{x^2 + 5x + 4} + \dfrac{10}{x+4} = \dfrac{4}{x+1}$

17. Simplify: $x + \dfrac{\dfrac{4}{x} - 1}{\dfrac{1}{x} - \dfrac{3}{x^2}}$

18. Write each fraction in terms of the LCM of the denominators.

$$\dfrac{x - 3}{x - 5}, \ \dfrac{x}{x^2 - 9x + 20}, \ \dfrac{1}{4 - x}$$

19. Solve: $\dfrac{6}{2x - 3} = \dfrac{5}{x + 5} + \dfrac{5}{2x^2 + 7x - 15}$

20. Simplify: $\dfrac{x^{n+1} + x}{x^{2n} - 1} \div \dfrac{x^{n+2} - x^2}{x^{2n} - 2x^n + 1}$

21. Simplify: $\dfrac{27x^3 - 8}{9x^3 + 6x^2 + 4x} \div \dfrac{9x^2 - 12x + 4}{9x^2 - 4}$

22. Simplify: $\dfrac{6x}{3x^2 - 7x + 2} - \dfrac{2}{3x - 1} + \dfrac{3x}{x - 2}$

23. Simplify: $\dfrac{x^3 - 27}{x^2 - 9}$

24. Simplify: $\dfrac{5}{3a^2b^3} + \dfrac{7}{8ab^4}$

25. Find the domain of $F(x) = \dfrac{x^2 - x}{3x^2 + 4}$.

26. Simplify: $\dfrac{12a^{4n} + 9a^{3n} - 6a^{2n}}{3a^{2n}}$

27. Simplify: $\dfrac{16 - x^2}{x^3 - 2x^2 - 8x}$

28. Simplify: $\dfrac{8x^3 - 27}{4x^2 - 9}$

29. Multiply: $\dfrac{16 - x^2}{6x + 12} \cdot \dfrac{x^2 + 5x + 6}{x^2 - 8x + 16}$

30. Divide: $\dfrac{x^{2n} - 5x^n + 4}{x^{2n} - 2x^n - 8} \div \dfrac{x^{2n} - 4x^n + 3}{x^{2n} + 8x^n + 12}$

31. Divide: $\dfrac{8x^3 - 64}{4x^3 + 4x^2 + x} \div \dfrac{x^2 + 2x + 4}{4x^2 - 1}$

32. Divide: $\dfrac{3 - x}{x^2 + 3x + 9} \div \dfrac{x^2 - 9}{x^3 - 27}$

33. Add: $\dfrac{3}{2a^3b^2} + \dfrac{5}{6a^2b}$

34. Simplify: $\dfrac{8}{9x^2 - 4} + \dfrac{5}{3x - 2} - \dfrac{4}{3x + 2}$

35. Simplify: $\dfrac{x - 6 + \dfrac{6}{x - 1}}{x + 3 - \dfrac{12}{x - 1}}$

36. Simplify: $\dfrac{x + \dfrac{3}{x - 4}}{3 + \dfrac{x}{x - 4}}$

37. Solve: $\dfrac{x + 2}{x - 3} = \dfrac{2x - 5}{x + 1}$

38. Solve: $\dfrac{5x}{2x - 3} + 4 = \dfrac{3}{2x - 3}$

39. Solve: $I = \dfrac{V}{R}$ for R.

40. Solve: $\dfrac{6}{x - 3} - \dfrac{1}{x + 3} = \dfrac{51}{x^2 - 9}$

41. Work The inlet pipe can fill a tub in 24 min. The drain pipe can empty the tub in 15 min. How long would it take to empty the tub with both pipes open?

42. Uniform Motion A bus and a cyclist leave a school at 8 A.M. and head for a stadium 90 mi away. The rate of the bus is three times the rate of the cyclist. The cyclist arrives 4 h after the bus. Find the rate of the bus.

43. Uniform Motion A helicopter travels 9 mi in the same amount of time that it takes an airplane to travel 10 mi. The rate of the airplane is 20 mph greater than the rate of the helicopter. Find the rate of the helicopter.

44. Education A student reads 2 pages of text in 5 min. At the same rate, how long will it take the student to read 150 pages?

45. Electronics The current (I) in an electric circuit varies inversely as the resistance (R). If the current in the circuit is 4 amps when the resistance is 50 ohms, find the current in the circuit when the resistance is 100 ohms.

46. Cartography On a certain map, 2.5 in. represents 10 mi. How many miles would be represented by 12 in.?

47. Safety The stopping distance (s) of a car varies directly as the square of the speed (v) of the car. For a car traveling at 50 mph, the stopping distance is 170 ft. Find the stopping distance of a car that is traveling at 65 mph.

48. Work An electrician requires 65 min to install a ceiling fan. The electrician and an apprentice working together take 40 min to install the fan. How long would it take the apprentice working alone to install the ceiling fan?

Chapter 6 Test

1. Solve: $\dfrac{3}{x+1} = \dfrac{2}{x}$

2. Simplify: $\dfrac{x^2+x-6}{x^2+7x+12} \div \dfrac{x^2-3x+2}{x^2+6x+8}$

3. Simplify: $\dfrac{2x-1}{x+2} - \dfrac{x}{x-3}$

4. Write each fraction in terms of the LCM of the denominators.

$\dfrac{x+1}{x^2+x-6}, \dfrac{2x}{x^2-9}$

5. Solve: $\dfrac{4x}{2x-1} = 2 - \dfrac{1}{2x-1}$

6. Simplify: $\dfrac{v^3-4v}{2v^2-5v+2}$

7. Simplify: $\dfrac{3x^2-12}{5x-15} \cdot \dfrac{2x^2-18}{x^2+5x+6}$

8. Determine the domain of $f(x) = \dfrac{3x^2-x+1}{x^2-9}$.

9. Simplify: $\dfrac{1 - \dfrac{1}{x} - \dfrac{12}{x^2}}{1 + \dfrac{6}{x} + \dfrac{9}{x^2}}$

10. Simplify: $\dfrac{1 - \dfrac{1}{x+2}}{1 - \dfrac{3}{x+4}}$

11. Simplify: $\dfrac{2x^2-x-3}{2x^2-5x+3} \div \dfrac{3x^2-x-4}{x^2-1}$

12. Solve: $\dfrac{4x}{x+1} - x = \dfrac{2}{x+1}$

13. Simplify: $\dfrac{2a^2-8a+8}{4+4a-3a^2}$

14. Solve $\dfrac{1}{r} = \dfrac{1}{2} - \dfrac{2}{t}$ for t.

15. Given $f(x) = \dfrac{3 - x^2}{x^3 - 2x^2 + 4}$, find $f(-1)$.

16. Simplify: $\dfrac{x + 2}{x^2 + 3x - 4} - \dfrac{2x}{x^2 - 1}$

17. Solve: $\dfrac{x + 1}{2x + 5} = \dfrac{x - 3}{x}$

18. Uniform Motion A cyclist travels 20 mi in the same amount of time as it takes a hiker to walk 6 mi. The rate of the cyclist is 7 mph greater than the rate of the hiker. Find the rate of the cyclist.

19. Electronics The electrical resistance (r) of a cable varies directly as its length (l) and inversely as the square of its diameter (d). If a cable 16,000 ft long and $\frac{1}{4}$ in. in diameter has a resistance of 3.2 ohms, what is the resistance of a cable that is 8000 ft long and $\frac{1}{2}$ in. in diameter?

20. Interior Design An interior designer uses 2 rolls of wallpaper for every 45 ft² of wall space in an office. At this rate, how many rolls of wallpaper are needed for an office that has 315 ft² of wall space?

21. Work One landscaper can till the soil for a lawn in 30 min, whereas it takes a second landscaper 15 min to do the same job. How long would it take to till the soil for the lawn with both landscapers working together?

22. Electronics The current (I) in an electric circuit is inversely proportional to the resistance (R). If the current is 0.25 amp when the resistance is 8 ohms, find the resistance when the current is 1.25 amps.

Cumulative Review Exercises

1. Simplify: $8 - 4[-3 - (-2)]^2 \div 5$

2. Solve: $\dfrac{2x - 3}{6} - \dfrac{x}{9} = \dfrac{x - 4}{3}$

3. Solve: $5 - |x - 4| = 2$

4. Find the domain of $\dfrac{x}{x - 3}$.

5. Given $P(x) = \dfrac{x - 1}{2x - 3}$, find $P(-2)$.

6. Write 0.000000035 in scientific notation.

7. Simplify: $\dfrac{(2a^{-2}b^3)^{-2}}{(4a)^{-1}}$

8. Solve: $x - 3(1 - 2x) \geq 1 - 4(2 - 2x)$
Write the solution set in interval notation.

9. Simplify: $(2a^2 - 3a + 1)(-2a^2)$

10. Factor: $2x^{2n} + 3x^n - 2$

11. Factor: $x^3y^3 - 27$

12. Simplify: $\dfrac{x^4 + x^3y - 6x^2y^2}{x^3 - 2x^2y}$

13. Find the equation of the line that contains the point $(-2, -1)$ and is parallel to the line $3x - 2y = 6$.

14. Solve: $8x^2 - 6x - 9 = 0$

15. Simplify: $\dfrac{4x^3 + 2x^2 - 10x + 1}{x - 2}$

16. Simplify: $\dfrac{16x^2 - 9y^2}{16x^2y - 12xy^2} \div \dfrac{4x^2 - xy - 3y^2}{12x^2y^2}$

17. Write each fraction in terms of the LCM of the denominators. $\dfrac{xy}{2x^2 + 2x}, \dfrac{2}{2x^4 - 2x^3 - 4x^2}$

18. Simplify: $\dfrac{5x}{3x^2 - x - 2} - \dfrac{2x}{x^2 - 1}$

19. Graph $-3x + 5y = -15$ by using the x- and y-intercepts.

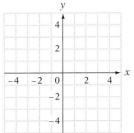

20. Graph the solution set: $\begin{array}{l} x + y \leq 3 \\ -2x + y > 4 \end{array}$

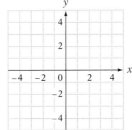

21. Evaluate the determinant:

$$\begin{vmatrix} 6 & 5 \\ 2 & -3 \end{vmatrix}$$

22. Simplify: $\dfrac{x - 4 + \dfrac{5}{x + 2}}{x + 2 - \dfrac{1}{x + 2}}$

23. Solve:
$$\begin{aligned} x + y + z &= 3 \\ -2x + y + 3z &= 2 \\ 2x - 4y + z &= -1 \end{aligned}$$

24. Solve: $|3x - 2| > 4$

25. Solve: $\dfrac{2}{x - 3} = \dfrac{5}{2x - 3}$

26. Solve: $\dfrac{3}{x^2 - 36} = \dfrac{2}{x - 6} - \dfrac{5}{x + 6}$

27. Solve $I = \dfrac{E}{R + r}$ for r.

28. Simplify: $(1 - x^{-1})^{-1}$

29. **Integers** The sum of two integers is fifteen. Five times the smaller integer is five more than twice the larger integer. Find the integers.

30. **Mixtures** How many pounds of almonds that cost $5.40 per pound must be mixed with 50 lb of peanuts that cost $2.60 per pound to make a mixture that costs $4.00 per pound?

31. **Elections** A pre-election survey showed that three out of five voters would vote in an election. At this rate, how many people would be expected to vote in a city of 125,000?

32. **Work** A new computer can work six times faster than an older computer. Working together, the computers can complete a job in 12 min. How long would it take the new computer working alone to complete the job?

33. **Uniform Motion** A plane can fly at a rate of 300 mph in calm air. Traveling with the wind, the plane flew 900 mi in the same amount of time as it took to fly 600 mi against the wind. Find the rate of the wind.

34. **Uniform Motion** Two people start from the same point on a circular exercise track that is 0.25 mi in circumference. The first person walks at 3 mph, and the second person jogs at 5 mph. After 1 h, how far apart (along the track) are the two people?

chapter

7 Exponents and Radicals

Thousands of man-made satellites have been launched into space and are now orbiting Earth. These satellites do hundreds of different tasks, from collecting data on our atmosphere to facilitating telecommunications. The speed at which a satellite orbits Earth is a function of its height above Earth's surface. The height at which it orbits Earth is dependent on the satellite's purpose. The relationship between a satellite's speed and its height above Earth can be modeled by a radical equation, such as the one used in **Exercise 26 on page 422.**

OBJECTIVES

Section 7.1

A To simplify expressions with rational exponents

B To write exponential expressions as radical expressions and to write radical expressions as exponential expressions

C To simplify radical expressions that are roots of perfect powers

Section 7.2

A To simplify radical expressions

B To add or subtract radical expressions

C To multiply radical expressions

D To divide radical expressions

Section 7.3

A To solve a radical equation

B To solve application problems

Section 7.4

A To simplify a complex number

B To add or subtract complex numbers

C To multiply complex numbers

D To divide complex numbers

Need help? For online student resources, such as section quizzes, visit this textbook's website at **math.college.hmco.com/students.**

Do these exercises to prepare for Chapter 7.

1. Complete: $48 = ? \cdot 3$

For Exercises 2 to 7, simplify.

2. 2^5

3. $6\left(\dfrac{3}{2}\right)$

4. $\dfrac{1}{2} - \dfrac{2}{3} + \dfrac{1}{4}$

5. $(3 - 7x) - (4 - 2x)$

6. $\dfrac{3x^5y^6}{12x^4y}$

7. $(3x - 2)^2$

For Exercises 8 and 9, multiply.

8. $(2 + 4x)(5 - 3x)$

9. $(6x - 1)(6x + 1)$

10. Solve: $x^2 - 14x - 5 = 10$

GO FIGURE • • •

You are planning a large dinner party. If you seat 5 people at each table, you end up with only 2 people at the last table. If you seat 3 people at each table, you have 9 people left over with no place to sit. There are less than 10 tables. How many guests are coming to the dinner party?

7.1 Rational Exponents and Radical Expressions

Objective A **To simplify expressions with rational exponents**

Point of Interest

Nicolas Chuquet (c. 1475), a French physician, wrote an algebra text in which he used a notation for expressions with fractional exponents. He wrote $R^2 6$ to mean $6^{1/2}$ and $R^3 15$ to mean $15^{1/3}$. This was an improvement over earlier notations that used words for these expressions.

In this section, the definition of an exponent is extended beyond integers so that any rational number can be used as an exponent. The definition is expressed in such a way that the Rules of Exponents hold true for rational exponents.

Consider the expression $(a^{1/n})^n$ for $a > 0$ and n a positive integer. Now simplify, assuming that the Rule for Simplifying the Power of an Exponential Expression is true.

$$(a^{1/n})^n = a^{\frac{1}{n} \cdot n} = a^1 = a$$

Because $(a^{1/n})^n = a$, the number $a^{1/n}$ is the number whose nth power is a.

If $a > 0$ and n is a positive number, then $a^{1/n}$ is called the nth root of a.

$25^{1/2} = 5$ because $(5)^2 = 25$.

$8^{1/3} = 2$ because $(2)^3 = 8$.

In the expression $a^{1/n}$, if a is a negative number and n is a positive even integer, then $a^{1/n}$ is not a real number.

$(-4)^{1/2}$ is not a real number, because there is no real number whose second power is -4.

When n is a positive odd integer, a can be a positive or a negative number.

$(-27)^{1/3} = -3$ because $(-3)^3 = -27$.

Integrating Technology

A calculator can be used to evaluate expressions with rational exponents. For example, to evaluate the expression at the right, press
(−) 27 ^ (1 ÷
3) ENTER . The display reads −3.

Using the definition of $a^{1/n}$ and the Rules of Exponents, it is possible to define any exponential expression that contains a rational exponent.

Rule for Rational Exponents

If m and n are positive integers and $a^{1/n}$ is a real number, then

$$a^{m/n} = (a^{1/n})^m$$

The expression $a^{m/n}$ can also be written $a^{m/n} = a^{m \cdot \frac{1}{n}} = (a^m)^{1/n}$. However, rewriting $a^{m/n}$ as $(a^m)^{1/n}$ is not as useful as rewriting it as $(a^{1/n})^m$. See the Take Note at the top of the next page.

As shown above, expressions that contain rational exponents do not always represent real numbers when the base of the exponential expression is a negative number. For this reason, **all variables in this chapter represent positive numbers unless otherwise stated.**

HOW TO Simplify: $27^{2/3}$

$27^{2/3} = (3^3)^{2/3}$ • Rewrite 27 as 3^3.

$= 3^{3(2/3)}$ • Multiply the exponents.

$= 3^2$ • Simplify.

$= 9$

HOW TO Simplify: $32^{-2/5}$

$32^{-2/5} = (2^5)^{-2/5}$ • Rewrite 32 as 2^5.

$= 2^{-2}$ • Multiply the exponents.

$= \dfrac{1}{2^2}$ • Use the Rule of Negative Exponents.

$= \dfrac{1}{4}$ • Simplify.

HOW TO Simplify: $a^{1/2} \cdot a^{2/3} \cdot a^{-1/4}$

$a^{1/2} \cdot a^{2/3} \cdot a^{-1/4} = a^{1/2 + 2/3 - 1/4}$ • Use the Rule for Multiplying Exponential Expressions.

$= a^{6/12 + 8/12 - 3/12}$

$= a^{11/12}$ • Simplify.

HOW TO Simplify: $(x^6 y^4)^{3/2}$

$(x^6 y^4)^{3/2} = x^{6(3/2)} y^{4(3/2)}$ • Use the Rule for Simplifying Powers of Products.

$= x^9 y^6$ • Simplify.

HOW TO Simplify: $\left(\dfrac{8a^3 b^{-4}}{64a^{-9} b^2}\right)^{2/3}$

$\left(\dfrac{8a^3 b^{-4}}{64a^{-9} b^2}\right)^{2/3} = \left(\dfrac{2^3 a^3 b^{-4}}{2^6 a^{-9} b^2}\right)^{2/3}$ • Rewrite 8 as 2^3 and 64 as 2^6.

$= (2^{-3} a^{12} b^{-6})^{2/3}$ • Use the Rule for Dividing Exponential Expressions.

$= 2^{-2} a^8 b^{-4}$ • Use the Rule for Simplifying Powers of Products.

$= \dfrac{a^8}{2^2 b^4} = \dfrac{a^8}{4b^4}$ • Use the Rule of Negative Exponents and simplify.

Example 1 Simplify: $64^{-2/3}$

Solution $64^{-2/3} = (2^6)^{-2/3} = 2^{-4}$ • $64 = 2^6$

$$= \frac{1}{2^4} = \frac{1}{16}$$

You Try It 1 Simplify: $16^{-3/4}$

Your solution

Example 2 Simplify: $(-49)^{3/2}$

Solution The base of the exponential expression is a negative number, while the denominator of the exponent is a positive even number.

Therefore, $(-49)^{3/2}$ is not a real number.

You Try It 2 Simplify: $(-81)^{3/4}$

Your solution

Example 3 Simplify: $(x^{1/2}y^{-3/2}z^{1/4})^{-3/2}$

Solution $(x^{1/2}y^{-3/2}z^{1/4})^{-3/2}$
$= x^{-3/4}y^{9/4}z^{-3/8}$ • Use the Rule for Simplifying
$= \dfrac{y^{9/4}}{x^{3/4}z^{3/8}}$ Powers of Products.

You Try It 3 Simplify: $(x^{3/4}y^{1/2}z^{-2/3})^{-4/3}$

Your solution

Example 4 Simplify: $\dfrac{x^{1/2}y^{-5/4}}{x^{-4/3}y^{1/3}}$

Solution $\dfrac{x^{1/2}y^{-5/4}}{x^{-4/3}y^{1/3}}$

$= x^{3/6-(-8/6)}y^{-15/12-4/12}$ • Use the Rule for Dividing Exponential Expressions.

$= x^{11/6}y^{-19/12} = \dfrac{x^{11/6}}{y^{19/12}}$

You Try It 4 Simplify: $\left(\dfrac{16a^{-2}b^{4/3}}{9a^4b^{-2/3}}\right)^{-1/2}$

Your solution

Solutions on pp. S21–S22

Objective B

To write exponential expressions as radical expressions and to write radical expressions as exponential expressions

Point of Interest

The radical sign was introduced in 1525 in a book by Christoff Rudolff called *Coss*. He modified the symbol to indicate square roots, cube roots, and fourth roots. The idea of using an index, as we do in our modern notation, did not occur until some years later.

Recall that $a^{1/n}$ is the nth root of a. The expression $\sqrt[n]{a}$ is another symbol for the nth root of a.

If a is a real number, then $a^{1/n} = \sqrt[n]{a}$.

In the expression $\sqrt[n]{a}$, the symbol $\sqrt{}$ is called a **radical**, n is the **index** of the radical, and a is the **radicand**. When $n = 2$, the radical expression represents a square root and the index 2 is usually not written.

An exponential expression with a rational exponent can be written as a radical expression.

> **Writing Exponential Expressions as Radical Expressions**
>
> If $a^{1/n}$ is a real number, then $a^{1/n} = \sqrt[n]{a}$ and $a^{m/n} = a^{m \cdot 1/n} = (a^m)^{1/n} = \sqrt[n]{a^m}$.

The expression $a^{m/n}$ can also be written $a^{m/n} = (a^{1/n})^m = (\sqrt[n]{a})^m$.

The exponential expression at the right has been written as a radical expression.

$$y^{2/3} = (y^2)^{1/3}$$
$$= \sqrt[3]{y^2}$$

The radical expressions at the right have been written as exponential expressions.

$$\sqrt[5]{x^6} = (x^6)^{1/5} = x^{6/5}$$
$$\sqrt{17} = (17)^{1/2} = 17^{1/2}$$

HOW TO Write $(5x)^{2/5}$ as a radical expression.

$(5x)^{2/5} = \sqrt[5]{(5x)^2}$ • The denominator of the rational exponent is the index of the radical. The numerator is the power of the radicand.

$= \sqrt[5]{25x^2}$ • Simplify.

HOW TO Write $\sqrt[3]{x^4}$ as an exponential expression with a rational exponent.

$\sqrt[3]{x^4} = (x^4)^{1/3}$ • The index of the radical is the denominator of the rational exponent. The power of the radicand is the numerator of the rational exponent.

$= x^{4/3}$ • Simplify.

HOW TO Write $\sqrt[3]{a^3 + b^3}$ as an exponential expression with a rational exponent.

$\sqrt[3]{a^3 + b^3} = (a^3 + b^3)^{1/3}$

Note that $(a^3 + b^3)^{1/3} \neq a + b$.

Example 5 Write $(3x)^{5/4}$ as a radical expression.

Solution $(3x)^{5/4} = \sqrt[4]{(3x)^5} = \sqrt[4]{243x^5}$

You Try It 5 Write $(2x^3)^{3/4}$ as a radical expression.

Your solution

Example 6 Write $-2x^{2/3}$ as a radical expression.

Solution $-2x^{2/3} = -2(x^2)^{1/3} = -2\sqrt[3]{x^2}$

You Try It 6 Write $-5a^{5/6}$ as a radical expression.

Your solution

Example 7 Write $\sqrt[4]{3a}$ as an exponential expression.

Solution $\sqrt[4]{3a} = (3a)^{1/4}$

You Try It 7 Write $\sqrt[3]{3ab}$ as an exponential expression.

Your solution

Example 8 Write $\sqrt{a^2 - b^2}$ as an exponential expression.

Solution $\sqrt{a^2 - b^2} = (a^2 - b^2)^{1/2}$

You Try It 8 Write $\sqrt[4]{x^4 + y^4}$ as an exponential expression.

Your solution

Solutions on p. S22

Objective C **To simplify radical expressions that are roots of perfect powers**

Every positive number has two square roots, one a positive number and one a negative number. For example, because $(5)^2 = 25$ and $(-5)^2 = 25$, there are two square roots of 25: 5 and -5.

The symbol $\sqrt{}$ is used to indicate the positive square root, or **principal square root.** To indicate the negative square root of a number, a negative sign is placed in front of the radical.

$\sqrt{25} = 5$

$-\sqrt{25} = -5$

The square root of zero is zero.

$\sqrt{0} = 0$

The square root of a negative number is not a real number, because the square of a real number must be positive.

$\sqrt{-25}$ is not a real number.

Note that

$$\sqrt{(-5)^2} = \sqrt{25} = 5 \text{ and } \sqrt{5^2} = \sqrt{25} = 5$$

This is true for all real numbers and is stated as the following result.

For any real number a, $\sqrt{a^2} = |a|$ and $-\sqrt{a^2} = -|a|$. If a is a positive real number, then $\sqrt{a^2} = a$ and $(\sqrt{a})^2 = a$.

Besides square roots, we can also determine cube roots, fourth roots, and so on.

$\sqrt[3]{8} = 2$, because $2^3 = 8$.

• The cube root of a positive number is positive.

$\sqrt[3]{-8} = -2$, because $(-2)^3 = -8$.

• The cube root of a negative number is negative.

$\sqrt[4]{625} = 5$, because $5^4 = 625$.

$\sqrt[5]{243} = 3$, because $3^5 = 243$.

The following properties hold true for finding the nth root of a real number.

If n is an even integer, then $\sqrt[n]{a^n} = |a|$ and $-\sqrt[n]{a^n} = -|a|$. If n is an odd integer, then $\sqrt[n]{a^n} = a$.

For example,

$$\sqrt[6]{y^6} = |y| \qquad -\sqrt[12]{x^{12}} = -|x| \qquad \sqrt[5]{b^5} = b$$

For the remainder of this chapter, we will assume that variable expressions inside a radical represent positive numbers. Therefore, it is not necessary to use the absolute value signs.

HOW TO Simplify: $\sqrt[4]{x^4 y^8}$

$\sqrt[4]{x^4 y^8} = (x^4 y^8)^{1/4}$

• The radicand is a perfect fourth power because the exponents on the variables are divisible by 4. Write the radical expressions as an exponential expression.

$= xy^2$

• Use the Rule for Simplifying Powers of Products.

HOW TO Simplify: $\sqrt[3]{125c^9d^6}$

$\sqrt[3]{125c^9d^6} = (5^3c^9d^6)^{1/3}$ • The radicand is a perfect cube because 125 is a perfect cube $(125 = 5^3)$ and all the exponents on the variables are divisible by 3.

$= 5c^3d^2$ • Use the Rule for Simplifying Powers of Products.

Note that **a variable expression is a perfect power if the exponents on the factors are evenly divisible by the index of the radical.**

The chart below shows roots of perfect powers. Knowledge of these roots is very helpful when simplifying radical expressions.

Square Roots		*Cube Roots*	*Fourth Roots*	*Fifth Roots*
$\sqrt{1} = 1$	$\sqrt{36} = 6$	$\sqrt[3]{1} = 1$	$\sqrt[4]{1} = 1$	$\sqrt[5]{1} = 1$
$\sqrt{4} = 2$	$\sqrt{49} = 7$	$\sqrt[3]{8} = 2$	$\sqrt[4]{16} = 2$	$\sqrt[5]{32} = 2$
$\sqrt{9} = 3$	$\sqrt{64} = 8$	$\sqrt[3]{27} = 3$	$\sqrt[4]{81} = 3$	$\sqrt[5]{243} = 3$
$\sqrt{16} = 4$	$\sqrt{81} = 9$	$\sqrt[3]{64} = 4$	$\sqrt[4]{256} = 4$	
$\sqrt{25} = 5$	$\sqrt{100} = 10$	$\sqrt[3]{125} = 5$	$\sqrt[4]{625} = 5$	

TAKE NOTE

From the chart, $\sqrt[5]{243} = 3$, which means that $3^5 = 243$. From this we know that $(-3)^5 = -243$, which means $\sqrt[5]{-243} = -3$.

HOW TO Simplify: $\sqrt[5]{-243x^5y^{15}}$

$\sqrt[5]{-243x^5y^{15}} = -3xy^3$ • From the chart, 243 is a perfect fifth power, and each exponent is divisible by 5. Therefore, the radicand is a perfect fifth power.

Example 9 Simplify: $\sqrt[3]{-125a^6b^9}$

Solution The radicand is a perfect cube.

$\sqrt[3]{-125a^6b^9} = -5a^2b^3$ • Divide each exponent by 3.

You Try It 9 Simplify: $\sqrt[3]{-8x^{12}y^3}$

Your solution

Example 10 Simplify: $-\sqrt[4]{16a^4b^8}$

Solution The radicand is a perfect fourth power.

$-\sqrt[4]{16a^4b^8} = -2ab^2$ • Divide each exponent by 4.

You Try It 10 Simplify: $-\sqrt[4]{81x^{12}y^8}$

Your solution

Solutions on p. S22

7.1 Exercises

Objective A **To simplify expressions with rational exponents**

For Exercises 1 to 80, simplify.

1. $8^{1/3}$

2. $16^{1/2}$

3. $9^{3/2}$

4. $25^{3/2}$

5. $27^{-2/3}$

6. $64^{-1/3}$

7. $32^{2/5}$

8. $16^{3/4}$

9. $(-25)^{5/2}$

10. $(-36)^{1/4}$

11. $\left(\dfrac{25}{49}\right)^{-3/2}$

12. $\left(\dfrac{8}{27}\right)^{-2/3}$

13. $x^{1/2}x^{1/2}$

14. $a^{1/3}a^{5/3}$

15. $y^{-1/4}y^{3/4}$

16. $x^{2/5} \cdot x^{-4/5}$

17. $x^{-2/3} \cdot x^{3/4}$

18. $x \cdot x^{-1/2}$

19. $a^{1/3} \cdot a^{3/4} \cdot a^{-1/2}$

20. $y^{-1/6} \cdot y^{2/3} \cdot y^{1/2}$

21. $\dfrac{a^{1/2}}{a^{3/2}}$

22. $\dfrac{b^{1/3}}{b^{4/3}}$

23. $\dfrac{y^{-3/4}}{y^{1/4}}$

24. $\dfrac{x^{-3/5}}{x^{1/5}}$

25. $\dfrac{y^{2/3}}{y^{-5/6}}$

26. $\dfrac{b^{3/4}}{b^{-3/2}}$

27. $(x^2)^{-1/2}$

28. $(a^8)^{-3/4}$

29. $(x^{-2/3})^6$

30. $(y^{-5/6})^{12}$

31. $(a^{-1/2})^{-2}$

32. $(b^{-2/3})^{-6}$

33. $(x^{-3/8})^{-4/5}$

34. $(y^{-3/2})^{-2/9}$

35. $(a^{1/2} \cdot a)^2$

36. $(b^{2/3} \cdot b^{1/6})^6$

37. $(x^{-1/2} \cdot x^{3/4})^{-2}$

38. $(a^{1/2} \cdot a^{-2})^3$

39. $(y^{-1/2} \cdot y^{2/3})^{2/3}$

40. $(b^{-2/3} \cdot b^{1/4})^{-4/3}$

41. $(x^8 y^2)^{1/2}$

42. $(a^3 b^9)^{2/3}$

43. $(x^4y^2z^6)^{3/2}$ **44.** $(a^8b^4c^4)^{3/4}$ **45.** $(x^{-3}y^6)^{-1/3}$ **46.** $(a^2b^{-6})^{-1/2}$

47. $(x^{-2}y^{1/3})^{-3/4}$ **48.** $(a^{-2/3}b^{2/3})^{3/2}$ **49.** $\left(\dfrac{x^{1/2}}{y^2}\right)^4$ **50.** $\left(\dfrac{b^{-3/4}}{a^{-1/2}}\right)^8$

51. $\dfrac{x^{1/4} \cdot x^{-1/2}}{x^{2/3}}$ **52.** $\dfrac{b^{1/2} \cdot b^{-3/4}}{b^{1/4}}$ **53.** $\left(\dfrac{y^{2/3} \cdot y^{-5/6}}{y^{1/9}}\right)^9$ **54.** $\left(\dfrac{a^{1/3} \cdot a^{-2/3}}{a^{1/2}}\right)^4$

55. $\left(\dfrac{b^2 \cdot b^{-3/4}}{b^{-1/2}}\right)^{-1/2}$ **56.** $\dfrac{(x^{-5/6} \cdot x^3)^{-2/3}}{x^{4/3}}$ **57.** $(a^{2/3}b^2)^6(a^3b^3)^{1/3}$ **58.** $(x^3y^{-1/2})^{-2}(x^{-3}y^2)^{1/6}$

59. $(16m^{-2}n^4)^{-1/2}(mn^{1/2})$ **60.** $(27m^3n^{-6})^{1/3}(m^{-1/3}n^{5/6})^6$

61. $\left(\dfrac{x^{1/2}y^{-3/4}}{y^{2/3}}\right)^{-6}$ **62.** $\left(\dfrac{x^{1/2}y^{-5/4}}{y^{-3/4}}\right)^{-4}$ **63.** $\left(\dfrac{2^{-6}b^{-3}}{a^{-1/2}}\right)^{-2/3}$ **64.** $\left(\dfrac{49c^{5/3}}{a^{-1/4}b^{5/6}}\right)^{-3/2}$

65. $y^{3/2}(y^{1/2} - y^{-1/2})$ **66.** $y^{3/5}(y^{2/5} + y^{-3/5})$ **67.** $a^{-1/4}(a^{5/4} - a^{9/4})$ **68.** $x^{4/3}(x^{2/3} + x^{-1/3})$

69. $x^n \cdot x^{3n}$ **70.** $a^{2n} \cdot a^{-5n}$ **71.** $x^n \cdot x^{n/2}$ **72.** $a^{n/2} \cdot a^{-n/3}$

73. $\dfrac{y^{n/2}}{y^{-n}}$ **74.** $\dfrac{b^{m/3}}{b^m}$ **75.** $(x^{2n})^n$ **76.** $(x^{5n})^{2n}$

77. $(x^{n/4}y^{n/8})^8$ **78.** $(x^{n/2}y^{n/3})^6$ **79.** $(x^{n/5}y^{n/10})^{20}$ **80.** $(x^{n/2}y^{n/5})^{10}$

Objective B **To write exponential expressions as radical expressions and to write radical expressions as exponential expressions**

For Exercises 81 to 96, rewrite the exponential expression as a radical expression.

81. $3^{1/4}$

82. $5^{1/2}$

83. $a^{3/2}$

84. $b^{4/3}$

85. $(2t)^{5/2}$

86. $(3x)^{2/3}$

87. $-2x^{2/3}$

88. $-3a^{2/5}$

89. $(a^2b)^{2/3}$

90. $(x^2y^3)^{3/4}$

91. $(a^2b^4)^{3/5}$

92. $(a^3b^7)^{3/2}$

93. $(4x-3)^{3/4}$

94. $(3x-2)^{1/3}$

95. $x^{-2/3}$

96. $b^{-3/4}$

For Exercises 97 to 112, rewrite the radical expression as an exponential expression.

97. $\sqrt{14}$

98. $\sqrt{7}$

99. $\sqrt[3]{x}$

100. $\sqrt[4]{y}$

101. $\sqrt[3]{x^4}$

102. $\sqrt[4]{a^3}$

103. $\sqrt[5]{b^3}$

104. $\sqrt[4]{b^5}$

105. $\sqrt[3]{2x^2}$

106. $\sqrt[5]{4y^7}$

107. $-\sqrt{3x^5}$

108. $-\sqrt[4]{4x^5}$

109. $3x\sqrt[3]{y^2}$

110. $2y\sqrt{x^3}$

111. $\sqrt{a^2-2}$

112. $\sqrt{3-y^2}$

Objective C **To simplify radical expressions that are roots of perfect powers**

For Exercises 113 to 136, simplify.

113. $\sqrt{x^{16}}$

114. $\sqrt{y^{14}}$

115. $-\sqrt{x^8}$

116. $-\sqrt{a^6}$

117. $\sqrt[3]{x^3y^9}$

118. $\sqrt[3]{a^6b^{12}}$

119. $-\sqrt[3]{x^{15}y^3}$

120. $-\sqrt[3]{a^9b^9}$

121. $\sqrt{16a^4b^{12}}$

122. $\sqrt{25x^8y^2}$

123. $\sqrt{-16x^4y^2}$

124. $\sqrt{-9a^4b^8}$

125. $\sqrt[3]{27x^9}$

126. $\sqrt[3]{8a^{21}b^6}$

127. $\sqrt[3]{-64x^9y^{12}}$

128. $\sqrt[3]{-27a^3b^{15}}$

129. $-\sqrt[4]{x^8y^{12}}$

130. $-\sqrt[4]{a^{16}b^4}$

131. $\sqrt[5]{x^{20}y^{10}}$

132. $\sqrt[5]{a^5b^{25}}$

133. $\sqrt[4]{81x^4y^{20}}$

134. $\sqrt[4]{16a^8b^{20}}$

135. $\sqrt[5]{32a^5b^{10}}$

136. $\sqrt[5]{-32x^{15}y^{20}}$

APPLYING THE CONCEPTS

137. Determine whether the following statements are true or false. If the statement is false, correct the right-hand side of the equation.

a. $\sqrt{(-2)^2} = -2$ **b.** $\sqrt[3]{(-3)^3} = -3$ **c.** $\sqrt[n]{a} = a^{1/n}$

d. $\sqrt[n]{a^n + b^n} = a + b$ **e.** $(a^{1/2} + b^{1/2})^2 = a + b$ **f.** $\sqrt[m]{a^n} = a^{mn}$

138. Simplify.

a. $\sqrt[3]{\sqrt{x^6}}$ **b.** $\sqrt[4]{\sqrt{a^8}}$ **c.** $\sqrt{\sqrt{81y^8}}$

d. $\sqrt{\sqrt[n]{a^{4n}}}$ **e.** $\sqrt[n]{\sqrt{b^{6n}}}$ **f.** $\sqrt{\sqrt[3]{x^{12}y^{24}}}$

139. If x is any real number, is $\sqrt{x^2} = x$ always true? Show why or why not.

7.2 Operations on Radical Expressions

Objective A **To simplify radical expressions**

Point of Interest

The Latin expression for irrational numbers was *numerus surdus*, which literally means "inaudible number." A prominent 16th-century mathematician wrote of irrational numbers, ". . . just as an infinite number is not a number, so an irrational number is not a true number, but lies hidden in some sort of cloud of infinity." In 1872, Richard Dedekind wrote a paper that established the first logical treatment of irrational numbers.

If a number is not a perfect power, its root can only be approximated; examples include $\sqrt{5}$ and $\sqrt[3]{3}$. These numbers are irrational numbers. Their decimal representations never terminate or repeat.

$$\sqrt{5} = 2.2360679\ldots \qquad \sqrt[3]{3} = 1.4422495\ldots$$

A radical expression is in simplest form when the radicand contains no factor that is a perfect power. The Product Property of Radicals is used to simplify radical expressions whose radicands are not perfect powers.

> **The Product Property of Radicals**
>
> If $\sqrt[n]{a}$ and $\sqrt[n]{b}$ are positive real numbers, then $\sqrt[n]{ab} = \sqrt[n]{a} \cdot \sqrt[n]{b}$ and $\sqrt[n]{a} \cdot \sqrt[n]{b} = \sqrt[n]{ab}$.

HOW TO Simplify: $\sqrt{48}$

$$\sqrt{48} = \sqrt{16 \cdot 3}$$

- Write the radicand as the product of a perfect square and a factor that does not contain a perfect square.

$$= \sqrt{16}\,\sqrt{3}$$

- Use the Product Property of Radicals to write the expression as a product.

$$= 4\sqrt{3}$$

- Simplify $\sqrt{16}$.

Note that 48 must be written as the product of a perfect square and *a factor that does not contain a perfect square.* Therefore, it would not be correct to rewrite $\sqrt{48}$ as $\sqrt{4 \cdot 12}$ and simplify the expression as shown at the right. Although 4 is a perfect square factor of 48, 12 contains a perfect square ($12 = 4 \cdot 3$) and thus $\sqrt{12}$ can be simplified further. Remember to find the largest perfect power that is a factor of the radicand.

$$\sqrt{48} = \sqrt{4 \cdot 12}$$
$$= \sqrt{4}\,\sqrt{12}$$
$$= 2\sqrt{12}$$

Not in simplest form

HOW TO Simplify: $\sqrt{18x^2y^3}$

$$\sqrt{18x^2y^3} = \sqrt{9x^2y^2 \cdot 2y}$$

- Write the radicand as the product of a perfect square and factors that do not contain a perfect square.

$$= \sqrt{9x^2y^2}\,\sqrt{2y}$$

- Use the Product Property of Radicals to write the expression as a product.

$$= 3xy\sqrt{2y}$$

- Simplify.

HOW TO Simplify: $\sqrt[3]{x^7}$

$\sqrt[3]{x^7} = \sqrt[3]{x^6 \cdot x}$

- Write the radicand as the product of a perfect cube and a factor that does not contain a perfect cube.

$= \sqrt[3]{x^6}\,\sqrt[3]{x}$

- Use the Product Property of Radicals to write the expression as a product.

$= x^2\,\sqrt[3]{x}$

- Simplify.

HOW TO Simplify: $\sqrt[4]{32x^7}$

$\sqrt[4]{32x^7} = \sqrt[4]{16x^4(2x^3)}$

- Write the radicand as the product of a perfect fourth power and factors that do not contain a perfect fourth power.

$= \sqrt[4]{16x^4}\,\sqrt[4]{2x^3}$

- Use the Product Property of Radicals to write the expression as a product.

$= 2x\,\sqrt[4]{2x^3}$

- Simplify.

Example 1 Simplify: $\sqrt[4]{x^9}$

Solution
$\sqrt[4]{x^9} = \sqrt[4]{x^8 \cdot x}$
$= \sqrt[4]{x^8}\,\sqrt[4]{x}$
$= x^2\,\sqrt[4]{x}$

- x^8 is a perfect fourth power.

You Try It 1 Simplify: $\sqrt[5]{x^7}$

Your solution

Example 2 Simplify: $\sqrt[3]{-27a^5b^{12}}$

Solution
$\sqrt[3]{-27a^5b^{12}}$
$= \sqrt[3]{-27a^3b^{12}(a^2)}$
$= \sqrt[3]{-27a^3b^{12}}\,\sqrt[3]{a^2}$
$= -3ab^4\,\sqrt[3]{a^2}$

- $-27a^3b^{12}$ is a perfect third power.

You Try It 2 Simplify: $\sqrt[3]{-64x^8y^{18}}$

Your solution

Solutions on p. S22

Objective B **To add or subtract radical expressions**

The Distributive Property is used to simplify the sum or difference of radical expressions that have the same radicand and the same index. For example,

$$3\sqrt{5} + 8\sqrt{5} = (3 + 8)\sqrt{5} = 11\sqrt{5}$$
$$2\sqrt[3]{3x} - 9\sqrt[3]{3x} = (2 - 9)\sqrt[3]{3x} = -7\sqrt[3]{3x}$$

Radical expressions that are in simplest form and have unlike radicands or different indices cannot be simplified by the Distributive Property. The expressions below cannot be simplified by the Distributive Property.

$$3\sqrt[4]{2} - 6\sqrt[3]{3} \qquad\qquad 2\sqrt[4]{4x} + 3\sqrt[3]{4x}$$

HOW TO Simplify: $3\sqrt{32x^2} - 2x\sqrt{2} + \sqrt{128x^2}$

$3\sqrt{32x^2} - 2x\sqrt{2} + \sqrt{128x^2}$

$= 3\sqrt{16x^2}\sqrt{2} - 2x\sqrt{2} + \sqrt{64x^2}\sqrt{2}$

$= 3 \cdot 4x\sqrt{2} - 2x\sqrt{2} + 8x\sqrt{2}$

$= 12x\sqrt{2} - 2x\sqrt{2} + 8x\sqrt{2}$

$= 18x\sqrt{2}$

- First simplify each term. Then combine like terms by using the Distributive Property.

Example 3

Simplify: $5b\sqrt[4]{32a^7b^5} - 2a\sqrt[4]{162a^3b^9}$

Solution

$5b\sqrt[4]{32a^7b^5} - 2a\sqrt[4]{162a^3b^9}$

$= 5b\sqrt[4]{16a^4b^4 \cdot 2a^3b} - 2a\sqrt[4]{81b^8 \cdot 2a^3b}$

$= 5b \cdot 2ab\sqrt[4]{2a^3b} - 2a \cdot 3b^2\sqrt[4]{2a^3b}$

$= 10ab^2\sqrt[4]{2a^3b} - 6ab^2\sqrt[4]{2a^3b}$

$= 4ab^2\sqrt[4]{2a^3b}$

You Try It 3

Simplify: $3xy\sqrt[3]{81x^5y} - \sqrt[3]{192x^8y^4}$

Your solution

Solution on p. S22

Objective C **To multiply radical expressions**

The Product Property of Radicals is used to multiply radical expressions with the same index.

$$\sqrt{3x} \cdot \sqrt{5y} = \sqrt{3x \cdot 5y} = \sqrt{15xy}$$

HOW TO Simplify: $\sqrt[3]{2a^5b}\,\sqrt[3]{16a^2b^2}$

$\sqrt[3]{2a^5b}\,\sqrt[3]{16a^2b^2} = \sqrt[3]{32a^7b^3}$

$= \sqrt[3]{8a^6b^3 \cdot 4a}$

$= 2a^2b\sqrt[3]{4a}$

- Use the Product Property of Radicals to multiply the radicands.
- Simplify.

HOW TO Simplify: $\sqrt{2x}(\sqrt{8x} - \sqrt{3})$

$\sqrt{2x}(\sqrt{8x} - \sqrt{3}) = \sqrt{2x}(\sqrt{8x}) - \sqrt{2x}(\sqrt{3})$

$= \sqrt{16x^2} - \sqrt{6x}$

$= 4x - \sqrt{6x}$

- Use the Distributive Property.
- Simplify.

HOW TO Simplify: $(2\sqrt{5} - 3)(3\sqrt{5} + 4)$

$(2\sqrt{5} - 3)(3\sqrt{5} + 4) = 6(\sqrt{5})^2 + 8\sqrt{5} - 9\sqrt{5} - 12$

$= 30 + 8\sqrt{5} - 9\sqrt{5} - 12$

$= 18 - \sqrt{5}$

- Use the FOIL method to multiply the numbers.

- Combine like terms.

HOW TO Simplify: $(4\sqrt{a} - \sqrt{b})(2\sqrt{a} + 5\sqrt{b})$

$$(4\sqrt{a} - \sqrt{b})(2\sqrt{a} + 5\sqrt{b})$$
$$= 8(\sqrt{a})^2 + 20\sqrt{ab} - 2\sqrt{ab} - 5(\sqrt{b})^2 \qquad \bullet \text{ Use the FOIL method.}$$
$$= 8a + 18\sqrt{ab} - 5b$$

The expressions $a + b$ and $a - b$ are **conjugates** of each other: binomial expressions that differ only in the sign of a term. Recall that $(a + b)(a - b) = a^2 - b^2$. This identity is used to simplify conjugate radical expressions.

HOW TO Simplify: $(\sqrt{11} - 3)(\sqrt{11} + 3)$

$$(\sqrt{11} - 3)(\sqrt{11} + 3) = (\sqrt{11})^2 - 3^2 = 11 - 9 \qquad \bullet \text{ The radical expressions}$$
$$= 2 \qquad\qquad\qquad\qquad\qquad \text{are conjugates.}$$

Example 4

Simplify: $\sqrt{3x}(\sqrt{27x^2} - \sqrt{3x})$

Solution
$\sqrt{3x}(\sqrt{27x^2} - \sqrt{3x}) = \sqrt{81x^3} - \sqrt{9x^2}$

$\quad = \sqrt{81x^3} - \sqrt{9x^2}$ • The Distributive Property

$\quad = \sqrt{81x^2 \cdot x} - \sqrt{9x^2}$ • Simplify each radical expression.

$\quad = 9x\sqrt{x} - 3x$

You Try It 4

Simplify: $\sqrt{5b}(\sqrt{3b} - \sqrt{10})$

Your solution

Example 5

Simplify: $(2\sqrt[3]{x} - 3)(3\sqrt[3]{x} - 4)$

Solution
$(2\sqrt[3]{x} - 3)(3\sqrt[3]{x} - 4)$

$\quad = 6\sqrt[3]{x^2} - 8\sqrt[3]{x} - 9\sqrt[3]{x} + 12$ • The FOIL method

$\quad = 6\sqrt[3]{x^2} - 17\sqrt[3]{x} + 12$

You Try It 5

Simplify: $(2\sqrt[3]{2x} - 3)(\sqrt[3]{2x} - 5)$

Your solution

Example 6

Simplify: $(2\sqrt{x} - \sqrt{2y})(2\sqrt{x} + \sqrt{2y})$

Solution
$(2\sqrt{x} - \sqrt{2y})(2\sqrt{x} + \sqrt{2y})$ • The expressions are conjugates.

$\quad = (2\sqrt{x})^2 - (\sqrt{2y})^2$

$\quad = 4x - 2y$

You Try It 6

Simplify: $(\sqrt{a} - 3\sqrt{y})(\sqrt{a} + 3\sqrt{y})$

Your solution

Solutions on p. S22

Objective D **To divide radical expressions**

The Quotient Property of Radicals is used to divide radical expressions with the same index.

> **The Quotient Property of Radicals**
>
> If $\sqrt[n]{a}$ and $\sqrt[n]{b}$ are real numbers, and $b \neq 0$, then
>
> $$\sqrt[n]{\frac{a}{b}} = \frac{\sqrt[n]{a}}{\sqrt[n]{b}} \quad \text{and} \quad \frac{\sqrt[n]{a}}{\sqrt[n]{b}} = \sqrt[n]{\frac{a}{b}}$$

HOW TO Simplify: $\sqrt[3]{\dfrac{81x^5}{y^6}}$

$\sqrt[3]{\dfrac{81x^5}{y^6}} = \dfrac{\sqrt[3]{81x^5}}{\sqrt[3]{y^6}}$

 • Use the Quotient Property of Radicals.

$= \dfrac{\sqrt[3]{27x^3 \cdot 3x^2}}{\sqrt[3]{y^6}} = \dfrac{3x\sqrt[3]{3x^2}}{y^2}$

 • Simplify each radical expression.

HOW TO Simplify: $\dfrac{\sqrt{5a^4b^7c^2}}{\sqrt{ab^3c}}$

$\dfrac{\sqrt{5a^4b^7c^2}}{\sqrt{ab^3c}} = \sqrt{\dfrac{5a^4b^7c^2}{ab^3c}}$

 • Use the Quotient Property of Radicals.

$= \sqrt{5a^3b^4c}$

 • Simplify the radicand.

$= \sqrt{a^2b^4 \cdot 5ac} = ab^2\sqrt{5ac}$

A radical expression is in simplest form when no radical remains in the denominator of the radical expression. The procedure used to remove a radical from the denominator is called **rationalizing the denominator.**

HOW TO Simplify: $\dfrac{5}{\sqrt{2}}$

$\dfrac{5}{\sqrt{2}} = \dfrac{5}{\sqrt{2}} \cdot 1 = \dfrac{5}{\sqrt{2}} \cdot \dfrac{\sqrt{2}}{\sqrt{2}}$

 • Multiply by $\dfrac{\sqrt{2}}{\sqrt{2}}$, which equals **1**.

$= \dfrac{5\sqrt{2}}{2}$

 • $\sqrt{2} \cdot \sqrt{2} = (\sqrt{2})^2 = 2$

TAKE NOTE

Multiplying by $\dfrac{\sqrt[3]{4x}}{\sqrt[3]{4x}}$ will not rationalize the denominator of $\dfrac{3}{\sqrt[3]{4x}}$.

$\dfrac{3}{\sqrt[3]{4x}} \cdot \dfrac{\sqrt[3]{4x}}{\sqrt[3]{4x}} = \dfrac{3\sqrt[3]{4x}}{\sqrt[3]{16x^2}}$

Because $16x^2$ is not a perfect cube, the denominator still contains a radical expression.

HOW TO Simplify: $\dfrac{3}{\sqrt[3]{4x}}$

$\dfrac{3}{\sqrt[3]{4x}} = \dfrac{3}{\sqrt[3]{4x}} \cdot \dfrac{\sqrt[3]{2x^2}}{\sqrt[3]{2x^2}}$

 • Because $4x \cdot 2x^2 = 8x^3$, a perfect cube, multiply the expression by $\dfrac{\sqrt[3]{2x^2}}{\sqrt[3]{2x^2}}$, which equals **1**.

$= \dfrac{3\sqrt[3]{2x^2}}{\sqrt[3]{8x^3}} = \dfrac{3\sqrt[3]{2x^2}}{2x}$

 • Simplify.

To simplify a fraction that has a square-root radical expression with two terms in the denominator, multiply the numerator and denominator by the conjugate of the denominator. Then simplify.

TAKE NOTE

Here is an example of using a conjugate to simplify a radical expression.

HOW TO Simplify: $\dfrac{\sqrt{x} - \sqrt{y}}{\sqrt{x} + \sqrt{y}}$

$$\dfrac{\sqrt{x} - \sqrt{y}}{\sqrt{x} + \sqrt{y}} = \dfrac{\sqrt{x} - \sqrt{y}}{\sqrt{x} + \sqrt{y}} \cdot \dfrac{\sqrt{x} - \sqrt{y}}{\sqrt{x} - \sqrt{y}}$$

• Multiply by $\dfrac{\sqrt{x} - \sqrt{y}}{\sqrt{x} - \sqrt{y}}$.

$$= \dfrac{(\sqrt{x})^2 - \sqrt{xy} - \sqrt{xy} + (\sqrt{y})^2}{(\sqrt{x})^2 - (\sqrt{y})^2} = \dfrac{x - 2\sqrt{xy} + y}{x - y}$$

• Simplify.

Example 7

Simplify: $\dfrac{5}{\sqrt{5x}}$

Solution

$$\dfrac{5}{\sqrt{5x}} = \dfrac{5}{\sqrt{5x}} \cdot \dfrac{\sqrt{5x}}{\sqrt{5x}} = \dfrac{5\sqrt{5x}}{(\sqrt{5x})^2}$$

• Rationalize the denominator.

$$= \dfrac{5\sqrt{5x}}{5x} = \dfrac{\sqrt{5x}}{x}$$

You Try It 7

Simplify: $\dfrac{y}{\sqrt{3y}}$

Your solution

Example 8

Simplify: $\dfrac{3x}{\sqrt[4]{2x}}$

Solution

$$\dfrac{3x}{\sqrt[4]{2x}} = \dfrac{3x}{\sqrt[4]{2x}} \cdot \dfrac{\sqrt[4]{8x^3}}{\sqrt[4]{8x^3}}$$

• Rationalize the denominator.

$$= \dfrac{3x\sqrt[4]{8x^3}}{\sqrt[4]{16x^4}} = \dfrac{3x\sqrt[4]{8x^3}}{2x}$$

$$= \dfrac{3\sqrt[4]{8x^3}}{2}$$

You Try It 8

Simplify: $\dfrac{3x}{\sqrt[3]{3x^2}}$

Your solution

Example 9

Simplify: $\dfrac{3}{5 - 2\sqrt{3}}$

Solution

$$\dfrac{3}{5 - 2\sqrt{3}} = \dfrac{3}{5 - 2\sqrt{3}} \cdot \dfrac{5 + 2\sqrt{3}}{5 + 2\sqrt{3}}$$

• Rationalize the denominator.

$$= \dfrac{15 + 6\sqrt{3}}{5^2 - (2\sqrt{3})^2}$$

$$= \dfrac{15 + 6\sqrt{3}}{25 - 12} = \dfrac{15 + 6\sqrt{3}}{13}$$

You Try It 9

Simplify: $\dfrac{3 + \sqrt{6}}{2 - \sqrt{6}}$

Your solution

Solutions on p. S22

7.2 Exercises

Objective A **To simplify radical expressions**

For Exercises 1 to 16, simplify.

1. $\sqrt{x^4 y^3 z^5}$

2. $\sqrt{x^3 y^6 z^9}$

3. $\sqrt{8a^3 b^8}$

4. $\sqrt{24a^9 b^6}$

5. $\sqrt{45x^2 y^3 z^5}$

6. $\sqrt{60xy^7 z^{12}}$

7. $\sqrt{-9x^3}$

8. $\sqrt{-x^2 y^5}$

9. $\sqrt[3]{a^{16} b^8}$

10. $\sqrt[3]{a^5 b^8}$

11. $\sqrt[3]{-125x^2 y^4}$

12. $\sqrt[3]{-216x^5 y^9}$

13. $\sqrt[3]{a^4 b^5 c^6}$

14. $\sqrt[3]{a^8 b^{11} c^{15}}$

15. $\sqrt[4]{16x^9 y^5}$

16. $\sqrt[4]{64x^8 y^{10}}$

Objective B **To add or subtract radical expressions**

For Exercises 17 to 42, simplify.

17. $2\sqrt{x} - 8\sqrt{x}$

18. $3\sqrt{y} + 12\sqrt{y}$

19. $\sqrt{8} - \sqrt{32}$

20. $\sqrt{27a} - \sqrt{8a}$

21. $\sqrt{18b} + \sqrt{75b}$

22. $2\sqrt{2x^3} + 4x\sqrt{8x}$

23. $3\sqrt{8x^2 y^3} - 2x\sqrt{32y^3}$

24. $2\sqrt{32x^2 y^3} - xy\sqrt{98y}$

25. $2a\sqrt{27ab^5} + 3b\sqrt{3a^3 b}$

26. $\sqrt[3]{128} + \sqrt[3]{250}$

27. $\sqrt[3]{16} - \sqrt[3]{54}$

28. $2\sqrt[3]{3a^4} - 3a\sqrt[3]{81a}$

29. $2b\sqrt[3]{16b^2} + \sqrt[3]{128b^5}$

30. $3\sqrt[3]{x^5 y^7} - 8xy\sqrt[3]{x^2 y^4}$

31. $3\sqrt[4]{32a^5} - a\sqrt[4]{162a}$

32. $2a\sqrt[4]{16ab^5} + 3b\sqrt[4]{256a^5b}$

33. $2\sqrt{50} - 3\sqrt{125} + \sqrt{98}$

34. $3\sqrt{108} - 2\sqrt{18} - 3\sqrt{48}$

35. $\sqrt{9b^3} - \sqrt{25b^3} + \sqrt{49b^3}$

36. $\sqrt{4x^7y^5} + 9x^2\sqrt{x^3y^5} - 5xy\sqrt{x^5y^3}$

37. $2x\sqrt{8xy^2} - 3y\sqrt{32x^3} + \sqrt{4x^3y^3}$

38. $5a\sqrt{3a^3b} + 2a^2\sqrt{27ab} - 4\sqrt{75a^5b}$

39. $\sqrt[3]{54xy^3} - 5\sqrt[3]{2xy^3} + y\sqrt[3]{128x}$

40. $2\sqrt[3]{24x^3y^4} + 4x\sqrt[3]{81y^4} - 3y\sqrt[3]{24x^3y}$

41. $2a\sqrt[4]{32b^5} - 3b\sqrt[4]{162a^4b} + \sqrt[4]{2a^4b^5}$

42. $6y\sqrt[4]{48x^5} - 2x\sqrt[4]{243xy^4} - 4\sqrt[4]{3x^5y^4}$

Objective C **To multiply radical expressions**

For Exercises 43 to 73, simplify.

43. $\sqrt{8}\ \sqrt{32}$

44. $\sqrt{14}\ \sqrt{35}$

45. $\sqrt[3]{4}\ \sqrt[3]{8}$

46. $\sqrt[3]{6}\ \sqrt[3]{36}$

47. $\sqrt{x^2y^5}\ \sqrt{xy}$

48. $\sqrt{a^3b}\ \sqrt{ab^4}$

49. $\sqrt{2x^2y}\ \sqrt{32xy}$

50. $\sqrt{5x^3y}\ \sqrt{10x^3y^4}$

51. $\sqrt[3]{x^2y}\ \sqrt[3]{16x^4y^2}$

52. $\sqrt[3]{4a^2b^3}\ \sqrt[3]{8ab^5}$

53. $\sqrt[4]{12ab^3}\ \sqrt[4]{4a^5b^2}$

54. $\sqrt[4]{36a^2b^4}\ \sqrt[4]{12a^5b^3}$

55. $\sqrt{3}\ (\sqrt{27} - \sqrt{3})$

56. $\sqrt{10}\ (\sqrt{10} - \sqrt{5})$

57. $\sqrt{x}\ (\sqrt{x} - \sqrt{2})$

58. $\sqrt{y}\ (\sqrt{y} - \sqrt{5})$

59. $\sqrt{2x}\ (\sqrt{8x} - \sqrt{32})$

60. $\sqrt{3a}\ (\sqrt{27a^2} - \sqrt{a})$

61. $(\sqrt{x} - 3)^2$

62. $(\sqrt{2x} + 4)^2$

63. $(4\sqrt{5} + 2)^2$

64. $2\sqrt{3x^2} \cdot 3\sqrt{12xy^3} \cdot \sqrt{6x^3y}$

65. $2\sqrt{14xy} \cdot 4\sqrt{7x^2y} \cdot 3\sqrt{8xy^2}$

66. $\sqrt[3]{8ab} \; \sqrt[3]{4a^2b^3} \; \sqrt[3]{9ab^4}$

67. $\sqrt[3]{2a^2b} \; \sqrt[3]{4a^3b^2} \; \sqrt[3]{8a^5b^6}$

68. $(\sqrt{2} - 3)(\sqrt{2} + 4)$

69. $(\sqrt{5} - 5)(2\sqrt{5} + 2)$

70. $(\sqrt{y} - 2)(\sqrt{y} + 2)$

71. $(\sqrt{x} - y)(\sqrt{x} + y)$

72. $(\sqrt{2x} - 3\sqrt{y})(\sqrt{2x} + 3\sqrt{y})$

73. $(2\sqrt{3x} - \sqrt{y})(2\sqrt{3x} + \sqrt{y})$

Objective D **To divide radical expressions**

74. When is a radical expression in simplest form?

75. Explain what it means to rationalize the denominator of a radical expression and how to do so.

For Exercises 76 to 112, simplify.

76. $\dfrac{\sqrt{32x^2}}{\sqrt{2x}}$

77. $\dfrac{\sqrt{60y^4}}{\sqrt{12y}}$

78. $\dfrac{\sqrt{42a^3b^5}}{\sqrt{14a^2b}}$

79. $\dfrac{\sqrt{65ab^4}}{\sqrt{5ab}}$

80. $\dfrac{1}{\sqrt{5}}$

81. $\dfrac{1}{\sqrt{2}}$

82. $\dfrac{1}{\sqrt{2x}}$

83. $\dfrac{2}{\sqrt{3y}}$

84. $\dfrac{5}{\sqrt{5x}}$

85. $\dfrac{9}{\sqrt{3a}}$

86. $\sqrt{\dfrac{x}{5}}$

87. $\sqrt{\dfrac{y}{2}}$

88. $\dfrac{3}{\sqrt[3]{2}}$

89. $\dfrac{5}{\sqrt[3]{9}}$

90. $\dfrac{3}{\sqrt[3]{4x^2}}$

91. $\dfrac{5}{\sqrt[3]{3y}}$

92. $\dfrac{\sqrt{40x^3y^2}}{\sqrt{80x^2y^3}}$

93. $\dfrac{\sqrt{15a^2b^5}}{\sqrt{30a^5b^3}}$

94. $\dfrac{\sqrt{24a^2b}}{\sqrt{18ab^4}}$

95. $\dfrac{\sqrt{12x^3y}}{\sqrt{20x^4y}}$

96. $\dfrac{5}{\sqrt{3}-2}$

97. $\dfrac{-2}{1-\sqrt{2}}$

98. $\dfrac{-3}{2-\sqrt{3}}$

99. $\dfrac{-4}{3-\sqrt{2}}$

100. $\dfrac{2}{\sqrt{5}+2}$

101. $\dfrac{5}{2-\sqrt{7}}$

102. $\dfrac{3}{\sqrt{y}-2}$

103. $\dfrac{-7}{\sqrt{x}-3}$

104. $\dfrac{\sqrt{2}-\sqrt{3}}{\sqrt{2}+\sqrt{3}}$

105. $\dfrac{\sqrt{3}+\sqrt{4}}{\sqrt{2}+\sqrt{3}}$

106. $\dfrac{2+3\sqrt{7}}{5-2\sqrt{7}}$

107. $\dfrac{2+3\sqrt{5}}{1-\sqrt{5}}$

108. $\dfrac{2\sqrt{3}-1}{3\sqrt{3}+2}$

109. $\dfrac{2\sqrt{a}-\sqrt{b}}{4\sqrt{a}+3\sqrt{b}}$

110. $\dfrac{2\sqrt{x}-4}{\sqrt{x}+2}$

111. $\dfrac{3\sqrt{y}-y}{\sqrt{y}+2y}$

112. $\dfrac{3\sqrt{x}-4\sqrt{y}}{3\sqrt{x}-2\sqrt{y}}$

APPLYING THE CONCEPTS

113. Determine whether the following statements are true or false. If the statement is false, correct the right-hand side of the equation.

a. $\sqrt[2]{3}\cdot\sqrt[3]{4}=\sqrt[5]{12}$
b. $\sqrt{3}\cdot\sqrt{3}=3$
c. $\sqrt[3]{x}\cdot\sqrt[3]{x}=x$

d. $\sqrt{x}+\sqrt{y}=\sqrt{x+y}$
e. $\sqrt[2]{2}+\sqrt[3]{3}=\sqrt[5]{2+3}$
f. $8\sqrt[5]{a}-2\sqrt[5]{a}=6\sqrt[5]{a}$

114. Multiply: $(\sqrt[3]{a}+\sqrt[3]{b})(\sqrt[3]{a^2}-\sqrt[3]{ab}+\sqrt[3]{b^2})$

115. Rewrite $\dfrac{\sqrt[4]{(a+b)^3}}{\sqrt{a+b}}$ as an expression with a single radical.

7.3 Solving Equations Containing Radical Expressions

Objective A **To solve a radical equation**

An equation that contains a variable expression in a radicand is a **radical equation.**

$$\left.\begin{array}{l}\sqrt[3]{2x-5}+x=7 \\ \sqrt{x+1}-\sqrt{x}=4\end{array}\right\} \begin{array}{l}\text{Radical} \\ \text{equations}\end{array}$$

The following property is used to solve a radical equation.

The Property of Raising Each Side of an Equation to a Power

If two numbers are equal, then the same powers of the numbers are equal.

If $a = b$, then $a^n = b^n$.

HOW TO Solve: $\sqrt{x-2}-6=0$

$$\sqrt{x-2}-6=0$$
$$\sqrt{x-2}=6 \qquad \bullet \text{ Isolate the radical by adding 6 to each side of the equation.}$$
$$(\sqrt{x-2})^2=6^2 \qquad \bullet \text{ Square each side of the equation.}$$
$$x-2=36 \qquad \bullet \text{ Simplify and solve for } x.$$
$$x=38$$

Check:

$$\begin{array}{c|c}\sqrt{x-2}-6=0 \\ \hline \sqrt{38-2}-6 & 0 \\ \sqrt{36}-6 & 0 \\ 6-6 & 0 \\ 0=0 \end{array}$$

38 checks as a solution. The solution is 38.

HOW TO Solve: $\sqrt[3]{x+2}=-3$

$$\sqrt[3]{x+2}=-3$$
$$(\sqrt[3]{x+2})^3=(-3)^3 \qquad \bullet \text{ Cube each side of the equation.}$$
$$x+2=-27 \qquad \bullet \text{ Solve the resulting equation.}$$
$$x=-29$$

Check:

$$\begin{array}{c|c}\sqrt[3]{x+2}=-3 \\ \hline \sqrt[3]{-29+2} & -3 \\ \sqrt[3]{-27} & -3 \\ -3=-3 \end{array}$$

-29 checks as a solution. The solution is -29.

Raising each side of an equation to an even power may result in an equation that has a solution that is not a solution of the original equation. This is called an **extraneous solution.** Here is an example:

> **HOW TO** Solve: $\sqrt{2x - 1} + \sqrt{x} = 2$

$$\sqrt{2x - 1} + \sqrt{x} = 2$$
$$\sqrt{2x - 1} = 2 - \sqrt{x}$$ • Solve for one of the radical expressions.
$$(\sqrt{2x - 1})^2 = (2 - \sqrt{x})^2$$ • Square each side. Recall that
$$2x - 1 = 4 - 4\sqrt{x} + x$$ $(a - b)^2 = a^2 - 2ab + b^2$.
$$x - 5 = -4\sqrt{x}$$
$$(x - 5)^2 = (-4\sqrt{x})^2$$

$$x^2 - 10x + 25 = 16x$$
$$x^2 - 26x + 25 = 0$$ • Solve the quadratic equation by factoring.
$$(x - 25)(x - 1) = 0$$
$$x = 25 \text{ or } x = 1$$

Check:

$\sqrt{2x - 1} + \sqrt{x} = 2$		$\sqrt{2x - 1} + \sqrt{x} = 2$	
$\sqrt{2(25) - 1} + \sqrt{25}$	2	$\sqrt{2(1) - 1} + \sqrt{1}$	2
$7 + 5$	2	$1 + 1$	2
$12 \ne 2$		$2 = 2$	

25 does not check as a solution. 1 checks as a solution. The solution is 1. Here 25 is an extraneous solution.

Example 1

Solve: $\sqrt{x - 1} + \sqrt{x + 4} = 5$

Solution
$$\sqrt{x - 1} + \sqrt{x + 4} = 5$$
$$\sqrt{x + 4} = 5 - \sqrt{x - 1}$$ • Subtract $\sqrt{x - 1}$.
$$(\sqrt{x + 4})^2 = (5 - \sqrt{x - 1})^2$$ • Square each side.
$$x + 4 = 25 - 10\sqrt{x - 1} + x - 1$$
$$-20 = -10\sqrt{x - 1}$$
$$2 = \sqrt{x - 1}$$
$$2^2 = (\sqrt{x - 1})^2$$ • Square each side.
$$4 = x - 1$$
$$5 = x$$

The solution checks. The solution is 5.

You Try It 1

Solve: $\sqrt{x} - \sqrt{x + 5} = 1$

Your solution

Example 2

Solve: $\sqrt[3]{3x - 1} = -4$

Solution
$$\sqrt[3]{3x - 1} = -4$$
$$(\sqrt[3]{3x - 1})^3 = (-4)^3$$ • Cube each side.
$$3x - 1 = -64$$
$$3x = -63$$
$$x = -21$$

The solution checks. The solution is -21.

You Try It 2

Solve: $\sqrt[4]{x - 8} = 3$

Your solution

Solutions on pp. S22–S23

Objective B **To solve application problems**

A right triangle contains one 90° angle. The side opposite the 90° angle is called the hypotenuse. The other two sides are called legs.

Pythagoras, a Greek mathematician, discovered that the square of the hypotenuse of a right triangle is equal to the sum of the squares of the two legs. Recall that this is called the Pythagorean Theorem.

*Pythagoras
(c. 580 B.C.–529 B.C.)*

Example 3

A ladder 20 ft long is leaning against a building. How high on the building will the ladder reach when the bottom of the ladder is 8 ft from the building? Round to the nearest tenth.

Strategy

To find the distance, use the Pythagorean Theorem. The hypotenuse is the length of the ladder. One leg is the distance from the bottom of the ladder to the base of the building. The distance along the building from the ground to the top of the ladder is the unknown leg.

Solution

$$c^2 = a^2 + b^2$$ • **Pythagorean Theorem**
$$20^2 = 8^2 + b^2$$ • **Replace** c **by 20 and** a **by 8.**
$$400 = 64 + b^2$$ • **Solve for** b.
$$336 = b^2$$

$$(336)^{1/2} = (b^2)^{1/2}$$ • **Raise each side to the** $\frac{1}{2}$ **power.**

$$\sqrt{336} = b$$ • $a^{1/2} = \sqrt{a}$
$$18.3 \approx b$$

The distance is 18.3 ft.

You Try It 3

Find the diagonal of a rectangle that is 6 cm long and 3 cm wide. Round to the nearest tenth.

Your strategy

Your solution

Solution on p. S23

Example 4

An object is dropped from a high building. Find the distance the object has fallen when its speed reaches 96 ft/s. Use the equation $v = \sqrt{64d}$, where v is the speed of the object in feet per second and d is the distance in feet.

Strategy

To find the distance the object has fallen, replace v in the equation with the given value and solve for d.

Solution

$$v = \sqrt{64d}$$
$$96 = \sqrt{64d}$$
$$(96)^2 = (\sqrt{64d})^2 \qquad \bullet \text{ Replace } v \text{ by } 96.$$
$$9216 = 64d \qquad \bullet \text{ Square each side.}$$
$$144 = d$$

The object has fallen 144 ft.

You Try It 4

How far would a submarine periscope have to be above the water to locate a ship 5.5 mi away? The equation for the distance in miles that the lookout can see is $d = \sqrt{1.5h}$, where h is the height in feet above the surface of the water. Round to the nearest hundredth.

Your strategy

Your solution

Example 5

Find the length of a pendulum that makes one swing in 1.5 s. The equation for the time of one swing is given by $T = 2\pi\sqrt{\dfrac{L}{32}}$, where T is the time in seconds and L is the length in feet. Use 3.14 for π. Round to the nearest hundredth.

Strategy

To find the length of the pendulum, replace T in the equation with the given value and solve for L.

Solution

$$T = 2\pi\sqrt{\frac{L}{32}}$$

$$1.5 = 2(3.14)\sqrt{\frac{L}{32}} \qquad \bullet \text{ Replace } T \text{ by } 1.5 \text{ and } \pi \text{ by } 3.14.$$

$$\frac{1.5}{2(3.14)} = \sqrt{\frac{L}{32}} \qquad \bullet \text{ Divide each side by } 2(3.14).$$

$$\left[\frac{1.5}{2(3.14)}\right]^2 = \left(\sqrt{\frac{L}{32}}\right)^2 \qquad \bullet \text{ Square each side.}$$

$$\left(\frac{1.5}{6.28}\right)^2 = \frac{L}{32} \qquad \bullet \text{ Solve for } L. \text{ Multiply each side by } 32.$$

$$1.83 \approx L$$

The length of the pendulum is 1.83 ft.

You Try It 5

Find the distance required for a car to reach a velocity of 88 ft/s when the acceleration is 22 ft/s². Use the equation $v = \sqrt{2as}$, where v is the velocity in feet per second, a is the acceleration, and s is the distance in feet.

Your strategy

Your solution

Solutions on p. S23

7.3 Exercises

To solve a radical equation

For Exercises 1 to 21, solve.

1. $\sqrt[3]{4x} = -2$

2. $\sqrt[3]{6x} = -3$

3. $\sqrt{3x - 2} = 5$

4. $\sqrt{3x + 9} - 12 = 0$

5. $\sqrt{4x - 3} - 5 = 0$

6. $\sqrt{4x - 2} = \sqrt{3x + 9}$

7. $\sqrt{2x + 4} = \sqrt{5x - 9}$

8. $\sqrt[3]{x - 2} = 3$

9. $\sqrt[3]{2x - 6} = 4$

10. $\sqrt[3]{3x - 9} = \sqrt[3]{2x + 12}$

11. $\sqrt[3]{x - 12} = \sqrt[3]{5x + 16}$

12. $\sqrt[4]{4x + 1} = 2$

13. $\sqrt[3]{2x - 3} + 5 = 2$

14. $\sqrt[3]{x - 4} + 7 = 5$

15. $\sqrt{x} + \sqrt{x - 5} = 5$

16. $\sqrt{x + 3} + \sqrt{x - 1} = 2$

17. $\sqrt{2x + 5} - \sqrt{2x} = 1$

18. $\sqrt{3x} - \sqrt{3x - 5} = 1$

19. $\sqrt{2x} - \sqrt{x - 1} = 1$

20. $\sqrt{2x - 5} + \sqrt{x + 1} = 3$

21. $\sqrt{2x + 2} + \sqrt{x} = 3$

To solve application problems

22. Physics An object is dropped from a bridge. Find the distance the object has fallen when its speed reaches 100 ft/s. Use the equation $v = \sqrt{64d}$, where v is the speed in feet per second and d is the distance in feet.

23. Physics The time it takes for an object to fall a certain distance is given by the equation $t = \sqrt{\dfrac{2d}{g}}$, where t is the time in seconds, d is the distance in feet, and g is the acceleration due to gravity. If an astronaut above the moon's surface drops an object, how far will it have fallen in 3 s? The acceleration on the moon's surface is 5.5 feet per second per second.

24. Physics The time it takes for an object to fall a certain distance is given by the equation $t = \sqrt{\dfrac{2d}{g}}$, where t is the time in seconds, d is the distance in feet, and g is the acceleration due to gravity. The acceleration due to gravity on Earth is 32 feet per second per second. If an object is dropped from an airplane, how far will it fall in 6 s?

25. **Television** High definition television (HDTV) gives consumers a wider viewing area, more like a film screen in a theater. A regular television with a 27-inch diagonal measurement has a screen 16.2 in. tall. An HDTV screen with the same 16.2-inch height would have a diagonal measuring 33 in. How many inches wider is the HDTV screen? Round to the nearest hundredth.

26. **Satellites** At what height above Earth's surface is a satellite in orbit if it is traveling at a speed of 7500 m/s? Use the equation $v = \sqrt{\dfrac{4 \times 10^{14}}{h + 6.4 \times 10^6}}$, where v is the speed of the satellite in meters per second and h is the height above Earth's surface in meters. Round to the nearest thousand.

27. **Pendulums** Find the length of a pendulum that makes one swing in 3 s. The equation for the time of one swing of a pendulum is $T = 2\pi\sqrt{\dfrac{L}{32}}$, where T is the time in seconds and L is the length in feet. Round to the nearest hundredth.

APPLYING THE CONCEPTS

28. Solve the following equations. Describe the solution by using the following terms: integer, rational number, irrational number, and real number. Note that more than one term may be used to describe the answer.
 a. $x^2 + 3 = 7$

 b. $x^{3/4} = 8$

29. Solve: $\sqrt{3x - 2} = \sqrt{2x - 3} + \sqrt{x - 1}$

30. Solve $a^2 + b^2 = c^2$ for a.

31. **Geometry** Solve $V = \dfrac{4}{3}\pi r^3$ for r.

32. **Geometry** Find the length of the side labeled x.

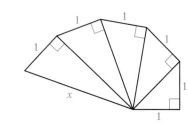

7.4 Complex Numbers

Objective A **To simplify a complex number**

The radical expression $\sqrt{-4}$ is not a real number, because there is no real number whose square is -4. However, the solution of an algebraic equation is sometimes the square root of a negative number.

For example, the equation $x^2 + 1 = 0$ does not have a real number solution, because there is no real number whose square is a negative number.

$$x^2 + 1 = 0$$
$$x^2 = -1$$

Around the 17th century, a new number, called an **imaginary number,** was defined so that a negative number would have a square root. The letter i was chosen to represent the number whose square is -1.

$$i^2 = -1$$

An imaginary number is defined in terms of i.

Point of Interest

The first written occurrence of an imaginary number was in a book published in 1545 by Hieronimo Cardan, where he wrote (in our modern notation) $5 + \sqrt{-15}$. He went on to say that the number "is as refined as it is useless." It was not until the 20th century that applications of complex numbers were found.

> **Definition of $\sqrt{-a}$**
>
> If a is a positive real number, then the principal square root of negative a is the imaginary number $i\sqrt{a}$.
>
> $$\sqrt{-a} = i\sqrt{a}$$

Here are some examples.

$$\sqrt{-16} = i\sqrt{16} = 4i$$
$$\sqrt{-12} = i\sqrt{12} = 2i\sqrt{3}$$
$$\sqrt{-21} = i\sqrt{21}$$
$$\sqrt{-1} = i\sqrt{1} = i$$

It is customary to write i in front of a radical to avoid confusing $\sqrt{a}\,i$ with \sqrt{ai}.

The real numbers and imaginary numbers make up the complex numbers.

> **Complex Number**
>
> A **complex number** is a number of the form $a + bi$, where a and b are real numbers and $i = \sqrt{-1}$. The number a is the **real part** of $a + bi$, and the number b is the **imaginary part.**

TAKE NOTE

The *imaginary part* of a complex number is a real number. As another example, the imaginary part of $6 - 8i$ is -8.

Examples of complex numbers are shown at the right.

Real Part	Imaginary Part
a +	bi
3 +	$2i$
8 −	$10i$

Complex numbers $a + bi$

┌─ Real Numbers $a + 0i$

└─ Imaginary Numbers $0 + bi$

A *real number* is a complex number in which $b = 0$.

An *imaginary number* is a complex number in which $a = 0$.

HOW TO Simplify: $\sqrt{20} - \sqrt{-50}$

$$\sqrt{20} - \sqrt{-50} = \sqrt{20} - i\sqrt{50}$$
$$= \sqrt{4 \cdot 5} - i\sqrt{25 \cdot 2}$$
$$= 2\sqrt{5} - 5i\sqrt{2}$$

- Write the complex number in the form $a + bi$.
- Use the Product Property of Radicals to simplify each radical.

Example 1

Simplify: $\sqrt{-80}$

Solution
$\sqrt{-80} = i\sqrt{80} = i\sqrt{16 \cdot 5} = 4i\sqrt{5}$

You Try It 1

Simplify: $\sqrt{-45}$

Your solution

Example 2

Simplify: $\sqrt{25} + \sqrt{-40}$

Solution
$\sqrt{25} + \sqrt{-40} = \sqrt{25} + i\sqrt{40}$
$= \sqrt{25} + i\sqrt{4 \cdot 10}$
$= 5 + 2i\sqrt{10}$

You Try It 2

Simplify: $\sqrt{98} - \sqrt{-60}$

Your solution

Solutions on p. S23

Objective B **To add or subtract complex numbers**

Addition and Subtraction of Complex Numbers

To add two complex numbers, add the real parts and add the imaginary parts. To subtract two complex numbers, subtract the real parts and subtract the imaginary parts.

$$(a + bi) + (c + di) = (a + c) + (b + d)i$$
$$(a + bi) - (c + di) = (a - c) + (b - d)i$$

HOW TO Simplify: $(3 - 7i) - (4 - 2i)$

$$(3 - 7i) - (4 - 2i) = (3 - 4) + [-7 - (-2)]i$$

- Subtract the real parts and subtract the imaginary parts of the complex numbers.

$$= -1 - 5i$$

HOW TO Simplify: $(3 + \sqrt{-12}) + (7 - \sqrt{-27})$

$(3 + \sqrt{-12}) + (7 - \sqrt{-27})$

$\quad = (3 + i\sqrt{12}) + (7 - i\sqrt{27})$

$\quad = (3 + i\sqrt{4 \cdot 3}) + (7 - i\sqrt{9 \cdot 3})$

$\quad = (3 + 2i\sqrt{3}) + (7 - 3i\sqrt{3})$

$\quad = 10 - i\sqrt{3}$

- Write each complex number in the form $a + bi$.
- Use the Product Property of Radicals to simplify each radical.

- Add the complex numbers.

Example 3

Simplify: $(3 + 2i) + (6 - 5i)$

Solution

$(3 + 2i) + (6 - 5i) = 9 - 3i$

You Try It 3

Simplify: $(-4 + 2i) - (6 - 8i)$

Your solution

Example 4

Simplify: $(9 - \sqrt{-8}) - (5 + \sqrt{-32})$

Solution

$(9 - \sqrt{-8}) - (5 + \sqrt{-32})$

$\quad = (9 - i\sqrt{8}) - (5 + i\sqrt{32})$

$\quad = (9 - i\sqrt{4 \cdot 2}) - (5 + i\sqrt{16 \cdot 2})$

$\quad = (9 - 2i\sqrt{2}) - (5 + 4i\sqrt{2})$

$\quad = 4 - 6i\sqrt{2}$

You Try It 4

Simplify: $(16 - \sqrt{-45}) - (3 + \sqrt{-20})$

Your solution

Example 5

Simplify: $(6 + 4i) + (-6 - 4i)$

Solution

$(6 + 4i) + (-6 - 4i) = 0 + 0i = 0$

This illustrates that the additive inverse of $a + bi$ is $-a - bi$.

You Try It 5

Simplify: $(3 - 2i) + (-3 + 2i)$

Your solution

Solutions on p. S23

Objective C **To multiply complex numbers**

When multiplying complex numbers, we often find that the term i^2 is a part of the product. Recall that $i^2 = -1$.

HOW TO Simplify: $2i \cdot 3i$

$2i \cdot 3i = 6i^2$

$\quad = 6(-1)$

$\quad = -6$

- Multiply the imaginary numbers.
- Replace i^2 by -1.
- Simplify.

HOW TO Simplify: $\sqrt{-6} \cdot \sqrt{-24}$

$$\sqrt{-6} \cdot \sqrt{-24} = i\sqrt{6} \cdot i\sqrt{24}$$

- Write each radical as the product of a real number and *i*.

$$= i^2\sqrt{144}$$

- Multiply the imaginary numbers.

$$= -\sqrt{144}$$

- Replace i^2 by -1.

$$= -12$$

- Simplify the radical expression.

Note from the last example that it would have been incorrect to multiply the radicands of the two radical expressions. To illustrate,

$$\sqrt{-6} \cdot \sqrt{-24} = \sqrt{(-6)(-24)} = \sqrt{144} = 12, \, not \, -12$$

The Product Property of Radicals does not hold true when both radicands are negative and the index is an even number.

HOW TO Simplify: $4i(3 - 2i)$

$$4i(3 - 2i) = 12i - 8i^2$$

- Use the Distributive Property to remove parentheses.

$$= 12i - 8(-1)$$

- Replace i^2 by -1.

$$= 8 + 12i$$

- Write the answer in the form $a + bi$.

The product of two complex numbers is defined as follows.

Multiplication of Complex Numbers

$(a + bi)(c + di) = (ac - bd) + (ad + bc)i$

One way to remember this rule is to think of the FOIL method.

HOW TO Simplify: $(2 + 4i)(3 - 5i)$

$$(2 + 4i)(3 - 5i) = 6 - 10i + 12i - 20i^2$$

- Use the FOIL method to find the product.

$$= 6 + 2i - 20i^2$$
$$= 6 + 2i - 20(-1)$$

- Replace i^2 by -1.

$$= 26 + 2i$$

- Write the answer in the form $a + bi$.

The conjugate of $a + bi$ is $a - bi$.
The product of conjugates, $(a + bi)(a - bi)$, is the real number $a^2 + b^2$.

$$(a + bi)(a - bi) = a^2 - b^2i^2$$
$$= a^2 - b^2(-1)$$
$$= a^2 + b^2$$

HOW TO Simplify: $(2 + 3i)(2 - 3i)$

$$(2 + 3i)(2 - 3i) = 2^2 + 3^2$$

- $a = 2, b = 3$.
 The product of the conjugates is $2^2 + 3^2$.

$$= 4 + 9$$
$$= 13$$

Note that the product of a complex number and its conjugate is a real number.

Example 6

Simplify: $(2i)(-5i)$

Solution

$(2i)(-5i) = -10i^2 = (-10)(-1) = 10$

You Try It 6

Simplify: $(-3i)(-10i)$

Your solution

Example 7

Simplify: $\sqrt{-10} \cdot \sqrt{-5}$

Solution

$\sqrt{-10} \cdot \sqrt{-5} = i\sqrt{10} \cdot i\sqrt{5}$
$= i^2\sqrt{50} = -\sqrt{25 \cdot 2} = -5\sqrt{2}$

You Try It 7

Simplify: $-\sqrt{-8} \cdot \sqrt{-5}$

Your solution

Example 8

Simplify: $3i(2 - 4i)$

Solution

$3i(2 - 4i) = 6i - 12i^2$ • **The Distributive Property**
$= 6i - 12(-1)$
$= 12 + 6i$

You Try It 8

Simplify: $-6i(3 + 4i)$

Your solution

Example 9

Simplify: $\sqrt{-8}(\sqrt{6} - \sqrt{-2})$

Solution

$\sqrt{-8}(\sqrt{6} - \sqrt{-2}) = i\sqrt{8}(\sqrt{6} - i\sqrt{2})$
$= i\sqrt{48} - i^2\sqrt{16}$
$= i\sqrt{16 \cdot 3} - (-1)\sqrt{16}$
$= 4i\sqrt{3} + 4 = 4 + 4i\sqrt{3}$

You Try It 9

Simplify: $\sqrt{-3}(\sqrt{27} - \sqrt{-6})$

Your solution

Example 10

Simplify: $(3 - 4i)(2 + 5i)$

Solution

$(3 - 4i)(2 + 5i) = 6 + 15i - 8i - 20i^2$ • **FOIL**
$= 6 + 7i - 20i^2$
$= 6 + 7i - 20(-1) = 26 + 7i$

You Try It 10

Simplify: $(4 - 3i)(2 - i)$

Your solution

Example 11

Simplify: $(4 + 5i)(4 - 5i)$

Solution

$(4 + 5i)(4 - 5i) = 4^2 + 5^2$ • **Conjugates**
$= 16 + 25 = 41$

You Try It 11

Simplify: $(3 + 6i)(3 - 6i)$

Your solution

Solutions on p. S23

Objective D **To divide complex numbers**

A rational expression containing one or more complex numbers is in simplest form when no imaginary number remains in the denominator.

HOW TO Simplify: $\dfrac{2 - 3i}{2i}$

$$\dfrac{2 - 3i}{2i} = \dfrac{2 - 3i}{2i} \cdot \dfrac{i}{i}$$

 • Multiply the numerator and denominator by $\dfrac{i}{i}$.

$$= \dfrac{2i - 3i^2}{2i^2}$$

$$= \dfrac{2i - 3(-1)}{2(-1)}$$

 • Replace i^2 by **−1**.

$$= \dfrac{3 + 2i}{-2} = -\dfrac{3}{2} - i$$

 • Simplify. Write the answer in the form $a + bi$.

HOW TO Simplify: $\dfrac{3 + 2i}{1 + i}$

$$\dfrac{3 + 2i}{1 + i} = \dfrac{3 + 2i}{1 + i} \cdot \dfrac{1 - i}{1 - i}$$

 • Multiply the numerator and denominator by the conjugate of **1** + *i*.

$$= \dfrac{3 - 3i + 2i - 2i^2}{1^2 + 1^2}$$

$$= \dfrac{3 - i - 2(-1)}{1 + 1}$$

 • Replace i^2 by **−1** and simplify.

$$= \dfrac{5 - i}{2} = \dfrac{5}{2} - \dfrac{1}{2}i$$

 • Write the answer in the form $a + bi$.

Example 12

Simplify: $\dfrac{5 + 4i}{3i}$

Solution

$$\dfrac{5 + 4i}{3i} = \dfrac{5 + 4i}{3i} \cdot \dfrac{i}{i} = \dfrac{5i + 4i^2}{3i^2}$$

$$= \dfrac{5i + 4(-1)}{3(-1)} = \dfrac{-4 + 5i}{-3} = \dfrac{4}{3} - \dfrac{5}{3}i$$

You Try It 12

Simplify: $\dfrac{2 - 3i}{4i}$

Your solution

Example 13

Simplify: $\dfrac{5 - 3i}{4 + 2i}$

Solution

$$\dfrac{5 - 3i}{4 + 2i} = \dfrac{5 - 3i}{4 + 2i} \cdot \dfrac{4 - 2i}{4 - 2i}$$

$$= \dfrac{20 - 10i - 12i + 6i^2}{4^2 + 2^2}$$

$$= \dfrac{20 - 22i + 6(-1)}{16 + 4}$$

$$= \dfrac{14 - 22i}{20} = \dfrac{7 - 11i}{10} = \dfrac{7}{10} - \dfrac{11}{10}i$$

You Try It 13

Simplify: $\dfrac{2 + 5i}{3 - 2i}$

Your solution

Solutions on p. S24

7.4 Exercises

Objective A To simplify a complex number

1. ✏️ What is an imaginary number? What is a complex number?

2. ✏️ Are all real numbers also complex numbers? Are all complex numbers also real numbers?

For Exercises 3 to 14, simplify.

3. $\sqrt{-4}$ **4.** $\sqrt{-64}$ **5.** $\sqrt{-98}$ **6.** $\sqrt{-72}$

7. $\sqrt{-27}$ **8.** $\sqrt{-75}$ **9.** $\sqrt{16} + \sqrt{-4}$ **10.** $\sqrt{25} + \sqrt{-9}$

11. $\sqrt{12} - \sqrt{-18}$ **12.** $\sqrt{60} - \sqrt{-48}$ **13.** $\sqrt{160} - \sqrt{-147}$ **14.** $\sqrt{96} - \sqrt{-125}$

Objective B To add or subtract complex numbers

For Exercises 15 to 24, simplify.

15. $(2 + 4i) + (6 - 5i)$ **16.** $(6 - 9i) + (4 + 2i)$

17. $(-2 - 4i) - (6 - 8i)$ **18.** $(3 - 5i) + (8 - 2i)$

19. $(8 - \sqrt{-4}) - (2 + \sqrt{-16})$ **20.** $(5 - \sqrt{-25}) - (11 - \sqrt{-36})$

21. $(12 - \sqrt{-50}) + (7 - \sqrt{-8})$ **22.** $(5 - \sqrt{-12}) - (9 + \sqrt{-108})$

23. $(\sqrt{8} + \sqrt{-18}) + (\sqrt{32} - \sqrt{-72})$ **24.** $(\sqrt{40} - \sqrt{-98}) - (\sqrt{90} + \sqrt{-32})$

Objective C To multiply complex numbers

For Exercises 25 to 42, simplify.

25. $(7i)(-9i)$ **26.** $(-6i)(-4i)$ **27.** $\sqrt{-2}\,\sqrt{-8}$

28. $\sqrt{-5}\,\sqrt{-45}$

29. $\sqrt{-3}\,\sqrt{-6}$

30. $\sqrt{-5}\,\sqrt{-10}$

31. $2i(6 + 2i)$

32. $-3i(4 - 5i)$

33. $\sqrt{-2}(\sqrt{8} + \sqrt{-2})$

34. $\sqrt{-3}(\sqrt{12} - \sqrt{-6})$

35. $(5 - 2i)(3 + i)$

36. $(2 - 4i)(2 - i)$

37. $(6 + 5i)(3 + 2i)$

38. $(4 - 7i)(2 + 3i)$

39. $(1 - i)\left(\dfrac{1}{2} + \dfrac{1}{2}i\right)$

40. $\left(\dfrac{4}{5} - \dfrac{2}{5}i\right)\left(1 + \dfrac{1}{2}i\right)$

41. $\left(\dfrac{6}{5} + \dfrac{3}{5}i\right)\left(\dfrac{2}{3} - \dfrac{1}{3}i\right)$

42. $(2 - i)\left(\dfrac{2}{5} + \dfrac{1}{5}i\right)$

Objective D **To divide complex numbers**

For Exercises 43 to 58, simplify.

43. $\dfrac{3}{i}$

44. $\dfrac{4}{5i}$

45. $\dfrac{2 - 3i}{-4i}$

46. $\dfrac{16 + 5i}{-3i}$

47. $\dfrac{4}{5 + i}$

48. $\dfrac{6}{5 + 2i}$

49. $\dfrac{2}{2 - i}$

50. $\dfrac{5}{4 - i}$

51. $\dfrac{1 - 3i}{3 + i}$

52. $\dfrac{2 + 12i}{5 + i}$

53. $\dfrac{\sqrt{-10}}{\sqrt{8} - \sqrt{-2}}$

54. $\dfrac{\sqrt{-2}}{\sqrt{12} - \sqrt{-8}}$

55. $\dfrac{2 - 3i}{3 + i}$

56. $\dfrac{3 + 5i}{1 - i}$

57. $\dfrac{5 + 3i}{3 - i}$

58. $\dfrac{3 - 2i}{2i + 3}$

APPLYING THE CONCEPTS

59. **a.** Is $3i$ a solution of $2x^2 + 18 = 0$?
b. Is $3 + i$ a solution of $x^2 - 6x + 10 = 0$?

60. Evaluate i^n for $n = 0, 1, 2, 3, 4, 5, 6,$ and 7. Make a conjecture about the value of i^n for any natural number. Using your conjecture, evaluate i^{76}.

Focus on Problem Solving

Another Look at Polya's Four-Step Process

Polya's four general steps to follow when attempting to solve a problem are to understand the problem, devise a plan, carry out the plan, and review the solution. (See the Focus on Problem Solving in the chapter entitled "Review of Real Numbers.") In the process of devising a plan (Step 2), it may be appropriate to write a mathematical expression or an equation. We will illustrate this with the following problem.

Number the letters of the alphabet in sequence from 1 to 26. (See the list at the left.) Find a word for which the product of the numerical values of the letters of the word equals 1,000,000. We will agree that a "word" is any sequence of letters that contains at least one vowel but is not necessarily in the dictionary.

A = 1
B = 2
C = 3
D = 4
E = 5
F = 6
G = 7
H = 8
I = 9
J = 10
K = 11
L = 12
M = 13
N = 14
O = 15
P = 16
Q = 17
R = 18
S = 19
T = 20
U = 21
V = 22
W = 23
X = 24
Y = 25
Z = 26

Understand the Problem

Consider REZB. The product of the values of the letters is $18 \cdot 5 \cdot 26 \cdot 2 = 4680$. This "word" is a sequence of letters with at least one vowel. However, the product of the numerical values of the letters is not 1,000,000. Thus this word does not solve our problem.

Devise a Plan

Actually, we should have known that the product of the values of the letters in REZB could not equal 1,000,000. The letter R has a factor of 9, and the letter Z has a factor of 13. Neither of these two numbers is a factor of 1,000,000. Consequently, R and Z cannot be letters in the word we are trying to find. This observation leads to an important consideration: Each of the letters that make up our word must be a factor of 1,000,000. To find these letters, consider the prime factorization of 1,000,000.

$$1,000,000 = 2^6 \cdot 5^6$$

Looking at the prime factorization, we note that only letters that contain 2 or 5 as factors are possible candidates. These letters are B, D, E, H, J, P, T, and Y. One additional point: Because 1 times any number is the number, the letter A can be part of any word we construct.

Our task is now to construct a word from these letters such that the product is 1,000,000. From the prime factorization above, we must have 2 as a factor six times and 5 as a factor six times.

Carry Out the Plan

We must construct a word with the characteristics described in our plan. Here is a possibility:

THEBEYE

Review the Solution

You should multiply the values of all the letters and verify that the product is 1,000,000. To ensure that you have an understanding of the problem, find other "words" that satisfy the conditions of the problem.

Projects and Group Activities

Solving Radical Equations with a Graphing Calculator

The radical equation $\sqrt{x-2} - 6 = 0$ was solved algebraically at the beginning of Section 7.3. To solve $\sqrt{x-2} - 6 = 0$ with a graphing calculator, use the left side of the equation and write an equation in two variables.

$$y = \sqrt{x-2} - 6$$

The graph of $y = \sqrt{x-2} - 6$ is shown below. The solution set of the equation $y = \sqrt{x-2} - 6$ is the set of ordered pairs (x, y) whose coordinates make the equation a true statement. The x-coordinate where $y = 0$ is the solution of the equation $\sqrt{x-2} - 6 = 0$. The solution is the x-intercept of the curve given by the equation $y = \sqrt{x-2} - 6$. The solution is 38.

The solution of the radical equation $\sqrt[3]{x-2} = -0.8$ is 1.488. A graphing calculator can be used to find any rational or irrational solution to a predetermined degree of accuracy. The graph of $y = \sqrt[3]{x-2} + 0.8$ is shown below. The graph intersects the x-axis at $x = 1.488$.

To graph the equation $y = \sqrt[3]{x-2} + 0.8$ on a TI-83, enter

Y= MATH 4 X,T,θ,n − 2) + .8 ZOOM 6

Enter 2nd CALC 2 to find the x-intercept of the equation.

You were to solve Exercises 1 to 21 in Section 7.3 by algebraic means. Try some of those exercises again, this time using a graphing calculator.

Use a graphing calculator to find the solutions of the equations in Exercises 1 to 3. Round to the nearest thousandth.

1. $\sqrt{x + 0.3} = 1.3$ **2.** $\sqrt[3]{x + 1.2} = -1.1$ **3.** $\sqrt[4]{3x - 1.5} = 1.4$

Integrating Technology

See the Keystroke Guide: *Graph* for instructions on using a graphing calculator to graph a function, and *Zero* for instructions on finding the x-intercepts of a function.

The Golden Rectangle

Take the time to look closely at each of the four rectangles shown at the left. Which rectangle seems to be the most pleasant to look at? Record the responses of all the students in your class. Which rectangle was named most frequently as being the most appealing?

Rectangle A

The length of rectangle A is three times the width. The ratio of the length to the width is 3 to 1.

Rectangle B

Rectangle B is a square. Its length is equal to its width. The ratio of the length to the width is 1 to 1.

Rectangle C

The ratio of the length of rectangle C to its width is 1.2 to 1.

Rectangle D

The shape of rectangle D is referred to as the golden rectangle. In a golden rectangle, the ratio of the length to the width is approximately 1.6 to 1.

Did most students in the class select the golden rectangle as the one most pleasant to look at? The early Greeks considered a golden rectangle to be the most pleasing to the eye. Therefore, they used it extensively in their art and architecture. The Parthenon in Athens, Greece, is an example of the use of the golden rectangle in Greek architecture. It was built about 435 B.C. as a temple to the goddess Athena.

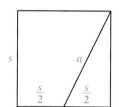

1. Construct a golden rectangle. Beginning with a square with each side of length s, draw a line segment from the midpoint of the base of the square to a vertex, as shown. Call this distance a. Make a rectangle whose width is that of the square and whose length is $\frac{s}{2} + a$.

2. Find the area of the rectangle in terms of s.

3. Find the exact ratio of the length to the width.

4. Find some examples of the golden rectangle in art and architecture. (*Note:* The World Wide Web is a wonderful source of information on the golden rectangle. Simply enter "golden rectangle" in a search engine. It will respond with hundreds of websites where you can learn about this rectangle's appearance in art, architecture, music, nature, and mathematics.)

Chapter 7 Summary

Key Words	**Examples**
$a^{1/n}$ is the *nth root of a*. [7.1A, p. 397]	$16^{1/2} = 4$ because $4^2 = 16$.
The expression $\sqrt[n]{a}$ is another symbol for the *nth* root of *a*. In the expression $\sqrt[n]{a}$, the symbol $\sqrt{}$ is called a *radical, n* is the *index* of the radical, and *a* is the *radicand*. [7.1B, p. 399]	$125^{1/3} = \sqrt[3]{125} = 5$ The index is 3, and the radicand is 125.
The symbol $\sqrt{}$ is used to indicate the positive square root, or *principle square root*, of a number. [7.1C, p. 401]	$\sqrt{16} = 4$ $-\sqrt{16} = -4$
The expressions $a + b$ and $a - b$ are called *conjugates* of each other. The product of conjugates of the form $(a + b)(a - b)$ is $a^2 - b^2$. [7.2C, p. 410]	$(x + 3)(x - 3) = x^2 - 3^2 = x^2 - 9$
The procedure used to remove a radical from the denominator of a radical expression is called *rationalizing the denominator*. [7.2D, p. 411]	$\dfrac{2}{1 - \sqrt{3}} = \dfrac{2}{1 - \sqrt{3}} \cdot \dfrac{1 + \sqrt{3}}{1 + \sqrt{3}}$ $= \dfrac{2(1 + \sqrt{3})}{(1 - \sqrt{3})(1 + \sqrt{3})}$ $= \dfrac{2 + 2\sqrt{3}}{1 - 3} = -1 - \sqrt{3}$
A *radical equation* is an equation that contains a variable expression in a radicand. [7.3A, p. 417]	$\sqrt{x - 2} - 3 = 6$ is a radical equation.
A *complex number* is a number of the form $a + bi$, where a and b are real numbers and $i = \sqrt{-1}$. For the complex number $a + bi$, a is the *real part* of the complex number and b is the *imaginary part* of the complex number. [7.4A, p. 423]	$3 + 2i$ is a complex number. 3 is the real part and 2 is the imaginary part of the complex number.

Essential Rules and Procedures	**Examples**
Rule for Rational Exponents [7.1A, p. 397] If m and n are positive integers and $a^{1/n}$ is a real number, then $a^{m/n} = (a^{1/n})^m$.	$8^{2/3} = (8^{1/3})^2 = 2^2 = 4$

Definition of *n*th Root of *a* [7.1B, p. 399]
If a is a real number, then $a^{1/n} = \sqrt[n]{a}$.

$x^{1/3} = \sqrt[3]{x}$

Writing Exponential Expressions as Radical Expressions
[7.1B, p. 399]
If $a^{1/n}$ is a real number, then $a^{m/n} = a^{m \cdot 1/n} = (a^m)^{1/n} = \sqrt[n]{a^m}$.
The expression $a^{m/n}$ can also be written $(\sqrt[n]{a})^m$.

$b^{3/4} = \sqrt[4]{b^3}$
$8^{2/3} = (\sqrt[3]{8})^2 = 2^2 = 4$

Product Property of Radicals [7.2A, p. 407]
If $\sqrt[n]{a}$ and $\sqrt[n]{b}$ are positive real numbers, then $\sqrt[n]{ab} = \sqrt[n]{a} \cdot \sqrt[n]{b}$
and $\sqrt[n]{a} \cdot \sqrt[n]{b} = \sqrt[n]{ab}$.

$\sqrt{9 \cdot 7} = \sqrt{9} \cdot \sqrt{7} = 3\sqrt{7}$

Quotient Property of Radicals [7.2D, p. 411]
If $\sqrt[n]{a}$ and $\sqrt[n]{b}$ are positive real numbers, and $b \neq 0$, then
$$\sqrt[n]{\frac{a}{b}} = \frac{\sqrt[n]{a}}{\sqrt[n]{b}} \text{ and } \frac{\sqrt[n]{a}}{\sqrt[n]{b}} = \sqrt[n]{\frac{a}{b}}.$$

$$\sqrt[3]{\frac{5}{27}} = \frac{\sqrt[3]{5}}{\sqrt[3]{27}} = \frac{\sqrt[3]{5}}{3}$$

Property of Raising Each Side of an Equation to a Power
[7.3A, p. 417]
If $a = b$, then $a^n = b^n$.

If $x = 4$, then $x^2 = 16$.

Pythagorean Theorem [7.3B, p. 419]
The square of the hypotenuse of a right triangle is equal to the
sum of the squares of the two legs.
$c^2 = a^2 + b^2$

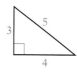

$5^2 = 3^2 + 4^2$

Definition of $\sqrt{-a}$ [7.4A, p. 423]
If a is a positive real number, then the principal square root of
negative a is the imaginary number $i\sqrt{a}$: $\sqrt{-a} = i\sqrt{a}$.

$\sqrt{-8} = i\sqrt{8} = 2i\sqrt{2}$

Addition and Subtraction of Complex Numbers
[7.4B, p. 424]
$(a + bi) + (c + di) = (a + c) + (b + d)i$
$(a + bi) - (c + di) = (a - c) + (b - d)i$

$(2 + 4i) + (3 + 6i)$
$\quad = (2 + 3) + (4 + 6)i$
$\quad = 5 + 10i$
$(4 + 3i) - (7 + 4i)$
$\quad = (4 - 7) + (3 - 4)i$
$\quad = -3 - i$

Multiplication of Complex Numbers [7.4C, p. 426]
$(a + bi)(c + di) = (ac - bd) + (ad + bc)i$
One way to remember this rule is to think of the **FOIL** method.

$(2 - 3i)(5 + 4i)$
$\quad = 10 + 8i - 15i - 12i^2$
$\quad = 10 - 7i - 12(-1)$
$\quad = 22 - 7i$

Chapter 7 Review Exercises

1. Simplify: $(16x^{-4}y^{12})^{1/4}(100x^6y^{-2})^{1/2}$

2. Solve: $\sqrt[4]{3x-5}=2$

3. Simplify: $(6-5i)(4+3i)$

4. Rewrite $7y\sqrt[3]{x^2}$ as an exponential expression.

5. Simplify: $(\sqrt{3}+8)(\sqrt{3}-2)$

6. Solve: $\sqrt{4x+9}+10=11$

7. Simplify: $\dfrac{x^{-3/2}}{x^{7/2}}$

8. Simplify: $\dfrac{8}{\sqrt{3y}}$

9. Simplify: $\sqrt[3]{-8a^6b^{12}}$

10. Simplify: $\sqrt{50a^4b^3}-ab\sqrt{18a^2b}$

11. Simplify: $\dfrac{x+2}{\sqrt{x}+\sqrt{2}}$

12. Simplify: $\dfrac{5+2i}{3i}$

13. Simplify: $\sqrt{18a^3b^6}$

14. Simplify:
$(\sqrt{50}+\sqrt{-72})-(\sqrt{162}-\sqrt{-8})$

15. Simplify: $3x\sqrt[3]{54x^8y^{10}}-2x^2y\sqrt[3]{16x^5y^7}$

16. Simplify: $\sqrt[3]{16x^4y}\,\sqrt[3]{4xy^5}$

17. Simplify: $i(3-7i)$

18. Rewrite $3x^{3/4}$ as a radical expression.

19. Simplify: $\sqrt[5]{-64a^8b^{12}}$

20. Simplify: $\dfrac{5 + 9i}{1 - i}$

21. Simplify: $\sqrt{-12}\,\sqrt{-6}$

22. Solve: $\sqrt{x - 5} + \sqrt{x + 6} = 11$

23. Simplify: $\sqrt[4]{81a^8b^{12}}$

24. Simplify: $\sqrt{-50}$

25. Simplify: $(-8 + 3i) - (4 - 7i)$

26. Simplify: $(5 - \sqrt{6})^2$

27. Simplify: $4x\sqrt{12x^2y} + \sqrt{3x^4y} - x^2\sqrt{27y}$

28. Simplify: $81^{-1/4}$

29. Simplify: $(a^{16})^{-5/8}$

30. Simplify: $-\sqrt{49x^6y^{16}}$

31. Rewrite $4a^{2/3}$ as a radical expression.

32. Rewrite $\sqrt[8]{b^5}$ as an exponential expression.

33. Simplify: $\sqrt[4]{x^6y^8z^{10}}$

34. Simplify: $\sqrt{54} + \sqrt{24}$

35. Simplify: $\sqrt{48x^5y} - x\sqrt{80x^2y}$

36. Simplify: $\sqrt{32}\,\sqrt{50}$

37. Simplify: $\sqrt{3x}(3 + \sqrt{3x})$

38. Simplify: $\dfrac{\sqrt{125x^6}}{\sqrt{5x^3}}$

39. Simplify: $\dfrac{\sqrt{x} + \sqrt{y}}{\sqrt{x} - \sqrt{y}}$

40. Simplify: $\sqrt{-36}$

41. Simplify: $\sqrt{49} - \sqrt{-16}$

42. Simplify: $\sqrt{200} + \sqrt{-12}$

43. Simplify: $(5 + 2i) + (4 - 3i)$

44. Simplify: $(9 - \sqrt{-16}) + (5 + \sqrt{-36})$

45. Simplify: $(3 - 9i) - 7$

46. Simplify: $(8i)(2i)$

47. Simplify: $\dfrac{-6}{i}$

48. Simplify: $\dfrac{7}{2 - i}$

49. Simplify: $\dfrac{\sqrt{16}}{\sqrt{4} - \sqrt{-4}}$

50. Solve: $\sqrt[3]{9x} = -6$

51. **Geometry** Find the width of a rectangle that has a diagonal of 13 in. and a length of 12 in.

52. **Energy** The velocity of the wind determines the amount of power generated by a windmill. A typical equation for this relationship is $v = 4.05 \sqrt[3]{P}$, where v is the velocity in miles per hour and P is the power in watts. Find the amount of power generated by a 20-mph wind. Round to the nearest whole number.

53. **Automotive Technology** Find the distance required for a car to reach a velocity of 88 ft/s when the acceleration is 16 ft/s². Use the equation $v = \sqrt{2as}$, where v is the velocity in feet per second, a is the acceleration, and s is the distance in feet.

54. **Home Maintenance** A 12-foot ladder is leaning against a house in preparation for washing the windows. How far from the house is the bottom of the ladder when the top of the ladder touches the house 10 ft above the ground? Round to the nearest hundredth.

Chapter 7 Test

1. Write $\frac{1}{2}\sqrt[4]{x^3}$ as an exponential expression.

2. Simplify:
 $\sqrt[3]{54x^7y^3} - x\sqrt[3]{128x^4y^3} - x^2\sqrt[3]{2xy^3}$

3. Write $3y^{2/5}$ as a radical expression.

4. Simplify: $(2 + 5i)(4 - 2i)$

5. Simplify: $(2\sqrt{x} + \sqrt{y})^2$

6. Simplify: $\dfrac{r^{2/3}\,r^{-1}}{r^{-1/2}}$

7. Solve: $\sqrt{x + 12} - \sqrt{x} = 2$

8. Simplify: $\sqrt[3]{8x^3y^6}$

9. Simplify: $\sqrt{3x}\,(\sqrt{x} - \sqrt{25x})$

10. Simplify: $(5 - 2i) - (8 - 4i)$

11. Simplify: $\sqrt{32x^4y^7}$

12. Simplify: $(2\sqrt{3} + 4)(3\sqrt{3} - 1)$

13. Simplify: $(2 + i) + (2 - i)(3 + 2i)$

14. Simplify: $\dfrac{4 - 2\sqrt{5}}{2 - \sqrt{5}}$

15. Simplify: $\sqrt{18a^3} + a\sqrt{50a}$

16. Simplify: $(\sqrt{a} - 3\sqrt{b})(2\sqrt{a} + 5\sqrt{b})$

17. Simplify: $\dfrac{(2x^{1/3}y^{-2/3})^6}{(x^{-4}y^8)^{1/4}}$

18. Simplify: $\dfrac{\sqrt{x}}{\sqrt{x} - \sqrt{y}}$

19. Simplify: $\dfrac{2 + 3i}{1 - 2i}$

20. Solve: $\sqrt[3]{2x - 2} + 4 = 2$

21. Simplify: $\left(\dfrac{4a^4}{b^2}\right)^{-3/2}$

22. Simplify: $\sqrt[3]{27a^4b^3c^7}$

23. Simplify: $\dfrac{\sqrt{32x^5y}}{\sqrt{2xy^3}}$

24. Simplify: $(\sqrt{-8})(\sqrt{-2})$

25. **Physics** An object is dropped from a high building. Find the distance the object has fallen when its speed reaches 192 ft/s. Use the equation $v = \sqrt{64d}$, where v is the speed of the object in feet per second and d is the distance in feet.

Cumulative Review Exercises

1. Identify the property that justifies the statement $(a + 2)b = ab + 2b$.

2. Given $f(x) = 3x^2 - 2x + 1$, evaluate $f(-3)$.

3. Solve: $5 - \dfrac{2}{3}x = 4$

4. Solve: $2[4 - 2(3 - 2x)] = 4(1 - x)$

5. Solve: $2 + |4 - 3x| = 5$

6. Solve: $6x - 3(2x + 2) > 3 - 3(x + 2)$
 Write the solution set in set-builder notation.

7. Solve: $|2x + 3| \le 9$

8. Factor: $81x^2 - y^2$

9. Factor: $x^5 + 2x^3 - 3x$

10. Find the equation of the line that passes through the points $(2, 3)$ and $(-1, 2)$.

11. Find the value of the determinant:
$$\begin{vmatrix} 1 & 2 & -3 \\ 0 & -1 & 2 \\ 3 & 1 & -2 \end{vmatrix}$$

12. Solve $P = \dfrac{R - C}{n}$ for C.

13. Simplify: $(2^{-1}x^2y^{-6})(2^{-1}y^{-4})^{-2}$

14. Simplify: $\dfrac{x^2y^3}{x^2 + 2x - 8} \cdot \dfrac{2x^2 - 7x + 6}{xy^4}$

15. Simplify: $\sqrt{40x^3} - x\sqrt{90x}$

16. Solve: $\dfrac{x}{x - 2} - 2x = \dfrac{-3}{x - 2}$

17. Graph $3x - 2y = -6$ and find the slope and y-intercept.

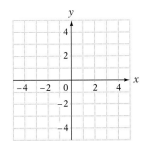

18. Graph the solution set of $3x + 2y \le 4$.

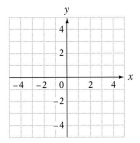

19. Simplify: $\dfrac{2i}{3 - i}$

20. Solve: $\sqrt[3]{3x - 4} + 5 = 1$

21. Simplify: $\dfrac{x}{2x - 3} + \dfrac{4}{x + 4}$

22. Solve by using Cramer's Rule:
$$2x - y = 4$$
$$-2x + 3y = 5$$

23. **Stamps** A collection of 30 stamps consists of 13¢ stamps and 18¢ stamps. The total value of the stamps is \$4.85. Find the number of 18¢ stamps.

24. **Uniform Motion** A sales executive traveled 25 mi by car and then an additional 625 mi by plane. The rate of the plane was five times greater than the rate of the car. The total time of the trip was 3 h. Find the rate of the plane.

25. **Astronomy** How long does it take light to travel to Earth from the moon when the moon is 232,500 mi from Earth? Light travels at a rate of 1.86×10^5 mi/s.

26. **Oceanography** How far would a submarine periscope have to be above the water to locate a ship 7 mi away? The equation for the distance in miles that the lookout can see is $d = \sqrt{1.5h}$, where h is the height in feet above the surface of the water. Round to the nearest tenth of a foot.

27. **Investments** The graph shows the amount invested and the annual income from an investment. Find the slope of the line between the two points shown on the graph. Then write a sentence that states the meaning of the slope.

8 Quadratic Equations

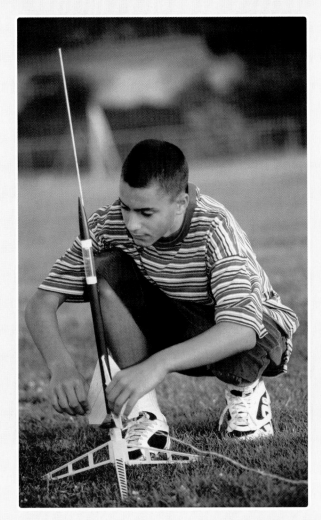

This model rocket, just like an airplane, is subject to the forces of weight and thrust as it moves through the air. The study of forces and their effect on the motion of objects traveling through the air is called aerodynamics. The motion of a model rocket through the air can be described and explained by the principals of physics, which were discovered by Sir Isaac Newton over 300 years ago. **Exercises 6 to 8 on page 477** use quadratic equations to mathematically model the rocket's movement, as well as the movements of other projectiles.

OBJECTIVES

Section 8.1

A To solve a quadratic equation by factoring

B To write a quadratic equation given its solutions

C To solve a quadratic equation by taking square roots

Section 8.2

A To solve a quadratic equation by completing the square

Section 8.3

A To solve a quadratic equation by using the quadratic formula

Section 8.4

A To solve an equation that is quadratic in form

B To solve a radical equation that is reducible to a quadratic equation

C To solve a fractional equation that is reducible to a quadratic equation

Section 8.5

A To solve a nonlinear inequality

Section 8.6

A To solve application problems

Need help? For online student resources, such as section quizzes, visit this textbook's website at **math.college.hmco.com/students.**

PREP TEST • • •

Do these exercises to prepare for Chapter 8.

1. Simplify: $\sqrt{18}$

2. Simplify: $\sqrt{-9}$

3. Simplify: $\dfrac{3x - 2}{x - 1} - 1$

4. Evaluate $b^2 - 4ac$ when $a = 2$, $b = -4$, and $c = 1$.

5. Is $4x^2 + 28x + 49$ a perfect square trinomial?

6. Factor: $4x^2 - 4x + 1$

7. Factor: $9x^2 - 4$

8. Graph $\{x \,|\, x < -1\} \cap \{x \,|\, x < 4\}$

$$\begin{array}{ccccccccccc} & & & & & & & & & & \\ \hline -5 & -4 & -3 & -2 & -1 & 0 & 1 & 2 & 3 & 4 & 5 \end{array}$$

9. Solve: $x(x - 1) = x + 15$

10. Solve: $\dfrac{4}{x - 3} = \dfrac{16}{x}$

GO FIGURE • • •

The numeral $0.AAA\ldots$ is a repeating decimal. Given that $A \neq B$ and $\sqrt{0.AAA\ldots} = \sqrt{B}$, find the value of A.

8.1 Solving Quadratic Equations by Factoring or by Taking Square Roots

Objective A To solve a quadratic equation by factoring

Recall that a *quadratic equation* is an equation of the form $ax^2 + bx + c = 0$, where a and b are coefficients, c is a constant, and $a \neq 0$.

Quadratic Equations
$$\begin{cases} 3x^2 - x + 2 = 0, & a = 3, & b = -1, & c = 2 \\ -x^2 + 4 = 0, & a = -1, & b = 0, & c = 4 \\ 6x^2 - 5x = 0, & a = 6, & b = -5, & c = 0 \end{cases}$$

A quadratic equation is in *standard form* when the polynomial is in descending order and equal to zero. Because the degree of the polynomial $ax^2 + bx + c$ is 2, a quadratic equation is also called a **second-degree equation.**

As we discussed earlier, quadratic equations sometimes can be solved by using the Principle of Zero Products. This method is reviewed here.

> **The Principle of Zero Products**
>
> If a and b are real numbers and $ab = 0$, then $a = 0$ or $b = 0$.

The Principle of Zero Products states that if the product of two factors is zero, then at least one of the factors must be zero.

TAKE NOTE

Recall that the steps involved in solving a quadratic equation by factoring are

1. Write the equation in standard form.
2. Factor.
3. Use the Principle of Zero Products to set each factor equal to 0.
4. Solve each equation.
5. Check the solutions.

HOW TO Solve by factoring: $3x^2 = 2 - 5x$

$$3x^2 = 2 - 5x$$
$$3x^2 + 5x - 2 = 0 \qquad \bullet \text{ Write the equation in standard form.}$$
$$(3x - 1)(x + 2) = 0 \qquad \bullet \text{ Factor.}$$
$$3x - 1 = 0 \qquad\qquad x + 2 = 0 \qquad \bullet \text{ Use the Principle of Zero Products to write two equations.}$$
$$3x = 1 \qquad\qquad\qquad x = -2 \qquad \bullet \text{ Solve each equation.}$$
$$x = \frac{1}{3}$$

$\frac{1}{3}$ and -2 check as solutions. The solutions are $\frac{1}{3}$ and -2.

TAKE NOTE

When a quadratic equation has two solutions that are the same number, the solution is called a **double root** of the equation. 3 is a double root of $x^2 - 6x = -9$.

HOW TO Solve by factoring: $x^2 - 6x = -9$

$$x^2 - 6x = -9$$
$$x^2 - 6x + 9 = 0 \qquad \bullet \text{ Write the equation in standard form.}$$
$$(x - 3)(x - 3) = 0 \qquad \bullet \text{ Factor.}$$
$$x - 3 = 0 \qquad\qquad x - 3 = 0 \qquad \bullet \text{ Use the Principle of Zero Products.}$$
$$x = 3 \qquad\qquad\qquad x = 3 \qquad \bullet \text{ Solve each equation.}$$

3 checks as a solution. The solution is 3.

Example 1

Solve by factoring: $2x(x - 3) = x + 4$

Solution

$$2x(x - 3) = x + 4$$
$$2x^2 - 6x = x + 4 \quad \bullet \text{ Write in standard form.}$$
$$2x^2 - 7x - 4 = 0$$
$$(2x + 1)(x - 4) = 0 \quad \bullet \text{ Solve by factoring.}$$
$$2x + 1 = 0 \qquad x - 4 = 0$$
$$2x = -1 \qquad x = 4$$
$$x = -\frac{1}{2}$$

The solutions are $-\frac{1}{2}$ and 4.

You Try It 1

Solve by factoring: $2x^2 = 7x - 3$

Your solution

Example 2

Solve for x by factoring: $x^2 - 4ax - 5a^2 = 0$

Solution

This is a literal equation. Solve for x in terms of a.

$$x^2 - 4ax - 5a^2 = 0$$
$$(x + a)(x - 5a) = 0$$
$$x + a = 0 \qquad x - 5a = 0$$
$$x = -a \qquad x = 5a$$

The solutions are $-a$ and $5a$.

You Try It 2

Solve for x by factoring: $x^2 - 3ax - 4a^2 = 0$

Your solution

Solutions on p. S24

Objective B **To write a quadratic equation given its solutions**

As shown below, the solutions of the equation $(x - r_1)(x - r_2) = 0$ are r_1 and r_2.

$$(x - r_1)(x - r_2) = 0$$

$$x - r_1 = 0 \qquad\qquad x - r_2 = 0$$
$$x = r_1 \qquad\qquad x = r_2$$

Check:

$(x - r_1)(x - r_2) = 0$	$(x - r_1)(x - r_2) = 0$
$(r_1 - r_1)(r_1 - r_2) \mid 0$	$(r_2 - r_1)(r_2 - r_2) \mid 0$
$0 \cdot (r_1 - r_2) \mid 0$	$(r_2 - r_1) \cdot 0 \mid 0$
$0 = 0$	$0 = 0$

Using the equation $(x - r_1)(x - r_2) = 0$ and the fact that r_1 and r_2 are solutions of this equation, it is possible to write a quadratic equation given its solutions.

HOW TO Write a quadratic equation that has solutions 4 and -5.

$$(x - r_1)(x - r_2) = 0$$
$$(x - 4)[x - (-5)] = 0 \quad \bullet \text{ Replace } r_1 \text{ by 4 and } r_2 \text{ by } -5.$$
$$(x - 4)(x + 5) = 0 \quad \bullet \text{ Simplify.}$$
$$x^2 + x - 20 = 0 \quad \bullet \text{ Multiply.}$$

HOW TO Write a quadratic equation with integer coefficients and solutions $\frac{2}{3}$ and $\frac{1}{2}$.

$$(x - r_1)(x - r_2) = 0$$

$$\left(x - \frac{2}{3}\right)\left(x - \frac{1}{2}\right) = 0$$ • Replace r_1 by $\frac{2}{3}$ and r_2 by $\frac{1}{2}$.

$$x^2 - \frac{7}{6}x + \frac{1}{3} = 0$$ • Multiply.

$$6\left(x^2 - \frac{7}{6}x + \frac{1}{3}\right) = 6 \cdot 0$$ • Multiply each side of the equation by the LCM of the denominators.

$$6x^2 - 7x + 2 = 0$$

Example 3

Write a quadratic equation with integer coefficients and solutions $\frac{1}{2}$ and -4.

Solution

$$(x - r_1)(x - r_2) = 0$$

$$\left(x - \frac{1}{2}\right)[x - (-4)] = 0 \quad \bullet \ r_1 = \frac{1}{2}, r_2 = -4$$

$$\left(x - \frac{1}{2}\right)(x + 4) = 0$$

$$x^2 + \frac{7}{2}x - 2 = 0$$

$$2\left(x^2 + \frac{7}{2}x - 2\right) = 2 \cdot 0$$

$$2x^2 + 7x - 4 = 0$$

You Try It 3

Write a quadratic equation with integer coefficients and solutions 3 and $-\frac{1}{2}$.

Your solution

Solution on p. S24

Objective C **To solve a quadratic equation by taking square roots**

The solution of the quadratic equation $x^2 = 16$ is shown at the right.

$$x^2 = 16$$
$$x^2 - 16 = 0$$
$$(x - 4)(x + 4) = 0$$
$$x - 4 = 0 \qquad x + 4 = 0$$
$$x = 4 \qquad\qquad x = -4$$

The solution can also be found by taking the square root of each side of the equation and writing the positive and negative square roots of the radicand. The notation ± 4 means $x = 4$ or $x = -4$.

$$x^2 = 16$$
$$\sqrt{x^2} = \sqrt{16}$$
$$x = \pm 4$$

The solutions are 4 and -4.

HOW TO Solve by taking square roots: $3x^2 = 54$

$$3x^2 = 54$$
$$x^2 = 18$$ • Solve for x^2.
$$\sqrt{x^2} = \sqrt{18}$$ • Take the square root of each side of the equation.
$$x = \pm\sqrt{18}$$ • Simplify.
$$x = \pm 3\sqrt{2}$$

The solutions are $3\sqrt{2}$ and $-3\sqrt{2}$. • $3\sqrt{2}$ and $-3\sqrt{2}$ check as solutions.

ΩSolving a quadratic equation by taking the square root of each side of the equation can lead to solutions that are complex numbers.

HOW TO Solve by taking square roots: $2x^2 + 18 = 0$

$$2x^2 + 18 = 0$$
$$2x^2 = -18 \qquad \bullet \text{ Solve for } x^2.$$
$$x^2 = -9$$
$$\sqrt{x^2} = \sqrt{-9} \qquad \bullet \text{ Take the square root of each side of the equation.}$$
$$x = \pm\sqrt{-9} \qquad \bullet \text{ Simplify.}$$
$$x = \pm 3i$$

Check:

$2x^2 + 18 = 0$		$2x^2 + 18 = 0$	
$2(3i)^2 + 18$	0	$2(-3i)^2 + 18$	0
$2(-9) + 18$	0	$2(-9) + 18$	0
$-18 + 18$	0	$-18 + 18$	0
	$0 = 0$		$0 = 0$

Study Tip

Always check the solution of an equation.

The solutions are $3i$ and $-3i$.

An equation containing the square of a binomial can be solved by taking square roots.

HOW TO Solve by taking square roots: $(x + 2)^2 - 24 = 0$

$$(x + 2)^2 - 24 = 0$$
$$(x + 2)^2 = 24 \qquad \bullet \text{ Solve for } (x + 2)^2.$$
$$\sqrt{(x + 2)^2} = \sqrt{24} \qquad \bullet \text{ Take the square root of each side of the equation. Then simplify.}$$
$$x + 2 = \pm\sqrt{24}$$
$$x + 2 = \pm 2\sqrt{6}$$

$$x + 2 = 2\sqrt{6} \qquad x + 2 = -2\sqrt{6} \qquad \bullet \text{ Solve for } x.$$
$$x = -2 + 2\sqrt{6} \qquad x = -2 - 2\sqrt{6}$$

The solutions are $-2 + 2\sqrt{6}$ and $-2 - 2\sqrt{6}$.

Example 4

Solve by taking square roots.
$3(x - 2)^2 + 12 = 0$

Solution

$$3(x - 2)^2 + 12 = 0$$
$$3(x - 2)^2 = -12 \qquad \bullet \text{ Solve for } (x - 2)^2.$$
$$(x - 2)^2 = -4$$
$$\sqrt{(x - 2)^2} = \sqrt{-4} \qquad \bullet \text{ Take the square root of each side of the equation.}$$
$$x - 2 = \pm\sqrt{-4}$$
$$x - 2 = \pm 2i$$

$$x - 2 = 2i \qquad x - 2 = -2i \qquad \bullet \text{ Solve for } x.$$
$$x = 2 + 2i \qquad x = 2 - 2i$$

The solutions are $2 + 2i$ and $2 - 2i$.

You Try It 4

Solve by taking square roots.
$2(x + 1)^2 - 24 = 0$

Your solution

Solution on p. S24

8.1 Exercises

Objective A **To solve a quadratic equation by factoring**

1. Explain why the restriction $a \neq 0$ is necessary in the definition of a quadratic equation.

2. What does the Principle of Zero Products state? How is it used to solve a quadratic equation?

For Exercises 3 to 6, write the quadratic equation in standard form with the coefficient of x^2 positive. Name the values of a, b, and c.

3. $2x^2 - 4x = 5$ **4.** $x^2 = 3x + 1$ **5.** $5x = 4x^2 + 6$ **6.** $3x^2 = 7$

For Exercises 7 to 37, solve by factoring.

7. $x^2 - 4x = 0$ **8.** $y^2 + 6y = 0$ **9.** $t^2 - 25 = 0$

10. $p^2 - 81 = 0$ **11.** $s^2 - s - 6 = 0$ **12.** $v^2 + 4v - 5 = 0$

13. $y^2 - 6y + 9 = 0$ **14.** $x^2 + 10x + 25 = 0$ **15.** $9z^2 - 18z = 0$

16. $4y^2 + 20y = 0$ **17.** $r^2 - 3r = 10$ **18.** $p^2 + 5p = 6$

19. $v^2 + 10 = 7v$ **20.** $t^2 - 16 = 15t$ **21.** $2x^2 - 9x - 18 = 0$

22. $3y^2 - 4y - 4 = 0$ **23.** $4z^2 - 9z + 2 = 0$ **24.** $2s^2 - 9s + 9 = 0$

25. $3w^2 + 11w = 4$ **26.** $2r^2 + r = 6$ **27.** $6x^2 = 23x + 18$

28. $6x^2 = 7x - 2$ **29.** $4 - 15u - 4u^2 = 0$ **30.** $3 - 2y - 8y^2 = 0$

31. $x + 18 = x(x - 6)$ **32.** $t + 24 = t(t + 6)$ **33.** $4s(s + 3) = s - 6$

34. $3v(v - 2) = 11v + 6$

35. $u^2 - 2u + 4 = (2u - 3)(u + 2)$

36. $(3v - 2)(2v + 1) = 3v^2 - 11v - 10$

37. $(3x - 4)(x + 4) = x^2 - 3x - 28$

For Exercises 38 to 52, solve for x by factoring.

38. $x^2 + 14ax + 48a^2 = 0$

39. $x^2 - 9bx + 14b^2 = 0$

40. $x^2 + 9xy - 36y^2 = 0$

41. $x^2 - 6cx - 7c^2 = 0$

42. $x^2 - ax - 20a^2 = 0$

43. $2x^2 + 3bx + b^2 = 0$

44. $3x^2 - 4cx + c^2 = 0$

45. $3x^2 - 14ax + 8a^2 = 0$

46. $3x^2 - 11xy + 6y^2 = 0$

47. $3x^2 - 8ax - 3a^2 = 0$

48. $3x^2 - 4bx - 4b^2 = 0$

49. $4x^2 + 8xy + 3y^2 = 0$

50. $6x^2 - 11cx + 3c^2 = 0$

51. $6x^2 + 11ax + 4a^2 = 0$

52. $12x^2 - 5xy - 2y^2 = 0$

Objective B **To write a quadratic equation given its solutions**

For Exercises 53 to 82, write a quadratic equation that has integer coefficients and has as solutions the given pair of numbers.

53. 2 and 5

54. 3 and 1

55. −2 and −4

56. −1 and −3

57. 6 and −1

58. −2 and 5

59. 3 and −3

60. 5 and −5

61. 4 and 4

62. 2 and 2

63. 0 and 5

64. 0 and −2

65. 0 and 3

66. 0 and -1

67. 3 and $\dfrac{1}{2}$

68. 2 and $\dfrac{2}{3}$

69. $-\dfrac{3}{4}$ and 2

70. $-\dfrac{1}{2}$ and 5

71. $-\dfrac{5}{3}$ and -2

72. $-\dfrac{3}{2}$ and -1

73. $-\dfrac{2}{3}$ and $\dfrac{2}{3}$

74. $-\dfrac{1}{2}$ and $\dfrac{1}{2}$

75. $\dfrac{1}{2}$ and $\dfrac{1}{3}$

76. $\dfrac{3}{4}$ and $\dfrac{2}{3}$

77. $\dfrac{6}{5}$ and $-\dfrac{1}{2}$

78. $\dfrac{3}{4}$ and $-\dfrac{3}{2}$

79. $-\dfrac{1}{4}$ and $-\dfrac{1}{2}$

80. $-\dfrac{5}{6}$ and $-\dfrac{2}{3}$

81. $\dfrac{3}{5}$ and $-\dfrac{1}{10}$

82. $\dfrac{7}{2}$ and $-\dfrac{1}{4}$

> **Objective C** **To solve a quadratic equation by taking square roots**

For Exercises 83 to 112, solve by taking square roots.

83. $y^2 = 49$

84. $x^2 = 64$

85. $z^2 = -4$

86. $v^2 = -16$

87. $s^2 - 4 = 0$

88. $r^2 - 36 = 0$

89. $4x^2 - 81 = 0$

90. $9x^2 - 16 = 0$

91. $y^2 + 49 = 0$

92. $z^2 + 16 = 0$

93. $v^2 - 48 = 0$

94. $s^2 - 32 = 0$

95. $r^2 - 75 = 0$

96. $u^2 - 54 = 0$

97. $z^2 + 18 = 0$

98. $t^2 + 27 = 0$

99. $(x - 1)^2 = 36$

100. $(x + 2)^2 = 25$

101. $3(y + 3)^2 = 27$

102. $4(s - 2)^2 = 36$

103. $5(z + 2)^2 = 125$

104. $2(y - 3)^2 = 18$

105. $\left(v - \dfrac{1}{2}\right)^2 = \dfrac{1}{4}$

106. $\left(r + \dfrac{2}{3}\right)^2 = \dfrac{1}{9}$

107. $(x + 5)^2 - 6 = 0$

108. $(t - 1)^2 - 15 = 0$

109. $(v - 3)^2 + 45 = 0$

110. $(x + 5)^2 + 32 = 0$

111. $\left(u + \dfrac{2}{3}\right)^2 - 18 = 0$

112. $\left(z - \dfrac{1}{2}\right)^2 - 20 = 0$

APPLYING THE CONCEPTS

For Exercises 113 to 120, write a quadratic equation that has as solutions the given pair of numbers.

113. $\sqrt{2}$ and $-\sqrt{2}$

114. $\sqrt{5}$ and $-\sqrt{5}$

115. i and $-i$

116. $2i$ and $-2i$

117. $2\sqrt{2}$ and $-2\sqrt{2}$

118. $3\sqrt{2}$ and $-3\sqrt{2}$

119. $i\sqrt{2}$ and $-i\sqrt{2}$

120. $2i\sqrt{3}$ and $-2i\sqrt{3}$

For Exercises 121 to 126, solve for x.

121. $4a^2x^2 = 36b^2,\ a > 0,\ b > 0$

122. $3y^2x^2 = 27z^2,\ y > 0,\ z > 0$

123. $(x + a)^2 - 4 = 0$

124. $(x - b)^2 - 1 = 0$

125. $(2x - 1)^2 = (2x + 3)^2$

126. $(x - 4)^2 = (x + 2)^2$

127. a. Show that the solutions of the equation $ax^2 + bx = 0$, $a > 0$, $b > 0$, are 0 and $-\dfrac{b}{a}$.

b. Show that the solutions of the equation $ax^2 + c = 0$, $a > 0$, $c > 0$, are $\dfrac{\sqrt{ca}}{a}i$ and $-\dfrac{\sqrt{ca}}{a}i$.

8.2 Solving Quadratic Equations by Completing the Square

Objective A **To solve a quadratic equation by completing the square**

Recall that a perfect-square trinomial is the square of a binomial.

Perfect-Square Trinomial		Square of a Binomial
$x^2 + 8x + 16$	$=$	$(x + 4)^2$
$x^2 - 10x + 25$	$=$	$(x - 5)^2$
$x^2 + 2ax + a^2$	$=$	$(x + a)^2$

For each perfect-square trinomial, the square of $\frac{1}{2}$ of the coefficient of x equals the constant term.

$$\left(\frac{1}{2}\ \text{coefficient of } x\right)^2 = \text{constant term}$$

$x^2 + 8x + 16,\qquad \left(\frac{1}{2} \cdot 8\right)^2 = 16$

$x^2 - 10x + 25,\qquad \left[\frac{1}{2}(-10)\right]^2 = 25$

$x^2 + 2ax + a^2,\qquad \left(\frac{1}{2} \cdot 2a\right)^2 = a^2$

Point of Interest

Early attempts to solve quadratic equations were primarily geometric. The Persian mathematician al-Khowarizmi (c. A.D. 800) essentially completed a square of $x^2 + 12x$ as follows.

This relationship can be used to write the constant term for a perfect-square trinomial. Adding to a binomial the constant term that makes it a perfect-square trinomial is called **completing the square.**

HOW TO Complete the square on $x^2 + 12x$. Write the resulting perfect-square trinomial as the square of a binomial.

$\left[\dfrac{1}{2}(12)\right]^2 = (6)^2 = 36$ • Find the constant term.

$x^2 + 12x + 36$ • Complete the square on $x^2 + 12x$ by adding the constant term.

$x^2 + 12x + 36 = (x + 6)^2$ • Write the resulting perfect-square trinomial as the square of a binomial.

HOW TO Complete the square on $z^2 - 3z$. Write the resulting perfect-square trinomial as the square of a binomial.

$\left[\dfrac{1}{2} \cdot (-3)\right]^2 = \left(-\dfrac{3}{2}\right)^2 = \dfrac{9}{4}$ • Find the constant term.

$z^2 - 3z + \dfrac{9}{4}$ • Complete the square on $z^2 - 3z$ by adding the constant term.

$z^2 - 3z + \dfrac{9}{4} = \left(z - \dfrac{3}{2}\right)^2$ • Write the resulting perfect-square trinomial as the square of a binomial.

Any quadratic equation can be solved by completing the square. Add to each side of the equation the term that completes the square. Rewrite the equation in the form $(x + a)^2 = b$. Then take the square root of each side of the equation.

HOW TO Solve by completing the square: $x^2 - 6x - 15 = 0$

$x^2 - 6x - 15 = 0$

$x^2 - 6x = 15$ • Add 15 to each side of the equation.

$x^2 - 6x + 9 = 15 + 9$ • Complete the square. Add $\left[\dfrac{1}{2}(-6)\right]^2 = (-3)^2 = 9$ to each side of the equation.

$(x - 3)^2 = 24$ • Factor the perfect-square trinomial.

$\sqrt{(x - 3)^2} = \sqrt{24}$ • Take the square root of each side of the equation.

$x - 3 = \pm\sqrt{24}$

$x - 3 = \pm 2\sqrt{6}$ • Solve for x.

$x - 3 = 2\sqrt{6}$ \qquad $x - 3 = -2\sqrt{6}$

$x = 3 + 2\sqrt{6}$ \qquad $x = 3 - 2\sqrt{6}$

Check:

$$\begin{array}{c|c} \quad x^2 - 6x - 15 = 0 \\ \hline (3 + 2\sqrt{6})^2 - 6(3 + 2\sqrt{6}) - 15 & 0 \\ 9 + 12\sqrt{6} + 24 - 18 - 12\sqrt{6} - 15 & 0 \\ 0 = 0 \end{array}$$

$$\begin{array}{c|c} \quad x^2 - 6x - 15 = 0 \\ \hline (3 - 2\sqrt{6})^2 - 6(3 - 2\sqrt{6}) - 15 & 0 \\ 9 - 12\sqrt{6} + 24 - 18 + 12\sqrt{6} - 15 & 0 \\ 0 = 0 \end{array}$$

The solutions are $3 + 2\sqrt{6}$ and $3 - 2\sqrt{6}$.

TAKE NOTE

The exact solutions of the equation $x^2 - 6x - 15 = 0$ are $3 + 2\sqrt{6}$ and $3 - 2\sqrt{6}$. 7.899 and -1.899 are approximate solutions of the equation.

In the example above, the solutions of the equation $x^2 - 6x - 15 = 0$ are $3 + 2\sqrt{6}$ and $3 - 2\sqrt{6}$. These are the exact solutions. However, in some situations it may be preferable to have decimal approximations of the solutions of a quadratic equation. Approximate solutions can be found by using a calculator and then rounding to the desired degree of accuracy.

$$3 + 2\sqrt{6} \approx 7.899 \text{ and } 3 - 2\sqrt{6} \approx -1.899$$

To the nearest thousandth, the approximate solutions of the equation $x^2 - 6x - 15 = 0$ are 7.899 and -1.899.

HOW TO Solve $2x^2 - x - 2 = 0$ by completing the square. Find the exact solutions and approximate the solutions to the nearest thousandth.

In order for us to complete the square on an expression, the coefficient of the squared term must be 1. After adding the constant term to each side of the equation, multiply each side of the equation by $\dfrac{1}{2}$.

$2x^2 - x - 2 = 0$

$2x^2 - x = 2$ • Add 2 to each side of the equation.

$\dfrac{1}{2}(2x^2 - x) = \dfrac{1}{2} \cdot 2$ • Multiply each side of the equation by $\dfrac{1}{2}$.

Integrating Technology

A graphing calculator can be used to check the proposed solution of a quadratic equation. For the equation at the right, enter the expression $2x^2 - x - 2$ in Y_1. Then evaluate the function for $\frac{1 + \sqrt{17}}{4}$ and $\frac{1 - \sqrt{17}}{4}$. The value of each should be zero. See the Keystroke Guide: *Evaluating Functions*.

$$x^2 - \frac{1}{2}x = 1$$

- The coefficient of x^2 is now 1.

$$x^2 - \frac{1}{2}x + \frac{1}{16} = 1 + \frac{1}{16}$$

- Complete the square. Add $\left[\frac{1}{2}\left(-\frac{1}{2}\right)\right]^2 = \left(-\frac{1}{4}\right)^2 = \frac{1}{16}$ to each side of the equation.

$$\left(x - \frac{1}{4}\right)^2 = \frac{17}{16}$$

- Factor the perfect-square trinomial.

$$\sqrt{\left(x - \frac{1}{4}\right)^2} = \sqrt{\frac{17}{16}}$$

- Take the square root of each side of the equation.

$$x - \frac{1}{4} = \pm \frac{\sqrt{17}}{4}$$

- Solve for x.

$$x - \frac{1}{4} = \frac{\sqrt{17}}{4} \qquad x - \frac{1}{4} = -\frac{\sqrt{17}}{4}$$

$$x = \frac{1}{4} + \frac{\sqrt{17}}{4} \qquad x = \frac{1}{4} - \frac{\sqrt{17}}{4}$$

The exact solutions are $\frac{1 + \sqrt{17}}{4}$ and $\frac{1 - \sqrt{17}}{4}$.

$$\frac{1 + \sqrt{17}}{4} \approx 1.281 \qquad \frac{1 - \sqrt{17}}{4} \approx -0.781$$

To the nearest thousandth, the solutions are 1.281 and -0.781.

Check of exact solutions:

$2x^2 - x - 2 = 0$	
$2\left(\dfrac{1 + \sqrt{17}}{4}\right)^2 - \dfrac{1 + \sqrt{17}}{4} - 2$	0
$2\left(\dfrac{18 + 2\sqrt{17}}{16}\right) - \dfrac{1 + \sqrt{17}}{4} - 2$	0
$\dfrac{18 + 2\sqrt{17}}{8} - \dfrac{2 + 2\sqrt{17}}{8} - 2$	0
$\dfrac{16}{8} - 2$	0
	$0 = 0$

$2x^2 - x - 2 = 0$	
$2\left(\dfrac{1 - \sqrt{17}}{4}\right)^2 - \dfrac{1 - \sqrt{17}}{4} - 2$	0
$2\left(\dfrac{18 - 2\sqrt{17}}{16}\right) - \dfrac{1 - \sqrt{17}}{4} - 2$	0
$\dfrac{18 - 2\sqrt{17}}{8} - \dfrac{2 - 2\sqrt{17}}{8} - 2$	0
$\dfrac{16}{8} - 2$	0
	$0 = 0$

The above example illustrates all the steps required in solving a quadratic equation by completing the square.

Study Tip

This is a new skill and one that is difficult for many students. Be sure to do all you need to do in order to be successful at solving equations by completing the square: Read through the introductory material, work through the HOW TO examples, study the paired Examples, do the You Try Its, and check your solutions against the ones in the back of the book. See *AIM for Success*, pages xxix and xxx.

Procedure for Solving a Quadratic Equation by Completing the Square

1. Write the equation in the form $ax^2 + bx = -c$.

2. Multiply both sides of the equation by $\frac{1}{a}$.

3. Complete the square on $x^2 + \frac{b}{a}x$. Add the number that completes the square to both sides of the equation.

4. Factor the perfect-square trinomial.

5. Take the square root of each side of the equation.

6. Solve the resulting equation for x.

7. Check the solutions.

Example 1

Solve by completing the square:
$4x^2 - 8x + 1 = 0$

Solution

$4x^2 - 8x + 1 = 0$

$\qquad 4x^2 - 8x = -1$ • Write in the form $ax^2 + bx = -c$.

$\dfrac{1}{4}(4x^2 - 8x) = \dfrac{1}{4}(-1)$ • Multiply both sides by $\dfrac{1}{a}$.

$\qquad x^2 - 2x = -\dfrac{1}{4}$

$x^2 - 2x + 1 = -\dfrac{1}{4} + 1$ • Complete the square.

$\qquad (x-1)^2 = \dfrac{3}{4}$ • Factor.

$\qquad \sqrt{(x-1)^2} = \sqrt{\dfrac{3}{4}}$ • Take square roots.

$\qquad x - 1 = \pm\dfrac{\sqrt{3}}{2}$

$x - 1 = \dfrac{\sqrt{3}}{2} \qquad\qquad x - 1 = -\dfrac{\sqrt{3}}{2}$ • Solve for x.

$\qquad x = 1 + \dfrac{\sqrt{3}}{2} \qquad\qquad x = 1 - \dfrac{\sqrt{3}}{2}$

$\qquad\quad = \dfrac{2 + \sqrt{3}}{2} \qquad\qquad\quad = \dfrac{2 - \sqrt{3}}{2}$

The solutions are $\dfrac{2 + \sqrt{3}}{2}$ and $\dfrac{2 - \sqrt{3}}{2}$.

Example 2

Solve by completing the square:
$x^2 + 4x + 5 = 0$

Solution

$x^2 + 4x + 5 = 0$

$\qquad x^2 + 4x = -5$

$x^2 + 4x + 4 = -5 + 4$ • Complete the square.

$\qquad (x + 2)^2 = -1$ • Factor.

$\qquad \sqrt{(x+2)^2} = \sqrt{-1}$ • Take square roots.

$\qquad x + 2 = \pm i$

$x + 2 = i \qquad\qquad x + 2 = -i$ • Solve for x.

$\qquad x = -2 + i \qquad\qquad x = -2 - i$

The solutions are $-2 + i$ and $-2 - i$.

You Try It 1

Solve by completing the square:
$4x^2 - 4x - 1 = 0$

Your solution

You Try It 2

Solve by completing the square:
$x^2 + 4x + 8 = 0$

Your solution

Solutions on pp. S24–S25

8.2 Exercises

Objective A **To solve a quadratic equation by completing the square**

For Exercises 1 to 48, solve by completing the square.

1. $x^2 - 4x - 5 = 0$

2. $y^2 + 6y + 5 = 0$

3. $v^2 + 8v - 9 = 0$

4. $w^2 - 2w - 24 = 0$

5. $z^2 - 6z + 9 = 0$

6. $u^2 + 10u + 25 = 0$

7. $r^2 + 4r - 7 = 0$

8. $s^2 + 6s - 1 = 0$

9. $x^2 - 6x + 7 = 0$

10. $y^2 + 8y + 13 = 0$

11. $z^2 - 2z + 2 = 0$

12. $t^2 - 4t + 8 = 0$

13. $s^2 - 5s - 24 = 0$

14. $v^2 + 7v - 44 = 0$

15. $x^2 + 5x - 36 = 0$

16. $y^2 - 9y + 20 = 0$

17. $p^2 - 3p + 1 = 0$

18. $r^2 - 5r - 2 = 0$

19. $t^2 - t - 1 = 0$

20. $u^2 - u - 7 = 0$

21. $y^2 - 6y = 4$

22. $w^2 + 4w = 2$

23. $x^2 = 8x - 15$

24. $z^2 = 4z - 3$

25. $v^2 = 4v - 13$

26. $x^2 = 2x - 17$

27. $p^2 + 6p = -13$

28. $x^2 + 4x = -20$

29. $y^2 - 2y = 17$

30. $x^2 + 10x = 7$

31. $z^2 = z + 4$

32. $r^2 = 3r - 1$

33. $x^2 + 13 = 2x$

34. $6v^2 - 7v = 3$

35. $4x^2 - 4x + 5 = 0$

36. $4t^2 - 4t + 17 = 0$

37. $9x^2 - 6x + 2 = 0$

38. $9y^2 - 12y + 13 = 0$

39. $2s^2 = 4s + 5$

40. $3u^2 = 6u + 1$

41. $2r^2 = 3 - r$

42. $2x^2 = 12 - 5x$

43. $y - 2 = (y - 3)(y + 2)$

44. $8s - 11 = (s - 4)(s - 2)$

45. $6t - 2 = (2t - 3)(t - 1)$

46. $2z + 9 = (2z + 3)(z + 2)$

47. $(x - 4)(x + 1) = x - 3$

48. $(y - 3)^2 = 2y + 10$

For Exercises 49 to 54, solve by completing the square. Approximate the solutions to the nearest thousandth.

49. $z^2 + 2z = 4$

50. $t^2 - 4t = 7$

51. $2x^2 = 4x - 1$

52. $3y^2 = 5y - 1$

53. $4z^2 + 2z = 1$

54. $4w^2 - 8w = 3$

APPLYING THE CONCEPTS

For Exercises 55 to 57, solve for x by completing the square.

55. $x^2 - ax - 2a^2 = 0$

56. $x^2 + 3ax - 4a^2 = 0$

57. $x^2 + 3ax - 10a^2 = 0$

58. **Sports** After a baseball is hit, the height h (in feet) of the ball above the ground t seconds after it is hit can be approximated by the equation $h = -16t^2 + 70t + 4$. Using this equation, determine when the ball will hit the ground. Round to the nearest hundredth. (*Hint:* The ball hits the ground when $h = 0$.)

59. **Sports** After a baseball is hit, there are two equations that can be considered. One gives the height h (in feet) the ball is above the ground t seconds after it is hit. The second is the distance s (in feet) the ball is from home plate t seconds after it is hit. A model of this situation is given by $h = -16t^2 + 70t + 4$ and $s = 44.5t$. Using this model, determine whether the ball will clear a fence 325 ft from home plate.

60. ✎ Explain how to complete the square on $x^2 + bx$.

8.3 Solving Quadratic Equations by Using the Quadratic Formula

Objective A **To solve a quadratic equation by using the quadratic formula**

A general formula known as the **quadratic formula** can be derived by applying the method of completing the square to the standard form of a quadratic equation. This formula can be used to solve any quadratic equation. The equation $ax^2 + bx + c = 0$ is solved by completing the square as follows.

$$ax^2 + bx + c = 0$$

Add the opposite of the constant term to each side of the equation.

$$ax^2 + bx + c + (-c) = 0 + (-c)$$

$$ax^2 + bx = -c$$

Multiply each side of the equation by the reciprocal of a, the coefficient of x^2.

$$\frac{1}{a}(ax^2 + bx) = \frac{1}{a}(-c)$$

$$x^2 + \frac{b}{a}x = -\frac{c}{a}$$

Complete the square by adding $\left(\frac{1}{2} \cdot \frac{b}{a}\right)^2 = \frac{b^2}{4a^2}$ to each side of the equation.

$$x^2 + \frac{b}{a}x + \frac{b^2}{4a^2} = \frac{b^2}{4a^2} - \frac{c}{a}$$

Simplify the right side of the equation.

$$x^2 + \frac{b}{a}x + \frac{b^2}{4a^2} = \frac{b^2}{4a^2} - \left(\frac{c}{a} \cdot \frac{4a}{4a}\right)$$

$$x^2 + \frac{b}{a}x + \frac{b^2}{4a^2} = \frac{b^2}{4a^2} - \frac{4ac}{4a^2}$$

$$x^2 + \frac{b}{a}x + \frac{b^2}{4a^2} = \frac{b^2 - 4ac}{4a^2}$$

Factor the perfect-square trinomial on the left side of the equation.

$$\left(x + \frac{b}{2a}\right)^2 = \frac{b^2 - 4ac}{4a^2}$$

Take the square root of each side of the equation.

$$\sqrt{\left(x + \frac{b}{2a}\right)^2} = \sqrt{\frac{b^2 - 4ac}{4a^2}}$$

$$x + \frac{b}{2a} = \pm\frac{\sqrt{b^2 - 4ac}}{2a}$$

Solve for x.

$$x + \frac{b}{2a} = \frac{\sqrt{b^2 - 4ac}}{2a} \qquad x + \frac{b}{2a} = -\frac{\sqrt{b^2 - 4ac}}{2a}$$

$$x = -\frac{b}{2a} + \frac{\sqrt{b^2 - 4ac}}{2a} \qquad x = -\frac{b}{2a} - \frac{\sqrt{b^2 - 4ac}}{2a}$$

$$= \frac{-b + \sqrt{b^2 - 4ac}}{2a} \qquad = \frac{-b - \sqrt{b^2 - 4ac}}{2a}$$

Point of Interest

Although mathematicians have studied quadratic equations since around 500 B.C., it was not until the 18th century that the formula was written as it is today. Of further note, the word *quadratic* has the same Latin root as the word *square*.

The Quadratic Formula

The solutions of $ax^2 + bx + c = 0$, $a \neq 0$, are

$$\frac{-b + \sqrt{b^2 - 4ac}}{2a} \quad \text{and} \quad \frac{-b - \sqrt{b^2 - 4ac}}{2a}$$

The quadratic formula is frequently written as $x = \dfrac{-b \pm \sqrt{b^2 - 4ac}}{2a}$.

HOW TO Solve by using the quadratic formula: $2x^2 + 5x + 3 = 0$

$$x = \frac{-b \pm \sqrt{b^2 - 4ac}}{2a}$$

$$= \frac{-(5) \pm \sqrt{(5)^2 - 4(2)(3)}}{2(2)}$$

$$= \frac{-5 \pm \sqrt{25 - 24}}{4}$$

$$= \frac{-5 \pm \sqrt{1}}{4} = \frac{-5 \pm 1}{4}$$

- The equation $2x^2 + 5x + 3 = 0$ is in standard form. $a = 2$, $b = 5$, $c = 3$
- Replace a, b, and c in the quadratic formula wth these values.

$$x = \frac{-5 + 1}{4} = \frac{-4}{4} = -1 \qquad x = \frac{-5 - 1}{4} = \frac{-6}{4} = -\frac{3}{2}$$

The solutions are -1 and $-\dfrac{3}{2}$.

HOW TO Solve $3x^2 = 4x + 6$ by using the quadratic formula. Find the exact solutions and approximate the solutions to the nearest thousandth.

$$3x^2 = 4x + 6$$
$$3x^2 - 4x - 6 = 0$$

$$x = \frac{-b \pm \sqrt{b^2 - 4ac}}{2a}$$

$$= \frac{-(-4) \pm \sqrt{(-4)^2 - 4(3)(-6)}}{2(3)}$$

$$= \frac{4 \pm \sqrt{16 - (-72)}}{6}$$

$$= \frac{4 \pm \sqrt{88}}{6} = \frac{4 \pm 2\sqrt{22}}{6}$$

$$= \frac{2(2 \pm \sqrt{22})}{2 \cdot 3} = \frac{2 \pm \sqrt{22}}{3}$$

- Write the equation in standard form. Subtract $4x$ and 6 from each side of the equation. $a = 3$, $b = -4$, $c = -6$
- Replace a, b, and c in the quadratic formula with these values.

Check:

$$3x^2 = 4x + 6$$

$$\begin{array}{c|c} 3\left(\dfrac{2 + \sqrt{22}}{3}\right)^2 & 4\left(\dfrac{2 + \sqrt{22}}{3}\right) + 6 \\ 3\left(\dfrac{4 + 4\sqrt{22} + 22}{9}\right) & \dfrac{8}{3} + \dfrac{4\sqrt{22}}{3} + \dfrac{18}{3} \\ 3\left(\dfrac{26 + 4\sqrt{22}}{9}\right) & \dfrac{26}{3} + \dfrac{4\sqrt{22}}{3} \\ \dfrac{26 + 4\sqrt{22}}{3} = & \dfrac{26 + 4\sqrt{22}}{3} \end{array}$$

$$3x^2 = 4x + 6$$

$$\begin{array}{c|c} 3\left(\dfrac{2 - \sqrt{22}}{3}\right)^2 & 4\left(\dfrac{2 - \sqrt{22}}{3}\right) + 6 \\ 3\left(\dfrac{4 - 4\sqrt{22} + 22}{9}\right) & \dfrac{8}{3} - \dfrac{4\sqrt{22}}{3} + \dfrac{18}{3} \\ 3\left(\dfrac{26 - 4\sqrt{22}}{9}\right) & \dfrac{26}{3} - \dfrac{4\sqrt{22}}{3} \\ \dfrac{26 - 4\sqrt{22}}{3} = & \dfrac{26 - 4\sqrt{22}}{3} \end{array}$$

The exact solutions are $\dfrac{2 + \sqrt{22}}{3}$ and $\dfrac{2 - \sqrt{22}}{3}$.

$$\frac{2 + \sqrt{22}}{3} \approx 2.230 \qquad \frac{2 - \sqrt{22}}{3} \approx -0.897$$

To the nearest thousandth, the solutions are 2.230 and -0.897.

Integrating Technology

See the Projects and Group Activities at the end of this chapter for instructions on using a graphing calculator to solve a quadratic equation. Instructions are also provided in the Keystroke Guide: *Zero*.

HOW TO Solve by using the quadratic formula: $4x^2 = 8x - 13$

$$4x^2 = 8x - 13$$

$$4x^2 - 8x + 13 = 0$$

$$x = \frac{-b \pm \sqrt{b^2 - 4ac}}{2a}$$

$$= \frac{-(-8) \pm \sqrt{(-8)^2 - 4 \cdot 4 \cdot 13}}{2 \cdot 4}$$

$$= \frac{8 \pm \sqrt{64 - 208}}{8} = \frac{8 \pm \sqrt{-144}}{8}$$

$$= \frac{8 \pm 12i}{8} = \frac{2 \pm 3i}{2}$$

- Write the equation in standard form.
- Use the quadratic formula.
- $a = 4, b = -8, c = 13$

The solutions are $1 + \frac{3}{2}i$ and $1 - \frac{3}{2}i$.

Of the three preceding examples, the first two had real number solutions; the last one had complex number solutions.

In the quadratic formula, the quantity $b^2 - 4ac$ is called the **discriminant.** When a, b, and c are real numbers, the discriminant determines whether a quadratic equation will have a double root, two real number solutions that are not equal, or two complex number solutions.

The Effect of the Discriminant on the Solutions of a Quadratic Equation

1. If $b^2 - 4ac = 0$, the equation has one real number solution, a double root.

2. If $b^2 - 4ac > 0$, the equation has two unequal real number solutions.

3. If $b^2 - 4ac < 0$, the equation has two complex number solutions.

HOW TO Use the discriminant to determine whether $x^2 - 4x - 5 = 0$ has one real number solution, two real number solutions, or two complex number solutions.

$$b^2 - 4ac$$

$$(-4)^2 - 4(1)(-5) = 16 + 20 = 36$$

$$36 > 0$$

- Evaluate the discriminant.
 $a = 1, b = -4, c = -5$

Because $b^2 - 4ac > 0$, the equation has two real number solutions.

Example 1 Solve by using the quadratic formula: $2x^2 - x + 5 = 0$

Solution

$$2x^2 - x + 5 = 0$$

$$a = 2, b = -1, c = 5$$

$$x = \frac{-b \pm \sqrt{b^2 - 4ac}}{2a}$$

$$= \frac{-(-1) \pm \sqrt{(-1)^2 - 4(2)(5)}}{2 \cdot 2}$$

$$= \frac{1 \pm \sqrt{1 - 40}}{4} = \frac{1 \pm \sqrt{-39}}{4}$$

$$= \frac{1 \pm i\sqrt{39}}{4}$$

The solutions are $\frac{1}{4} + \frac{\sqrt{39}}{4}i$ and $\frac{1}{4} - \frac{\sqrt{39}}{4}i$.

You Try It 1 Solve by using the quadratic formula: $x^2 - 2x + 10 = 0$

Your solution

Solution on p. S25

Example 2

Solve by using the quadratic formula:
$2x^2 = (x - 2)(x - 3)$

Solution

$2x^2 = (x - 2)(x - 3)$

$2x^2 = x^2 - 5x + 6$ • **Write in standard form.**

$x^2 + 5x - 6 = 0$

$a = 1, b = 5, c = -6$

$x = \dfrac{-b \pm \sqrt{b^2 - 4ac}}{2a}$

$ = \dfrac{-5 \pm \sqrt{5^2 - 4(1)(-6)}}{2 \cdot 1}$

$ = \dfrac{-5 \pm \sqrt{25 + 24}}{2} = \dfrac{-5 \pm \sqrt{49}}{2}$

$ = \dfrac{-5 \pm 7}{2}$

$x = \dfrac{-5 + 7}{2} \qquad x = \dfrac{-5 - 7}{2}$

$ = \dfrac{2}{2} = 1 \qquad = \dfrac{-12}{2} = -6$

The solutions are 1 and -6.

You Try It 2

Solve by using the quadratic formula:
$4x^2 = 4x - 1$

Your solution

Example 3

Use the discriminant to determine whether $4x^2 - 2x + 5 = 0$ has one real number solution, two real number solutions, or two complex number solutions.

Solution

$a = 4, b = -2, c = 5$

$b^2 - 4ac = (-2)^2 - 4(4)(5)$

$ = 4 - 80$

$ = -76$

$-76 < 0$

Because the discriminant is less than zero, the equation has two complex number solutions.

You Try It 3

Use the discriminant to determine whether $3x^2 - x - 1 = 0$ has one real number solution, two real number solutions, or two complex number solutions.

Your solution

Solutions on p. S25

8.3 Exercises

Objective A **To solve a quadratic equation by using the quadratic formula**

1. Write the quadratic formula. What does each variable in the formula represent?

2. Write the expression that appears under the radical symbol in the quadratic formula. What is this quantity called? What can it be used to determine?

For Exercises 3 to 29, solve by using the quadratic formula.

3. $x^2 - 3x - 10 = 0$

4. $z^2 - 4z - 8 = 0$

5. $y^2 + 5y - 36 = 0$

6. $z^2 - 3z - 40 = 0$

7. $w^2 = 8w + 72$

8. $t^2 = 2t + 35$

9. $v^2 = 24 - 5v$

10. $x^2 = 18 - 7x$

11. $2y^2 + 5y - 1 = 0$

12. $4p^2 - 7p + 1 = 0$

13. $8s^2 = 10s + 3$

14. $12t^2 = 5t + 2$

15. $x^2 = 14x - 4$

16. $v^2 = 12v - 24$

17. $2z^2 - 2z - 1 = 0$

18. $6w^2 = 9w - 1$

19. $z^2 + 2z + 2 = 0$

20. $p^2 - 4p + 5 = 0$

21. $y^2 - 2y + 5 = 0$

22. $x^2 + 6x + 13 = 0$

23. $s^2 - 4s + 13 = 0$

24. $t^2 - 6t + 10 = 0$

25. $2w^2 - 2w - 5 = 0$

26. $2v^2 + 8v + 3 = 0$

27. $2x^2 + 6x + 5 = 0$

28. $2y^2 + 2y + 13 = 0$

29. $4t^2 - 6t + 9 = 0$

For Exercises 30 to 35, solve by using the quadratic formula. Approximate the solutions to the nearest thousandth.

30. $x^2 - 6x - 6 = 0$

31. $p^2 - 8p + 3 = 0$

32. $r^2 - 2r = 4$

33. $w^2 + 4w = 1$

34. $3t^2 = 7t + 1$

35. $2y^2 = y + 5$

For Exercises 36 to 41, use the discriminant to determine whether the quadratic equation has one real number solution, two real number solutions, or two complex number solutions.

36. $2z^2 - z + 5 = 0$

37. $3y^2 + y + 1 = 0$

38. $9x^2 - 12x + 4 = 0$

39. $4x^2 + 20x + 25 = 0$

40. $2v^2 - 3v - 1 = 0$

41. $3w^2 + 3w - 2 = 0$

APPLYING THE CONCEPTS

42. Sports The height h, in feet, of a baseball above the ground t seconds after it is hit by a Little Leaguer can be approximated by the equation $h = -0.01t^2 + 2t + 3.5$. Does the ball reach a height of 100 ft?

3.5 ft

43. Sports The height h, in feet, of an arrow shot upward can be approximated by the equation $h = 128t - 16t^2$, where t is the time in seconds. Does the arrow reach a height of 275 ft?

44. Name the three methods of solving a quadratic equation that have been discussed. What are the advantages and disadvantages of each?

For what values of p do the quadratic equations in Exercises 45 and 46 have two real number solutions that are not equal? Write the answer in set-builder notation.

45. $x^2 - 6x + p = 0$

46. $x^2 + 10x + p = 0$

For what values of p do the quadratic equations in Exercises 47 and 48 have two complex number solutions? Write the answer in interval notation.

47. $x^2 - 2x + p = 0$

48. $x^2 + 4x + p = 0$

49. Find all values of x that satisfy the equation $x^2 + ix + 2 = 0$.

50. Show that the equation $x^2 + bx - 1 = 0$, $b \in$ real numbers, always has real number solutions regardless of the value of b.

51. Show that the equation $2x^2 + bx - 2 = 0$, $b \in$ real numbers, always has real number solutions regardless of the value of b.

8.4 Solving Equations That Are Reducible to Quadratic Equations

Objective A **To solve an equation that is quadratic in form**

Certain equations that are not quadratic can be expressed in quadratic form by making suitable substitutions. An equation is **quadratic in form** if it can be written as $au^2 + bu + c = 0$.

The equation $x^4 - 4x^2 - 5 = 0$ is quadratic in form.

$$x^4 - 4x^2 - 5 = 0$$
$$(x^2)^2 - 4(x^2) - 5 = 0$$
$$u^2 - 4u - 5 = 0 \qquad \bullet \text{ Let } x^2 = u.$$

The equation $y - y^{1/2} - 6 = 0$ is quadratic in form.

$$y - y^{1/2} - 6 = 0$$
$$(y^{1/2})^2 - (y^{1/2}) - 6 = 0$$
$$u^2 - u - 6 = 0 \qquad \bullet \text{ Let } y^{1/2} = u.$$

The key to recognizing equations that are quadratic in form is as follows. When the equation is written in standard form, the exponent on one variable term is $\frac{1}{2}$ the exponent on the other variable term.

HOW TO Solve: $z + 7z^{1/2} - 18 = 0$

$$z + 7z^{1/2} - 18 = 0 \qquad \bullet \text{ The equation is quadratic in form.}$$
$$(z^{1/2})^2 + 7(z^{1/2}) - 18 = 0$$
$$u^2 + 7u - 18 = 0 \qquad \bullet \text{ Let } z^{1/2} = u.$$

$$(u - 2)(u + 9) = 0 \qquad \bullet \text{ Solve by factoring.}$$

$$\begin{array}{ll} u - 2 = 0 & u + 9 = 0 \\ u = 2 & u = -9 \end{array}$$

$$\begin{array}{ll} z^{1/2} = 2 & z^{1/2} = -9 \\ \sqrt{z} = 2 & \sqrt{z} = -9 \end{array} \qquad \bullet \text{ Replace } u \text{ by } z^{1/2}.$$

$$\begin{array}{ll} (\sqrt{z})^2 = 2^2 & (\sqrt{z})^2 = (-9)^2 \\ z = 4 & z = 81 \end{array} \qquad \bullet \text{ Solve for } z.$$

Check each solution.

Check:

$$\begin{array}{c|c} z + 7z^{1/2} - 18 = 0 & \\ \hline 4 + 7(4)^{1/2} - 18 & 0 \\ 4 + 7 \cdot 2 - 18 & 0 \\ 4 + 14 - 18 & 0 \\ & 0 = 0 \end{array}$$

$$\begin{array}{c|c} z + 7z^{1/2} - 18 = 0 & \\ \hline 81 + 7(81)^{1/2} - 18 & 0 \\ 81 + 7 \cdot 9 - 18 & 0 \\ 81 + 63 - 18 & 0 \\ & 126 \neq 0 \end{array}$$

4 checks as a solution, but 81 does not check as a solution.

The solution is 4.

TAKE NOTE

When each side of an equation is squared, the resulting equation may have a solution that is not a solution of the original equation.

Example 1 Solve: $x^4 + x^2 - 12 = 0$

You Try It 1 Solve: $x - 5x^{1/2} + 6 = 0$

Solution

$$x^4 + x^2 - 12 = 0$$
$$(x^2)^2 + (x^2) - 12 = 0$$
$$u^2 + u - 12 = 0$$
$$(u - 3)(u + 4) = 0$$

$$u - 3 = 0 \qquad u + 4 = 0$$
$$u = 3 \qquad\quad u = -4$$

Replace u by x^2.

$$x^2 = 3 \qquad\qquad x^2 = -4$$
$$\sqrt{x^2} = \sqrt{3} \qquad \sqrt{x^2} = \sqrt{-4}$$
$$x = \pm\sqrt{3} \qquad\quad x = \pm 2i$$

The solutions are $\sqrt{3}$, $-\sqrt{3}$, $2i$, and $-2i$.

Your solution

Solution on p. S25

Objective B **To solve a radical equation that is reducible to a quadratic equation**

Certain equations containing radicals can be expressed as quadratic equations.

HOW TO Solve: $\sqrt{x + 2} + 4 = x$

$$\sqrt{x + 2} + 4 = x$$

$$\sqrt{x + 2} = x - 4$$ • Solve for the radical expression.

$$(\sqrt{x + 2})^2 = (x - 4)^2$$ • Square each side of the equation.

$$x + 2 = x^2 - 8x + 16$$ • Simplify.

$$0 = x^2 - 9x + 14$$ • Write the equation in standard form.

$$0 = (x - 7)(x - 2)$$ • Solve for x.

$$x - 7 = 0 \qquad x - 2 = 0$$
$$x = 7 \qquad\quad x = 2$$

Check each solution.

Check:
$$\frac{\sqrt{x + 2} + 4 = x}{\begin{array}{c|c} \sqrt{7 + 2} + 4 & 7 \\ \sqrt{9} + 4 & 7 \\ 3 + 4 & 7 \\ 7 = 7 \end{array}}$$

$$\frac{\sqrt{x + 2} + 4 = x}{\begin{array}{c|c} \sqrt{2 + 2} + 4 & 2 \\ \sqrt{4} + 4 & 2 \\ 2 + 4 & 2 \\ 6 \ne 2 \end{array}}$$

7 checks as a solution, but 2 does not check as a solution.

The solution is 7.

Example 2

Solve : $\sqrt{7y - 3} + 3 = 2y$

Solution

$\sqrt{7y - 3} + 3 = 2y$

 • Solve for the radical.

$\sqrt{7y - 3} = 2y - 3$

$(\sqrt{7y - 3})^2 = (2y - 3)^2$ • Square each side.

$7y - 3 = 4y^2 - 12y + 9$ • Write in standard form.

$0 = 4y^2 - 19y + 12$

$0 = (4y - 3)(y - 4)$ • Solve by factoring.

$4y - 3 = 0 \qquad y - 4 = 0$

$\quad\; 4y = 3 \qquad\quad\; y = 4$

$\quad\;\; y = \dfrac{3}{4}$

4 checks as a solution.

$\dfrac{3}{4}$ does not check as a solution.

The solution is 4.

You Try It 2

Solve: $\sqrt{2x + 1} + x = 7$

Your solution

Example 3

Solve: $\sqrt{2y + 1} - \sqrt{y} = 1$

Solution

$\sqrt{2y + 1} - \sqrt{y} = 1$

Solve for one of the radical expressions.

$\sqrt{2y + 1} = \sqrt{y} + 1$

$(\sqrt{2y + 1})^2 = (\sqrt{y} + 1)^2$ • Square each side.

$2y + 1 = y + 2\sqrt{y} + 1$

$\quad\;\; y = 2\sqrt{y}$

$\quad\;\; y^2 = (2\sqrt{y})^2$ • Square each side.

$\quad\;\; y^2 = 4y$

$y^2 - 4y = 0$

$y(y - 4) = 0$

$y = 0 \qquad y - 4 = 0$

$\qquad\qquad\quad y = 4$

0 and 4 check as solutions.

The solutions are 0 and 4.

You Try It 3

Solve: $\sqrt{2x - 1} + \sqrt{x} = 2$

Your solution

Solutions on p. S25

Objective C **To solve a fractional equation that is reducible to a quadratic equation**

After each side of a fractional equation has been multiplied by the LCM of the denominators, the resulting equation may be a quadratic equation.

HOW TO Solve: $\dfrac{1}{r} + \dfrac{1}{r+1} = \dfrac{3}{2}$

$$\frac{1}{r} + \frac{1}{r+1} = \frac{3}{2}$$

$$2r(r+1)\left(\frac{1}{r} + \frac{1}{r+1}\right) = 2r(r+1) \cdot \frac{3}{2}$$ • Multiply each side of the equation by the LCM of the denominators.

$$2(r+1) + 2r = r(r+1) \cdot 3$$
$$2r + 2 + 2r = 3r(r+1)$$
$$4r + 2 = 3r^2 + 3r$$
$$0 = 3r^2 - r - 2$$ • Write the equation in standard form.
$$0 = (3r+2)(r-1)$$ • Solve for r by factoring.

$$3r + 2 = 0 \qquad r - 1 = 0$$
$$3r = -2 \qquad r = 1$$
$$r = -\frac{2}{3}$$

$-\dfrac{2}{3}$ and 1 check as solutions.

The solutions are $-\dfrac{2}{3}$ and 1.

Example 4

Solve: $\dfrac{9}{x-3} = 2x + 1$

Solution

$$\frac{9}{x-3} = 2x + 1$$

$$(x-3)\frac{9}{x-3} = (x-3)(2x+1)$$ • Clear denominators.

$$9 = 2x^2 - 5x - 3$$ • Write in standard form.

$$0 = 2x^2 - 5x - 12$$
$$0 = (2x+3)(x-4)$$ • Solve by factoring.

$$2x + 3 = 0 \qquad x - 4 = 0$$
$$2x = -3 \qquad x = 4$$
$$x = -\frac{3}{2}$$

The solutions are $-\dfrac{3}{2}$ and 4.

You Try It 4

Solve: $3y + \dfrac{25}{3y-2} = -8$

Your solution

Solution on p. S26

8.4 Exercises

Objective A **To solve an equation that is quadratic in form**

For Exercises 1 to 18, solve.

1. $x^4 - 13x^2 + 36 = 0$

2. $y^4 - 5y^2 + 4 = 0$

3. $z^4 - 6z^2 + 8 = 0$

4. $t^4 - 12t^2 + 27 = 0$

5. $p - 3p^{1/2} + 2 = 0$

6. $v - 7v^{1/2} + 12 = 0$

7. $x - x^{1/2} - 12 = 0$

8. $w - 2w^{1/2} - 15 = 0$

9. $z^4 + 3z^2 - 4 = 0$

10. $y^4 + 5y^2 - 36 = 0$

11. $x^4 + 12x^2 - 64 = 0$

12. $x^4 - 81 = 0$

13. $p + 2p^{1/2} - 24 = 0$

14. $v + 3v^{1/2} - 4 = 0$

15. $y^{2/3} - 9y^{1/3} + 8 = 0$

16. $z^{2/3} - z^{1/3} - 6 = 0$

17. $9w^4 - 13w^2 + 4 = 0$

18. $4y^4 - 7y^2 - 36 = 0$

Objective B **To solve a radical equation that is reducible to a quadratic equation**

For Exercises 19 to 36, solve.

19. $\sqrt{x + 1} + x = 5$

20. $\sqrt{x - 4} + x = 6$

21. $x = \sqrt{x} + 6$

22. $\sqrt{2y - 1} = y - 2$

23. $\sqrt{3w + 3} = w + 1$

24. $\sqrt{2s + 1} = s - 1$

25. $\sqrt{4y + 1} - y = 1$

26. $\sqrt{3s + 4} + 2s = 12$

27. $\sqrt{10x + 5} - 2x = 1$

28. $\sqrt{t + 8} = 2t + 1$

29. $\sqrt{p + 11} = 1 - p$

30. $x - 7 = \sqrt{x - 5}$

31. $\sqrt{x - 1} - \sqrt{x} = -1$

32. $\sqrt{y} + 1 = \sqrt{y + 5}$

33. $\sqrt{2x - 1} = 1 - \sqrt{x - 1}$

34. $\sqrt{x + 6} + \sqrt{x + 2} = 2$

35. $\sqrt{t + 3} + \sqrt{2t + 7} = 1$

36. $\sqrt{5 - 2x} = \sqrt{2 - x} + 1$

Objective C **To solve a fractional equation that is reducible to a quadratic equation**

For Exercises 37 to 51, solve.

37. $x = \dfrac{10}{x - 9}$

38. $z = \dfrac{5}{z - 4}$

39. $\dfrac{t}{t + 1} = \dfrac{-2}{t - 1}$

40. $\dfrac{2v}{v - 1} = \dfrac{5}{v + 2}$

41. $\dfrac{y - 1}{y + 2} + y = 1$

42. $\dfrac{2p - 1}{p - 2} + p = 8$

43. $\dfrac{3r + 2}{r + 2} - 2r = 1$

44. $\dfrac{2v + 3}{v + 4} + 3v = 4$

45. $\dfrac{2}{2x + 1} + \dfrac{1}{x} = 3$

46. $\dfrac{3}{s} - \dfrac{2}{2s - 1} = 1$

47. $\dfrac{16}{z - 2} + \dfrac{16}{z + 2} = 6$

48. $\dfrac{2}{y + 1} + \dfrac{1}{y - 1} = 1$

49. $\dfrac{t}{t - 2} + \dfrac{2}{t - 1} = 4$

50. $\dfrac{4t + 1}{t + 4} + \dfrac{3t - 1}{t + 1} = 2$

51. $\dfrac{5}{2p - 1} + \dfrac{4}{p + 1} = 2$

APPLYING THE CONCEPTS

52. Solve: $(x^2 - 7)^{1/2} = (x - 1)^{1/2}$

53. Solve: $(\sqrt{x} - 2)^2 - 5\sqrt{x} + 14 = 0$ (*Hint:* Let $u = \sqrt{x} - 2$.)

54. Solve: $(\sqrt{x} + 3)^2 - 4\sqrt{x} - 17 = 0$ (*Hint:* Let $u = \sqrt{x} + 3$.)

55. **Mathematics** The fourth power of a number is twenty-five less than ten times the square of the number. Find the number.

8.5 Quadratic Inequalities and Rational Inequalities

Objective A **To solve a nonlinear inequality**

A **quadratic inequality** is one that can be written in the form $ax^2 + bx + c < 0$ or $ax^2 + bx + c > 0$, where $a \neq 0$. The symbols \leq and \geq can also be used. The solution set of a quadratic inequality can be found by solving a compound inequality.

To solve $x^2 - 3x - 10 > 0$, first factor the trinomial.

$$x^2 - 3x - 10 > 0$$
$$(x + 2)(x - 5) > 0$$

There are two cases for which the product of the factors will be positive: (1) both factors are positive, or (2) both factors are negative.

(1) $x + 2 > 0$ and $x - 5 > 0$
(2) $x + 2 < 0$ and $x - 5 < 0$

Solve each pair of compound inequalities.

(1) $x + 2 > 0$ and $x - 5 > 0$
$x > -2$ $x > 5$
$\{x \,|\, x > -2\} \cap \{x \,|\, x > 5\} = \{x \,|\, x > 5\}$

(2) $x + 2 < 0$ and $x - 5 < 0$
$x < -2$ $x < 5$
$\{x \,|\, x < -2\} \cap \{x \,|\, x < 5\} = \{x \,|\, x < -2\}$

Because the two cases for which the product will be positive are connected by *or*, the solution set is the union of the solution sets of the individual inequalities.

$$\{x \,|\, x > 5\} \cup \{x \,|\, x < -2\} = \{x \,|\, x > 5 \text{ or } x < -2\}$$

Although the solution set of any quadratic inequality can be found by using the method outlined above, a graphical method is often easier to use.

HOW TO Solve and graph the solution set of $x^2 - x - 6 < 0$.

Factor the trinomial.

$$x^2 - x - 6 < 0$$
$$(x - 3)(x + 2) < 0$$

On a number line, draw lines indicating the numbers that make each factor equal to zero.

$x - 3 = 0 \qquad x + 2 = 0$
$ x = 3 \qquad x = -2$

TAKE NOTE

For each factor, choose a number in each region. For example: when $x = -4$, $x - 3$ is negative; when $x = 1$, $x - 3$ is negative; and when $x = 4$, $x - 3$ is positive. When $x = -4$, $x + 2$ is negative; when $x = 1$, $x + 2$ is positive; and when $x = 4$, $x + 2$ is positive.

For each factor, place plus signs above the number line for those regions where the factor is positive and minus signs where the factor is negative.

Because $x^2 - x - 6 < 0$, the solution set will be the regions where one factor is positive and the other factor is negative.

Write the solution set.

$$\{x \,|\, -2 < x < 3\}$$

The graph of the solution set of $x^2 - x - 6 < 0$ is shown at the right.

HOW TO Solve and graph the solution set of $(x - 2)(x + 1)(x - 4) > 0$.

On a number line, identify for each factor the regions where the factor is positive and those where the factor is negative.

There are two regions where the product of the three factors is positive.

Write the solution set.

$\{x \mid -1 < x < 2 \text{ or } x > 4\}$

The graph of the solution set of $(x - 2)(x + 1)(x - 4) > 0$ is shown at the right.

HOW TO Solve: $\dfrac{2x - 5}{x - 4} \le 1$

Rewrite the inequality so that 0 appears on the right side of the inequality.

$$\dfrac{2x - 5}{x - 4} \le 1$$

$$\dfrac{2x - 5}{x - 4} - 1 \le 0$$

Simplify.

$$\dfrac{2x - 5}{x - 4} - \dfrac{x - 4}{x - 4} \le 0$$

$$\dfrac{x - 1}{x - 4} \le 0$$

On a number line, identify for each factor of the numerator and each factor of the denominator the regions where the factor is positive and those where the factor is negative.

The region where the quotient of the two factors is negative is between 1 and 4.

Write the solution set. $\{x \mid 1 \le x < 4\}$

Note that 1 is part of the solution set but 4 is not because the denominator of the rational expression is zero when $x = 4$.

Example 1

Solve and graph the solution set of
$2x^2 - x - 3 \ge 0$.

Solution

$$2x^2 - x - 3 \ge 0$$
$$(2x - 3)(x + 1) \ge 0$$

$$\left\{ x \mid x \le -1 \text{ or } x \ge \dfrac{3}{2} \right\}$$

You Try It 1

Solve and graph the solution set of
$2x^2 - x - 10 \le 0$.

Your solution

Solution on p. S26

8.5 Exercises

Objective A **To solve a nonlinear inequality**

1. If $(x - 3)(x - 5) > 0$, what must be true of the values of $x - 3$ and $x - 5$?

2. For the inequality $\frac{x - 2}{x - 3} \leq 1$, which of the values 1, 2, and 3 is not a possible element of the solution set? Why?

For Exercises 3 to 18, solve and graph the solution set.

3. $(x - 4)(x + 2) > 0$

4. $(x + 1)(x - 3) > 0$

5. $x^2 - 3x + 2 \geq 0$

6. $x^2 + 5x + 6 > 0$

7. $x^2 - x - 12 < 0$

8. $x^2 + x - 20 < 0$

9. $(x - 1)(x + 2)(x - 3) < 0$

10. $(x + 4)(x - 2)(x + 1) > 0$

11. $(x + 4)(x - 2)(x - 1) \geq 0$

12. $(x - 1)(x + 5)(x - 2) \leq 0$

13. $\dfrac{x - 4}{x + 2} > 0$

14. $\dfrac{x + 2}{x - 3} > 0$

15. $\dfrac{x - 3}{x + 1} \leq 0$

16. $\dfrac{x - 1}{x} > 0$

17. $\dfrac{(x-1)(x+2)}{x-3} \le 0$

18. $\dfrac{(x+3)(x-1)}{x-2} \ge 0$

For Exercises 19 to 33, solve.

19. $x^2 - 16 > 0$

20. $x^2 - 4 \ge 0$

21. $x^2 - 9x \le 36$

22. $x^2 + 4x > 21$

23. $4x^2 - 8x + 3 < 0$

24. $2x^2 + 11x + 12 \ge 0$

25. $\dfrac{3}{x-1} < 2$

26. $\dfrac{x}{(x-1)(x+2)} \ge 0$

27. $\dfrac{x-2}{(x+1)(x-1)} \le 0$

28. $\dfrac{1}{x} < 2$

29. $\dfrac{x}{2x-1} \ge 1$

30. $\dfrac{x}{2x-3} \le 1$

31. $\dfrac{x}{2-x} \le -3$

32. $\dfrac{3}{x-2} > \dfrac{2}{x+2}$

33. $\dfrac{3}{x-5} > \dfrac{1}{x+1}$

APPLYING THE CONCEPTS

For Exercises 34 to 39, graph the solution set.

34. $(x+2)(x-3)(x+1)(x+4) > 0$

35. $(x-1)(x+3)(x-2)(x-4) \ge 0$

36. $(x^2 + 2x - 8)(x^2 - 2x - 3) < 0$

37. $(x^2 + 2x - 3)(x^2 + 3x + 2) \ge 0$

38. $(x^2 + 1)(x^2 - 3x + 2) > 0$

39. $\dfrac{x^2(3-x)(2x+1)}{(x+4)(x+2)} \ge 0$

8.6 Applications of Quadratic Equations

Objective A **To solve application problems**

The application problems in this section are similar to problems solved earlier in the text. Each of the strategies for the problems in this section will result in a quadratic equation.

> **HOW TO** A small pipe takes 16 min longer to empty a tank than does a larger pipe. Working together, the pipes can empty the tank in 6 min. How long would it take each pipe working alone to empty the tank?

Empties tank in t minutes

Empties tank in $(t + 16)$ minutes

Empties $\frac{1}{t}$ of the tank each minute

Empties $\frac{1}{t+16}$ of the tank each minute

Strategy for Solving an Application Problem

1. Determine the type of problem. Is it a uniform motion problem, a geometry problem, an integer problem, or a work problem?

The problem is a work problem.

2. Choose a variable to represent the unknown quantity. Write numerical or variable expressions for all the remaining quantities. These results can be recorded in a table.

The unknown time of the larger pipe: t
The unknown time of the smaller pipe: $t + 16$

	Rate of Work	·	*Time Worked*	=	*Part of Task Completed*
Larger pipe	$\frac{1}{t}$	·	6	=	$\frac{6}{t}$
Smaller pipe	$\frac{1}{t+16}$	·	6	=	$\frac{6}{t+16}$

3. Determine how the quantities are related.

$$\frac{6}{t} + \frac{6}{t+16} = 1$$ • The sum of the parts of the task completed must equal 1.

$$t(t+16)\left(\frac{6}{t} + \frac{6}{t+16}\right) = t(t+16) \cdot 1$$
$$(t+16)6 + 6t = t^2 + 16t$$
$$6t + 96 + 6t = t^2 + 16t$$
$$0 = t^2 + 4t - 96$$
$$0 = (t+12)(t-8)$$

$t + 12 = 0 \qquad t - 8 = 0$
$\quad\quad t = -12 \qquad\quad t = 8$

Because time cannot be negative, the solution $t = -12$ is not possible.

The time for the smaller pipe is $t + 16$.

$t + 16 = 8 + 16 = 24$ • Replace *t* by 8 and evaluate.

The larger pipe requires 8 min to empty the tank.
The smaller pipe requires 24 min to empty the tank.

Example 1

In 8 h, two campers rowed 15 mi down a river and then rowed back to their campsite. The rate of the river's current was 1 mph. Find the rate at which the campers rowed.

Strategy
- This is a uniform motion problem.
- Unknown rowing rate of the campers: r

	Distance	Rate	Time
Down river	15	$r + 1$	$\dfrac{15}{r + 1}$
Up river	15	$r - 1$	$\dfrac{15}{r - 1}$

- The total time of the trip was 8 h.

Solution

$$\frac{15}{r + 1} + \frac{15}{r - 1} = 8$$

$$(r + 1)(r - 1)\left(\frac{15}{r + 1} + \frac{15}{r - 1}\right) = (r + 1)(r - 1)8$$

$$(r - 1)15 + (r + 1)15 = (r^2 - 1)8$$

$$15r - 15 + 15r + 15 = 8r^2 - 8$$

$$30r = 8r^2 - 8$$

$$0 = 8r^2 - 30r - 8$$

$$0 = 2(4r^2 - 15r - 4)$$

$$0 = 2(4r + 1)(r - 4)$$

$$
\begin{array}{ll}
4r + 1 = 0 & r - 4 = 0 \\
4r = -1 & r = 4 \\
r = -\dfrac{1}{4} &
\end{array}
$$

The solution $r = -\dfrac{1}{4}$ is not possible, because the rate cannot be a negative number.

The rowing rate was 4 mph.

You Try It 1

The length of a rectangle is 3 m more than the width. The area is 54 m². Find the length of the rectangle.

Your strategy

Your solution

Solution on p. S26

8.6 Exercises

Objective A **To solve application problems**

1. **Geometry** The length of the base of a triangle is 1 cm less than five times the height of the triangle. The area of the triangle is 21 cm². Find the height of the triangle and the length of the base of the triangle.

2. **Geometry** The length of a rectangle is 2 ft less than three times the width of the rectangle. The area of the rectangle is 65 ft². Find the length and width of the rectangle.

3. **Geography** The state of Colorado is almost perfectly rectangular, with its north border 111 mi longer than its west border. If the state encompasses 104,000 mi², estimate the dimensions of Colorado. Round to the nearest mile.

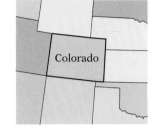

4. **Geometry** A square piece of cardboard is formed into a box by cutting 10-centimeter squares from each of the four corners and then folding up the sides, as shown in the figure. If the volume, V, of the box is to be 49,000 cm³, what size square piece of cardboard is needed? Recall that $V = LWH$.

10 cm
10 cm

5. **Safety** A car with good tire tread can stop in less distance than a car with poor tread. The formula for the stopping distance d, in feet, of a car with good tread on dry cement is approximated by $d = 0.04v^2 + 0.5v$, where v is the speed of the car. If the driver must be able to stop within 60 ft, what is the maximum safe speed, to the nearest mile per hour, of the car?

6. **Rockets** A model rocket is launched with an initial velocity of 200 ft/s. The height h, in feet, of the rocket t seconds after the launch is given by $h = -16t^2 + 200t$. How many seconds after the launch will the rocket be 300 ft above the ground? Round to the nearest hundredth of a second.

7. **Physics** The height of a projectile fired upward is given by the formula $s = v_0 t - 16t^2$, where s is the height in feet, v_0 is the initial velocity, and t is the time in seconds. Find the time for a projectile to return to Earth if it has an initial velocity of 200 ft/s.

8. **Physics** The height of a projectile fired upward is given by the formula $s = v_0 t - 16t^2$, where s is the height in feet, v_0 is the initial velocity, and t is the time in seconds. Find the time for a projectile to reach a height of 64 ft if it has an initial velocity of 128 ft/s. Round to the nearest hundredth of a second.

9. **Safety** In Germany, there is no speed limit on some portions of the autobahn (highway). Other portions have a speed limit of 180 km/h (approximately 112 mph). The distance d, in meters, required to stop a car traveling at v kilometers per hour is $d = 0.019v^2 + 0.69v$. Approximate, to the nearest tenth, the maximum speed a driver can be traveling and still be able to stop within 150 m.

German Autobahn System

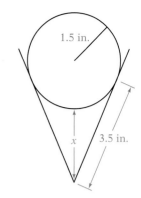

10. **Ice Cream** A perfectly spherical scoop of mint chocolate chip ice cream is placed in a cone, as shown in the figure. How far is the bottom of the scoop of ice cream from the bottom of the cone? Round to the nearest tenth. (*Hint:* A line segment from the center of the ice cream to the point at which the ice cream touches the cone is perpendicular to the edge of the cone.)

11. **Tanks** A small pipe can fill a tank in 6 min more time than it takes a larger pipe to fill the same tank. Working together, both pipes can fill the tank in 4 min. How long would it take each pipe working alone to fill the tank?

12. **Uniform Motion** A cruise ship made a trip of 100 mi in 8 h. The ship traveled the first 40 mi at a constant rate before increasing its speed by 5 mph. Then it traveled another 60 mi at the increased speed. Find the rate of the cruise ship for the first 40 mi.

13. **Uniform Motion** The Concorde's speed in calm air was 1320 mph. Flying with the wind, the Concorde could fly from New York to London, a distance of approximately 4000 mi, in 0.5 h less than the time required to make the return trip. Find the rate of the wind to the nearest mile per hour.

14. **Uniform Motion** A car travels 120 mi. A second car, traveling 10 mph faster than the first car, makes the same trip in 1 h less time. Find the speed of each car.

15. **Uniform Motion** For a portion of the Green River in Utah, the rate of the river's current is 4 mph. A tour guide can row 5 mi down this river and back in 3 h. Find the rowing rate of the guide in calm water.

16. **Construction** The height h, in feet, of an arch is given by the equation $h(x) = -\frac{3}{64}x^2 + 27$, where $|x|$ is the distance in feet from the center of the arch.

 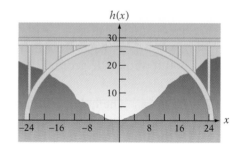

 a. What is the maximum height of the arch?
 b. What is the height of the arch 8 ft to the right of the center?
 c. How far from the center is the arch 8 ft tall? Round to the nearest hundredth.

APPLYING THE CONCEPTS

17. **Geometry** The surface area of the ice cream cone shown at the right is given by $A = \pi r^2 + \pi rs$, where r is the radius of the circular top of the cone and s is the slant height of the cone. If the surface area of the cone is 11.25π in^2 and the slant height is 6 in., find the radius of the cone.

18. **Physics** Using Torricelli's Principle, it can be shown that the depth d of a liquid in a bottle with a hole of area 0.5 cm^2 in its side can be approximated by $d = 0.0034t^2 - 0.52518t + 20$, where t is the time since a stopper was removed from the hole. When will the depth be 10 cm? Round to the nearest tenth of a second.

Focus on Problem Solving

Inductive and Deductive Reasoning

Consider the following sums of odd positive integers.

$1 + 3 = 4 = 2^2$ Sum of the first two odd numbers is 2^2.

$1 + 3 + 5 = 9 = 3^2$ Sum of the first three odd numbers is 3^2.

$1 + 3 + 5 + 7 = 16 = 4^2$ Sum of the first four odd numbers is 4^2.

$1 + 3 + 5 + 7 + 9 = 25 = 5^2$ Sum of the first five odd numbers is 5^2.

1. Make a conjecture about the value of the sum

$$1 + 3 + 5 + 7 + 9 + 11 + 13 + 15$$

without adding the numbers.

If the pattern continues, a possible conjecture is that the sum of the first eight odd positive integers is $8^2 = 64$. By adding the numbers, you can verify that this is correct. Inferring that the pattern established by a few cases is valid for all cases is an example of **inductive reasoning.**

2. Use inductive reasoning to find the next figure in the sequence below.

The fact that a pattern appears to be true does not *prove* that it is true for all cases. For example, consider the polynomial $n^2 - n + 41$. If we begin substituting positive integer values for n and evaluating the expression, we produce the following table.

n	$n^2 - n + 41$	
1	$1^2 - 1 + 41 = 41$	41 is a prime number.
2	$2^2 - 2 + 41 = 43$	43 is a prime number.
3	$3^2 - 3 + 41 = 47$	47 is a prime number.
4	$4^2 - 4 + 41 = 53$	53 is a prime number.

Even if we try a number like 30, we have $30^2 - 30 + 41 = 911$, and 911 is a prime number. Thus it appears that the value of this polynomial is a prime number for any value of n. However, the conjecture is not true. For instance, if $n = 41$, we have $41^2 - 41 + 41 = 41^2 = 1681$, which is not a prime number because it is divisible by 41. This illustrates that inductive reasoning may lead to incorrect conclusions and that an inductive proof must be available to prove conjectures. There is such a proof, called *mathematical induction,* that you may study in a future math course.

Now consider the true statement that the sum of the measures of the interior angles of a triangle is 180°. The figure at the left is a triangle; therefore, the sum of the measures of the interior angles must be 180°. This is an example of *deductive reasoning.* **Deductive reasoning** uses a rule or statement of fact to reach a conclusion.

$\angle A + \angle B + \angle C = 180°$

3. Use deductive reasoning to complete the following sentence. All even numbers are divisible by 2. Because 14,386 is an even number, . . .

4. a. Use the pattern given below to find the three missing terms in
___ ___ ___ 64.

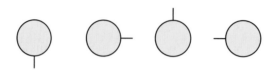

A 2 3 4 B 6 7 8 C 10 11 12 D 14 15 16...

 b. Is this inductive or deductive reasoning?

5. a. Draw the next figure in the sequence:

 b. Is this inductive or deductive reasoning?

6. a. If ◊◊◊◊◊ = ‡‡‡ and ‡‡‡ = ∇∇∇∇, how many ◊'s equal ∇∇∇∇∇∇∇∇?
 b. Is this inductive or deductive reasoning?

7. a. If ♠♠ = ♦♦♦♦♦♦, and ♦♦♦ = ♣♣, and ♣♣♣♣ = ♥, how many ♥'s equal ♠♠♠♠♠♠?
 b. Is this inductive or deductive reasoning?

For Exercises 8 to 10, determine whether inductive or deductive reasoning is being used.

8. The Atlanta Braves have won eight games in a row. Therefore, the Atlanta Braves will win their next game.

9. All quadrilaterals have four sides. A square is a quadrilateral. Therefore, a square has four sides.

10. Every English setter likes to hunt. Duke is an English setter, so Duke likes to hunt.

Projects and Group Activities

Using a Graphing Calculator to Solve a Quadratic Equation

Recall that an *x*-intercept of the graph of an equation is a point at which the graph crosses the *x*-axis. For the graph in Figure 1, the *x*-intercepts are $(-2, 0)$ and $(3, 0)$.

Recall also that to find the *x*-intercept of a graph, set $y = 0$ and then solve for x. For the equation in Figure 1, if we set $y = 0$, the resulting equation is $0 = x^2 - x - 6$, which is a quadratic equation. Solving this equation by factoring, we have

$$0 = x^2 - x - 6$$
$$0 = (x + 2)(x - 3)$$

$$x + 2 = 0 \qquad\qquad x - 3 = 0$$
$$x = -2 \qquad\qquad x = 3$$

Figure 1

Thus the solutions of the equation are the x-coordinates of the x-intercepts of the graph.

Figure 2

This connection between the solutions of an equation and the x-intercepts of a graph allows us to find approximations of the real number solutions of an equation graphically. For example, to use a TI-83 to approximate the solutions of $2x^2 + 5x - 1 = 0$ graphically, use the following keystrokes to graph $y = 2x^2 + 5x - 1$ and find the decimal approximations of the x-coordinates of the x-intercepts. Use the window shown in the graph in Figure 2.

Use the arrow keys to move to the left of the leftmost x-intercept. Press ENTER. Use the arrow keys to move to a point just to the right of the leftmost x-intercept. Press ENTER twice. The x-coordinate at the bottom of the screen is the approximation of one solution of the equation. To find the other solution, use the same procedure but move the cursor first to the left and then to the right of the rightmost x-intercept.

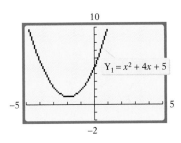

Figure 3

Attempting to find the solutions of an equation graphically will not necessarily yield all the solutions. Because the x-coordinates of the x-intercepts of a graph are *real* numbers, only real number solutions can be found. For instance, consider the equation $x^2 + 4x + 5 = 0$. The graph of $y = x^2 + 4x + 5$ is shown in Figure 3. Note that the graph has no x-intercepts and that consequently it has no real number solutions. However, $x^2 + 4x + 5 = 0$ does have complex number solutions that can be obtained by using the quadratic formula. They are $-2 + i$ and $-2 - i$.

Use a graphing calculator to approximate the solutions of the equations in Exercises 1 to 4.

1. $x^2 + 6x + 1 = 0$

2. $-x^2 + x + 3 = 0$

3. $2x^2 + 4x - 1 = 0$

4. $x^2 + 4x + 5 = 0$

Chapter 8 Summary

Key Words

A *quadratic equation* is an equation of the form $ax^2 + bx + c = 0$, where $a \neq 0$. A quadratic equation is also called a *second-degree equation*. A quadratic equation is in *standard form* when the polynomial is in descending order and equal to zero. [8.1A, p. 445]

When a quadratic equation has two solutions that are the same number, the solution is called a *double root* of the equation. [8.1A, p. 445]

Examples

$3x^2 + 4x - 7 = 0$, $x^2 - 1 = 0$, and $4x^2 + 8x = 0$ are quadratic equations. They are all written in standard form.

$$x^2 - 4x + 4 = 0$$
$$(x - 2)(x - 2) = 0$$
$$x - 2 = 0 \qquad x - 2 = 0$$
$$x = 2 \qquad x = 2$$
2 is a double root.

For an equation of the form $ax^2 + bx + c = 0$, the quantity $b^2 - 4ac$ is called the *discriminant*. [8.3A, p. 461]

$2x^2 - 3x + 5 = 0$
$a = 2, b = -3, c = 5$
$b^2 - 4ac = (-3)^2 - 4(2)(5)$
$\qquad\qquad = 9 - 40 = -31$

An equation is *quadratic in form* if it can be written as $au^2 + bu + c = 0$. [8.4A, p. 465]

$6x^4 - 5x^2 - 4 = 0$
$6(x^2)^2 - 5(x^2) - 4 = 0$
$6u^2 - 5u - 4 = 0$
The equation is quadratic in form.

A *quadratic inequality* is one that can be written in the form $ax^2 + bx + c < 0$ or $ax^2 + bx + c > 0$, where $a \neq 0$. The symbols \leq and \geq can also be used. [8.5A, p. 471]

$3x^2 + 5x - 8 \leq 0$ is a quadratic inequality.

Essential Rules and Procedures

Examples

Principle of Zero Products [8.1A, p. 445]
If a and b are real numbers and $ab = 0$, then $a = 0$ or $b = 0$.

$(x - 3)(x + 4) = 0$
$x - 3 = 0 \qquad x + 4 = 0$
$\qquad x = 3 \qquad\qquad x = -4$

To Solve a Quadratic Equation by Factoring [8.1A, p. 445]
1. Write the equation in standard form.
2. Factor the polynomial.
3. Use the Principle of Zero Products to set each factor equal to 0.
4. Solve each equation.
5. Check the solutions.

$x^2 + 2x = 35$
$x^2 + 2x - 35 = 0$
$(x - 5)(x + 7) = 0$
$x - 5 = 0 \qquad x + 7 = 0$
$\qquad x = 5 \qquad\qquad x = -7$

To Write a Quadratic Equation Given Its Solutions
[8.1B, p. 446]
Use the equation $(x - r_1)(x - r_2) = 0$. Replace r_1 with one solution and r_2 with the other solution. Then multiply the two factors.

Write a quadratic equation that has solutions -3 and 6.
$(x - r_1)(x - r_2) = 0$
$[x - (-3)](x - 6) = 0$
$(x + 3)(x - 6) = 0$
$x^2 - 3x - 18 = 0$

To Solve a Quadratic Equation by Taking Square Roots
[8.1C, pp. 447, 448]
1. Solve for x^2 or for $(x + a)^2$.
2. Take the square root of each side of the equation.
3. Simplify.
4. Check the solutions.

$$(x + 2)^2 - 9 = 0$$
$$(x + 2)^2 = 9$$
$$\sqrt{(x + 2)^2} = \sqrt{9}$$
$$x + 2 = \pm\sqrt{9}$$
$$x + 2 = \pm 3$$
$$x + 2 = 3 \qquad x + 2 = -3$$
$$x = 1 \qquad\qquad x = -5$$

To Complete the Square [8.2A, p. 453]

Add to a binomial of the form $x^2 + bx$ the square of $\frac{1}{2}$ of the coefficient of x, making it a perfect-square trinomial.

To complete the square on $x^2 - 8x$, add $\left[\frac{1}{2}(-8)\right]^2 = 16$: $x^2 - 8x + 16$.

To Solve a Quadratic Equation by Completing the Square
[8.2A, p. 455]
1. Write the equation in the form $ax^2 + bx = -c$.
2. Multiply both sides of the equation by $\frac{1}{a}$.
3. Complete the square on $x^2 + \frac{b}{a}x$. Add the number that completes the square to both sides of the equation.
4. Factor the perfect-square trinomial.
5. Take the square root of each side of the equation.
6. Solve the resulting equation for x.
7. Check the solutions.

$$x^2 + 4x - 1 = 0$$
$$x^2 + 4x = 1$$
$$x^2 + 4x + 4 = 1 + 4$$
$$(x + 2)^2 = 5$$
$$\sqrt{(x + 2)^2} = \sqrt{5}$$
$$x + 2 = \pm\sqrt{5}$$
$$x + 2 = \sqrt{5} \qquad x + 2 = -\sqrt{5}$$
$$x = -2 + \sqrt{5} \qquad x = -2 - \sqrt{5}$$

Quadratic Formula [8.3A, p. 459]

The solutions of $ax^2 + bx + c = 0$, $a \neq 0$, are $x = \frac{-b \pm \sqrt{b^2 - 4ac}}{2a}$.

$$2x^2 - 3x + 4 = 0$$
$$a = 2, b = -3, c = 4$$
$$x = \frac{-(-3) \pm \sqrt{(-3)^2 - 4(2)(4)}}{2(2)}$$
$$= \frac{3 \pm \sqrt{-23}}{4} = \frac{3 \pm i\sqrt{23}}{4}$$
$$= \frac{3}{4} \pm \frac{\sqrt{23}}{4}i$$

The Effect of the Discriminant on the Solutions of a Quadratic Equation [8.3A, p. 461]
1. If $b^2 - 4ac = 0$, the equation has one real number solution, a double root.
2. If $b^2 - 4ac > 0$, the equation has two unequal real number solutions.
3. If $b^2 - 4ac < 0$, the equation has two complex number solutions.

$x^2 + 8x + 16 = 0$ has a double root because $b^2 - 4ac = 8^2 - 4(1)(16) = 0$.
$2x^2 + 3x - 5 = 0$ has two unequal real number solutions because $b^2 - 4ac = 3^2 - 4(2)(-5) = 49$.
$3x^2 + 2x + 4 = 0$ has two complex number solutions because $b^2 - 4ac = 2^2 - 4(3)(4) = -44$.

Chapter 8 Review Exercises

1. Solve by factoring: $2x^2 - 3x = 0$

2. Solve for x by factoring: $6x^2 + 9cx = 6c^2$

3. Solve by taking square roots:
 $x^2 = 48$

4. Solve by taking square roots:
 $\left(x + \dfrac{1}{2}\right)^2 + 4 = 0$

5. Solve by completing the square:
 $x^2 + 4x + 3 = 0$

6. Solve by completing the square:
 $7x^2 - 14x + 3 = 0$

7. Solve by using the quadratic formula:
 $12x^2 - 25x + 12 = 0$

8. Solve by using the quadratic formula:
 $x^2 - x + 8 = 0$

9. Write a quadratic equation that has integer coefficients and has solutions 0 and -3.

10. Write a quadratic equation that has integer coefficients and has solutions $\dfrac{3}{4}$ and $-\dfrac{2}{3}$.

11. Solve by completing the square:
 $x^2 - 2x + 8 = 0$

12. Solve by completing the square:
 $(x - 2)(x + 3) = x - 10$

13. Solve by using the quadratic formula:
 $3x(x - 3) = 2x - 4$

14. Use the discriminant to determine whether $3x^2 - 5x + 1 = 0$ has one real number solution, two real number solutions, or two complex number solutions.

15. Solve: $(x + 3)(2x - 5) < 0$

16. Solve: $(x - 2)(x + 4)(2x + 3) \leq 0$

17. Solve: $x^{2/3} + x^{1/3} - 12 = 0$

18. Solve: $2(x - 1) + 3\sqrt{x - 1} - 2 = 0$

19. Solve: $3x = \dfrac{9}{x - 2}$

20. Solve: $\dfrac{3x + 7}{x + 2} + x = 3$

21. Solve and graph the solution set:
$\dfrac{x - 2}{2x - 3} \geq 0$

22. Solve and graph the solution set:
$\dfrac{(2x - 1)(x + 3)}{x - 4} \leq 0$

23. Solve: $x = \sqrt{x} + 2$

24. Solve: $2x = \sqrt{5x + 24} + 3$

25. Solve: $\dfrac{x - 2}{2x + 3} - \dfrac{x - 4}{x} = 2$

26. Solve: $1 - \dfrac{x + 4}{2 - x} = \dfrac{x - 3}{x + 2}$

27. Write a quadratic equation that has integer coefficients and has solutions $\dfrac{1}{3}$ and -3.

28. Solve by factoring: $2x^2 + 9x = 5$

29. Solve: $2(x + 1)^2 - 36 = 0$

30. Solve by using the quadratic formula: $x^2 + 6x + 10 = 0$

31. Solve: $\dfrac{2}{x - 4} + 3 = \dfrac{x}{2x - 3}$

32. Solve: $x^4 - 28x^2 + 75 = 0$

33. Solve: $\sqrt{2x - 1} + \sqrt{2x} = 3$

34. Solve: $2x^{2/3} + 3x^{1/3} - 2 = 0$

35. Solve: $\sqrt{3x - 2} + 4 = 3x$

36. Solve by completing the square:
$x^2 - 10x + 7 = 0$

37. Solve: $\dfrac{2x}{x - 4} + \dfrac{6}{x + 1} = 11$

38. Solve by using the quadratic formula:
$9x^2 - 3x = 1$

39. Solve: $2x = 4 - 3\sqrt{x - 1}$

40. Solve: $1 - \dfrac{x + 3}{3 - x} = \dfrac{x - 4}{x + 3}$

41. Use the discriminant to determine whether $2x^2 - 5x = 6$ has one real number solution, two real number solutions, or two complex number solutions.

42. Solve: $x^2 - 3x \le 10$

43. **Sports** To prepare for an upcoming race, a sculling crew rowed 16 mi down a river and back in 6 h. If the rate of the river's current is 2 mph, find the sculling crew's rate of rowing in calm water.

44. **Geometry** The length of a rectangle is two more than twice the width. The area of the rectangle is 60 cm². Find the length and width of the rectangle.

$2W + 2$

W

45. **Integers** The sum of the squares of three consecutive even integers is fifty-six. Find the three integers.

46. **Computers** An older computer requires 12 min longer to print the payroll than does a newer computer. Together the computers can print the payroll in 8 min. Find the time for the new computer working alone to print the payroll.

47. **Uniform Motion** A car travels 200 mi. A second car, making the same trip, travels 10 mph faster than the first car and makes the trip in 1 h less time. Find the speed of each car.

Chapter 8 Test

1. Solve by factoring: $3x^2 + 10x = 8$

2. Solve by factoring: $6x^2 - 5x - 6 = 0$

3. Write a quadratic equation that has integer coefficients and has solutions 3 and -3.

4. Write a quadratic equation that has integer coefficients and has solutions $\frac{1}{2}$ and -4.

5. Solve by taking square roots:
$3(x - 2)^2 - 24 = 0$

6. Solve by completing the square:
$x^2 - 6x - 2 = 0$

7. Solve by completing the square:
$3x^2 - 6x = 2$

8. Solve by using the quadratic formula:
$2x^2 - 2x = 1$

9. Solve by using the quadratic formula:
$x^2 + 4x + 12 = 0$

10. Use the discriminant to determine whether $3x^2 - 4x = 1$ has one real number solution, two real number solutions, or two complex number solutions.

11. Use the discriminant to determine whether $x^2 - 6x = -15$ has one real number solution, two real number solutions, or two complex number solutions.

12. Solve: $2x + 7x^{1/2} - 4 = 0$

13. Solve: $x^4 - 4x^2 + 3 = 0$

14. Solve: $\sqrt{2x + 1} + 5 = 2x$

15. Solve: $\sqrt{x - 2} = \sqrt{x} - 2$

16. Solve: $\dfrac{2x}{x - 3} + \dfrac{5}{x - 1} = 1$

17. Solve and graph the solution set of $(x - 2)(x + 4)(x - 4) < 0$.

<--+---+---+---+---+---+---+---+---+---+---+-->
$-5 \; -4 \; -3 \; -2 \; -1 \;\; 0 \;\; 1 \;\; 2 \;\; 3 \;\; 4 \;\; 5$

18. Solve and graph the solution set of $\dfrac{2x - 3}{x + 4} \le 0$.

<--+---+---+---+---+---+---+---+---+---+---+-->
$-5 \; -4 \; -3 \; -2 \; -1 \;\; 0 \;\; 1 \;\; 2 \;\; 3 \;\; 4 \;\; 5$

19. **Sports** A basketball player shoots at a basket that is 25 ft away. The height h, in feet, of the ball above the ground after t seconds is given by $h = -16t^2 + 32t + 6.5$. How many seconds after the ball is released does it hit the basket? *Note:* the basket is 10 ft off the ground. Round to the nearest hundredth.

20. **Uniform Motion** The rate of a river's current is 2 mph. A canoe was rowed 6 mi down the river and back in 4 h. Find the rowing rate in calm water.

Cumulative Review Exercises

1. Evaluate $2a^2 - b^2 \div c^2$ when $a = 3$, $b = -4$, and $c = -2$.

2. Solve: $\dfrac{2x - 3}{4} - \dfrac{x + 4}{6} = \dfrac{3x - 2}{8}$

3. Find the slope of the line containing the points $(3, -4)$ and $(-1, 2)$.

4. Find the equation of the line that contains the point $(1, 2)$ and is parallel to the line $x - y = 1$.

5. Factor: $-3x^3y + 6x^2y^2 - 9xy^3$

6. Factor: $6x^2 - 7x - 20$

7. Factor: $a^nx + a^ny - 2x - 2y$

8. Divide: $(3x^3 - 13x^2 + 10) \div (3x - 4)$

9. Simplify: $\dfrac{x^2 + 2x + 1}{8x^2 + 8x} \cdot \dfrac{4x^3 - 4x^2}{x^2 - 1}$

10. Find the distance between the points $(-2, 3)$ and $(2, 5)$.

11. Solve $S = \dfrac{n}{2}(a + b)$ for b.

12. Simplify: $-2i(7 - 4i)$

13. Simplify: $a^{-1/2}(a^{1/2} - a^{3/2})$

14. Simplify: $\dfrac{\sqrt[3]{8x^4y^5}}{\sqrt[3]{16xy^6}}$

15. Solve: $\dfrac{x}{x + 2} - \dfrac{4x}{x + 3} = 1$

16. Solve: $\dfrac{x}{2x + 3} - \dfrac{3}{4x^2 - 9} = \dfrac{x}{2x - 3}$

17. Solve: $x^4 - 6x^2 + 8 = 0$

18. Solve: $\sqrt{3x + 1} - 1 = x$

19. Solve: $|3x - 2| < 8$

20. Find the x- and y-intercepts of the graph of $6x - 5y = 15$.

21. Graph the solution set:
$$x + y \leq 3$$
$$2x - y < 4$$

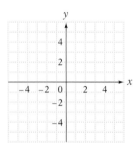

22. Solve by using Cramer's Rule.
$$x + y + z = 2$$
$$-x + 2y - 3z = -9$$
$$x - 2y - 2z = -1$$

23. Given $f(x) = \dfrac{2x - 3}{x^2 - 1}$, find $f(-2)$.

24. Find the domain of the function
$$f(x) = \dfrac{x - 2}{x^2 - 2x - 15}.$$

25. Solve and graph the solution set of $x^3 + x^2 - 6x < 0$.

26. Solve and graph the solution set of
$$\dfrac{(x - 1)(x - 5)}{x + 3} \geq 0.$$

27. Mechanics A piston rod for an automobile is $9\dfrac{3}{8}$ in. long with a tolerance of $\dfrac{1}{64}$ in. Find the lower and upper limits of the length of the piston rod.

28. Geometry The length of the base of a triangle is $(x + 8)$ ft. The height is $(2x - 4)$ ft. Find the area of the triangle in terms of the variable x.

29. Use the discriminant to determine whether $2x^2 + 4x + 3 = 0$ has one real number solution, two real number solutions, or two complex number solutions.

30. Depreciation The graph shows the relationship between the cost of a building and the depreciation allowed for income tax purposes. Find the slope of the line between the two points shown on the graph. Write a sentence that states the meaning of the slope.

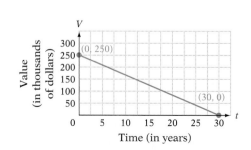

chapter

9 Functions and Relations

Due to constantly changing airfares and schedules, thousands of available vacation packages, and the vast amount of travel information on the Internet, travel planning can be frustrating and time-consuming. For assistance, tourists and businesspeople often turn to travel agents, who assess their needs and make the appropriate travel arrangements for them. Like any business, the travel agency's main goal is to earn a profit. Profit is the difference between a company's revenue (the total amount of money the company earns by selling its products or services) and its costs (the total amount of money the company spends in doing business). The **Projects and Group Activities on pages 532 and 533** use quadratic equations to determine the maximum profit companies can earn.

OBJECTIVES

Section 9.1

A To graph a quadratic function
B To find the *x*-intercepts of a parabola
C To find the minimum or maximum of a quadratic function
D To solve application problems

Section 9.2

A To graph functions

Section 9.3

A To perform operations on functions
B To find the composition of two functions

Section 9.4

A To determine whether a function is one-to-one
B To find the inverse of a function

Need help? For online student resources, such as section quizzes, visit this textbook's website at **math.college.hmco.com/students.**

Do these exercises to prepare for Chapter 9.

1. Evaluate $-\dfrac{b}{2a}$ for $b = -4$ and $a = 2$.

2. Given $y = -x^2 + 2x + 1$, find the value of y when $x = -2$.

3. Given $f(x) = x^2 - 3x + 2$, find $f(-4)$.

4. Evaluate $p(r) = r^2 - 5$ when $r = 2 + h$.

5. Solve: $0 = 3x^2 - 7x - 6$

6. Solve by using the quadratic formula: $0 = x^2 - 4x + 1$

7. Solve $x = 2y + 4$ for y.

8. Find the domain and range of the relation $\{(-2, 4), (3, 5), (4, 6), (6, 5)\}$. Is the relation a function?

9. What value is excluded from the domain of $f(x) = \dfrac{3}{x - 8}$?

10. Graph: $x = -2$

GO FIGURE ● ● ●

Each time the two hands of a certain standard, analog, 12-hour clock form a 180° angle, a bell chimes once. From noon today until noon tomorrow, how many chimes will be heard?

9.1 Properties of Quadratic Functions

Objective A To graph a quadratic function

Recall that a linear function is one that can be expressed by the equation $f(x) = mx + b$. The graph of a linear function has certain characteristics. It is a straight line with slope m and y-intercept $(0, b)$. A **quadratic function** is one that can be expressed by the equation $f(x) = ax^2 + bx + c$, $a \neq 0$. The graph of this function, called a **parabola,** also has certain characteristics. The graph of a quadratic function can be drawn by finding ordered pairs that belong to the function.

TAKE NOTE

Sometimes the value of the independent variable is called the **input** because it is *put in* place of the independent variable. The result of evaluating the function is called the **output.**

An **input/output table** shows the results of evaluating a function for various values of the independent variable. An input/output table for $f(x) = x^2 - 2x - 3$ is shown at the right.

HOW TO Graph $f(x) = x^2 - 2x - 3$.

By evaluating the function for various values of x, find enough ordered pairs to determine the shape of the graph.

x	$f(x) = x^2 - 2x - 3$	$f(x)$	(x, y)
-2	$f(-2) = (-2)^2 - 2(-2) - 3$	5	$(-2, 5)$
-1	$f(-1) = (-1)^2 - 2(-1) - 3$	0	$(-1, 0)$
0	$f(0) = (0)^2 - 2(0) - 3$	-3	$(0, -3)$
1	$f(1) = (1)^2 - 2(1) - 3$	-4	$(1, -4)$
2	$f(2) = (2)^2 - 2(2) - 3$	-3	$(2, -3)$
3	$f(3) = (3)^2 - 2(3) - 3$	0	$(3, 0)$
4	$f(4) = (4)^2 - 2(4) - 3$	5	$(4, 5)$

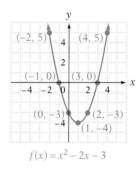

$f(x) = x^2 - 2x - 3$

Because the value of $f(x) = x^2 - 2x - 3$ is a real number for all values of x, the domain of f is all real numbers. From the graph, it appears that no value of y is less than -4. Thus the range is $\{y \,|\, y \geq -4\}$. The range can also be determined algebraically, as shown below, by completing the square.

TAKE NOTE

In completing the square, 1 is both added and subtracted. Because $1 - 1 = 0$, the expression $x^2 - 2x - 3$ is not changed. Note that

$(x - 1)^2 - 4$
$= (x^2 - 2x + 1) - 4$
$= x^2 - 2x - 3$

which is the original expression.

$f(x) = x^2 - 2x - 3$
$\quad = (x^2 - 2x) - 3$ • Group the variable terms.
$\quad = (x^2 - 2x + 1) - 1 - 3$ • Complete the square on $x^2 - 2x$. Add and subtract $\left[\dfrac{1}{2}(-2)\right]^2 = 1$ to and from $x^2 - 2x$.
$\quad = (x - 1)^2 - 4$ • Factor and combine like terms.

Because the square of a real number is always nonnegative, we have

$(x - 1)^2 \geq 0$
$(x - 1)^2 - 4 \geq -4$ • Subtract 4 from each side of the inequality.
$f(x) \geq -4$ • $f(x) = (x - 1)^2 - 4$
$y \geq -4$

From the last inequality, the range is $\{y \,|\, y \geq -4\}$.

$a > 0$

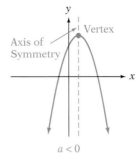

$a < 0$

In general, the graph of $f(x) = ax^2 + bx + c$, $a \neq 0$, resembles a "cup" shape as shown at the left. When $a > 0$, the parabola opens up and the **vertex of the parabola** is the point with the smallest y-coordinate. When $a < 0$, the parabola opens down and the vertex is the point with the largest y-coordinate. The **axis of symmetry of a parabola** is the vertical line that passes through the vertex of the parabola and is parallel to the y-axis. To understand the axis of symmetry, think of folding the graph along that vertical line. The two portions of the graph will match up.

The vertex and axis of symmetry of a parabola can be found by completing the square.

HOW TO Find the vertex and axis of symmetry of the graph of $F(x) = x^2 + 4x + 3$.

To find the coordinates of the vertex, complete the square.

$$F(x) = x^2 + 4x + 3$$
$$= (x^2 + 4x) + 3 \qquad \bullet \text{ Group the variable terms.}$$
$$= (x^2 + 4x + 4) - 4 + 3 \qquad \bullet \text{ Complete the square on } x^2 + 4x. \text{ Add and}$$
$$\text{subtract } \left[\frac{1}{2}(4)\right]^2 = 4 \text{ to and from } x^2 + 4x.$$
$$= (x + 2)^2 - 1 \qquad \bullet \text{ Factor and combine like terms.}$$

Because a, the coefficient of x^2, is positive ($a = 1$), the parabola opens up and the vertex is the point with the smallest y-coordinate. Because $(x + 2)^2 \geq 0$ for all values of x, the smallest y-coordinate occurs when $(x + 2)^2 = 0$. The quantity $(x + 2)^2$ is equal to zero when $x = -2$. Therefore, the x-coordinate of the vertex is -2.

To find the y-coordinate of the vertex, evaluate the function at $x = -2$.

$$F(x) = (x + 2)^2 - 1$$
$$F(-2) = (-2 + 2)^2 - 1 = -1$$

The y-coordinate of the vertex is -1.

From the results above, the coordinates of the vertex are $(-2, -1)$.

The axis of symmetry is the vertical line that passes through the vertex. The equation of the vertical line that passes through the point $(-2, -1)$ is $x = -2$. The axis of symmetry is the line $x = -2$.

$F(x) = x^2 + 4x + 3$

By following the process illustrated in the preceding example and completing the square on $f(x) = ax^2 + bx + c$, we can find a formula for the coordinates of the vertex and the equation of the axis of symmetry of a parabola.

Vertex and Axis of Symmetry of a Parabola

Let $f(x) = ax^2 + bx + c$ be the equation of a parabola. The coordinates of the vertex are $\left(-\dfrac{b}{2a}, f\left(-\dfrac{b}{2a}\right)\right)$. The equation of the axis of symmetry is $x = -\dfrac{b}{2a}$.

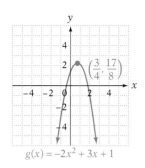

$g(x) = -2x^2 + 3x + 1$

HOW TO Find the vertex of the parabola whose equation is $g(x) = -2x^2 + 3x + 1$. Then graph the equation.

x-coordinate of the vertex: $-\dfrac{b}{2a} = -\dfrac{3}{2(-2)} = \dfrac{3}{4}$

• From the equation $g(x) = -2x^2 + 3x + 1$, $a = -2$, $b = 3$.

y-coordinate of the vertex: $g(x) = -2x^2 + 3x + 1$

$$g\left(\frac{3}{4}\right) = -2\left(\frac{3}{4}\right)^2 + 3\left(\frac{3}{4}\right) + 1$$

$$g\left(\frac{3}{4}\right) = \frac{17}{8}$$

• Evaluate the function at the value of the x-coordinate of the vertex.

The vertex is $\left(\dfrac{3}{4}, \dfrac{17}{8}\right)$.

Because a is negative, the graph opens down. Find a few ordered pairs that belong to the function and then sketch the graph, as shown at the left.

Once the y-coordinate of the vertex is known, the range of the function can be determined. Here, the graph of g opens down, so the y-coordinate of the vertex is the largest value of y. Therefore, the range of g is $\left\{y \mid y \leq \dfrac{17}{8}\right\}$. The domain is $\{x \mid x \in \text{real numbers}\}$.

TAKE NOTE

Once the coordinates of the vertex are found, the range of the quadratic function can be determined.

Example 1 Find the vertex and axis of symmetry of the parabola whose equation is $y = -x^2 + 4x + 1$. Then graph the equation.

You Try It 1 Find the vertex and axis of symmetry of the parabola whose equation is $y = 4x^2 + 4x + 1$. Then graph the equation.

Solution x-coordinate of the vertex:

$$-\frac{b}{2a} = -\frac{4}{2(-1)} = 2$$

y-coordinate of vertex:
$$y = -x^2 + 4x + 1$$
$$= -(2)^2 + 4(2) + 1$$
$$= 5$$

Vertex: (2, 5)
Axis of symmetry: $x = 2$

Your solution

Example 2 Find the domain and range of $f(x) = 0.5x^2 - 3$. Then graph the equation.

You Try It 2 Find the domain and range of $f(x) = -x^2 - 2x - 1$. Then graph the equation.

Solution x-coordinate of the vertex:

$$-\frac{b}{2a} = -\frac{0}{2(0.5)} = 0$$

y-coordinate of vertex:
$$f(x) = 0.5x^2 - 3$$
$$f(0) = 0.5(0)^2 - 3$$
$$= -3$$

Vertex: (0, −3)
The domain is
$\{x \mid x \in \text{real numbers}\}$.
The range is $\{y \mid y \geq -3\}$.

Your solution

Solutions on p. S26

Objective B **To find the *x*-intercepts of a parabola**

Recall that a point at which a graph crosses the *x*- or *y*-axis is called an *intercept* of the graph. The *x*-intercepts of the graph of an equation occur when $y = 0$; the *y*-intercepts occur when $x = 0$.

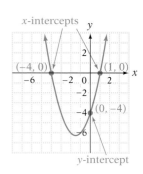

The graph of $y = x^2 + 3x - 4$ is shown at the right. The points whose coordinates are $(-4, 0)$ and $(1, 0)$ are the *x*-intercepts of the graph. The point whose coordinates are $(0, -4)$ is the *y*-intercept of the graph.

HOW TO Find the *x*-intercepts of the parabola whose equation is $y = 4x^2 - 4x + 1$.

To find the *x*-intercepts, let $y = 0$ and then solve for *x*.

$y = 4x^2 - 4x + 1$
$0 = 4x^2 - 4x + 1$ • Let *y* = 0.

$0 = (2x - 1)(2x - 1)$ • Solve for *x* by factoring and using the Principle of Zero Products.

$$2x - 1 = 0 \qquad\qquad 2x - 1 = 0$$
$$2x = 1 \qquad\qquad\quad 2x = 1$$
$$x = \frac{1}{2} \qquad\qquad\quad x = \frac{1}{2}$$

The *x*-intercept is $\left(\frac{1}{2}, 0\right)$.

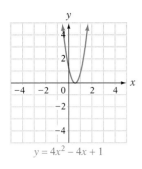

$y = 4x^2 - 4x + 1$

Integrating Technology

See the Keystroke Guide: *Zero* for instructions on using a graphing calculator to find the *x*-intercepts of the graph of a parabola.

In the preceding example, the parabola has only one *x*-intercept. In this case, the parabola is said to be *tangent* to the *x*-axis at $x = \frac{1}{2}$.

HOW TO Find the *x*-intercepts of $y = 2x^2 - x - 6$.

To find the *x*-intercepts, let $y = 0$ and solve for *x*.

$y = 2x^2 - x - 6$
$0 = 2x^2 - x - 6$ • Let *y* = 0.
$0 = (2x + 3)(x - 2)$ • Solve for *x* by factoring and using the Principle of Zero Products.

$$2x + 3 = 0 \qquad\qquad x - 2 = 0$$
$$x = -\frac{3}{2} \qquad\qquad\quad x = 2$$

The *x*-intercepts are $\left(-\frac{3}{2}, 0\right)$ and $(2, 0)$.

TAKE NOTE

A zero of a function is the *x*-coordinate of the *x*-intercept of the graph of the function. Because the *x*-intercepts of the graph of $f(x) = 2x^2 - x - 6$ are $\left(-\frac{3}{2}, 0\right)$ and $(2, 0)$, the zeros are $-\frac{3}{2}$ and 2.

If the equation above, $y = 2x^2 - x - 6$, is written in functional notation as $f(x) = 2x^2 - x - 6$, then to find the *x*-intercepts you would let $f(x) = 0$ and solve for *x*. A value of *x* for which $f(x) = 0$ is a **zero of a function.** Thus $-\frac{3}{2}$ and 2 are *zeros* of $f(x) = 2x^2 - x - 6$.

Integrating Technology

See the Keystroke Guide: *Zero* for instructions on using a graphing calculator to find the zeros of a function.

HOW TO Find the zeros of $f(x) = x^2 - 2x - 1$.

To find the zeros, let $f(x) = 0$ and solve for x.

$$f(x) = x^2 - 2x - 1$$
$$0 = x^2 - 2x - 1$$
$$x = \frac{-b \pm \sqrt{b^2 - 4ac}}{2a}$$
$$= \frac{-(-2) \pm \sqrt{(-2)^2 - 4(1)(-1)}}{2(1)}$$
$$= \frac{2 \pm \sqrt{4 + 4}}{2} = \frac{2 \pm \sqrt{8}}{2}$$
$$= \frac{2 \pm 2\sqrt{2}}{2}$$
$$= 1 \pm \sqrt{2}$$

- Because $x^2 - 2x - 1$ does not easily factor, use the quadratic formula to solve for x.
- $a = 1$, $b = -2$, $c = -1$

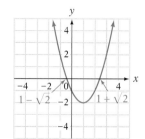

TAKE NOTE

The graph of $f(x) = x^2 - 2x - 1$ is shown at the right. Note that the zeros are the x-intercepts of the graph of f.

The zeros of the function are $1 - \sqrt{2}$ and $1 + \sqrt{2}$.

The preceding examples suggest that there is a relationship among the x-intercepts of the graph of a function, the zeros of a function, and the solutions of an equation. In fact, those three concepts are different ways of discussing the same number. The choice depends on the focus of the discussion. If we are discussing graphing, then the intercept is our focus; if we are discussing functions, then the zero of the function is our focus; and if we are discussing equations, the solution of the equation is our focus.

The graph of a parabola may not have x-intercepts. The graph of $y = -x^2 + 2x - 2$ is shown at the right. Note that the graph does not pass through the x-axis and thus there are no x-intercepts. This means there are no real number zeros of $f(x) = -x^2 + 2x - 2$ and there are no real number solutions of $-x^2 + 2x - 2 = 0$.

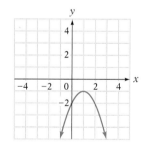

When we apply the quadratic formula, we find that the solutions of the equation $-x^2 + 2x - 2 = 0$ are $1 - i$ and $1 + i$. Thus the zeros of $f(x) = -x^2 + 2x - 2$ are the complex numbers $1 - i$ and $1 + i$.

Study Tip

The paragraph at the right begins with the word "Recall." This signals that the content refers to material presented earlier in the text. The ideas presented here will be more meaningful if you return to the word *discriminant* on page 461 and review the concepts presented there.

Recall that the *discriminant* of $ax^2 + bx + c$ is the expression $b^2 - 4ac$ and that this expression can be used to determine whether $ax^2 + bx + c = 0$ has zero, one, or two real number solutions. Because there is a connection between the solutions of $ax^2 + bx + c = 0$ and the x-intercepts of the graph of $y = ax^2 + bx + c$, the discriminant can be used to determine the number of x-intercepts of a parabola.

The Effect of the Discriminant on the Number of x-Intercepts of a Parabola

1. If $b^2 - 4ac = 0$, the parabola has one x-intercept.
2. If $b^2 - 4ac > 0$, the parabola has two x-intercepts.
3. If $b^2 - 4ac < 0$, the parabola has no x-intercepts.

> **HOW TO** Use the discriminant to determine the number of x-intercepts of the parabola whose equation is $y = 2x^2 - x + 2$.
>
> $b^2 - 4ac$ • Evaluate the discriminant.
> $(-1)^2 - 4(2)(2) = 1 - 16 = -15$ • $a = 2, b = -1, c = 2$
> $-15 < 0$
>
> The discriminant is less than zero. Therefore, the parabola has no x-intercepts.

Example 3

Find the x-intercepts of $y = 2x^2 - 5x + 2$.

Solution
$y = 2x^2 - 5x + 2$
$0 = 2x^2 - 5x + 2$ • Let $y = 0$.
$0 = (2x - 1)(x - 2)$ • Solve for x by factoring.

$2x - 1 = 0 \quad x - 2 = 0$
$\quad\quad 2x = 1 \quad\quad\quad x = 2$
$\quad\quad\quad x = \dfrac{1}{2}$

The x-intercepts are $\left(\dfrac{1}{2}, 0\right)$ and $(2, 0)$.

You Try It 3

Find the x-intercepts of $y = x^2 + 3x + 4$.

Your solution

Example 4

Find the zeros of $f(x) = x^2 + 4x + 5$.

Solution
$f(x) = x^2 + 4x + 5$
$\quad 0 = x^2 + 4x + 5$ • Let $f(x) = 0$.

$x = \dfrac{-b \pm \sqrt{b^2 - 4ac}}{2a}$ • Use the quadratic formula.

$\quad = \dfrac{-4 \pm \sqrt{4^2 - 4(1)(5)}}{2(1)}$ • $a = 1, b = 4, c = 5$

$\quad = \dfrac{-4 \pm \sqrt{16 - 20}}{2} = \dfrac{-4 \pm \sqrt{-4}}{2}$

$\quad = \dfrac{-4 \pm 2i}{2} = -2 \pm i$

The zeros of the function are $-2 + i$ and $-2 - i$.

You Try It 4

Find the zeros of $g(x) = x^2 - x + 6$.

Your solution

Example 5

Use the discriminant to determine the number of x-intercepts of $y = x^2 - 6x + 9$.

Solution
$b^2 - 4ac$ • Evaluate the discriminant.
$(-6)^2 - 4(1)(9)$ • $a = 1, b = -6, c = 9$
$\quad = 36 - 36 = 0$
The discriminant is equal to zero.
The parabola has one x-intercept.

You Try It 5

Use the discriminant to determine the number of x-intercepts of $y = x^2 - x - 6$.

Your solution

Solutions on pp. S26–S27

Objective C **To find the minimum or maximum of a quadratic function**

The graph of $f(x) = x^2 - 2x + 3$ is shown at the right. Because a is positive, the parabola opens up. The vertex of the parabola is the lowest point on the parabola. It is the point that has the minimum y-coordinate. This point represents the **minimum value of the function.**

Vertex (1, 2) Minimum y-coordinate = 2

The graph of $f(x) = -x^2 + 2x + 1$ is shown at the right. Because a is negative, the parabola opens down. The vertex of the parabola is the highest point on the parabola. It is the point that has the maximum y-coordinate. This point represents the **maximum value of the function.**

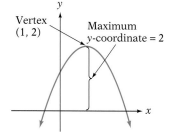

Vertex (1, 2) Maximum y-coordinate = 2

To find the minimum or maximum value of a quadratic function, first find the x-coordinate of the vertex. Then evaluate the function at that value.

Integrating Technology

See the Projects and Group Activities at the end of this chapter for instructions on using a graphing calculator to find the minimum or maximum value of a function.

HOW TO Find the maximum or minimum value of the function $f(x) = -2x^2 + 4x + 3$.

$$x = -\frac{b}{2a} = -\frac{4}{2(-2)} = 1$$

• Find the x-coordinate of the vertex. $a = -2$, $b = 4$

$$f(x) = -2x^2 + 4x + 3$$
$$f(1) = -2(1)^2 + 4(1) + 3$$
$$= -2 + 4 + 3$$
$$= 5$$

• Evaluate the function at $x = 1$.

The maximum value of the function is 5.

• Because $a < 0$, the graph of f opens down. Therefore, the function has a maximum value.

Example 6

Find the maximum or minimum value of $f(x) = 2x^2 - 3x + 1$.

Solution

$$x = -\frac{b}{2a} = -\frac{-3}{2(2)} = \frac{3}{4}$$

• The x-coordinate of the vertex

$$f(x) = 2x^2 - 3x + 1$$
$$f\left(\frac{3}{4}\right) = 2\left(\frac{3}{4}\right)^2 - 3\left(\frac{3}{4}\right) + 1$$

• $x = \frac{3}{4}$

$$= \frac{9}{8} - \frac{9}{4} + 1 = -\frac{1}{8}$$

Because a is positive, the graph opens up. The function has a minimum value.

The minimum value of the function is $-\frac{1}{8}$.

You Try It 6

Find the maximum or minimum value of $f(x) = -3x^2 + 4x - 1$.

Your solution

Solution on p. S27

Objective D **To solve application problems**

HOW TO A carpenter is forming a rectangular floor for a storage shed. The perimeter of the rectangle is 44 ft. What dimensions of the rectangle would give the floor a maximum area? What is the maximum area?

We are given the perimeter of the rectangle, and we want to find the dimensions of the rectangle that will yield the maximum area for the floor. Use the equation for the perimeter of a rectangle.

$$P = 2L + 2W$$
$$44 = 2L + 2W \qquad \bullet \ P = 44$$
$$22 = L + W \qquad \bullet \ \text{Divide both sides of the equation by 2.}$$
$$22 - L = W \qquad \bullet \ \text{Solve the equation for } W.$$

Now use the equation for the area of a rectangle. Use substitution to express the area in terms of L.

$$A = LW$$
$$A = L(22 - L) \qquad \bullet \ \text{From the equation above, } W = 22 - L. \text{ Substitute}$$
$$\qquad\qquad\qquad\qquad\qquad 22 - L \text{ for } W.$$
$$A = 22L - L^2 \qquad \bullet \ \text{The area of the rectangle is } 22L - L^2.$$

To find the length of the rectangle, find the L-coordinate of the vertex of the function $f(L) = -L^2 + 22L$.

$$L = -\frac{b}{2a} = -\frac{22}{2(-1)} = 11 \qquad \bullet \ \text{For the equation } f(L) = -L^2 + 22L,$$
$$\qquad\qquad\qquad\qquad\qquad\qquad\qquad a = -1 \text{ and } b = 22.$$

The length of the rectangle is 11 ft.

To find the width, replace L in $22 - L$ by the L-coordinate of the vertex and evaluate.

$$W = 22 - L$$
$$W = 22 - 11 = 11 \qquad \bullet \ \text{Replace } L \text{ by 11 and evaluate.}$$

The width of the rectangle is 11 ft.

The dimensions of the rectangle that would give the floor a maximum area are 11 ft by 11 ft.

To find the maximum area of the floor, evaluate $f(L) = -L^2 + 22L$ at the L-coordinate of the vertex.

$$f(L) = -L^2 + 22L$$
$$f(11) = -(11)^2 + 22(11) \qquad \bullet \ \text{Evaluate the function at 11.}$$
$$= -121 + 242 = 121$$

The maximum area of the floor is 121 ft².

Point of Interest

Calculus is a branch of mathematics that demonstrates, among other things, how to find the maximum or minimum of functions other than quadratic functions. These are very important problems in applied mathematics. For instance, an automotive engineer wants to design a car whose shape will *minimize* the effects of air flow. The same engineer tries to *maximize* the efficiency of the car's engine. Similarly, an economist may try to determine what business practices will *minimize* cost and *maximize* profit.

The graph of the function $f(L) = -L^2 + 22L$ is shown at the right. Note that the vertex of the parabola is $(11, 121)$. For any value of L less than 11, the area of the floor will be less than 121 ft². For any value of L greater than 11, the area of the floor will be less than 121 ft². The maximum value of the function is 121, and the maximum value occurs when $L = 11$.

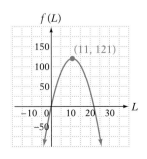

Example 7

A mining company has determined that the cost c, in dollars per ton, of mining a mineral is given by $c(x) = 0.2x^2 - 2x + 12$, where x is the number of tons of the mineral mined. Find the number of tons of the mineral that should be mined to minimize the cost. What is the minimum cost per ton?

Strategy

• To find the number of tons of the mineral that should be mined to minimize the cost, find the x-coordinate of the vertex.
• To find the minimum cost, evaluate $c(x)$ at the x-coordinate of the vertex.

Solution

$c(x) = 0.2x^2 - 2x + 12$ • $a = 0.2$, $b = -2$, $c = 12$

$x = -\dfrac{b}{2a} = -\dfrac{(-2)}{2(0.2)} = 5$ • The x-coordinate of the vertex

To minimize the cost, 5 tons of the mineral should be mined.

$c(x) = 0.2x^2 - 2x + 12$
$c(5) = 0.2(5)^2 - 2(5) + 12$ • $x = 5$
$\quad\ \ = 5 - 10 + 12$
$\quad\ \ = 7$

The minimum cost is $7 per ton.

Note: The graph of the function $c(x) = 0.2x^2 - 2x + 12$ is shown below. The vertex of the parabola is $(5, 7)$. For any value of x less than 5, the cost per ton is greater than $7. For any value of x greater than 5, the cost per ton is greater than $7. 7 is the minimum value of the function, and the minimum value occurs when $x = 5$.

You Try It 7

The height s, in feet, of a ball thrown straight up is given by $s(t) = -16t^2 + 64t$, where t is the time in seconds. Find the time it takes the ball to reach its maximum height. What is the maximum height?

Your strategy

Your solution

Solution on p. S27

Example 8

Find two numbers whose difference is 10 and whose product is a minimum. What is the minimum product of the two numbers?

You Try It 8

A rectangular fence is being constructed along a stream to enclose a picnic area. If there are 100 ft of fence available, what dimensions of the rectangle will produce the maximum area for picnicking?

Strategy

• Let x represent one number. Because the difference between the two numbers is 10,

$$x + 10$$

represents the other number.
[*Note:* $(x + 10) - (x) = 10$]
Then their product is represented by

$$x(x + 10) = x^2 + 10x$$

• To find one of the two numbers, find the x-coordinate of the vertex of
$f(x) = x^2 + 10x$.
• To find the other number, replace x in $x + 10$ by the x-coordinate of the vertex and evaluate.
• To find the minimum product, evaluate the function at the x-coordinate of the vertex.

Your strategy

Solution

$f(x) = x^2 + 10x$ • $a = 1, b = 10, c = 0$

$x = -\dfrac{b}{2a} = -\dfrac{10}{2(1)} = -5$ • One number is -5.

$\begin{aligned} &x + 10 \\ &-5 + 10 = 5 \end{aligned}$ • The other number is 5.

The numbers are -5 and 5.

$\begin{aligned} f(x) &= x^2 + 10x \\ f(-5) &= (-5)^2 + 10(-5) \\ &= 25 - 50 \\ &= -25 \end{aligned}$

The minimum product of the two numbers is -25.

Your solution

Solution on p. S27

9.1 Exercises

Objective A **To graph a quadratic function**

1. What is a quadratic function?

2. Describe the graph of a parabola.

3. What is the vertex of a parabola?

4. What is the axis of symmetry of the graph of a parabola?

5. The axis of symmetry of a parabola is the line $x = -5$. What is the x-coordinate of the vertex of the parabola?

6. The axis of symmetry of a parabola is the line $x = 8$. What is the x-coordinate of the vertex of the parabola?

7. The vertex of a parabola is $(7, -9)$. What is the equation of the axis of symmetry of the parabola?

8. The vertex of a parabola is $(-4, 10)$. What is the equation of the axis of symmetry of the parabola?

For Exercises 9 to 23, find the vertex and axis of symmetry of the parabola given by each equation. Then graph the equation.

9. $y = x^2 - 2x - 4$

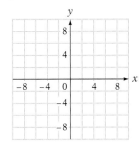

10. $y = x^2 + 4x - 4$

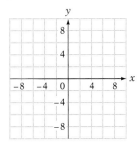

11. $y = -x^2 + 2x - 3$

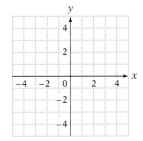

12. $y = -x^2 + 4x - 5$

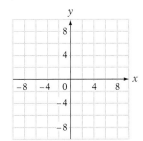

13. $f(x) = x^2 - x - 6$

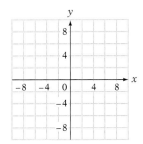

14. $G(x) = x^2 - x - 2$

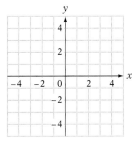

15. $F(x) = x^2 - 3x + 2$

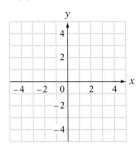

16. $y = 2x^2 - 4x + 1$

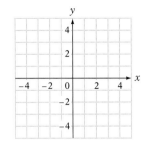

17. $y = -2x^2 + 6x$

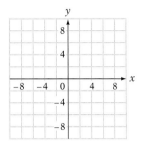

18. $y = \dfrac{1}{2}x^2 + 4$

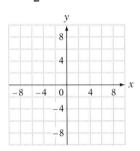

19. $y = -\dfrac{1}{4}x^2 - 1$

20. $h(x) = \dfrac{1}{2}x^2 - x + 1$

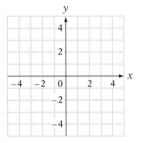

21. $P(x) = -\dfrac{1}{2}x^2 + 2x - 3$

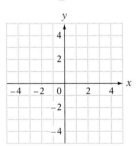

22. $y = \dfrac{1}{2}x^2 + 2x - 6$

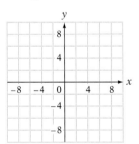

23. $y = -\dfrac{1}{2}x^2 + x - 3$

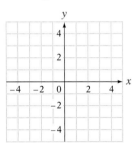

For Exercises 24 to 29, state the domain and range of the function.

24. $f(x) = 2x^2 - 4x - 5$

25. $f(x) = 2x^2 + 8x + 3$

26. $f(x) = -2x^2 - 3x + 2$

27. $f(x) = -x^2 + 6x - 9$

28. $f(x) = -x^2 - 4x - 5$

29. $f(x) = x^2 + 4x - 3$

Objective B To find the *x*-intercepts of a parabola

30. **a.** What is an x-intercept of the graph of a parabola?
b. How many x-intercepts can the graph of a parabola have?

31. **a.** What is the y-intercept of the graph of a parabola?
b. How many y-intercepts can the graph of a parabola have?

For Exercises 32 to 43, find the x-intercepts of the parabola given by each equation.

32. $y = x^2 - 4$

33. $y = x^2 - 9$

34. $y = 2x^2 - 4x$

35. $y = 3x^2 + 6x$

36. $y = x^2 - x - 2$

37. $y = x^2 - 2x - 8$

38. $y = 2x^2 - x - 1$

39. $y = 2x^2 - 5x - 3$

40. $y = x^2 + 2x - 1$

41. $y = x^2 + 4x - 3$

42. $y = x^2 + 6x + 10$

43. $y = -x^2 - 4x - 5$

For Exercises 44 to 55, find the zeros of the function.

44. $f(x) = x^2 + 3x + 2$

45. $f(x) = x^2 - 6x + 9$

46. $f(x) = -x^2 + 4x - 5$

47. $f(x) = -x^2 + 3x + 8$

48. $f(x) = 2x^2 - 3x$

49. $f(x) = -3x^2 + 4x$

50. $f(x) = 2x^2 - 4$

51. $f(x) = 3x^2 + 6$

52. $f(x) = 2x^2 + 3x + 2$

53. $f(x) = 3x^2 - x + 4$

54. $f(x) = -3x^2 - 4x + 1$

55. $f(x) = -2x^2 + x + 5$

For Exercises 56 to 67, use the discriminant to determine the number of x-intercepts of the graph.

56. $y = 2x^2 + 2x - 1$

57. $y = -x^2 - x + 3$

58. $y = x^2 - 8x + 16$

59. $y = x^2 - 10x + 25$ **60.** $y = -3x^2 - x - 2$ **61.** $y = -2x^2 + x - 1$

62. $y = -2x^2 + x + 1$ **63.** $y = 4x^2 - x - 2$ **64.** $y = 2x^2 + x + 1$

65. $y = 2x^2 + x + 4$ **66.** $y = -3x^2 + 2x - 8$ **67.** $y = 4x^2 + 2x - 5$

68. The zeros of the function $f(x) = x^2 - 2x - 3$ are -1 and 3. What are the x-intercepts of the graph of the equation $y = x^2 - 2x - 3$?

69. The zeros of the function $f(x) = x^2 - x - 20$ are -4 and 5. What are the x-intercepts of the graph of the equation $y = x^2 - x - 20$?

Objective C **To find the minimum or maximum of a quadratic function**

70. What is the minimum or maximum value of a quadratic function?

71. Describe how to find the minimum or maximum value of a quadratic function.

72. Does the function have a minimum or a maximum value?
 a. $f(x) = -x^2 + 6x - 1$ **b.** $f(x) = 2x^2 - 4$ **c.** $f(x) = -5x^2 + x$

73. Does the function have a minimum or a maximum value?
 a. $f(x) = 3x^2 - 2x + 4$ **b.** $f(x) = -x^2 + 9$ **c.** $f(x) = 6x^2 - 3x$

For Exercises 74 to 85, find the minimum or maximum value of the quadratic function.

74. $f(x) = x^2 - 2x + 3$ **75.** $f(x) = 2x^2 + 4x$ **76.** $f(x) = -2x^2 + 4x - 3$

77. $f(x) = -2x^2 + 4x - 5$ **78.** $f(x) = -2x^2 - 3x + 4$ **79.** $f(x) = -2x^2 - 3x$

80. $f(x) = 2x^2 + 3x - 8$ **81.** $f(x) = 3x^2 + 3x - 2$ **82.** $f(x) = -3x^2 + x - 6$

83. $f(x) = -x^2 - x + 2$ **84.** $f(x) = x^2 - 5x + 3$ **85.** $f(x) = 3x^2 + 5x + 2$

86. Which of the following parabolas has the greatest minimum value?
 a. $y = x^2 - 2x - 3$ **b.** $y = x^2 - 10x + 20$ **c.** $y = 3x^2 - 6$

87. Which of the following parabolas has the greatest maximum value?
 a. $y = -2x^2 + 2x - 1$ **b.** $y = -x^2 + 8x - 2$ **c.** $y = -4x^2 + 3$

88. The vertex of a parabola that opens up is $(-4, 7)$. Does the function have a maximum or a minimum value? What is the maximum or minimum value of the function?

89. The vertex of a parabola that opens down is $(3, -5)$. Does the function have a maximum or a minimum value? What is the maximum or minimum value of the function?

Objective D **To solve application problems**

90. Physics The height s, in feet, of a rock thrown upward at an initial speed of 64 ft/s from a cliff 50 ft above an ocean beach is given by the function $s(t) = -16t^2 + 64t + 50$, where t is the time in seconds. Find the maximum height above the beach that the rock will attain.

50 ft

91. Business A manufacturer of microwave ovens believes that the revenue R, in dollars, the company receives is related to the price P, in dollars, of an oven by the function $R(P) = 125P - 0.25P^2$. What price will give the maximum revenue?

92. Business A tour operator believes that the profit P, in dollars, from selling x tickets is given by $P(x) = 40x - 0.25x^2$. Using this model, what is the maximum profit the tour operator can expect?

93. The Olympics An event in the Summer Olympics is 10-meter springboard diving. In this event, the height s, in meters, of a diver above the water t seconds after jumping is given by $s(t) = -4.9t^2 + 7.8t + 10$. What is the maximum height that the diver will be above the water? Round to the nearest tenth.

94. Chemistry A pool is treated with a chemical to reduce the amount of algae in the pool. The amount of algae in the pool t days after the treatment can be approximated by the function $A(t) = 40t^2 - 400t + 500$. How many days after the treatment will the pool have the least amount of algae?

95. Construction The suspension cable that supports a small footbridge hangs in the shape of a parabola. The height h, in feet, of the cable above the bridge is given by the function $h(x) = 0.25x^2 - 0.8x + 25$, where x is the distance in feet from one end of the bridge. What is the minimum height of the cable above the bridge?

96. **Telescopes** An equation that models the thickness h, in inches, of the mirror in the telescope at the Palomar Mountain Observatory is given by $h(x) = 0.000379x^2 - 0.0758x + 24$, where x is measured in inches from the edge of the mirror. Find the minimum thickness of the mirror.

97. Fire Science The height s, in feet, of water squirting from a certain fire hose nozzle is given by the equation $s(x) = -\frac{1}{30}x^2 + 2x + 5$, where x is the horizontal distance, in feet, from the nozzle. How high on a building 40 ft from the fire hose will the water land?

98. **Fountains** The Buckingham Fountain in Chicago shoots water from a nozzle at the base of the fountain. The height h, in feet, of the water above the ground t seconds after it leaves the nozzle is given by the function $h(t) = -16t^2 + 90t + 15$. What is the maximum height of the water? Round to the nearest tenth.

99. Safety On wet concrete the stopping distance s, in feet, of a car traveling v miles per hour is given by the function $s(v) = 0.055v^2 + 1.1v$. What is the maximum speed at which a car could be traveling and still stop at a stop sign 44 ft away?

100. Sports Some football fields are built in a parabolic mound shape so that water will drain off the field. A model for the shape of such a field is given by $h(x) = -0.00023475x^2 + 0.0375x$, where h is the height of the field, in feet, at a distance of x feet from the sideline. What is the maximum height? Round to the nearest tenth.

101. Mathematics Find two numbers whose sum is 20 and whose product is a maximum.

102. Mathematics Find two numbers whose difference is 14 and whose product is a minimum.

103. Ranching A rancher has 200 ft of fencing to build a rectangular corral alongside an existing fence. Determine the dimensions of the corral that will maximize the enclosed area.

APPLYING THE CONCEPTS

104. One root of the quadratic equation $2x^2 - 5x + k = 0$ is 4. What is the other root?

105. What is the value of k if the vertex of the parabola $y = x^2 - 8x + k$ is a point on the x-axis?

106. The roots of the function $f(x) = mx^2 + nx + 1$ are -2 and 3. What are the roots of the function $g(x) = nx^2 + mx - 1$?

9.2 Graphs of Functions

Objective A **To graph functions**

The graphs of the polynomial functions $L(x) = mx + b$ (a straight line) and $Q(x) = ax^2 + bx + c$, $a \neq 0$ (a parabola) were discussed in previous objectives. The graphs of other functions can also be drawn by finding ordered pairs that belong to the function, plotting the points that correspond to the ordered pairs, and then drawing a curve through the points.

HOW TO Graph: $F(x) = x^3$

Select several values of x and evaluate the function.

x	$F(x) = x^3$	$F(x)$	(x, y)
-2	$(-2)^3$	-8	$(-2, -8)$
-1	$(-1)^3$	-1	$(-1, -1)$
0	0^3	0	$(0, 0)$
1	1^3	1	$(1, 1)$
2	2^3	8	$(2, 8)$

Plot the ordered pairs and draw a graph through the points.

HOW TO Graph: $g(x) = x^3 - 4x + 5$

Select several values of x and evaluate the function.

x	$g(x) = x^3 - 4x + 5$	$g(x)$	(x, y)
-3	$(-3)^3 - 4(-3) + 5$	-10	$(-3, -10)$
-2	$(-2)^3 - 4(-2) + 5$	5	$(-2, 5)$
-1	$(-1)^3 - 4(-1) + 5$	8	$(-1, 8)$
0	$(0)^3 - 4(0) + 5$	5	$(0, 5)$
1	$(1)^3 - 4(1) + 5$	2	$(1, 2)$
2	$(2)^3 - 4(2) + 5$	5	$(2, 5)$

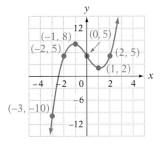

Plot the ordered pairs and draw a graph through the points.

Note from the graphs of the two cubic functions above that the shapes of the graphs can be quite different. The following graphs of typical cubic polynomial functions show the general shape of a cubic polynomial.

As the degree of a polynomial increases, the graph of the polynomial function can change significantly. In such cases, it may be necessary to plot many points before an accurate graph can be drawn. Only polynomials of degree 3 are considered here.

Integrating Technology

See the Keystroke Guide: *Graph* for instructions on using a graphing calculator to graph a function.

HOW TO Graph: $f(x) = |x + 2|$

This is an **absolute-value function.**

| x | $f(x) = |x + 2|$ | $f(x)$ | (x, y) |
|-----|------------------|--------|----------|
| -3 | $|-3 + 2|$ | 1 | $(-3, 1)$ |
| -2 | $|-2 + 2|$ | 0 | $(-2, 0)$ |
| -1 | $|-1 + 2|$ | 1 | $(-1, 1)$ |
| 0 | $|0 + 2|$ | 2 | $(0, 2)$ |
| 1 | $|1 + 2|$ | 3 | $(1, 3)$ |
| 2 | $|2 + 2|$ | 4 | $(2, 4)$ |

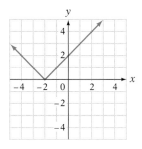

In general, the graph of the absolute value of a linear polynomial is V-shaped.

HOW TO Graph: $R(x) = \sqrt{2x - 4}$

This is a **radical function.** Because the square root of a negative number is not a real number, the domain of this function requires that $2x - 4 \geq 0$. Solve this inequality for x.

$$2x - 4 \geq 0$$
$$2x \geq 4$$
$$x \geq 2$$

The domain is $\{x \mid x \geq 2\}$. This means that only values of x that are greater than or equal to **2** can be chosen as values at which to evaluate the function. In this case, some of the y-coordinates must be approximated.

x	$R(x) = \sqrt{2x - 4}$	$R(x)$	(x, y)
2	$\sqrt{2(2) - 4}$	0	$(2, 0)$
3	$\sqrt{2(3) - 4}$	1.41	$(3, 1.41)$
4	$\sqrt{2(4) - 4}$	2	$(4, 2)$
5	$\sqrt{2(5) - 4}$	2.45	$(5, 2.45)$
6	$\sqrt{2(6) - 4}$	2.83	$(6, 2.83)$

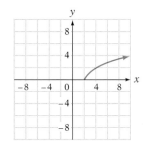

Recall that a function is a special type of relation, one for which no two ordered pairs have the same first coordinate. Graphically, this means that the graph of a function cannot pass through two points that have the same x-coordinate and different y-coordinates. For instance, the graph at the right is not the graph of a function because there are ordered pairs with the same x-coordinate and different y-coordinates.

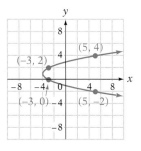

The last graph illustrates a general statement that can be made about whether a graph defines a function. It is called the **vertical-line test.**

Vertical-Line Test

A graph defines a function if any vertical line intersects the graph at no more than one point.

For example, the graph of a nonvertical straight line is the graph of a function. Any vertical line intersects the graph no more than once. The graph of a circle, however, is not the graph of a function. There are vertical lines that intersect the graph at more than one point.

 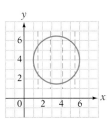

Some practical applications can be modeled by a graph that is not the graph of a function. Below is an example of such an application.

One of the causes of smog is an inversion layer of air where temperatures at higher altitudes are warmer than those at lower altitudes. The graph at the right shows the altitudes at which various temperatures were recorded. As shown by the dashed lines in the graph, there are two altitudes at which the temperature was 25°C. This means that there are two ordered pairs (shown in the graph) with the same first coordinate but different second coordinates. The graph does not define a function.

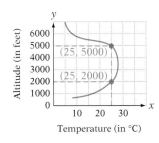

When a graph does define a function, the domain and range can be estimated from the graph.

TAKE NOTE

To determine the domain, think of collapsing the graph onto the x-axis and determining the interval on the x-axis that the graph covers. To determine the range, think of collapsing the graph onto the y-axis and determining the interval on the y-axis that the graph covers.

HOW TO Determine the domain and range of the function given by the graph at the right. Write the answer in set-builder notation.

The solid dots on the graph indicate its beginning and ending points.

The domain is the set of x-coordinates.
Domain: $\{x \mid 1 \le x \le 6\}$

The range is the set of y-coordinates.
Range: $\{y \mid 2 \le y \le 5\}$

TAKE NOTE

In the first example, the domain and range are given in set-builder notation. In the second example, the domain and range are given in interval notation. Either method may be used.

HOW TO Determine the domain and range of the function given by the graph at the right. Write the answer in interval notation.

The arrows on the graph indicate that the graph continues in the same manner.

The domain is the set of x-coordinates.
Domain: $(-\infty, \infty)$

The range is the set of y-coordinates.
Range: $[-4, 4]$

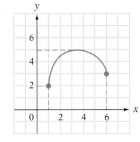

Example 1

Use the vertical-line test to determine whether the graph shown is the graph of a function.

Solution

A vertical line intersects the graph more than once. The graph is not the graph of a function.

You Try It 1

Use the vertical-line test to determine whether the graph shown is the graph of a function.

Your solution

Example 2

Graph and write in set-builder notation the domain and range of the function defined by $f(x) = x^3 - 3x$.

Solution

Domain:
$\{x \mid x \in \text{real numbers}\}$
Range:
$\{y \mid y \in \text{real numbers}\}$

You Try It 2

Graph and write in set-builder notation the domain and range of the function defined by $f(x) = -\frac{1}{2}x^3 + 2x$.

Your solution

Example 3

Graph and write in interval notation the domain and range of the function defined by $f(x) = |x| + 2$.

Solution

Domain: $(-\infty, \infty)$
Range: $[2, \infty)$

You Try It 3

Graph and write in interval notation the domain and range of the function defined by $f(x) = |x - 3|$.

Your solution

Example 4

Graph and write in set-builder notation the domain and range of the function defined by $f(x) = \sqrt{2 - x}$.

Solution

Domain: $\{x \mid x \leq 2\}$
Range: $\{y \mid y \geq 0\}$

You Try It 4

Graph and write in set-builder notation the domain and range of the function defined by $f(x) = -\sqrt{x - 1}$.

Your solution

Solutions on pp. S27–S28

9.2 Exercises

Objective A **To graph functions**

For Exercises 1 to 6, use the vertical-line test to determine whether the graph is the graph of a function.

1.

2.

3.

4.

5.

6.

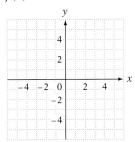

For Exercises 7 to 18, graph the function and state its domain and range in set-builder notation.

7. $f(x) = 3|2 - x|$

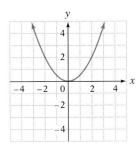

8. $f(x) = x^3 - 1$

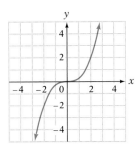

9. $f(x) = 1 - x^3$

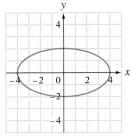

10. $f(x) = \sqrt{1 + x}$

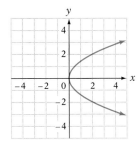

11. $f(x) = \sqrt{4 - x}$

12. $f(x) = |2x - 1|$

13. $f(x) = x^3 + 4x^2 + 4x$

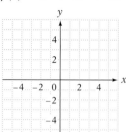

14. $f(x) = x^3 - x^2 - x + 1$

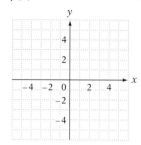

15. $f(x) = -\sqrt{x + 2}$

16. $f(x) = -\sqrt{x - 3}$

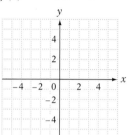

17. $f(x) = |2x + 2|$

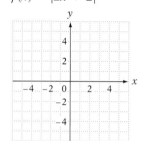

18. $f(x) = 2|x + 1|$

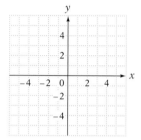

APPLYING THE CONCEPTS

19. If $f(x) = \sqrt{x - 2}$ and $f(a) = 4$, find a.

20. If $f(x) = \sqrt{x + 5}$ and $f(a) = 3$, find a.

21. $f(a, b)$ = the sum of a and b
$g(a, b)$ = the product of a and b
Find $f(2, 5) + g(2, 5)$.

22. The graph of the function f is shown at the right. For this function, which of the following are true?
a. $f(4) = 1$ **b.** $f(0) = 3$ **c.** $f(-3) = 2$

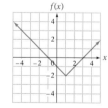

23. Let $f(x)$ be the digit in the xth decimal place of the repeating decimal $0.\overline{387}$. For example, $f(3) = 7$ because 7 is the digit in the third decimal place. Find $f(14)$.

24. Given $f(x) = (x + 1)(x - 1)$, for what values of x is $f(x)$ negative? Write your answer in set-builder notation.

25. Given $f(x) = (x + 2)(x - 2)$, for what values of x is $f(x)$ negative? Write your answer in set-builder notation.

26. Given $f(x) = -|x + 3|$, for what value of x is $f(x)$ greatest?

27. Given $f(x) = |2x - 2|$, for what value of x is $f(x)$ smallest?

9.3 Algebra of Functions

Objective A **To perform operations on functions**

The operations of addition, subtraction, multiplication, and division of functions are defined as follows.

Operations on Functions

If f and g are functions and x is an element of the domain of each function, then

$(f + g)(x) = f(x) + g(x)$ 　　　$(f \cdot g)(x) = f(x) \cdot g(x)$

$(f - g)(x) = f(x) - g(x)$ 　　　$\left(\dfrac{f}{g}\right)(x) = \dfrac{f(x)}{g(x)}$, $g(x) \neq 0$

HOW TO　Given $f(x) = x^2 + 1$ and $g(x) = 3x - 2$, find $(f + g)(3)$ and $(f \cdot g)(-1)$.

$$\begin{aligned}
(f + g)(3) &= f(3) + g(3) \\
&= [(3)^2 + 1] + [3(3) - 2] \\
&= 10 + 7 = 17 \\
(f \cdot g)(-1) &= f(-1) \cdot g(-1) \\
&= [(-1)^2 + 1] \cdot [3(-1) - 2] \\
&= 2 \cdot (-5) = -10
\end{aligned}$$

Consider the functions f and g from the last example. Let $S(x)$ be the sum of the two functions. Then

$$\begin{aligned}
S(x) &= (f + g)(x) = f(x) + g(x) \\
&= [x^2 + 1] + [3x - 2] \\
S(x) &= x^2 + 3x - 1
\end{aligned}$$

- The definition of addition of functions
- $f(x) = x^2 + 1$, $g(x) = 3x - 2$

Now evaluate $S(3)$.

$$\begin{aligned}
S(x) &= x^2 + 3x - 1 \\
S(3) &= (3)^2 + 3(3) - 1 \\
&= 9 + 9 - 1 \\
&= 17 = (f + g)(3)
\end{aligned}$$

Note that $S(3) = 17$ and $(f + g)(3) = 17$. This shows that adding $f(x) + g(x)$ and then evaluating is the same as evaluating $f(x)$ and $g(x)$ and then adding. The same is true for the other operations on functions. For instance, let $P(x)$ be the product of the functions f and g. Then

$$\begin{aligned}
P(x) &= (f \cdot g)(x) = f(x) \cdot g(x) \\
&= (x^2 + 1)(3x - 2) \\
&= 3x^3 - 2x^2 + 3x - 2 \\
P(-1) &= 3(-1)^3 - 2(-1)^2 + 3(-1) - 2 \\
&= -3 - 2 - 3 - 2 \\
&= -10
\end{aligned}$$

- Note that $P(-1) = -10$ and $(f \cdot g)(-1) = -10$.

HOW TO Given $f(x) = 2x^2 - 5x + 3$ and $g(x) = x^2 - 1$, find $\left(\dfrac{f}{g}\right)(1)$.

$$\left(\frac{f}{g}\right)(1) = \frac{f(1)}{g(1)}$$

$$= \frac{2(1)^2 - 5(1) + 3}{(1)^2 - 1} = \frac{0}{0} \qquad \bullet \text{ Not a real number}$$

Because $\dfrac{0}{0}$ is not defined, the expression $\left(\dfrac{f}{g}\right)(1)$ cannot be evaluated.

Example 1

Given $f(x) = x^2 - x + 1$ and $g(x) = x^3 - 4$, find $(f - g)(3)$.

Solution

$$(f - g)(3) = f(3) - g(3)$$
$$= (3^2 - 3 + 1) - (3^3 - 4)$$
$$= 7 - 23$$
$$= -16$$

$(f - g)(3) = -16$

You Try It 1

Given $f(x) = x^2 + 2x$ and $g(x) = 5x - 2$, find $(f + g)(-2)$.

Your solution

Example 2

Given $f(x) = x^2 + 2$ and $g(x) = 2x + 3$, find $(f \cdot g)(-2)$.

Solution

$$(f \cdot g)(-2) = f(-2) \cdot g(-2)$$
$$= [(-2)^2 + 2] \cdot [2(-2) + 3]$$
$$= 6(-1)$$
$$= -6$$

$(f \cdot g)(-2) = -6$

You Try It 2

Given $f(x) = 4 - x^2$ and $g(x) = 3x - 4$, find $(f \cdot g)(3)$.

Your solution

Example 3

Given $f(x) = x^2 + 4x + 4$ and $g(x) = x^3 - 2$, find $\left(\dfrac{f}{g}\right)(3)$.

Solution

$$\left(\frac{f}{g}\right)(3) = \frac{f(3)}{g(3)}$$

$$= \frac{3^2 + 4(3) + 4}{3^3 - 2}$$

$$= \frac{25}{25}$$

$$= 1$$

$\left(\dfrac{f}{g}\right)(3) = 1$

You Try It 3

Given $f(x) = x^2 - 4$ and $g(x) = x^2 + 2x + 1$, find $\left(\dfrac{f}{g}\right)(4)$.

Your solution

Solutions on p. S28

Objective B **To find the composition of two functions**

Composition of functions is another way in which functions can be combined. This method of combining functions uses the output of one function as an input for a second function.

Suppose a forest fire is started by lightning striking a tree, and the spread of the fire can be approximated by a circle whose radius r, in feet, is given by $r(t) = 24\sqrt{t}$, where t is the number of hours after the tree is struck by the lightning. The area of the fire is the area of a circle and is given by the formula $A(r) = \pi r^2$. Because the area of the fire depends on the radius of the circle and the radius depends on the time since the tree was struck, there is a relationship between the area of the fire and time. This relationship can be found by evaluating the formula for the area of a circle using $r(t) = 24\sqrt{t}$.

$$A(r) = \pi r^2$$
$$A[r(t)] = \pi [r(t)]^2 \quad \bullet \text{ Replace } r \text{ by } r(t).$$
$$= \pi [24\sqrt{t}]^2 \quad \bullet \; r(t) = 24\sqrt{t}.$$
$$= 576\pi t \quad \bullet \text{ Simplify.}$$

The result is the function $A(t) = 576\pi t$, which gives the area of the fire in terms of the time since the lightning struck. For instance, when $t = 3$, we have

$$A(t) = 576\pi t$$
$$A(3) = 576\pi (3)$$
$$\approx 5429$$

Three hours after the lightning strikes, the area of the fire is approximately 5429 ft².

The function produced above, in which one function was evaluated with another function, is referred to as the *composition* of A with r. The notation $A \circ r$ is used to denote this composition of functions. That is,

$$(A \circ r)(t) = 576\pi t$$

Definition of the Composition of Two Functions

Let f and g be two functions such that $g(x)$ is in the domain of f for all x in the domain of g. Then the **composition of the two functions**, denoted by $f \circ g$, is the function whose value at x is given by $(f \circ g)(x) = f[g(x)]$.

The function defined by $(f \circ g)(x)$ is also called the **composite** of f and g and represents a **composite function.** We read $(f \circ g)(x)$ or $f[g(x)]$ as "f of g of x."

The function machine at the right illustrates the composition of the two functions $g(x) = x^2$ and $f(x) = 2x$. Note that a composite function combines two functions. First one function pairs an input with an output. Then that output is used as the input for a second function, which in turn produces a final output.

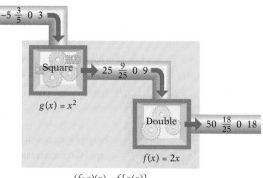

$(f \circ g)(x) = f[g(x)]$

Consider $f(x) = 3x - 1$ and $g(x) = x^2 + 1$. The expression $(f \circ g)(-2)$ or, equivalently, $f[g(-2)]$ means to evaluate the function f at $g(-2)$.

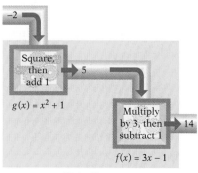

$f[g(-2)] = 14$

$$g(x) = x^2 + 1$$
$$g(-2) = (-2)^2 + 1 \qquad \bullet \text{ Evaluate } g \text{ at } -2.$$
$$g(-2) = 5$$

$$f(x) = 3x - 1$$
$$f(5) = 3(5) - 1 = 14 \qquad \bullet \text{ Evaluate } f \text{ at } g(-2) = 5.$$

If we apply our function machine analogy, the composition of functions looks something like the figure at the left.

We can find a general expression for $f[g(x)]$ by evaluating f at $g(x)$. For instance, using $f(x) = 3x - 1$ and $g(x) = x^2 + 1$, we have

$$f(x) = 3x - 1$$
$$f[g(x)] = 3[g(x)] - 1 \qquad \bullet \text{ Replace } x \text{ by } g(x).$$
$$= 3[x^2 + 1] - 1 \qquad \bullet \text{ Replace } g(x) \text{ by } x^2 + 1.$$
$$= 3x^2 + 2 \qquad \bullet \text{ Simplify.}$$

The requirement in the definition of the composition of two functions that $g(x)$ be in the domain of f for all x in the domain of g is important. For instance, let

$$f(x) = \frac{1}{x - 1} \qquad \text{and} \qquad g(x) = 3x - 5$$

When $x = 2$,

$$g(2) = 3(2) - 5 = 1$$
$$f[g(2)] = f(1) = \frac{1}{1 - 1} = \frac{1}{0} \qquad \bullet \text{ This is not a real number.}$$

In this case, $g(2)$ is not in the domain of f. Thus the composition is not defined at 2.

HOW TO Given $f(x) = x^3 - x + 1$ and $g(x) = 2x^2 - 10$, evaluate $(g \circ f)(2)$.

$$f(2) = (2)^3 - (2) + 1 = 7$$
$$(g \circ f)(2) = g[f(2)]$$
$$= g(7)$$
$$= 2(7)^2 - 10 = 88$$

> **HOW TO** Given $f(x) = 3x - 2$ and $g(x) = x^2 - 2x$, find $(f \circ g)(x)$.
>
> $$(f \circ g)(x) = f[g(x)]$$
> $$= 3(x^2 - 2x) - 2$$
> $$= 3x^2 - 6x - 2$$

When we evaluate compositions of functions, the order in which the functions are applied is important. In the two diagrams below, the order of the *square* function, $g(x) = x^2$, and the *double* function, $f(x) = 2x$, is interchanged. Note that the final outputs are different. Therefore, $(f \circ g)(x) \neq (g \circ f)(x)$.

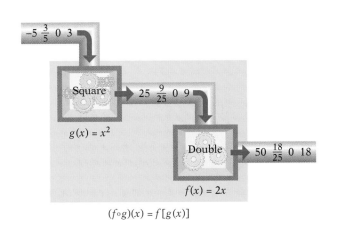

$(f \circ g)(x) = f[g(x)]$

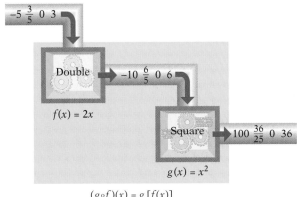

$(g \circ f)(x) = g[f(x)]$

Example 4

Given $f(x) = x^2 - x$ and $g(x) = 3x - 2$, find $f[g(3)]$.

Solution

$g(x) = 3x - 2$

$g(3) = 3(3) - 2 = 9 - 2 = 7$ • Evaluate *g* at 3.

$f(x) = x^2 - x$ • Evaluate *f* at $g(3) = 7$.

$f[g(3)] = f(7) = 7^2 - 7 = 42$

You Try It 4

Given $f(x) = 1 - 2x$ and $g(x) = x^2$, find $f[g(-1)]$.

Your solution

Example 5

Given $s(t) = t^2 + 3t - 1$ and $v(t) = 2t + 1$, determine $s[v(t)]$.

Solution

$s(t) = t^2 + 3t - 1$ • Evaluate *s* at *v*(*t*).

$s[v(t)] = (2t + 1)^2 + 3(2t + 1) - 1$

$= (4t^2 + 4t + 1) + 6t + 3 - 1$

$= 4t^2 + 10t + 3$

You Try It 5

Given $L(s) = s + 1$ and $M(s) = s^3 + 1$, determine $M[L(s)]$.

Your solution

Solutions on p. S28

9.3 Exercises

To perform operations on functions

For Exercises 1 to 9, let $f(x) = 2x^2 - 3$ and $g(x) = -2x + 4$. Find:

1. $f(2) - g(2)$

2. $f(3) - g(3)$

3. $f(0) + g(0)$

4. $f(1) + g(1)$

5. $(f \cdot g)(2)$

6. $(f \cdot g)(-1)$

7. $\left(\dfrac{f}{g}\right)(4)$

8. $\left(\dfrac{f}{g}\right)(-1)$

9. $\left(\dfrac{g}{f}\right)(-3)$

For Exercises 10 to 18, let $f(x) = 2x^2 + 3x - 1$ and $g(x) = 2x - 4$. Find:

10. $f(-3) + g(-3)$

11. $f(1) + g(1)$

12. $f(-2) - g(-2)$

13. $f(4) - g(4)$

14. $(f \cdot g)(-2)$

15. $(f \cdot g)(1)$

16. $\left(\dfrac{f}{g}\right)(2)$

17. $\left(\dfrac{f}{g}\right)(-3)$

18. $(f \cdot g)\left(\dfrac{1}{2}\right)$

For Exercises 19 to 21, let $f(x) = x^2 + 3x - 5$ and $g(x) = x^3 - 2x + 3$. Find:

19. $f(2) - g(2)$

20. $(f \cdot g)(-3)$

21. $\left(\dfrac{f}{g}\right)(-2)$

22. ✎ Explain the meaning of the notation $f[g(3)]$.

23. ✎ Explain the meaning of the expression $(f \circ g)(-2)$.

To find the composition of two functions

Given $f(x) = 2x - 3$ and $g(x) = 4x - 1$, evaluate the composite functions in Exercises 24 to 29.

24. $f[g(0)]$

25. $g[f(0)]$

26. $f[g(2)]$

27. $g[f(-2)]$

28. $f[g(x)]$

29. $g[f(x)]$

Given $h(x) = 2x + 4$ and $f(x) = \frac{1}{2}x + 2$, evaluate the composite functions in Exercises 30 to 35.

30. $h[f(0)]$ **31.** $f[h(0)]$ **32.** $h[f(2)]$

33. $f[h(-1)]$ **34.** $h[f(x)]$ **35.** $f[h(x)]$

Given $g(x) = x^2 + 3$ and $h(x) = x - 2$, evaluate the composite functions in Exercises 36 to 41.

36. $g[h(0)]$ **37.** $h[g(0)]$ **38.** $g[h(4)]$

39. $h[g(-2)]$ **40.** $g[h(x)]$ **41.** $h[g(x)]$

Given $f(x) = x^2 + x + 1$ and $h(x) = 3x + 2$, evaluate the composite functions in Exercises 42 to 47.

42. $f[h(0)]$ **43.** $h[f(0)]$ **44.** $f[h(-1)]$

45. $h[f(-2)]$ **46.** $f[h(x)]$ **47.** $h[f(x)]$

Given $f(x) = x - 2$ and $g(x) = x^3$, evaluate the composite functions in Exercises 48 to 53.

48. $f[g(2)]$ **49.** $f[g(-1)]$ **50.** $g[f(2)]$

51. $g[f(-1)]$ **52.** $f[g(x)]$ **53.** $g[f(x)]$

54. **Oil Spills** Suppose the spread of an oil leak from a tanker can be approximated by a circle with the tanker at its center and radius r, in feet. The radius of the spill t hours after the beginning of the leak is given by $r(t) = 45t$.
 a. Find the area of the spill as a function of time.
 b. What is the area of the spill after 3 h? Round to the nearest whole number.

55. **Manufacturing** Suppose the manufacturing cost, in dollars, per digital camera is given by the function $M(x) = \dfrac{50x + 10{,}000}{x}$. A camera store will sell the cameras by marking up the manufacturing cost per camera, $M(x)$, by 60%.
 a. Express the selling price of a camera as a function of the number of cameras to be manufactured. That is, find $S \circ M$.
 b. Find $(S \circ M)(5000)$.
 c. Explain the meaning of the answer to part **b**.

56. Manufacturing The number of electric scooters e that a factory can produce per day is a function of the number of hours h it operates and is given by $e(h) = 250h$ for $0 \le h \le 10$. The daily cost c to manufacture e electric scooters is given by the function $c(e) = 0.05e^2 + 60e + 1000$.

 a. Find $(c \circ e)(h)$.

 b. Evaluate $(c \circ e)(10)$.

 c. Write a sentence that explains the meaning of the answer to part **b**.

57. Automobile Rebates A car dealership offers a \$1500 rebate and a 10% discount off the price of a new car. Let p be the sticker price of a new car on the dealer's lot, r the price after the rebate, and d the discounted price. Then $r(p) = p - 1500$ and $d(p) = 0.90p$.

 a. Write a composite function for the dealer taking the rebate first and then the discount.

 b. Write a composite function for the dealer taking the discount first and then the rebate.

 c. Which composite function would you prefer the dealer to use when you buy a new car?

APPLYING THE CONCEPTS

The graphs of f and g are shown at the right. Use the graphs to determine the values of the composite functions in Exercises 58 to 63.

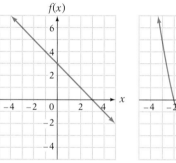

58. $f[g(-1)]$ **59.** $g[f(1)]$

60. $(g \circ f)(2)$ **61.** $(f \circ g)(3)$

62. $g[f(3)]$ **63.** $f[g(0)]$

For Exercises 64 to 69, let $g(x) = x^2 - 1$. Find:

64. $g(2 + h)$ **65.** $g(3 + h) - g(3)$ **66.** $g(-1 + h) - g(-1)$

67. $\dfrac{g(1 + h) - g(1)}{h}$ **68.** $\dfrac{g(-2 + h) - g(-2)}{h}$ **69.** $\dfrac{g(a + h) - g(a)}{h}$

For Exercises 70 to 75, let $f(x) = 2x$, $g(x) = 3x - 1$, and $h(x) = x - 2$. Find:

70. $f(g[h(2)])$ **71.** $g(h[f(1)])$ **72.** $h(g[f(-1)])$

73. $f(h[g(0)])$ **74.** $f(g[h(x)])$ **75.** $g(f[h(x)])$

9.4 One-to-One and Inverse Functions

Objective A **To determine whether a function is one-to-one**

Recall that a function is a set of ordered pairs in which no two ordered pairs that have the same first coordinate have different second coordinates. This means that given any x, there is only one y that can be paired with that x. A **one-to-one function** satisfies the additional condition that given any y, there is only one x that can be paired with the given y. One-to-one functions are commonly written as 1–1.

> **One-to-One Function**
>
> A function f is a 1–1 function if for any a and b in the domain of f, $f(a) = f(b)$ implies that $a = b$.

This definition states that if the y-coordinates of an ordered pair are equal, $f(a) = f(b)$, then the x-coordinates must be equal, $a = b$.

The function defined by $f(x) = 2x + 1$ is a 1–1 function. To show this, determine $f(a)$ and $f(b)$. Then form the equation $f(a) = f(b)$.

$$f(a) = 2a + 1 \qquad f(b) = 2b + 1$$

$$f(a) = f(b)$$
$$2a + 1 = 2b + 1$$
$$2a = 2b \qquad \bullet \text{ Subtract 1 from each side of the equation.}$$
$$a = b \qquad \bullet \text{ Divide each side of the equation by 2.}$$

Because $f(a) = f(b)$ implies that $a = b$, the function is a 1–1 function.

Consider the function defined by $g(x) = x^2 - x$. Evaluate the function at -2 and 3.

$$g(-2) = (-2)^2 - (-2) = 6 \qquad g(3) = 3^2 - 3 = 6$$

From this evaluation, $g(-2) = 6$ and $g(3) = 6$, but $-2 \neq 3$. Thus g is not a 1–1 function.

The graphs of $f(x) = 2x + 1$ and $g(x) = x^2 - x$ are shown below. Note that a horizontal line intersects the graph of f at no more than one point. However, a horizontal line intersects the graph of g at more than one point.

$f(x) = 2x + 1$

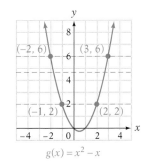

$g(x) = x^2 - x$

Looking at the graph of f on the previous page, note that for each y-coordinate there is only one x-coordinate. Thus f is a 1–1 function. From the graph of g, however, there are *two* x-coordinates for a given y-coordinate. For instance, $(-2, 6)$ and $(3, 6)$ are the coordinates of two points on the graph for which the y-coordinates are the same and the x-coordinates are different. Therefore, g is not a 1–1 function.

> **Horizontal-Line Test**
>
> The graph of a function represents the graph of a 1–1 function if any horizontal line intersects the graph at no more than one point.

Example 1	**You Try It 1**
Determine whether the graph is the graph of a 1–1 function.	Determine whether the graph is the graph of a 1–1 function.
Solution	**Your solution**
Because a horizontal line intersects the graph more than once, the graph is not the graph of a 1–1 function.	

Solution on p. S28

Objective B **To find the inverse of a function**

The **inverse of a function** is the set of ordered pairs formed by reversing the coordinates of each ordered pair of the function.

For example, the set of ordered pairs of the function defined by $f(x) = 2x$ with domain $\{-2, -1, 0, 1, 2\}$ is $\{(-2, -4), (-1, -2), (0, 0), (1, 2), (2, 4)\}$. The set of ordered pairs of the inverse function is $\{(-4, -2), (-2, -1), (0, 0), (2, 1), (4, 2)\}$.

From the ordered pairs of f, we have

$$\text{Domain} = \{-2, -1, 0, 1, 2\} \quad \text{and} \quad \text{range} = \{-4, -2, 0, 2, 4\}$$

From the ordered pairs of the inverse function, we have

$$\text{Domain} = \{-4, -2, 0, 2, 4\} \quad \text{and} \quad \text{range} = \{-2, -1, 0, 1, 2\}$$

Note that the domain of the inverse function is the range of the original function, and the range of the inverse function is the domain of the original function.

Now consider the function defined by $g(x) = x^2$ with domain $\{-2, -1, 0, 1, 2\}$. The set of ordered pairs of this function is $\{(-2, 4), (-1, 1), (0, 0), (1, 1), (2, 4)\}$. Reversing the ordered pairs gives $\{(4, -2), (1, -1), (0, 0), (1, 1), (4, 2)\}$. These ordered pairs do not satisfy the condition of a function, because there are ordered pairs with the same first coordinate and different second coordinates. This example illustrates that not all functions have an inverse function.

The graphs of $f(x) = 2x$ and $g(x) = x^2$, with the set of real numbers as the domain, are shown below.

$f(x) = 2x$

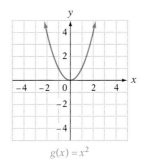

$g(x) = x^2$

By the horizontal-line test, f is a 1–1 function but g is not.

> **Condition for an Inverse Function**
>
> A function f has an inverse function if and only if f is a 1–1 function.

The symbol f^{-1} is used to denote the inverse of the function f. The symbol $f^{-1}(x)$ is read "f inverse of x."

$f^{-1}(x)$ is *not* the reciprocal of $f(x)$ but is, rather, the notation for the inverse of a 1–1 function.

To find the inverse of a function, interchange x and y. Then solve for y.

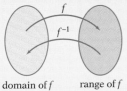
HOW TO Find the inverse of the function defined by $f(x) = 3x + 6$.

$$f(x) = 3x + 6$$
$$y = 3x + 6$$ • Replace $f(x)$ by y.
$$x = 3y + 6$$ • Interchange x and y.
$$x - 6 = 3y$$ • Solve for y.
$$\frac{1}{3}x - 2 = y$$

$$f^{-1}(x) = \frac{1}{3}x - 2$$ • Replace y by $f^{-1}(x)$.

The inverse of the function $f(x) = 3x + 6$ is $f^{-1}(x) = \frac{1}{3}x - 2$.

The fact that the ordered pairs of the inverse of a function are the reverse of those of the original function has a graphical interpretation. In the graph in the middle, the points from the figure on the left are plotted with the coordinates reversed. The inverse function is graphed by drawing a smooth curve through those points, as shown in the figure on the right.

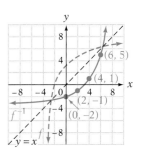

Note that the dashed graph of $y = x$ is shown in the figure on the bottom right of the preceding page. If two functions are inverses of each other, their graphs are mirror images with respect to the graph of the line $y = x$.

The composition of a function and its inverse has a special property.

Composition of Inverse Functions Property

$$f^{-1}[f(x)] = x \quad \text{and} \quad f[f^{-1}(x)] = x$$

This property can be used to determine whether two functions are inverses of each other.

HOW TO Are $f(x) = 2x - 4$ and $g(x) = \frac{1}{2}x + 2$ inverses of each other?

To determine whether the functions are inverses, use the Composition of Inverse Functions Property.

$$f[g(x)] = 2\left(\frac{1}{2}x + 2\right) - 4 \qquad\qquad g[f(x)] = \frac{1}{2}(2x - 4) + 2$$
$$= x + 4 - 4 \qquad\qquad\qquad\qquad = x - 2 + 2$$
$$= x \qquad\qquad\qquad\qquad\qquad\qquad = x$$

Because $f[g(x)] = x$ and $g[f(x)] = x$, the functions are inverses of each other.

Example 2

Find the inverse of the function defined by $f(x) = 2x - 3$.

Solution

$$f(x) = 2x - 3$$
$$y = 2x - 3 \qquad \bullet \text{ Replace } f(x) \text{ by } y.$$
$$x = 2y - 3 \qquad \bullet \text{ Interchange } x \text{ and } y.$$
$$x + 3 = 2y \qquad \bullet \text{ Solve for } y.$$
$$\frac{x}{2} + \frac{3}{2} = y$$
$$f^{-1}(x) = \frac{x}{2} + \frac{3}{2} \qquad \bullet \text{ Replace } y \text{ by } f^{-1}(x).$$

The inverse of the function is given by $f^{-1}(x) = \frac{x}{2} + \frac{3}{2}$.

You Try It 2

Find the inverse of the function defined by $f(x) = \frac{1}{2}x + 4$.

Your solution

Example 3

Are $f(x) = 3x - 6$ and $g(x) = \frac{1}{3}x + 2$ inverses of each other?

Solution

$$f[g(x)] = 3\left(\frac{1}{3}x + 2\right) - 6 = x + 6 - 6 = x$$

$$g[f(x)] = \frac{1}{3}(3x - 6) + 2 = x - 2 + 2 = x$$

Yes, the functions are inverses of each other.

You Try It 3

Are $f(x) = 2x - 6$ and $g(x) = \frac{1}{2}x - 3$ inverses of each other?

Your solution

Solutions on p. S28

9.4 Exercises

Objective A **To determine whether a function is one-to-one**

1. ✏ What is a 1–1 function?

2. ✏ What is the horizontal-line test?

For Exercises 3 to 14, determine whether the graph represents the graph of a 1–1 function.

3.

4.

5.

6.

7.

8.

9.

10.

11.

12.

13.

14.

Objective B **To find the inverse of a function**

15. How are the ordered pairs of the inverse of a function related to the function?

16. Why is it that not all functions have an inverse function?

For Exercises 17 to 24, find the inverse of the function. If the function does not have an inverse function, write "no inverse."

17. $\{(1, 0), (2, 3), (3, 8), (4, 15)\}$

18. $\{(1, 0), (2, 1), (-1, 0), (-2, 0)\}$

19. $\{(3, 5), (-3, -5), (2, 5), (-2, -5)\}$

20. $\{(-5, -5), (-3, -1), (-1, 3), (1, 7)\}$

21. $\{(0, -2), (-1, 5), (3, 3), (-4, 6)\}$

22. $\{(-2, -2), (0, 0), (2, 2), (4, 4)\}$

23. $\{(-2, -3), (-1, 3), (0, 3), (1, 3)\}$

24. $\{(2, 0), (1, 0), (3, 0), (4, 0)\}$

For Exercises 25 to 42, find $f^{-1}(x)$.

25. $f(x) = 4x - 8$

26. $f(x) = 3x + 6$

27. $f(x) = 2x + 4$

28. $f(x) = x - 5$

29. $f(x) = \dfrac{1}{2}x - 1$

30. $f(x) = \dfrac{1}{3}x + 2$

31. $f(x) = -2x + 2$

32. $f(x) = -3x - 9$

33. $f(x) = \dfrac{2}{3}x + 4$

34. $f(x) = \dfrac{3}{4}x - 4$

35. $f(x) = -\dfrac{1}{3}x + 1$

36. $f(x) = -\dfrac{1}{2}x + 2$

37. $f(x) = 2x - 5$ **38.** $f(x) = 3x + 4$ **39.** $f(x) = 5x - 2$

40. $f(x) = 4x - 2$ **41.** $f(x) = 6x - 3$ **42.** $f(x) = -8x + 4$

For Exercises 43 to 45, let $f(x) = 3x - 5$. Find:

43. $f^{-1}(0)$ **44.** $f^{-1}(2)$ **45.** $f^{-1}(4)$

For Exercises 46 to 48, state whether the graph is the graph of a function. If it is the graph of a function, does the function have an inverse?

46.

47.

48.
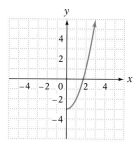

For Exercises 49 to 56, use the Composition of Inverse Functions Property to determine whether the functions are inverses.

49. $f(x) = 4x; g(x) = \dfrac{x}{4}$ **50.** $g(x) = x + 5; h(x) = x - 5$

51. $f(x) = 3x; h(x) = \dfrac{1}{3x}$ **52.** $h(x) = x + 2; g(x) = 2 - x$

53. $g(x) = 3x + 2; f(x) = \dfrac{1}{3}x - \dfrac{2}{3}$ **54.** $h(x) = 4x - 1; f(x) = \dfrac{1}{4}x + \dfrac{1}{4}$

55. $f(x) = \dfrac{1}{2}x - \dfrac{3}{2}; g(x) = 2x + 3$ **56.** $g(x) = -\dfrac{1}{2}x - \dfrac{1}{2}; h(x) = -2x + 1$

APPLYING THE CONCEPTS

Given the graph of the 1–1 function in each of Exercises 57 to 62, draw the graph of the inverse of the function by using the technique shown in Objective B of this section.

57.

58.

59.

60.

61.

62.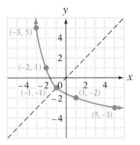

Each of the tables in Exercises 63 and 64 defines a function. Is the inverse of the function a function? Explain your answer.

63. **Grading Scale**

Score	Grade
90–100	A
80–89	B
70–79	C
60–69	D
0–59	F

64. **First-Class Postage Rates**

Weight	Cost
$0 < w < 1$	$.37
$1 < w \leq 2$	$.60
$2 < w \leq 3$	$.83
$3 < w \leq 4$	$1.06

Given that f is a 1–1 function and that $f(0) = 1$, $f(3) = -1$, and $f(5) = -3$, find the inverse functions in Exercises 65 to 67.

65. $f^{-1}(-3)$ **66.** $f^{-1}(-1)$ **67.** $f^{-1}(1)$

Given that f is a 1–1 function and that $f(-3) = 3$, $f(-4) = 7$, and $f(0) = 8$, find the inverse functions in Exercises 68 to 70.

68. $f^{-1}(3)$ **69.** $f^{-1}(7)$ **70.** $f^{-1}(8)$

71. Is the inverse of a constant function a function? Explain your answer.

72. The graphs of all functions given by $f(x) = mx + b$, $m \neq 0$, are straight lines. Are all of these functions 1–1 functions? If so, explain why. If not, give an example of a linear function that is not 1–1.

Focus on Problem Solving

Proof in Mathematics

Mathematics is based on the concept of proof, which in turn is based on principles of logic. It was Aristotle (384 B.C. to 322 B.C.) who took the logical principles that had been developed over time and set them in a structured form. From this structured form, several methods of proof were developed.

$$\text{A true} \xrightarrow{\text{implies}} \text{B true} \xrightarrow{\text{implies}} \text{C true}$$
$$\text{A true} \xrightarrow{\text{implies}} \text{C true}$$

A deductive proof is one that proceeds in a carefully ordered sequence of steps. It is the type of proof you may have seen in a geometry class. Deductive proofs can be modeled as shown at the left.

Here is an example of a deductive proof. It proves that the product of a negative number and a positive number is a negative number. The proof is based on the fact that every real number has exactly one additive inverse. For instance, the additive inverse of 4 is -4 and the additive inverse of $-\frac{2}{3}$ is $\frac{2}{3}$.

Theorem: If a and b are positive real numbers, then $(-a)b = -(ab)$.

Proof:
$$\begin{aligned} (-a)b + ab &= (-a + a)b \\ &= 0 \cdot b \\ &= 0 \end{aligned}$$

- Distributive Property
- Inverse Property of Addition
- Multiplication Property of Zero

Since $(-a)b + ab = 0$, $(-a)b$ is the additive inverse of ab. But $-(ab)$ is also the additive inverse of ab and because every number has exactly one additive inverse, $(-a)b = -(ab)$.

Here is another example of a deductive proof. It is a proof of part of the Pythagorean Theorem.

Theorem: If a and b are the lengths of the legs of a right triangle and c is the length of the hypotenuse, then $a^2 + b^2 = c^2$.

Proof: Let ΔDBA be a right triangle and let BC be the altitude to the hypotenuse of ΔDBA, as shown at the left. Then ΔBCD and ΔBCA are right triangles.

1. ΔDBA is a right triangle with height BC.

 Given.

2. ΔDBC is similar to ΔBAC and ΔDBA is similar to ΔBCA.

 Theorem from geometry: the altitude to the hypotenuse of a right triangle forms triangles that are similar to each other.

3. $\dfrac{a}{b} = \dfrac{x}{a}$

 The ratios of corresponding sides of similar triangles are equal.

4. $a^2 = bx$

 Multiply each side by ab.

5. $\dfrac{x + b}{c} = \dfrac{c}{b}$

 The ratios of corresponding sides of similar triangles are equal.

6. $bx + b^2 = c^2$

 Multiply each side by cb.

7. $a^2 + b^2 = c^2$

 Substitution Property from (**4**).

Because the first statement of the proof is true and each subsequent statement of the proof is true, the first statement implies that the last statement is true. Therefore, we have proved: If a and b are the lengths of the legs of a right triangle and c is the length of the hypotenuse, then $a^2 + b^2 = c^2$.

Projects and Group Activities

Finding the Maximum or Minimum of a Function Using a Graphing Calculator

Recall that for a quadratic function, the maximum or minimum value of the function is the y-coordinate of the vertex of the graph of the function. We determined algebraically that the coordinates of the vertex are $\left(-\dfrac{b}{2a}, f\left(-\dfrac{b}{2a}\right)\right)$. It is also possible to use a graphing calculator to approximate a maximum or minimum value. The method used to find the value depends on the calculator model. However, it is generally true for most calculators that the point at which the function has a maximum or minimum must be shown on the screen. If it is not, you must select a different viewing window.

For a TI-83 graphing calculator, press 2nd CALC 3 to determine a minimum value or 2nd CALC 4 to determine a maximum value. Move the cursor to a point on the curve that is to the left of the minimum (or maximum). Press ENTER. Move the cursor to a point on the curve that is to the right of the minimum (or maximum). Press ENTER twice. The minimum (or maximum) value is shown as the y-coordinate at the bottom of the screen.

1. Find the maximum or minimum value of $f(x) = -x^2 - 3x + 2$ by using a graphing calculator.

2. Find the maximum or minimum value of $f(x) = x^2 - 4x - 2$ by using a graphing calculator.

3. Find the maximum or minimum value of $f(x) = -\sqrt{2}x^2 - \pi x + 2$ by using a graphing calculator.

4. Find the maximum or minimum value of $f(x) = \dfrac{x^2}{\sqrt{5}} - \sqrt{7}x - \sqrt{11}$ by using a graphing calculator.

5. **Business** A manufacturer of camera lenses estimated that the average monthly cost C of production is given by the function $C(x) = 0.1x^2 - 20x + 2000$, where x is the number of lenses produced each month. Find the number of lenses the company should produce in order to minimize the average cost.

6. **Compensation** The net annual income I, in dollars, of a family physician can be modeled by the equation $I(x) = -290(x - 48)^2 + 148{,}000$, where x is the age of the physician and $27 \le x \le 70$.
 a. Find the age at which the physician's income will be a maximum.
 b. Find the maximum income.

Business Applications of Maximum and Minimum Values of Quadratic Functions

A company's revenue, R, is the total amount of money the company earned by selling its products. The cost, C, is the total amount of money the company spent to manufacture and sell its products. A company's profit, P, is the difference between the revenue and the cost.

$$P = R - C$$

A company's revenue and cost may be represented by equations.

HOW TO The owners of a travel agency are selling tickets for a tour. The travel agency must pay a cost per person of $10. A maximum of 180 people can be accommodated on the tour. For every $.25 increase in the price of a ticket over the $10 cost, the owners estimate that they will sell 1 ticket less. Therefore, the price per ticket can be represented by

$$p = 10 + 0.25(180 - x)$$
$$= 55 - 0.25x$$

where x represents the number of tickets sold. Determine the maximum profit and the cost per ticket that yields the maximum profit.

Solution

The cost is represented by $10x$ and the revenue is represented by $x(55 - 0.25x)$, where x represents the number of tickets sold.

Use the equation $P = R - C$ to represent the profit.

$$P = R - C$$
$$P = x(55 - 0.25x) - 10x$$
$$P = -0.25x^2 + 45x$$

The graph of $P = -0.25x^2 + 45x$ is a parabola that opens down. Find the x-coordinates of the vertex.

$$x = -\frac{b}{2a} = -\frac{45}{2(-0.25)} = 90$$

To find the maximum profit, evaluate $P = -0.25x^2 + 45x$ at 90.

$$P = -0.25x^2 + 45x$$
$$P = -0.25(90^2) + 45(90)$$
$$P = 2025$$

To find the price per ticket, evaluate $p = 55 - 0.25x$ at 90.

$$p = 55 - 0.25x$$
$$p = 55 - 0.25(90)$$
$$p = 32.50$$

The travel agency can expect a maximum profit of $2025 when 90 people take the tour at a ticket price of $32.50 per person.

The graph of the profit function is shown at the left.

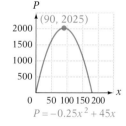

$P = -0.25x^2 + 45x$

1. **Business** A company's total monthly cost, in dollars, for producing and selling n DVDs per month is $C = 25n + 4000$. The company's revenue, in dollars, from selling all n DVDs is $R = 275n - 0.2n^2$.
 a. How many DVDs must the company produce and sell each month in order to make the maximum monthly profit?
 b. What is the maximum monthly profit?

2. **Business** A shipping company has determined that its cost to deliver x packages per week is $1000 + 6x$. The price per package that the company charges when it sends x packages per week is $24 - 0.02x$. Determine the company's maximum weekly profit and the price per package that yields the maximum profit.

Chapter 9 Summary

Key Words	**Examples**

A *quadratic function* is one that can be expressed by the equation $f(x) = ax^2 + bx + c, a \neq 0$. [9.1A, p. 493]

$f(x) = 4x^2 + 3x - 5$ is a quadratic function.

The value of the independent variable is sometimes called the *input*. The result of evaluating the function is called the *output*. An *input/output* table shows the result of evaluating a function for various values of the independent variable. [9.1A, p. 493]

$f(x) = x^2 + 4x - 6$

x	y
-2	-10
-1	-9
0	-6
1	-1
2	6

The graph of a quadratic function is a *parabola*. When $a > 0$, the parabola opens up and the *vertex* of the parabola is the point with the smallest y-coordinate. When $a < 0$, the parabola opens down and the *vertex* of the parabola is the point with the largest y-coordinate. The *axis of symmetry* is the vertical line that passes through the vertex of the parabola and is parallel to the y-axis. [9.1A, pp. 493, 494]

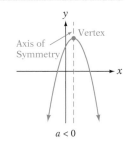

A point at which a graph crosses the x-axis is called an *x-intercept* of the graph. The x-intercepts occur when $y = 0$. [9.1B, p. 496]

$$x^2 - 5x + 6 = 0$$
$$(x - 2)(x - 3) = 0$$
$$x - 2 = 0 \qquad x - 3 = 0$$
$$x = 2 \qquad\quad x = 3$$

The x-intercepts are $(2, 0)$ and $(3, 0)$.

A value of x for which $f(x) = 0$ is a *zero* of the function. [9.1B, p. 496]

$$f(x) = x^2 - 7x + 10$$
$$0 = x^2 - 7x + 10$$
$$0 = (x - 2)(x - 5)$$
$$x - 2 = 0 \qquad x - 5 = 0$$
$$x = 2 \qquad\quad x = 5$$

2 and 5 are zeros of the function.

The graph of a quadratic function has a *minimum value* if $a > 0$ and a *maximum value* if $a < 0$. [9.1C, p. 499]

The graph of $f(x) = 3x^2 - x + 4$ has a minimum value.
The graph of $f(x) = -x^2 - 6x + 5$ has a maximum value.

A function is a *one-to-one function* if, for any a and b in the domain of f, $f(a) = f(b)$ implies $a = b$. This means that given any y, there is only one x that can be paired with the given y. One-to-one functions are commonly written as 1–1. [9.4A, p. 523]

A linear function is a 1–1 function. A quadratic function is not a 1–1 function.

The *inverse of a function* is the set of ordered pairs formed by reversing the coordinates of each ordered pair of the function. [9.4B, p. 524]

The inverse of the function $\{(1, 2), (2, 4), (3, 6), (4, 8), (5, 10)\}$ is $\{(2, 1), (4, 2), (6, 3), (8, 4), (10, 5)\}$.

Essential Rules and Procedures

Examples

Vertex and Axis of Symmetry of a Parabola [9.1A, p. 494]
Let $f(x) = ax^2 + bx + c$ be the equation of the parabola. The coordinates of the vertex are $\left(-\dfrac{b}{2a}, f\left(-\dfrac{b}{2a} \right) \right)$. The equation of the axis of symmetry is $x = -\dfrac{b}{2a}$.

$f(x) = x^2 - 2x - 4$
$a = 1, b = -2$
$-\dfrac{b}{2a} = -\dfrac{-2}{2(1)} = 1$
$f(1) = 1^2 - 2(1) - 4 = -5$

The coordinates of the vertex are $(1, -5)$.
The axis of symmetry is $x = 1$.

Effect of the Discriminant on the Number of *x*-Intercepts of a Parabola [9.1B, p. 497]
1. If $b^2 - 4ac = 0$, the parabola has one *x*-intercept.
2. If $b^2 - 4ac > 0$, the parabola has two *x*-intercepts.
3. If $b^2 - 4ac < 0$, the parabola has no *x*-intercepts.

$x^2 + 8x + 16 = 0$ has one *x*-intercept because $b^2 - 4ac = 8^2 - 4(1)(16) = 0$.

$2x^2 + 3x - 5 = 0$ has two *x*-intercepts because $b^2 - 4ac = 3^2 - 4(2)(-5) = 49$.

$3x^2 + 2x + 4 = 0$ has no *x*-intercepts because $b^2 - 4ac = 2^2 - 4(3)(4) = -44$.

To Find the Minimum or Maximum Value of a Quadratic Function [9.1C, p. 499]
First find the *x*-coordinate of the vertex. Then evaluate the function at that value.

$f(x) = x^2 - 2x - 4$
$a = 1, b = -2$
$-\dfrac{b}{2a} = -\dfrac{-2}{2(1)} = 1$
$f(1) = 1^2 - 2(1) - 4 = -5$
$a > 0$. The minimum value of the function is -5.

Vertical-Line Test [9.2A, p. 510]

A graph defines a function if any vertical line intersects the
graph at no more than one point.

A function

Not a function

Operations on Functions [9.3A, p. 515]

If f and g are functions and x is an element of the domain of
each function, then

$(f + g)(x) = f(x) + g(x)$
$(f - g)(x) = f(x) - g(x)$
$(f \cdot g)(x) = f(x) \cdot g(x)$
$\left(\dfrac{f}{g}\right)(x) = \dfrac{f(x)}{g(x)}, g(x) \neq 0$

Given $f(x) = x + 2$ and $g(x) = 2x$, then
$(f + g)(4) = f(4) + g(4)$
$\qquad = (4 + 2) + 2(4) = 6 + 8$
$\qquad = 14$
$(f - g)(4) = f(4) - g(4)$
$\qquad = (4 + 2) - 2(4) = 6 - 8$
$\qquad = -2$
$(f \cdot g)(4) = f(4) \cdot g(4)$
$\qquad = (4 + 2) \cdot 2(4) = 6 \cdot 8$
$\qquad = 48$
$\left(\dfrac{f}{g}\right)(4) = \dfrac{f(4)}{g(4)} = \dfrac{4 + 2}{2(4)} = \dfrac{6}{8} = \dfrac{3}{4}$

Definition of the Composition of Two Functions
[9.3B, p. 517]

The composition of two functions f and g, symbolized by $f \circ g$,
is the function whose value at x is given by $(f \circ g)(x) = f[g(x)]$.

Given $f(x) = x - 4$ and $g(x) = 4x$, then
$(f \circ g)(2) = f[g(2)]$
$\qquad = f(8) \qquad$ because $g(2) = 8$
$\qquad = 8 - 4 = 4$

Horizontal-Line Test [9.4A, p. 524]

The graph of a function represents the graph of a 1–1 function
if any horizontal line intersects the graph at no more than one
point.

A 1-1
function

Not a 1-1
function

Condition for an Inverse Function [9.4B, p. 525]

A function f has an inverse function if and only if f is a 1–1
function.

The function $f(x) = x^2$ does not have
an inverse function. When $y = 4$,
$x = 2$ or -2; therefore, the function
$f(x)$ is not a 1–1 function.

Composition of Inverse Functions Property [9.4B, p. 526]

$f^{-1}[f(x)] = x$ and $f[f^{-1}(x)] = x$.

$f(x) = 2x - 3 \qquad f^{-1}(x) = \dfrac{1}{2}x + \dfrac{3}{2}$

$f^{-1}[f(x)] = \dfrac{1}{2}(2x - 3) + \dfrac{3}{2} = x$

$f[f^{-1}(x)] = 2\left(\dfrac{1}{2}x + \dfrac{3}{2}\right) - 3 = x$

Chapter 9 Review Exercises

1. Is the graph below the graph of a function?

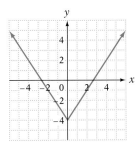

2. Is the graph below the graph of a 1–1 function?

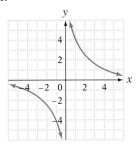

3. Graph $f(x) = 3x^3 - 2$. State the domain and range.

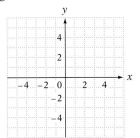

4. Graph $f(x) = \sqrt{x + 4}$. State the domain and range.

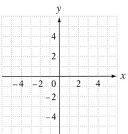

5. Use the discriminant to determine the number of x-intercepts of $y = -3x^2 + 4x + 6$.

6. Find the x-intercepts of $y = 3x^2 + 9x$.

7. Find the zeros of $f(x) = 3x^2 + 2x + 2$.

8. Find the maximum value of $f(x) = -2x^2 + 4x + 1$.

9. Find the minimum value of $f(x) = x^2 - 7x + 8$.

10. Given $f(x) = x^2 + 4$ and $g(x) = 4x - 1$, find $f[g(0)]$.

11. Given $f(x) = 6x + 8$ and $g(x) = 4x + 2$, find $g[f(-1)]$.

12. Given $f(x) = 3x^2 - 4$ and $g(x) = 2x + 1$, find $f[g(x)]$.

13. Are the functions given by $f(x) = -\frac{1}{4}x + \frac{5}{4}$ and $g(x) = -4x + 5$ inverses of each other?

14. Graph $f(x) = x^2 + 2x - 4$. State the domain and range.

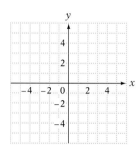

15. Find the vertex and axis of symmetry of the parabola whose equation is $y = x^2 - 2x + 3$. Then graph the equation.

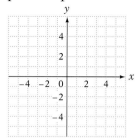

16. Graph $f(x) = |x| - 3$. State the domain and range.

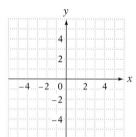

17. Is the graph below the graph of a 1–1 function?

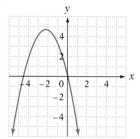

For Exercises 18 to 21, use $f(x) = x^2 + 2x - 3$ and $g(x) = x^2 - 2$.

18. Evaluate: $(f + g)(2)$

19. Evaluate: $(f - g)(-4)$

20. Evaluate: $(f \cdot g)(-4)$

21. Evaluate: $\left(\dfrac{f}{g}\right)(3)$

22. Given $f(x) = 2x^2 + x - 5$ and $g(x) = 3x - 1$, find $g[f(x)]$.

23. Find the inverse of $f(x) = -6x + 4$.

24. Find the inverse of $f(x) = \dfrac{2}{3}x - 12$.

25. Find the inverse of $f(x) = \dfrac{1}{2}x + 8$.

26. **Geometry** The perimeter of a rectangle is 28 ft. What dimensions would give the rectangle a maximum area?

Chapter 9 Test

1. Find the zeros of the function $f(x) = 2x^2 - 3x + 4$.

2. Find the x-intercepts of $y = x^2 + 3x - 8$.

3. Use the discriminant to determine the number of zeros of the parabola $y = 3x^2 + 2x - 4$.

4. Given $f(x) = x^2 + 2x - 3$ and $g(x) = x^3 - 1$, find $(f - g)(2)$.

5. Given $f(x) = x^3 + 1$ and $g(x) = 2x - 3$, find $(f \cdot g)(-3)$.

6. Given $f(x) = 4x - 5$ and $g(x) = x^2 + 3x + 4$, find $\left(\dfrac{f}{g}\right)(-2)$.

7. Given $f(x) = x^2 + 4$ and $g(x) = 2x^2 + 2x + 1$, find $(f - g)(-4)$.

8. Given $f(x) = 4x + 2$ and $g(x) = \dfrac{x}{x + 1}$, find $f[g(3)]$.

9. Given $f(x) = 2x^2 - 7$ and $g(x) = x - 1$, find $f[g(x)]$.

10. Find the maximum value of the function $f(x) = -x^2 + 8x - 7$.

11. Find the inverse of the function $f(x) = 4x - 2$.

12. Find the inverse of the function $f(x) = \frac{1}{4}x - 4$.

13. Find the inverse of the function {(2, 6), (3, 5), (4, 4), (5, 3)}.

14. Are the functions $f(x) = \frac{1}{2}x + 2$ and $g(x) = 2x - 4$ inverses of each other?

15. Find the maximum product of two numbers whose sum is 20.

16. Graph $f(x) = -\sqrt{3 - x}$. State the domain and range.

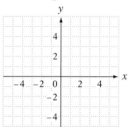

17. Graph $f(x) = \left|\frac{1}{2}x\right| - 2$. State the domain and range.

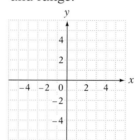

18. Graph $f(x) = x^3 - 3x + 2$. State the domain and range.

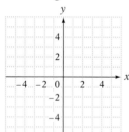

19. Determine whether the graph is the graph of a 1–1 function.

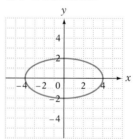

20. Geometry The perimeter of a rectangle is 200 cm. What dimensions would give the rectangle a maximum area? What is the maximum area?

Cumulative Review Exercises

1. Evaluate $-3a + \left|\dfrac{3b - ab}{3b - c}\right|$ when $a = 2$, $b = 2$, and $c = -2$.

2. Graph $\{x \mid x < -3\} \cap \{x \mid x > -4\}$.

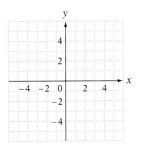

3. Solve: $\dfrac{3x - 1}{6} - \dfrac{5 - x}{4} = \dfrac{5}{6}$

4. Solve: $4x - 2 < -10$ or $3x - 1 > 8$
Write the solution set in set-builder notation.

5. Solve: $|8 - 2x| \geq 0$

6. Simplify: $\left(\dfrac{3a^3b}{2a}\right)^2 \left(\dfrac{a^2}{-3b^2}\right)^3$

7. Simplify: $(x - 4)(2x^2 + 4x - 1)$

8. Solve by using the addition method:
$6x - 2y = -3$
$4x + y = 5$

9. Factor: $x^3y + x^2y^2 - 6xy^3$

10. Solve: $(b + 2)(b - 5) = 2b + 14$

11. Solve: $x^2 - 2x > 15$

12. Simplify: $\dfrac{x^2 + 4x - 5}{2x^2 - 3x + 1} - \dfrac{x}{2x - 1}$

13. Solve: $\dfrac{5}{x^2 + 7x + 12} = \dfrac{9}{x + 4} - \dfrac{2}{x + 3}$

14. Simplify: $\dfrac{4 - 6i}{2i}$

15. Graph $f(x) = \dfrac{1}{4}x^2$. Find the vertex and axis of symmetry.

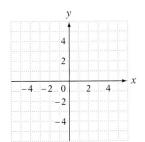

16. Graph the solution set of $3x - 4y \geq 8$.

17. Find the equation of the line containing the points $(-3, 4)$ and $(2, -6)$.

18. Find the equation of the line that contains the point $(-3, 1)$ and is perpendicular to the line $2x - 3y = 6$.

19. Solve: $3x^2 = 3x - 1$

20. Solve: $\sqrt{8x + 1} = 2x - 1$

21. Find the minimum value of the function $f(x) = 2x^2 - 3$.

22. Find the range of $f(x) = |3x - 4|$ if the domain is $\{0, 1, 2, 3\}$.

23. Is this set of ordered pairs a function? $\{(-3, 0), (-2, 0), (-1, 1), (0, 1)\}$

24. Solve: $\sqrt[3]{5x - 2} = 2$

25. Given $g(x) = 3x - 5$ and $h(x) = \frac{1}{2}x + 4$, find $g[h(2)]$.

26. Find the inverse of the function given by $f(x) = -3x + 9$.

27. **Mixtures** Find the cost per pound of a tea mixture made from 30 lb of tea costing $4.50 per pound and 45 lb of tea costing $3.60 per pound.

28. **Mixtures** How many pounds of an 80% copper alloy must be mixed with 50 lb of a 20% copper alloy to make an alloy that is 40% copper?

29. **Mixtures** Six ounces of an insecticide are mixed with 16 gal of water to make a spray for spraying an orange grove. How much additional insecticide is required if it is to be mixed with 28 gal of water?

30. **Tanks** A large pipe can fill a tank in 8 min less time than it takes a smaller pipe to fill the same tank. Working together, both pipes can fill the tank in 3 min. How long would it take the larger pipe, working alone, to fill the tank?

31. **Physics** The distance, d, that a spring stretches varies directly as the force, f, used to stretch the spring. If a force of 50 lb can stretch a spring 30 in., how far can a force of 40 lb stretch the spring?

32. **Music** The frequency of vibration, f, in a pipe in an open pipe organ varies inversely as the length, L, of the pipe. If the air in a pipe 2 m long vibrates 60 times per minute, find the frequency in a pipe that is 1.5 m long.

10 Exponential and Logarithmic Functions

This archaeopteryx skeleton is an exciting find for archaeologists. The archaeopteryx is thought to be from the Upper Jurassic period, and is considered by some scientists to be the most primitive of all birds. To determine exactly how old the bones are, archaeologists will use the carbon-dating method, which involves an exponential function. Instruments are used to detect the amount of carbon-14 left in an object. Since carbon-14 occurs naturally in living things and gradually decays after death, the amount of carbon-14 remaining can reveal the object's age. **The problem on page 574** gives an example of the carbon dating process.

Need help? For online student resources, such as section quizzes, visit this textbook's website at **math.college.hmco.com/students.**

OBJECTIVES

Section 10.1

A To evaluate an exponential function
B To graph an exponential function

Section 10.2

A To find the logarithm of a number
B To use the Properties of Logarithms to simplify expressions containing logarithms
C To use the Change-of-Base Formula

Section 10.3

A To graph a logarithmic function

Section 10.4

A To solve an exponential equation
B To solve a logarithmic equation

Section 10.5

A To solve application problems

Do these exercises to prepare for Chapter 10.

1. Simplify: 3^{-2}

2. Simplify: $\left(\dfrac{1}{2}\right)^{-4}$

3. Complete: $\dfrac{1}{8} = 2^{?}$

4. Evaluate $f(x) = x^4 + x^3$ for $x = -1$ and $x = 3$.

5. Solve: $3x + 7 = x - 5$

6. Solve: $16 = x^2 - 6x$

7. Evaluate $A(1 + i)^n$ for $A = 5000$, $i = 0.04$, and $n = 6$. Round to the nearest hundredth.

8. Graph: $f(x) = x^2 - 1$

GO FIGURE • • •

What is the ones digit of $1 + 9 + 9^2 + 9^3 + 9^4 + \cdots + 9^{2000} + 9^{2001}$?

10.1 Exponential Functions

Objective A **To evaluate an exponential function**

The growth of a $500 savings account that earns 5% annual interest compounded daily is shown in the graph at the right. In approximately 14 years, the savings account contains approximately $1000, twice the initial amount. The growth of this savings account is an example of an exponential function.

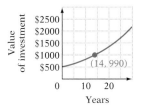

The pressure of the atmosphere at a certain height is shown in the graph at the right. This is another example of an exponential function. From the graph, we read that the air pressure is approximately 6.5 lb/in² at an altitude of 20,000 ft.

Definition of an Exponential Function

The **exponential function** with base *b* is defined by

$$f(x) = b^x$$

where *b* > 0, *b* ≠ 1, and *x* is any real number.

In the definition of an exponential function, *b*, the base, is required to be positive. If the base were a negative number, the value of the function would be a complex number for some values of *x*. For instance, the value of $f(x) = (-4)^x$ when $x = \frac{1}{2}$ is $f\left(\frac{1}{2}\right) = (-4)^{1/2} = \sqrt{-4} = 2i$. To avoid complex number values of a function, the base of the exponential function is always a positive number.

Integrating Technology

A graphing calculator can be used to evaluate an exponential expression for an irrational number.

HOW TO Evaluate $f(x) = 2^x$ at $x = 3$ and $x = -2$.

$f(3) = 2^3 = 8$ • **Substitute 3 for *x* and simplify.**

$f(-2) = 2^{-2} = \dfrac{1}{2^2} = \dfrac{1}{4}$ • **Substitute −2 for *x* and simplify.**

To evaluate an exponential expression for an irrational number such as $\sqrt{2}$, we obtain an approximation to the value of the function by approximating the irrational number. For instance, the value of $f(x) = 4^x$ when $x = \sqrt{2}$ can be approximated by using an approximation of $\sqrt{2}$.

$$f(\sqrt{2}) = 4^{\sqrt{2}} \approx 4^{1.4142} \approx 7.1029$$

Because $f(x) = b^x$ ($b > 0$, $b \neq 1$) can be evaluated at both rational and irrational numbers, the domain of f is all real numbers. And because $b^x > 0$ for all values of x, the range of f is the positive real numbers.

A frequently used base in applications of exponential functions is an irrational number designated by e. **The number e is approximately 2.71828183.** It is an irrational number, so it has a nonterminating, nonrepeating decimal representation.

> **Natural Exponential Function**
>
> The function defined by $f(x) = e^x$ is called the **natural exponential function.**

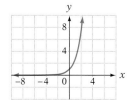

The e^x key on a calculator can be used to evaluate the natural exponential function. The graph of $y = e^x$ is shown at the left.

Example 1

Evaluate $f(x) = \left(\dfrac{1}{2}\right)^x$ at $x = 2$ and $x = -3$.

Solution

$$f(x) = \left(\dfrac{1}{2}\right)^x$$

$$f(2) = \left(\dfrac{1}{2}\right)^2 = \dfrac{1}{4} \qquad \bullet \ x = 2$$

$$f(-3) = \left(\dfrac{1}{2}\right)^{-3} = 2^3 = 8 \qquad \bullet \ x = -3$$

You Try It 1

Evaluate $f(x) = \left(\dfrac{2}{3}\right)^x$ at $x = 3$ and $x = -2$.

Your solution

Example 2

Evaluate $f(x) = 2^{3x-1}$ at $x = 1$ and $x = -1$.

Solution

$$f(x) = 2^{3x-1}$$

$$f(1) = 2^{3(1)-1} = 2^2 = 4 \qquad \bullet \ x = 1$$

$$f(-1) = 2^{3(-1)-1} = 2^{-4} = \dfrac{1}{2^4} = \dfrac{1}{16} \qquad \bullet \ x = -1$$

You Try It 2

Evaluate $f(x) = 2^{2x+1}$ at $x = 0$ and $x = -2$.

Your solution

Example 3

Evaluate $f(x) = e^{2x}$ at $x = 1$ and $x = -1$. Round to the nearest ten-thousandth.

Solution

$$f(x) = e^{2x}$$

$$f(1) = e^{2 \cdot 1} = e^2 \approx 7.3891 \qquad \bullet \ x = 1$$

$$f(-1) = e^{2(-1)} = e^{-2} \approx 0.1353 \qquad \bullet \ x = -1$$

You Try It 3

Evaluate $f(x) = e^{2x-1}$ at $x = 2$ and $x = -2$. Round to the nearest ten-thousandth.

Your solution

Solutions on p. S29

Objective B **To graph an exponential function**

Integrating Technology

See the Keystroke Guide: *Graph* for instructions on using a graphing calculator to graph functions.

Some properties of an exponential function can be seen from its graph.

HOW TO Graph $f(x) = 2^x$.

Think of this as the equation $y = 2^x$.

Choose values of x and find the corresponding values of y. The results can be recorded in a table.

Graph the ordered pairs on a rectangular coordinate system.

Connect the points with a smooth curve.

x	$f(x) = y$
-2	$2^{-2} = \dfrac{1}{4}$
-1	$2^{-1} = \dfrac{1}{2}$
0	$2^0 = 1$
1	$2^1 = 2$
2	$2^2 = 4$
3	$2^3 = 8$

Note that any vertical line would intersect the graph at only one point. Therefore, by the vertical-line test, the graph of $f(x) = 2^x$ is the graph of a function. Also note that any horizontal line would intersect the graph at only one point. Therefore, the graph of $f(x) = 2^x$ is the graph of a one-to-one function.

HOW TO Graph $f(x) = \left(\dfrac{1}{2}\right)^x$.

Think of this as the equation $y = \left(\dfrac{1}{2}\right)^x$.

Choose values of x and find the corresponding values of y.

Graph the ordered pairs on a rectangular coordinate system.

Connect the points with a smooth curve.

x	$f(x) = y$
-3	$\left(\dfrac{1}{2}\right)^{-3} = 8$
-2	$\left(\dfrac{1}{2}\right)^{-2} = 4$
-1	$\left(\dfrac{1}{2}\right)^{-1} = 2$
0	$\left(\dfrac{1}{2}\right)^{0} = 1$
1	$\left(\dfrac{1}{2}\right)^{1} = \dfrac{1}{2}$
2	$\left(\dfrac{1}{2}\right)^{2} = \dfrac{1}{4}$

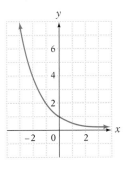

Applying the vertical-line and horizontal-line tests reveals that the graph of $f(x) = \left(\dfrac{1}{2}\right)^x$ is also the graph of a one-to-one function.

HOW TO Graph $f(x) = 2^{-x}$.

Think of this as the equation $y = 2^{-x}$.

Choose values of x and find the corresponding values of y.

Graph the ordered pairs on a rectangular coordinate system.

Connect the points with a smooth curve.

x	y
-3	8
-2	4
-1	2
0	1
1	$\dfrac{1}{2}$
2	$\dfrac{1}{4}$

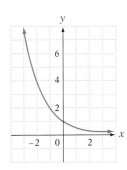

TAKE NOTE

Note that because $2^{-x} = (2^{-1})^x = \left(\dfrac{1}{2}\right)^x$, the graphs of $f(x) = 2^{-x}$ and $f(x) = \left(\dfrac{1}{2}\right)^x$ are the same.

Example 4 Graph: $f(x) = 3^{\frac{1}{2}x - 1}$

Solution

x	y
-2	$\dfrac{1}{9}$
0	$\dfrac{1}{3}$
2	1
4	3

You Try It 4 Graph: $f(x) = 2^{-\frac{1}{2}x}$

Your solution

Example 5 Graph: $f(x) = 2^x - 1$

Solution

x	y
-2	$-\dfrac{3}{4}$
-1	$-\dfrac{1}{2}$
0	0
1	1
2	3
3	7

You Try It 5 Graph: $f(x) = 2^x + 1$

Your solution

Example 6 Graph: $f(x) = \left(\dfrac{1}{3}\right)^x - 2$

Solution

x	y
-2	7
-1	1
0	-1
1	$-\dfrac{5}{3}$
2	$-\dfrac{17}{9}$

You Try It 6 Graph: $f(x) = 2^{-x} + 2$

Your solution

Example 7 Graph: $f(x) = 2^{-\frac{1}{2}x} - 1$

Solution

x	y
-6	7
-4	3
-2	1
0	0
2	$-\dfrac{1}{2}$
4	$-\dfrac{3}{4}$

You Try It 7 Graph: $f(x) = \left(\dfrac{1}{2}\right)^{-\frac{1}{2}x} + 2$

Your solution

Solutions on p. S29

10.1 Exercises

Objective A **To evaluate an exponential function**

1. What is an exponential function?

2. What is the natural exponential function?

3. Which of the following cannot be the base of an exponential function?
 a. 7 **b.** $\frac{1}{4}$ **c.** -5 **d.** 0.01

4. Which of the following cannot be the base of an exponential function?
 a. 0.9 **b.** 476 **c.** 8 **d.** $-\frac{1}{2}$

5. Given $f(x) = 3^x$, evaluate the following.
 a. $f(2)$ **b.** $f(0)$ **c.** $f(-2)$

6. Given $H(x) = 2^x$, evaluate the following.
 a. $H(-3)$ **b.** $H(0)$ **c.** $H(2)$

7. Given $g(x) = 2^{x+1}$, evaluate the following.
 a. $g(3)$ **b.** $g(1)$ **c.** $g(-3)$

8. Given $F(x) = 3^{x-2}$, evaluate the following.
 a. $F(-4)$ **b.** $F(-1)$ **c.** $F(0)$

9. Given $P(x) = \left(\frac{1}{2}\right)^{2x}$, evaluate the following.
 a. $P(0)$ **b.** $P\left(\frac{3}{2}\right)$ **c.** $P(-2)$

10. Given $R(t) = \left(\frac{1}{3}\right)^{3t}$, evaluate the following.
 a. $R\left(-\frac{1}{3}\right)$ **b.** $R(1)$ **c.** $R(-2)$

11. Given $G(x) = e^{x/2}$, evaluate the following. Round to the nearest ten-thousandth.
 a. $G(4)$ **b.** $G(-2)$ **c.** $G\left(\frac{1}{2}\right)$

12. Given $f(x) = e^{2x}$, evaluate the following. Round to the nearest ten-thousandth.
 a. $f(-2)$ **b.** $f\left(-\frac{2}{3}\right)$ **c.** $f(2)$

13. Given $H(r) = e^{-r+3}$, evaluate the following. Round to the nearest ten-thousandth.
 a. $H(-1)$ **b.** $H(3)$ **c.** $H(5)$

14. Given $P(t) = e^{-\frac{1}{2}t}$, evaluate the following. Round to the nearest ten-thousandth.
 a. $P(-3)$ **b.** $P(4)$ **c.** $P\left(\frac{1}{2}\right)$

15. Given $F(x) = 2^{x^2}$, evaluate the following.

 a. $F(2)$ **b.** $F(-2)$ **c.** $F\left(\dfrac{3}{4}\right)$

16. Given $Q(x) = 2^{-x^2}$, evaluate the following.

 a. $Q(3)$ **b.** $Q(-1)$ **c.** $Q(-2)$

17. Given $f(x) = e^{-x^2/2}$, evaluate the following. Round to the nearest ten-thousandth.

 a. $f(-2)$ **b.** $f(2)$ **c.** $f(-3)$

18. Given $f(x) = e^{-2x} + 1$, evaluate the following. Round to the nearest ten-thousandth.

 a. $f(-1)$ **b.** $f(3)$ **c.** $f(-2)$

Objective B **To graph an exponential function**

Graph the functions in Exercises 19 to 30.

19. $f(x) = 3^x$

20. $f(x) = 3^{-x}$

21. $f(x) = 2^{x+1}$

22. $f(x) = 2^{x-1}$

23. $f(x) = \left(\dfrac{1}{3}\right)^x$

24. $f(x) = \left(\dfrac{2}{3}\right)^x$

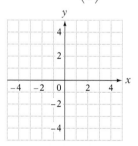

25. $f(x) = 2^{-x} + 1$

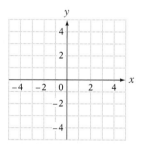

26. $f(x) = 2^x - 3$

27. $f(x) = \left(\dfrac{1}{3}\right)^{-x}$

28. $f(x) = \left(\dfrac{3}{2}\right)^{-x}$

29. $f(x) = \left(\dfrac{1}{2}\right)^{-x} + 2$

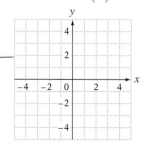

30. $f(x) = \left(\dfrac{1}{2}\right)^x - 1$

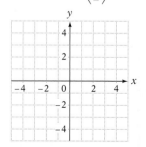

31. Which of the following functions have the same graph?

 a. $f(x) = 3^x$ **b.** $f(x) = \left(\dfrac{1}{3}\right)^x$ **c.** $f(x) = x^3$ **d.** $f(x) = 3^{-x}$

32. Which of the following functions have the same graph?

 a. $f(x) = x^4$ **b.** $f(x) = 4^{-x}$ **c.** $f(x) = 4^x$ **d.** $f(x) = \left(\dfrac{1}{4}\right)^x$

33. Graph $f(x) = 3^x$ and $f(x) = 3^{-x}$ and find the point of intersection of the two graphs.

34. Graph $f(x) = 2^{x+1}$ and $f(x) = 2^{-x+1}$ and find the point of intersection of the two graphs.

35. Graph $f(x) = \left(\dfrac{1}{3}\right)^x$. What are the x- and y-intercepts of the graph of the function?

36. Graph $f(x) = \left(\dfrac{1}{3}\right)^{-x}$. What are the x- and y-intercepts of the graph of the function?

APPLYING THE CONCEPTS

 Use a graphing calculator to graph the functions in Exercises 37 to 39.

37. $P(x) = \left(\sqrt{3}\right)^x$ **38.** $Q(x) = \left(\sqrt{3}\right)^{-x}$ **39.** $f(x) = \pi^x$

 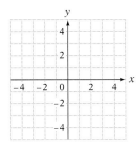

40. Evaluate $\left(1 + \dfrac{1}{n}\right)^n$ for $n = 100$, 1000, $10{,}000$, and $100{,}000$ and compare the results with the value of e, the base of the natural exponential function. On the basis of your evaluation, complete the following sentence:

As n increases, $\left(1 + \dfrac{1}{n}\right)^n$ becomes closer to _____.

41. **Physics** If air resistance is ignored, the speed v, in feet per second, of an object t seconds after it has been dropped is given by $v = 32t$. However, if air resistance is considered, then the speed depends on the mass (and on other things). For a certain mass, the speed t seconds after it has been dropped is given by $v = 32(1 - e^{-t})$.

 a. Graph this equation. *Suggestion:* Use Xmin = 0, Xmax = 5.5, Ymin = 0, Ymax = 40, and Yscl = 5.

 b. The point whose approximate coordinates are (2, 27.7) is on this graph. Write a sentence that explains the meaning of these coordinates.

Introduction to Logarithms

Objective A

To find the logarithm of a number

Suppose a bacteria colony that originally contained 1000 bacteria doubled in size every hour. Then the table at the right would show the number of bacteria in that colony after 1, 2, and 3 h.

Time (in hours)	Number of Bacteria
0	1000
1	2000
2	4000
3	8000

The exponential function $A = 1000(2^t)$, where A is the number of bacteria in the colony at time t, is a model of the growth of the colony. For instance, when $t = 3$ hours, we have

$$A = 1000(2^t)$$

$$A = 1000(2^3) \qquad \text{• Replace } t \text{ by 3.}$$

$$A = 1000(8) = 8000$$

After 3 h there are 8000 bacteria in the colony.

Now we ask, "How long will it take for there to be 32,000 bacteria in the colony?" To answer the question, we must solve the *exponential equation* $32{,}000 = 1000(2^t)$. By trial and error, we find that when $t = 5$,

$$A = 1000(2^t)$$

$$A = 1000(2^5) \qquad \text{• Replace } t \text{ by 5.}$$

$$A = 1000(32) = 32{,}000$$

After 5 h there will be 32,000 bacteria in the colony.

Now suppose we want to know how long it takes before the colony reaches 50,000 bacteria. To answer that question, we must find t so that $50{,}000 = 1000(2^t)$. Using trial and error again, we find that

$$1000(2^5) = 32{,}000 \qquad \text{and} \qquad 1000(2^6) = 64{,}000$$

Because 50,000 is between 32,000 and 64,000, we conclude that t is between 5 and 6 h. If we try $t = 5.5$ (halfway between 5 and 6), then

$$A = 1000(2^t)$$

$$A = 1000(2^{5.5}) \qquad \text{• Replace } t \text{ by 5.5.}$$

$$A \approx 1000(45.25) = 45{,}250$$

In 5.5 h, there are approximately 45,250 bacteria in the colony. Because this is less than 50,000, the value of t must be a little greater than 5.5.

We could continue to use trial and error to find the correct value of t, but it would be more efficient if we could just solve the exponential equation $50{,}000 = 1000(2^t)$ for t. If we follow the procedures for solving equations that were discussed earlier in the text, we have

$$50{,}000 = 1000(2^t)$$
$$50 = 2^t \qquad \bullet \text{ Divide each side of the equation by 1000.}$$

To proceed to the next step, it would be helpful to have a function that would find the power of 2 that produces 50.

Around the mid-sixteenth century, mathematicians created such a function, which we now call a *logarithmic function*. We write the solution of $50 = 2^t$ as $t = \log_2 50$. This is read "t equals the logarithm base 2 of 50." and it means "t equals the power of 2 that produces 50." When logarithms were first introduced, tables were used to find a numerical value of t. Today, a calculator is used. Using a calculator, we can approximate the value of t as 5.644. This means that $2^{5.644} \approx 50$.

TAKE NOTE

Note that earlier, when we tried $t = 5.5$, we stated that the actual value of t must be greater than 5.5 and that 5.644 is a little greater than 5.5.

The equivalence of the expressions $50 = 2^t$ and $t = \log_2 50$ are described in the following definition of **logarithm.**

TAKE NOTE

Read $\log_b x$ as "the logarithm of x, base b" or "log base b of x."

Definition of Logarithm

For $x > 0$, $b > 0$, $b \neq 1$, $y = \log_b x$ is equivalent to $x = b^y$.

The table at the right shows equivalent statements written in both exponential and logarithmic form.

Exponential Form	Logarithmic Form
$2^4 = 16$	$\log_2 16 = 4$
$\left(\dfrac{2}{3}\right)^2 = \dfrac{4}{9}$	$\log_{2/3}\left(\dfrac{4}{9}\right) = 2$
$10^{-1} = 0.1$	$\log_{10}(0.1) = -1$

Study Tip

Be sure you can rewrite an exponential equation as a logarithmic equation and can rewrite a logarithmic equation as an exponential equation. This relationship is very important.

HOW TO Write $\log_3 81 = 4$ in exponential form.

$\log_3 81 = 4$ is equivalent to $3^4 = 81$.

HOW TO Write $10^{-2} = 0.01$ in logarithmic form.

$10^{-2} = 0.01$ is equivalent to $\log_{10}(0.01) = -2$.

It is important to note that the exponential function is a 1–1 function and thus has an inverse function. **The inverse function of the exponential function is called a logarithm.**

The 1–1 property of exponential functions can be used to evaluate some logarithms.

> **1–1 Property of Exponential Functions**
> For $b > 0$, $b \neq 1$, if $b^u = b^v$, then $u = v$.

HOW TO Evaluate $\log_2 8$.

$\log_2 8 = x$ • Write an equation.

$8 = 2^x$ • Write the equation in its equivalent exponential form.

$2^3 = 2^x$ • Write 8 as 2^3.

$3 = x$ • Use the 1–1 Property of Exponential Functions.

$\log_2 8 = 3$

HOW TO Solve $\log_4 x = -2$ for x.

$\log_4 x = -2$

$4^{-2} = x$ • Write the equation in its equivalent exponential form.

$\dfrac{1}{16} = x$ • Simplify the negative exponent.

The solution is $\dfrac{1}{16}$.

Integrating Technology

The logarithms of most numbers are irrational numbers. Therefore, the value displayed by a calculator is an approximation.

Logarithms to the base 10 are called **common logarithms.** Usually the base, 10, is omitted when writing the common logarithm of a number. Therefore, $\log_{10} x$ is written $\log x$. To find the common logarithm of most numbers, a calculator is necessary. A calculator was used to find the value of $\log 384$, shown below.

$$\log 384 \approx 2.5843312$$

When e (the base of the natural exponential function) is used as the base of a logarithm, the logarithm is referred to as the **natural logarithm** and is abbreviated $\ln x$. This is read "el en x." Use a calculator to approximate natural logarithms.

$$\ln 23 \approx 3.135494216$$

Example 1

Evaluate: $\log_3 \dfrac{1}{9}$

Solution

$\log_3 \dfrac{1}{9} = x$ • Write an equation.

$\dfrac{1}{9} = 3^x$ • Write the equivalent exponential form.

$3^{-2} = 3^x$ • $\dfrac{1}{9} = 3^{-2}$

$-2 = x$ • The bases are the same. The exponents are equal.

$\log_3 \dfrac{1}{9} = -2$

You Try It 1

Evaluate: $\log_4 64$

Your solution

Solution on p. S29

Example 2 Solve for x: $\log_5 x = 2$

Solution
$$\log_5 x = 2$$
$$5^2 = x \qquad \bullet \text{ Write the equivalent}$$
$$25 = x \qquad\quad \text{exponential form.}$$

The solution is 25.

You Try It 2 Solve for x: $\log_2 x = -4$

Your solution

Example 3 Solve $\log x = -1.5$ for x. Round to the nearest ten-thousandth.

Solution
$$\log x = -1.5$$
$$10^{-1.5} = x \qquad \bullet \text{ Write the equivalent}$$
$$\qquad\qquad\qquad \text{exponential form.}$$
$$0.0316 \approx x \qquad \bullet \text{ Use a calculator.}$$

You Try It 3 Solve $\ln x = 3$ for x. Round to the nearest ten-thousandth.

Your solution

Solutions on p. S29

Objective B **To use the Properties of Logarithms to simplify expressions containing logarithms**

Because a logarithm is a special kind of exponent, the Properties of Logarithms are similar to the Properties of Exponents.

The property of logarithms that states that the logarithm of the product of two numbers equals the sum of the logarithms of the two numbers is similar to the property of exponents that states that to multiply two exponential expressions with the same base, we add the exponents.

TAKE NOTE

Pay close attention to this property. Note, for instance, that this property states that $\log_3(4 \cdot p) = \log_3 4 + \log_3 p$. It also states that $\log_5 9 + \log_5 z = \log_5(9z)$. It does *not* state any relationship regarding the expression $\log_b(x + y)$. **This expression cannot be simplified.**

> **The Logarithm Property of the Product of Two Numbers**
>
> For any positive real numbers x, y, and b, $b \neq 1$, $\log_b(xy) = \log_b x + \log_b y$.

A proof of this property can be found in the Appendix.

HOW TO Write $\log_b(6z)$ in expanded form.

$$\log_b(6z) = \log_b 6 + \log_b z \qquad \bullet \text{ Use the Logarithm Property of Products.}$$

HOW TO Write $\log_b 12 + \log_b r$ as a single logarithm.

$$\log_b 12 + \log_b r = \log_b(12r) \qquad \bullet \text{ Use the Logarithm Property of Products.}$$

The Logarithm Property of Products can be extended to include the logarithm of the product of more than two factors. For instance,

$$\log_b(xyz) = \log_b x + \log_b y + \log_b z$$

$$\log_b(7rt) = \log_b 7 + \log_b r + \log_b t$$

A second property of logarithms involves the logarithm of the quotient of two numbers. This property of logarithms is also based on the facts that a logarithm is an exponent and that to divide two exponential expressions with the same base, we subtract the exponents.

> **The Logarithm Property of the Quotient of Two Numbers**
>
> For any positive real numbers x, y, and b, $b \neq 1$,
>
> $$\log_b \frac{x}{y} = \log_b x - \log_b y.$$

A proof of this property can be found in the Appendix.

HOW TO Write $\log_b \dfrac{p}{8}$ in expanded form.

$\log_b \dfrac{p}{8} = \log_b p - \log_b 8$ • Use the Logarithm Property of Quotients.

HOW TO Write $\log_b y - \log_b v$ as a single logarithm.

$\log_b y - \log_b v = \log_b \dfrac{y}{v}$ • Use the Logarithm Property of Quotients.

A third property of logarithms is used to simplify powers of a number.

Point of Interest

Logarithms were developed independently by Jobst Burgi (1552–1632) and John Napier (1550–1617) as a means of simplifying the calculations of astronomers. The idea was to devise a method by which two numbers could be multiplied by performing additions. Napier is usually given credit for logarithms because he published his results first.

John Napier

In Napier's original work, the logarithm of 10,000,000 was 0. After this work was published, Napier, in discussions with Henry Briggs (1561–1631), decided that tables of logarithms would be easier to use if the logarithm of 1 were 0. Napier died before new tables could be determined, and Briggs took on the task. His table consisted of logarithms accurate to 30 decimal places, all accomplished without a calculator!

The logarithms Briggs calculated are the common logarithms mentioned earlier.

> **The Logarithm Property of the Power of a Number**
>
> For any positive real numbers x and b, $b \neq 1$, and for any real number r,
> $\log_b x^r = r \log_b x.$

A proof of this property can be found in the Appendix.

HOW TO Rewrite $\log_b x^3$ in terms of $\log_b x$.

$\log_b x^3 = 3 \log_b x$ • Use the Logarithm Property of Powers.

HOW TO Rewrite $\dfrac{2}{3} \log_b x$ with a coefficient of 1.

$\dfrac{2}{3} \log_b x = \log_b x^{2/3}$ • Use the Logarithm Property of Powers.

The following table summarizes the properties of logarithms that we have discussed, along with three other properties.

> **Summary of the Properties of Logarithms**
>
> Let x, y, and b be positive real numbers with $b \neq 1$. Then
>
> | Product Property | $\log_b(x \cdot y) = \log_b x + \log_b y$ |
> | Quotient Property | $\log_b \dfrac{x}{y} = \log_b x - \log_b y$ |
> | Power Property | $\log_b x^r = r \log_b x$, r a real number |
> | Logarithm of One | $\log_b 1 = 0$ |
> | Inverse Property | $b^{\log_b x} = x$ and $\log_b b^x = x$ |
> | 1–1 Property | If $\log_b x = \log_b y$, then $x = y$. |

HOW TO Write $\log_b \frac{xy}{z}$ in expanded form.

$\log_b \dfrac{xy}{z} = \log_b (xy) - \log_b z$ • Use the Logarithm Property of Quotients.

$\qquad = \log_b x + \log_b y - \log_b z$ • Use the Logarithm Property of Products.

HOW TO Write $\log_b \frac{x^2}{y^3}$ in expanded form.

$\log_b \dfrac{x^2}{y^3} = \log_b x^2 - \log_b y^3$ • Use the Logarithm Property of Quotients.

$\qquad = 2 \log_b x - 3 \log_b y$ • Use the Logarithm Property of Powers.

HOW TO Write $2 \log_b x + 4 \log_b y$ as a single logarithm with a coefficient of 1.

$2 \log_b x + 4 \log_b y = \log_b x^2 + \log_b y^4$ • Use the Logarithm Property of Powers.

$\qquad = \log_b x^2 y^4$ • Use the Logarithm Property of Products.

Example 4

Write $\log \sqrt{x^3 y}$ in expanded form.

Solution

$\log \sqrt{x^3 y} = \log (x^3 y)^{1/2} = \dfrac{1}{2} \log (x^3 y)$ • Power Property

$\qquad = \dfrac{1}{2} (\log x^3 + \log y)$ • Product Property

$\qquad = \dfrac{1}{2} (3 \log x + \log y)$ • Power Property

$\qquad = \dfrac{3}{2} \log x + \dfrac{1}{2} \log y$ • Distributive Property

You Try It 4

Write $\log_8 \sqrt[3]{xy^2}$ in expanded form.

Your solution

Example 5

Write $\dfrac{1}{2} (\log_3 x - 3 \log_3 y + \log_3 z)$ as a single logarithm with a coefficient of 1.

Solution

$\dfrac{1}{2} (\log_3 x - 3 \log_3 y + \log_3 z)$

$= \dfrac{1}{2} (\log_3 x - \log_3 y^3 + \log_3 z)$

$= \dfrac{1}{2} \left(\log_3 \dfrac{x}{y^3} + \log_3 z \right)$

$= \dfrac{1}{2} \left(\log_3 \dfrac{xz}{y^3} \right) = \log_3 \left(\dfrac{xz}{y^3} \right)^{1/2} = \log_3 \sqrt{\dfrac{xz}{y^3}}$

You Try It 5

Write $\dfrac{1}{3} (\log_4 x - 2 \log_4 y + \log_4 z)$ as a single logarithm with a coefficient of 1.

Your solution

Example 6

Find $\log_4 4^7$.

Solution

$\log_4 4^7 = 7$ • Inverse Property

You Try It 6

Find $\log_9 1$.

Your solution

Solutions on pp. S29–S30

Objective C **To use the Change-of-Base Formula**

Although only common logarithms and natural logarithms are programmed into calculators, the logarithms for other positive bases can be found.

> **HOW TO** Evaluate $\log_5 22$.
>
> $\log_5 22 = x$ • Write an equation.
> $5^x = 22$ • Write the equation in its equivalent exponential form.
> $\log 5^x = \log 22$ • Apply the common logarithm to each side of the equation.
> $x \log 5 = \log 22$ • Use the Power Property of Logarithms.
> $x = \dfrac{\log 22}{\log 5}$ • Exact answer
> $x \approx 1.92057$ • Approximate answer
> $\log_5 22 \approx 1.92057$

In the third step above, the natural logarithm, instead of the common logarithm, could have been applied to each side of the equation. The same result would have been obtained.

Using a procedure similar to the one used to evaluate $\log_5 22$, we can derive a formula for changing bases.

> **Change-of-Base Formula**
>
> $$\log_a N = \frac{\log_b N}{\log_b a}$$

> **HOW TO** Evaluate $\log_2 14$.
>
> $\log_2 14 = \dfrac{\log 14}{\log 2}$ • Use the Change-of-Base Formula with $N = 14$, $a = 2$, $b = 10$.
> ≈ 3.80735

In the last example, common logarithms were used. Here is the same example using natural logarithms. Note that the answers are the same.

$$\log_2 14 = \frac{\ln 14}{\ln 2} \approx 3.80735$$

Example 7

Evaluate $\log_8 0.137$ by using natural logarithms.

Solution

$\log_8 0.137 = \dfrac{\ln 0.137}{\ln 8} \approx -0.95592$

You Try It 7

Evaluate $\log_3 0.834$ by using natural logarithms.

Your solution

Example 8

Evaluate $\log_2 90.813$ by using common logarithms.

Solution

$\log_2 90.813 = \dfrac{\log 90.813}{\log 2} \approx 6.50483$

You Try It 8

Evaluate $\log_7 6.45$ by using common logarithms.

Your solution

Solutions on p. S30

10.2 Exercises

Objective A **To find the logarithm of a number**

1. **a.** What is a common logarithm?
 b. How is the common logarithm of $4z$ written?

2. **a.** What is a natural logarithm?
 b. How is the natural logarithm of $3x$ written?

For Exercises 3 to 10, write the exponential expression in logarithmic form.

3. $5^2 = 25$

4. $10^3 = 1000$

5. $4^{-2} = \dfrac{1}{16}$

6. $3^{-3} = \dfrac{1}{27}$

7. $10^y = x$

8. $e^y = x$

9. $a^x = w$

10. $b^y = c$

For Exercises 11 to 18, write the logarithmic expression in exponential form.

11. $\log_3 9 = 2$

12. $\log_2 32 = 5$

13. $\log 0.01 = -2$

14. $\log_5 \dfrac{1}{5} = -1$

15. $\ln x = y$

16. $\log x = y$

17. $\log_b u = v$

18. $\log_c x = y$

Evaluate the expressions in Exercises 19 to 30.

19. $\log_3 81$

20. $\log_7 49$

21. $\log_2 128$

22. $\log_5 125$

23. $\log 100$

24. $\log 0.001$

25. $\ln e^3$

26. $\ln e^2$

27. $\log_8 1$

28. $\log_3 243$

29. $\log_5 625$

30. $\log_2 64$

For Exercises 31 to 38, solve for x.

31. $\log_3 x = 2$

32. $\log_5 x = 1$

33. $\log_4 x = 3$

34. $\log_2 x = 6$

35. $\log_7 x = -1$

36. $\log_8 x = -2$

37. $\log_6 x = 0$

38. $\log_4 x = 0$

For Exercises 39 to 46, solve for x. Round to the nearest hundredth.

39. $\log x = 2.5$ **40.** $\log x = 3.2$ **41.** $\log x = -1.75$ **42.** $\log x = -2.1$

43. $\ln x = 2$ **44.** $\ln x = 1.4$ **45.** $\ln x = -\dfrac{1}{2}$ **46.** $\ln x = -1.7$

Objective B **To use the Properties of Logarithms to simplify expressions containing logarithms**

47. What is the Product Property of Logarithms?

48. What is the Quotient Property of Logarithms?

For Exercises 49 to 76, express as a single logarithm with a coefficient of 1.

49. $\log_3 x^3 + \log_3 y^2$ **50.** $\log_7 x + \log_7 z^2$ **51.** $\ln x^4 - \ln y^2$

52. $\ln x^2 - \ln y$ **53.** $3 \log_7 x$ **54.** $4 \log_8 y$

55. $3 \ln x + 4 \ln y$ **56.** $2 \ln x - 5 \ln y$ **57.** $2(\log_4 x + \log_4 y)$

58. $3(\log_5 r + \log_5 t)$ **59.** $2 \log_3 x - \log_3 y + 2 \log_3 z$ **60.** $4 \log_5 r - 3 \log_5 s + \log_5 t$

61. $\ln x - (2 \ln y + \ln z)$ **62.** $2 \log_b x - 3(\log_b y + \log_b z)$

63. $\dfrac{1}{2}(\log_6 x - \log_6 y)$ **64.** $\dfrac{1}{3}(\log_8 x - \log_8 y)$

65. $2(\log_4 s - 2 \log_4 t + \log_4 r)$ **66.** $3(\log_9 x + 2 \log_9 y - 2 \log_9 z)$

67. $\ln x - 2(\ln y + \ln z)$ **68.** $\ln t - 3(\ln u + \ln v)$

69. $3 \log_2 t - 2(\log_2 r - \log_2 v)$ **70.** $2 \log_{10} x - 3(\log_{10} y - \log_{10} z)$

71. $\dfrac{1}{2}(3\log_4 x - 2\log_4 y + \log_4 z)$

72. $\dfrac{1}{3}(4\log_5 t - 5\log_5 u - 7\log_5 v)$

73. $\dfrac{1}{2}(\ln x - 3\ln y)$

74. $\dfrac{1}{3}\ln a + \dfrac{2}{3}\ln b$

75. $\dfrac{1}{2}\log_2 x - \dfrac{2}{3}\log_2 y + \dfrac{1}{2}\log_2 z$

76. $\dfrac{2}{3}\log_3 x + \dfrac{1}{3}\log_3 y - \dfrac{1}{2}\log_3 z$

For Exercises 77 to 100, write the logarithm in expanded form.

77. $\log_8(xz)$

78. $\log_7(rt)$

79. $\log_3 x^5$

80. $\log_2 y^7$

81. $\log_b \dfrac{r}{s}$

82. $\log_c \dfrac{z}{4}$

83. $\log_3(x^2 y^6)$

84. $\log_4(t^4 u^2)$

85. $\log_7 \dfrac{u^3}{v^4}$

86. $\log_{10} \dfrac{s^5}{t^2}$

87. $\log_2(rs)^2$

88. $\log_3(x^2 y)^3$

89. $\ln(x^2 yz)$

90. $\ln(xy^2 z^3)$

91. $\log_5 \dfrac{xy^2}{z^4}$

92. $\log_b \dfrac{r^2 s}{t^3}$

93. $\log_8 \dfrac{x^2}{yz^2}$

94. $\log_9 \dfrac{x}{y^2 z^3}$

95. $\log_4 \sqrt{x^3 y}$

96. $\log_3 \sqrt{x^5 y^3}$

97. $\log_7 \sqrt{\dfrac{x^3}{y}}$

98. $\log_b \sqrt[3]{\dfrac{r^2}{t}}$

99. $\log_3 \dfrac{t}{\sqrt{x}}$

100. $\log_4 \dfrac{x}{\sqrt{y^2 z}}$

Objective C **To use the Change-of-Base Formula**

Evaluate the expressions in Exercises 101 to 124. Round to the nearest ten-thousandth.

101. $\log_{10} 7$

102. $\log_{10} 9$

103. $\log_{10} \dfrac{3}{5}$

104. $\log_{10} \dfrac{13}{3}$

105. $\ln 4$

106. $\ln 6$

107. $\ln \dfrac{17}{6}$

108. $\ln \dfrac{13}{17}$

109. $\log_8 6$

110. $\log_4 8$

111. $\log_5 30$

112. $\log_6 28$

113. $\log_3 0.5$

114. $\log_5 0.6$

115. $\log_7 1.7$

116. $\log_6 3.2$

117. $\log_5 15$

118. $\log_3 25$

119. $\log_{12} 120$

120. $\log_9 90$

121. $\log_4 2.55$

122. $\log_8 6.42$

123. $\log_5 67$

124. $\log_8 35$

APPLYING THE CONCEPTS

125. For each of the following, answer True or False. Assume all variables represent positive numbers.

a. $\log_3(-9) = -2$

b. $x^y = z$ and $\log_x z = y$ are equivalent equations.

c. $\log(x^{-1}) = \dfrac{1}{\log x}$

d. $\log \dfrac{x}{y} = \log x - \log y$

e. $\log(x \cdot y) = \log x \cdot \log y$

f. If $\log x = \log y$, then $x = y$.

126. Complete each statement using the equation $\log_a b = c$.

a. $a^c =$ _____

b. $\log_a(a^c) =$ _____

10.3 Graphs of Logarithmic Functions

Objective A **To graph a logarithmic function**

The graph of a logarithmic function can be drawn by using the relationship between the exponential and logarithmic functions.

HOW TO Graph: $f(x) = \log_2 x$

Think of $f(x) = \log_2 x$ as the equation $y = \log_2 x$.

Write the equivalent exponential equation.

$$f(x) = \log_2 x$$
$$y = \log_2 x$$
$$x = 2^y$$

Because the equation is solved for x in terms of y, it is easier to choose values of y and find the corresponding values of x. The results can be recorded in a table.

Graph the ordered pairs on a rectangular coordinate system.

Connect the points with a smooth curve.

x	y
$\frac{1}{4}$	-2
$\frac{1}{2}$	-1
1	0
2	1
4	2

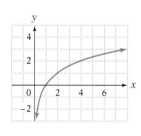

Applying the vertical-line and horizontal-line tests reveals that $f(x) = \log_2 x$ is a one-to-one function.

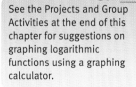

HOW TO Graph: $f(x) = \log_2 x + 1$

Think of $f(x) = \log_2 x + 1$ as the equation $y = \log_2 x + 1$.

Solve for $\log_2 x$.

Write the equivalent exponential equation.

$$f(x) = \log_2 x + 1$$
$$y = \log_2 x + 1$$
$$y - 1 = \log_2 x$$
$$2^{y-1} = x$$

Choose values of y and find the corresponding values of x.

Graph the ordered pairs on a rectangular coordinate system.

Connect the points with a smooth curve.

x	y
$\frac{1}{4}$	-1
$\frac{1}{2}$	0
1	1
2	2
4	3

Example 1 Graph: $f(x) = \log_3 x$

Solution

$f(x) = \log_3 x$

$y = \log_3 x$ · $f(x) = y$

$3^y = x$ · Write the equivalent exponential equation.

x	y
$\frac{1}{9}$	-2
$\frac{1}{3}$	-1
1	0
3	1

You Try It 1 Graph: $f(x) = \log_2(x - 1)$

Your solution

Example 2 Graph: $f(x) = 2\log_3 x$

Solution

$f(x) = 2\log_3 x$

$y = 2\log_3 x$ · $f(x) = y$

$\dfrac{y}{2} = \log_3 x$ · Divide both sides by 2.

$3^{y/2} = x$ · Write the equivalent exponential equation.

x	y
$\frac{1}{9}$	-4
$\frac{1}{3}$	-2
1	0
3	2

You Try It 2 Graph: $f(x) = \log_3(2x)$

Your solution

Example 3 Graph: $f(x) = -\log_2(x - 2)$

Solution

$f(x) = -\log_2(x - 2)$

$y = -\log_2(x - 2)$ · $f(x) = y$

$-y = \log_2(x - 2)$ · Multiply both sides by -1.

$2^{-y} = x - 2$ · Write the equivalent exponential equation.

$2^{-y} + 2 = x$

x	y
6	-2
4	-1
3	0
$\frac{5}{2}$	1
$\frac{9}{4}$	2
$\frac{17}{8}$	3

You Try It 3 Graph: $f(x) = -\log_3(x + 1)$

Your solution

Solutions on p. S30

10.3 Exercises

Objective A **To graph a logarithmic function**

1. Is the function $f(x) = \log x$ a 1–1 function? Why or why not?

2. Name two characteristics of the graph of $y = \log_b x$, $b > 1$.

3. What is the relationship between the graphs of $x = 3^y$ and $y = \log_3 x$?

4. What is the relationship between the graphs of $y = 3^x$ and $y = \log_3 x$?

Graph the functions in Exercises 5 to 16.

5. $f(x) = \log_4 x$

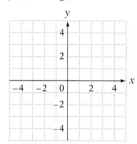

6. $f(x) = \log_2(x + 1)$

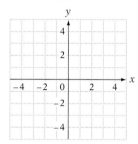

7. $f(x) = \log_3(2x - 1)$

8. $f(x) = \log_2\left(\dfrac{1}{2}x\right)$

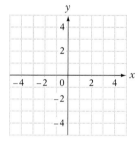

9. $f(x) = 3\log_2 x$

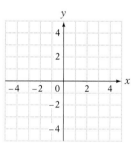

10. $f(x) = \dfrac{1}{2}\log_2 x$

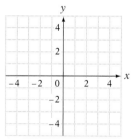

11. $f(x) = -\log_2 x$

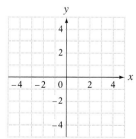

12. $f(x) = -\log_3 x$

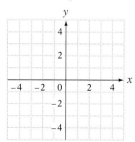

13. $f(x) = \log_2(x - 1)$

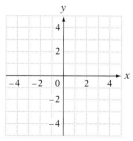

14. $f(x) = \log_3(2 - x)$

15. $f(x) = -\log_2(x - 1)$

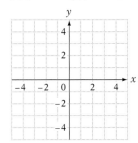

16. $f(x) = -\log_2(1 - x)$

APPLYING THE CONCEPTS

 Use a graphing calculator to graph the functions in Exercises 17 to 22.

17. $f(x) = x - \log_2(1 - x)$

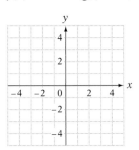

18. $f(x) = -\dfrac{1}{2}\log_2 x - 1$

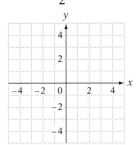

19. $f(x) = \dfrac{x}{2} - 2\log_2(x + 1)$

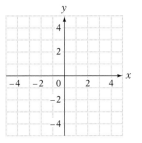

20. $f(x) = x + \log_3(2 - x)$

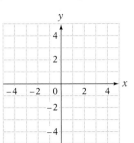

21. $f(x) = x^2 - 10\ln(x - 1)$

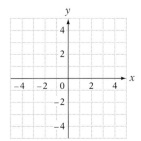

22. $f(x) = \dfrac{x}{3} - 3\log_2(x + 3)$

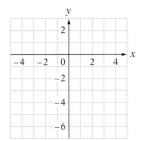

23. **Astronomy** Astronomers use the *distance modulus* of a star as a method of determining the star's distance from Earth. The formula is $M = 5\log s - 5$, where M is the distance modulus and s is the star's distance from the Earth in parsecs. (One parsec $\approx 1.9 \times 10^{13}$ mi)

 a. Graph the equation.

 b. The point with coordinates (25.1, 2) is on the graph. Write a sentence that describes the meaning of this ordered pair.

24. **Typing** Without practice, the proficiency of a typist decreases. The equation $s = 60 - 7\ln(t + 1)$, where s is the typing speed in words per minute and t is the number of months without typing, approximates this decrease.

 a. Graph the equation.

 b. The point with coordinates (4, 49) is on the graph. Write a sentence that describes the meaning of this ordered pair.

10.4 Solving Exponential and Logarithmic Equations

Objective A **To solve an exponential equation**

An **exponential equation** is one in which a variable occurs in the exponent. The equations at the right are exponential equations.

$$6^{2x+1} = 6^{3x-2}$$

$$4^x = 3$$

$$2^{x+1} = 7$$

An exponential equation in which each side of the equation can be expressed in terms of the same base can be solved by using the 1–1 Property of Exponential Functions. Recall that the 1–1 Property of Exponential Functions states that for $b > 0$, $b \neq 1$,

if $b^x = b^y$, then $x = y$.

In the two examples below, this property is used in solving exponential equations.

HOW TO Solve: $10^{3x+5} = 10^{x-3}$

$$10^{3x+5} = 10^{x-3}$$
$$3x + 5 = x - 3$$

- Use the 1–1 Property of Exponential Functions to equate the exponents.

$$2x + 5 = -3$$
$$2x = -8$$
$$x = -4$$

- Solve the resulting equation.

Check:
$$\begin{array}{c|c} \multicolumn{2}{c}{10^{3x+5} = 10^{x-3}} \\ \hline 10^{3(-4)+5} & 10^{-4-3} \\ 10^{-12+5} & 10^{-7} \\ 10^{-7} &= 10^{-7} \end{array}$$

The solution is -4.

HOW TO Solve: $9^{x+1} = 27^{x-1}$

$$9^{x+1} = 27^{x-1}$$
$$(3^2)^{x+1} = (3^3)^{x-1}$$

- $3^2 = 9$; $3^3 = 27$

$$3^{2x+2} = 3^{3x-3}$$
$$2x + 2 = 3x - 3$$

- Use the 1–1 Property of Exponential Functions to equate the exponents.

$$2 = x - 3$$
$$5 = x$$

- Solve for *x*.

Check:
$$\begin{array}{c|c} \multicolumn{2}{c}{9^{x+1} = 27^{x-1}} \\ \hline 9^{5+1} & 27^{5-1} \\ 9^6 & 27^4 \\ 531{,}441 &= 531{,}441 \end{array}$$

The solution is 5.

Content:

OK writing the real thing.

Writing the real markdown now:

Fine, writing all text directly.



Below is the full transcription content of page 568:

I really need to just write it. Here:

Writing now for real, no more loops.

Producing the markdown body now.

Here is the body:

Writing page 568 content now:

Objective B **To solve a logarithmic equation**

A logarithmic equation can be solved by using the Properties of Logarithms. Here are two examples.

> **HOW TO** Solve: $\log_9 x + \log_9 (x - 8) = 1$
>
> $\log_9 x + \log_9 (x - 8) = 1$
> $\qquad \log_9 [x(x - 8)] = 1$
>
> - Use the Logarithm Property of Products to rewrite the left side of the equation.
>
> $\qquad\qquad 9^1 = x(x - 8)$
> $\qquad\qquad\ 9 = x^2 - 8x$
> $\qquad\qquad\ 0 = x^2 - 8x - 9$
> $\qquad\qquad\ 0 = (x - 9)(x + 1)$
>
> - Write the equation in exponential form.
> - Write in standard form.
> - Factor and use the Principle of Zero Products.
>
> $x - 9 = 0 \qquad x + 1 = 0$
> $\quad\ \ x = 9 \qquad\quad\ x = -1$

Replacing x by 9 in the original equation reveals that 9 checks as a solution. Replacing x by -1 in the original equation results in the expression $\log_9 (-1)$. Because the logarithm of a negative number is not a real number, -1 does not check as a solution.

The solution of the equation is 9.

> **HOW TO** Solve: $\log_3 6 - \log_3 (2x + 3) = \log_3 (x + 1)$
>
> $\log_3 6 - \log_3 (2x + 3) = \log_3 (x + 1)$
> $\qquad\quad \log_3 \dfrac{6}{2x + 3} = \log_3 (x + 1)$
>
> - Use the Quotient Property of Logarithms to rewrite the left side of the equation.
>
> $\qquad\qquad \dfrac{6}{2x + 3} = x + 1$
>
> - Use the 1–1 Property of Logarithms.
>
> $\qquad\qquad\qquad 6 = (2x + 3)(x + 1)$
> $\qquad\qquad\qquad 6 = 2x^2 + 5x + 3$
> $\qquad\qquad\qquad 0 = 2x^2 + 5x - 3$
> $\qquad\qquad\qquad 0 = (2x - 1)(x + 3)$
>
> - Write in standard form.
> - Solve for x.
>
> $2x - 1 = 0 \qquad x + 3 = 0$
> $\qquad\ \ x = \dfrac{1}{2} \qquad\quad x = -3$

Replacing x by $\dfrac{1}{2}$ in the original equation reveals that $\dfrac{1}{2}$ checks as a solution. Replacing x by -3 in the original equation results in the expression $\log_3 (-2)$. Because the logarithm of a negative number is not a real number, -3 does not check as a solution.

The solution of the equation is $\dfrac{1}{2}$.

Example 3

Solve for x: $\log_3(2x - 1) = 2$

Solution

$\log_3(2x - 1) = 2$

$\quad\quad\quad 3^2 = 2x - 1$ • Write in exponential form.

$\quad\quad\quad\quad 9 = 2x - 1$

$\quad\quad\quad 10 = 2x$

$\quad\quad\quad\quad 5 = x$

The solution is 5.

You Try It 3

Solve for x: $\log_4(x^2 - 3x) = 1$

Your solution

Example 4

Solve for x: $\log_2 x - \log_2(x - 1) = \log_2 2$

Solution

$\log_2 x - \log_2(x - 1) = \log_2 2$

$\log_2\left(\dfrac{x}{x - 1}\right) = \log_2 2$ • Quotient Property

$\dfrac{x}{x - 1} = 2$ • 1–1 Property of Logarithms

$(x - 1)\left(\dfrac{x}{x - 1}\right) = (x - 1)2$

$\quad\quad\quad\quad x = 2x - 2$

$\quad\quad\quad -x = -2$

$\quad\quad\quad\quad x = 2$

The solution is 2.

You Try It 4

Solve for x: $\log_3 x + \log_3(x + 3) = \log_3 4$

Your solution

Example 5

Solve for x:
$\log_2(3x + 8) = \log_2(2x + 2) + \log_2(x - 2)$

Solution

$\log_2(3x + 8) = \log_2(2x + 2) + \log_2(x - 2)$

$\log_2(3x + 8) = \log_2[(2x + 2)(x - 2)]$

$\log_2(3x + 8) = \log_2(2x^2 - 2x - 4)$

$3x + 8 = 2x^2 - 2x - 4$ • 1–1 Property of Logarithms

$0 = 2x^2 - 5x - 12$

$0 = (2x + 3)(x - 4)$ • Solve by factoring.

$2x + 3 = 0 \quad\quad\quad x - 4 = 0$

$x = -\dfrac{3}{2} \quad\quad\quad\quad x = 4$

$-\dfrac{3}{2}$ does not check as a solution;

4 checks as a solution. The solution is 4.

You Try It 5

Solve for x: $\log_3 x + \log_3(x + 6) = 3$

Your solution

Solutions on p. S30

10.4 Exercises

Objective A To solve an exponential equation

1. What is an exponential equation?

2. **a.** What does the 1–1 Property of Exponential Functions state?
 b. Provide an example of when you would use this property.

For Exercises 3 to 26, solve for x. For Exercises 15 to 26, round to the nearest ten-thousandth.

3. $5^{4x-1} = 5^{x-2}$

4. $7^{4x-3} = 7^{2x+1}$

5. $8^{x-4} = 8^{5x+8}$

6. $10^{4x-5} = 10^{x+4}$

7. $9^x = 3^{x+1}$

8. $2^{x-1} = 4^x$

9. $8^{x+2} = 16^x$

10. $9^{3x} = 81^{x-4}$

11. $16^{2-x} = 32^{2x}$

12. $27^{2x-3} = 81^{4-x}$

13. $25^{3-x} = 125^{2x-1}$

14. $8^{4x-7} = 64^{x-3}$

15. $5^x = 6$

16. $7^x = 10$

17. $e^x = 3$

18. $e^x = 2$

19. $10^x = 21$

20. $10^x = 37$

21. $2^{-x} = 7$

22. $3^{-x} = 14$

23. $2^{x-1} = 6$

24. $4^{x+1} = 9$

25. $3^{2x-1} = 4$

26. $4^{-x+2} = 12$

Objective B To solve a logarithmic equation

27. What is a logarithmic equation?

28. What does the 1–1 Property of Logarithms state?

For Exercises 29 to 46, solve for x.

29. $\log_2(2x - 3) = 3$

30. $\log_4(3x + 1) = 2$

31. $\log_2(x^2 + 2x) = 3$

32. $\log_3(x^2 + 6x) = 3$

33. $\log_5 \dfrac{2x}{x - 1} = 1$

34. $\log_6 \dfrac{3x}{x + 1} = 1$

35. $\log x = \log (1 - x)$

36. $\ln(3x - 2) = \ln(x + 1)$

37. $\ln 5 = \ln(4x - 13)$

38. $\log_3(x - 2) = \log_3(2x)$

39. $\ln(3x + 2) = 4$

40. $\ln(2x + 3) = -1$

41. $\log_2(8x) - \log_2(x^2 - 1) = \log_2 3$

42. $\log_5(3x) - \log_5(x^2 - 1) = \log_5 2$

43. $\log_9 x + \log_9(2x - 3) = \log_9 2$

44. $\log_6 x + \log_6(3x - 5) = \log_6 2$

45. $\log_8(6x) = \log_8 2 + \log_8(x - 4)$

46. $\log_7(5x) = \log_7 3 + \log_7(2x + 1)$

APPLYING THE CONCEPTS

For Exercises 47 to 52, solve for x. Round to the nearest ten-thousandth.

47. $8^{\frac{x}{2}} = 6$

48. $4^{\frac{x}{3}} = 2$

49. $5^{\frac{3x}{2}} = 7$

50. $9^{\frac{2x}{3}} = 8$

51. $1.2^{\frac{x}{2} - 1} = 1.4$

52. $5.6^{\frac{x}{3} + 1} = 7.8$

53. Physics A model for the distance s, in feet, that an object experiencing air resistance will fall in t seconds is given by $s = 312.5 \ln \dfrac{e^{0.32t} + e^{-0.32t}}{2}$.

 a. Graph this equation. *Suggestion:* Use Xmin = 0, Xmax = 4.5, Ymin = 0, Ymax = 140, and Yscl = 20.
 b. Determine, to the nearest hundredth of a second, the time it takes the object to travel 100 ft.

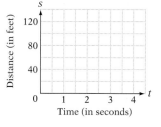

54. Physics A model for the distance s, in feet, that an object experiencing air resistance will fall in t seconds is given by $s = 78 \ln \dfrac{e^{0.8t} + e^{-0.8t}}{2}$.

 a. Graph this equation. *Suggestion:* Use Xmin = 0, Xmax = 4.5, Ymin = 0, Ymax = 140, and Yscl = 20.
 b. Determine, to the nearest hundredth of a second, the time it takes the object to travel 125 ft.

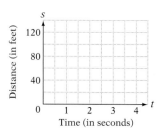

55. The following "proof" shows that $0.5 < 0.25$. Explain the error.

$$1 < 2$$
$$1 \cdot \log 0.5 < 2 \cdot \log 0.5$$
$$\log 0.5 < \log (0.5)^2$$
$$0.5 < (0.5)^2$$
$$0.5 < 0.25$$

10.5 Applications of Exponential and Logarithmic Functions

Objective A **To solve application problems**

A biologist places one single-celled bacterium in a culture, and each hour that particular species of bacteria divides into two bacteria. After one hour there will be two bacteria. After two hours, each of these two bacteria will divide and there will be four bacteria. After three hours, each of the four bacteria will divide and there will be eight bacteria.

Point of Interest

C. Northcote Parkinson

Parkinson's Law, named after C. Northcote Parkinson, is sometimes stated as "A job will expand to fill the time alloted for the job." However, Parkinson actually said that in any new government administration, administrative employees will be added at the rate of about 5% to 6% per year. This is an example of exponential growth and means that a staff of 500 will grow to approximately 630 by the end of a 4-year term.

The table at the right shows the number of bacteria in the culture after various intervals of time t, in hours. Values in this table could also be found by using the exponential equation $N = 2^t$.

Time, t	Number of Bacteria, N
0	1
1	2
2	4
3	8
4	16

The equation $N = 2^t$ is an example of an **exponential growth equation.** In general, any equation that can be written in the form $A = A_0 b^{kt}$, where A is the size at time t, A_0 is the initial size, $b > 1$, and k is a positive real number, is an exponential growth equation. These equations are important not only in population growth studies but also in physics, chemistry, psychology, and economics.

Recall that interest is the amount of money that one pays (or receives) when borrowing (or investing) money. **Compound interest** is interest that is computed not only on the original principal but also on the interest already earned. The compound interest formula is an exponential equation.

The **compound interest formula** is $A = P(1 + i)^n$, where P is the original value of an investment, i is the interest rate per compounding period, n is the total number of compounding periods, and A is the value of the investment after n periods.

HOW TO An investment broker deposits $1000 into an account that earns 8% annual interest compounded quarterly. What is the value of the investment after 3 years?

$i = \dfrac{8\%}{4} = \dfrac{0.08}{4} = 0.02$ • Find i, the interest rate per quarter. The quarterly rate is the annual rate divided by 4, the number of quarters in 1 year.

$n = 4 \cdot 3 = 12$ • Find n, the number of compounding periods. The investment is compounded quarterly, 4 times a year, for 3 years.

$A = P(1 + i)^n$ • Use the compound interest formula.

$A = 1000(1 + 0.02)^{12}$ • Replace P, i, and n by their values.

$A \approx 1268$ • Solve for A.

The value of the investment after 3 years is approximately $1268.

 Exponential decay offers another example of an exponential equation. One of the most common illustrations of exponential decay is the decay of a radioactive substance. For instance, tritium, a radioactive nucleus of hydrogen that has been used in luminous watch dials, has a half-life of approximately 12 years. This means that one-half of any given amount of tritium will disintegrate in 12 years.

The table at the right indicates the amount of an initial 10-microgram sample of tritium that remains after various intervals of time, t, in years. Values in this table could also be found by using the exponential equation $A = 10(0.5)^{t/12}$.

Time, t	Amount, A
0	10
12	5
24	2.5
36	1.25
48	0.625

The equation $A = 10(0.5)^{t/12}$ is an example of an **exponential decay equation.** Comparing this equation to the exponential growth equation, note that for exponential growth, the base of the exponential equation is greater than 1, whereas for exponential decay, the base is between 0 and 1.

A method by which an archaeologist can measure the age of a bone is called carbon dating. Carbon dating is based on a radioactive isotope of carbon called carbon-14, which has a half-life of approximately 5570 years. The exponential decay equation is given by $A = A_0(0.5)^{t/5570}$, where A_0 is the original amount of carbon-14 present in the bone, t is the age of the bone in years, and A is the amount of carbon-14 present after t years.

HOW TO A bone that originally contained 100 mg of carbon-14 now has 70 mg of carbon-14. What is the approximate age of the bone?

$A = A_0(0.5)^{t/5570}$ • Use the exponential decay equation.

$70 = 100(0.5)^{t/5570}$ • Replace A by 70 and A_0 by 100 and solve for t.

$0.7 = (0.5)^{t/5570}$ • Divide each side by 100.

$\log 0.7 = \log(0.5)^{t/5570}$ • Take the common logarithm of each side of the equation.

$\log 0.7 = \dfrac{t}{5570} \log 0.5$ • Power Property

$\dfrac{5570 \log 0.7}{\log 0.5} = t$ • Multiply by 5570, and divide by log 0.5.

$2866 \approx t$

The bone is approximately 2866 years old.

 A chemist measures the acidity or alkalinity of a solution by measuring the concentration of hydrogen ions, H^+, in the solution using the formula $pH = -\log(H^+)$. A neutral solution such as distilled water has a pH of 7, acids have a pH less than 7, and alkaline solutions (also called basic solutions) have a pH greater than 7.

HOW TO Find the pH of orange juice that has a hydrogen ion concentration, H^+, of 2.9×10^{-4}. Round to the nearest tenth.

$pH = -\log(H^+)$

$= -\log(2.9 \times 10^{-4})$ • $H^+ = 2.9 \times 10^{-4}$

≈ 3.5376

The pH of the orange juice is approximately 3.5.

Point of Interest

Søren Sørensen

The pH scale was created by the Danish biochemist Søren Sørensen in 1909 to measure the acidity of water used in the brewing of beer. pH is an abbreviation for *pondus hydrogenii*, which translates as "potential hydrogen."

Logarithmic functions are used to scale very large or very small numbers into numbers that are easier to comprehend. For instance, the *Richter scale magnitude* of an earthquake uses a logarithmic function to convert the intensity of shock waves I into a number M, which for most earthquakes is in the range of 0 to 10. The intensity I of an earthquake is often given in terms of the constant I_0, where I_0 is the intensity of the smallest earthquake, called a *zero-level earthquake*, that can be measured on a seismograph near the earthquake's epicenter. An earthquake with an intensity I has a Richter scale magnitude of $M = \log \dfrac{I}{I_0}$, where I_0 is the measure of the intensity of a zero-level earthquake.

> **HOW TO** Find the Richter scale magnitude of the 2003 Amazonas, Brazil, earthquake that had an intensity I of $12{,}589{,}254 I_0$. Round to the nearest tenth.
>
> $M = \log \dfrac{I}{I_0}$
>
> $M = \log \dfrac{12{,}589{,}254 I_0}{I_0}$ • $I = 12{,}589{,}254 I_0$
>
> $M = \log 12{,}589{,}254$ • Divide the numerator and denominator by I_0.
>
> $M \approx 7.1$ • Evaluate $\log 12{,}589{,}254$.
>
> The 2003 Amazonas, Brazil, earthquake had a Richter scale magnitude of 7.1.

If you know the Richter scale magnitude of an earthquake, you can determine the intensity of the earthquake.

> **HOW TO** Find the intensity of the 2003 Colima, Mexico, earthquake that measured 7.6 on the Richter scale. Write the answer in terms of I_0. Round to the nearest thousand.
>
> $M = \log \dfrac{I}{I_0}$
>
> $7.6 = \log \dfrac{I}{I_0}$ • Replace M by 7.6.
>
> $10^{7.6} = \dfrac{I}{I_0}$ • Write in exponential form.
>
> $10^{7.6} I_0 = I$ • Multiply both sides by I_0.
>
> $39{,}810{,}717 I_0 \approx I$ • Evaluate $10^{7.6}$.
>
> The 2003 Colima, Mexico, earthquake had an intensity that was approximately 39,811,000 times the intensity of a zero-level earthquake.

Point of Interest

Charles F. Richter

The Richter scale was created by seismologist Charles F. Richter in 1935. Note that a tenfold increase in the intensity level of an earthquake increases the Richter scale magnitude of the earthquake by only 1.

The percent of light that will pass through a substance is given by $\log P = -kd$, where P is the percent of light passing through the substance, k is a constant depending on the substance, and d is the thickness of the substance in meters.

> **HOW TO** For certain parts of the ocean, $k = 0.03$. Using this value, at what depth will the percent of light be 50% of the light at the surface of the ocean? Round to the nearest meter.
>
> $\log P = -kd$
>
> $\log(0.5) = -0.03d$ • Replace P by 0.5 (50%) and k by 0.03.
>
> $\dfrac{\log(0.5)}{-0.03} = d$ • Solve for d.
>
> $10.0343 \approx d$
>
> At a depth of about 10 m, the light will be 50% of the light at the surface.

Example 1

An investment of \$3000 is placed into an account that earns 12% annual interest compounded monthly. In approximately how many years will the investment be worth twice the original amount?

Strategy

To find the time, solve the compound interest formula for n. Use $A = 6000$, $P = 3000$, and $i = \frac{12\%}{12} = \frac{0.12}{12} = 0.01$.

Solution

$$A = P(1 + i)^n$$
$$6000 = 3000(1 + 0.01)^n$$
$$6000 = 3000(1.01)^n$$
$$2 = (1.01)^n$$
$$\log 2 = \log(1.01)^n \quad \bullet \text{ Take the log of each side.}$$
$$\log 2 = n \log 1.01 \quad \bullet \text{ Power Property}$$
$$\frac{\log 2}{\log 1.01} = n \quad \bullet \text{ Divide both sides by log 1.01.}$$
$$70 \approx n$$

70 months ÷ 12 ≈ 5.8 years

In approximately 6 years, the investment will be worth \$6000.

You Try It 1

Find the hydrogen ion concentration, H^+, of vinegar that has a pH of 2.9.

Your strategy

Your solution

Example 2

The number of words per minute that a student can type will increase with practice and can be approximated by the equation $N = 100[1 - (0.9)^t]$, where N is the number of words typed per minute after t days of instruction. Find the number of words a student will type per minute after 8 days of instruction.

Strategy

To find the number of words per minute, replace t in the equation by its given value and solve for N.

Solution
$$N = 100[1 - (0.9)^t]$$
$$= 100[1 - (0.9)^8] \quad \bullet \ t = 8$$
$$\approx 56.95$$

After 8 days of instruction, a student will type approximately 57 words per minute.

You Try It 2

On April 29, 2003, an earthquake measuring 4.6 on the Richter scale struck Fort Payne, Alabama. Find the intensity of the quake in terms of I_0.

Your strategy

Your solution

Solutions on p. S31

10.5 Exercises

Objective A **To solve application problems**

Compound Interest For Exercises 1 to 4, use the compound interest formula $A = P(1 + i)^n$, where P is the original value of an investment, i is the interest rate per compounding period, n is the total number of compounding periods, and A is the value of the investment after n periods.

1. An investment broker deposits $1000 into an account that earns 8% annual interest compounded quarterly. What is the value of the investment after 2 years? Round to the nearest dollar.

2. A financial advisor recommends that a client deposit $2500 into a fund that earns 7.5% annual interest compounded monthly. What will be the value of the investment after 3 years? Round to the nearest cent.

3. To save for college tuition, the parents of a preschooler invest $5000 in a bond fund that earns 6% annual interest compounded monthly. In approximately how many years will the investment be worth $15,000?

4. A hospital administrator deposits $10,000 into an account that earns 6% annual interest compounded monthly. In approximately how many years will the investment be worth $15,000?

Radioactivity For Exercises 5 to 8, use the exponential decay equation $A = A_0\left(\dfrac{1}{2}\right)^{t/k}$, where A is the amount of a radioactive material present after time t, k is the half-life of the radioactive substance, and A_0 is the original amount of the radioactive substance. Round to the nearest tenth.

5. An isotope of technetium is used to prepare images of internal body organs. This isotope has a half-life of approximately 6 h. A patient is injected with 30 mg of this isotope.
 a. What is the technetium level in the patient after 3 h?
 b. How long (in hours) will it take for the technetium level to reach 20 mg?

6. Iodine-131 is an isotope that is used to study the functioning of the thyroid gland. This isotope has a half-life of approximately 8 days. A patient is given an injection that contains 8 micrograms of iodine-131.
 a. What is the iodine level in the patient after 5 days?
 b. How long (in days) will it take for the iodine level to reach 5 micrograms?

7. A sample of promethium-147 (used in some luminous paints) weighs 25 mg. One year later, the sample weighs 18.95 mg. What is the half-life of promethium-147, in years?

8. Francium-223 is a very rare radioactive isotope discovered in 1939 by Marguerite Percy. A 3-microgram sample of francium-223 decays to 2.54 micrograms in 5 min. What is the half-life of francium-223, in minutes?

9. **Earth Science** Earth's atmospheric pressure changes as you rise above its surface. At an altitude of h kilometers, where $0 < h < 80$, the pressure P in newtons per square centimeter is approximately modeled by the equation $P(h) = 10.13e^{-0.116h}$.

 a. What is the approximate pressure at 40 km above Earth's surface?

 b. What is the approximate pressure on Earth's surface?

 c. Does atmospheric pressure increase or decrease as you rise above Earth's surface?

10. **Demography** The U.S. Census Bureau provides information about various segments of the population in the United States. The following table gives the number of people, in millions, age 80 and older at the beginning of each decade from 1910 to 2000. An equation that approximately models the data is $y = 0.18808(1.0365)^x$, where x is the number of years since 1900 and y is the population, in millions, of people age 80 and over.

Year	1910	1920	1930	1940	1950	1960	1970	1980	1990	2000
Number of people age 80 and over (in millions)	0.3	0.4	0.5	0.8	1.1	1.6	2.3	2.9	3.9	9.3

 a. According to the model, what is the predicted population of this age group in the year 2020? Round to the nearest tenth of a million. (*Hint:* You will need to determine the x-value for the year 2020.)

 b. In what year does this model predict that the population of this age group will be 15 million? Round to the nearest year.

Chemistry For Exercises 11 and 12, use the equation $pH = -\log(H^+)$, where H^+ is the hydrogen ion concentration of a solution. Round to the nearest tenth.

11. Find the pH of milk, which has a hydrogen ion concentration of 3.97×10^{-7}.

12. Find the pH of a baking soda solution for which the hydrogen ion concentration is 3.98×10^{-9}.

Light For Exercises 13 and 14, use the equation $\log P = -kd$, which gives the relationship between the percent P, as a decimal, of light passing through a substance of thickness d, in meters.

13. The value of k for a swimming pool is approximately 0.05. At what depth, in meters, will the percent of light be 75% of the light at the surface of the pool?

14. The constant k for a piece of blue stained glass is 20. What percent of light will pass through a piece of this glass that is 0.005 m thick?

Sound For Exercises 15 and 16, use the equation $D = 10(\log I + 16)$, where D is the number of decibels of a sound and I is the power of the sound measured in watts. Round to the nearest whole number.

15. Find the number of decibels of normal conversation. The power of the sound of normal conversation is approximately 3.2×10^{-10} watts.

16. The loudest sound made by any animal is made by the blue whale and can be heard more than 500 mi away. The power of the sound is 630 watts. Find the number of decibels of sound emitted by the blue whale.

17. **Sports** During the 1980s and 1990s, the average time T to play a Major League baseball game increased each year. If the year 1981 is represented by $x = 1$, then the function $T(x) = 149.57 + 7.63 \ln x$ approximates the time T, in minutes, to play a Major League baseball game for the years $x = 1$ to $x = 19$. By how many minutes did the average time of a Major League baseball game increase from 1981 to 1999? Round to the nearest tenth.

18. **Postage** In 1962, the cost of a first-class postage stamp was \$.04. In 2004, the cost was \$.37. The increase in cost can be modeled by the equation $C = 0.04e^{0.057t}$, where C is the cost and t is the number of years after 1962. According to this model, in what year did a first-class postage stamp cost \$.22?

19. **Chemistry** The intensity I of an x-ray after it has passed through a material that is x centimeters thick is given by $I = I_0e^{-kx}$, where I_0 is the initial intensity of the x-ray and k is a number that depends on the material. The constant k for copper is 3.2. Find the thickness of copper that is needed so that the intensity of an x-ray after passing through the copper is 25% of the original intensity. Round to the nearest tenth.

20. **Natural Resources** One model for the time it will take for the world's oil supply to be depleted is given by the equation $T = 14.29 \ln (0.00411r + 1)$, where r is the estimated world oil reserves in billions of barrels and T is the time, in years, before that amount of oil is depleted. Use this equation to determine how many barrels of oil are necessary to meet the world demand for 20 years. Round to the nearest tenth.

Seismology For Exercises 21 to 24, use the Richter scale equation $M = \log \dfrac{I}{I_0}$, where M is the magnitude of an earthquake, I is the intensity of the shock waves, and I_0 is the measure of the intensity of a zero-level earthquake.

21. On May 21, 2003, an earthquake struck Northern Algeria. The earthquake had an intensity of $I = 6,309,573I_0$. Find the Richter scale magnitude of the earthquake. Round to the nearest tenth.

22. The earthquake on November 17, 2003, in the Aleutian Islands of Moska had an intensity of $I = 63{,}095{,}734I_0$. Find the Richter scale magnitude of the earthquake. Round to the nearest tenth.

23. An earthquake in Japan on March 2, 1933 measured 8.9 on the Richter scale. Find the intensity of the earthquake in terms of I_0. Round to the nearest whole number.

24. An earthquake that occurred in China in 1978 measured 8.2 on the Richter scale. Find the intensity of the earthquake in terms of I_0. Round to the nearest whole number.

Seismology Shown at the right is a *seismogram*, which is used to measure the magnitude of an earthquake. The magnitude is determined by the amplitude A of a shock wave and the difference in time t, in seconds, between the occurrences of two types of waves called *primary waves* and *secondary waves*. As you can see on the graph, a primary wave is abbreviated *p-wave* and a secondary wave is abbreviated *s-wave*. The amplitude A of a wave is one-half the difference between its highest and lowest points. For this graph, A is 23 mm. The equation is $M = \log A + 3 \log 8t - 2.92$. Use this information for Exercises 25 to 27.

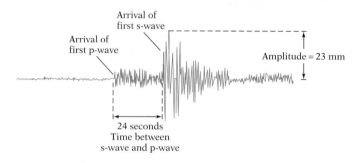

25. Determine the magnitude of the earthquake for the seismogram given in the figure. Round to the nearest tenth.

26. Find the magnitude of an earthquake that has a seismogram with an amplitude of 30 mm and for which t is 21 s.

27. Find the magnitude of an earthquake that has a seismogram with an amplitude of 28 mm and for which t is 28 s.

APPLYING THE CONCEPTS

28. **Investments** The value of an investment in an account that earns an annual interest rate of 7% compounded daily grows according to the equation $A = A_0\left(1 + \dfrac{0.07}{365}\right)^{365t}$, where A_0 is the original value of an investment and t is the time in years. Find the time for the investment to double in value. Round to the nearest year.

29. **Investments** Some banks now use continuous compounding of an amount invested. In this case, the equation that relates the value of an initial investment of P dollars after t years earning an annual interest rate r is given by $A = Pe^{rt}$. Using this equation, find the value after 5 years of an investment of \$2500 in an account that earns 5% annual interest.

Focus on Problem Solving

Proof by Contradiction

The four-step plan for solving problems that we have used before is restated here.

1. Understand the problem.
2. Devise a plan.
3. Carry out the plan.
4. Review the solution.

One of the techniques that can be used in the second step is a method called *proof by contradiction*. In this method you assume that the conditions of the problem you are trying to solve can be met and then show that your assumption leads to a condition you already know is not true.

To illustrate this method, suppose we try to prove that $\sqrt{2}$ is a rational number. We begin by recalling that a rational number is one that can be written as the quotient of integers. Therefore, let a and b be two integers with no common factors. If $\sqrt{2}$ is a rational number, then

$$\sqrt{2} = \frac{a}{b}$$

$$(\sqrt{2})^2 = \left(\frac{a}{b}\right)^2 \qquad \bullet \text{ Square each side.}$$

$$2 = \frac{a^2}{b^2}$$

$$2b^2 = a^2 \qquad \bullet \text{ Multiply each side by } b^2.$$

From the last equation, a^2 is an even number. Because a^2 is an even number, a is an even number. Now divide each side of the last equation by 2.

$$2b^2 = a^2$$

$$2b^2 = a \cdot a$$

$$b^2 = a \cdot x \qquad \bullet \; x = \frac{a}{2}$$

Because a is an even number, $a \cdot x$ is an even number. Because $a \cdot x$ is an even number, b^2 is an even number, and this in turn means that b is an even number. This result, however, contradicts the assumption that a and b are two integers with no common factors. Because this assumption is now known not to be true, we conclude that our original assumption, that $\sqrt{2}$ is a rational number, is false. This proves that $\sqrt{2}$ is an irrational number.

Try a proof by contradiction for the following problem: "Is it possible to write numbers using each of the digits 0, 1, 2, 3, 4, 5, 6, 7, 8, and 9 exactly once such that the sum of the numbers is exactly 100?"* Here are some suggestions. First note that the sum of the 10 digits is 45. This means that some of the digits used must be tens digits. Let x be the sum of those digits.

a. What is the sum of the remaining units digits?

b. Express "the sum of the units digits and the tens digits equals 100" as an equation.

c. Solve the equation for x.

d. Explain why this result means that it is impossible to satisfy both conditions of the problem.

*Problem adapted from G. Polya, *How to Solve It: A New Aspect of Mathematical Method.* Copyright © 1947 by G. Polya. Reprinted by permission of Princeton University Press.

Projects and Group Activities

Solving Exponential and Logarithmic Equations Using a Graphing Calculator

A graphing calculator can be used to draw the graphs of logarithmic functions. Note that there are two logarithmic keys, **LOG** and **LN** , on a graphing calculator. The first key gives the values of common logarithms (base 10), and the second gives the values of natural logarithms (base e).

To graph $y = \ln x$, press the **Y=** key. Clear any equations already entered. Press **LN** **X,T,θ,n** **)** **GRAPH** . The graph is shown at the left with a viewing window of Xmin = -1, Xmax = 8, Ymin = -5, and Ymax = 5.

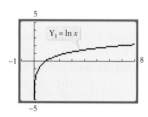

Some exponential and logarithmic equations cannot be solved algebraically. In these cases, a graphical approach may be appropriate. Here is an example.

HOW TO Solve $\ln (2x + 4) = x^2$ for x. Round to the nearest hundredth.

Rewrite the equation by subtracting x^2 from each side.

$$\ln (2x + 4) = x^2$$
$$\ln (2x + 4) - x^2 = 0$$

The zeros of $f(x) = \ln (2x + 4) - x^2$ are the solutions of $\ln (2x + 4) - x^2 = 0$.

Graph f and use the zero feature of the graphing calculator to estimate the solutions to the nearest hundredth.

The zeros are approximately -0.89 and 1.38. The solutions are -0.89 and 1.38.

For Exercises 1 to 8, solve for x by graphing. Round to the nearest hundredth.

1. $2^x = 2x + 4$

2. $3^x = -x - 1$

3. $e^x = -2x - 2$

4. $e^x = 3x + 4$

5. $\log (2x - 1) = -x + 3$

6. $\log (x + 4) = -2x + 1$

7. $\ln (x + 2) = x^2 - 3$

8. $\ln x = -x^2 + 1$

To graph a logarithmic function with a base other than base 10 or base e, the Change-of-Base Formula is used. Here is the Change-of-Base Formula, discussed earlier:

$$\log_a N = \frac{\log_b N}{\log_b a}$$

To graph $y = \log_2 x$, first change from base 2 logarithms to the equivalent base 10 logarithms or natural logarithms. Base 10 is shown here.

$$y = \log_2 x = \frac{\log_{10} x}{\log_{10} 2}$$ • $a = 2, b = 10, N = x$

Then graph the equivalent equation $y = \dfrac{\log x}{\log 2}$. The graph is shown at the left with a viewing window of Xmin = -1, Xmax = 8, Ymin = -5, and Ymax = 5.

Credit Reports and FICO® Scores[1]

When a consumer applies for a loan, the lender generally wants to know the consumer's credit history. For this, the lender turns to a credit reporting agency. These agencies maintain files on millions of borrowers. They provide the lender with a credit report, which lists information such as the types of credit the consumer uses, the length of time the consumer's credit accounts have been open, the amounts owed by the consumer, and whether the consumer has paid his or her bills on time. Along with the credit report, the lender can buy a credit score based on the information in the report. The credit score gives the lender a quick measure of the consumer's credit risk, or how likely the consumer is to repay a debt. It answers the lender's question: "If I lend this person money (or give this person a credit card), what is the probability that I will be paid back in a timely manner?"

The most widely used credit bureau score is the FICO® score, so named because scores are produced from software developed by Fair Isaac Corporation (FICO). FICO scores range from 300 to 850. The higher the score, the lower the predicted credit risk for lenders.

The following graph shows the delinquency rate, or credit risk, associated with ranges of FICO scores. For example, the delinquency rate of consumers in the 550–599 range of scores is 51%. This means that within the next 2 years, for every 100 borrowers in this range, approximately 51 will default on a loan, file for bankruptcy, or fail to pay a credit card bill within 90 days of the due date.

Because the deliquency rate depends on the FICO score, we can let the independent variable x represent the FICO score and the dependent variable y represent the delinquency rate. We will use the middle of each range for the values of x. The resulting ordered pairs are recorded in the table at the right.

x	y
400	87
525	71
575	51
625	31
675	15
725	5
775	2
825	1

[1]Source: **www.myfico.com**

1. Use your graphing calculator to find a logarithmic equation that approximates these data. Here are instructions for the TI-83:

 Press STAT ENTER. This will bring up a table into which you can enter data. Enter the values of x in L1. Enter the values of y in L2.

 Press STAT again, press the right arrow key to highlight CALC, and arrow down to LnReg (ln for the natural logarithm). Press ENTER twice. The values for a and b in the equation $y = a + b \ln x$ will appear on the screen.

2. Use your equation to predict the delinquency rate of a consumer with a score of 500. Round to the nearest whole number.

3. The equation pairs a delinquency rate of 36% with what score?

4. Use the graph on page 583 to determine the highest FICO score that will result in a delinquency rate of 71% or more.

5. Use the graph on page 583 to determine the lowest FICO score that will result in a delinquency rate of 5% or lower.

6. Use the Internet to find the name of at least one of the major credit reporting agencies.

Chapter 10 Summary

Key Words	Examples
A function of the form $f(x) = b^x$, where b is a positive real number not equal to 1, is an *exponential function*. The number b is the *base* of the exponential function. [10.1A, p. 545]	$f(x) = 3^x$ is an exponential function. 3 is the base of the function.
The function defined by $f(x) = e^x$ is called the *natural exponential function*. [10.1A, p. 546]	$f(x) = 2e^{x-1}$ is a natural exponential function. e is an irrational number approximately equal to 2.71828183.
Because the exponential function is a 1–1 function, it has an inverse function that is called a *logarithm*. The definition of logarithm is: For $x > 0$, $b > 0$, $b \neq 1$, $y = \log_b x$ is equivalent to $x = b^y$. [10.2A, p. 553]	$\log_2 8 = 3$ is equivalent to $8 = 2^3$.
Logarithms with base 10 are called *common logarithms*. We usually omit the base, 10, when writing the common logarithm of a number. [10.2A, p. 554]	$\log_{10} 100 = 2$ is usually written $\log 100 = 2$.
When e (the base of the natural exponential function) is used as a base of a logarithm, the logarithm is referred to as the *natural logarithm* and is abbreviated $\ln x$. [10.2A, p. 554]	$\log_e 100 \approx 4.61$ is usually written $\ln 100 \approx 4.61$.

An *exponential equation* is one in which a variable occurs in the exponent. [10.4A, p. 567]

$2^x = 12$ is an exponential equation.

An *exponential growth equation* is an equation that can be written in the form $A = A_0 b^{kt}$, where A is the size at time t, A_0 is the initial size, $b > 1$, and k is a positive real number. In an *exponential decay equation*, the base is between 0 and 1. [10.5A, pp. 573–574]

$P = 1000(1.03)^n$ is an exponential growth equation.
$A = 10(0.5)^x$ is an exponential decay equation.

Essential Rules and Procedures

Examples

1–1 Property of Exponential Functions [10.2A, p. 554]
For $b > 0$, $b \neq 1$, if $b^u = b^v$, then $u = v$.

If $b^x = b^5$, then $x = 5$.

Logarithm Property of the Product of Two Numbers
[10.2B, p. 555]
For any positive real numbers x, y, and b, $b \neq 1$,
$\log_b(xy) = \log_b x + \log_b y$.

$\log_b(3x) = \log_b 3 + \log_b x$

Logarithm Property of the Quotient of Two Numbers
[10.2B, p. 556]
For any positive real numbers x, y, and b, $b \neq 1$,

$\log_b \dfrac{x}{y} = \log_b x - \log_b y$.

$\log_b \dfrac{x}{20} = \log_b x - \log_b 20$

Logarithm Property of the Power of a Number
[10.2B, p. 556]
For any positive real numbers x and b, $b \neq 1$, and for any real number r, $\log_b x^r = r \log_b x$.

$\log_b x^5 = 5 \log_b x$

Logarithm Property of One [10.2B, p. 556]
For any positive real numbers x, y, and b, $b \neq 1$, $\log_b 1 = 0$.

$\log_6 1 = 0$

Inverse Property of Logarithms [10.2B, p. 556]
For any positive real numbers x, y, and b, $b \neq 1$, $b^{\log_b x} = x$ and $\log_b b^x = x$.

$\log_3 3^4 = 4$

1–1 Property of Logarithms [10.2B, p. 556]
For any positive real numbers x, y, and b, $b \neq 1$, if $\log_b x = \log_b y$, then $x = y$.

If $\log_5(x - 2) = \log_5 3$, then $x - 2 = 3$.

Change-of-Base Formula [10.2C, p. 558]

$\log_a N = \dfrac{\log_b N}{\log_b a}$

$\log_3 12 = \dfrac{\log 12}{\log 3}$ $\log_6 16 = \dfrac{\ln 16}{\ln 6}$

Chapter 10 Review Exercises

1. Evaluate $f(x) = e^{x-2}$ at $x = 2$.

2. Write $\log_5 25 = 2$ in exponential form.

3. Graph: $f(x) = 3^{-x} + 2$

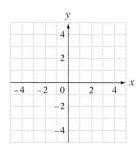

4. Graph: $f(x) = \log_3(x - 1)$

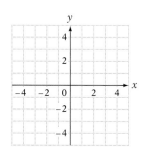

5. Write $\log_3 \sqrt[5]{x^2 y^4}$ in expanded form.

6. Write $2 \log_3 x - 5 \log_3 y$ as a single logarithm with a coefficient of 1.

7. Solve: $27^{2x+4} = 81^{x-3}$

8. Solve: $\log_5 \dfrac{7x + 2}{3x} = 1$

9. Find $\log_6 22$. Round to the nearest ten-thousandth.

10. Solve: $\log_2 x = 5$

11. Solve: $\log_3(x + 2) = 4$

12. Solve: $\log_{10} x = 3$

13. Write $\dfrac{1}{3}(\log_7 x + 4 \log_7 y)$ as a single logarithm with a coefficient of 1.

14. Write $\log_8 \sqrt{\dfrac{x^5}{y^3}}$ in expanded form.

15. Write $2^5 = 32$ in logarithmic form.

16. Find $\log_3 1.6$. Round to the nearest ten-thousandth.

17. Solve $3^{x+2} = 5$. Round to the nearest thousandth.

18. Evaluate $f(x) = \left(\frac{2}{3}\right)^{x+2}$ at $x = -3$.

19. Solve: $\log_8(x + 2) - \log_8 x = \log_8 4$

20. Solve: $\log_6(2x) = \log_6 2 + \log_6(3x - 4)$

21. Graph: $f(x) = \left(\frac{2}{3}\right)^{x+1}$

22. Graph: $f(x) = \log_2(2x - 1)$

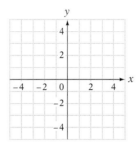

23. Evaluate: $\log_6 36$

24. Write $\frac{1}{3}(\log_2 x - \log_2 y)$ as a single logarithm with a coefficient of 1.

25. Solve for x: $9^{2x} = 3^{x+3}$

26. Evaluate $f(x) = \left(\frac{3}{5}\right)^x$ at $x = 0$.

27. Solve for x: $\log_5 x = -1$

28. Write $3^4 = 81$ in logarithmic form.

29. Solve for x: $\log x + \log(x - 2) = \log 15$

30. Write $\log_5 \sqrt[3]{x^2 y}$ in expanded form.

31. Solve for x: $9^{2x-5} = 9^{x-3}$

32. Evaluate $f(x) = 7^{x+2}$ at $x = -3$.

33. Evaluate: $\log_2 16$

34. Solve for x: $\log_6 x = \log_6 2 + \log_6 (2x - 3)$

35. Evaluate $\log_2 5$. Round to the nearest ten-thousandth.

36. Solve for x: $4^x = 8^{x-1}$

37. Solve for x: $\log_5 x = 4$

38. Write $3 \log_b x - 7 \log_b y$ as a single logarithm with a coefficient of 1.

39. Evaluate $f(x) = 5^{-x-1}$ at $x = -2$.

40. Solve $5^{x-2} = 7$ for x. Round to the nearest ten-thousandth.

41. Investments Use the compound interest formula $P = A(1 + i)^n$, where A is the original value of an investment, i is the interest rate per compounding period, n is the number of compounding periods, and P is the value of the investment after n periods, to find the value of an investment after 2 years. The amount of the investment is \$4000, and it is invested at 8% compounded monthly. Round to the nearest dollar.

42. **Seismology** An earthquake in Japan in September, 2003, had an intensity of $I = 199{,}526{,}232 I_0$. Find the Richter scale magnitude of the earthquake. Use the Richter scale equation $M = \log \dfrac{I}{I_0}$, where M is the magnitude of an earthquake, I is the intensity of the shock waves, and I_0 is the measure of the intensity of a zero-level earthquake. Round to the nearest tenth.

43. **Radioactivity** Use the exponential decay equation $A = A_0 \left(\dfrac{1}{2}\right)^{t/k}$, where A is the amount of a radioactive material present after time t, k is the half-life of the radioactive material, and A_0 is the original amount of radioactive material, to find the half-life of a material that decays from 25 mg to 15 mg in 20 days. Round to the nearest whole number.

44. **Sound** The number of decibels, D, of a sound can be given by the equation $D = 10(\log I + 16)$, where I is the power of the sound measured in watts. Find the number of decibels of sound emitted from a busy street corner for which the power of the sound is 5×10^{-6} watts.

Chapter 10 Test

1. Evaluate $f(x) = \left(\dfrac{2}{3}\right)^x$ at $x = 0$.

2. Evaluate $f(x) = 3^{x+1}$ at $x = -2$.

3. Graph: $f(x) = 2^x - 3$

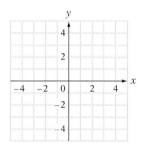

4. Graph: $f(x) = 2^x + 2$

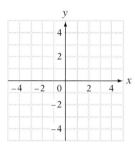

5. Evaluate: $\log_4 16$

6. Solve for x: $\log_3 x = -2$

7. Graph: $f(x) = \log_2(2x)$

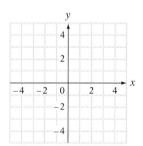

8. Graph: $f(x) = \log_3(x + 1)$

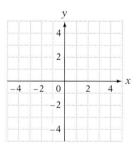

9. Write $\log_6 \sqrt{xy^3}$ in expanded form.

10. Write $\dfrac{1}{2}(\log_3 x - \log_3 y)$ as a single logarithm with a coefficient of 1.

11. Write $\ln \dfrac{x}{\sqrt{z}}$ in expanded form.

12. Write $3 \ln x - \ln y - \dfrac{1}{2} \ln z$ as a single logarithm with a coefficient of 1.

13. Solve for x: $3^{7x+1} = 3^{4x-5}$

14. Solve for x: $8^x = 2^{x-6}$

15. Solve for x: $3^x = 17$

16. Solve for x: $\log x + \log(x - 4) = \log 12$

17. Solve for x: $\log_6 x + \log_6(x - 1) = 1$

18. Find $\log_5 9$.

19. Find $\log_3 19$.

20. **Radioactivity** Use the exponential decay equation $A = A_0 \left(\dfrac{1}{2}\right)^{t/k}$, where A is the amount of a radioactive material present after time t, k is the half-life of the material, and A_0 is the original amount of radioactive material, to find the half-life of a material that decays from 10 mg to 9 mg in 5 h. Round to the nearest whole number.

Cumulative Review Exercises

1. Solve: $4 - 2[x - 3(2 - 3x) - 4x] = 2x$

2. Find the equation of the line that contains the point $(2, -2)$ and is parallel to the line $2x - y = 5$.

3. Factor: $4x^{2n} + 7x^n + 3$

4. Simplify: $\dfrac{1 - \dfrac{5}{x} + \dfrac{6}{x^2}}{1 + \dfrac{1}{x} - \dfrac{6}{x^2}}$

5. Simplify: $\dfrac{\sqrt{xy}}{\sqrt{x} - \sqrt{y}}$

6. Solve by completing the square: $x^2 - 4x - 6 = 0$

7. Write a quadratic equation that has integer coefficients and has solutions $\dfrac{1}{3}$ and -3.

8. Graph the solution set: $2x - y < 3$
$$x + y < 1$$

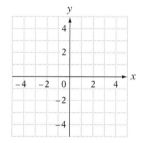

9. Solve by the addition method:
$$3x - y + z = 3$$
$$x + y + 4z = 7$$
$$3x - 2y + 3z = 8$$

10. Simplify: $\dfrac{x - 4}{2 - x} - \dfrac{1 - 6x}{2x^2 - 7x + 6}$

11. Solve: $x^2 + 4x - 5 \le 0$
Write the solution set in set-builder notation.

12. Solve: $|2x - 5| \le 3$

13. Graph: $f(x) = \left(\dfrac{1}{2}\right)^x + 1$

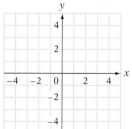

14. Graph: $f(x) = \log_2 x - 1$

15. Evaluate the function $f(x) = 2^{-x-1}$ at $x = -3$.

16. Solve for x: $\log_5 x = 3$

17. Write $3 \log_b x - 5 \log_b y$ as a single logarithm with a coefficient of 1.

18. Find $\log_3 7$.

19. Solve for x: $4^{5x-2} = 4^{3x+2}$

20. Solve for x: $\log x + \log(2x + 3) = \log 2$

21. **Banking** A bank offers two types of business checking accounts. One account has a charge of $5 per month plus 2 cents per check. The second account has a charge of $2 per month plus 8 cents per check. How many checks can a customer who has the second type of account write if it is to cost the customer less than the first type of checking account?

22. **Mixtures** Find the cost per pound of a mixture made from 16 lb of chocolate that costs $4.00 per pound and 24 lb of chocolate that costs $2.50 per pound.

23. **Uniform Motion** A plane can fly at a rate of 225 mph in calm air. Traveling with the wind, the plane flew 1000 mi in the same amount of time that it took to fly 800 mi against the wind. Find the rate of the wind.

24. **Physics** The distance, d, that a spring stretches varies directly as the force, f, used to stretch the spring. If a force of 20 lb stretches a spring 6 in., how far will a force of 34 lb stretch the spring?

25. **Carpentry** A carpenter purchased 80 ft of redwood and 140 ft of fir for a total cost of $67. A second purchase, at the same prices, included 140 ft of redwood and 100 ft of fir for a total cost of $81. Find the cost of redwood and of fir.

26. **Investments** The compound interest formula is $A = P(1 + i)^n$, where P is the original value of an investment, i is the interest rate per compounding period, n is the total number of compounding periods, and A is the value of the investment after n periods. Use the compound interest formula to find how many years it will take for an investment of $5000 to double in value. The investment earns 7% annual interest and is compounded semiannually.

11

Conic Sections

When creating a parabolic-shaped mirror for use in a telescope, such as the one shown in this photo, there is little room for error. The multimillion dollar Hubble Telescope made international headlines when it was discovered that the outer edge of its primary mirror was ground too flat by one fiftieth the thickness of a human hair! This flaw (which was later corrected) caused Hubble to transmit fuzzy images because the light from viewed objects was scattering instead of reflecting to a single point. This single point is called the *focus*, as explained in **Exercises 25 and 26 on page 600.**

Need help? For online student resources, such as section quizzes, visit this textbook's website at **math.college.hmco.com/students.**

OBJECTIVES

Section 11.1

A To graph a parabola

Section 11.2

A To find the equation of a circle and then graph the circle

B To write the equation of a circle in standard form

Section 11.3

A To graph an ellipse with center at the origin

B To graph a hyperbola with center at the origin

Section 11.4

A To solve a nonlinear system of equations

Section 11.5

A To graph the solution set of a quadratic inequality in two variables

B To graph the solution set of a nonlinear system of inequalities

PREP TEST • • •

Do these exercises to prepare for Chapter 11.

1. Find the distance between the points whose coordinates are $(-2, 3)$ and $(4, -1)$. Round to the nearest hundredth.

2. Complete the square on $x^2 - 8x$. Write the resulting perfect-square trinomial as the square of a binomial.

3. Solve $\dfrac{x^2}{16} + \dfrac{y^2}{9} = 1$ for x when $y = 3$ and when $y = 0$.

4. Solve by the substitution method:
$$7x + 4y = 3$$
$$y = x - 2$$

5. Solve by the addition method:
$$4x - y = 9$$
$$2x + 3y = -13$$

6. Find the axis of symmetry and the vertex of the graph of $y = x^2 - 4x + 2$.

7. Graph: $f(x) = -2x^2 + 4x$

8. Graph the solution set of $5x - 2y > 10$.

9. Graph the solution set:
$$x + 2y \leq 4$$
$$x - y \leq 2$$

GO FIGURE • • •

What is the minimum number of times a paper must be folded in half so that its thickness is at least the height of the Empire State Building? Assume that the paper can be folded in half indefinitely and that the paper is 0.01 in. thick. The height of the Empire State Building is 1250 ft.

11.1 The Parabola

Objective A ## To graph a parabola

The **conic sections** are curves that can be constructed from the intersection of a plane and a right circular cone. The *parabola*, which was introduced earlier, is one of these curves. Here we will review some of that previous discussion and look at equations of parabolas that were not discussed before.

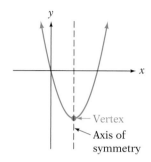

The four conic sections (parabola, circle, ellipse, and hyperbola) are obtained by slicing a cone with planes of various orientations.

Every parabola has an axis of symmetry and a vertex that is on the axis of symmetry. To understand the axis of symmetry, think of folding the paper along that axis. The two halves of the curve will match up.

The graph of the equation $y = ax^2 + bx + c, a \neq 0$, is a parabola with the axis of symmetry parallel to the y-axis. The parabola opens up when $a > 0$ and opens down when $a < 0$. When the parabola opens up, the vertex is the lowest point on the parabola. When the parabola opens down, the vertex is the highest point on the parabola.

The coordinates of the vertex can be found by completing the square.

HOW TO Find the vertex of the parabola whose equation is $y = x^2 - 4x + 5$.

$y = x^2 - 4x + 5$ • **Group the terms involving x.**
$y = (x^2 - 4x) + 5$ • **Complete the square on $x^2 - 4x$. Note that 4**
$y = (x^2 - 4x + 4) - 4 + 5$ **is added and subtracted. Because $4 - 4 = 0$,**
 the equation is not changed.
$y = (x - 2)^2 + 1$ • **Factor the trinomial and combine like terms.**

The coefficient of x^2 is positive, so the parabola opens up. The vertex is the lowest point on the parabola, or the point that has the least y-coordinate.

Because $(x - 2)^2 \geq 0$ for all x, the least y-coordinate occurs when $(x - 2)^2 = 0$, which occurs when $x = 2$. This means the x-coordinate of the vertex is 2.

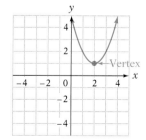

To find the y-coordinate of the vertex, replace x in $y = (x - 2)^2 + 1$ by 2 and solve for y.

$y = (x - 2)^2 + 1$
$ = (2 - 2)^2 + 1 = 1$

The vertex is (2, 1).

By following the procedure of the last example and completing the square on the equation $y = ax^2 + bx + c$, we find that the **x-coordinate of the vertex** is $-\dfrac{b}{2a}$.
The y-coordinate of the vertex can then be determined by substituting this value of x into $y = ax^2 + bx + c$ and solving for y.

Because the axis of symmetry is parallel to the y-axis and passes through the vertex, the equation of the axis of symmetry is $x = -\dfrac{b}{2a}$.

Point of Interest

Golden Gate Bridge

The suspension cables for some bridges, such as the Golden Gate bridge, hang in the shape of a parabola. Parabolic shapes are also used for mirrors in telescopes and in certain antenna designs.

HOW TO Find the axis of symmetry and vertex of the parabola whose equation is $y = -3x^2 + 6x + 1$. Then sketch its graph.

x-coordinate: $-\dfrac{b}{2a} = -\dfrac{6}{2(-3)} = 1$

- Find the x-coordinate of the vertex and the axis of symmetry. $a = -3$, $b = 6$

The x-coordinate of the vertex is 1.
The axis of symmetry is the line $x = 1$.

To find the y-coordinate of the vertex, replace x by 1 and solve for y.

$y = -3x^2 + 6x + 1$
$ = -3(1)^2 + 6(1) + 1 = 4$

The vertex is $(1, 4)$.

Because a is negative, the parabola opens down.

Find a few ordered pairs and use symmetry to sketch the graph.

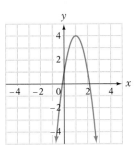

Integrating Technology

The instructions in the Keystroke Guide: *Min and Max* can be used to find the vertex of a parabola.

HOW TO Find the axis of symmetry and vertex of the parabola whose equation is $y = x^2 - 2$. Then sketch its graph.

x-coordinate: $-\dfrac{b}{2a} = -\dfrac{0}{2(1)} = 0$

- Find the x-coordinate of the vertex and the axis of symmetry. $a = 1$, $b = 0$

The x-coordinate of the vertex is 0.
The axis of symmetry is the line $x = 0$.

To find the y-coordinate of the vertex, replace x by 0 and solve for y.

$y = x^2 - 2$
$ = 0^2 - 2 = -2$

The vertex is $(0, -2)$.

Because a is positive, the parabola opens up.

Find a few ordered pairs and use symmetry to sketch the graph.

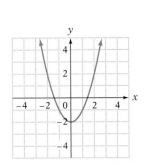

Study Tip

Express in your own words the difference between an equation of a parabola that opens up or down and one that opens right or left. Doing so will help you to remember the differences.

The graph of an equation of the form $x = ay^2 + by + c$, $a \neq 0$, is also a parabola. In this case, the parabola opens to the right when a is positive and opens to the left when a is negative.

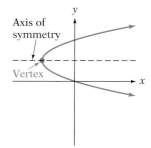

For a parabola of this form, the **y-coordinate of the vertex** is $-\dfrac{b}{2a}$. The axis of symmetry is the line $y = -\dfrac{b}{2a}$.

Using the vertical-line test, the graph of a parabola of this form is not the graph of a function. The graph of $x = ay^2 + by + c$ is a relation.

HOW TO　Find the axis of symmetry and vertex of the parabola whose equation is $x = 2y^2 - 8y + 5$. Then sketch its graph.

y-coordinate: $-\dfrac{b}{2a} = -\dfrac{-8}{2(2)} = 2$　• Find the y-coordinate of the vertex and the axis of symmetry. $a = 2$, $b = -8$

The y-coordinate of the vertex is 2.
The axis of symmetry is the line $y = 2$.

To find the x-coordinate of the vertex, replace y by 2 and solve for x.

$$x = 2y^2 - 8y + 5$$
$$= 2(2)^2 - 8(2) + 5 = -3$$

The vertex is $(-3, 2)$.

Since a is positive, the parabola opens to the right.

Find a few ordered pairs and use symmetry to sketch the graph.

Point of Interest

Hypatia

Hypatia (c. 340–415) is considered the first prominent woman mathematician. She lectured in mathematics and philosophy at the Museum in Alexandria, at that time the most distinguished place of learning in the world. One of the topics on which Hypatia lectured was conic sections. One historian has claimed that with the death (actually the murder) of Hypatia, "the long and glorious history of Greek mathematics was at an end."

HOW TO　Find the axis of symmetry and vertex of the parabola whose equation is $x = -2y^2 - 4y - 3$. Then sketch its graph.

y-coordinate: $-\dfrac{b}{2a} = -\dfrac{-4}{2(-2)} = -1$　• Find the y-coordinate of the vertex and the axis of symmetry. $a = -2$, $b = -4$

The y-coordinate of the vertex is -1.
The axis of symmetry is the line $y = -1$.

To find the x-coordinate of the vertex, replace y by -1 and solve for x.

$$x = -2y^2 - 4y - 3$$
$$= -2(-1)^2 - 4(-1) - 3 = -1$$

The vertex is $(-1, -1)$.

Because a is negative, the parabola opens to the left.

Find a few ordered pairs and use symmetry to sketch the graph.

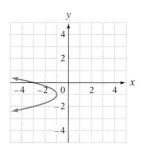

Example 1

Find the axis of symmetry and vertex of the parabola whose equation is $y = x^2 - 4x + 3$. Then sketch its graph.

Solution

$$-\frac{b}{2a} = -\frac{-4}{2(1)} = 2$$

Axis of symmetry:
$\quad x = 2$

$y^2 = x^2 - 4x + 3$
$y = 2^2 - 4(2) + 3$
$\quad = -1$

Vertex: $(2, -1)$

You Try It 1

Find the axis of symmetry and vertex of the parabola whose equation is $y = x^2 + 2x + 1$. Then sketch its graph.

Your solution

Example 2

Find the axis of symmetry and vertex of the parabola whose equation is $x = 2y^2 - 4y + 1$. Then sketch its graph.

Solution

$$-\frac{b}{2a} = -\frac{-4}{2(2)} = 1$$

Axis of symmetry:
$\quad y = 1$

$x = 2y^2 - 4y + 1$
$x = 2(1)^2 - 4(1) + 1$
$\quad = -1$

Vertex: $(-1, 1)$

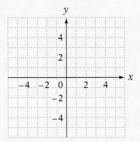

You Try It 2

Find the axis of symmetry and vertex of the parabola whose equation is $x = -y^2 - 2y + 2$. Then sketch its graph.

Your solution

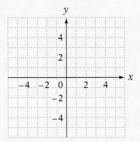

Example 3

Find the axis of symmetry and vertex of the parabola whose equation is $y = x^2 + 1$. Then sketch its graph.

Solution

$$-\frac{b}{2a} = -\frac{0}{2(1)} = 0$$

Axis of symmetry:
$\quad x = 0$

$y = x^2 + 1$
$y = 0^2 + 1$
$\quad = 1$

Vertex: $(0, 1)$

You Try It 3

Find the axis of symmetry and vertex of the parabola whose equation is $y = x^2 - 2x - 1$. Then sketch its graph.

Your solution

Solutions on p. S31

11.1 Exercises

Objective A **To graph a parabola**

a. For Exercises 1 to 6, state whether the axis of symmetry is a vertical or horizontal line. **b.** State the direction in which the parabola opens.

1. $y = 3x^2 - 4x + 7$

2. $y = -x^2 + 5x - 2$

3. $x = y^2 + 2y - 8$

4. $x = -3y^2 - y + 9$

5. $x = -\dfrac{1}{2}y^2 - 4y - 7$

6. $y = \dfrac{1}{4}x^2 + 6x - 1$

For Exercises 7 to 24, find the axis of symmetry and vertex of the parabola given by the equation. Then sketch its graph.

7. $x = y^2 - 3y - 4$

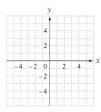

8. $y = x^2 - 2$

9. $y = x^2 + 2$

10. $x = -\dfrac{1}{2}y^2 + 4$

11. $x = -\dfrac{1}{4}y^2 - 1$

12. $x = \dfrac{1}{2}y^2 - y + 1$

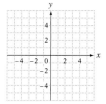

13. $x = -\dfrac{1}{2}y^2 + 2y - 3$

14. $y = -\dfrac{1}{2}x^2 + 2x + 6$

15. $y = \dfrac{1}{2}x^2 + x - 3$

16. $x = y^2 + 6y + 5$

17. $x = y^2 - y - 6$

18. $x = -y^2 + 2y - 3$

19. $y = 2x^2 + 4x - 5$

20. $y = 2x^2 - x - 5$

21. $y = 2x^2 - x - 3$

22. $y = x^2 - 5x + 4$

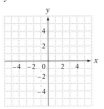

23. $y = x^2 + 5x + 6$

24. $x = y^2 - 2y - 5$

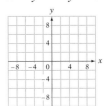

APPLYING THE CONCEPTS

Science Parabolas have a unique property that is important in the design of telescopes and antennas. If a parabola has a mirrored surface, then all light rays parallel to the axis of symmetry of the parabola are reflected to a single point called the **focus** of the parabola. The location of this point is p units from the vertex on the axis of symmetry. The value of p is given by $p = \dfrac{1}{4a}$, where $y = ax^2$ is the equation of a parabola with vertex at the origin. For the graph of $y = \dfrac{1}{4}x^2$ shown at the right, the coordinates of the focus are $(0, 1)$. For Exercises 25 and 26, find the coordinates of the focus of the parabola given by the equation.

25. $y = 2x^2$

26. $y = \dfrac{1}{10}x^2$

Parallel rays of light are reflected to the focus

$a = \dfrac{1}{4}$

$p = \dfrac{1}{4a} = \dfrac{1}{4(1/4)} = 1$

11.2 The Circle

Objective A **To find the equation of a circle and then graph the circle**

A **circle** is a conic section formed by the intersection of a cone and a plane parallel to the base of the cone.

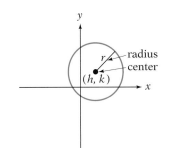

TAKE NOTE

As the angle of the plane that intersects the cone changes, different conic sections are formed. For a parabola, the plane was *parallel to the side* of the cone. For a circle, the plane is *parallel to the base* of the cone.

A circle can be defined as all points (x, y) in the plane that are a fixed distance from a given point (h, k), called the **center.** The fixed distance is the **radius** of the circle.

Using the distance formula, the equation of a circle can be determined.

Let (h, k) be the coordinates of the center of the circle, let r be the radius, and let (x, y) be any point on the circle. Then, by the distance formula,

$$r = \sqrt{(x - h)^2 + (y - k)^2}$$

Squaring each side of the equation gives the equation of a circle.

$$r^2 = [\sqrt{(x - h)^2 + (y - k)^2}\,]^2$$
$$r^2 = (x - h)^2 + (y - k)^2$$

The Standard Form of the Equation of a Circle

Let r be the radius of a circle and let (h, k) be the coordinates of the center of the circle. Then the equation of the circle is given by

$$(x - h)^2 + (y - k)^2 = r^2$$

HOW TO Sketch a graph of $(x - 1)^2 + (y + 2)^2 = 9$.

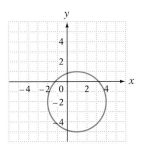

- The standard form of the equation is
 $(x - 1)^2 + [y - (-2)]^2 = 3^2$.
 Center: (1, −2) Radius: 3

HOW TO Find the equation of the circle with radius 4 and center $(-1, 2)$. Then sketch its graph.

$$(x - h)^2 + (y - k)^2 = r^2$$

• Use the standard form of the equation of a circle.

$$[x - (-1)]^2 + (y - 2)^2 = 4^2$$
$$(x + 1)^2 + (y - 2)^2 = 16$$

• Replace r by 4, h by -1, and k by 2.

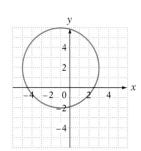

• Sketch the graph by drawing a circle with center $(-1, 2)$ and radius 4.

TAKE NOTE

Applying the vertical-line test reveals that the graph of a circle is not the graph of a function. The graph of a circle is the graph of a relation.

HOW TO Find the equation of the circle that passes through the point whose coordinates are $(3, -4)$ and whose center is $(2, 1)$.

A point on the circle is known and the center is known. Use the distance formula to find the radius of the circle.

$$r = \sqrt{(x_1 - x_2)^2 + (y_1 - y_2)^2}$$
$$= \sqrt{(3 - 2)^2 + (-4 - 1)^2}$$
$$= \sqrt{1^2 + (-5)^2}$$
$$= \sqrt{1 + 25}$$
$$= \sqrt{26}$$

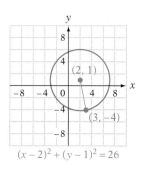

$$(x - 2)^2 + (y - 1)^2 = 26$$

Using $r = \sqrt{26}$ and $(h, k) = (2, 1)$, write the equation of the circle.
Note that $r^2 = (\sqrt{26})^2 = 26$.

$$(x - h)^2 + (y - k)^2 = r^2$$
$$(x - 2)^2 + (y - 1)^2 = 26$$

Example 1

Sketch a graph of $(x + 2)^2 + (y - 1)^2 = 4$.

Solution
Center: $(-2, 1)$
Radius: $\sqrt{4} = 2$

You Try It 1

Sketch a graph of $(x - 2)^2 + (y + 3)^2 = 9$.

Your solution

Solution on p. S31

Example 2

Find the equation of the circle with radius 5 and center $(-1, 3)$. Then sketch its graph.

Solution

$$(x - h)^2 + (y - k)^2 = r^2$$
$$[x - (-1)]^2 + (y - 3)^2 = 5^2 \quad \bullet \ h = -1, k = 3, r = 5$$
$$(x + 1)^2 + (y - 3)^2 = 25$$

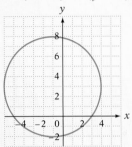

You Try It 2

Find the equation of the circle with radius 4 and center $(2, -3)$. Then sketch its graph.

Your solution

Solution on p. S31

Objective B **To write the equation of a circle in standard form**

S t u d y T i p

See Objective 8.2A to review the process of completing the square.

The equation of a circle can also be expressed as $x^2 + y^2 + ax + by + c = 0$. This is called the **general form of the equation of a circle.** An equation in general form can be written in standard form by completing the square.

HOW TO Find the center and the radius of a circle whose equation is given by $x^2 + y^2 + 4x + 2y + 1 = 0$. Then sketch its graph.

Write the equation in standard form by completing the square on x and y.

$$x^2 + y^2 + 4x + 2y + 1 = 0$$
$$x^2 + y^2 + 4x + 2y = -1$$

$$(x^2 + 4x) + (y^2 + 2y) = -1$$

$$(x^2 + 4x + 4) + (y^2 + 2y + 1) = -1 + 4 + 1$$

$$(x + 2)^2 + (y + 1)^2 = 4$$

- **Subtract the constant term from each side of the equation.**
- **Rewrite the equation by grouping x terms and y terms.**
- **Complete the square on $x^2 + 4x$ and $y^2 + 2y$. Add 4 and 1 to both sides.**
- **Factor the trinomials.**

The center is $(-2, -1)$; the radius is 2.

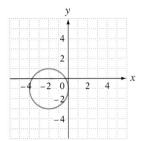

Example 3

Write the equation of the circle $x^2 + y^2 - 4x + 2y + 4 = 0$ in standard form. Then sketch its graph.

Solution

$$x^2 + y^2 - 4x + 2y + 4 = 0$$
$$(x^2 - 4x) + (y^2 + 2y) = -4$$
$$(x^2 - 4x + 4) + (y^2 + 2y + 1) = -4 + 4 + 1$$
$$(x - 2)^2 + (y + 1)^2 = 1$$

Center: $(2, -1)$; Radius: 1

You Try It 3

Write the equation of the circle $x^2 + y^2 - 2x - 15 = 0$ in standard form. Then sketch its graph.

Your solution

Example 4

Write the equation of the circle $x^2 + y^2 + 3x - 2y - 1 = 0$ in standard form. Then sketch its graph.

Solution

$$x^2 + y^2 + 3x - 2y - 1 = 0$$
$$x^2 + y^2 + 3x - 2y = 1$$
$$(x^2 + 3x) + (y^2 - 2y) = 1$$
$$\left(x^2 + 3x + \frac{9}{4}\right) + (y^2 - 2y + 1) = 1 + \frac{9}{4} + 1$$
$$\left(x + \frac{3}{2}\right)^2 + (y - 1)^2 = \frac{17}{4}$$

Center: $\left(-\frac{3}{2}, 1\right)$; Radius: $\sqrt{\frac{17}{4}} = \frac{\sqrt{17}}{2} \approx 2$

You Try It 4

Write the equation of the circle $x^2 + y^2 - 4x + 8y + 15 = 0$ in standard form. Then sketch its graph.

Your solution

Solutions on pp. S31–S32

11.2 Exercises

Objective A **To find the equation of a circle and then graph the circle**

For Exercises 1 to 6, sketch a graph of the circle given by the equation.

1. $(x - 2)^2 + (y + 2)^2 = 9$

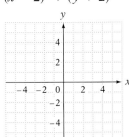

2. $(x + 2)^2 + (y - 3)^2 = 16$

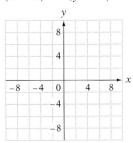

3. $(x + 3)^2 + (y - 1)^2 = 25$

4. $(x - 2)^2 + (y + 3)^2 = 4$

5. $(x + 2)^2 + (y + 2)^2 = 4$

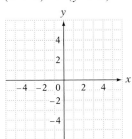

6. $(x - 1)^2 + (y - 2)^2 = 25$

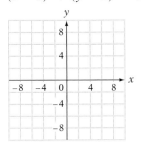

7. Find the equation of the circle with radius 2 and center $(2, -1)$. Then sketch its graph.

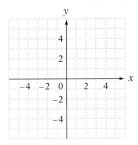

8. Find the equation of the circle with radius 3 and center $(-1, -2)$. Then sketch its graph.

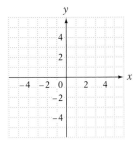

9. Find the equation of the circle that passes through the point $(1, 2)$ and whose center is $(-1, 1)$. Then sketch its graph.

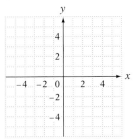

10. Find the equation of the circle that passes through the point $(-1, 3)$ and whose center is $(-2, 1)$. Then sketch its graph.

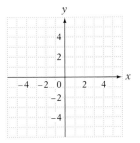

| Objective B | **To write the equation of a circle in standard form** |

For Exercises 11 to 16, write the equation of the circle in standard form. Then sketch its graph.

11. $x^2 + y^2 - 2x + 4y - 20 = 0$

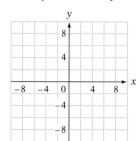

12. $x^2 + y^2 - 4x + 8y + 4 = 0$

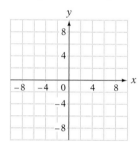

13. $x^2 + y^2 + 6x + 8y + 9 = 0$

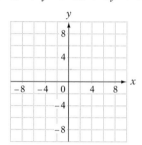

14. $x^2 + y^2 - 6x + 10y + 25 = 0$

15. $x^2 + y^2 - 2x + 2y - 23 = 0$

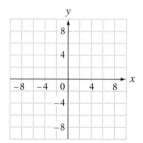

16. $x^2 + y^2 - 6x + 4y + 9 = 0$

APPLYING THE CONCEPTS

17. Find the equation of a circle that has center $(4, 0)$ and passes through the origin.

18. Find the equation of a circle that has a diameter with endpoints $(-2, 4)$ and $(2, -2)$.

19. **Geometry** The radius of a sphere is 12 in. What is the radius of the circle that is formed by the intersection of the sphere and a plane that is 6 in. from the center of the sphere?

20. Find the equation of the circle with center at $(3, 3)$ if the circle is tangent to the x-axis.

21. Is $x^2 + y^2 + 4x + 8y + 24 = 0$ the equation of a circle? If not, explain why not. If so, find the radius and the coordinates of the center.

11.3 The Ellipse and the Hyperbola

Objective A To graph an ellipse with center at the origin

The orbits of the planets around the sun are "oval" shaped. This oval shape can be described as an **ellipse,** which is another of the conic sections.

 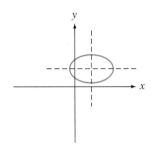

There are two axes of symmetry for an ellipse. The intersection of these two axes is the **center** of the ellipse.

An ellipse with center at the origin is shown at the right. Note that there are two x-intercepts and two y-intercepts.

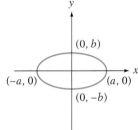

Point of Interest

The word *ellipse* comes from the Greek word *ellipsis*, which means "deficient." The method by which the early Greeks analyzed the conics caused a certain area in the construction of the ellipse to be less than another area (deficient). The word *ellipsis* in English, meaning "omission," has the same Greek root as the word *ellipse.*

The Standard Form of the Equation of an Ellipse with Center at the Origin

The equation of an ellipse with center at the origin is $\dfrac{x^2}{a^2} + \dfrac{y^2}{b^2} = 1$.

The x-intercepts are $(a, 0)$ and $(-a, 0)$. The y-intercepts are $(0, b)$ and $(0, -b)$.

By finding the x- and y-intercepts of an ellipse and using the fact that the ellipse is oval-shaped, we can sketch a graph of the ellipse.

HOW TO Sketch the graph of the ellipse whose equation is $\dfrac{x^2}{9} + \dfrac{y^2}{4} = 1$.

The x-intercepts are $(3, 0)$ and $(-3, 0)$.
The y-intercepts are $(0, 2)$ and $(0, -2)$.

• $a^2 = 9,\ b^2 = 4$

• Use the intercepts and symmetry to sketch the graph of the ellipse.

Integrating Technology

See the Projects and Group Activities at the end of this chapter for instructions on graphing conic sections using a graphing calculator.

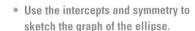

Using the vertical-line test, we find that the graph of an ellipse is not the graph of a function. The graph of an ellipse is the graph of a relation.

> **HOW TO** Sketch a graph of the ellipse whose equation is $\frac{x^2}{16} + \frac{y^2}{16} = 1$.

The x-intercepts are $(4, 0)$ and $(-4, 0)$.
The y-intercepts are $(0, 4)$ and $(0, -4)$.

- $a^2 = 16$, $b^2 = 16$

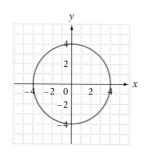

- Use the intercepts and symmetry to sketch the graph of the ellipse.

The graph in this last example is the graph of a circle. A circle is a special case of an ellipse. It occurs when $a^2 = b^2$ in the equation $\frac{x^2}{a^2} + \frac{y^2}{b^2} = 1$.

Example 1

Sketch a graph of the ellipse whose equation is $\frac{x^2}{9} + \frac{y^2}{16} = 1$.

Solution
$a^2 = 9$, $b^2 = 16$
x-intercepts:
$(3, 0)$ and $(-3, 0)$

y-intercepts:
$(0, 4)$ and $(0, -4)$

You Try It 1

Sketch a graph of the ellipse whose equation is $\frac{x^2}{4} + \frac{y^2}{25} = 1$.

Your solution

Example 2

Sketch a graph of the ellipse whose equation is $\frac{x^2}{16} + \frac{y^2}{12} = 1$.

Solution
$a^2 = 16$, $b^2 = 12$
x-intercepts:
$(4, 0)$ and $(-4, 0)$

y-intercepts:
$(0, 2\sqrt{3})$
and $(0, -2\sqrt{3})$
$(2\sqrt{3} \approx 3.5)$

You Try It 2

Sketch a graph of the ellipse whose equation is $\frac{x^2}{18} + \frac{y^2}{9} = 1$.

Your solution

Solutions on p. S32

Objective B **To graph a hyperbola with center at the origin**

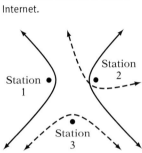

A **hyperbola** is a conic section that can be formed by the intersection of a cone and a plane perpendicular to the base of the cone.

 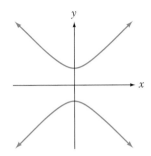

The hyperbola has two **vertices** and an **axis of symmetry** that passes through the vertices. The **center** of a hyperbola is the point halfway between the two vertices.

The graphs at the right show two possible graphs of a hyperbola with center at the origin.

In the first graph, an axis of symmetry is the x-axis, and the vertices are x-intercepts.

In the second graph, an axis of symmetry is the y-axis, and the vertices are y-intercepts.

Note that in either case, the graph of a hyperbola is not the graph of a function. The graph of a hyperbola is the graph of a relation.

The Standard Form of the Equation of a Hyperbola with Center at the Origin

The equation of a hyperbola for which an axis of symmetry is the x-axis is $\dfrac{x^2}{a^2} - \dfrac{y^2}{b^2} = 1$. The vertices are $(a, 0)$ and $(-a, 0)$.

The equation of a hyperbola for which an axis of symmetry is the y-axis is $\dfrac{y^2}{b^2} - \dfrac{x^2}{a^2} = 1$. The vertices are $(0, b)$ and $(0, -b)$.

To sketch a hyperbola, it is helpful to draw two lines that are "approached" by the hyperbola. These two lines are called **asymptotes.** As a point on the hyperbola gets farther from the origin, the hyperbola "gets closer to" the asymptotes.

Because the asymptotes are straight lines, their equations are linear equations. The equations of the asymptotes for a hyperbola with center at the origin are $y = \dfrac{b}{a}x$ and $y = -\dfrac{b}{a}x$.

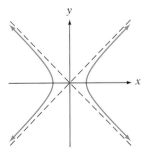

S t u d y T i p
You now have completed the lessons on the four conic sections. You need to be able to recognize the equation of each. To test yourself, you might do the exercises in the Chapter 11 Review Exercises.

HOW TO Sketch a graph of the hyperbola whose equation is $\dfrac{y^2}{9} - \dfrac{x^2}{4} = 1$.

An axis of symmetry is the y-axis.
The vertices are $(0, 3)$ and $(0, -3)$.

- $b^2 = 9$, $a^2 = 4$

The asymptotes are $y = \dfrac{3}{2}x$ and $y = -\dfrac{3}{2}x$.

- Locate the vertices. Sketch the asymptotes. Use symmetry and the fact that the hyperbola will approach the asymptotes to sketch the graph.

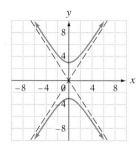

Example 3

Sketch a graph of the hyperbola whose equation is $\dfrac{x^2}{16} - \dfrac{y^2}{4} = 1$.

Solution
$a^2 = 16$, $b^2 = 4$
Axis of symmetry:
x-axis

Vertices:
$(4, 0)$ and $(-4, 0)$

Asymptotes:
$y = \dfrac{1}{2}x$ and $y = -\dfrac{1}{2}x$

You Try It 3

Sketch a graph of the hyperbola whose equation is $\dfrac{x^2}{9} - \dfrac{y^2}{25} = 1$.

Your solution

Example 4

Sketch a graph of the hyperbola whose equation is $\dfrac{y^2}{16} - \dfrac{x^2}{25} = 1$.

Solution
$a^2 = 25$, $b^2 = 16$
Axis of symmetry:
y-axis

Vertices:
$(0, 4)$ and $(0, -4)$

Asymptotes:
$y = \dfrac{4}{5}x$ and $y = -\dfrac{4}{5}x$

You Try It 4

Sketch a graph of the hyperbola whose equation is $\dfrac{y^2}{9} - \dfrac{x^2}{9} = 1$.

Your solution

Solutions on p. S32

11.3 Exercises

Objective A **To graph an ellipse with center at the origin**

For Exercises 1 to 9, sketch a graph of the ellipse given by the equation.

1. $\dfrac{x^2}{4} + \dfrac{y^2}{9} = 1$

2. $\dfrac{x^2}{25} + \dfrac{y^2}{16} = 1$

3. $\dfrac{x^2}{25} + \dfrac{y^2}{9} = 1$

4. $\dfrac{x^2}{16} + \dfrac{y^2}{9} = 1$

5. $\dfrac{x^2}{36} + \dfrac{y^2}{16} = 1$

6. $\dfrac{x^2}{49} + \dfrac{y^2}{64} = 1$

7. $\dfrac{x^2}{16} + \dfrac{y^2}{49} = 1$

8. $\dfrac{x^2}{25} + \dfrac{y^2}{36} = 1$

9. $\dfrac{x^2}{4} + \dfrac{y^2}{25} = 1$

Objective B **To graph a hyperobla with center at the origin**

For Exercises 10 to 21, sketch a graph of the hyperbola given by the equation.

10. $\dfrac{x^2}{9} - \dfrac{y^2}{16} = 1$

11. $\dfrac{x^2}{25} - \dfrac{y^2}{4} = 1$

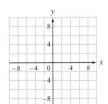

12. $\dfrac{y^2}{16} - \dfrac{x^2}{9} = 1$

13. $\dfrac{y^2}{16} - \dfrac{x^2}{25} = 1$

14. $\dfrac{x^2}{16} - \dfrac{y^2}{4} = 1$

15. $\dfrac{x^2}{9} - \dfrac{y^2}{49} = 1$

16. $\dfrac{y^2}{25} - \dfrac{x^2}{9} = 1$

17. $\dfrac{y^2}{4} - \dfrac{x^2}{16} = 1$

18. $\dfrac{x^2}{4} - \dfrac{y^2}{25} = 1$

19. $\dfrac{x^2}{36} - \dfrac{y^2}{9} = 1$

20. $\dfrac{y^2}{9} - \dfrac{x^2}{36} = 1$

21. $\dfrac{y^2}{25} - \dfrac{x^2}{4} = 1$

APPLYING THE CONCEPTS

a. For Exercises 22 to 27, write the equation in standard form and identify the graph. **b.** Graph the equation.

22. $4x^2 + 9y^2 = 36$

23. $16x^2 + 25y^2 = 400$

24. $9x^2 - 25y^2 = 225$

25. $25y^2 - 4x^2 = -100$

26. $9y^2 - 16x^2 = 144$

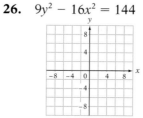

27. $4y^2 - x^2 = 36$

11.4 Solving Nonlinear Systems of Equations

Objective A **To solve a nonlinear system of equations**

A **nonlinear system of equations** is a system in which one or more of the equations are not linear equations. A nonlinear system of equations can be solved by using either the substitution method or the addition method.

> **HOW TO** Solve: $2x - y = 4$ (1)
> $y^2 = 4x$ (2)
>
> When a nonlinear system of equations contains a linear equation, the substitution method is used.
>
> Solve Equation (1) for y.
>
> $2x - y = 4$
> $-y = -2x + 4$
> $y = 2x - 4$
>
> Substitute $2x - 4$ for y into equation (2).
>
> $y^2 = 4x$
> $(2x - 4)^2 = 4x$ • $y = 2x - 4$
> $4x^2 - 16x + 16 = 4x$ • Solve for x.
> $4x^2 - 20x + 16 = 0$
>
> $4(x^2 - 5x + 4) = 0$
> $4(x - 4)(x - 1) = 0$
>
> $x - 4 = 0 \qquad x - 1 = 0$
> $x = 4 \qquad x = 1$
>
> Substitute the values of x into the equation $y = 2x - 4$ and solve for y.
>
> $y = 2x - 4 \qquad\qquad y = 2x - 4$
> $y = 2(4) - 4$ • $x = 4$ $\qquad y = 2(1) - 4$ • $x = 1$
> $y = 4 \qquad\qquad\qquad y = -2$
>
> The solutions are $(4, 4)$ and $(1, -2)$.

The graph of this system of equations is shown at the right. The graph of the line intersects the parabola at two points. The coordinates of these points are the solutions of the system of equations.

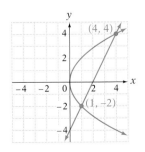

HOW TO Solve: $x^2 + y^2 = 4$ (1)
$\qquad\qquad\qquad\quad y = x + 3$ (2)

Use the substitution method.

Substitute the expression for y into Equation (1).

$$x^2 + y^2 = 4$$
$$x^2 + (x + 3)^2 = 4 \qquad \bullet\ y = x + 3$$
$$x^2 + x^2 + 6x + 9 = 4 \qquad \bullet\ \text{Solve for } x.$$
$$2x^2 + 6x + 5 = 0$$

Consider the discriminant of $2x^2 + 6x + 5 = 0$.

$$b^2 - 4ac$$
$$6^2 - 4 \cdot 2 \cdot 5 = 36 - 40 = -4 \qquad \bullet\ a = 2, b = 6, c = 5$$

Because the discriminant is less than zero, the equation $2x^2 + 6x + 5 = 0$ has two complex number solutions. Therefore, the system of equations has no real number solutions.

The graphs of the equations of this system are shown at the right. Note that the two graphs do not intersect.

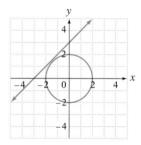

HOW TO Solve: $4x^2 + y^2 = 16$ (1)
$\qquad\qquad\qquad\quad x^2 + y^2 = 4$ (2)

Use the addition method.

$$4x^2 + y^2 = 16$$
$$\underline{-x^2 - y^2 = -4} \qquad \bullet\ \text{Multiply Equation (2) by } -1.$$
$$3x^2 = 12 \qquad\quad \bullet\ \text{Add the equations.}$$

Solve for x.

$$x^2 = 4$$
$$x = \pm 2$$

Substitute the values of x into Equation (2) and solve for y.

$$x^2 + y^2 = 4 \qquad\qquad\qquad x^2 + y^2 = 4$$
$$2^2 + y^2 = 4 \quad \bullet\ x = 2 \qquad (-2)^2 + y^2 = 4 \quad \bullet\ x = -2$$
$$y^2 = 0 \qquad\qquad\qquad\qquad y^2 = 0$$
$$y = 0 \qquad\qquad\qquad\qquad y = 0$$

The solutions are $(2, 0)$ and $(-2, 0)$.

The graphs of the equations in this system are shown at the right. Note that the graphs intersect at the points whose coordinates are $(2, 0)$ and $(-2, 0)$.

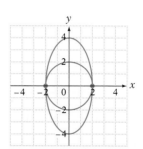

HOW TO Solve: $2x^2 - y^2 = 1$ (1)
$x^2 + 2y^2 = 18$ (2)

Use the addition method.

$4x^2 - 2y^2 = 2$ • Multiply Equation (1) by 2.
$\underline{x^2 + 2y^2 = 18}$
$\qquad 5x^2 = 20$ • Add the equations.
$\qquad\; x^2 = 4$ • Solve for x.
$\qquad\;\; x = \pm 2$

Substitute the values of x into one of the equations and solve for y. Equation (2) is used here.

$x^2 + 2y^2 = 18$ $x^2 + 2y^2 = 18$
$2^2 + 2y^2 = 18$ • $x = 2$ $(-2)^2 + 2y^2 = 18$ • $x = -2$
$\quad\; 2y^2 = 14$ $\qquad\; 2y^2 = 14$
$\qquad y^2 = 7$ $\qquad\; y^2 = 7$
$\qquad\; y = \pm\sqrt{7}$ $\qquad\;\; y = \pm\sqrt{7}$

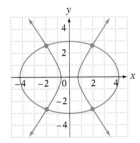

The solutions are $(2, \sqrt{7})$, $(2, -\sqrt{7})$, $(-2, \sqrt{7})$, and $(-2, -\sqrt{7})$.

The graphs of the equations in this system are shown at the left. Note that there are four points of intersection.

Note from the preceding three examples that the number of points at which the graphs of the equations of the system intersect is the same as the number of real number solutions of the system of equations. Here are three more examples.

$x^2 + y^2 = 4$
$\qquad y = x^2 + 2$

The graphs intersect at one point.
The system of equations has one solution.

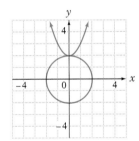

$y = x^2$
$y = -x + 2$

The graphs intersect at two points.
The system of equations has two solutions.

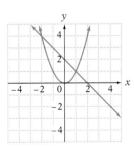

$(x + 2)^2 + (y - 2)^2 = 4$
$\qquad\qquad x = y^2$

The graphs do not intersect.
The system of equations has no solution.

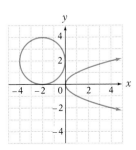

Example 1 Solve: (1) $y = 2x^2 - 3x - 1$
(2) $y = x^2 - 2x + 5$

You Try It 1 Solve: $y = 2x^2 + x - 3$
$y = 2x^2 - 2x + 9$

Solution Use the substitution method.

$$y = x^2 - 2x + 5$$
$$2x^2 - 3x - 1 = x^2 - 2x + 5$$
$$x^2 - x - 6 = 0$$
$$(x - 3)(x + 2) = 0$$

$$x - 3 = 0 \qquad x + 2 = 0$$
$$x = 3 \qquad x = -2$$

Substitute each value of x into Equation (1).

$$y = 2x^2 - 3x - 1$$
$$y = 2(3)^2 - 3(3) - 1$$
$$y = 18 - 9 - 1$$
$$y = 8$$

$$y = 2x^2 - 3x - 1$$
$$y = 2(-2)^2 - 3(-2) - 1$$
$$y = 8 + 6 - 1$$
$$y = 13$$

The solutions are $(3, 8)$ and $(-2, 13)$.

Your solution

Example 2 Solve: (1) $3x^2 - 2y^2 = 26$
(2) $x^2 - y^2 = 5$

You Try It 2 Solve: $x^2 - y^2 = 10$
$x^2 + y^2 = 8$

Solution Use the addition method.
Multiply Equation (2) by -2.

$$-2x^2 + 2y^2 = -10$$
$$3x^2 - 2y^2 = 26$$
$$x^2 = 16$$
$$x = \pm\sqrt{16} = \pm 4$$

Substitute each value of x into Equation (2).

$$x^2 - y^2 = 5 \qquad\qquad x^2 - y^2 = 5$$
$$4^2 - y^2 = 5 \qquad\qquad (-4)^2 - y^2 = 5$$
$$16 - y^2 = 5 \qquad\qquad 16 - y^2 = 5$$
$$-y^2 = -11 \qquad\qquad -y^2 = -11$$
$$y^2 = 11 \qquad\qquad y^2 = 11$$
$$y = \pm\sqrt{11} \qquad\qquad y = \pm\sqrt{11}$$

The solutions are $(4, \sqrt{11})$, $(4, -\sqrt{11})$, $(-4, \sqrt{11})$, and $(-4, -\sqrt{11})$.

Your solution

Solutions on p. S32

11.4 Exercises

Objective A To solve a nonlinear system of equations

For Exercises 1 to 33, solve.

1. $y = x^2 - x - 1$
 $y = 2x + 9$

2. $y = x^2 - 3x + 1$
 $y = x + 6$

3. $y^2 = -x + 3$
 $x - y = 1$

4. $y^2 = 4x$
 $x - y = -1$

5. $y^2 = 2x$
 $x + 2y = -2$

6. $y^2 = 2x$
 $x - y = 4$

7. $x^2 + 2y^2 = 12$
 $2x - y = 2$

8. $x^2 + 4y^2 = 37$
 $x - y = -4$

9. $x^2 + y^2 = 13$
 $x + y = 5$

10. $x^2 + y^2 = 16$
 $x - 2y = -4$

11. $4x^2 + y^2 = 12$
 $y = 4x^2$

12. $2x^2 + y^2 = 6$
 $y = 2x^2$

13. $y = x^2 - 2x - 3$
 $y = x - 6$

14. $y = x^2 + 4x + 5$
 $y = -x - 3$

15. $3x^2 - y^2 = -1$
 $x^2 + 4y^2 = 17$

16. $x^2 + y^2 = 10$
 $x^2 + 9y^2 = 18$

17. $2x^2 + 3y^2 = 30$
 $x^2 + y^2 = 13$

18. $x^2 + y^2 = 61$
 $x^2 - y^2 = 11$

19. $y = 2x^2 - x + 1$
 $y = x^2 - x + 5$

20. $y = -x^2 + x - 1$
 $y = x^2 + 2x - 2$

21. $2x^2 + 3y^2 = 24$
 $x^2 - y^2 = 7$

22. $2x^2 + 3y^2 = 21$
$\quad\ \ x^2 + 2y^2 = 12$

23. $\quad x^2 + y^2 = 36$
$\quad 4x^2 + 9y^2 = 36$

24. $2x^2 + 3y^2 = 12$
$\quad\ \ x^2 - y^2 = 25$

25. $11x^2 - 2y^2 = 4$
$\quad\ \ 3x^2 + y^2 = 15$

26. $x^2 + 4y^2 = 25$
$\quad x^2 - y^2 = 5$

27. $2x^2 - y^2 = 7$
$\quad 2x - y = 5$

28. $3x^2 + 4y^2 = 7$
$\quad\ \ x - 2y = -3$

29. $y = 3x^2 + x - 4$
$\quad y = 3x^2 - 8x + 5$

30. $y = 2x^2 + 3x + 1$
$\quad y = 2x^2 + 9x + 7$

31. $\quad\ \ x = y + 3$
$\quad x^2 + y^2 = 5$

32. $\quad x - y = -6$
$\quad x^2 + y^2 = 4$

33. $\quad y = x^2 + 4x + 4$
$\quad x + 2y = 4$

APPLYING THE CONCEPTS

34. Give the maximum number of points in which the following pairs of graphs can intersect.
a. A line and an ellipse.
b. A line and a parabola.
c. Two ellipses with centers at the origin.
d. Two different circles with centers at the origin.
e. Two hyperbolas with centers at the origin.

 For Exercises 35 to 40, solve by graphing. Round the solutions of the systems to the nearest thousandth.

35. $\quad y = 2^x$
$\quad x + y = 3$

36. $\quad y = 3^{-x}$
$\quad x^2 + y^2 = 9$

37. $\quad y = \log_2 x$
$\quad \dfrac{x^2}{9} + \dfrac{y^2}{1} = 1$

38. $\quad\ \ y = \log_3 x$
$\quad x^2 + y^2 = 4$

39. $\quad y = -\log_3 x$
$\quad x + y = 4$

40. $\quad y = \left(\dfrac{1}{2}\right)^x$
$\quad \dfrac{x^2}{9} + \dfrac{y^2}{4} = 1$

11.5 Quadratic Inequalities and Systems of Inequalities

Objective A To graph the solution set of a quadratic inequality in two variables

The **graph of a quadratic inequality in two variables** is a region of the plane that is bounded by one of the conic sections (parabola, circle, ellipse, or hyperbola). When graphing an inequality of this type, first replace the inequality symbol with an equals sign. Graph the resulting conic using a dashed curve when the original inequality is less than ($<$) or greater than ($>$). Use a solid curve when the original inequality is \leq or \geq. Use the point (0, 0) to determine which region of the plane to shade. If (0, 0) is a solution of the inequality, then shade the region of the plane containing (0, 0). If not, shade the other portion of the plane.

> **HOW TO** Graph the solution set: $x^2 + y^2 > 9$
>
> Change the inequality to an equality.
>
> $x^2 + y^2 = 9$
>
> This is the equation of a circle with center (0, 0) and radius 3.
>
> The original inequality is $>$, so the graph is drawn as a dashed circle.
>
> Substitute the point (0, 0) in the inequality. Because $0^2 + 0^2 > 9$ is not true, the point (0, 0) should not be in the shaded region.

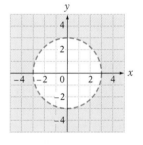

Not all inequalities have a solution set that can be graphed by shading a region of the plane. For example, the inequality

$$\frac{x^2}{9} + \frac{y^2}{4} < -1$$

has no ordered-pair solutions. The solution set is the empty set, and no region of the plane is shaded.

> **HOW TO** Graph the solution set: $y \leq x^2 + 2x + 2$
>
> Change the inequality to an equality.
>
> $y = x^2 + 2x + 2$
>
> This is the equation of a parabola that opens up. The vertex is $(-1, 1)$ and the axis of symmetry is the line $x = -1$.
>
> The original inequality is \leq, so the graph is drawn as a solid curve.
>
> Substitute the point (0, 0) into the inequality. Because $0 \leq 0^2 + 2(0) + 2$ is true, the point (0, 0) should be in the shaded region.

TAKE NOTE

Any point that is not on the graph of the parabola can be used as a test point. For example, the point (2, 1) can be used.

$y \leq x^2 + 2x + 2$
$1 \leq 2^2 + 2(2) + 2$
$1 \leq 10$

This is a true inequality. The point (2, 1) should be in the shaded region.

> **HOW TO** Graph the solution set: $\dfrac{y^2}{9} - \dfrac{x^2}{4} \geq 1$
>
> Write the inequality as an equality.
>
> $$\dfrac{y^2}{9} - \dfrac{x^2}{4} = 1$$
>
> This is the equation of a hyperbola. The vertices are $(0, -3)$ and $(0, 3)$. The equations of the asymptotes are $y = \dfrac{3}{2}x$ and $y = -\dfrac{3}{2}x$.
>
> The original inequality is \geq, so the graph is drawn as a solid curve.
>
> Substitute the point $(0, 0)$ into the inequality. Because $\dfrac{0^2}{9} - \dfrac{0^2}{4} \geq 1$ is not true, the point $(0, 0)$ should not be in the shaded region.

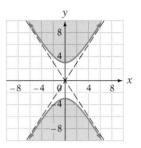

Example 1

Graph the solution set: $\dfrac{x^2}{9} + \dfrac{y^2}{16} \leq 1$

Solution
Graph the ellipse
$\dfrac{x^2}{9} + \dfrac{y^2}{16} = 1$ as a
solid curve.

Shade the region of the plane that includes $(0, 0)$.

You Try It 1

Graph the solution set: $\dfrac{x^2}{9} - \dfrac{y^2}{16} \leq 1$

Your solution

Example 2

Graph the solution set: $x > y^2 - 4y + 2$

Solution
Graph the parabola
$x = y^2 - 4y + 2$
as a dashed curve.

Shade the region that does not include $(0, 0)$.

You Try It 2

Graph the solution set: $\dfrac{x^2}{16} + \dfrac{y^2}{4} > 1$

Your solution

Solutions on pp. S32–S33

Objective B **To graph the solution set of a nonlinear system of inequalities**

Recall that the solution set of a system of inequalities is the intersection of the solution sets of the individual inequalities. To graph the solution set of a system of inequalities, first graph the solution set of each inequality. The solution set of the system of inequalities is the region of the plane represented by the intersection of the two shaded regions.

HOW TO Graph the solution set: $x^2 + y^2 \leq 16$
$y \geq x^2$

Graph the solution set of each inequality.

$x^2 + y^2 = 16$ is the equation of a circle with center at the origin and a radius of 4. Because the original inequality is \leq, use a solid curve.

$y = x^2$ is the equation of a parabola that opens up. The vertex is $(0, 0)$. Because the original inequality is \geq, use a solid curve.

Substitute the point $(0, 0)$ into the inequality $x^2 + y^2 \leq 16$. $0^2 + 0^2 \leq 16$ is a true statement. Shade inside the circle.

Substitute $(2, 0)$ into the inequality $y \geq x^2$. [We cannot use $(0, 0)$, a point on the parabola.] $0 > 4$ is a false statement. Shade inside the parabola.

The solution is the intersection of the two shaded regions.

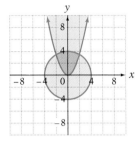

TAKE NOTE

Other points can be used as test points. For example, the point $(6, 2)$ can be used for the circle.

$x^2 + y^2 \leq 16$
$6^2 + 2^2 \leq 16$
$40 \leq 16$

This is false. The point $(6, 2)$ should not be in the shaded region.

Example 3

Graph the solution set: $\dfrac{x^2}{9} + \dfrac{y^2}{4} \geq 1$
$\dfrac{x^2}{4} - \dfrac{y^2}{9} \geq 1$

Solution

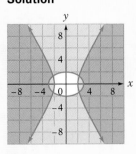

- Draw the ellipse
 $\dfrac{x^2}{9} + \dfrac{y^2}{4} = 1$ as a solid
 curve. Shade outside the ellipse.
- Draw the hyperbola
 $\dfrac{x^2}{4} - \dfrac{y^2}{9} = 1$ as a solid
 curve. $(0, 0)$ should not be in the shaded region.

You Try It 3

Graph the solution set: $x^2 + y^2 < 16$
$y^2 > x$

Your solution

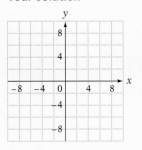

Solution on p. S33

Example 4

Graph the solution set: $y > x^2$
$\quad\quad\quad\quad\quad\quad\quad\quad\quad\quad y < x + 2$

Solution

- Draw the parabola $y = x^2$ as a dashed curve. Shade inside the parabola.
- Draw the line $y = x + 2$ as a dashed line. Shade below the line.

You Try It 4

Graph the solution set: $\dfrac{x^2}{4} + \dfrac{y^2}{9} \le 1$
$\quad\quad\quad\quad\quad\quad\quad\quad\quad\quad x > y^2 - 2$

Your solution

Example 5

Graph the solution set: $\dfrac{x^2}{9} - \dfrac{y^2}{16} > 1$
$\quad\quad\quad\quad\quad\quad\quad\quad\quad\quad x^2 + y^2 < 4$

Solution

- Draw the hyperbola $\dfrac{x^2}{9} - \dfrac{y^2}{16} = 1$ as a dashed curve. $(0, 0)$ should not be in the shaded region.
- Draw the circle $x^2 + y^2 = 4$ as a dashed curve. Shade inside the circle.

The solution sets of the two inequalities do not intersect. This system of inequalities has no real number solution.

You Try It 5

Graph the solution set: $\dfrac{x^2}{16} + \dfrac{y^2}{25} \ge 1$
$\quad\quad\quad\quad\quad\quad\quad\quad\quad\quad x^2 + y^2 < 9$

Your solution

Solutions on p. S33

11.5 Exercises

Objective A **To graph the solution set of a quadratic inequality in two variables**

1. How is the graph of the solution set of a nonlinear inequality that uses ≤ or ≥ different from the graph of the solution set of a nonlinear inequality that uses < or >?

2. How can you determine which region to shade when graphing the solution set of a nonlinear inequality?

For Exercises 3 to 26, graph the solution set.

3. $y \le x^2 - 4x + 3$

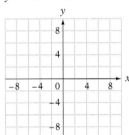

4. $y < x^2 - 2x - 3$

5. $(x - 1)^2 + (y + 2)^2 \le 9$

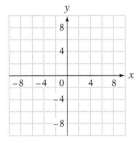

6. $(x + 2)^2 + (y - 3)^2 > 4$

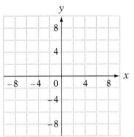

7. $(x + 3)^2 + (y - 2)^2 \ge 9$

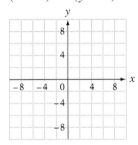

8. $(x - 2)^2 + (y + 1)^2 \le 16$

9. $\dfrac{x^2}{16} + \dfrac{y^2}{25} < 1$

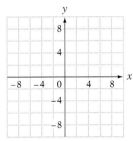

10. $\dfrac{x^2}{9} + \dfrac{y^2}{4} \ge 1$

11. $\dfrac{x^2}{25} - \dfrac{y^2}{9} \le 1$

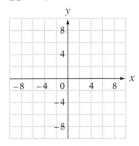

12. $\dfrac{y^2}{25} - \dfrac{x^2}{36} > 1$

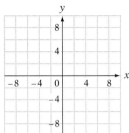

13. $\dfrac{x^2}{4} + \dfrac{y^2}{16} \geq 1$

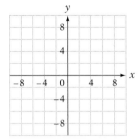

14. $\dfrac{x^2}{4} - \dfrac{y^2}{16} \leq 1$

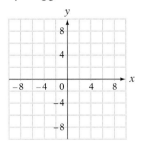

15. $y \leq x^2 - 2x + 3$

16. $x \leq y^2 + 2y + 1$

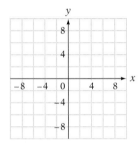

17. $\dfrac{y^2}{9} - \dfrac{x^2}{16} \leq 1$

18. $\dfrac{x^2}{16} - \dfrac{y^2}{4} < 1$

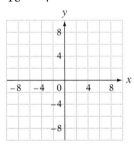

19. $\dfrac{x^2}{9} + \dfrac{y^2}{1} \leq 1$

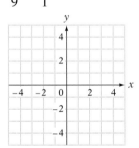

20. $\dfrac{x^2}{16} + \dfrac{y^2}{4} > 1$

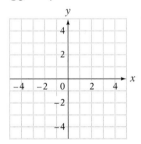

21. $(x - 1)^2 + (x + 3)^2 \leq 25$

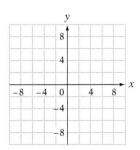

22. $(x + 1)^2 + (y - 2)^2 \geq 16$

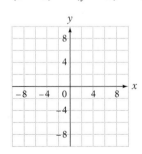

23. $\dfrac{y^2}{25} - \dfrac{x^2}{4} \leq 1$

24. $\dfrac{x^2}{9} - \dfrac{y^2}{25} \geq 1$

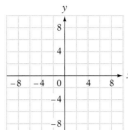

25. $\dfrac{x^2}{25} + \dfrac{y^2}{9} \leq 1$

26. $\dfrac{x^2}{36} + \dfrac{y^2}{4} \leq 1$

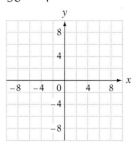

Objective B **To graph the solution set of a nonlinear system of inequalities**

For Exercises 27 to 38, graph the solution set.

27. $y \leq (x - 2)^2$
$y + x > 4$

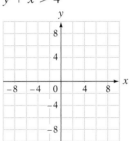

28. $x^2 + y^2 < 1$
$x + y \geq 4$

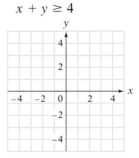

29. $x^2 + y^2 < 16$
$y > x + 1$

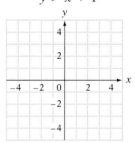

30. $y > x^2 - 4$
$y < x - 2$

31. $\dfrac{x^2}{4} + \dfrac{y^2}{16} \leq 1$

$y \leq -\dfrac{1}{2}x + 2$

32. $\dfrac{y^2}{4} - \dfrac{x^2}{25} \geq 1$

$y \leq \dfrac{2}{3}x + 4$

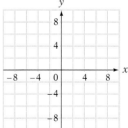

33. $x \geq y^2 - 3y + 2$
$y \geq 2x - 2$

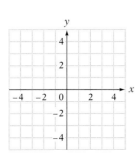

34. $x^2 + y^2 \leq 25$

$y \leq -\dfrac{1}{3}x + 2$

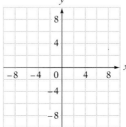

35. $x^2 + y^2 < 25$
$\dfrac{x^2}{9} + \dfrac{y^2}{36} < 1$

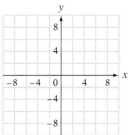

36. $\dfrac{x^2}{9} - \dfrac{y^2}{4} < 1$

$\dfrac{x^2}{25} + \dfrac{y^2}{9} < 1$

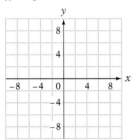

37. $x^2 + y^2 > 4$

$x^2 + y^2 < 25$

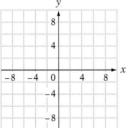

38. $\dfrac{x^2}{25} + \dfrac{y^2}{16} \le 1$

$\dfrac{x^2}{4} + \dfrac{y^2}{4} \ge 1$

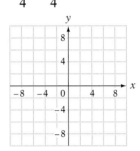

APPLYING THE CONCEPTS

For Exercises 39 to 47, graph the solution set.

39. $y > x^2 - 3$

$y < x + 3$

$x \le 0$

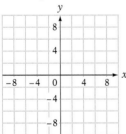

40. $x^2 + y^2 \le 25$

$y > x + 1$

$x \ge 0$

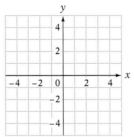

41. $x^2 + y^2 < 3$

$x > y^2 - 1$

$y \ge 0$

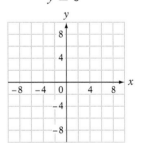

42. $\dfrac{x^2}{4} - \dfrac{y^2}{25} \le 1$

$\dfrac{x^2}{4} + \dfrac{y^2}{4} \le 1$

$y \ge 0$

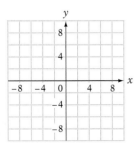

43. $\dfrac{x^2}{16} + \dfrac{y^2}{4} \le 1$

$x^2 + y^2 \le 4$

$x \ge 0$

$y \le 0$

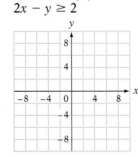

44. $\dfrac{x^2}{4} + \dfrac{y^2}{25} \le 1$

$x > y^2 - 4$

$x \le 0$

$y \ge 0$

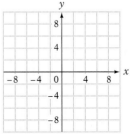

45. $y > 2^x$

$x + y < 4$

46. $y < \left(\dfrac{1}{2}\right)^x$

$2x - y \ge 2$

47. $y \ge \log_2 x$

$x^2 + y^2 < 9$

Focus on Problem Solving

Using a Variety of Problem-Solving Techniques

We have examined several problem-solving strategies throughout the text. See if you can apply those techniques to Exercises 1 to 6.

1. Eight coins look exactly alike, but one is lighter than the others. Explain how the different coin can be found in two weighings on a balance scale.

2. For the sequence of numbers 1, 1, 2, 3, 5, 8, 13, . . . , identify a possible pattern and then use that pattern to determine the next number in the sequence.

3. Arrange the numbers 1, 2, 3, 4, 5, 6, 7, 8, and 9 in the squares at the right such that the sum of any row, column, or diagonal is 15. (*Suggestion:* Note that 1, 5, 9; 2, 5, 8; 3, 5, 7; and 4, 5, 6 all add to 15. Because 5 is part of each sum, this suggests that 5 should be placed in the center of the squares.)

4. A restaurant charges $12.00 for a pizza that has a diameter of 9 in. Determine the selling price of a pizza with a diameter of 18 in. such that the selling price per square inch is the same as for the 9-inch pizza.

5. You have a balance scale and weights of 1 g, 4 g, 8 g, and 16 g. Using only these weights, can you weigh something that weighs each of the following?
 a. 3 g **b.** 7 g **c.** 11 g **d.** 13 g **e.** 19 g

6. Can the checkerboard at the right be covered with dominos (which look like ▭) such that every square on the board is covered by a domino? Why or why not? (*Note:* The dominos cannot overlap.)

Projects and Group Activities

The Eccentricity and Foci of an Ellipse

The graph of an ellipse can be long and thin, or it can have a shape that is very close to a circle. The **eccentricity,** *e*, of an ellipse is a measure of its "roundness."

The shapes of ellipses with various eccentricities are shown below.

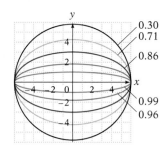

Planet	Eccentricity
Mercury	0.206
Venus	0.007
Earth	0.017
Mars	0.093
Jupiter	0.049
Saturn	0.051
Uranus	0.046
Neptune	0.005
Pluto	0.250

1. Based on the eccentricities of the ellipses shown above, complete the sentence. "As the eccentricity of an ellipse gets closer to 1, the ellipses get *flatter/rounder.*"

2. The planets travel around the sun in elliptical orbits. The eccentricities of the orbits of the planets are shown in the table at the left. Which planet has the most nearly circular orbit?

For an ellipse given by the equation $\frac{x^2}{a^2} + \frac{y^2}{b^2} = 1$, $a > b$, a formula for eccentricity is $e = \frac{\sqrt{a^2 - b^2}}{a}$. Use this formula to find the eccentricity of the ellipses in Exercises 3 and 4. If necessary, round to the nearest hundredth.

3. $\dfrac{x^2}{25} + \dfrac{y^2}{16} = 1$

4. $\dfrac{x^2}{9} + \dfrac{y^2}{4} = 1$

Ellipses have a reflective property that has been used in the design of some buildings. The **foci** of an ellipse are two points on the longer axis, called the **major axis,** of the ellipse.

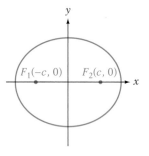

If light or sound emanates from one focus, it is reflected to the other focus. This phenomenon results in what are called "whispering galleries." The rotunda of the Capitol Building in Washington, D.C. is a whispering gallery. A whisper spoken by a person at one focus can be heard clearly by a person at the other focus.

Foci

The foci are c units from the center of an ellipse, where $c = \sqrt{a^2 - b^2}$ for an ellipse whose equation is $\frac{x^2}{a^2} + \frac{y^2}{b^2} = 1$, $a > b$.

5. Find the foci for an ellipse whose equation is $\dfrac{x^2}{169} + \dfrac{y^2}{144} = 1$.

Graphing Conic Sections Using a Graphing Calculator

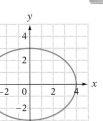

Consider the graph of the ellipse $\frac{x^2}{16} + \frac{y^2}{9} = 1$ shown at the left. Because a vertical line can intersect the graph at more than one point, the graph is not the graph of a function. And because the graph is not the graph of a function, the equation does not represent a function. Consequently, the equation cannot be entered into a graphing calculator to be graphed. However, by solving the equation for y, we have

$$\frac{y^2}{9} = 1 - \frac{x^2}{16}$$

$$y^2 = 9\left(1 - \frac{x^2}{16}\right)$$

$$y = \pm 3\sqrt{1 - \frac{x^2}{16}}$$

There are two solutions for y, which can be written

$$y_1 = 3\sqrt{1 - \frac{x^2}{16}} \quad \text{and} \quad y_2 = -3\sqrt{1 - \frac{x^2}{16}}$$

Each of these equations is the equation of a function and therefore can be entered into a graphing calculator. The graph of $y_1 = 3\sqrt{1 - \frac{x^2}{16}}$ is shown in blue below, and the graph of $y_2 = -3\sqrt{1 - \frac{x^2}{16}}$ is shown in red. Note that together, the graphs are the graph of an ellipse.

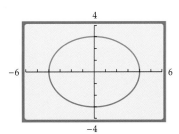

A similar technique can be used to graph a hyperbola. To graph $\frac{x^2}{16} - \frac{y^2}{4} = 1$, solve for y. Then graph each equation, $y_1 = 2\sqrt{\frac{x^2}{16} - 1}$ and $y_2 = -2\sqrt{\frac{x^2}{16} - 1}$.

 For Exercises 1 to 4, solve each equation for y. Then graph using a graphing calculator.

1. $\dfrac{x^2}{25} + \dfrac{y^2}{49} = 1$ **2.** $\dfrac{x^2}{4} + \dfrac{y^2}{64} = 1$ **3.** $\dfrac{x^2}{16} - \dfrac{y^2}{4} = 1$ **4.** $\dfrac{x^2}{9} - \dfrac{y^2}{36} = 1$

Chapter 11 Summary

Key Words	**Examples**
Conic sections are curves that can be constructed from the intersection of a plane and a right circular cone. [11.1A, p. 595]	The four conic sections presented in this chapter are the parabola, the circle, the ellipse, and the hyperbola.
The *asymptotes* of a hyperbola are the two straight lines that are "approached" by the hyperbola. As the graph of the hyperbola gets farther from the origin, the hyberbola "gets closer to" the asymptotes. [11.3B, p .609]	
A *nonlinear system of equations* is a system in which one or more of the equations are not linear equations. [11.4A, p. 613]	$x^2 + y^2 = 16$ $y = x^2 - 2$
The *graph of a quadratic inequality in two variables* is a region of the plane that is bounded by one of the conic sections. [11.5A, p. 619]	$y \geq x^2 - 2x - 4$

Essential Rules and Procedures

Examples

Equations of a Parabola [11.1A, pp. 595–597]
$y = ax^2 + bx + c, a \neq 0$
When $a > 0$, the parabola opens up.
When $a < 0$, the parabola opens down.

The x-coordinate of the vertex is $-\dfrac{b}{2a}$.

The axis of symmetry is the line $x = -\dfrac{b}{2a}$.

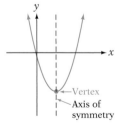

$x = ay^2 + by + c, a \neq 0$
When $a > 0$, the parabola opens to the right.
When $a < 0$, the parabola opens to the left.

The y-coordinate of the vertex is $-\dfrac{b}{2a}$.

The axis of symmetry is the line $y = -\dfrac{b}{2a}$.

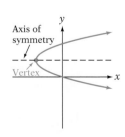

Standard Form of the Equation of a Circle [11.2A, p. 601]

$(x - h)^2 + (y - k)^2 = r^2$

The center of the circle is (h, k), and the radius is r.

$(x - 2)^2 + (y + 1)^2 = 9$
$(h, k) = (2, -1)$
$r = 3$

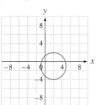

General Form of the Equation of a Circle [11.2B, p. 603]

The equation of a circle can also be expressed as
$x^2 + y^2 + ax + by + c = 0$.

$x^2 + y^2 + 4x + 2y + 1 = 0$ is the equation of a circle in general form.

Equation of an Ellipse [11.3A, p. 607]

$\dfrac{x^2}{a^2} + \dfrac{y^2}{b^2} = 1$

The x-intercepts are $(a, 0)$ and $(-a, 0)$.
The y-intercepts are $(0, b)$ and $(0, -b)$.

$\dfrac{x^2}{9} + \dfrac{y^2}{1} = 1$

x-intercepts:
$(3, 0)$ and $(-3, 0)$

y-intercepts:
$(0, 1)$ and $(0, -1)$

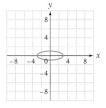

Equation of a Hyperbola [11.3B, p. 609]

$\dfrac{x^2}{a^2} - \dfrac{y^2}{b^2} = 1$

An axis of symmetry is the x-axis.
The vertices are $(a, 0)$ and $(-a, 0)$.

The equations of the asymptotes are $y = \dfrac{b}{a}x$ and $y = -\dfrac{b}{a}x$.

$\dfrac{x^2}{9} - \dfrac{y^2}{4} = 1$

Vertices:
$(3, 0)$, and $(-3, 0)$
Asymptotes:
$y = \dfrac{2}{3}x$ and
$y = -\dfrac{2}{3}x$

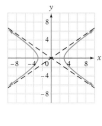

$\dfrac{y^2}{b^2} - \dfrac{x^2}{a^2} = 1$

An axis of symmetry is the y-axis.
The vertices are $(0, b)$ and $(0, -b)$.

The equations of the asymptotes are $y = \dfrac{b}{a}x$ and $y = -\dfrac{b}{a}x$.

$\dfrac{y^2}{1} - \dfrac{x^2}{4} = 1$

Vertices:
$(0, 1)$, and $(0, -1)$
Asymptotes:
$y = \dfrac{1}{2}x$ and
$y = -\dfrac{1}{2}x$

Chapter 11 Review Exercises

1. Sketch a graph of $y = -2x^2 + x - 2$.

2. Sketch a graph of $\dfrac{x^2}{1} + \dfrac{y^2}{9} = 1$.

 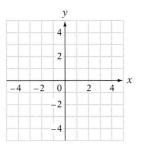

3. Graph the solution set: $\dfrac{x^2}{9} - \dfrac{y^2}{16} < 1$

 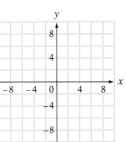

4. Sketch a graph of $(x + 3)^2 + (y + 1)^2 = 1$.

 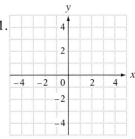

5. Find the axis of symmetry and vertex of the parabola $y = x^2 - 4x + 8$.

6. Solve: $y^2 = 2x^2 - 3x + 6$
 $ y^2 = 2x^2 + 5x - 2$

7. Graph the solution set:
 $\dfrac{x^2}{25} + \dfrac{y^2}{16} \le 1$
 $\dfrac{y^2}{4} - \dfrac{x^2}{4} \ge 1$

 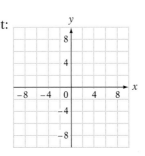

8. Sketch a graph of $\dfrac{x^2}{25} - \dfrac{y^2}{1} = 1$.

9. Find the equation of the circle that passes through the point whose coordinates are $(2, -1)$ and whose center is $(-1, 2)$.

10. Find the axis of symmetry and vertex of the parabola $y = -x^2 + 7x - 8$.

11. Solve: $x = 2y^2 - 3y + 1$
 $\ 3x - 2y = 0$

12. Find the equation of the circle that passes through the point $(4, 6)$ and whose center is $(0, -3)$.

13. Find the equation of the circle with radius 6 and center $(-1, 5)$.

14. Solve: $2x^2 + y^2 = 19$
 $\ 3x^2 - y^2 = 6$

15. Write the equation $x^2 + y^2 + 4x - 2y = 4$ in standard form.

16. Solve: $y = x^2 + 5x - 6$
$y = x - 10$

17. Graph the solution set:
$$\frac{x^2}{16} + \frac{y^2}{4} > 1$$

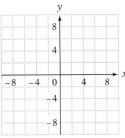

18. Sketch a graph of
$$\frac{y^2}{16} - \frac{x^2}{9} = 1.$$

19. Graph the solution set:
$(x - 2)^2 + (y + 1)^2 \leq 16$

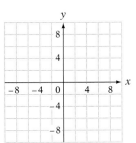

20. Sketch a graph of
$x^2 + (y - 2)^2 = 9.$

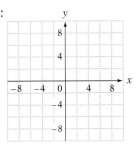

21. Sketch a graph of
$$\frac{x^2}{25} + \frac{y^2}{9} = 1.$$

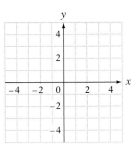

22. Graph the solution set:
$y \geq -x^2 - 2x + 3$

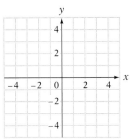

23. Graph the solution set:
$$\frac{x^2}{16} + \frac{y^2}{4} < 1$$
$$x^2 + y^2 > 9$$

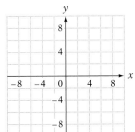

24. Graph the solution set:
$y \geq x^2 - 4x + 2$
$$y \leq \frac{1}{3}x - 1$$

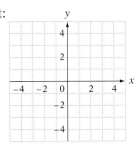

25. Sketch a graph of
$x = 2y^2 - 6y + 5.$

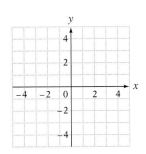

26. Graph the solution set:
$$\frac{x^2}{9} + \frac{y^2}{1} \geq 1$$
$$\frac{x^2}{4} - \frac{y^2}{1} \leq 1$$

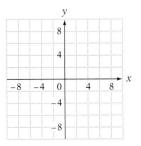

27. Sketch a graph of
$y = x^2 - 4x - 1$.

28. Sketch a graph of
$x = y^2 - 1$.

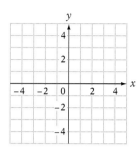

29. Find the equation of the circle with radius 5 and center $(3, -4)$.

30. Write the equation of the circle $x^2 + y^2 - 6x + 4y - 23 = 0$ in standard form.

31. Sketch a graph of the circle $(x + 1)^2 + (y - 4)^2 = 36$.

32. Sketch a graph of the circle $x^2 + y^2 - 8x + 4y + 16 = 0$.

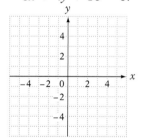

33. Sketch a graph of
$\dfrac{x^2}{36} + \dfrac{y^2}{16} = 1$.

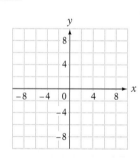

34. Sketch a graph of
$\dfrac{x^2}{1} + \dfrac{y^2}{25} = 1$.

35. Sketch a graph of
$\dfrac{x^2}{9} - \dfrac{y^2}{36} = 1$.

36. Sketch a graph of
$\dfrac{y^2}{36} - \dfrac{x^2}{4} = 1$.

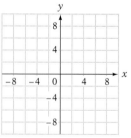

37. Solve: $y = x^2 - 3x - 4$
$y = x + 1$

38. Graph the solution set: $x \le y^2 + 2y - 3$

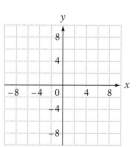

Chapter 11 Test

1. Find the equation of the circle with radius 4 and center $(-3, -3)$.

2. Find the equation of the circle that passes through the point $(2, 5)$ and whose center is $(-2, 1)$.

3. Sketch a graph of $\dfrac{y^2}{25} - \dfrac{x^2}{16} = 1$.

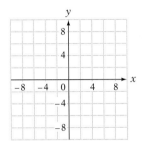

4. Graph the solution set: $x^2 + y^2 < 36$
$$x + y > 4$$

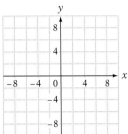

5. Find the axis of symmetry of the parabola $y = -x^2 + 6x - 5$.

6. Solve: $x^2 - y^2 = 24$
$$2x^2 + 5y^2 = 55$$

7. Find the vertex of the parabola $y = -x^2 + 3x - 2$.

8. Sketch a graph of $x = y^2 - y - 2$.

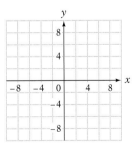

9. Sketch a graph of $\dfrac{x^2}{16} + \dfrac{y^2}{4} = 1$.

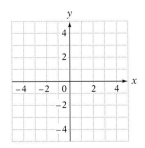

10. Graph the solution set: $\dfrac{x^2}{25} + \dfrac{y^2}{4} \le 1$

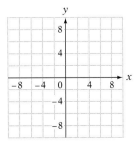

11. Find the equation of the circle with radius 3 and center $(-2, 4)$.

12. Solve: $x = 3y^2 + 2y - 4$
$\qquad\quad x = y^2 - 5y$

13. Solve: $x^2 + 2y^2 = 4$
$\qquad\quad\; x + y = 2$

14. Find the equation of the circle that passes through the point $(2, 4)$ and whose center is $(-1, -3)$.

15. Graph the solution set: $\dfrac{x^2}{25} - \dfrac{y^2}{16} \geq 1$
$\qquad\qquad\qquad\qquad\quad x^2 + y^2 \leq 9$

16. Write the equation $x^2 + y^2 - 4x + 2y + 1 = 0$ in standard form and then sketch its graph.

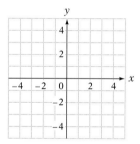

17. Sketch a graph of $y = \dfrac{1}{2}x^2 + x - 4$.

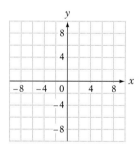

18. Sketch a graph of $(x - 2)^2 + (y + 1)^2 = 9$.

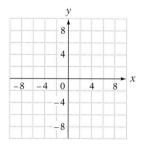

19. Sketch a graph of $\dfrac{x^2}{9} - \dfrac{y^2}{4} = 1$.

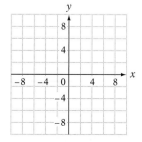

20. Graph the solution set: $\dfrac{x^2}{16} - \dfrac{y^2}{25} < 1$

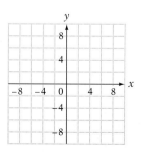

Cumulative Review Exercises

1. Graph the solution set.
$\{x \mid x < 4\} \cap \{x \mid x > 2\}$

$$\xleftarrow{\quad}\underset{-5\ -4\ -3\ -2\ -1\ \ 0\ \ 1\ \ 2\ \ 3\ \ 4\ \ 5}{\rule{0pt}{0pt}}\xrightarrow{\quad}$$

2. Solve: $\dfrac{5x - 2}{3} - \dfrac{1 - x}{5} = \dfrac{x + 4}{10}$

3. Solve: $4 + |3x + 2| < 6$

4. Find the equation of the line that contains the point $(2, -3)$ and has slope $-\dfrac{3}{2}$.

5. Find the equation of the line that contains the point $(4, -2)$ and is perpendicular to the line $y = -x + 5$.

6. Multiply: $x^{2n}(x^{2n} + 2x^{n} - 3x)$

7. Factor: $(x - 1)^3 - y^3$

8. Solve: $\dfrac{3x - 2}{x + 4} \le 1$

9. Simplify: $\dfrac{ax - bx}{ax + ay - bx - by}$

10. Simplify: $\dfrac{x - 4}{3x - 2} - \dfrac{1 + x}{3x^2 + x - 2}$

11. Solve: $\dfrac{6x}{2x - 3} - \dfrac{1}{2x - 3} = 7$

12. Graph the solution set: $5x + 2y > 10$

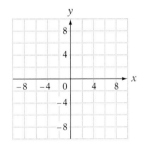

13. Simplify: $\left(\dfrac{12a^2b^2}{a^{-3}b^{-4}} \right)^{-1} \left(\dfrac{ab}{4^{-1}a^{-2}b^4} \right)^{2}$

14. Write $2\sqrt[4]{x^3}$ as an exponential expression.

15. Simplify: $\sqrt{18} - \sqrt{-25}$

16. Solve: $2x^2 + 2x - 3 = 0$

17. Solve: $x^2 + y^2 = 20$
$\quad\quad\quad\ x^2 - y^2 = 12$

18. Solve: $x - \sqrt{2x - 3} = 3$

19. Evaluate the function $f(x) = -x^2 + 3x - 2$ at $x = -3$.

20. Find the inverse of the function $f(x) = 4x + 8$.

21. Find the maximum value of the function $f(x) = -2x^2 + 4x - 2$.

22. Find the maximum product of two numbers whose sum is 40.

23. Find the distance between the points $(2, 4)$ and $(-1, 0)$.

24. Find the equation of the circle that passes through the point $(3, 1)$ and whose center is $(-1, 2)$.

25. Graph: $x = y^2 - 2y + 3$

26. Graph: $\dfrac{x^2}{25} + \dfrac{y^2}{4} = 1$

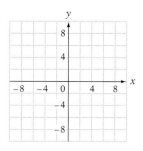

27. Graph: $\dfrac{y^2}{4} - \dfrac{x^2}{25} = 1$

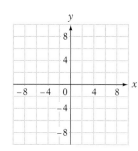

28. Graph the solution set:
$$(x - 1)^2 + y^2 \le 25$$
$$y^2 < x$$

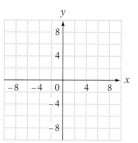

29. **The Arts** Tickets for a school play sold for $12.00 for each adult and $4.50 for each child. The total receipts for the 192 tickets sold were $1479. Find the number of adult tickets sold.

30. **Uniform Motion** A motorcycle travels 180 mi in the same amount of time that a car travels 144 mi. The rate of the motorcycle is 12 mph greater than the rate of the car. Find the rate of the motorcycle.

31. **Sports** The rate of a river's current is 1.5 mph. A rowing crew can row 12 mi downstream and 12 mi back in 6 h. Find the rowing rate of the crew in calm water.

32. **Mechanics** The speed, v, of a gear varies inversely as the number of teeth, t. If a gear that has 36 teeth makes 30 revolutions per minute, how many revolutions per minute will a gear that has 60 teeth make?

chapter

12 Sequences and Series

Stores use lasers, like the one shown here, to scan each product sold. The laser reads a code—on groceries, a Universal Product Code (UPC), and on books, an International Standard Book Number (ISBN)—and interprets the numbers identifying the product. Each product has a different numerical pattern, which is a series of numbers that is divisible by a particular number. The **Projects and Group Activities on pages 670 and 671** explain how ISBNs and UPCs are chosen.

Need help? For online student resources, such as section quizzes, visit this textbook's website at **math.college.hmco.com/students.**

Do these exercises to prepare for Chapter 12.

1. Simplify: $[3(1) - 2] + [3(2) - 2] + [3(3) - 2]$

2. Evaluate $f(n) = \dfrac{n}{n + 2}$ for $n = 6$.

3. Evaluate $a_1 + (n - 1)d$ for $a_1 = 2$, $n = 5$, and $d = 4$.

4. Evaluate $a_1 r^{n - 1}$ for $a_1 = -3$, $r = -2$, and $n = 6$.

5. Evaluate $\dfrac{a_1(1 - r^n)}{1 - r}$ for $a_1 = -2$, $r = -4$, and $n = 5$.

6. Simplify: $\dfrac{\frac{4}{10}}{1 - \frac{1}{10}}$

7. Expand: $(x + y)^2$

8. Expand: $(x + y)^3$

GO FIGURE • • •

What is the next number in the following sequence?

1 32 81 64 25 6

12.1 Introduction to Sequences and Series

Objective A **To write the terms of a sequence**

An investor deposits \$100 in an account that earns 10% interest compounded annually. The amount of interest earned each year can be determined by using the compound interest formula.

The amount of interest earned in each of the first 4 years of the investment is shown at the right.

Year	1	2	3	4
Interest Earned	\$10	\$11	\$12.10	\$13.31

The list of numbers 10, 11, 12.10, 13.31 is called a *sequence*. A **sequence** is an ordered list of numbers. The list 10, 11, 12.10, 13.31 is ordered because the position of a number in this list indicates the year in which that amount of interest was earned.

Each of the numbers of a sequence is called a **term** of the sequence.

For the sequence at the right, the first term is 2, the second term is 4, the third term is 6, and the fourth term is 8.

$2, 4, 6, 8, \ldots$

Examples of other sequences are shown at the right. These sequences are separated into two groups. A **finite sequence** contains a finite number of terms. An **infinite sequence** contains an infinite number of terms.

$$\left. \begin{matrix} 1, 1, 2, 3, 5, 8 \\ 1, 2, 3, 4, 5, 6, 7, 8 \\ 1, -1, 1, -1 \end{matrix} \right\} \text{Finite Sequences}$$

$$\left. \begin{matrix} 1, 3, 5, 7, \ldots \\ 1, \dfrac{1}{2}, \dfrac{1}{4}, \dfrac{1}{8}, \ldots \\ 1, 1, 2, 3, 5, 8, \ldots \end{matrix} \right\} \text{Infinite Sequences}$$

A general sequence is shown at the right. The first term is a_1, the second term is a_2, the third term is a_3, and the nth term, also called the **general term** of the sequence, is a_n. Note that each term of the sequence is paired with a natural number.

$a_1, a_2, a_3, \ldots, a_n, \ldots$

Frequently a sequence has a definite pattern that can be expressed by a formula.

TAKE NOTE

Note that a sequence is a function whose domain is the set of natural numbers.

Each term of the sequence shown at the right is paired with a natural number by the formula $a_n = 3n$. The first term, a_1, is 3. The second term, a_2, is 6. The third term, a_3, is 9. The nth term, a_n, is $3n$.

$a_n = 3n$

$a_1, \quad a_2, \quad a_3, \quad \ldots, a_n, \ldots$

$3(1), 3(2), 3(3), \ldots, 3(n), \ldots$

$3, \quad 6, \quad 9, \quad \ldots, 3n, \ldots$

HOW TO Write the first three terms of the sequence whose nth term is given by the formula $a_n = 4n$.

$a_n = 4n$

$a_1 = 4(1) = 4$ • Replace n by 1.

$a_2 = 4(2) = 8$ • Replace n by 2.

$a_3 = 4(3) = 12$ • Replace n by 3.

The first term is 4, the second term is 8, and the third term is 12.

HOW TO Find the fifth and seventh terms of the sequence whose nth term is given by the formula $a_n = \dfrac{1}{n^2}$.

$$a_n = \frac{1}{n^2}$$

$$a_5 = \frac{1}{5^2} = \frac{1}{25}$$ • Replace n by 5.

$$a_7 = \frac{1}{7^2} = \frac{1}{49}$$ • Replace n by 7.

The fifth term is $\dfrac{1}{25}$ and the seventh term is $\dfrac{1}{49}$.

Example 1

Write the first three terms of the sequence whose nth term is given by the formula $a_n = 2n - 1$.

Solution
$a_n = 2n - 1$
$a_1 = 2(1) - 1 = 1$ The first term is 1.
$a_2 = 2(2) - 1 = 3$ The second term is 3.
$a_3 = 2(3) - 1 = 5$ The third term is 5.

You Try It 1

Write the first four terms of the sequence whose nth term is given by the formula $a_n = n(n + 1)$.

Your solution

Example 2

Find the eighth and tenth terms of the sequence whose nth term is given by the formula $a_n = \dfrac{n}{n + 1}$.

Solution

$$a_n = \frac{n}{n + 1}$$

$$a_8 = \frac{8}{8 + 1} = \frac{8}{9}$$ The eighth term is $\dfrac{8}{9}$.

$$a_{10} = \frac{10}{10 + 1} = \frac{10}{11}$$ The tenth term is $\dfrac{10}{11}$.

You Try It 2

Find the sixth and ninth terms of the sequence whose nth term is given by the formula $a_n = \dfrac{1}{n(n + 2)}$.

Your solution

Solutions on p. S33

Objective B **To find the sum of a series**

In the last objective, the sequence 10, 11, 12.10, 13.31 was shown to represent the amount of interest earned in each of 4 years of an investment.

10, 11, 12.10, 13.31

The sum of the terms of this sequence represents the total interest earned by the investment over the four-year period.

$10 + 11 + 12.10 + 13.31 = 46.41$

The total interest earned over the four-year period is $46.41.

The indicated sum of the terms of a sequence is called a **series.** Given the sequence 10, 11, 12.10, 13.31, we can write the series $10 + 11 + 12.10 + 13.31$.

S_n is used to designate the sum of the first n terms of a sequence.

For the preceding example, the sums of the series S_1, S_2, S_3, and S_4 represent the total interest earned for 1, 2, 3, and 4 years, respectively.

$$S_1 = 10 \qquad\qquad = 10$$
$$S_2 = 10 + 11 \qquad = 21$$
$$S_3 = 10 + 11 + 12.10 \qquad = 33.10$$
$$S_4 = 10 + 11 + 12.10 + 13.31 = 46.41$$

For the general sequence a_1, a_2, a_3, ... , a_n, the series S_1, S_2, S_3, and S_n are shown at the right.

$$S_1 = a_1$$
$$S_2 = a_1 + a_2$$
$$S_3 = a_1 + a_2 + a_3$$
$$S_n = a_1 + a_2 + a_3 + \cdots + a_n$$

It is convenient to represent a series in a compact form called **summation notation,** or **sigma notation.** The Greek letter sigma, Σ, is used to indicate a sum.

The first four terms of the sequence whose nth term is given by the formula $a_n = 2n$ are 2, 4, 6, and 8. The corresponding series is shown at the right written in summation notation; it is read "the sum from 1 to 4 of $2n$." The letter n is called the **index** of the summation.

$$\sum_{n=1}^{4} 2n$$

To write the terms of the series, replace n by the consecutive integers from 1 to 4.

$$\sum_{n=1}^{4} 2n = 2(1) + 2(2) + 2(3) + 2(4)$$

The series is $2 + 4 + 6 + 8$.

$$= 2 + 4 + 6 + 8$$

The sum of the series is 20.

$$= 20$$

HOW TO Evaluate the series $\displaystyle\sum_{n=1}^{6} n^2$.

$$\sum_{n=1}^{6} n^2 = 1^2 + 2^2 + 3^2 + 4^2 + 5^2 + 6^2 \qquad \bullet \text{ Replace } n \text{ by 1, 2, 3, 4, 5, and 6.}$$

$$= 1 + 4 + 9 + 16 + 25 + 36 \qquad \bullet \text{ Simplify.}$$

$$= 91$$

TAKE NOTE

The placement of the parentheses in this example is important. For instance,

$$\sum_{i=1}^{3} 2i - 1$$
$$= 2(1) + 2(2) + 2(3) - 1$$
$$= 2 + 4 + 6 - 1 = 11$$

This is *not* the same result as the sum of

$$\sum_{i=1}^{3} (2i - 1).$$

HOW TO Evaluate the series $\displaystyle\sum_{i=1}^{3} (2i - 1)$.

$$\sum_{i=1}^{3} (2i - 1) = [2(1) - 1] + [2(2) - 1] + [2(3) - 1] \qquad \bullet \text{ Replace } i \text{ by 1, 2, and 3.}$$

$$= 1 + 3 + 5 \qquad \bullet \text{ Simplify.}$$

$$= 9$$

HOW TO Evaluate the series $\displaystyle\sum_{n=3}^{6} \frac{1}{2}n$.

$\displaystyle\sum_{n=3}^{6} \frac{1}{2}n = \frac{1}{2}(3) + \frac{1}{2}(4) + \frac{1}{2}(5) + \frac{1}{2}(6)$ • Replace n by 3, 4, 5, and 6.

$\qquad\qquad = \frac{3}{2} + 2 + \frac{5}{2} + 3$ • Simplify.

$\qquad\qquad = 9$

HOW TO Write $\displaystyle\sum_{i=1}^{5} x^i$ in expanded form.

This is a variable series.

$\displaystyle\sum_{i=1}^{5} x^i = x + x^2 + x^3 + x^4 + x^5$ • Replace i by 1, 2, 3, 4, and 5.

Example 3

Evaluate the series $\displaystyle\sum_{n=1}^{4} (7 - n)$.

Solution

$\displaystyle\sum_{n=1}^{4} (7 - n)$

$\quad = (7 - 1) + (7 - 2) + (7 - 3) + (7 - 4)$
$\quad = 6 + 5 + 4 + 3$
$\quad = 18$

You Try It 3

Evaluate the series $\displaystyle\sum_{i=3}^{6} (i^2 - 2)$.

Your solution

Example 4

Write $\displaystyle\sum_{i=1}^{5} ix$ in expanded form.

Solution

$\displaystyle\sum_{i=1}^{5} ix = 1 \cdot x + 2 \cdot x + 3 \cdot x + 4 \cdot x + 5 \cdot x$

$\qquad = x + 2x + 3x + 4x + 5x$

You Try It 4

Write $\displaystyle\sum_{n=1}^{4} \frac{x}{n}$ in expanded form.

Your solution

Solutions on p. S33

12.1 Exercises

Objective A **To write the terms of a sequence**

1. ✎ What is a sequence?

2. ✎ What is the difference between a finite sequence and an infinite sequence?

3. Name the third term in the sequence $2, 5, 8, 11, 14, \ldots$

4. Name the fourth term in the sequence $1, 2, 4, 8, 16, 32, \ldots$

For Exercises 5 to 16, write the first four terms of the sequence whose nth term is given by the formula.

5. $a_n = n + 1$

6. $a_n = n - 1$

7. $a_n = 2n + 1$

8. $a_n = 3n - 1$

9. $a_n = 2 - 2n$

10. $a_n = 1 - 2n$

11. $a_n = 2^n$

12. $a_n = 3^n$

13. $a_n = n^2 + 1$

14. $a_n = n - \dfrac{1}{n}$

15. $a_n = n^2 - \dfrac{1}{n}$

16. $a_n = (-1)^{n+1} n$

For Exercises 17 to 28, find the indicated term of the sequence whose nth term is given by the formula.

17. $a_n = 3n + 4; a_{12}$

18. $a_n = 2n - 5; a_{10}$

19. $a_n = n(n - 1); a_{11}$

20. $a_n = \dfrac{n}{n + 1}; a_{12}$

21. $a_n = (-1)^{n-1} n^2; a_{15}$

22. $a_n = (-1)^{n-1}(n - 1); a_{25}$

23. $a_n = \left(\dfrac{1}{2}\right)^n; a_8$

24. $a_n = \left(\dfrac{2}{3}\right)^n; a_5$

25. $a_n = (n + 2)(n + 3); a_{17}$

26. $a_n = (n + 4)(n + 1); a_7$

27. $a_n = \dfrac{(-1)^{2n-1}}{n^2}; a_6$

28. $a_n = \dfrac{(-1)^{2n}}{n + 4}; a_{16}$

Objective B **To find the sum of a series**

For Exercises 29 to 40, find the sum of the series.

29. $\displaystyle\sum_{n=1}^{5} (2n + 3)$

30. $\displaystyle\sum_{i=1}^{7} (i + 2)$

31. $\displaystyle\sum_{i=1}^{4} 2i$

32. $\displaystyle\sum_{n=1}^{7} n$

33. $\displaystyle\sum_{i=1}^{6} i^2$

34. $\displaystyle\sum_{i=1}^{5} (i^2 + 1)$

35. $\displaystyle\sum_{n=1}^{6} (-1)^n$

36. $\displaystyle\sum_{n=1}^{4} \frac{1}{2n}$

37. $\displaystyle\sum_{i=3}^{6} i^3$

38. $\displaystyle\sum_{i=1}^{4} (-1)^{i-1}(i + 1)$

39. $\displaystyle\sum_{n=3}^{5} \frac{(-1)^{n-1}}{n - 2}$

40. $\displaystyle\sum_{n=4}^{7} \frac{(-1)^{n-1}}{n - 3}$

For Exercises 41 to 49, write the series in expanded form.

41. $\displaystyle\sum_{n=1}^{5} 2x^n$

42. $\displaystyle\sum_{n=1}^{4} \frac{2n}{x}$

43. $\displaystyle\sum_{i=1}^{5} \frac{x^i}{i}$

44. $\displaystyle\sum_{i=2}^{4} \frac{x^i}{2i - 1}$

45. $\displaystyle\sum_{n=1}^{5} x^{2n}$

46. $\displaystyle\sum_{n=1}^{4} x^{2n-1}$

47. $\displaystyle\sum_{i=1}^{4} \frac{x^i}{i^2}$

48. $\displaystyle\sum_{n=1}^{4} nx^{n-1}$

49. $\displaystyle\sum_{i=1}^{5} x^{-i}$

APPLYING THE CONCEPTS

50. In the first box below, $\frac{1}{2}$ of the box is shaded. In successive boxes the sums $\frac{1}{2} + \frac{1}{4}$ and $\frac{1}{2} + \frac{1}{4} + \frac{1}{8}$ are shown. Can you identify the infinite sum $\frac{1}{2} + \frac{1}{4} + \frac{1}{8} + \frac{1}{16} + \cdots$?

51. Rewrite $1 + \frac{1}{2} + \frac{1}{3} + \cdots + \frac{1}{n}$ using sigma notation.

52. Indicate whether the following statements are true or false.
a. No two terms of a sequence can be equal.
b. The sum of an infinite number of terms is an infinite number.

53. Write a paragraph about the history of the sigma notation used for summation.

54. Write a paragraph explaining the Fibonacci Sequence.

Fibonacci

12.2 Arithmetic Sequences and Series

Objective A To find the *n*th term of an arithmetic sequence

A company's expenses for training a new employee are quite high. To encourage employees to continue their employment with the company, a company that has a six-month training program offers a starting salary of $2400 per month and then a $200-per-month pay increase each month during the training period.

The sequence at the right shows the employee's monthly salaries during the training period. Each term of the sequence is found by adding $200 to the preceding term.

Month	1	2	3	4	5	6
Salary	2400	2600	2800	3000	3200	3400

The sequence 2400, 2600, 2800, 3000, 3200, 3400 is called an *arithmetic sequence*. An **arithmetic sequence,** or **arithmetic progression,** is one in which the difference between any two consecutive terms is constant. The difference between consecutive terms is called the **common difference** of the sequence.

Each of the sequences shown at the right is an arithmetic sequence. To find the common difference of an arithmetic sequence, subtract the first term from the second term.

2, 7, 12, 17, 22, ... Common difference: 5

3, 1, -1, -3, -5, ... Common difference: -2

$1, \frac{3}{2}, 2, \frac{5}{2}, 3, \frac{7}{2}$ Common difference: $\frac{1}{2}$

Consider an arithmetic sequence in which the first term is a_1 and the common difference is d. Adding the common difference to each successive term of the arithmetic sequence yields a formula for the *n*th term.

The first term is a_1. $a_1 = a_1$

To find the second term, add the common difference d to the first term. $a_2 = a_1 + d$

To find the third term, add the common difference d to the second term. $a_3 = a_2 + d = (a_1 + d) + d$
$a_3 = a_1 + 2d$

To find the fourth term, add the common difference d to the third term. $a_4 = a_3 + d = (a_1 + 2d) + d$
$a_4 = a_1 + 3d$

Note the relationship between the term number and the number that multiplies d. The multiplier of d is 1 less than the term number. $a_n = a_1 + (n - 1)d$

> **The Formula for the *n*th Term of an Arithmetic Sequence**
>
> The *n*th term of an arithmetic sequence with a common difference of d is given by
> $a_n = a_1 + (n - 1)d$.

HOW TO Find the 27th term of the arithmetic sequence -4, -1, 2, 5, 8,

$d = a_2 - a_1 = -1 - (-4) = 3$ • Find the common difference.

$a_n = a_1 + (n - 1)d$
$a_{27} = -4 + (27 - 1)3$
$a_{27} = -4 + (26)3$
$a_{27} = -4 + 78$
$a_{27} = 74$

• Use the Formula for the nth Term of an Arithmetic Sequence to find the 27th term.
$n = 27$, $a_1 = -4$, $d = 3$

HOW TO Find the formula for the nth term of the arithmetic sequence -5, -2, 1, 4,

$d = a_2 - a_1 = -2 - (-5) = 3$ • Find the common difference.

$a_n = a_1 + (n - 1)d$
$a_n = -5 + (n - 1)3$
$a_n = -5 + 3n - 3$
$a_n = 3n - 8$

• Use the Formula for the nth Term of an Arithmetic Sequence. $a_1 = -5$, $d = 3$

HOW TO Find the formula for the nth term of the arithmetic sequence 2, -3, -8, -13,

$d = a_2 - a_1 = -3 - 2 = -5$ • Find the common difference.

$a_n = a_1 + (n - 1)d$
$a_n = 2 + (n - 1)(-5)$
$a_n = 2 - 5n + 5$
$a_n = -5n + 7$

• Use the Formula for the nth Term of an Arithmetic Sequence. $a_1 = 2$, $d = -5$

Example 1

Find the 10th term of the arithmetic sequence 1, 6, 11, 16,

Solution
$d = a_2 - a_1 = 6 - 1 = 5$ • Find d.

$a_n = a_1 + (n - 1)d$
$a_{10} = 1 + (10 - 1)5 = 1 + (9)5$ • Find a_{10}.
$\quad = 1 + 45 = 46$

You Try It 1

Find the 15th term of the arithmetic sequence 9, 3, -3, -9,

Your solution

Example 2

Find the number of terms in the finite arithmetic sequence -3, 1, 5, ..., 41.

Solution
$d = a_2 - a_1 = 1 - (-3) = 4$ • Find d.

$a_n = a_1 + (n - 1)d$
$41 = -3 + (n - 1)4$ • $a_n = 41$, $d = 4$
$41 = -3 + 4n - 4$
$41 = -7 + 4n$
$48 = 4n$
$12 = n$

There are 12 terms in the sequence.

You Try It 2

Find the number of terms in the finite arithmetic sequence 7, 9, 11, ..., 59.

Your solution

Solutions on pp. S33–S34

Objective B **To find the sum of an arithmetic series**

The indicated sum of the terms of an arithmetic sequence is called an **arithmetic series.** The sum of an arithmetic series can be found by using a formula. A proof of this formula appears in the Appendix.

Point of Interest

This formula was proved in *Aryabhatiya,* which was written by Aryabhata around 499. The book is the earliest known Indian mathematical work by an identifiable author. Although the proof of the formula appears in that text, the formula was known before Aryabhata's time.

The Formula for the Sum of n Terms of an Arithmetic Series

Let a_1 be the first term of a finite arithmetic sequence, let n be the number of terms, and let a_n be the last term of the sequence.

Then the sum of the series S_n is given by $S_n = \dfrac{n}{2}(a_1 + a_n)$.

HOW TO Find the sum of the first 10 terms of the arithmetic sequence 2, 4, 6, 8,

$d = a_2 - a_1 = 4 - 2 = 2$ • Find the common difference.

$a_n = a_1 + (n - 1)d$
$a_{10} = 2 + (10 - 1)2$
$a_{10} = 2 + (9)2 = 20$

• Use the Formula for the nth Term of an Arithmetic Sequence to find the 10th term.
$n = 10$, $a_1 = 2$, $d = 2$

$S_n = \dfrac{n}{2}(a_1 + a_n)$

• Use the Formula for the Sum of n Terms of an Arithmetic Series. $n = 10$, $a_1 = 2$, $a_n = 20$

$S_{10} = \dfrac{10}{2}(2 + 20) = 5(22) = 110$

HOW TO Find the sum of the arithmetic series $\displaystyle\sum_{n=1}^{25} (3n + 1)$.

$a_n = 3n + 1$
$a_1 = 3(1) + 1 = 4$ • Find the 1st term.
$a_{25} = 3(25) + 1 = 76$ • Find the 25th term.

$S_n = \dfrac{n}{2}(a_1 + a_n)$

• Use the Formula for the Sum of n Terms of an Arithmetic Series. $n = 25$, $a_1 = 4$, $a_n = 76$

$S_{25} = \dfrac{25}{2}(4 + 76) = \dfrac{25}{2}(80)$

$S_{25} = 1000$

Example 3

Find the sum of the first 20 terms of the arithmetic sequence 3, 8, 13, 18,

Solution

$d = a_2 - a_1 = 8 - 3 = 5$ • Find d.

$a_n = a_1 + (n - 1)d$
$a_{20} = 3 + (20 - 1)5 = 3 + (19)5$ • Find a_{20}.
$= 3 + 95 = 98$

$S_n = \dfrac{n}{2}(a_1 + a_n)$ • Use the formula.

$S_{20} = \dfrac{20}{2}(3 + 98) = 10(101) = 1010$

You Try It 3

Find the sum of the first 25 terms of the arithmetic sequence −4, −2, 0, 2, 4,

Your solution

Solution on p. S34

Find the sum of the arithmetic series
$$\sum_{n=1}^{15}(2n + 3).$$

Solution

$a_n = 2n + 3$
$a_1 = 2(1) + 3 = 5$ • Find a_1.
$a_{15} = 2(15) + 3 = 33$ • Find a_{15}.

$S_n = \dfrac{n}{2}(a_1 + a_n)$ • Use the formula.

$S_{15} = \dfrac{15}{2}(5 + 33) = \dfrac{15}{2}(38) = 285$

Find the sum of the arithmetic series
$$\sum_{n=1}^{18}(3n - 2).$$

Your solution

Solution on p. S34

To solve application problems

The distance a ball rolls down a ramp each second is given by an arithmetic sequence. The distance in feet traveled by the ball during the nth second is given by $2n - 1$. Find the distance the ball will travel during the 10th second.

Strategy

To find the distance:

• Find the 1st and 2nd terms.
• Find the common difference of the arithmetic sequence.
• Use the Formula for the nth Term of an Arithmetic Sequence.

Solution

$a_n = 2n - 1$ • The given sequence
$a_1 = 2(1) - 1 = 1$ • Find a_1.
$a_2 = 2(2) - 1 = 3$ • Find a_2.
$d = a_2 - a_1 = 3 - 1 = 2$ • Find d.

$a_n = a_1 + (n - 1)d$ • Use the formula.
$a_{10} = 1 + (10 - 1)2 = 1 + (9)2$
$\quad\;\, = 1 + 18 = 19$

The ball will travel 19 ft during the 10th second.

A contest offers 20 prizes. The first prize is $10,000, and each successive prize is $300 less than the preceding prize. What is the value of the 20th-place prize? What is the total amount of prize money being awarded?

Your strategy

Your solution

Solution on p. S34

12.2 Exercises

Objective A **To find the nth term of an arithmetic sequence**

For Exercises 1 to 9, find the indicated term of the arithmetic sequence.

1. 1, 11, 21, ...; a_{15}

2. 3, 8, 13, ...; a_{20}

3. $-6, -2, 2, ...; a_{15}$

4. $-7, -2, 3, ...; a_{14}$

5. $2, \dfrac{5}{2}, 3, ...; a_{31}$

6. $1, \dfrac{5}{4}, \dfrac{3}{2}, ...; a_{17}$

7. $-4, -\dfrac{5}{2}, -1, ...; a_{12}$

8. $-\dfrac{5}{3}, -1, -\dfrac{1}{3}, ...; a_{22}$

9. 8, 5, 2, ...; a_{40}

For Exercises 10 to 15, find the formula for the nth term of the arithmetic sequence.

10. 1, 2, 3, ...

11. 1, 4, 7, ...

12. 6, 2, -2, ...

13. 3, 0, -3, ...

14. $2, \dfrac{7}{2}, 5, ...$

15. 7, 4.5, 2, ...

For Exercises 16 to 24, find the number of terms in the finite arithmetic sequence.

16. 1, 5, 9, ..., 81

17. 3, 8, 13, ..., 98

18. 2, 0, -2, ..., -56

19. 1, -3, -7, ..., -75

20. $\dfrac{5}{2}, 3, \dfrac{7}{2}, ..., 13$

21. $\dfrac{7}{3}, \dfrac{13}{3}, \dfrac{19}{3}, ..., \dfrac{79}{3}$

22. 1, 0.75, 0.50, ..., -4

23. 3.5, 2, 0.5, ..., -25

24. $-3.4, -2.8, -2.2, ..., 11$

Objective B **To find the sum of an arithmetic series**

For Exercises 25 to 30, find the sum of the indicated number of terms of the arithmetic sequence.

25. 1, 3, 5, ...; $n = 50$

26. 2, 4, 6, ...; $n = 25$

27. 20, 18, 16, ...; $n = 40$

28. 25, 20, 15, ...; $n = 22$

29. $\dfrac{1}{2}, 1, \dfrac{3}{2}, ...; n = 27$

30. $2, \dfrac{11}{4}, \dfrac{7}{2}, ...; n = 10$

For Exercises 31 to 36, find the sum of the arithmetic series.

31. $\displaystyle\sum_{i=1}^{15} (3i - 1)$

32. $\displaystyle\sum_{i=1}^{15} (3i + 4)$

33. $\displaystyle\sum_{n=1}^{17} \left(\frac{1}{2}n + 1\right)$

34. $\displaystyle\sum_{n=1}^{10} (1 - 4n)$

35. $\displaystyle\sum_{i=1}^{15} (4 - 2i)$

36. $\displaystyle\sum_{n=1}^{10} (5 - n)$

Objective C **To solve application problems**

37. Exercise An exercise program calls for walking 10 min each day for a week. Each week thereafter, the amount of time spent walking increases by 5 min per day. In how many weeks will a person be walking 60 min each day?

38. Product Displays A display of cans in a grocery store consists of 24 cans in the bottom row, 21 cans in the next row, and so on in an arithmetic sequence. The top row has 3 cans. Find the number of cans in the display.

39. Physics An object that is dropped from a low-flying airplane will fall 16 ft the first second, 48 ft the next second, 80 ft the third second, and so on in an arithmetic sequence. Find the distance that the object will fall during the fifth second. (Assume that air resistance is negligible.)

40. The Arts The loge seating section in a concert hall consists of 27 rows of chairs. There are 73 seats in the first row, 79 seats in the second row, 85 seats in the third row, and so on in an arithmetic sequence. How many seats are in the loge seating section?

41. Compensation The salary schedule for a software engineer apprentice program is $1500 for the first month and a $300-per-month salary increase for the next 7 months. Find the monthly salary during the eighth month. Find the total salary for the eight-month period.

APPLYING THE CONCEPTS

42. State whether the following statements are true or false.
 a. The sum of the successive terms of an arithmetic sequence always increases in value.
 b. Each successive term in an arithmetic series is found by adding a fixed number.

43. Write $\displaystyle\sum_{i=1}^{2} \log 2i$ as a single logarithm with a coefficient of 1.

44. Write a formula for the nth term of the sequence of the odd natural numbers.

12.3 Geometric Sequences and Series

Objective A To find the nth term of a geometric sequence

An ore sample contains 20 mg of a radioactive material with a half-life of one week. The amount of the radioactive material that the sample contains at the beginning of each week can be determined by using an exponential decay equation.

The sequence at the right represents the amount in the sample at the beginning of each week. Each term of the sequence is found by multiplying the preceding term by $\frac{1}{2}$.

Week	1	2	3	4	5
Amount	20	10	5	2.5	1.25

The sequence 20, 10, 5, 2.5, 1.25 is called a *geometric sequence*. A **geometric sequence,** or **geometric progression,** is one in which each successive term of the sequence is the same nonzero constant multiple of the preceding term. The common multiple is called the **common ratio** of the sequence.

> **TAKE NOTE**
>
> Geometric sequences are different from arithmetic sequences. For a geometric sequence, every two successive terms have the same *ratio*. For an arithmetic sequence, every two successive terms have the same *difference*.

Each of the sequences shown at the right is a geometric sequence. To find the common ratio of a geometric sequence, divide the second term of the sequence by the first term.

3, 6, 12, 24, 48, ... Common ratio: 2

4, -12, 36, -108, 324, ... Common ratio: -3

6, 4, $\frac{8}{3}$, $\frac{16}{9}$, $\frac{32}{27}$, ... Common ratio: $\frac{2}{3}$

Consider a geometric sequence in which the first term is a_1 and the common ratio is r. Multiplying each successive term of the geometric sequence by the common ratio yields a formula for the nth term.

The first term is a_1. $a_1 = a_1$

To find the second term, multiply the first term by the common ratio r. $a_2 = a_1 r$

To find the third term, multiply the second term by the common ratio r. $a_3 = (a_1 r)r$
$a_3 = a_1 r^2$

To find the fourth term, multiply the third term by the common ratio r. $a_4 = (a_1 r^2)r$
$a_4 = a_1 r^3$

Note the relationship between the term number and the number that is the exponent on r. The exponent on r is 1 less than the term number. $a_n = a_1 r^{n-1}$

> **The Formula for the nth Term of a Geometric Sequence**
>
> The nth term of a geometric sequence with first term a_1 and common ratio r is given by $a_n = a_1 r^{n-1}$.

Integrating Technology

See the Keystroke Guide: *Sequences and Series* for instructions on using a graphing calculator to display the terms of a sequence.

HOW TO Find the 6th term of the geometric sequence 3, 6, 12,

$$r = \frac{a_2}{a_1} = \frac{6}{3} = 2$$
 • Find the common ratio.

$$a_n = a_1 r^{n-1}$$
 • Use the Formula for the nth Term of a Geometric Sequence
$$a_6 = 3(2)^{6-1}$$
 to find the 6th term. $n = 6$, $a_1 = 3$, $r = 2$
$$a_6 = 3(2)^5 = 3(32)$$
$$a_6 = 96$$

HOW TO Find a_3 for the geometric sequence 8, a_2, a_3, -27,

$$a_n = a_1 r^{n-1}$$
 • Use the Formula for the nth Term of a Geometric
$$a_4 = a_1 r^{4-1}$$
 Sequence to find the common ratio. The 1st and
$$-27 = 8r^3$$
 4th terms are known. Substitute 4 for n.
 • $a_4 = -27$, $a_1 = 8$

$$-\frac{27}{8} = r^3$$
 • Solve for r^3.

$$-\frac{3}{2} = r$$
 • Take the cube root of each side of the equation.

$$a_n = a_1 r^{n-1}$$
 • Use the Formula for the nth Term of a
$$a_3 = 8\left(-\frac{3}{2}\right)^{3-1} = 8\left(-\frac{3}{2}\right)^2 = 8\left(\frac{9}{4}\right)$$
 Geometric Sequence to find a_3.
 $n = 3$, $a_1 = 8$, $r = -\frac{3}{2}$
$$a_3 = 18$$

Example 1

Find the 7th term of the geometric sequence 3, -6, 12,

Solution

$$r = \frac{a_2}{a_1} = \frac{-6}{3} = -2$$ • Find r. $a_2 = -6$, $a_1 = 3$
$$a_n = a_1 r^{n-1}$$
 • Use the formula.
$$a_7 = 3(-2)^{7-1} = 3(-2)^6 = 3(64)$$ • $n = 7$, $r = -2$
$$a_7 = 192$$

You Try It 1

Find the 5th term of the geometric sequence 5, 2, $\frac{4}{5}$,

Your solution

Solution on p. S34

Example 2

Find a_2 for the geometric sequence 2, a_2, a_3, 54,

Solution

$a_n = a_1 r^{n-1}$ • Use the formula.

$a_4 = 2r^{4-1}$ • $n = 4$, $a_1 = 2$

$54 = 2r^{4-1}$ • $a_4 = 54$

$54 = 2r^3$

$27 = r^3$

$3 = r$ • The common ratio is 3.

$a_n = a_1 r^{n-1}$

$a_2 = 2(3)^{2-1} = 2(3)$ • $n = 2$, $r = 3$

$a_2 = 6$

You Try It 2

Find a_3 for the geometric sequence 3, a_2, a_3, −192,

Your solution

Solution on p. S34

Objective B **To find the sum of a finite geometric series**

The indicated sum of the terms of a geometric sequence is called a **geometric series.** The sum of a geometric series can be found by a formula. A proof of this formula appears in the Appendix.

Point of Interest

Geometric series are used extensively in the mathematics of finance. Finite geometric series are used to calculate loan balances and monthly payments for amortized loans.

> **The Formula for the Sum of *n* Terms of a Finite Geometric Series**
>
> Let a_1 be the first term of a finite geometric sequence, let n be the number of terms, and let r be the common ratio. Then the sum of the series S_n is given by $S_n = \dfrac{a_1(1 - r^n)}{1 - r}$.

HOW TO Find the sum of the terms of the geometric sequence 2, 8, 32, 128, 512.

$r = \dfrac{a_2}{a_1} = \dfrac{8}{2} = 4$ • Find the common ratio.

$S_n = \dfrac{a_1(1 - r^n)}{1 - r}$ • Use the Formula for the Sum of *n* Terms of a Finite Geometric Series. $n = 5$, $a_1 = 2$, $r = 4$

$S_5 = \dfrac{2(1 - 4^5)}{1 - 4} = \dfrac{2(1 - 1024)}{-3}$

$S_5 = \dfrac{2(-1023)}{-3} = \dfrac{-2046}{-3}$

$S_5 = 682$

HOW TO Find the sum of the geometric series $\sum_{n=1}^{10} (-20)(-2)^{n-1}$.

$a_n = (-20)(-2)^{n-1}$

$a_1 = (-20)(-2)^{1-1} = (-20)(-2)^0$ • Find the 1st term.
 $= (-20)(1) = -20$

$a_2 = (-20)(-2)^{2-1} = (-20)(-2)^1$ • Find the 2nd term.
 $= (-20)(-2) = 40$

$r = \dfrac{a_2}{a_1} = \dfrac{40}{-20} = -2$ • Find the common ratio.

$S_n = \dfrac{a_1(1 - r^n)}{1 - r}$ • Use the Formula for the Sum of n Terms of a Finite Geometric Series. $n = 10$, $a_1 = -20$, $r = -2$

$S_{10} = \dfrac{-20[1 - (-2)^{10}]}{1 - (-2)}$

$S_{10} = \dfrac{-20(1 - 1024)}{3}$

$S_{10} = \dfrac{-20(-1023)}{3} = \dfrac{20{,}460}{3}$

$S_{10} = 6820$

Example 3

Find the sum of the terms of the geometric sequence 3, 6, 12, 24, 48, 96.

Solution

$r = \dfrac{a_2}{a_1} = \dfrac{6}{3} = 2$ • Find r. $a_2 = 6$, $a_1 = 3$

$S_n = \dfrac{a_1(1 - r^n)}{1 - r}$ • Use the formula.

$S_6 = \dfrac{3(1 - 2^6)}{1 - 2} = \dfrac{3(1 - 64)}{-1} = -3(-63) = 189$

You Try It 3

Find the sum of the terms of the geometric sequence $1, -\dfrac{1}{3}, \dfrac{1}{9}, -\dfrac{1}{27}$.

Your solution

Example 4

Find the sum of the geometric series $\sum_{n=1}^{4} 4^n$.

Solution

$a_n = 4^n$

$a_1 = 4^1 = 4$ • Find a_1.
$a_2 = 4^2 = 16$ • Find a_2.

$r = \dfrac{a_2}{a_1} = \dfrac{16}{4} = 4$ • Find r.

$S_n = \dfrac{a_1(1 - r^n)}{1 - r}$ • Use the formula.

$S_4 = \dfrac{4(1 - 4^4)}{1 - 4} = \dfrac{4(1 - 256)}{-3} = \dfrac{4(-255)}{-3} = \dfrac{-1020}{-3}$

$= 340$

You Try It 4

Find the sum of the geometric series $\sum_{n=1}^{5} \left(\dfrac{1}{2}\right)^n$.

Your solution

Solutions on pp. S34–S35

Objective C **To find the sum of an infinite geometric series**

When the absolute value of the common ratio of a geometric sequence is less than 1, $|r| < 1$, then as n becomes larger, r^n becomes closer to zero.

Examples of geometric sequences for which $|r| < 1$ are shown at the right. Note that as the number of terms increases, the value of the last term listed gets closer to zero.

$$1, \frac{1}{3}, \frac{1}{9}, \frac{1}{27}, \frac{1}{81}, \frac{1}{243}, \cdots$$

$$1, -\frac{1}{2}, \frac{1}{4}, -\frac{1}{8}, \frac{1}{16}, -\frac{1}{32}, \cdots$$

The indicated sum of the terms of an infinite geometric sequence is called an **infinite geometric series.**

An example of an infinite geometric series is shown at the right. The first term is 1. The common ratio is $\frac{1}{3}$.

$$1 + \frac{1}{3} + \frac{1}{9} + \frac{1}{27} + \frac{1}{81} + \frac{1}{243} + \cdots$$

The sums of the first 5, 7, 12, and 15 terms, along with the values of r^n, are shown at the right. Note that as n increases, the sum of the terms gets closer to 1.5 and the value of r^n gets closer to zero.

n	S_n	r^n
5	1.4938272	0.0041152
7	1.4993141	0.0004572
12	1.4999972	0.0000019
15	1.4999999	0.0000001

Using the Formula for the Sum of n Terms of a Finite Geometric Series and the fact that r^n approaches zero when $|r| < 1$ and n increases, we can find a formula for the sum of an infinite geometric series.

The sum of the first n terms of a geometric series is shown at the right. If $|r| < 1$, then r^n can be made very close to zero by using larger and larger values of n. Therefore, the sum of the first n terms is approximately $\frac{a_1}{1 - r}$.

approximately zero

$$S_n = \frac{a_1(1 - r^n)}{1 - r}$$

$$S_n \approx \frac{a_1(1 - 0)}{1 - r} = \frac{a_1}{1 - r}$$

The Formula for the Sum of an Infinite Geometric Series

The sum of an infinite geometric series in which $|r| < 1$ and a_1 is the first term is given by $S = \dfrac{a_1}{1 - r}$.

When $|r| > 1$, the infinite geometric series does not have a sum. For example, the sum of the infinite geometric series $1 + 2 + 4 + 8 + \cdots$ increases without limit.

> **HOW TO** Find the sum of the terms of the infinite geometric sequence $1, -\frac{1}{2}, \frac{1}{4}, -\frac{1}{8}, \ldots$.
>
> The common ratio is $-\frac{1}{2}$. $\left|-\frac{1}{2}\right| < 1$.
>
> $$S = \frac{a_1}{1-r} = \frac{1}{1-\left(-\frac{1}{2}\right)} = \frac{1}{\frac{3}{2}} = \frac{2}{3}$$ • Use the Formula for the Sum of an Infinite Geometric Series.

The sum of an infinite geometric series can be applied to nonterminating, repeating decimals.

The repeating decimal shown below has been rewritten as an infinite geometric series with first term $\frac{3}{10}$ and common ratio $\frac{1}{10}$.

$$0.33\overline{3} = 0.3 + 0.03 + 0.003 + \cdots$$
$$= \frac{3}{10} + \frac{3}{100} + \frac{3}{1000} + \cdots$$

The sum of the series can be calculated by using the Formula for the Sum of an Infinite Geometric Series.

$$S = \frac{a_1}{1-r} = \frac{\frac{3}{10}}{1-\frac{1}{10}} = \frac{\frac{3}{10}}{\frac{9}{10}} = \frac{3}{9} = \frac{1}{3}$$ • $a_1 = \frac{3}{10}, r = \frac{1}{10}$

This method can be used to find an equivalent numerical fraction for any repeating decimal.

> **HOW TO** Find an equivalent fraction for $0.12\overline{2}$.
>
> Write the decimal as an infinite geometric series. Note that the geometric series does not begin with the first term. It begins with $\frac{2}{100}$. The common ratio is $\frac{1}{10}$.
>
> $$0.122\overline{2} = 0.1 + 0.02 + 0.002 + 0.0002 + \cdots$$
> $$= \frac{1}{10} + \underbrace{\frac{2}{100} + \frac{2}{1000} + \frac{2}{10{,}000} + \cdots}_{\text{Geometric Series}}$$
>
> Use the Formula for the Sum of an Infinite Geometric Series.
>
> $$S = \frac{a_1}{1-r} = \frac{\frac{2}{100}}{1-\frac{1}{10}} = \frac{\frac{2}{100}}{\frac{9}{10}} = \frac{2}{90}$$ • $a_1 = \frac{2}{100}, r = \frac{1}{10}$
>
> Add $\frac{1}{10}$ to the sum of the geometric series.
>
> $$0.122\overline{2} = \frac{1}{10} + \frac{2}{90} = \frac{11}{90}$$
>
> An equivalent fraction for $0.122\overline{2}$ is $\frac{11}{90}$.

Example 5

Find the sum of the terms of the infinite geometric sequence $2, \frac{1}{2}, \frac{1}{8}, \ldots$.

Solution

$r = \frac{a_2}{a_1} = \frac{\frac{1}{2}}{2} = \frac{1}{4}$ • Find r. $a_2 = \frac{1}{2}$, $a_1 = 2$

$S = \frac{a_1}{1-r} = \frac{2}{1-\frac{1}{4}} = \frac{2}{\frac{3}{4}} = \frac{8}{3}$ • Use the formula.

You Try It 5

Find the sum of the terms of the infinite geometric sequence $3, -2, \frac{4}{3}, -\frac{8}{9}, \ldots$.

Your solution

Example 6

Find an equivalent fraction for $0.3\overline{6}$.

Solution

$0.3\overline{6} = 0.36 + 0.0036 + 0.000036 + \cdots$

$= \frac{36}{100} + \frac{36}{10,000} + \frac{36}{1,000,000} + \cdots$

$S = \frac{a_1}{1-r} = \frac{\frac{36}{100}}{1-\frac{1}{100}} = \frac{\frac{36}{100}}{\frac{99}{100}} = \frac{36}{99} = \frac{4}{11}$

An equivalent fraction is $\frac{4}{11}$.

You Try It 6

Find an equivalent fraction for $0.\overline{66}$.

Your solution

Example 7

Find an equivalent fraction for $0.2\overline{345}$.

Solution

$0.2\overline{345} = 0.23 + 0.0045 + 0.000045 + \cdots$

$= \frac{23}{100} + \frac{45}{10,000} + \frac{45}{1,000,000} + \cdots$

$S = \frac{a_1}{1-r} = \frac{\frac{45}{10,000}}{1-\frac{1}{100}} = \frac{\frac{45}{10,000}}{\frac{99}{100}} = \frac{45}{9900}$

$0.2\overline{345} = \frac{23}{100} + \frac{45}{9900} = \frac{2277}{9900} + \frac{45}{9900}$

$= \frac{2322}{9900} = \frac{129}{550}$

An equivalent fraction is $\frac{129}{550}$.

You Try It 7

Find an equivalent fraction for $0.3\overline{421}$.

Your solution

Solutions on p. S35

Objective D **To solve application problems**

Example 8

On the first swing, the length of the arc through which a pendulum swings is 16 in. The length of each successive swing is $\frac{7}{8}$ the length of the preceding swing. Find the length of the arc on the fifth swing. Round to the nearest tenth.

Strategy

To find the length of the arc on the fifth swing, use the Formula for the nth Term of a Geometric Sequence.

Solution

$n = 5, a_1 = 16, r = \dfrac{7}{8}$

$a_n = a_1 r^{n-1}$

$a_5 = 16\left(\dfrac{7}{8}\right)^{5-1} = 16\left(\dfrac{7}{8}\right)^4 = 16\left(\dfrac{2401}{4096}\right)$

$\quad = \dfrac{2401}{256} \approx 9.4$

The length of the arc on the fifth swing is 9.4 in.

You Try It 8

You start a chain letter and send it to three friends. Each of the three friends sends the letter to three other friends, and the sequence is repeated. Assuming that no one breaks the chain, find how many letters will have been mailed from the first through the sixth mailings.

Your strategy

Your solution

Example 9

Assume that the value of a mutual fund will increase at a rate of 7% per year. If $25,000 is invested on January 1, 2008, find the value of the mutual fund on December 31, 2017. Round to the nearest cent.

Strategy

To find the value of the investment, use the Formula for the nth Term of a Geometric Sequence, $a_n = a_1 r^{n-1}$.

Solution

$n = 11, a_1 = 25,000, r = 1.07$

$a_n = a_1 r^{n-1}$
$a_{11} = 25,000(1.07)^{11-1} = 25,000(1.07)^{10}$
$\quad \approx 49,178.78$

The value of the investment after 10 years is $49,178.78.

You Try It 9

Assume that the value of a mutual fund will increase at a rate of 8% per year. If $20,000 is invested on January 1, 2008, find the value of the mutual fund on December 31, 2017. Round to the nearest cent.

Your strategy

Your solution

Solutions on p. S35

12.3 Exercises

Objective A To find the *n*th term of a geometric sequence

1. ✏️ Explain the difference between an arithmetic sequence and a geometric sequence.

2. ✏️ Explain how to find the common ratio of a geometric sequence.

For Exercises 3 to 8, find the indicated term of the geometric sequence.

3. $2, 8, 32, \ldots; a_9$

4. $4, 3, \dfrac{9}{4}, \ldots; a_8$

5. $6, -4, \dfrac{8}{3}, \ldots; a_7$

6. $-5, 15, -45, \ldots; a_7$

7. $-\dfrac{1}{16}, \dfrac{1}{8}, -\dfrac{1}{4}, \ldots; a_{10}$

8. $\dfrac{1}{27}, \dfrac{1}{9}, \dfrac{1}{3}, \ldots; a_7$

For Exercises 9 to 14, find a_2 and a_3 for the geometric sequence.

9. $9, a_2, a_3, \dfrac{8}{3}, \ldots$

10. $8, a_2, a_3, \dfrac{27}{8}, \ldots$

11. $3, a_2, a_3, -\dfrac{8}{9}, \ldots$

12. $6, a_2, a_3, -48, \ldots$

13. $-3, a_2, a_3, 192, \ldots$

14. $5, a_2, a_3, 625, \ldots$

Objective B To find the sum of a finite geometric series

For Exercises 15 to 18, find the sum of the indicated number of terms of the geometric sequence.

15. $2, 6, 18, \ldots; n = 7$

16. $-4, 12, -36, \ldots; n = 7$

17. $12, 9, \dfrac{27}{4}, \ldots; n = 5$

18. $3, 3\sqrt{2}, 6, \ldots; n = 12$

For Exercises 19 to 22, find the sum of the geometric series.

19. $\displaystyle\sum_{i=1}^{5} (2)^i$

20. $\displaystyle\sum_{n=1}^{6} \left(\dfrac{3}{2}\right)^n$

21. $\displaystyle\sum_{i=1}^{5} \left(\dfrac{1}{3}\right)^i$

22. $\displaystyle\sum_{n=1}^{6} \left(\dfrac{2}{3}\right)^n$

Objective C To find the sum of an infinite geometric series

For Exercises 23 to 26, find the sum of the infinite geometric series.

23. $3 + 2 + \dfrac{4}{3} + \cdots$

24. $2 - \dfrac{1}{4} + \dfrac{1}{32} + \cdots$

25. $\dfrac{7}{10} + \dfrac{7}{100} + \dfrac{7}{1000} + \cdots$

26. $\dfrac{5}{100} + \dfrac{5}{10,000} + \dfrac{5}{1,000,000} + \cdots$

For Exercises 27 to 34, find an equivalent fraction for the repeating decimal.

27. $0.88\overline{8}$

28. $0.55\overline{5}$

29. $0.22\overline{2}$

30. $0.99\overline{9}$

31. $0.45\overline{45}$

32. $0.18\overline{18}$

33. $0.166\overline{6}$

34. $0.833\overline{3}$

Objective D **To solve application problems**

35. Radioactivity A laboratory ore sample contains 400 mg of a radioactive material with a half-life of 1 h. Find the amount of radioactive material in the sample at the beginning of the fourth hour.

36. Pendulum On the first swing, the length of the arc through which a pendulum swings is 20 in. The length of each successive swing is $\dfrac{4}{5}$ the length of the preceding swing. What is the total distance the pendulum has traveled during four swings? Round to the nearest tenth of an inch.

37. Sports To test the "bounce" of a tennis ball, the ball is dropped from a height of 10 ft. The ball bounces to 75% of its previous height with each bounce. How high does the ball bounce on the sixth bounce? Round to the nearest tenth of a foot.

38. Real Estate A real estate broker estimates that a certain piece of land will increase in value at a rate of 15% each year. If the original value of the land is $25,000, what will be its value in 10 years?

39. Compensation Suppose an employee receives a wage of 1¢ on the first day of work, 2¢ the second day, 4¢ the third day, and so on in a geometric sequence. Find the total amount of money earned for working 30 days.

APPLYING THE CONCEPTS

40. State whether the following statements are true or false.
 a. The sum of an infinite geometric series is infinity.
 b. The sum of a geometric series always increases in value.
 c. $1, 8, 27, \ldots, k^3, \ldots$ is a geometric sequence.
 d. $\sqrt{1}, \sqrt{2}, \sqrt{3}, \sqrt{4}, \ldots, \sqrt{n}, \ldots$ is a geometric sequence.

41. The third term of a geometric sequence is 3, and the sixth term is $\dfrac{1}{9}$. Find the first term.

42. Given $a_n = 162$, $r = -3$, and $S_n = 122$ for a geometric sequence, find a_1 and n.

12.4 Binomial Expansions

Objective A To expand $(a + b)^n$

By carefully observing the series for each expansion of the binomial $(a + b)^n$ shown below, it is possible to identify some interesting patterns.

$(a + b)^1 = a + b$
$(a + b)^2 = a^2 + 2ab + b^2$
$(a + b)^3 = a^3 + 3a^2b + 3ab^2 + b^3$
$(a + b)^4 = a^4 + 4a^3b + 6a^2b^2 + 4ab^3 + b^4$
$(a + b)^5 = a^5 + 5a^4b + 10a^3b^2 + 10a^2b^3 + 5ab^4 + b^5$

PATTERNS FOR THE VARIABLE PART

1. The first term is a^n. The exponent on a decreases by 1 for each successive term.

2. The exponent on b increases by 1 for each successive term. The last term is b^n.

3. The degree of each term is n.

HOW TO Write the variable parts of the terms in the expansion of $(a + b)^6$.

$a^6, a^5b, a^4b^2, a^3b^3, a^2b^4, ab^5, b^6$

• The first term is a^6. For each successive term, the exponent on a decreases by 1, and the exponent on b increases by 1. The last term is b^6.

The variable parts in the general expansion of $(a + b)^n$ are

$a^n, a^{n-1}b, a^{n-2}b^2, \ldots, a^{n-r}b^r, \ldots, ab^{n-1}, b^n$

A pattern for the coefficients of the terms of the expanded binomial can be found by writing the coefficients in a triangular array known as **Pascal's Triangle.**

Each row begins and ends with the number 1. Any other number in a row is the sum of the two closest numbers above it. For example, $4 + 6 = 10$.

	For $(a + b)^1$:				1		1				
	For $(a + b)^2$:			1		2		1			
	For $(a + b)^3$:		1		3		3		1		
	For $(a + b)^4$:	1		4		6		4		1	
	For $(a + b)^5$:	1	5		10		10		5		1

HOW TO Write the sixth row of Pascal's Triangle.

To write the sixth row, first write the numbers of the fifth row. The first and last numbers of the sixth row are 1. Each of the other numbers of the sixth row can be obtained by finding the sum of the two closest numbers above it in the fifth row.

These numbers will be the coefficients of the terms in the expansion of $(a + b)^6$.

Point of Interest

Blaise Pascal

Blaise Pascal (1623–1662) is given credit for this triangle, which he first published in 1654. In that publication, the triangle looked like

1	2	3	4	5...	
1	1	1	1	1	1...
2	1	2	3	4...	
3	1	3	6...		
4	1	4...			
5	1				

Thus the triangle was rotated 45° from the way it is shown today.

The first European publication of the triangle is attributed to Peter Apianus in 1527. However, there are versions of it in a Chinese text by Yang Hui that dates from 1250. In that text, Hui demonstrated how to find the third, fourth, fifth, and sixth roots of a number by using the triangle.

Using the numbers of the sixth row of Pascal's Triangle for the coefficients and using the pattern for the variable part of each term, we can write the expanded form of $(a + b)^6$ as follows:

$$(a + b)^6 = a^6 + 6a^5b + 15a^4b^2 + 20a^3b^3 + 15a^2b^4 + 6ab^5 + b^6$$

Although Pascal's Triangle can be used to find the coefficients for the expanded form of the power of any binomial, this method is inconvenient when the power of the binomial is large. An alternative method for determining these coefficients is based on the concept of **_n_ factorial.**

n Factorial

n! (which is read "*n* factorial") is the product of the first *n* consecutive natural numbers. 0! is defined to be 1.

$$n! = n(n - 1)(n - 2) \cdots 3 \cdot 2 \cdot 1$$

1!, 5!, and 7! are shown at the right.

$1! = 1$
$5! = 5 \cdot 4 \cdot 3 \cdot 2 \cdot 1 = 120$
$7! = 7 \cdot 6 \cdot 5 \cdot 4 \cdot 3 \cdot 2 \cdot 1 = 5040$

HOW TO Evaluate: 6!

$6! = 6 \cdot 5 \cdot 4 \cdot 3 \cdot 2 \cdot 1 = 720$ • Write the factorial as a product. Then simplify.

HOW TO Evaluate: $\dfrac{7!}{4!\,3!}$

$\dfrac{7!}{4!\,3!} = \dfrac{7 \cdot 6 \cdot 5 \cdot 4 \cdot 3 \cdot 2 \cdot 1}{(4 \cdot 3 \cdot 2 \cdot 1)(3 \cdot 2 \cdot 1)} = 35$ • Write each factorial as a product. Then simplify.

The coefficients in a binomial expansion can be given in terms of factorials.

Note that in the expansion of $(a + b)^5$, the coefficient of a^2b^3 can be given by $\dfrac{5!}{2!\,3!}$. The numerator is the factorial of the power of the binomial. The denominator is the product of the factorials of the exponents on a and b.

$(a + b)^5 =$
$a^5 + 5a^4b + 10a^3b^2 + 10a^2b^3 + 5ab^4 + b^5$

$\dfrac{5!}{2!\,3!} = \dfrac{5 \cdot 4 \cdot 3 \cdot 2 \cdot 1}{(2 \cdot 1)(3 \cdot 2 \cdot 1)} = 10$

Binomial Coefficients

The coefficients of $a^{n-r}b^r$ in the expansion of $(a + b)^n$ are given by $\dfrac{n!}{(n - r)!\,r!}$. The symbol $\dbinom{n}{r}$ is used to express this quotient.

$$\dbinom{n}{r} = \dfrac{n!}{(n - r)!\,r!}$$

HOW TO Evaluate: $\dbinom{8}{5}$

$\dbinom{8}{5} = \dfrac{8!}{(8 - 5)!\,5!} = \dfrac{8!}{3!\,5!}$ • Write the quotient of the factorials.

$= \dfrac{8 \cdot 7 \cdot 6 \cdot 5 \cdot 4 \cdot 3 \cdot 2 \cdot 1}{(3 \cdot 2 \cdot 1)(5 \cdot 4 \cdot 3 \cdot 2 \cdot 1)} = 56$ • Simplify.

HOW TO Evaluate: $\dbinom{7}{0}$

$\dbinom{7}{0} = \dfrac{7!}{(7-0)!\,0!} = \dfrac{7!}{7!\,0!}$ • Write the quotient of the factorials.

$= \dfrac{7 \cdot 6 \cdot 5 \cdot 4 \cdot 3 \cdot 2 \cdot 1}{(7 \cdot 6 \cdot 5 \cdot 4 \cdot 3 \cdot 2 \cdot 1)(1)} = 1$ • Recall that $0! = 1$.

Using factorials and the pattern for the variable part of each term, we can write a formula for any natural number power of a binomial.

The Binomial Expansion Formula

$(a + b)^n = \dbinom{n}{0}a^n + \dbinom{n}{1}a^{n-1}b + \dbinom{n}{2}a^{n-2}b^2 + \cdots +$

$\dbinom{n}{r}a^{n-r}b^r + \cdots + \dbinom{n}{n}b^n$

The Binomial Expansion Formula is used below to expand $(a + b)^5$.

$(a + b)^5 = \dbinom{5}{0}a^5 + \dbinom{5}{1}a^4b + \dbinom{5}{2}a^3b^2 + \dbinom{5}{3}a^2b^3 + \dbinom{5}{4}ab^4 + \dbinom{5}{5}b^5$

$= a^5 + 5a^4b + 10a^3b^2 + 10a^2b^3 + 5ab^4 + b^5$

HOW TO Write $(x - 2)^4$ in expanded form.

Use the Binomial Expansion Formula. Then simplify each term.

$(x - 2)^4 = \dbinom{4}{0}x^4 + \dbinom{4}{1}x^3(-2) + \dbinom{4}{2}x^2(-2)^2 + \dbinom{4}{3}x(-2)^3 + \dbinom{4}{4}(-2)^4$

$= 1x^4 + 4x^3(-2) + 6x^2(4) + 4x(-8) + 1(16)$

$= x^4 - 8x^3 + 24x^2 - 32x + 16$

HOW TO Write $(4x + 3y)^3$ in expanded form.

Use the Binomial Expansion Formula. Then simplify each term.

$(4x + 3y)^3 = \dbinom{3}{0}(4x)^3 + \dbinom{3}{1}(4x)^2(3y) + \dbinom{3}{2}(4x)(3y)^2 + \dbinom{3}{3}(3y)^3$

$= 1(64x^3) + 3(16x^2)(3y) + 3(4x)(9y^2) + 1(27y^3)$

$= 64x^3 + 144x^2y + 108xy^2 + 27y^3$

The Binomial Theorem can also be used to write any term of a binomial expansion. The formula for the rth term is given below. First note that in the expansion of $(a + b)^5$, the exponent on b is 1 less than the term number.

$(a + b)^5 = a^5 + 5a^4b + 10a^3b^2 + 10a^2b^3 + 5ab^4 + b^5$

The Formula for the rth Term in a Binomial Expansion

The rth term of $(a + b)^n$ is $\dbinom{n}{r-1}a^{n-r+1}b^{r-1}$.

HOW TO Find the 4th term in the expansion of $(x + 3)^7$.

Use the Formula for the rth Term in a Binomial Expansion. $r = 4$

$$\binom{n}{r-1}a^{n-r+1}b^{r-1}$$

$$\binom{7}{4-1}x^{7-4+1}(3)^{4-1} = \binom{7}{3}x^4(3)^3 = 35x^4(27)$$

$$= 945x^4$$

HOW TO Find the 12th term in the expansion of $(2x - 1)^{14}$.

Use the Formula for the rth Term in a Binomial Expansion. $r = 12$

$$\binom{14}{12-1}(2x)^{14-12+1}(-1)^{12-1} = \binom{14}{11}(2x)^3(-1)^{11} = 364(8x^3)(-1)$$

$$= -2912x^3$$

Example 1

Write $(2x + y)^3$ in expanded form.

Solution

$(2x + y)^3$

$$= \binom{3}{0}(2x)^3 + \binom{3}{1}(2x)^2(y) + \binom{3}{2}(2x)(y)^2 + \binom{3}{3}(y)^3$$

$$= 1(8x^3) + 3(4x^2)(y) + 3(2x)(y^2) + 1(y^3)$$

$$= 8x^3 + 12x^2y + 6xy^2 + y^3$$

You Try It 1

Write $(3m - n)^4$ in expanded form.

Your solution

Example 2

Find the first three terms in the expansion of $(x + 3)^{15}$.

Solution

$(x + 3)^{15}$

$$= \binom{15}{0}x^{15} + \binom{15}{1}x^{14}(3) + \binom{15}{2}x^{13}(3)^2 + \cdots$$

$$= 1(x^{15}) + 15x^{14}(3) + 105x^{13}(9) + \cdots$$

$$= x^{15} + 45x^{14} + 945x^{13} + \cdots$$

You Try It 2

Find the first three terms in the expansion of $(y - 2)^{10}$.

Your solution

Example 3

Find the 4th term in the expansion of $(5x - y)^6$.

Solution

$$\binom{n}{r-1}a^{n-r+1}b^{r-1}$$

$$\binom{6}{4-1}(5x)^{6-4+1}(-y)^{4-1} \quad \bullet \; n = 6, a = 5x, b = -y, \; r = 4$$

$$= \binom{6}{3}(5x)^3(-y)^3$$

$$= 20(125x^3)(-y^3)$$

$$= -2500x^3y^3$$

You Try It 3

Find the 3rd term in the expansion of $(t - 2s)^7$.

Your solution

Solutions on pp. S35–S36

12.4 Exercises

Objective A To expand $(a + b)^n$

Evaluate the expressions in Exercises 1 to 24.

1. $3!$

2. $4!$

3. $8!$

4. $9!$

5. $0!$

6. $1!$

7. $\dfrac{5!}{2! \, 3!}$

8. $\dfrac{8!}{5! \, 3!}$

9. $\dfrac{6!}{6! \, 0!}$

10. $\dfrac{10!}{10! \, 0!}$

11. $\dfrac{9!}{6! \, 3!}$

12. $\dfrac{10!}{2! \, 8!}$

13. $\dbinom{7}{2}$

14. $\dbinom{8}{6}$

15. $\dbinom{10}{2}$

16. $\dbinom{9}{6}$

17. $\dbinom{9}{0}$

18. $\dbinom{10}{10}$

19. $\dbinom{6}{3}$

20. $\dbinom{7}{6}$

21. $\dbinom{11}{1}$

22. $\dbinom{13}{1}$

23. $\dbinom{4}{2}$

24. $\dbinom{8}{4}$

Write the expressions in Exercises 25 to 32 in expanded form.

25. $(x + y)^4$

26. $(r - s)^3$

27. $(x - y)^5$

28. $(y - 3)^4$

29. $(2m + 1)^4$

30. $(2x + 3y)^3$

31. $(2r - 3)^5$

32. $(x + 3y)^4$

For Exercises 33 to 42, find the first three terms in the expansion.

33. $(a + b)^{10}$

34. $(a + b)^9$

35. $(a - b)^{11}$

36. $(a - b)^{12}$

37. $(2x + y)^8$

38. $(x + 3y)^9$

39. $(4x - 3y)^8$

40. $(2x - 5)^7$

41. $\left(x + \dfrac{1}{x}\right)^7$

42. $\left(x - \dfrac{1}{x}\right)^8$

For Exercises 43 to 50, find the indicated term in the expansion.

43. $(2x - 1)^7$; 4th term

44. $(x + 4)^5$; 3rd term

45. $(x^2 - y^2)^6$; 2nd term

46. $(x^2 + y^2)^7$; 6th term

47. $(y - 1)^9$; 5th term

48. $(x - 2)^8$; 8th term

49. $\left(n + \dfrac{1}{n}\right)^5$; 2nd term

50. $\left(x + \dfrac{1}{2}\right)^6$; 3rd term

APPLYING THE CONCEPTS

51. State whether the following statements are true or false.
 a. $0! \cdot 4! = 0$
 b. $\dfrac{4!}{0!}$ is undefined.
 c. In the expansion of $(a + b)^8$, the exponent on a for the fifth term is 5.
 d. $n!$ is the sum of the first n positive numbers.
 e. There are n terms in the expansion of $(a + b)^n$.
 f. The sum of the exponents in each term of the expansion of $(a + b)^n$ is n.

52. Write a summary of the history of the development of Pascal's Triangle. Include some of the properties of the triangle.

Focus on Problem Solving

Forming Negations A **statement** is a declarative sentence that is either true or false but not both true and false. Here are a few examples of statements.

> Denver is the capital of Colorado.
> The word *dog* has four letters.
> Florida is a state in the United States.
> In the year 2020, the President of the United States will be a woman.

Problem solving sometimes requires that you form the **negation** of a statement. If a statement is true, its negation must be false. If a statement is false, its negation must be true. For instance,

Babe Ruth

> "Henry Aaron broke Babe Ruth's home run record" is a true sentence. Accordingly, the sentence "Henry Aaron did not break Babe Ruth's home run record" is false.

> "Dogs have twelve paws" is a false sentence, so the sentence "Dogs do not have twelve paws" is a true sentence.

To form the negation of a statement requires a precise understanding of the statement being negated. Here are some phrases that demand special attention.

Statement	Meaning
At least seven	Greater than or equal to seven
More than seven	More than seven (does not include 7)
At most seven	Less than or equal to seven
Less than (or *fewer than*) seven	Less than seven (does not include 7)

Consider the sentence "At least three people scored over 90 on the last test." The negation of that sentence is "Fewer than three people scored over 90 on the last test."

For Exercises 1 to 4, write the negation of the given sentence.

1. There was at least one error on the test.

2. The temperature was less than 30 degrees.

3. The mountain is more than 5000 ft tall.

4. There are at most five vacancies for the field trip to New York.

When the word *all, no* (or *none*), or *some* occurs in a sentence, forming the negation requires careful thought. Consider the sentence "All roads are paved with cement." This sentence is not true because there are some roads—dirt roads, for example—that are not paved with cement. Because the sentence is false, its negation must be true. You might be tempted to write "All roads are not paved with cement" as the negation, but that sentence is not true because some roads are paved with cement. The correct negation is "*Some* roads are *not* paved with cement."

Now consider the sentence "Some fish live in aquariums." Because this sentence is true, the negation must be false. Writing "Some fish do not live in aquariums" as the negation is not correct because that sentence is true. The correct negation is "*All* fish do *not* live in aquariums."

The sentence "No houses have basements" is false because there is at least one house with a basement. Because this sentence is false, its negation must be true. The negation is "Some houses do have basements."

Statement	*Negation*
All *A* are *B*.	Some *A* are not *B*.
No *A* are *B*.	Some *A* are *B*.
Some *A* are *B*.	No *A* are *B*.
Some *A* are not *B*.	All *A* are *B*.

Write the negation of each sentence in Exercises 5 to 12.

5. All trees are tall.

6. All cats chase mice.

7. Some flowers have red blooms.

8. No golfers like tennis.

9. Some students do not like math.

10. No honest people are politicians.

11. All cars have power steering.

12. Some televisions are black and white.

Projects and Group Activities

ISBN and UPC Numbers

Every book that is cataloged for the Library of Congress must have an ISBN (International Standard Book Number). An ISBN is a 10-digit number of the form $a_1\text{-}a_2a_3a_4\text{-}a_5a_6a_7a_8a_9\text{-}c$. For instance, the ISBN for the fourth edition of the *American Heritage Dictionary* is 0-395-82517-2. The first number, 0, indicates that the book is written in English. The next three numbers, 395, indicate the publisher (Houghton Mifflin Company). The next five numbers, 82517, identify the book (*American Heritage Dictionary*), and the last digit, *c*, is called a *check digit*. This digit is chosen such that the following sum is divisible by 11.

$$a_1(10) + a_2(9) + a_3(8) + a_4(7) + a_5(6) + a_6(5) + a_7(4) + a_8(3) + a_9(2) + c$$

For the *American Heritage Dictionary*,

$$0(10) + 3(9) + 9(8) + 5(7) + 8(6) + 2(5) + 5(4) + 1(3) + 7(2) + c$$
$$= 229 + c$$

ISBN 0-395-82517-2

The last digit of the ISBN is chosen as 2 because $229 + 2 = 231$ and $231 \div 11 = 21$ remainder 0. The value of *c* could be any number from 0 to 10. The number 10 is coded as an X.

One purpose of the ISBN method of coding books is to ensure that orders placed for books are accurately filled. For instance, suppose a clerk sends an order for the *American Heritage Dictionary* and inadvertently enters the number 0-395-82571-2 (the 7 and the 1 have been transposed). Now

$$0(10) + 3(9) + 9(8) + 5(7) + 8(6) + 2(5) + 5(4) + 7(3) + 1(2) + 2 = 237$$

and if we divide 237 by 11, the remainder is not zero. Thus an error has been made in the ISBN.

1. Determine the ISBN check digit for the book *A Brief History of Time* by Stephen Hawking. The first nine digits of the ISBN are 0-553-05340-?

2. Is 0-395-12370-4 a possible ISBN?

3. Check the ISBN of this intermediate algebra textbook.

Another coding scheme that is closely related to the ISBN is the UPC (Universal Product Code). This number is placed on many items and is particularly useful in grocery stores. A check-out clerk passes the number by a scanner, which reads the number and records the price on the cash register.

The UPC is a 12-digit number. The first six digits identify the manufacturer. The next five digits are a product identification number given by the manufacturer to identify the product. The final digit is a check digit. The check digit is chosen such that the following sum is divisible by 10.

$$a_1(3) + a_2 + a_3(3) + a_4 + a_5(3) + a_6 + a_7(3) + a_8 + a_9(3) + a_{10} + a_{11}(3) + c$$

The UPC for the TI-83 Plus calculator is 0-33317-19865-8. Using the expression above, we find that

$$0(3) + 3 + 3(3) + 3 + 1(3) + 7 + 1(3) + 9 + 8(3) + 6 + 5(3) + c = 82 + c$$

Because $82 + 8$ is a multiple of 10, the check digit is 8.

4. A new product a company has developed has been assigned the UPC 0-21443-32912-?. Determine the check digit for the UPC.

5. Is 1-32342-65933-9 a valid UPC?

6. Check the UPC of a product in your home.

Proof of Purchase TI-83 Plus

0 33317 19865 8

Chapter 12 Summary

Key Words

A *sequence* is an ordered list of numbers. Each of the numbers of a sequence is called a *term* of the sequence. For a general sequence, the first term is a_1, the second term is a_2, the third term is a_3, and the nth term, also called the *general term* of the sequence, is a_n. [12.1A, p. 641]

Examples

$1, \dfrac{1}{3}, \dfrac{1}{9}, \dfrac{1}{27}$ is a sequence.

The terms are $1, \dfrac{1}{3}, \dfrac{1}{9}$, and $\dfrac{1}{27}$.

A *finite sequence* contains a finite number of terms. [12.1A, p. 641]

2, 4, 6, 8 is a finite sequence.

An *infinite sequence* contains an infinite number of terms. [12.1A, p. 641]

1, 3, 5, 7, ... is an infinite sequence.

The indicated sum of the terms of a sequence is called a *series*. S_n is used to designate the sum of the first n terms of a sequence. [12.1B, p. 643]

Given the sequence 1, 5, 9, 13, the series is $1 + 5 + 9 + 13$.

Summation notation, or *sigma notation*, is used to represent a series in a compact form. The Greek capital letter sigma, Σ, is used to indicate the sum. [12.1B, p. 643]

$$\sum_{n=1}^{4} 2n = 2(1) + 2(2) + 2(3) + 2(4)$$
$$= 20$$

An *arithmetic sequence*, or *arithmetic progression*, is one in which the difference between any two consecutive terms is the same constant. The difference between consecutive terms is called the *common difference* of the sequence. [12.2A, p. 647]

3, 9, 15, 21, ... is an arithmetic sequence. $9 - 3 = 6$; 6 is the common diference.

The indicated sum of the terms of an arithmetic sequence is an *arithmetic series*. [12.2B, p. 649]

1, 2, 3, 4, 5 is an arithmetic sequence. The indicated sum is $1 + 2 + 3 + 4 + 5$.

A *geometric sequence*, or *geometric progression*, is one in which each successive term of the sequence is the same nonzero constant multiple of the preceding term. The common multiple is called the *common ratio* of the sequence. [12.3A, p. 653]

9, 3, 1, $\frac{1}{3}$, ... is a geometric sequence. $\frac{3}{9} = \frac{1}{3}$; $\frac{1}{3}$ is the common ratio.

The indicated sum of the terms of a geometric sequence is a *geometric series*. [12.3B, p. 655]

5, 10, 20, 40 is a geometric sequence. The indicated sum is $5 + 10 + 20 + 40$.

n factorial, written $n!$, is the product of the first n consecutive natural number. 0! is defined to be 1. [12.4A, p. 664]

$5! = 5 \cdot 4 \cdot 3 \cdot 2 \cdot 1 = 120$

Essential Rules and Procedures

Examples

Formula for the *n*th Term of an Arithmetic Sequence
[12.2A, p. 647]
The *n*th term of an arithmetic sequence with a common difference of d is given by $a_n = a_1 + (n - 1)d$.

4, 7, 10, 13, ...
$d = 7 - 4 = 3$; $a_1 = 4$
$a_{10} = 4 + (10 - 1)3 = 4 + 27 = 31$

Formula for the Sum of *n* Terms of an Arithmetic Series
[12.2B, p. 649]
Let a_1 be the first term of a finite arithmetic sequence, n be the number of terms, and a_n be the last term of the sequence. Then the sum of the series S_n is given by $S_n = \frac{n}{2}(a_1 + a_n)$.

$5, 8, 11, 14, \ldots \qquad d = 3, a_{12} = 38$

$S_{12} = \frac{12}{2}(5 + 38) = 258$

Formula for the *n*th Term of a Geometric Sequence
[12.3A, p. 654]
The nth term of a geometric sequence with first term a_1 and common ratio r is given by $a_n = a_1 r^{n-1}$.

$2, 6, 18, 54, \ldots \qquad a_1 = 2, r = 3$

$a_{10} = 2(3)^{10-1} = 2(3)^9 = 39{,}366$

Formula for the Sum of *n* Terms of a Finite Geometric Series [12.3B, p. 655]
Let a_1 be the first term of a finite geometric sequence, n be the number of terms, and r be the common ratio. Then the sum of the series S_n is given by $S_n = \frac{a_1(1 - r^n)}{1 - r}$.

$1, 4, 16, 64, \ldots \qquad a_1 = 1, r = 4$

$S_8 = \frac{1(1 - 4^8)}{1 - 4} = \frac{1 - 65{,}536}{-3}$

$= 21{,}845$

Formula for the Sum of an Infinite Geometric Series
[12.3C, p. 657]
The sum of an infinite geometric series in which $|r| < 1$ and a_1 is the first term is given by $S = \frac{a_1}{1 - r}$.

$2, 1, \frac{1}{2}, \frac{1}{4}, \ldots \qquad a_1 = 2, r = \frac{1}{2}$

$S = \frac{2}{1 - \frac{1}{2}} = \frac{2}{\frac{1}{2}} = 4$

Binomials Coefficients [12.4A, p. 664]
The coefficients of $a^{n-r}b^r$ in the expansion of $(a + b)^n$ are given by $\binom{n}{r}$, where $\binom{n}{r} = \frac{n!}{(n - r)!\, r!}$.

$\binom{6}{2} = \frac{6!}{(6 - 2)!\, 2!} = \frac{6!}{4!\, 2!} = 15$

Binomial Expansion Formula [12.4A, p. 665]

$(a + b)^n = \binom{n}{0}a^n + \binom{n}{1}a^{n-1}b + \binom{n}{2}a^{n-2}b^2 + \cdots$

$+ \binom{n}{r}a^{n-r}b^r + \cdots + \binom{n}{n}b^n$

$(x + y)^4$

$= \binom{4}{0}x^4 + \binom{4}{1}x^3y + \binom{4}{2}x^2y^2$

$+ \binom{4}{3}xy^3 + \binom{4}{4}y^4$

$= x^4 + 4x^3y + 6x^2y^2 + 4xy^3 + y^4$

Formula for the *r*th Term in a Binomial Expansion
[12.4A, p. 665]

The rth term of $(a + b)^n$ is $\binom{n}{r - 1}a^{n-r+1}b^{r-1}$.

The sixth term of $(2x - y)^9$ is

$\binom{9}{5}(2x)^{9-6+1}(-y)^{6-1} = -2016x^4y^5.$

Chapter 12 Review Exercises

1. Write $\displaystyle\sum_{i=1}^{4} 2x^{i-1}$ in expanded form.

2. Write the 10th term of the sequence whose nth term is given by the formula $a_n = 3n - 2$.

3. Find an equivalent fraction for $0.63\overline{3}$.

4. Find the sum of the infinite geometric series $\dfrac{3}{10} + \dfrac{3}{100} + \dfrac{3}{1000} + \cdots$.

5. Find the sum of the arithmetic series $\displaystyle\sum_{i=1}^{30}(4i - 1)$.

6. Find the sum of the series $\displaystyle\sum_{n=1}^{5}(3n - 2)$.

7. Find an equivalent fraction for $0.23\overline{23}$.

8. Evaluate: $\dbinom{9}{3}$

9. Find the sum of the geometric series $\displaystyle\sum_{n=1}^{5} 2(3)^n$.

10. Find the number of terms in the finite arithmetic sequence $8, 2, -4, \ldots, -118$.

11. Find the sum of the geometric series $\displaystyle\sum_{n=1}^{8}\left(\dfrac{1}{2}\right)^n$. Round to the nearest thousandth.

12. Find the 5th term in the expansion of $(3x - y)^8$.

13. Write the 5th term of the sequence whose nth term is given by the formula $a_n = \dfrac{(-1)^{2n-1}n}{n^2 + 2}$.

14. Find the sum of the first seven terms of the geometric sequence $5, 5\sqrt{5}, 25, \ldots$. Round to the nearest whole number.

15. Find the formula for the nth term of the arithmetic sequence $-7, -2, 3, \ldots$.

16. Find the 8th term in the expansion of $(x - 2y)^{11}$.

17. Find the 12th term of the geometric sequence $1, \sqrt{3}, 3, \ldots$.

18. Find the sum of the first 40 terms of the arithmetic sequence $11, 13, 15, \ldots$.

19. Evaluate: $\begin{pmatrix} 12 \\ 9 \end{pmatrix}$

20. Find the sum of the series $\sum_{n=1}^{4} \frac{(-1)^{n-1}n}{n+1}$.

21. Write $\sum_{i=1}^{5} \frac{(2x)^i}{i}$ in expanded form.

22. Find the 8th term of the geometric sequence $3, 1, \frac{1}{3}, \ldots$.

23. Write $(x - 3y^2)^5$ in expanded form.

24. Find the sum of the infinite geometric series $4 - 1 + \frac{1}{4} - \cdots$.

25. Find the sum of the infinite geometric series $2 + \frac{4}{3} + \frac{8}{9} + \cdots$.

26. Evaluate: $\frac{12!}{5!\,8!}$

27. Find the 20th term of the arithmetic sequence $1, 5, 9, \ldots$.

28. Write $\sum_{i=1}^{6} 4x^i$ in expanded form.

29. Find the number of terms in the finite arithmetic sequence $-3, -7, -11, \ldots, -59$.

30. Find the 7th term of the geometric sequence $5, 5\sqrt{3}, 15, \ldots$.

31. Find the sum of the terms of the infinite geometric sequence $3, 2, \frac{4}{3}, \ldots$.

32. Write the 17th term of the sequence whose nth term is given by the formula $a_n = \frac{10}{n+3}$.

33. Find the 10th term of the arithmetic sequence $-10, -4, 2, \ldots$.

34. Find the sum of the first 18 terms of the arithmetic sequence $-19, -12, -5, \ldots$.

35. Find the sum of the first 6 terms of the geometric sequence 8, −16, 32,

36. Evaluate: $\dfrac{7!}{5!\,2!}$

37. Find the 7th term in the expansion of $(3x + y)^9$.

38. Find the sum of the series $\displaystyle\sum_{n=1}^{5} (4n - 3)$.

39. Write the 8th term in the sequence whose nth term is given by the formula $a_n = \dfrac{n + 2}{2n}$.

40. Find the formula for the nth term of the arithmetic sequence −4, 0, 4,

41. Find the 5th term of the geometric sequence $10, -2, \dfrac{2}{5}, \ldots$.

42. Find an equivalent fraction for $0.36\overline{6}$.

43. Find the 37th term of the arithmetic sequence −37, −32, −27,

44. Find the sum of the first five terms of the geometric sequence 1, 4, 16,

45. Exercise An exercise program calls for walking 15 min each day for a week. Each week thereafter, the amount of time spent walking increases by 3 min per day. In how many weeks will a person be walking 60 min each day?

46. Temperature The temperature of a hot water spa is 102°F. Each hour the temperature is 5% lower than during the preceding hour. Find the temperature of the spa after 8 h. Round to the nearest tenth.

47. Compensation The salary schedule for an apprentice electrician is $2400 for the first month and an $80-per-month salary increase for the next eight months. Find the total salary for the nine-month period.

48. Radioactivity A laboratory radioactive sample contains 200 mg of a material with a half-life of 1 h. Find the amount of radioactive material in the sample at the beginning of the seventh hour.

Chapter 12 Test

1. Find the sum of the series $\displaystyle\sum_{n=1}^{4} (3n + 1)$.

2. Evaluate: $\dbinom{9}{6}$

3. Find the 7th term of the geometric sequence $4, 4\sqrt{2}, 8, \ldots$.

4. Write the 14th term of the sequence whose nth term is given by the formula $a_n = \dfrac{8}{n + 2}$.

5. Write in expanded form: $\displaystyle\sum_{i=1}^{4} 3x^i$

6. Find the sum of the first 18 terms of the arithmetic sequence $-25, -19, -13, \ldots$.

7. Find the sum of the terms of the infinite geometric sequence $4, 3, \dfrac{9}{4}, \ldots$.

8. Find the number of terms in the finite arithmetic sequence $-5, -8, -11, \ldots, -50$.

9. Find an equivalent fraction for $0.2\overline{33}$.

10. Evaluate: $\dfrac{8!}{4!\,4!}$

11. Find the 5th term of the geometric sequence $6, 2, \dfrac{2}{3}, \ldots$.

12. Find the 4th term in the expansion of $(x - 2y)^7$.

13. Find the 35th term of the arithmetic sequence $-13, -16, -19, \ldots$.

14. Find the formula for the nth term of the arithmetic sequence $12, 9, 6, \ldots$.

15. Find the sum of the first five terms of the geometric sequence $-6, 12, -24, \ldots$.

16. Find the 6th and 7th terms of the sequence whose nth term is given by the formula $a_n = \dfrac{n+1}{n}$.

17. Find the sum of the first six terms of the geometric sequence $1, \dfrac{3}{2}, \dfrac{9}{4}, \ldots$.

18. Find the sum of the first 21 terms of the arithmetic sequence $5, 12, 19, \ldots$.

19. **Inventory** An inventory of supplies for a yarn manufacturer indicated that 7500 skeins of yarn were in stock on January 1. On February 1, and on the first of the month for each successive month, the manufacturer sent 550 skeins of yarn to retail outlets. How many skeins were in stock after the shipment on October 1?

20. **Radioactivity** An ore sample contains 320 mg of a radioactive substance with a half-life of one day. Find the amount of radioactive material in the sample at the beginning of the fifth day.

Cumulative Review Exercises

1. Graph: $3x - 2y = -4$

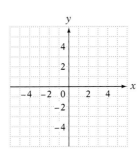

2. Factor: $2x^6 + 16$

3. Simplify: $\dfrac{4x^2}{x^2 + x - 2} - \dfrac{3x - 2}{x + 2}$

4. Given $f(x) = 2x^2 - 3x$, find $f(-2)$.

5. Simplify: $\sqrt{2y}\,(\sqrt{8xy} - \sqrt{y})$

6. Solve by using the quadratic formula: $2x^2 - x + 7 = 0$

7. Solve: $5 - \sqrt{x} = \sqrt{x + 5}$

8. Find the equation of the circle that passes through the point $(4, 2)$ and whose center is $(-1, -1)$.

9. Solve by the addition method:
$3x - 3y = 2$
$6x - 4y = 5$

10. Evaluate the determinant:
$\begin{vmatrix} -3 & 1 \\ 4 & 2 \end{vmatrix}$

11. Solve: $2x - 1 > 3$ or $1 - 3x > 7$
Write the solution set in set-builder notation.

12. Graph: $2x - 3y < 9$

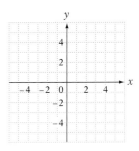

13. Write $\log_5 \sqrt{\dfrac{x}{y}}$ in expanded form.

14. Solve for x: $4^x = 8^{x-1}$

15. Write the 5th and 6th terms of the sequence whose nth term is given by the formula $a_n = n(n - 1)$.

16. Find the sum of the series $\displaystyle\sum_{n=1}^{7} (-1)^{n-1}(n + 2)$.

17. Find the 33rd term of the arithmetic sequence $-7, -10, -13, \ldots$.

18. Find the sum of the terms of the infinite geometric sequence $3, -2, \frac{4}{3}, \ldots$.

19. Find an equivalent fraction for $0.4\overline{6}$.

20. Find the 6th term in the expansion of $(2x + y)^6$.

21. **Mixtures** How many ounces of pure water must be added to 200 oz of an 8% salt solution to make a 5% salt solution?

22. **Computers** A new computer can complete a payroll in 16 min less time than it takes an older computer to complete the same payroll. Working together, both computers can complete the payroll in 15 min. How long would it take each computer working alone to complete the payroll?

23. **Uniform Motion** A boat traveling with the current went 15 mi in 2 h. Against the current it took 3 h to travel the same distance. Find the rate of the boat in calm water and the rate of the current.

24. **Radioactivity** An 80-milligram sample of a radioactive material decays to 55 mg in 30 days. Use the exponential decay equation $A = A_0\left(\frac{1}{2}\right)^{t/k}$, where A is the amount of radioactive material present after time t, k is the half-life of the material, and A_0 is the original amount of radioactive material, to find the half-life of the 80-milligram sample. Round to the nearest whole number.

25. **The Arts** A theater-in-the-round has 62 seats in the first row, 74 seats in the second row, 86 seats in the third row, and so on in an arithmetic sequence. Find the total number of seats in the theater if there are 12 rows of seats.

26. **Sports** To test the "bounce" of a ball, the ball is dropped from a height of 8 ft. The ball bounces to 80% of its previous height with each bounce. How high does the ball bounce on the fifth bounce? Round to the nearest tenth.

Final Exam

1. Simplify: $12 - 8[3 - (-2)]^2 \div 5 - 3$

2. Evaluate $\dfrac{a^2 - b^2}{a - b}$ when $a = 3$ and $b = -4$.

3. Simplify: $5 - 2[3x - 7(2 - x) - 5x]$

4. Solve: $\dfrac{3}{4}x - 2 = 4$

5. Solve: $\dfrac{2 - 4x}{3} - \dfrac{x - 6}{12} = \dfrac{5x - 2}{6}$

6. Solve: $8 - |5 - 3x| = 1$

7. Graph $2x - 3y = 9$ using the x- and y-intercepts.

8. Find the equation of the line containing the points $(3, -2)$ and $(1, 4)$.

9. Find the equation of the line that contains the point $(-2, 1)$ and is perpendicular to the line $3x - 2y = 6$.

10. Simplify: $2a[5 - a(2 - 3a) - 2a] + 3a^2$

11. Factor: $8 - x^3y^3$

12. Factor: $x - y - x^3 + x^2y$

13. Simplify: $(2x^3 - 7x^2 + 4) \div (2x - 3)$

14. Simplify: $\dfrac{x^2 - 3x}{2x^2 - 3x - 5} \div \dfrac{4x - 12}{4x^2 - 4}$

15. Simplify: $\dfrac{x - 2}{x + 2} - \dfrac{x + 3}{x - 3}$

16. Simplify: $\dfrac{\dfrac{3}{x} + \dfrac{1}{x + 4}}{\dfrac{1}{x} + \dfrac{3}{x + 4}}$

17. Solve: $\dfrac{5}{x - 2} - \dfrac{5}{x^2 - 4} = \dfrac{1}{x + 2}$

18. Solve $a_n = a_1 + (n - 1)d$ for d.

19. Simplify: $\left(\dfrac{4x^2y^{-1}}{3x^{-1}y}\right)^{-2}\left(\dfrac{2x^{-1}y^2}{9x^{-2}y^2}\right)^3$

20. Simplify: $\left(\dfrac{3x^{2/3}y^{1/2}}{6x^2y^{4/3}}\right)^6$

21. Simplify: $x\sqrt{18x^2y^3} - y\sqrt{50x^4y}$

22. Simplify: $\dfrac{\sqrt{16x^5y^4}}{\sqrt{32xy^7}}$

23. Simplify: $\dfrac{3}{2 + i}$

24. Write a quadratic equation that has integer coefficients and has solutions $-\dfrac{1}{2}$ and 2.

25. Solve by using the quadratic formula:
$2x^2 - 3x - 1 = 0$

26. Solve: $x^{2/3} - x^{1/3} - 6 = 0$

27. Solve: $\dfrac{2}{x} - \dfrac{2}{2x + 3} = 1$

28. Graph: $f(x) = -x^2 + 4$

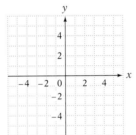

29. Graph: $\dfrac{x^2}{16} + \dfrac{y^2}{4} = 1$

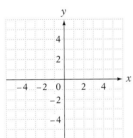

30. Find the inverse of the function
$f(x) = \dfrac{2}{3}x - 4.$

31. Solve by the addition method:
$3x - 2y = 1$
$5x - 3y = 3$

32. Evaluate the determinant:
$\begin{vmatrix} 3 & 4 \\ -1 & 2 \end{vmatrix}$

33. Solve: $x^2 - y^2 = 4$
$x + y = 1$

34. Solve: $2 - 3x < 6$ and $2x + 1 > 4$
Write the solution set in interval notation.

35. Solve: $|2x + 5| < 3$

36. Graph the solution set: $3x + 2y > 6$

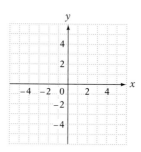

37. Graph: $f(x) = \log_2(x + 1)$

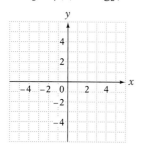

38. Write $2(\log_2 a - \log_2 b)$ as a single logarithm with a coefficient of 1.

39. Solve for x: $\log_3 x - \log_3(x - 3) = \log_3 2$

40. Write $\displaystyle\sum_{i=1}^{5} 2y^i$ in expanded form.

41. Find an equivalent fraction for $0.5\overline{1}$.

42. Find the 3rd term in the expansion of $(x - 2y)^9$.

43. Education An average score of 70–79 in a history class receives a C grade. A student has scores of 64, 58, 82, and 77 on four history tests. Find the range of scores on the fifth test that will give the student a C grade for the course.

44. Uniform Motion A jogger and a cyclist set out at 8 A.M. from the same point headed in the same direction. The average speed of the cyclist is two and a half times the average speed of the jogger. In 2 h, the cyclist is 24 mi ahead of the jogger. How far did the cyclist ride in that time?

45. Investments You have a total of $12,000 invested in two simple interest accounts. On one account, a money market fund, the annual simple interest rate is 8.5%. On the other account, a tax free bond fund, the annual simple interest rate is 6.4%. The total annual interest earned by the two accounts is $936. How much do you have invested in each account?

46. Geometry The length of a rectangle is 1 ft less than three times the width. The area of the rectangle is 140 ft². Find the length and width of the rectangle.

47. The Stock Market Three hundred shares of a utility stock earn a yearly dividend of $486. How many additional shares of the utility stock would give a total dividend income of $810?

48. Uniform Motion An account executive traveled 45 mi by car and then an additional 1050 mi by plane. The rate of the plane was seven times the rate of the car. The total time for the trip was $3\frac{1}{4}$ h. Find the rate of the plane.

49. Physics An object is dropped from the top of a building. Find the distance the object has fallen when its speed reaches 75 ft/s. Use the equation $v = \sqrt{64d}$, where v is the speed of the object and d is the distance in feet. Round to the nearest whole number.

50. Uniform Motion A small plane made a trip of 660 mi in 5 h. The plane traveled the first 360 mi at a constant rate before increasing its speed by 30 mph. Then it traveled another 300 mi at the increased speed. Find the rate of the plane for the first 360 mi.

51. Light The intensity, L, of a light source is inversely proportional to the square of the distance, d, from the source. If the intensity is 8 foot-candles at a distance of 20 ft, what is the intensity when the distance is 4 ft?

52. Uniform Motion A motorboat traveling with the current can go 30 mi in 2 h. Against the current, it takes 3 h to go the same distance. Find the rate of the motorboat in calm water and the rate of the current.

53. Investments An investor deposits $4000 into an account that earns 9% annual interest compounded monthly. Use the compound interest formula $P = A(1 + i)^n$, where A is the original value of the investment, i is the interest rate per compounding period, n is the total number of compounding periods, and P is the value of the investment after n periods, to find the value of the investment after 2 years. Round to the nearest cent.

54. Real Estate Assume that the average value of a home increases 6% per year. How much will a house costing $80,000 be worth in 20 years? Round to the nearest dollar.

Appendix A

Keystroke Guide for the TI-83 and TI-83 Plus

Basic Operations

Numerical calculations are performed on the **home screen.** You can always return to the home screen by pressing (2nd) QUIT. Pressing (CLEAR) erases the home screen.

To evaluate the expression $-2(3 + 5) - 8 \div 4$, use the following keystrokes.

 (-) 2 (3 + 5) − 8 ÷ 4 (ENTER)

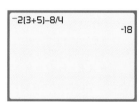

Note: There is a difference between the key to enter a negative number, (-), and the key for subtraction, (−). You cannot use these keys interchangeably.

The (2nd) key is used to access the commands in gold writing above a key. For instance, to evaluate the $\sqrt{49}$, press (2nd) √ 49) (ENTER).

The (ALPHA) key is used to place a letter on the screen. One reason to do this is to store a value of a variable. The following keystrokes give A the value of 5.

5 (STO▸) (ALPHA) A (ENTER)

This value is now available in calculations. For instance, we can find the value of $3a^2$ by using the following keystrokes: 3 (ALPHA) A (x²). To display the value of the variable on the screen, press (2nd) RCL (ALPHA) A.

Note: When you use the (ALPHA) key, only capital letters are available on the TI-83 calculator.

Complex Numbers

To perform operations on complex numbers, first press (MODE) and then use the arrow keys to select a+bi. Then press (ENTER) (2nd) QUIT.

Addition of complex numbers To add $(3 + 4i) + (2 - 7i)$, use the keystrokes

(3 + 4 (2nd) i) +

(2 − 7 (2nd) i) (ENTER).

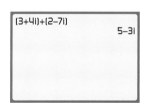

Division of complex numbers. To divide $\frac{26 + 2i}{2 + 4i}$, use the keystrokes (26 + 2 (2nd) i) ÷

(2 + 4 (2nd) i) (ENTER).

Note: Operations for subtraction and multiplication are similar.

TAKE NOTE

The descriptions in the margins (for example, Basic Operations and Complex Numbers) are the same as those used in the text and are arranged alphabetically.

Additional operations on complex numbers can be found by selecting CPX under the MATH key.

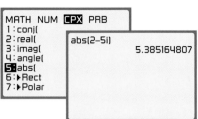

To find the absolute value of $2 - 5i$, press MATH (scroll to CPX) (scroll to abs) ENTER (2 – 5 2nd i) ENTER .

Evaluating Functions

There are various methods of evaluating a function but all methods require that the expression be entered as one of the ten functions Y₁ to Y₀. To evaluate $f(x) = \frac{x^2}{x-1}$ when $x = -3$, enter the expression into, for instance, Y₁, and then press VARS ▶ 11 ((-) 3) ENTER .

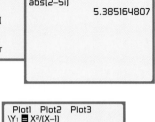

Note: If you try to evaluate a function at a number that is not in the domain of the function, you will get an error message. For instance, 1 is not in the domain of $f(x) = \frac{x^2}{x-1}$. If we try to evaluate the function at 1, the error screen at the right appears.

Evaluating Variable Expressions

To evaluate a variable expression, first store the values of each variable. Then enter the variable expression. For instance, to evaluate $s^2 + 2sl$ when $s = 4$ and $l = 5$, use the following keystrokes.

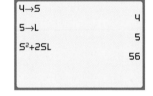

4 STO▶ ALPHA S ENTER 5 STO▶ ALPHA L ENTER ALPHA S
x^2 + 2 ALPHA S ALPHA L ENTER

Graph

To graph a function, use the Y= key to enter the expression for the function, select a suitable viewing window, and then press GRAPH . For instance, to graph $f(x) = 0.1x^3 - 2x - 1$ in the standard viewing window, use the following keystrokes.

Y= 0.1 X,T,θ,n ^ 3 – 2 X,T,θ,n – 1 ZOOM (scroll to 6) ENTER

Note: For the keystrokes above, you do not have to scroll to 6. Alternatively, use ZOOM 6. This will select the standard viewing window and automatically start the graph. Use the WINDOW key to create a custom window for a graph.

Graphing Inequalities

To illustrate this feature, we will graph $y \leq 2x - 1$. Enter $2x - 1$ into Y₁. Because $y \leq 2x - 1$, we want to shade below the graph. Move the cursor to the left of Y₁ and press ENTER three times. Press GRAPH .

Note: To shade above the graph, move the cursor to the left of Y₁ and press ENTER two times. An inequality with the symbol ≤ or ≥ should be graphed with a solid line, and an inequality with the symbol < or > should be graphed with a dashed line. However, the graph of a linear inequality on a graphing calculator does not distinguish between a solid line and a dashed line.

To graph the solution set of a system of inequalities, solve each inequality for y and graph each inequality. The solution set is the intersection of the two inequalities. The solution set of $\begin{array}{l} 3x + 2y > 10 \\ 4x - 3y \le 5 \end{array}$ is shown at the right.

Intersect The INTERSECT feature is used to solve a system of equations. To illustrate this feature, we will use the system of equations $\begin{array}{l} 2x - 3y = 13 \\ 3x + 4y = -6 \end{array}$.

Note: Some equations can be solved by this method. See the section "Solve an equation" below. Also, this method is used to find a number in the domain of a function for a given number in the range. See the section "Find a domain element."

Solve each of the equations in the system of equations for y. In this case, we have $y = \frac{2}{3}x - \frac{13}{3}$ and $y = -\frac{3}{4}x - \frac{3}{2}$.

Use the Y-editor to enter $\frac{2}{3}x - \frac{13}{3}$ into Y₁ and $-\frac{3}{4}x - \frac{3}{2}$ into Y₂. Graph the two functions in the standard viewing window. (If the window does not show the point of intersection of the two graphs, adjust the window until you can see the point of intersection.)

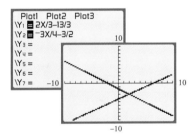

Press 2nd CALC (scroll to 5, intersect) ENTER.

Alternatively, you can just press 2nd CALC 5.

First curve? is shown at the bottom of the screen and identifies one of the two graphs on the screen. Press ENTER.

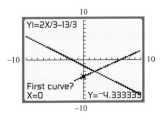

Second curve? is shown at the bottom of the screen and identifies the second of the two graphs on the screen. Press ENTER.

Guess? shown at the bottom of the screen asks you to use the left or right arrow key to move the cursor to the *approximate* location of the point of intersection. (If there are two or more points of intersection, it does not matter which one you choose first.) Press ENTER.

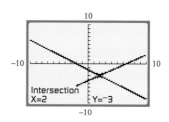

The solution of the system of equations is $(2, -3)$.

Solve an equation To illustrate the steps involved, we will solve the equation $2x + 4 = -3x - 1$. The idea is to write the equation as the system of equations $\begin{array}{l} y = 2x + 4 \\ y = -3x - 1 \end{array}$ and then use the steps for solving a system of equations.

Use the Y-editor to enter the left and right sides of the equation into Y_1 and Y_2. Graph the two functions and then follow the steps for Intersect.

The solution is -1, the x-coordinate of the point of intersection.

Find a domain element For this example, we will find a number in the domain of $f(x) = -\frac{2}{3}x + 2$ that corresponds to 4 in the range of the function. This is like solving the system of equations $y = -\frac{2}{3}x + 2$ and $y = 4$.

Use the Y= editor to enter the expression for the function in Y_1 and the desired output, 4, in Y_2. Graph the two functions and then follow the steps for Intersect.

The point of intersection is $(-3, 4)$. The number -3 in the domain of f produces an output of 4 in the range of f.

Math Pressing **MATH** gives you access to many built-in functions. The following keystrokes will convert 0.125 to a fraction: .125 **MATH** 1 **ENTER** .

Additional built-in functions under **MATH** can be found by pressing **MATH** ▶. For instance, to evaluate $-|-25|$, press **(-)** **MATH** ▶ 1 **(-)** 25 **)** **ENTER** .

See your owner's manual for assistance with other functions under the **MATH** key.

Min and Max The local minimum and the local maximum values of a function are calculated by accessing the CALC menu. For this demonstration, we will find the minimum value and the maximum value of $f(x) = 0.2x^3 + 0.3x^2 - 3.6x + 2$.

Enter the function into Y_1. Press 2nd CALC (scroll to 3 for minimum of the function) ENTER.

Alternatively, you can just press 2nd CALC 3.

Left Bound? shown at the bottom of the screen asks you to use the left or right arrow key to move the cursor to the *left* of the minimum. Press ENTER.

Right Bound? shown at the bottom of the screen asks you to use the left or right arrow key to move the cursor to the *right* of the minimum. Press ENTER.

Guess? shown at the bottom of the screen asks you to use the left or right arrow key to move the cursor to the *approximate* location of the minimum. Press ENTER.

The minimum value of the function is the *y*-coordinate. For this example, the minimum value of the function is -2.4.

The *x*-coordinate for the minimum is 2. However, because of rounding errors in the calculation, it is shown as a number close to 2.

To find the maximum value of the function, follow the same steps as above except select maximum under the CALC menu. The screens for this calculation are shown below.

The maximum value of the function is 10.1.

Radical To evaluate a square-root expression, press 2nd $\sqrt{\ }$.
Expressions

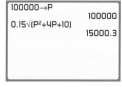

For instance, to evaluate $0.15\sqrt{p^2 + 4p + 10}$ when $p = 100,000$, first store 100,000 in P. Then press 0.15 2nd $\sqrt{\ }$ ALPHA P x² + 4 ALPHA P + 10) ENTER.

To evaluate a radical expression other than a square root, access $\sqrt[x]{\ }$ by pressing MATH. For instance, to evaluate $\sqrt[4]{67}$, press 4 (the index of the radical) MATH (scroll to 5) ENTER 67 ENTER.

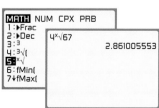

Scientific Notation

To enter a number in scientific notation, use EE. For instance, to find $\frac{3.45 \times 10^{-12}}{1.5 \times 10^{25}}$, press 3.45 (2nd) EE (-) 12 (÷) 1.5 (2nd) EE 25 (ENTER). The answer is 2.3×10^{-37}.

Sequences and Series

The terms of a sequence and the sum of a series can be calculated by using the (2nd) LIST feature.

Store a sequence A sequence is stored in one of the lists L_1 through L_6. For instance, to store the sequence 1, 3, 5, 7, 9 in L_1, use the following keystrokes.

Display the terms of a sequence The terms of a sequence are displayed by using the function seq(expression, variable, begin, end, increment). For instance, to display the 3rd through 8th terms of the sequence given by $a_n = n^2 + 6$, enter the following keystrokes.

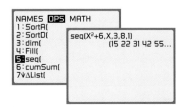

The keystrokes (STO▸) (2nd) L1 (ENTER) store the terms of the sequence in L_1. This is not necessary but is sometimes helpful if additional work will be done with that sequence.

Find a sequence of partial sums To find a sequence of partial sums, use the cumSum(function. For instance, to find the sequence of partial sums for 2, 4, 6, 8, 10, use the following keystrokes.

If a sequence is stored as a list in L_1, then the sequence of partial sums can be calculated by pressing (2nd) LIST (scroll to 6 [or press 6]) (ENTER) (2nd) L1 (.) (ENTER).

Find the sum of a series The sum of a series is calculated using sum<list, start, end>. For instance, to find $\sum_{n=3}^{6} (n^2 + 2)$, enter the following keystrokes.

Table There are three steps in creating an input/output table for a function. First use the editor to input the function. The second step is setting up the table, and the third step is displaying the table.

To set up the table, press **2nd** TBLSET. **TblStart** is the first value of the independent variable in the input/output table. △**Tbl** is the difference between successive values. Setting this to 1 means that, for this table, the input values are $-2, -1, 0, 1, 2. \ldots$ If △**Tbl** $= 0.5$, then the input values are $-2, -1.5, -1, -0.5, 0, 0.5, \ldots$

Indpnt is the independent variable. When this is set to **Auto**, values of the independent variable are automatically entered into the table. **Depend** is the dependent variable. When this is set to **Auto**, values of the dependent variable are automatically entered into the table.

To display the table, press **2nd** TABLE. An input/output table for $f(x) = x^2 - 1$ is shown at the right.

Once the table is on the screen, the up and down arrow keys can be used to display more values in the table. For the table at the right, we used the up arrow key to move to $x = -7$.

An input/output table for any given input can be created by selecting **Ask** for the independent variable. The table at the right shows an input/output table for $f(x) = \dfrac{4x}{x-2}$ for selected values of x. Note the word **ERROR** when 2 was entered. This occurred because f is not defined when $x = 2$.

Note: Using the table feature in **Ask** mode is the same as evaluating a function for given values of the independent variable. For instance, from the table at the right, we have $f(4) = 8$.

Test The TEST feature has many uses, one of which is to graph the solution set of a linear inequality in one variable. To illustrate this feature, we will graph the solution set of $x - 1 < 4$. Press **Y=** **X,T,θ,n** **−** 1 **2nd** TEST (scroll to 5) **ENTER** 4 **GRAPH**.

Trace Once a graph is drawn, pressing **TRACE** will place a cursor on the screen, and the coordinates of the point below the cursor are shown at the bottom of the screen. Use the left and right arrow keys to move the cursor along the graph. For the graph at the right, we have $f(4.8) = 3.4592$, where $f(x) = 0.1x^3 - 2x + 2$ is shown at the top left of the screen.

In TRACE mode, you can evaluate a function at any value of the independent variable that is within Xmin and Xmax. To do this, first graph the function. Now press **TRACE** (the value of x) **ENTER**. For the graph at the left below, we used $x = -3.5$. If a value of x is chosen outside the window, an error message is displayed.

 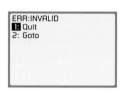

In the example above where we entered -3.5 for x, the value of the function was calculated as 4.7125. This means that $f(-3.5) = 4.7125$. The keystrokes **2nd** QUIT **VARS** ◄ 11 **MATH** 1 **ENTER** will convert the decimal value to a fraction.

When the TRACE feature is used with two or more graphs, the up and down arrow keys are used to move between the graphs. The graphs below are for the functions $f(x) = 0.1x^3 - 2x + 2$ and $g(x) = 2x - 3$. By using the up and down arrows, we can place the cursor on either graph. The right and left arrows are used to move along the graph.

Window The viewing window for a graph is controlled by pressing **WINDOW**. Xmin and Xmax are the minimum value and maximum value, respectively, of the independent variable shown on the graph. Xscl is the distance between tic marks on the x-axis. Ymin and Ymax are the minimum value and maximum value, respectively, of the dependent variable shown on the graph. Yscl is the distance between tic marks on the y-axis. Leave Xres as 1.

Note: In the standard viewing window, the distance between tic marks on the x-axis is different from the distance between tic marks on the y-axis. This will distort a graph. A more accurate picture of a graph can be created by using a square viewing window. See ZOOM.

Y= The **Y=** editor is used to enter the expression for a function. There are ten possible functions, labeled Y_1 to Y_0, that can be active at any one time. For instance, to enter $f(x) = x^2 + 3x - 2$ as Y_1, use the following keystrokes.

Y= **X,T,θ,n** **x²** **+** 3 **X,T,θ,n** **−** 2

Note: If an expression is already entered for Y₁, place the cursor anywhere on that expression and press [CLEAR].

To enter $s = \frac{2v - 1}{v^3 - 3}$ into Y₂, place the cursor to the right of the equals sign for Y₂. Then press [(] 2 [X,T,θ,n] [−] 1 [)] [÷] [(] [X,T,θ,n] [^] 3 [−] 3 [)].

> Note: When we enter an equation, the independent variable, v in the expression above, is entered using [X,T,θ,n]. The dependent variable, s in the expression above, is one of Y₁ to Y₀. Also note the use of parentheses to ensure the correct order of operations.

Observe the black rectangle that covers the equals sign for the two examples we have shown. This rectangle means that the function is "active." If we were to press [GRAPH], then the graph of both functions would appear. You can make a function inactive by using the arrow keys to move the cursor over the equals sign of that function and then pressing [ENTER]. This will remove the black rectangle. We have done that for Y₂, as shown at the right. Now if [GRAPH] is pressed, only Y₁ will be graphed.

It is also possible to control the appearance of the graph by moving the cursor on the [Y=] screen to the left of any Y. With the cursor in this position, pressing [ENTER] will change the appearance of the graph. The options are shown at the right.

Zero The ZERO feature of a graphing calculator is used for various calculations: to find the x-intercepts of a function, to solve some equations, and to find the zero of a function.

x-intercepts To illustrate the procedure for finding x-intercepts, we will use $f(x) = x^2 + x - 2$.

First, use the Y-editor to enter the expression for the function and then graph the function in the standard viewing window. (It may be necessary to adjust this window so that the intercepts are visible.) Once the graph is displayed, use the keystrokes below to find the x-intercepts of the graph of the function.

Press [2nd] CALC (scroll to 2 for zero of the function) [ENTER].

Alternatively, you can just press [2nd] CALC 2.

Left Bound? shown at the bottom of the screen asks you to use the left or right arrow key to move the cursor to the *left* of the desired x-intercept. Press [ENTER].

Right Bound? shown at the bottom of the screen asks you to use the left or right arrow key to move the cursor to the *right* of the desired x-intercept. Press [ENTER].

Guess? shown at the bottom of the screen asks you to use the left or right arrow key to move the cursor to the *approximate* location of the desired *x*-intercept. Press ⏎.

The *x*-coordinate of an *x*-intercept is -2. Therefore, an *x*-intercept is $(-2, 0)$.

To find the other *x*-intercept, follow the same steps as above. The screens for this calculation are shown below.

A second *x*-intercept is $(1, 0)$.

Solve an equation To use the ZERO feature to solve an equation, first rewrite the equation with all terms on one side. For instance, one way to solve the equation $x^3 - x + 1 = -2x + 3$ is first to rewrite it as $x^3 + x - 2 = 0$. Enter $x^3 + x - 2$ into Y₁ and then follow the steps for finding *x*-intercepts.

Find the real zeros of a function To find the real zeros of a function, follow the steps for finding *x*-intercepts.

Zoom Pressing ZOOM allows you to select some preset viewing windows. This key also gives you access to ZBox, Zoom In, and Zoom Out. These functions enable you to redraw a selected portion of a graph in a new window. Some windows used frequently in this text are shown below.

Appendix B

Proofs and Tables

Proofs of Logarithmic Properties

In each of the following proofs of logarithmic properties, it is assumed that the Properties of Exponents are true for all real number exponents.

The Logarithm Property of the Product of Two Numbers

For any positive real numbers x, y, and b, $b \neq 1$, $\log_b xy = \log_b x + \log_b y$.

Proof: Let $\log_b x = m$ and $\log_b y = n$.

Write each equation in its equivalent exponential form.
Use substitution and the Properties of Exponents.

Write the equation in its equivalent logarithmic form.
Substitute $\log_b x$ for m and $\log_b y$ for n.

$$x = b^m \qquad y = b^n$$
$$xy = b^m b^n$$
$$xy = b^{m+n}$$
$$\log_b xy = m + n$$
$$\log_b xy = \log_b x + \log_b y$$

The Logarithm Property of the Quotient of Two Numbers

For any positive real numbers x, y, and b, $b \neq 1$, $\log_b \dfrac{x}{y} = \log_b x - \log_b y$.

Proof: Let $\log_b x = m$ and $\log_b y = n$.
Write each equation in its equivalent exponential form.

Use substitution and the Properties of Exponents.

Write the equation in its equivalent logarithmic form.

Substitute $\log_b x$ for m and $\log_b y$ for n.

$$x = b^m \qquad y = b^n$$
$$\frac{x}{y} = \frac{b^m}{b^n}$$
$$\frac{x}{y} = b^{m-n}$$
$$\log_b \frac{x}{y} = m - n$$
$$\log_b \frac{x}{y} = \log_b x - \log_b y$$

The Logarithm Property of the Power of a Number

For any real numbers x, r, and b, $b \neq 1$, $\log_b x^r = r \log_b x$.

Proof: Let $\log_b x = m$.
Write the equation in its equivalent exponential form.
Raise both sides to the r power.

Write the equation in its equivalent logarithmic form.
Substitute $\log_b x$ for m.

$$x = b^m$$
$$x^r = (b^m)^r$$
$$x^r = b^{mr}$$
$$\log_b x^r = mr$$
$$\log_b x^r = r \log_b x$$

Proof of the Formula for the Sum of n Terms of a Geometric Series

Theorem: The sum of the first n terms of a geometric sequence whose nth term is ar^{n-1} is given by $S_n = \dfrac{a(1 - r^n)}{1 - r}$.

Proof: Let S_n represent the sum of n terms of the sequence.
Multiply each side of the equation by r.
Subtract the two equations.
Assuming $r \neq 1$, solve for S_n.

$$S_n = a + ar + ar^2 + \cdots + ar^{n-2} + ar^{n-1}$$
$$rS_n = ar + ar^2 + ar^3 + \cdots + ar^{n-1} + ar^n$$
$$S_n - rS_n = a - ar^n$$
$$(1 - r)S_n = a(1 - r^n)$$
$$S_n = \frac{a(1 - r^n)}{1 - r}$$

Proof of the Formula for the Sum of n Terms of an Arithmetic Series

Each term of the arithmetic sequence shown at the right was found by adding 3 to the previous term.

$$2, 5, 8,...,17, 20$$

Each term of the reverse arithmetic sequence can be found by subtracting 3 from the previous term.

$$20, 17, 14,...,5, 2$$

This idea is used in the following proof.

Theorem: The sum of the first n terms of an arithmetic sequence for which a_1 is the first term, n is the number of terms, a_n is the last term, and d is the common difference is given by $S_n = \dfrac{n}{2}(a_1 + a_n)$.

Proof: Let S_n represent the sum of the sequence.

$$S_n = a_1 + (a_1 + d) + (a_1 + 2d) + \cdots + a_n$$

Write the terms of the sum of the sequence in reverse order. The sum will be the same.

$$S_n = a_n + (a_n - d) + (a_n - 2d) + \cdots + a_1$$

Add the two equations.

$$2S_n = (a_1 + a_n) + (a_1 + a_n) + (a_1 + a_n) + \cdots + (a_1 + a_n)$$

Simplify the right side of the equation by using the fact that there are n terms in the sequence.

$$2S_n = n(a_1 + a_n)$$

Solve for S_n.

$$S_n = \dfrac{n}{2}(a_1 + a_n)$$

Table of Symbols

Symbol	Meaning
$+$	add
$-$	subtract
$\cdot, \times, (a)(b)$	multiply
$\dfrac{a}{b}, \div$	divide
$(\,)$	parentheses, a grouping symbol
$[\,]$	brackets, a grouping symbol
π	pi, a number approximately equal to $\dfrac{22}{7}$ or 3.14
$-a$	the opposite, or additive inverse, of a
$\dfrac{1}{a}$	the reciprocal, or multiplicative inverse, of a
$=$	is equal to
\approx	is approximately equal to
\neq	is not equal to

Symbol	Meaning		
$<$	is less than		
\leq	is less than or equal to		
$>$	is greater than		
\geq	is greater than or equal to		
(a, b)	an ordered pair whose first component is a and whose second component is b		
$^\circ$	degree (for angles)		
\sqrt{a}	the principal square root of a		
$\varnothing, \{\}$	the empty set		
$	a	$	the absolute value of a
\cup	union of two sets		
\cap	intersection of two sets		
\in	is an element of (for sets)		
\notin	is not an element of (for sets)		

Table of Measurement Abbreviations

U.S. Customary System

Length		**Capacity**		**Weight**		**Area**	
in.	inches	oz	fluid ounces	oz	ounces	in²	square inches
ft	feet	c	cups	lb	pounds	ft²	square feet
yd	yards	qt	quarts				
mi	miles	gal	gallons				

Metric System

Length		**Capacity**		**Weight/Mass**		**Area**	
mm	millimeter (0.001 m)	ml	milliliter (0.001 L)	mg	milligram (0.001 g)	cm²	square centimeters
cm	centimeter (0.01 m)	cl	centiliter (0.01 L)	cg	centigram (0.01 g)	m²	square meters
dm	decimeter (0.1 m)	dl	deciliter (0.1 L)	dg	decigram (0.1 g)		
m	meter	L	liter	g	gram		
dam	decameter (10 m)	dal	decaliter (10 L)	dag	decagram (10 g)		
hm	hectometer (100 m)	hl	hectoliter (100 L)	hg	hectogram (100 g)		
km	kilometer (1000 m)	kl	kiloliter (1000 L)	kg	kilogram (1000 g)		

Time

h	hours	min	minutes	s	seconds

Table of Properties

Properties of Real Numbers

The Associative Property of Addition

If a, b, and c are real numbers, then

$(a + b) + c = a + (b + c)$.

The Commutative Property of Addition

If a and b are real numbers, then

$a + b = b + a$.

The Addition Property of Zero

If a is a real number, then

$a + 0 = 0 + a = a$.

The Multiplication Property of Zero

If a is a real number, then

$a \cdot 0 = 0 \cdot a = 0$.

The Inverse Property of Addition

If a is a real number, then

$a + (-a) = (-a) + a = 0$.

The Associative Property of Multiplication

If a, b, and c are real numbers, then

$(a \cdot b) \cdot c = a \cdot (b \cdot c)$.

The Commutative Property of Multiplication

If a and b are real numbers, then

$a \cdot b = b \cdot a$.

The Multiplication Property of One

If a is a real number, then

$a \cdot 1 = 1 \cdot a = a$.

The Inverse Property of Multiplication

If a is a real number and $a \neq 0$, then

$a \cdot \dfrac{1}{a} = \dfrac{1}{a} \cdot a = 1$.

Distributive Property

If a, b, and c are real numbers, then

$a(b + c) = ab + ac$.

Properties of Equations

Addition Property of Equations

If $a = b$, then $a + c = b + c$.

Multiplication Property of Equations

If $a = b$ and $c \neq 0$, then $a \cdot c = b \cdot c$.

Properties of Inequalities

Addition Property of Inequalities

If $a > b$, then $a + c > b + c$.

If $a < b$, then $a + c < b + c$.

Multiplication Property of Inequalities

If $a > b$ and $c > 0$, then $ac > bc$.

If $a < b$ and $c > 0$, then $ac < bc$.

If $a > b$ and $c < 0$, then $ac < bc$.

If $a < b$ and $c < 0$, then $ac > bc$.

Properties of Exponents

If m and n are integers, then $x^m \cdot x^n = x^{m+n}$.

If m and n are integers, then $(x^m)^n = x^{mn}$.

If $x \neq 0$, then $x^0 = 1$.

If m and n are integers and $x \neq 0$, then $\dfrac{x^m}{x^n} = x^{m-n}$.

If m, n, and p are integers, then $(x^m \cdot y^n)^p = x^{mp}y^{np}$.

If n is a positive integer and $x \neq 0$, then

$x^{-n} = \dfrac{1}{x^n}$ and $\dfrac{1}{x^{-n}} = x^n$.

If m, n, and p are integers and $y \neq 0$, then $\left(\dfrac{x^m}{y^n}\right)^p = \dfrac{x^{mp}}{y^{np}}$.

Principle of Zero Products

If $a \cdot b = 0$, then $a = 0$ or $b = 0$.

Properties of Radical Expressions

If a and b are positive real numbers, then $\sqrt{ab} = \sqrt{a}\sqrt{b}$.

If a and b are positive real numbers, then $\sqrt{\dfrac{a}{b}} = \dfrac{\sqrt{a}}{\sqrt{b}}$.

Property of Squaring Both Sides of an Equation

If a and b are real numbers and $a = b$, then $a^2 = b^2$.

Properties of Logarithms

If x, y, and b are positive real numbers and $b \neq 1$, then

$\log_b(xy) = \log_b x + \log_b y$.

If x, y, and b are positive real numbers and $b \neq 1$, then

$\log_b \dfrac{x}{y} = \log_b x - \log_b y$.

If x and b are positive real numbers, $b \neq 1$, and r is any real number, then $\log_b x^r = r \log_b x$.

If x and b are positive real numbers and $b \neq 1$, then

$\log_b b^x = x$.

Table of Algebraic and Geometric Formulas

Slope of a line

$$m = \frac{y_2 - y_1}{x_2 - x_1}, x_1 \neq x_2$$

Point-slope formula for a line

$$y - y_1 = m(x - x_1)$$

Quadratic Formula

$$x = \frac{-b \pm \sqrt{b^2 - 4ac}}{2a}$$

discriminant $= b^2 - 4ac$

Perimeter and Area of a Triangle, and Sum of the Measures of the Angles

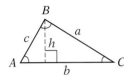

$$P = a + b + c$$

$$A = \frac{1}{2}bh$$

$$A + B + C = 180°$$

Pythagorean Theorem

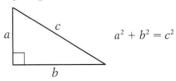

$$a^2 + b^2 = c^2$$

Perimeter and Area of a Rectangle

$$P = 2L + 2W$$

$$A = LW$$

Perimeter and Area of a Square

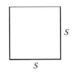

$$P = 4s$$

$$A = s^2$$

Area of a Trapezoid

$$A = \frac{1}{2}h(b_1 + b_2)$$

Circumference and Area of a Circle

$$C = 2\pi r$$

$$A = \pi r^2$$

Volume and Surface Area of a Rectangular Solid

$$V = LWH$$

$$SA = 2LW + 2LH + 2WH$$

Volume and Surface Area of a Sphere

$$V = \frac{4}{3}\pi r^3$$

$$SA = 4\pi r^2$$

Volume and Surface Area of a Right Circular Cylinder

$$V = \pi r^2 h$$

$$SA = 2\pi r^2 + 2\pi rh$$

Volume and Surface Area of a Right Circular Cone

$$V = \frac{1}{3}\pi r^2 h$$

$$SA = \pi r^2 + \pi rl$$

Solutions to Chapter 1 "You Try It"

SECTION 1.1

You Try It 1 Replace z by each of the elements of the set and determine whether the inequality is true.
$$z > -5$$
$$-10 > -5 \quad \text{False}$$
$$-5 > -5 \quad \text{False}$$
$$6 > -5 \quad \text{True}$$
The inequality is true for 6.

You Try It 2 **a.** Replace d in $-d$ by each element of the set and determine the value of the expression.
$$-d$$
$$-(-11) = 11$$
$$-(0) = 0$$
$$-(8) = -8$$
b. Replace d in $|d|$ by each element of the set and determine the value of the expression.
$$|d|$$
$$|-11| = 11$$
$$|0| = 0$$
$$|8| = 8$$

You Try It 3 $\{-5, -4, -3, -2, -1\}$

You Try It 4 $\{x \mid x \geq 15, x \in \text{whole numbers}\}$

You Try It 5 $C \cup D = \{1, 3, 5, 7, 9, 11, 13, 17\}$

You Try It 6 $E \cap F = \varnothing$

You Try It 7 Draw a parenthesis at 0 to indicate that 0 is not in the set. Draw a solid line to the left of 0.

$$\overleftarrow{\quad\underset{-5\,-4\,-3\,-2\,-1\ \ 0\ \ 1\ \ 2\ \ 3\ \ 4\ \ 5}{\mid\ \mid\ \mid\ \mid\ \mid\)\ \mid\ \mid\ \mid\ \mid\ \mid}\quad}$$

You Try It 8 This is the set of real numbers less than -2 *or* greater than -1.

$$\overleftarrow{\quad\underset{-5\,-4\,-3\,-2\,-1\ \ 0\ \ 1\ \ 2\ \ 3\ \ 4\ \ 5}{\mid\ \mid\)\ (\ \mid\ \mid\ \mid\ \mid\ \mid\ \mid\ \mid}\quad}$$

You Try It 9 This is the set of real numbers less than 1 *and* greater than -3.

$$\overleftarrow{\quad\underset{-5\,-4\,-3\,-2\,-1\ \ 0\ \ 1\ \ 2\ \ 3\ \ 4\ \ 5}{\mid\ \mid\ (\ \mid\ \mid\ \mid\)\ \mid\ \mid\ \mid\ \mid}\quad}$$

You Try It 10 **a.** The set $\{x \mid x < -1\}$ is the numbers less than -1. In interval notation, this is written $(-\infty, -1)$.
b. The set $\{x \mid -2 \leq x < 4\}$ is the numbers greater than or equal to -2 and less than 4. In interval notation, this is written $[-2, 4)$.

You Try It 11 **a.** The ∞ symbol indicates that the interval extends forever. Therefore, $(3, \infty)$ is the numbers greater than 3. In set-builder notation, this is written $\{x \mid x > 3\}$.
b. $(-4, 1]$ is the numbers greater than -4 and less than or equal to 1. In set-builder notation, this is written $\{x \mid -4 < x \leq 1\}$.

You Try It 12 Draw a bracket at -2 to show that it is in the set. The ∞ symbol indicates that the set extends forever. Draw a solid line to the right of -2.

$$\overleftarrow{\quad\underset{-5\,-4\,-3\,-2\,-1\ \ 0\ \ 1\ \ 2\ \ 3\ \ 4\ \ 5}{\mid\ \mid\ \mid\ [\ \mid\ \mid\ \mid\ \mid\ \mid\ \mid\ \mid}\quad}$$

SECTION 1.2

You Try It 1 $6 - (-8) - 10$
$$= 6 + 8 + (-10) \quad \bullet \text{ Rewrite}$$
$$= 14 + (-10) \qquad\quad \text{subtraction as}$$
$$= 4 \qquad\qquad\qquad \text{addition of the}$$
opposite. Then add.

You Try It 2 $12(-3)|-6|$
$$= 12(-3)(6) \quad \bullet \text{ Find the absolute}$$
$$= (-36)(6) \qquad \text{value of } -6.$$
$$= -216 \qquad\quad \text{Then multiply.}$$

You Try It 3
$$-\frac{28}{|-14|} = -\frac{28}{14}$$
$$= -2$$

You Try It 4

$$\frac{5}{6} - \frac{3}{8} - \frac{7}{9} = \frac{5}{6} + \frac{-3}{8} + \frac{-7}{9}$$
$$= \frac{5}{6} \cdot \frac{12}{12} + \frac{-3}{8} \cdot \frac{9}{9} + \frac{-7}{9} \cdot \frac{8}{8}$$
$$= \frac{60}{72} + \frac{-27}{72} + \frac{-56}{72}$$
$$= \frac{60 + (-27) + (-56)}{72} = \frac{-23}{72}$$
$$= -\frac{23}{72}$$

You Try It 5

$$\frac{5}{8} \div \left(-\frac{15}{40}\right) = -\frac{5}{8} \cdot \frac{40}{15}$$

$$= -\frac{5 \cdot 40}{8 \cdot 15} = -\frac{5}{3}$$

You Try It 6 **a.** $-2^5 = -(2 \cdot 2 \cdot 2 \cdot 2 \cdot 2) = -32$
b. $(-5)^2 = (-5)(-5) = 25$

You Try It 7

$$\left(\frac{2}{3}\right)^4 = \left(\frac{2}{3}\right)\left(\frac{2}{3}\right)\left(\frac{2}{3}\right)\left(\frac{2}{3}\right)$$

$$= \frac{2 \cdot 2 \cdot 2 \cdot 2}{3 \cdot 3 \cdot 3 \cdot 3}$$

$$= \frac{16}{81}$$

You Try It 8

$$-2\left[3 - 3^3 + \frac{16 + 2}{1 - 10}\right]$$

$$= -2\left[3 - 3^3 + \frac{18}{-9}\right] \quad \bullet \text{ Inside brackets}$$

$$= -2[3 - 3^3 + (-2)]$$

$$= -2[3 - 27 + (-2)]$$

$$= -2[-24 + (-2)]$$

$$= -2[-26]$$

$$= 52 \qquad\qquad \bullet \text{ Multiplication}$$

You Try It 9

$$\frac{1}{3} + \frac{5}{8} \div \frac{15}{16} - \frac{7}{12} = \frac{1}{3} + \frac{5}{8} \cdot \frac{16}{15} - \frac{7}{12} \quad \bullet \text{ Division and}$$
$$\text{multiplication}$$

$$= \frac{1}{3} + \frac{2}{3} - \frac{7}{12}$$

$$= 1 - \frac{7}{12} \qquad\qquad \bullet \text{ Addition and}$$
$$\text{subtraction}$$

$$= \frac{5}{12}$$

You Try It 10

$$\frac{11}{12} - \frac{\dfrac{5}{4}}{2 - \dfrac{7}{2}} \cdot \frac{3}{4} = \frac{11}{12} - \frac{\dfrac{5}{4}}{-\dfrac{3}{2}} \cdot \frac{3}{4} \quad \bullet \text{ Simplify the}$$
$$\text{complex fraction.}$$

$$= \frac{11}{12} \cdot \left[\frac{5}{4} \cdot \left(-\frac{2}{3}\right)\right] \cdot \frac{3}{4}$$

$$= \frac{11}{12} - \left(-\frac{5}{6}\right) \cdot \frac{3}{4}$$

$$= \frac{11}{12} - \left(-\frac{5}{8}\right) \qquad \bullet \text{ Multiplication}$$

$$= \frac{37}{24} \qquad\qquad \bullet \text{ Subtraction}$$

SECTION 1.3

You Try It 1 $3x + (-3x) = 0$

You Try It 2 The Associative Property of Addition

You Try It 3

$-2x^2 + 3(4xy - z)$
$-2(-3)^2 + 3[4(-3)(-1) - 2]$

$$= -2(-3)^2 + 3[12 - 2]$$

$$= -2(-3)^2 + 3(10)$$

$$= -2(9) + 3(10)$$

$$= -18 + 30$$

$$= 12$$

You Try It 4

$2x - y|4x^2 - y^2|$
$2(-2) - (-6)|4(-2)^2 - (-6)^2|$

$$= 2(-2) - (-6)|4 \cdot 4 - 36|$$

$$= 2(-2) - (-6)|16 - 36|$$

$$= 2(-2) - (-6)|-20|$$

$$= 2(-2) - (-6)(20)$$

$$= -4 - (-120)$$

$$= -4 + 120$$

$$= 116$$

You Try It 5

$9 - 2(3y - 4) + 5y$
$= 9 - 6y + 8 + 5y \quad \bullet \text{ The Order of Operations}$
$= -y + 17 \qquad\qquad \text{ Agreement requires}$
$\text{multiplication before subtraction.}$

You Try It 6

$6z - 3(5 - 3z) + 5(2z - 3)$
$= 6z - 15 + 9z + 10z - 15 \quad \bullet \text{ Distributive Property}$
$= 25z - 30 \qquad\qquad\qquad \bullet \text{ Combine like terms.}$

SECTION 1.4

You Try It 1 twice x divided by the difference between x and 7

$$\frac{2x}{x - 7}$$

You Try It 2 the product of negative 3 and the square of d

$$-3d^2$$

You Try It 3 The smaller number is x.
The larger number is $16 - x$.
the difference between twice the smaller number and the larger number
$2x - (16 - x) = 2x - 16 + x$
$\qquad\qquad\qquad\quad = 3x - 16$

You Try It 4 Let the number be x.
the difference between 14 and the sum of a number and 7

$14 - (x + 7) = 14 - x - 7$
$\qquad\qquad\qquad = -x + 7$

You Try It 5 the height of the Empire State Building: h
the length of *Destiny*: $h + 56$

You Try It 6 the depth of the shallow end: d
the depth of the deep end: $2d + 2$

You Try It 7 the total credit card bill: x
the cost of restaurant meals: $\dfrac{1}{4}x$

Solutions to Chapter 2 "You Try It"

SECTION 2.1

You Try It 1
$$x + 4 = -3$$
$$x + 4 - 4 = -3 - 4 \quad \bullet \text{ Subtract 4 from}$$
$$x = -7 \qquad\qquad \text{each side.}$$

The solution is -7.

You Try It 2
$$-3x = 18$$
$$\frac{-3x}{-3} = \frac{18}{-3} \quad \bullet \text{ Divide each side by } -3.$$
$$x = -6$$

The solution is -6.

You Try It 3
$$6x - 5 - 3x = 14 - 5x$$
$$3x - 5 = 14 - 5x \qquad \bullet \text{ Combine like terms.}$$
$$3x - 5 + 5x = 14 - 5x + 5x \quad \bullet \text{ Add } 5x \text{ to each side.}$$
$$8x - 5 = 14$$
$$8x - 5 + 5 = 14 + 5 \qquad \bullet \text{ Add 5 to each side.}$$
$$8x = 19$$
$$\frac{8x}{8} = \frac{19}{8} \qquad\qquad \bullet \text{ Divide each side by 8.}$$
$$x = \frac{19}{8}$$

The solution is $\dfrac{19}{8}$.

You Try It 4
$$6(5 - x) - 12 = 2x - 3(4 + x)$$
$$30 - 6x - 12 = 2x - 12 - 3x$$
$$18 - 6x = -x - 12$$
$$18 - 6x + x = -x - 12 + x$$
$$18 - 5x = -12$$
$$18 - 5x - 18 = -12 - 18$$
$$-5x = -30$$
$$\frac{-5x}{-5} = \frac{-30}{-5}$$
$$x = 6$$

The solution is 6.

You Try It 5
$$S = C - rC$$
$$S - C = -rC \qquad \bullet \text{ Subtract } C \text{ from each side.}$$
$$\frac{S - C}{-C} = r \qquad \bullet \text{ Divide each side by } -C.$$
$$r = -\frac{S - C}{C} = \frac{C - S}{C}$$

SECTION 2.2

You Try It 1

Strategy

- The first number: n
 The second number: $2n$
 The third number: $4n - 3$
- The sum of the numbers is 81.
 $n + 2n + (4n - 3) = 81$

Solution

$$n + 2n + (4n - 3) = 81$$
$$7n - 3 = 81 \qquad \bullet \text{ Combine like terms.}$$
$$7n = 84 \qquad \bullet \text{ Add 3 to each side.}$$
$$n = 12 \qquad \bullet \text{ Divide each side by 7.}$$
$$2n = 2(12) = 24$$
$$4n - 3 = 4(12) - 3 = 48 - 3 = 45$$

The numbers are 12, 24, and 45.

You Try It 2

Strategy

- First odd integer: n
 Second odd integer: $n + 2$
 Third odd integer: $n + 4$
- Three times the sum of the first two integers is ten more than the product of the third integer and four.
 $3[n + (n + 2)] = (n + 4)4 + 10$

Solution

$$3[n + (n + 2)] = (n + 4)4 + 10$$
$$3[2n + 2] = 4n + 16 + 10 \quad \bullet \text{ Add like terms and use the Distributive Property.}$$
$$6n + 6 = 4n + 26 \qquad \bullet \text{ Use the Distributive Property and add like terms.}$$
$$2n + 6 = 26 \qquad \bullet \text{ Subtract } 4n \text{ from each side.}$$
$$2n = 20 \qquad \bullet \text{ Subtract 6 from each side.}$$
$$n = 10 \qquad \bullet \text{ Divide each side by 2.}$$

Because 10 is not an odd integer, there is no solution.

You Try It 3

Strategy

- Number of 3¢ stamps: x
 Number of 10¢ stamps: $2x + 2$
 Number of 15¢ stamps: $3x$

Stamps	Number	Value	Total Value
3¢	x	3	$3x$
10¢	$2x + 2$	10	$10(2x + 2)$
15¢	$3x$	15	$45x$

- The sum of the total values of each type of stamp equals the total value of all the stamps (156 cents).
 $3x + 10(2x + 2) + 45x = 156$

Solution

$$3x + 10(2x + 2) + 45x = 156$$
$$3x + 20x + 20 + 45x = 156 \qquad \bullet \text{ Distributive Property}$$
$$68x + 20 = 156 \qquad \bullet \text{ Combine like terms.}$$
$$68x = 136 \qquad \bullet \text{ Subtract 20 from each side.}$$
$$x = 2 \qquad \bullet \text{ Divide each side by 68.}$$
$$3x = 3(2) = 6$$

There are six 15¢ stamps in the collection.

SECTION 2.3

You Try It 1

Strategy

- Pounds of $3.00 hamburger: x
 Pounds of $1.80 hamburger: $75 - x$
 Pounds of the $2.20 mixture: 75

	Amount	Cost	Value
$3.00 hamburger	x	3.00	$3.00x$
$1.80 hamburger	$75 - x$	1.80	$1.80(75 - x)$
Mixture	75	2.20	$75(2.20)$

- The sum of the values before mixing equals the value after mixing.
 $3.00x + 1.80(75 - x) = 75(2.20)$

Solution

$$3.00x + 1.80(75 - x) = 75(2.20)$$
$$3x + 135 - 1.80x = 165$$
$$1.2x + 135 = 165$$
$$1.2x = 30$$
$$x = 25$$
$$75 - x = 75 - 25 = 50$$

The mixture must contain 25 lb of the $3.00 hamburger and 50 lb of the $1.80 hamburger.

You Try It 2

Strategy

- Pounds of 22% hamburger: x
 Pounds of 12% hamburger: $80 - x$
 Pounds of 18% hamburger: 80

	Amount	Percent	Quantity
22%	x	0.22	$0.22x$
12%	$80 - x$	0.12	$0.12(80 - x)$
18%	80	0.18	$0.18(80)$

- The sum of the quantities before mixing is equal to the quantity after mixing.
 $0.22x + 0.12(80 - x) = 0.18(80)$

Solution

$$0.22x + 0.12(80 - x) = 0.18(80)$$
$$0.22x + 9.6 - 0.12x = 14.4$$
$$0.10x + 9.6 = 14.4$$
$$0.10x = 4.8 \qquad \bullet \text{ Subtract 9.6 from each side.}$$
$$x = 48 \qquad \bullet \text{ Divide each side by 0.10.}$$
$$80 - x = 80 - 48 = 32$$

The butcher needs 48 lb of the hamburger that is 22% fat and 32 lb of the hamburger that is 12% fat.

You Try It 3

Strategy

- Rate of the second plane: r
 Rate of the first plane: $r + 30$

	Rate	Time	Distance
1st plane	$r + 30$	4	$4(r + 30)$
2nd plane	r	4	$4r$

- The total distance traveled by the two planes is 1160 mi.
 $4(r + 30) + 4r = 1160$

Solution

$$4(r + 30) + 4r = 1160$$
$$4r + 120 + 4r = 1160 \qquad \bullet \text{ Distributive Property}$$
$$8r + 120 = 1160 \qquad \bullet \text{ Combine like terms.}$$
$$8r = 1040 \qquad \bullet \text{ Subtract 120 from each side.}$$
$$r = 130 \qquad \bullet \text{ Divide each side by 8.}$$
$$r + 30 = 130 + 30 = 160$$

The first plane is traveling 160 mph. The second plane is traveling 130 mph.

SECTION 2.4

You Try It 1

$$2x - 1 < 6x + 7$$
$$-4x - 1 < 7 \qquad \bullet \text{ Subtract } 6x \text{ from each side.}$$
$$-4x < 8 \qquad \bullet \text{ Add 1 to each side.}$$
$$\frac{-4x}{-4} > \frac{8}{-4} \qquad \bullet \text{ Divide each side by } -4.$$
$$x > -2$$
$$\{x \,|\, x > -2\}$$

Number line from −5 to 5 with open parenthesis at −2 and arrow extending right.

You Try It 2
$$5x - 2 \le 4 - 3(x - 2)$$
$$5x - 2 \le 4 - 3x + 6$$
$$5x - 2 \le 10 - 3x$$
$$8x - 2 \le 10$$
$$8x \le 12$$
$$\frac{8x}{8} \le \frac{12}{8}$$
$$x \le \frac{3}{2}$$
$$\left(-\infty, \frac{3}{2}\right]$$

You Try It 3
$$-2 \le 5x + 3 \le 13$$
$$-2 - 3 \le 5x + 3 - 3 \le 13 - 3 \quad \bullet \text{ Subtract 3 from each of the three parts.}$$
$$-5 \le 5x \le 10$$
$$\frac{-5}{5} \le \frac{5x}{5} \le \frac{10}{5} \quad \bullet \text{ Divide each of the three parts by 5.}$$
$$-1 \le x \le 2$$
$$[-1, 2]$$

You Try It 4
$$2 - 3x > 11 \quad \text{or} \quad 5 + 2x > 7$$
$$-3x > 9 \qquad\qquad 2x > 2$$
$$x < -3 \qquad\qquad x > 1$$
$$\{x \,|\, x < -3\} \qquad \{x \,|\, x > 1\}$$
$$\{x \,|\, x < -3\} \cup \{x \,|\, x > 1\}$$
$$= \{x \,|\, x < -3 \text{ or } x > 1\}$$

You Try It 5

Strategy To find the maximum height, substitute the given values in the inequality $\frac{1}{2}bh < A$ and solve.

Solution
$$\frac{1}{2}bh < A$$
$$\frac{1}{2}(12)(x + 2) < 50$$
$$6(x + 2) < 50$$
$$6x + 12 < 50$$
$$6x < 38$$
$$x < \frac{19}{3}$$

The largest integer less than $\frac{19}{3}$ is 6.

$$x + 2 = 6 + 2 = 8$$

The maximum height of the triangle is 8 in.

You Try It 6

Strategy To find the range of scores, write and solve an inequality using N to represent the score on the last test.

Solution
$$80 \le \frac{72 + 94 + 83 + 70 + N}{5} \le 89$$
$$80 \le \frac{319 + N}{5} \le 89$$
$$5 \cdot 80 \le 5\left(\frac{319 + N}{5}\right) \le 5 \cdot 89$$
$$400 \le 319 + N \le 445$$
$$400 - 319 \le 319 + N - 319 \le 445 - 319$$
$$81 \le N \le 126$$

Because 100 is the maximum score, the range of scores to receive a B grade is $81 \le N \le 100$.

SECTION 2.5

You Try It 1
$$|2x - 3| = 5$$

$$2x - 3 = 5 \qquad 2x - 3 = -5$$
$$2x = 8 \qquad 2x = -2 \qquad \bullet \text{ Add 3.}$$
$$x = 4 \qquad x = -1 \qquad \bullet \text{ Divide by 2.}$$

The solutions are 4 and -1.

You Try It 2 $|x - 3| = -2$

There is no solution to this equation because the absolute value of a number must be nonnegative.

You Try It 3
$$5 - |3x + 5| = 3$$
$$-|3x + 5| = -2 \qquad \bullet \text{ Subtract 5.}$$
$$|3x + 5| = 2 \qquad \bullet \text{ Multiply by } -1.$$
$$3x + 5 = 2 \qquad 3x + 5 = -2$$
$$3x = -3 \qquad 3x = -7$$
$$x = -1 \qquad x = -\frac{7}{3}$$

The solutions are -1 and $-\frac{7}{3}$.

You Try It 4 $|3x + 2| < 8$

$$-8 < 3x + 2 < 8$$
$$-8 - 2 < 3x + 2 - 2 < 8 - 2$$
$$-10 < 3x < 6$$
$$\frac{-10}{3} < \frac{3x}{3} < \frac{6}{3}$$
$$-\frac{10}{3} < x < 2$$
$$\left\{ x \,\middle|\, -\frac{10}{3} < x < 2 \right\}$$

You Try It 7
$|5x + 3| > 8$

$5x + 3 < -8$	or	$5x + 3 > 8$
$5x < -11$		$5x > 5$
$x < -\dfrac{11}{5}$		$x > 1$

$$\left\{ x \,\middle|\, x < -\frac{11}{5} \right\} \qquad \{x \,|\, x > 1\}$$
$$\left\{ x \,\middle|\, x < -\frac{11}{5} \right\} \cup \{x \,|\, x > 1\}$$
$$= \left\{ x \,\middle|\, x < -\frac{11}{5} \text{ or } x > 1 \right\}$$

You Try It 5 $|3x - 7| < 0$

The absolute value of a number must be nonnegative.

The solution set is the empty set.

\varnothing

You Try It 8

Strategy

Let b represent the diameter of the bushing, T the tolerance, and d the lower and upper limits of the diameter. Solve the absolute value inequality $|d - b| \leq T$ for d.

Solution

$$|d - b| \leq T$$
$$|d - 2.55| \leq 0.003$$
$$-0.003 \leq d - 2.55 \leq 0.003$$
$$-0.003 + 2.55 \leq d - 2.55 + 2.55 \leq 0.003 + 2.55$$
$$2.547 \leq d \leq 2.553$$

The lower and upper limits of the diameter of the bushing are 2.547 in. and 2.553 in.

You Try It 6 $|2x + 7| \geq -1$

The absolute value of a number is nonnegative.

The solution set is the set of real numbers.

Solutions to Chapter 3 "You Try It"

SECTION 3.1

You Try It 1 $y = \dfrac{3x}{x + 1}$

Replace x by -2 and solve for y.

$$y = \frac{3(-2)}{-2 + 1} = \frac{-6}{-1} = 6$$

The ordered-pair solution is $(-2, 6)$.

You Try It 2 $y = |x + 1|$

x	y
-3	2
-2	1
-1	0
0	1
1	2

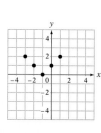

You Try It 3 $(x_1, y_1) = (5, -2)$ $(x_2, y_2) = (-4, 3)$

$$d = \sqrt{(x_1 - x_2)^2 + (y_1 - y_2)^2}$$
$$= \sqrt{[5 - (-4)]^2 + (-2 - 3)^2}$$
$$= \sqrt{9^2 + (-5)^2}$$
$$= \sqrt{81 + 25}$$
$$= \sqrt{106} \approx 10.30$$

You Try It 4 $(x_1, y_1) = (-3, -5)$ $(x_2, y_2) = (-2, 3)$

$$x_m = \frac{x_1 + x_2}{2} \qquad y_m = \frac{y_1 + y_2}{2}$$
$$= \frac{-3 + (-2)}{2} \qquad = \frac{-5 + 3}{2}$$
$$= -\frac{5}{2} \qquad = -1$$

The midpoint is $\left(-\dfrac{5}{2}, -1 \right)$.

You Try It 5

Strategy To draw a scatter diagram:
- Draw a coordinate grid with the horizontal axis representing the year and the vertical axis representing the number of deaths.
- Graph the ordered pairs $(1998, 14)$, $(1999, 28)$, $(2000, 17)$, $(2001, 18)$, and $(2002, 23)$.

Solution

SECTION 3.2

You Try It 1 Domain: $\{-1, 3, 4, 6\}$
Range: $\{5\}$

- The domain is the set of first coordinates.

You Try It 2 $G(x) = \dfrac{3x}{x + 2}$

$G(-4) = \dfrac{3(-4)}{-4 + 2} = \dfrac{-12}{-2} = 6$

You Try It 3 $f(x) = x^2 - 11$
$f(3h) = (3h)^2 - 11$
$\quad\quad = 9h^2 - 11$

You Try It 4 $h(z) = 3z + 1$

$h(0) = 3(0) + 1 = 1$

$h\left(\dfrac{1}{3}\right) = 3\left(\dfrac{1}{3}\right) + 1 = 2$

$h\left(\dfrac{2}{3}\right) = 3\left(\dfrac{2}{3}\right) + 1 = 3$

$h(1) = 3(1) + 1 = 4$

The range is $\{1, 2, 3, 4\}$.

You Try It 5 $f(x) = \dfrac{2}{x - 5}$

For $x = 5, 5 - 5 = 0$.

$f(5) = \dfrac{2}{5 - 5} = \dfrac{2}{0}$, which is not a real number.

5 is excluded from the domain of the function.

SECTION 3.3

You Try It 1

You Try It 2

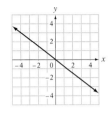

You Try It 3 $-3x + 2y = 4$

$2y = 3x + 4$

$y = \dfrac{3}{2}x + 2$

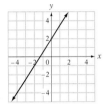

You Try It 4 $y - 3 = 0$
$\quad\quad\quad y = 3$

- The graph of $y = 3$ goes through the point $(0, 3)$. .

You Try It 5

x-intercept:		y-intercept:	
$3x - y = 2$		$3x - y = 2$	
$3x - 0 = 2$	• Let	$3(0) - y = 2$	• Let
$3x = 2$	$y = 0$.	$-y = 2$	$x = 0$.
$x = \dfrac{2}{3}$		$y = -2$	

x-intercept: $\left(\dfrac{2}{3}, 0\right)$ y-intercept: $(0, -2)$

You Try It 6 *x*-intercept: *y*-intercept:

$$y = \frac{1}{4}x + 1 \qquad (0, b)$$
$$b = 1$$
$$0 = \frac{1}{4}x + 1 \qquad (0, 1)$$

$$-\frac{1}{4}x = 1$$

$$x = -4$$

$$(-4, 0)$$

You Try It 4 $2x + 3y = 6$
$$3y = -2x + 6$$
$$y = -\frac{2}{3}x + 2$$
$$m = -\frac{2}{3} = \frac{-2}{3}$$
y-intercept $= (0, 2)$

You Try It 7

... (wait)

You Try It 7

● Graph $h = \frac{3}{4}L + 50$.

When $L = 20$, $h = 65$. When $L = 40$, $h = 80$.

The ordered pair (32, 74) means that a person with a stride of 32 in. is 74 in. tall.

You Try It 5 $(x_1, y_1) = (-3, -2)$
$$m = 3$$

SECTION 3.4

You Try It 1 Let $P_1 = (4, -3)$ and $P_2 = (2, 7)$.

$$m = \frac{y_2 - y_1}{x_2 - x_1} = \frac{7 - (-3)}{2 - 4} = \frac{10}{-2} = -5$$

The slope is -5.

You Try It 2 Let $P_1 = (6, -1)$ and $P_2 = (6, 7)$.
$$m = \frac{y_2 - y_1}{x_2 - x_1} = \frac{7 - (-1)}{6 - 6} = \frac{8}{0}$$

Division by zero is not defined.

The slope of the line is undefined.

You Try It 3 $P_1 = (5, 25{,}000), P_2 = (2, 55{,}000)$

$$m = \frac{y_2 - y_1}{x_2 - x_1}$$

$$= \frac{55{,}000 - 25{,}000}{2 - 5}$$ ● $(x_1, y_1) = (2, 55{,}000),$
$(x_2, y_2) = (5, 25{,}000)$

$$= \frac{30{,}000}{-3}$$

$$= -10{,}000$$

A slope of $-10{,}000$ means that the value of the recycling truck is decreasing by \$10,000 per year.

SECTION 3.5

You Try It 1 $m = -\frac{1}{3}$ $(x_1, y_1) = (-3, -2)$

$$y - y_1 = m(x - x_1)$$

$$y - (-2) = -\frac{1}{3}[x - (-3)]$$

$$y + 2 = -\frac{1}{3}(x + 3)$$

$$y + 2 = -\frac{1}{3}x - 1$$

$$y = -\frac{1}{3}x - 3$$

The equation of the line is $y = -\frac{1}{3}x - 3$.

You Try It 2 $m = -3$ $(x_1, y_1) = (4, -3)$

$$y - y_1 = m(x - x_1)$$
$$y - (-3) = -3(x - 4)$$
$$y + 3 = -3x + 12$$
$$y = -3x + 9$$

The equation of the line is $y = -3x + 9$.

You Try It 3 Let $(x_1, y_1) = (2, 0)$ and $(x_2, y_2) = (5, 3)$.

$$m = \frac{y_2 - y_1}{x_2 - x_1} = \frac{3 - 0}{5 - 2} = \frac{3}{3} = 1$$

$$y - y_1 = m(x - x_1)$$
$$y - 0 = 1(x - 2)$$
$$y = 1(x - 2)$$
$$y = x - 2$$

The equation of the line is $y = x - 2$.

You Try It 4 Let $(x_1, y_1) = (2, 3)$ and
$(x_2, y_2) = (-5, 3)$.

$$m = \frac{y_2 - y_1}{x_2 - x_1} = \frac{3 - 3}{-5 - 2} = \frac{0}{-7} = 0$$

The line has zero slope.
The line is a horizontal line.
All points on the line have an
ordinate of 3.
The equation of the line is $y = 3$.

You Try It 5

Strategy • Select the independent and
dependent variables. The function
is to predict the Celsius
temperature, so that quantity is the
dependent variable, y. The
Fahrenheit temperature is the
independent variable, x.
• From the given data, two ordered
pairs are (212, 100) and (32, 0).
Use these ordered pairs to
determine the linear function.

Solution Let $(x_1, y_1) = (32, 0)$ and
$(x_2, y_2) = (212, 100)$.

$$m = \frac{y_2 - y_1}{x_2 - x_1} = \frac{100 - 0}{212 - 32} = \frac{100}{180} = \frac{5}{9}$$
$$y - y_1 = m(x - x_1)$$
$$y - 0 = \frac{5}{9}(x - 32)$$
$$y = \frac{5}{9}(x - 32), \text{ or } C = \frac{5}{9}(F - 32).$$

The linear function is
$f(F) = \frac{5}{9}(F - 32).$

SECTION 3.6

You Try It 1
$$m_1 = \frac{1 - (-3)}{7 - (-2)} = \frac{4}{9}$$

 • $(x_1, y_1) = (-2, -3),$
 $(x_2, y_2) = (7, 1)$

$$m_2 = \frac{-5 - 1}{6 - 4} = \frac{-6}{2} = -3$$

 • $(x_1, y_1) = (4, 1),$
 $(x_2, y_2) = (6, -5)$

$$m_1 \cdot m_2 = \frac{4}{9}(-3) = -\frac{4}{3}$$

No, the lines are not perpendicular.

You Try It 2
$$5x + 2y = 2$$
$$2y = -5x + 2$$
$$y = -\frac{5}{2}x + 1$$
$$m_1 = -\frac{5}{2}$$
$$5x + 2y = -6$$
$$2y = -5x - 6$$
$$y = -\frac{5}{2}x - 3$$
$$m_2 = -\frac{5}{2}$$
$$m_1 = m_2 = -\frac{5}{2}$$

Yes, the lines are parallel.

You Try It 3
$$x - 4y = 3$$
$$-4y = -x + 3$$
$$y = \frac{1}{4}x - \frac{3}{4}$$
$$m_1 = \frac{1}{4}$$
$$m_1 \cdot m_2 = -1$$
$$\frac{1}{4} \cdot m_2 = -1$$
$$m_2 = -4$$
$$y - y_1 = m(x - x_1)$$
$$y - 2 = -4[x - (-2)] \quad \bullet \ (x_1, y_1) = (-2, 2)$$
$$y - 2 = -4(x + 2)$$
$$y - 2 = -4x - 8$$
$$y = -4x - 6$$

The equation of the line is
$y = -4x - 6.$

SECTION 3.7

You Try It 1
$$x + 3y > 6$$
$$3y > -x + 6$$
$$y > -\frac{1}{3}x + 2$$

You Try It 2 $y < 2$

Solutions to Chapter 4 "You Try It"

SECTION 4.1

You Try It 1

- Find the point of intersection of the graphs of the equations.

The solution is $(-1, 2)$.

You Try It 2

- Graph the two equations.

The lines are parallel and therefore do not intersect. The system of equations has no solution. The system of equations is inconsistent.

You Try It 3

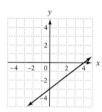

- Graph the two equations.

The two equations represent the same line. The system of equations is dependent. The solutions are the ordered pairs $\left(x, \frac{3}{4}x - 3\right)$.

You Try It 4

(1) $3x - y = 3$
(2) $6x + 3y = -4$

Solve Equation (1) for y.

$3x - y = 3$
$-y = -3x + 3$
$y = 3x - 3$

Substitute into Equation (2).

$6x + 3y = -4$
$6x + 3(3x - 3) = -4$
$6x + 9x - 9 = -4$
$15x - 9 = -4$
$15x = 5$
$x = \dfrac{5}{15} = \dfrac{1}{3}$

Substitute the value of x into Equation (1).

$3x - y = 3$
$3\left(\dfrac{1}{3}\right) - y = 3$
$1 - y = 3$
$-y = 2$
$y = -2$

The solution is $\left(\dfrac{1}{3}, -2\right)$.

You Try It 5

(1) $y = 2x - 3$
(2) $3x - 2y = 6$
 $3x - 2y = 6$
$3x - 2(2x - 3) = 6$
 $3x - 4x + 6 = 6$
 $-x + 6 = 6$
 $-x = 0$
 $x = 0$

Substitute the value of x into Equation (1).

$y = 2x - 3$
$y = 2(0) - 3$
$y = 0 - 3$
$y = -3$

The solution is $(0, -3)$.

- Graph the two equations.

You Try It 6

(1) $6x - 3y = 6$
(2) $2x - y = 2$

Solve Equation (2) for y.

$2x - y = 2$
$-y = -2x + 2$
$y = 2x - 2$

Substitute into Equation (1).

$6x - 3y = 6$
$6x - 3(2x - 2) = 6$
$6x - 6x + 6 = 6$
$6 = 6$

The system of equations is dependent. The solutions are the ordered pairs $(x, 2x - 2)$.

● Graph the two equations.

You Try It 7

Strategy ● Amount invested at 4.2%: x
Amount invested at 6%: y

	Principal	Rate	Interest
Amount at 4.2%	x	0.042	$0.042x$
Amount at 6%	y	0.06	$0.06y$

● The total investment is $13,600. The two accounts earn the same interest.

Solution $x + y = 13,600$
$0.042x = 0.06y$
$x = \dfrac{10}{7}y$

Substitute $\dfrac{10}{7}y$ for x in
$x + y = 13,600$ and solve for y.

$\dfrac{10}{7}y + y = 13,600$

$\dfrac{17}{7}y = 13,600$

$y = 5600$

$x + 5600 = 13,600$
$x = 8000$

$8000 must be invested at 4.2% and $5600 invested at 6%.

SECTION 4.2

You Try It 1 (1) $2x + 5y = 6$
(2) $3x - 2y = 6x + 2$

Write Equation (2) in the form
$Ax + By = C$.

$3x - 2y = 6x + 2$
$-3x - 2y = 2$

Solve the system: $2x + 5y = 6$
$-3x - 2y = 2$

Eliminate y.
$2(2x + 5y) = 2(6)$
$5(-3x - 2y) = 5(2)$

$4x + 10y = 12$
$-15x - 10y = 10$
$\overline{-11x = 22}$ ● Add the equations.
$x = -2$ ● Solve for x.

Replace x in Equation (1).
$2x + 5y = 6$
$2(-2) + 5y = 6$
$-4 + 5y = 6$
$5y = 10$
$y = 2$

The solution is $(-2, 2)$.

You Try It 2 $2x + y = 5$
$4x + 2y = 6$

Eliminate y.
$-2(2x + y) = -2(5)$
$4x + 2y = 6$

$-4x - 2y = -10$
$4x + 2y = 6$
$\overline{0x + 0y = -4}$ ● Add the equations.
$0 = -4$

This is not a true equation.
The system is inconsistent and therefore has no solution.

You Try It 3 (1) $x - y + z = 6$
(2) $2x + 3y - z = 1$
(3) $x + 2y + 2z = 5$

Eliminate z. Add Equations (1) and (2).
$x - y + z = 6$
$2x + 3y - z = 1$
$\overline{3x + 2y = 7}$ ● Equation (4)

Multiply Equation (2) by 2 and add to Equation (3).
$4x + 6y - 2z = 2$
$x + 2y + 2z = 5$
$\overline{5x + 8y = 7}$ ● Equation (5)

Solve the system of two equations.
(4) $3x + 2y = 7$
(5) $5x + 8y = 7$

Multiply Equation (4) by -4 and add to Equation (5).
$-12x - 8y = -28$
$5x + 8y = 7$
$\overline{-7x = -21}$
$x = 3$

(Continued)

(Continued)

Replace x by 3 in Equation (4).

$$3x + 2y = 7$$
$$3(3) + 2y = 7$$
$$9 + 2y = 7$$
$$2y = -2$$
$$y = -1$$

Replace x by 3 and y by -1 in Equation (1).

$$x - y + z = 6$$
$$3 - (-1) + z = 6$$
$$4 + z = 6$$
$$z = 2$$

The solution is $(3, -1, 2)$.

SECTION 4.3

You Try It 1

$$\begin{vmatrix} -1 & -4 \\ 3 & -5 \end{vmatrix} = -1(-5) - (-4)(3)$$
$$= 5 + 12 = 17$$

The value of the determinant is 17.

You Try It 2 Expand by cofactors of the first row.

$$\begin{vmatrix} 1 & 4 & -2 \\ 3 & 1 & 1 \\ 0 & -2 & 2 \end{vmatrix}$$

$$= 1\begin{vmatrix} 1 & 1 \\ -2 & 2 \end{vmatrix} - 4\begin{vmatrix} 3 & 1 \\ 0 & 2 \end{vmatrix} + (-2)\begin{vmatrix} 3 & 1 \\ 0 & -2 \end{vmatrix}$$
$$= 1(2 + 2) - 4(6 - 0) - 2(-6 - 0)$$
$$= 4 - 24 + 12$$
$$= -8$$

The value of the determinant is -8.

You Try It 3

$$\begin{vmatrix} 3 & -2 & 0 \\ 1 & 4 & 2 \\ -2 & 1 & 3 \end{vmatrix}$$

$$= 3\begin{vmatrix} 4 & 2 \\ 1 & 3 \end{vmatrix} - (-2)\begin{vmatrix} 1 & 2 \\ -2 & 3 \end{vmatrix} + 0\begin{vmatrix} 1 & 4 \\ -2 & 1 \end{vmatrix}$$
$$= 3(12 - 2) + 2(3 + 4) + 0$$
$$= 3(10) + 2(7)$$
$$= 30 + 14$$
$$= 44$$

The value of the determinant is 44.

You Try It 4

$$3x - y = 4$$
$$6x - 2y = 5$$

$$D = \begin{vmatrix} 3 & -1 \\ 6 & -2 \end{vmatrix} = 0$$

Because $D = 0$, $\dfrac{D_x}{D}$ is undefined.

Therefore, the system of equations is dependent or inconsistent. That is, the system is not independent.

You Try It 5

$$2x - y + z = -1$$
$$3x + 2y - z = 3$$
$$x + 3y + z = -2$$

$$D = \begin{vmatrix} 2 & -1 & 1 \\ 3 & 2 & -1 \\ 1 & 3 & 1 \end{vmatrix} = 21$$

$$D_x = \begin{vmatrix} -1 & -1 & 1 \\ 3 & 2 & -1 \\ -2 & 3 & 1 \end{vmatrix} = 9$$

$$D_y = \begin{vmatrix} 2 & -1 & 1 \\ 3 & 3 & -1 \\ 1 & -2 & 1 \end{vmatrix} = -3$$

$$D_z = \begin{vmatrix} 2 & -1 & -1 \\ 3 & 2 & 3 \\ 1 & 3 & -2 \end{vmatrix} = -42$$

$$x = \frac{D_x}{D} = \frac{9}{21} = \frac{3}{7}$$

$$y = \frac{D_y}{D} = \frac{-3}{21} = -\frac{1}{7}$$

$$z = \frac{D_z}{D} = \frac{-42}{21} = -2$$

The solution is $\left(\dfrac{3}{7}, -\dfrac{1}{7}, -2\right)$.

SECTION 4.4

You Try It 1

Strategy

- Rate of the rowing team in calm water: t
 Rate of the current: c

	Rate	Time	Distance
With current	$t + c$	2	$2(t + c)$
Against current	$t - c$	2	$2(t - c)$

- The distance traveled with the current is 18 mi.
 The distance traveled against the current is 10 mi.

$$2(t + c) = 18$$
$$2(t - c) = 10$$

Solution

$$2(t + c) = 18 \qquad \frac{1}{2} \cdot 2(t + c) = \frac{1}{2} \cdot 18$$

$$2(t - c) = 10 \qquad \frac{1}{2} \cdot 2(t - c) = \frac{1}{2} \cdot 10$$

$$t + c = 9$$
$$\underline{t - c = 5}$$
$$2t = 14$$
$$t = 7$$

$$t + c = 9$$
$$7 + c = 9 \qquad \bullet \text{ Substitute 7 for } t.$$
$$c = 2$$

The rate of the rowing team in calm water is 7 mph.
The rate of the current is 2 mph.

You Try It 2

Strategy

- Number of dimes: d
 Number of nickels: n
 Number of quarters: q
- There are 19 coins in a bank that contains only nickels, dimes, and quarters. ($n + d + q = 19$)
 The value of the coins is \$2. ($5n + 10d + 25q = 200$)
 There are twice as many nickels as dimes. ($n = 2d$)

Solution

(1) $n + d + q = 19$
(2) $n = 2d$
(3) $5n + 10d + 25q = 200$

Solve the system of equations. Substitute $2d$ for n in Equation (1) and Equation (3).

$$2d + d + q = 19$$
$$5(2d) + 10d + 25q = 200$$

(4) $3d + q = 19$
(5) $20d + 25q = 200$

Solve the system of equations in two variables by multiplying Equation (4) by -25 and then adding it to Equation (5).

$$-75d - 25q = -475$$
$$\underline{20d + 25q = 200}$$
$$-55d = -275$$
$$d = 5$$

Substituting the value of d into Equation (2), we have $n = 10$. Substituting the values of d and n into Equation (1), we have $q = 4$.

There are 10 nickels, 5 dimes, and 4 quarters in the bank.

SECTION 4.5

You Try It 1

Shade above the solid line $y = 2x - 3$.
Shade above the dotted line $y = -3x$.

The solution set of the system is the intersection of the solution sets of the individual inequalities.

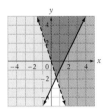

You Try It 2

$$3x + 4y > 12$$
$$4y > -3x + 12$$
$$y > -\frac{3}{4}x + 3$$

Shade above the dotted line

$$y = -\frac{3}{4}x + 3.$$

Shade below the dotted line

$$y = \frac{3}{4}x - 1.$$

The solution set of the system is the intersection of the solution sets of the individual inequalities.

Solutions to Chapter 5 "You Try It"

SECTION 5.1

You Try It 1
$$(-3a^2b^4)(-2ab^3)^4 = (-3a^2b^4)[(-2)^4a^4b^{12}]$$
$$= (-3a^2b^4)(16a^4b^{12})$$
$$= -48a^6b^{16}$$

You Try It 2 $(y^{n-3})^2 = y^{(n-3)2}$ • **Multiply the exponents.**
 $= y^{2n-6}$

You Try It 3 $[(ab^3)^3]^4 = [a^3b^9]^4$
 $= a^{12}b^{36}$

You Try It 4
$$\frac{20r^{-2}t^{-5}}{-16r^{-3}s^{-2}} = -\frac{4 \cdot 5r^{-2-(-3)}s^2t^{-5}}{4 \cdot 4}$$
$$= -\frac{5rs^2}{4t^5}$$

You Try It 5
$$\frac{(9u^{-6}v^4)^{-1}}{(6u^{-3}v^{-2})^{-2}} = \frac{9^{-1}u^6v^{-4}}{6^{-2}u^6v^4}$$
$$= 9^{-1} \cdot 6^2u^0v^{-8}$$
$$= \frac{36}{9v^8}$$
$$= \frac{4}{v^8}$$

You Try It 6

$$\frac{a^{2n+1}}{a^{n+3}} = a^{2n+1-(n+3)}$$

• Subtract the exponents.

$$= a^{2n+1-n-3}$$

$$= a^{n-2}$$

You Try It 7 $942,000,000 = 9.42 \times 10^8$

You Try It 8 $2.7 \times 10^{-5} = 0.000027$

You Try It 9

$$\frac{5,600,000 \times 0.000000081}{900 \times 0.000000028}$$

$$= \frac{5.6 \times 10^6 \times 8.1 \times 10^{-8}}{9 \times 10^2 \times 2.8 \times 10^{-8}}$$

$$= \frac{(5.6)(8.1) \times 10^{6+(-8)-2-(-8)}}{(9)(2.8)}$$

$$= 1.8 \times 10^4 = 18,000$$

You Try It 10

Strategy To find the number of arithmetic operations:
• Find the reciprocal of 1×10^{-7}, which is the number of operations performed in 1 s.
• Write the number of seconds in 1 min (60) in scientific notation.
• Multiply the number of arithmetic operations per second by the number of seconds in 1 min.

Solution $\dfrac{1}{1 \times 10^{-7}} = 10^7$

$60 = 6 \times 10$

$6 \times 10 \times 10^7$

6×10^8

The computer can perform 6×10^8 operations in 1 min.

SECTION 5.2

You Try It 1
$R(x) = -2x^4 - 5x^3 + 2x - 8$
$R(2) = -2(2)^4 - 5(2)^3 + 2(2) - 8$ • Replace x by 2.
$\quad = -2(16) - 5(8) + 4 - 8$ Simplify.
$\quad = -32 - 40 + 4 - 8$
$\quad = -76$

You Try It 2 The leading coefficient is -3, the constant term is -12, and the degree is 4.

You Try It 3 **a.** Yes, this is a polynomial function.
b. No, this is not a polynomial function. A polynomial function does not have a variable expression raised to a negative power.

c. No, this is not a polynomial function. A polynomial function does not have a variable expression within a radical.

You Try It 4

x	y
-4	5
-3	0
-2	-3
-1	-4
0	-3
1	0
2	5

You Try It 5

x	y
-3	28
-2	9
-1	2
0	1
1	0
2	-7
3	-26

You Try It 6

$$\begin{aligned}-3x^2 - \ \ 4x + \ \ 9 \\ -5x^2 - \ \ 7x + \ \ 1 \\ \hline -8x^2 - 11x + 10\end{aligned}$$

You Try It 7 Add the additive inverse of $6x^2 + 3x - 7$ to $-5x^2 + 2x - 3$.

$$\begin{aligned}-5x^2 + 2x - 3 \\ -6x^2 - 3x + 7 \\ \hline -11x^2 - \ \ x + 4\end{aligned}$$

You Try It 8
$S(x) = (4x^3 - 3x^2 + 2) + (-2x^2 + 2x - 3)$
$\quad = 4x^3 - 5x^2 + 2x - 1$

$S(-1) = 4(-1)^3 - 5(-1)^2 + 2(-1) - 1$
$\quad = 4(-1) - 5(1) - 2 - 1$
$\quad = -4 - 5 - 2 - 1$
$\quad = -12$

You Try It 9
$D(x) = P(x) - R(x)$

$D(x) = (5x^{2n} - 3x^n - 7) - (-2x^{2n} - 5x^n + 8)$
$\quad = (5x^{2n} - 3x^n - 7) + (2x^{2n} + 5x^n - 8)$
$\quad = 7x^{2n} + 2x^n - 15$

SECTION 5.3

You Try It 1
$(2b^2 - 7b - 8)(-5b)$
$\quad = 2b^2(-5b) - 7b(-5b) - 8(-5b)$ • Use the
$\quad = -10b^3 + 35b^2 + 40b$ Distributive Property.

You Try It 2 $x^2 - 2x[x - x(4x - 5) + x^2]$
$$= x^2 - 2x[x - 4x^2 + 5x + x^2]$$
$$= x^2 - 2x[6x - 3x^2]$$
$$= x^2 - 12x^2 + 6x^3$$
$$= 6x^3 - 11x^2$$

You Try It 3
$y^{n+3}(y^{n-2} - 3y^2 + 2)$
$$= y^{n+3}(y^{n-2}) - (y^{n+3})(3y^2) + (y^{n+3})(2) \quad \bullet \text{ Use the}$$
$$= y^{n+3+(n-2)} - 3y^{n+3+2} + 2y^{n+3} \qquad\qquad \textbf{Distributive}$$
$$= y^{2n+1} - 3y^{n+5} + 2y^{n+3} \qquad\qquad\qquad \textbf{Property.}$$

You Try It 4
$$\begin{array}{r} -2b^2 + 5b - 4 \\ -3b + 2 \\ \hline -4b^2 + 10b - 8 \\ 6b^3 - 15b^2 + 12b \\ \hline 6b^3 - 19b^2 + 22b - 8 \end{array}$$
$\bullet \ 2(-2b^2 + 5b - 4)$
$\bullet \ -3b(-2b^2 + 5b - 4)$

You Try It 5
$(3x - 4)(2x - 3) = 6x^2 - 9x - 8x + 12 \quad \bullet \text{ FOIL}$
$$= 6x^2 - 17x + 12$$

You Try It 6
$(2x^n + y^n)(x^n - 4y^n)$
$$= 2x^{2n} - 8x^n y^n + x^n y^n - 4y^{2n} \quad \bullet \text{ FOIL}$$
$$= 2x^{2n} - 7x^n y^n - 4y^{2n}$$

You Try It 7 $(3x - 7)(3x + 7)$ \bullet **The sum and differ-**
$$= 9x^2 - 49 \qquad\qquad \textbf{ence of two terms}$$

You Try It 8 $(2x^n + 3)(2x^n - 3)$ \bullet **The sum and differ-**
$$= 4x^{2n} - 9 \qquad\qquad \textbf{ence of two terms}$$

You Try It 9 $(3x - 4y)^2$ \bullet **The square**
$$= 9x^2 - 24xy + 16y^2 \qquad \textbf{of a binomial}$$

You Try It 10 $(2x^n - 8)^2$ \bullet **The square**
$$= 4x^{2n} - 32x^n + 64 \qquad \textbf{of a binomial}$$

You Try It 11

Strategy To find the area, replace the variables b and h in the equation $A = \frac{1}{2}bh$ by the given values and solve for A.

Solution $A = \frac{1}{2}bh$

$A = \frac{1}{2}(2x + 6)(x - 4)$

$A = (x + 3)(x - 4)$

$A = x^2 - 4x + 3x - 12 \quad \bullet \text{ FOIL}$

$A = x^2 - x - 12$

The area is $(x^2 - x - 12)$ ft^2.

You Try It 12

Strategy To find the volume, subtract the volume of the small rectangular solid from the volume of the large rectangular solid.

Large rectangular solid:
Length $= L_1 = 12x$
Width $= W_1 = 7x + 2$
Height $= H_1 = 5x - 4$
Small rectangular solid:
Length $= L_2 = 12x$
Width $= W_2 = x$
Height $= H_2 = 2x$

Solution
V = Volume of large rectangular solid $-$ volume of small rectangular solid
$V = (L_1 \cdot W_1 \cdot H_1) - (L_2 \cdot W_2 \cdot H_2)$
$V = (12x)(7x + 2)(5x - 4) - (12x)(x)(2x)$
$\quad = (84x^2 + 24x)(5x - 4) - (12x^2)(2x)$
$\quad = (420x^3 - 336x^2 + 120x^2 - 96x) - 24x^3$
$\quad = 396x^3 - 216x^2 - 96x$

The volume is $(396x^3 - 216x^2 - 96x)$ ft^3.

You Try It 13

Strategy To find the area, replace the variable r in the equation $A = \pi r^2$ by the given value and solve for A.

Solution $A = \pi r^2$
$A \approx 3.14(2x + 3)^2$
$\quad = 3.14(4x^2 + 12x + 9)$
$\quad = 12.56x^2 + 37.68x + 28.26$

The area is $(12.56x^2 + 37.68x + 28.26)$ cm^2.

SECTION 5.4

You Try It 1
$$\frac{4x^3y + 8x^2y^2 - 4xy^3}{2xy} = \frac{4x^3y}{2xy} + \frac{8x^2y^2}{2xy} - \frac{4xy^3}{2xy}$$
$$= 2x^2 + 4xy - 2y^2$$
Check: $2xy(2x^2 + 4xy - 2y^2) = 4x^3y + 8x^2y^2 - 4xy^3$

You Try It 2
$$\begin{array}{r} 5x - 1 \\ 3x + 4 \overline{)15x^2 + 17x - 20} \\ \underline{15x^2 + 20x} \\ -3x - 20 \\ \underline{-3x - 4} \\ -16 \end{array}$$

$$\frac{15x^2 + 17x - 20}{3x + 4} = 5x - 1 - \frac{16}{3x + 4}$$

You Try It 3

$$3x - 1\overline{)3x^3 + 8x^2 - 6x + 2}$$

with quotient $x^2 + 3x - 1$

$$\underline{3x^3 - x^2}$$
$$9x^2 - 6x$$
$$\underline{9x^2 - 3x}$$
$$-3x + 2$$
$$\underline{-3x + 1}$$
$$1$$

$$\frac{3x^3 + 8x^2 - 6x + 2}{3x - 1} = x^2 + 3x - 1 + \frac{1}{3x - 1}$$

You Try It 4

$$x^2 - 3x + 2\overline{)3x^4 - 11x^3 + 16x^2 - 16x + 8}$$

with quotient $3x^2 - 2x + 4$

$$\underline{3x^4 - 9x^3 + 6x^2}$$
$$-2x^3 + 10x^2 - 16x$$
$$\underline{-2x^3 + 6x^2 - 4x}$$
$$4x^2 - 12x + 8$$
$$\underline{4x^2 - 12x + 8}$$
$$0$$

$$\frac{3x^4 - 11x^3 + 16x^2 - 16x + 8}{x^2 - 3x + 2} = 3x^2 - 2x + 4$$

You Try It 5

-2	6	8	-5
		-12	8
	6	-4	3

$$(6x^2 + 8x - 5) \div (x + 2)$$

$$= 6x - 4 + \frac{3}{x + 2}$$

You Try It 6

2	5	-12	-8	16
		10	-4	-24
	5	-2	-12	-8

$$(5x^3 - 12x^2 - 8x + 16) \div (x - 2)$$

$$= 5x^2 - 2x - 12 - \frac{8}{x - 2}$$

You Try It 7

3	2	-3	-8	0	-2
		6	9	3	9
	2	3	1	3	7

$$(2x^4 - 3x^3 - 8x^2 - 2) \div (x - 3)$$

$$= 2x^3 + 3x^2 + x + 3 + \frac{7}{x - 3}$$

You Try It 8

2	2	-3	-5
		4	2
	2	1	-3

$$P(2) = -3$$

You Try It 9

-3	2	-5	0	7
		-6	33	-99
	2	-11	33	-92

$$P(-3) = -92$$

SECTION 5.5

You Try It 1 The GCF of $3x^3y$, $6x^2y^2$, and $3xy^3$ is $3xy$.

$$3x^3y - 6x^2y^2 - 3xy^3 = 3xy(x^2 - 2xy - y^2)$$

You Try It 2 The GCF of $6t^{2n}$ and $9t^n$ is $3t^n$.

$$6t^{2n} - 9t^n = 3t^n(2t^n - 3)$$

You Try It 3
$$3(6x - 7y) - 2x^2(6x - 7y)$$
$$= (6x - 7y)(3 - 2x^2)$$

• The common binomial factor is $(6x - 7y)$.

You Try It 4 $4a^2 - 6a - 6ax + 9x$
$$= (4a^2 - 6a) - (6ax - 9x)$$
$$= 2a(2a - 3) - 3x(2a - 3)$$
$$= (2a - 3)(2a - 3x)$$

You Try It 5 $x^2 - x - 20 = (x + 4)(x - 5)$

You Try It 6 $x^2 + 5xy + 6y^2 = (x + 2y)(x + 3y)$

You Try It 7 $4x^2 + 15x - 4 = (x + 4)(4x - 1)$

You Try It 8 $10x^2 + 39x + 14 = (2x + 7)(5x + 2)$

You Try It 9
The GCF of $3a^3b^3$, $3a^2b^2$, and $60ab$ is $3ab$.

$$3a^3b^3 + 3a^2b^2 - 60ab = 3ab(a^2b^2 + ab - 20)$$
$$= 3ab(ab + 5)(ab - 4)$$

You Try It 10
We are looking for two polynomials that, when multiplied, equal $3x^2 - 17x - 6$.
We must factor $3x^2 - 17x - 6$.
$$3x^2 - 17x - 6 = (3x + 1)(x - 6)$$
$$f(x) = 3x + 1 \text{ and } g(x) = x - 6.$$

SECTION 5.6

You Try It 1
$$x^2 - 36y^4 = x^2 - (6y^2)^2$$
$$= (x + 6y^2)(x - 6y^2).$$

• Difference of two squares

You Try It 2
$$9x^2 + 12x + 4 = (3x + 2)^2$$

• Perfect-square trinomial

You Try It 3

$(a + b)^2 - (a - b)^2$ • **Difference of two squares**

$\quad = [(a + b) + (a - b)][(a + b) - (a - b)]$

$\quad = (a + b + a - b)(a + b - a + b)$

$\quad = (2a)(2b) = 4ab$

You Try It 4

$a^3b^3 - 27 = (ab)^3 - 3^3$ • **Difference of two cubes**

$\quad = (ab - 3)(a^2b^2 + 3ab + 9)$

You Try It 5

$8x^3 + y^3z^3 = (2x)^3 + (yz)^3$ • **Sum of two cubes**

$\quad = (2x + yz)(4x^2 - 2xyz + y^2z^2)$

You Try It 6

$(x - y)^3 + (x + y)^3$ • **Sum of two cubes**

$\quad = [(x - y) + (x + y)]$

$\quad\quad \times [(x - y)^2 - (x - y)(x + y) + (x + y)^2]$

$\quad = 2x[x^2 - 2xy + y^2 - (x^2 - y^2)$

$\quad\quad + x^2 + 2xy + y^2]$

$\quad = 2x(x^2 - 2xy + y^2 - x^2 + y^2 + x^2$

$\quad\quad + 2xy + y^2)$

$\quad = 2x(x^2 + 3y^2)$

You Try It 7 Let $u = x^2$.

$$3x^4 + 4x^2 - 4 = 3u^2 + 4u - 4$$
$$= (u + 2)(3u - 2)$$
$$= (x^2 + 2)(3x^2 - 2)$$

You Try It 8

$18x^3 - 6x^2 - 60x = 6x(3x^2 - x - 10)$ • **GCF**

$\quad = 6x(3x + 5)(x - 2)$

You Try It 9 $4x - 4y - x^3 + x^2y$ • **Factor by grouping.**

$\quad = (4x - 4y) - (x^3 - x^2y)$

$\quad = 4(x - y) - x^2(x - y)$

$\quad = (x - y)(4 - x^2)$

$\quad = (x - y)(2 + x)(2 - x)$

You Try It 10

$x^{4n} - x^{2n}y^{2n} = x^{2n+2n} - x^{2n}y^{2n}$ • **GCF**

$\quad = x^{2n}(x^{2n} - y^{2n})$

$\quad = x^{2n}[(x^n)^2 - (y^n)^2]$ • **Difference of**

$\quad = x^{2n}(x^n + y^n)(x^n - y^n)$ **two squares**

You Try It 11

$ax^5 - ax^2y^6 = ax^2(x^3 - y^6)$ • **GCF**

$\quad = ax^2(x - y^2)(x^2 + xy^2 + y^4)$ • **Difference of two cubes**

SECTION 5.7

You Try It 1

$(x + 4)(x - 1) = 14$

$\quad x^2 + 3x - 4 = 14$ • **FOIL**

$\quad x^2 + 3x - 18 = 0$ • **Write in standard form.**

$(x + 6)(x - 3) = 0$ • **Factor the left side.**

$x + 6 = 0 \quad\quad x - 3 = 0$ • **Principle of**

$\quad x = -6 \quad\quad\quad x = 3$ **Zero Products**

The solutions are -6 and 3.

You Try It 2

Strategy Draw a diagram. Then use the formula for the area of a triangle.

Solution $A = \dfrac{1}{2}bh$

$$54 = \dfrac{1}{2}x(x + 3)$$
$$108 = x(x + 3)$$
$$0 = x^2 + 3x - 108$$
$$0 = (x - 9)(x + 12)$$

$x - 9 = 0 \quad\quad x + 12 = 0$

$\quad x = 9 \quad\quad\quad\quad x = -12$

$x + 3 = 9 + 3 = 12$

The base is 9 cm; the height is 12 cm.

Solutions to Chapter 6 "You Try It"

SECTION 6.1

You Try It 1 $f(x) = \dfrac{3 - 5x}{x^2 + 5x + 6}$

$\quad f(2) = \dfrac{3 - 5(2)}{2^2 + 5(2) + 6}$ • **Replace x by 2.**

$\quad\quad = \dfrac{3 - 10}{4 + 10 + 6}$

$\quad\quad = \dfrac{-7}{20}$

$\quad\quad = -\dfrac{7}{20}$

You Try It 2 Set the denominator equal to zero.

 Then solve for x.

$2x^2 - 7x + 3 = 0$ • **Solve the**

$(2x - 1)(x - 3) = 0$ **quadratic**

$2x - 1 = 0 \quad\quad x - 3 = 0$ **equation by**

$\quad x = \dfrac{1}{2} \quad\quad\quad x = 3$ **factoring.**

The domain is $\left\{x \mid x \neq \dfrac{1}{2}, 3\right\}$.

You Try It 3 $\dfrac{6x^4 - 24x^3}{12x^3 - 48x^2} = \dfrac{6x^3(x - 4)}{12x^2(x - 4)}$ • **Factor out the GCF.**

$\quad\quad = \dfrac{6x^3 \overset{1}{\cancel{(x - 4)}}}{12x^2 \underset{1}{\cancel{(x - 4)}}} = \dfrac{x}{2}$

You Try It 4 $\dfrac{21a^3b - 14a^3b^2}{7a^2b} = \dfrac{7a^3b(3 - 2b)}{7a^2b}$ • **Factor out the GCF.**

$\quad\quad = a(3 - 2b)$

You Try It 5

$$\frac{20x - 15x^2}{15x^3 - 5x^2 - 20x} = \frac{5x(4 - 3x)}{5x(3x^2 - x - 4)}$$

$$= \frac{5x(4 - 3x)}{5x(3x - 4)(x + 1)}$$

$$= \frac{5x\overset{-1}{\cancel{(4 - 3x)}}}{5x\underset{1}{\cancel{(3x - 4)}}(x + 1)}$$

$$= -\frac{1}{x + 1}$$

You Try It 6

$$\frac{x^{2n} + x^n - 12}{x^{2n} - 3x^n} = \frac{(x^n + 4)(x^n - 3)}{x^n(x^n - 3)}$$

$$= \frac{(x^n + 4)\overset{1}{\cancel{(x^n - 3)}}}{x^n\underset{1}{\cancel{(x^n - 3)}}}$$

$$= \frac{x^n + 4}{x^n}$$

You Try It 7

$$\frac{12 + 5x - 3x^2}{x^2 + 2x - 15} \cdot \frac{2x^2 + x - 45}{3x^2 + 4x}$$

$$= \frac{(4 + 3x)(3 - x)}{(x + 5)(x - 3)} \cdot \frac{(2x - 9)(x + 5)}{x(3x + 4)}$$

$$= \frac{(4 + 3x)(3 - x)(2x - 9)(x + 5)}{(x + 5)(x - 3) \cdot x(3x + 4)}$$

$$= \frac{(4 + 3x)\overset{-1}{\cancel{(3 - x)}}(2x - 9)\overset{1}{\cancel{(x + 5)}}}{\underset{1}{\cancel{(x + 5)}}\underset{1}{\cancel{(x - 3)}} \cdot x\underset{1}{\cancel{(3x + 4)}}} = -\frac{2x - 9}{x}$$

You Try It 8

$$\frac{2x^2 - 13x + 20}{x^2 - 16} \cdot \frac{2x^2 + 9x + 4}{6x^2 - 7x - 5}$$

$$= \frac{(2x - 5)(x - 4)}{(x - 4)(x + 4)} \cdot \frac{(2x + 1)(x + 4)}{(3x - 5)(2x + 1)}$$

$$= \frac{(2x - 5)(x - 4) \cdot (2x + 1)(x + 4)}{(x - 4)(x + 4) \cdot (3x - 5)(2x + 1)}$$

$$= \frac{(2x - 5)\overset{1}{\cancel{(x - 4)}}\overset{1}{\cancel{(2x + 1)}}\overset{1}{\cancel{(x + 4)}}}{\underset{1}{\cancel{(x - 4)}}\underset{1}{\cancel{(x + 4)}}(3x - 5)\underset{1}{\cancel{(2x + 1)}}} = \frac{2x - 5}{3x - 5}$$

You Try It 9

$$\frac{7a^3b^7}{15x^2y^3} \div \frac{21a^5b^2}{20x^4y}$$

$$= \frac{7a^3b^7}{15x^2y^3} \cdot \frac{20x^4y}{21a^5b^2}$$ • Rewrite division as multiplication by the reciprocal.

$$= \frac{7a^3b^7 \cdot 20x^4y}{15x^2y^3 \cdot 21a^5b^2}$$

$$= \frac{4b^5x^2}{9a^2y^2}$$

You Try It 10

$$\frac{6x^2 - 3xy}{10ab^4} \div \frac{16x^2y^2 - 8xy^3}{15a^2b^2}$$

$$= \frac{6x^2 - 3xy}{10ab^4} \cdot \frac{15a^2b^2}{16x^2y^2 - 8xy^3}$$ • Rewrite division as multiplication by the reciprocal.

$$= \frac{3x(2x - y)}{10ab^4} \cdot \frac{15a^2b^2}{8xy^2(2x - y)}$$

$$= \frac{3x(2x - y) \cdot 15a^2b^2}{10ab^4 \cdot 8xy^2(2x - y)}$$

$$= \frac{3x\overset{1}{\cancel{(2x - y)}} \cdot 15a^2b^2}{10ab^4 \cdot 8xy^2\underset{1}{\cancel{(2x - y)}}} = \frac{9a}{16b^2y^2}$$

You Try It 11

$$\frac{6x^2 - 7x + 2}{3x^2 + x - 2} \div \frac{4x^2 - 8x + 3}{5x^2 + x - 4}$$

$$= \frac{6x^2 - 7x + 2}{3x^2 + x - 2} \cdot \frac{5x^2 + x - 4}{4x^2 - 8x + 3}$$

$$= \frac{(2x - 1)(3x - 2)}{(x + 1)(3x - 2)} \cdot \frac{(x + 1)(5x - 4)}{(2x - 1)(2x - 3)}$$

$$= \frac{(2x - 1)(3x - 2)(x + 1)(5x - 4)}{(x + 1)(3x - 2)(2x - 1)(2x - 3)}$$

$$= \frac{\overset{1}{\cancel{(2x - 1)}}\overset{1}{\cancel{(3x - 2)}}\overset{1}{\cancel{(x + 1)}}(5x - 4)}{\underset{1}{\cancel{(x + 1)}}\underset{1}{\cancel{(3x - 2)}}\underset{1}{\cancel{(2x - 1)}}(2x - 3)} = \frac{5x - 4}{2x - 3}$$

SECTION 6.2

You Try It 1

The LCM is $(2x - 5)(x + 4)$.

$$\frac{2x}{2x - 5} = \frac{2x}{2x - 5} \cdot \frac{x + 4}{x + 4} = \frac{2x^2 + 8x}{(2x - 5)(x + 4)}$$

$$\frac{3}{x + 4} = \frac{3}{x + 4} \cdot \frac{2x - 5}{2x - 5} = \frac{6x - 15}{(2x - 5)(x + 4)}$$

You Try It 2

$2x^2 - 11x + 15 = (x - 3)(2x - 5)$; $x^2 - 3x = x(x - 3)$

The LCM is $x(x - 3)(2x - 5)$.

$$\frac{3x}{2x^2 - 11x + 15} = \frac{3x}{(x - 3)(2x - 5)} \cdot \frac{x}{x} = \frac{3x^2}{x(x - 3)(2x - 5)}$$

$$\frac{x - 2}{x^2 - 3x} = \frac{x - 2}{x(x - 3)} \cdot \frac{2x - 5}{2x - 5} = \frac{2x^2 - 9x + 10}{x(x - 3)(2x - 5)}$$

You Try It 3

$2x - x^2 = x(2 - x) = -x(x - 2)$;
$3x^2 - 5x - 2 = (x - 2)(3x + 1)$

The LCM is $x(x - 2)(3x + 1)$.

$$\frac{2x - 7}{2x - x^2} = -\frac{2x - 7}{x(x - 2)} \cdot \frac{3x + 1}{3x + 1} = -\frac{6x^2 - 19x - 7}{x(x - 2)(3x + 1)}$$

$$\frac{3x - 2}{3x^2 - 5x - 2} = \frac{3x - 2}{(x - 2)(3x + 1)} \cdot \frac{x}{x} = \frac{3x^2 - 2x}{x(x - 2)(3x + 1)}$$

You Try It 4

The LCM is ab.

$$\frac{2}{b} - \frac{1}{a} + \frac{4}{ab} = \frac{2}{b} \cdot \frac{a}{a} - \frac{1}{a} \cdot \frac{b}{b} + \frac{4}{ab}$$

$$= \frac{2a}{ab} - \frac{b}{ab} + \frac{4}{ab} = \frac{2a - b + 4}{ab}$$

You Try It 5

The LCM is $a(a - 5)(a + 5)$.

$$\frac{a - 3}{a^2 - 5a} + \frac{a - 9}{a^2 - 25}$$

$$= \frac{a - 3}{a(a - 5)} \cdot \frac{a + 5}{a + 5} + \frac{a - 9}{(a - 5)(a + 5)} \cdot \frac{a}{a}$$

$$= \frac{(a - 3)(a + 5) + a(a - 9)}{a(a - 5)(a + 5)}$$

$$= \frac{(a^2 + 2a - 15) + (a^2 - 9a)}{a(a - 5)(a + 5)}$$

$$= \frac{a^2 + 2a - 15 + a^2 - 9a}{a(a - 5)(a + 5)}$$

$$= \frac{2a^2 - 7a - 15}{a(a - 5)(a + 5)} = \frac{(2a + 3)(a - 5)}{a(a - 5)(a + 5)}$$

$$= \frac{(2a + 3)\overset{1}{\cancel{(a - 5)}}}{a\underset{1}{\cancel{(a - 5)}}(a + 5)} = \frac{2a + 3}{a(a + 5)}$$

You Try It 6

The LCM is $(x - 4)(x + 1)$.

$$\frac{2x}{x - 4} - \frac{x - 1}{x + 1} + \frac{2}{x^2 - 3x - 4}$$

$$= \frac{2x}{x - 4} \cdot \frac{x + 1}{x + 1} - \frac{x - 1}{x + 1} \cdot \frac{x - 4}{x - 4} + \frac{2}{(x - 4)(x + 1)}$$

$$= \frac{2x(x + 1) - (x - 1)(x - 4) + 2}{(x - 4)(x + 1)}$$

$$= \frac{(2x^2 + 2x) - (x^2 - 5x + 4) + 2}{(x - 4)(x + 1)}$$

$$= \frac{x^2 + 7x - 2}{(x - 4)(x + 1)}$$

SECTION 6.3

You Try It 1

The LCM is $x - 3$.

$$\frac{2x + 5 + \dfrac{14}{x - 3}}{4x + 16 + \dfrac{49}{x - 3}} = \frac{2x + 5 + \dfrac{14}{x - 3}}{4x + 16 + \dfrac{49}{x - 3}} \cdot \frac{x - 3}{x - 3}$$

$$= \frac{(2x + 5)(x - 3) + \dfrac{14}{x - 3}(x - 3)}{(4x + 16)(x - 3) + \dfrac{49}{x - 3}(x - 3)}$$

$$= \frac{2x^2 - x - 15 + 14}{4x^2 + 4x - 48 + 49} = \frac{2x^2 - x - 1}{4x^2 + 4x + 1}$$

$$= \frac{(2x + 1)(x - 1)}{(2x + 1)(2x + 1)} = \frac{\overset{1}{\cancel{(2x + 1)}}(x - 1)}{\underset{1}{\cancel{(2x + 1)}}(2x + 1)} = \frac{x - 1}{2x + 1}$$

You Try It 2

The LCM of the denominators is x.

$$2 - \frac{1}{2 - \dfrac{1}{x}} = 2 - \frac{1}{2 - \dfrac{1}{x}} \cdot \frac{x}{x}$$

$$= 2 - \frac{1 \cdot x}{2 \cdot x - \dfrac{1}{x} \cdot x} = 2 - \frac{x}{2x - 1}$$

The LCM of the denominators is $2x - 1$.

$$2 - \frac{x}{2x - 1} = 2 \cdot \frac{2x - 1}{2x - 1} - \frac{x}{2x - 1}$$

$$= \frac{4x - 2}{2x - 1} - \frac{x}{2x - 1}$$

$$= \frac{4x - 2 - x}{2x - 1} = \frac{3x - 2}{2x - 1}$$

SECTION 6.4

You Try It 1

$$\frac{5}{x - 2} = \frac{3}{4}$$

$$\frac{5}{x - 2} \cdot 4(x - 2) = \frac{3}{4} \cdot 4(x - 2)$$

• Multiply each side by the LCM of $x - 2$ and 4.

$$5 \cdot 4 = 3(x - 2)$$

$$20 = 3x - 6$$

$$26 = 3x$$

$$\frac{26}{3} = x$$

The solution is $\dfrac{26}{3}$.

You Try It 2

$$\frac{5}{2x - 3} = \frac{-2}{x + 1}$$

$$\frac{5}{2x - 3}(x + 1)(2x - 3) = \frac{-2}{x + 1}(x + 1)(2x - 3)$$

$$5(x + 1) = -2(2x - 3)$$

$$5x + 5 = -4x + 6$$

$$9x + 5 = 6$$

$$9x = 1$$

$$x = \frac{1}{9}$$

The solution is $\dfrac{1}{9}$.

You Try It 3

Strategy To find the cost, write and solve a proportion using x to represent the cost.

Solution
$$\frac{2}{5.80} = \frac{15}{x}$$

$$\frac{2}{5.80} \cdot x(5.80) = \frac{15}{x} \cdot x(5.80)$$

$$2x = 15(5.80)$$

$$2x = 87$$

$$x = 43.50$$

The cost of 15 lb of cashews is $43.50.

SECTION 6.5

You Try It 1
$$\frac{x}{x-2} + x = \frac{6}{x-2}$$

$$(x-2)\left(\frac{x}{x-2} + x\right) = (x-2)\left(\frac{6}{x-2}\right)$$

$$(x-2)\left(\frac{x}{x-2}\right) + (x-2)x = (x-2)\left(\frac{6}{x-2}\right)$$

$$x + x^2 - 2x = 6$$

$$x^2 - x = 6$$

$$x^2 - x - 6 = 0$$

$$(x-3)(x+2) = 0$$

$$x = 3 \qquad x = -2$$

-2 and 3 check as solutions.

The solutions are -2 and 3.

You Try It 2
$$\frac{5}{x-2} = 2x + \frac{3x-1}{x-2}$$

$$(x-2)\left(\frac{5}{x-2}\right) = (x-2)\left(2x + \frac{3x-1}{x-2}\right)$$

$$5 = (x-2)(2x) + 3x - 1$$

$$5 = 2x^2 - 4x + 3x - 1$$

$$0 = 2x^2 - x - 6$$

$$0 = (2x+3)(x-2)$$

$$2x + 3 = 0 \qquad x - 2 = 0$$

$$x = -\frac{3}{2} \qquad x = 2$$

2 does not check as a solution.

The solution is $-\dfrac{3}{2}$.

You Try It 3
$$\frac{P_1V_1}{T_1} = \frac{P_2V_2}{T_2}$$

$$T_1T_2\left(\frac{P_1V_1}{T_1}\right) = T_1T_2\left(\frac{P_2V_2}{T_2}\right)$$

 • **Multiply by the LCM of T_1 and T_2.**

$$T_2P_1V_1 = T_1P_2V_2$$

$$\frac{T_2P_1V_1}{P_2V_2} = \frac{T_1P_2V_2}{P_2V_2}$$

$$\frac{T_2P_1V_1}{P_2V_2} = T_1$$

You Try It 4

Strategy • Time required for the small pipe to fill the tank: x

	Rate	Time	Part
Large pipe	$\dfrac{1}{9}$	6	$\dfrac{6}{9}$
Small pipe	$\dfrac{1}{x}$	6	$\dfrac{6}{x}$

 • The sum of the part of the task completed by the large pipe and the part of the task completed by the small pipe is 1.

Solution
$$\frac{6}{9} + \frac{6}{x} = 1$$

$$\frac{2}{3} + \frac{6}{x} = 1$$
 • **Simplify $\dfrac{6}{9}$.**

$$3x\left(\frac{2}{3} + \frac{6}{x}\right) = 3x \cdot 1$$
 • **Multiply by the LCM of 3 and x.**

$$2x + 18 = 3x$$

$$18 = x$$

The small pipe working alone will fill the tank in 18 h.

You Try It 5

Strategy
 • Rate of the wind: r

	Distance	Rate	Time
With wind	700	$150 + r$	$\dfrac{700}{150 + r}$
Against wind	500	$150 - r$	$\dfrac{500}{150 - r}$

 • The time flying with the wind equals the time flying against the wind.

Solution

$$\frac{700}{150 + r} = \frac{500}{150 - r}$$

$$(150 + r)(150 - r)\left(\frac{700}{150 + r}\right) = (150 + r)(150 - r)\left(\frac{500}{150 - r}\right)$$

$$(150 - r)700 = (150 + r)500$$

$$105{,}000 - 700r = 75{,}000 + 500r$$

$$30{,}000 = 1200r$$

$$25 = r$$

The rate of the wind is 25 mph.

SECTION 6.6

You Try It 1

Strategy To find the distance:

- Write the basic direct variation equation, replace the variables by the given values, and solve for k.
- Write the direct variation equation, replacing k by its value. Substitute 5 for t and solve for s.

Solution $s = kt^2$ • **Direct variation equation**
$64 = k(2)^2$ • **Replace s by 64 and t by 2.**
$64 = k \cdot 4$
$16 = k$

$s = 16t^2 = 16(5)^2 = 400$ • $k = 16, t = 5$

The object will fall 400 ft in 5 s.

You Try It 2

Strategy To find the resistance:

- Write the basic inverse variation equation, replace the variables by the given values, and solve for k.
- Write the inverse variation equation, replacing k by its value. Substitute 0.02 for d and solve for R.

Solution

$$R = \frac{k}{d^2}$$ • **Inverse variation equation**

$$0.5 = \frac{k}{(0.01)^2}$$ • **Replace R by 0.5 and d by 0.01.**

$$0.5 = \frac{k}{0.0001}$$

$$0.00005 = k$$

$$R = \frac{0.00005}{d^2} = \frac{0.00005}{(0.02)^2} = 0.125$$ • $k = 0.00005, \; d = 0.02$

The resistance is 0.125 ohm.

You Try It 3

Strategy • Write the basic combined variation equation, replace the variables by the given values, and solve for k.
- Write the combined variation equation, replacing k by its value and substituting 4 for W, 8 for d, and 16 for L. Solve for s.

Solution

$$s = \frac{kWd^2}{L}$$ • **Combined variation equation**

$$1200 = \frac{k(2)(12)^2}{12}$$ • **Replace s by 1200, W by 2, d by 12, and L by 12.**

$$1200 = 24k$$

$$50 = k$$

$$s = \frac{50Wd^2}{L}$$ • **Replace k by 50 in the combined variation equation.**

$$= \frac{50(4)8^2}{16}$$ • **Replace W by 4, d by 8, and L by 16.**

$$= 800$$

The strength of the beam is 800 lb.

Solutions to Chapter 7 "You Try It"

SECTION 7.1

You Try It 1 $16^{-3/4} = (2^4)^{-3/4}$ • $16 = 2^4$
$$= 2^{-3}$$
$$= \frac{1}{2^3} = \frac{1}{8}$$

You Try It 2 $(-81)^{3/4}$
The base of the exponential expression is negative, and the denominator of the exponent is a positive even number.

Therefore, $(-81)^{3/4}$ is not a real number.

You Try It 3 $(x^{3/4}y^{1/2}z^{-2/3})^{-4/3} = x^{-1}y^{-2/3}z^{8/9}$
$$= \frac{z^{8/9}}{xy^{2/3}}$$

You Try It 4

$$\left(\frac{16a^{-2}b^{4/3}}{9a^4b^{-2/3}}\right)^{-1/2} = \left(\frac{2^4a^{-6}b^2}{3^2}\right)^{-1/2}$$
 • Use the Rule for Dividing Exponential Expressions.

$$= \frac{2^{-2}a^3b^{-1}}{3^{-1}}$$
 • Use the Rule for Simplifying Powers of Products.

$$= \frac{3a^3}{2^2b} = \frac{3a^3}{4b}$$

You Try It 5
$$(2x^3)^{3/4} = \sqrt[4]{(2x^3)^3}$$
$$= \sqrt[4]{8x^9}$$

You Try It 6
$$-5a^{5/6} = -5(a^5)^{1/6}$$
$$= -5\sqrt[6]{a^5}$$

You Try It 7
$$\sqrt[3]{3ab} = (3ab)^{1/3}$$

You Try It 8
$$\sqrt[4]{x^4 + y^4} = (x^4 + y^4)^{1/4}$$

You Try It 9
$$\sqrt[3]{-8x^{12}y^3} = -2x^4y$$

You Try It 10
$$-\sqrt[4]{81x^{12}y^8} = -3x^3y^2$$

SECTION 7.2

You Try It 1
$$\sqrt[5]{x^7} = \sqrt[5]{x^5 \cdot x^2}$$
$$= \sqrt[5]{x^5}\sqrt[5]{x^2}$$
 • x^5 is a perfect fifth power.
$$= x\sqrt[5]{x^2}$$

You Try It 2
$$\sqrt[3]{-64x^8y^{18}} = \sqrt[3]{-64x^6y^{18}(x^2)}$$
$$= \sqrt[3]{-64x^6y^{18}}\sqrt[3]{x^2}$$
 • $-64x^6y^{18}$ is a perfect third power.
$$= -4x^2y^6\sqrt[3]{x^2}$$

You Try It 3
$$3xy\sqrt[3]{81x^5y} - \sqrt[3]{192x^8y^4}$$
$$= 3xy\sqrt[3]{27x^3 \cdot 3x^2y} - \sqrt[3]{64x^6y^3 \cdot 3x^2y}$$
$$= 3xy\sqrt[3]{27x^3}\sqrt[3]{3x^2y} - \sqrt[3]{64x^6y^3}\sqrt[3]{3x^2y}$$
$$= 3xy \cdot 3x\sqrt[3]{3x^2y} - 4x^2y\sqrt[3]{3x^2y}$$
$$= 9x^2y\sqrt[3]{3x^2y} - 4x^2y\sqrt[3]{3x^2y} = 5x^2y\sqrt[3]{3x^2y}$$

You Try It 4
$$\sqrt{5b}(\sqrt{3b} - \sqrt{10})$$
$$= \sqrt{15b^2} - \sqrt{50b}$$
 • The Distributive Property
$$= \sqrt{b^2 \cdot 15} - \sqrt{25 \cdot 2b}$$
 • Simplify each radical expression.
$$= \sqrt{b^2}\sqrt{15} - \sqrt{25}\sqrt{2b}$$
$$= b\sqrt{15} - 5\sqrt{2b}$$

You Try It 5
$$(2\sqrt[3]{2x} - 3)(\sqrt[3]{2x} - 5)$$
$$= 2\sqrt[3]{4x^2} - 10\sqrt[3]{2x} - 3\sqrt[3]{2x} + 15$$
 • The FOIL method
$$= 2\sqrt[3]{4x^2} - 13\sqrt[3]{2x} + 15$$

You Try It 6
$$(\sqrt{a} - 3\sqrt{y})(\sqrt{a} + 3\sqrt{y})$$
$$= (\sqrt{a})^2 - (3\sqrt{y})^2$$
 • The expressions are conjugates.
$$= a - 9y$$

You Try It 7
$$\frac{y}{\sqrt{3y}} = \frac{y}{\sqrt{3y}} \cdot \frac{\sqrt{3y}}{\sqrt{3y}} = \frac{y\sqrt{3y}}{\sqrt{9y^2}}$$
 • Rationalize the denominator.
$$= \frac{y\sqrt{3y}}{3y} = \frac{\sqrt{3y}}{3}$$

You Try It 8
$$\frac{3x}{\sqrt[3]{3x^2}} = \frac{3x}{\sqrt[3]{3x^2}} \cdot \frac{\sqrt[3]{9x}}{\sqrt[3]{9x}} = \frac{3x\sqrt[3]{9x}}{\sqrt[3]{27x^3}}$$
 • Rationalize the denominator.
$$= \frac{3x\sqrt[3]{9x}}{3x} = \sqrt[3]{9x}$$

You Try It 9
$$\frac{3 + \sqrt{6}}{2 - \sqrt{6}} = \frac{3 + \sqrt{6}}{2 - \sqrt{6}} \cdot \frac{2 + \sqrt{6}}{2 + \sqrt{6}}$$
 • Rationalize the denominator.
$$= \frac{6 + 3\sqrt{6} + 2\sqrt{6} + (\sqrt{6})^2}{2^2 - (\sqrt{6})^2}$$
$$= \frac{6 + 5\sqrt{6} + 6}{4 - 6} = \frac{12 + 5\sqrt{6}}{-2}$$
$$= -\frac{12 + 5\sqrt{6}}{2}$$

SECTION 7.3

You Try It 1
$$\sqrt{x} - \sqrt{x + 5} = 1$$
 • Add $\sqrt{x + 5}$ to each side.
$$\sqrt{x} = 1 + \sqrt{x + 5}$$
$$(\sqrt{x})^2 = (1 + \sqrt{x + 5})^2$$
 • Square each side.
$$x = 1 + 2\sqrt{x + 5} + x + 5$$
$$-6 = 2\sqrt{x + 5}$$
$$-3 = \sqrt{x + 5}$$
$$(-3)^2 = (\sqrt{x + 5})^2$$
 • Square each side.
$$9 = x + 5$$
$$4 = x$$

4 does not check as a solution. The equation has no solution.

You Try It 2
$$\sqrt[4]{x - 8} = 3$$
$$(\sqrt[4]{x - 8})^4 = 3^4$$
 • Raise each side to the fourth power.
$$x - 8 = 81$$
$$x = 89$$

Check:

$$\dfrac{\sqrt[4]{x-8}=3}{\sqrt[4]{89-8}\ \Big|\ 3}$$
$$\sqrt[4]{81}\ \Big|\ 3$$
$$3=3$$

The solution is 89.

You Try It 3

Strategy To find the diagonal, use the Pythagorean Theorem. One leg is the length of the rectangle. The second leg is the width of the rectangle. The hypotenuse is the diagonal of the rectangle.

Solution

$$c^2 = a^2 + b^2 \qquad \bullet \text{ Pythagorean Theorem}$$
$$c^2 = (6)^2 + (3)^2 \qquad \bullet \text{ Replace } a \text{ by 6 and } b \text{ by 3.}$$
$$c^2 = 36 + 9 \qquad \bullet \text{ Solve for } c.$$
$$c^2 = 45$$

$$(c^2)^{1/2} = (45)^{1/2} \qquad \bullet \text{ Raise each side to the } \dfrac{1}{2} \text{ power.}$$

$$c = \sqrt{45} \qquad \bullet\ a^{1/2} = \sqrt{a}$$
$$c \approx 6.7$$

The diagonal is 6.7 cm.

You Try It 4

Strategy To find the height, replace d in the equation with the given value and solve for h.

Solution

$$d = \sqrt{1.5h}$$
$$5.5 = \sqrt{1.5h} \qquad \bullet \text{ Replace } d \text{ by 5.5.}$$
$$(5.5)^2 = (\sqrt{1.5h})^2 \qquad \bullet \text{ Square each side.}$$
$$30.25 = 1.5h$$
$$20.17 \approx h$$

The periscope must be approximately 20.17 ft above the water.

You Try It 5

Strategy To find the distance, replace the variables v and a in the equation by their given values and solve for s.

Solution

$$v = \sqrt{2as}$$
$$88 = \sqrt{2 \cdot 22s} \qquad \bullet \text{ Replace } v \text{ by 88}$$
$$88 = \sqrt{44s} \qquad\qquad \text{ and } a \text{ by 22.}$$
$$(88)^2 = (\sqrt{44s})^2 \qquad \bullet \text{ Square each side.}$$
$$7744 = 44s$$
$$176 = s$$

The distance required is 176 ft.

SECTION 7.4

You Try It 1 $\sqrt{-45} = i\sqrt{45} = i\sqrt{9 \cdot 5} = 3i\sqrt{5}$

You Try It 2 $\sqrt{98} - \sqrt{-60} = \sqrt{98} - i\sqrt{60}$
$$= \sqrt{49 \cdot 2} - i\sqrt{4 \cdot 15}$$
$$= 7\sqrt{2} - 2i\sqrt{15}$$

You Try It 3 $(-4 + 2i) - (6 - 8i) = -10 + 10i$

You Try It 4 $(16 - \sqrt{-45}) - (3 + \sqrt{-20})$
$$= (16 - i\sqrt{45}) - (3 + i\sqrt{20})$$
$$= (16 - i\sqrt{9 \cdot 5}) - (3 + i\sqrt{4 \cdot 5})$$
$$= (16 - 3i\sqrt{5}) - (3 + 2i\sqrt{5})$$
$$= 13 - 5i\sqrt{5}$$

You Try It 5 $(3 - 2i) + (-3 + 2i) = 0 + 0i = 0$

You Try It 6 $(-3i)(-10i) = 30i^2 = 30(-1) = -30$

You Try It 7
$$-\sqrt{-8} \cdot \sqrt{-5} = -i\sqrt{8} \cdot i\sqrt{5} = -i^2\sqrt{40}$$
$$= -(-1)\sqrt{40} = \sqrt{4 \cdot 10} = 2\sqrt{10}$$

You Try It 8
$$-6i(3 + 4i) = -18i - 24i^2 \qquad \bullet \text{ The Distributive}$$
$$= -18i - 24(-1) \qquad\qquad \text{ Property}$$
$$= 24 - 18i$$

You Try It 9
$$\sqrt{-3}(\sqrt{27} - \sqrt{-6}) = i\sqrt{3}(\sqrt{27} - i\sqrt{6})$$
$$= i\sqrt{81} - i^2\sqrt{18}$$
$$= i\sqrt{81} - (-1)\sqrt{9 \cdot 2}$$
$$= 9i + 3\sqrt{2}$$
$$= 3\sqrt{2} + 9i$$

You Try It 10
$$(4 - 3i)(2 - i) = 8 - 4i - 6i + 3i^2 \qquad \bullet \text{ FOIL}$$
$$= 8 - 10i + 3i^2$$
$$= 8 - 10i + 3(-1)$$
$$= 5 - 10i$$

You Try It 11 $(3 + 6i)(3 - 6i) = 3^2 + 6^2 \quad \bullet \text{ Conjugates}$
$$= 9 + 36$$
$$= 45$$

You Try It 12

$$\frac{2 - 3i}{4i} = \frac{2 - 3i}{4i} \cdot \frac{i}{i}$$

$$= \frac{2i - 3i^2}{4i^2}$$

$$= \frac{2i - 3(-1)}{4(-1)}$$

$$= \frac{3 + 2i}{-4} = -\frac{3}{4} - \frac{1}{2}i$$

You Try It 13

$$\frac{2 + 5i}{3 - 2i} = \frac{2 + 5i}{3 - 2i} \cdot \frac{3 + 2i}{3 + 2i} = \frac{6 + 4i + 15i + 10i^2}{3^2 + 2^2}$$

$$= \frac{6 + 19i + 10(-1)}{9 + 4} = \frac{-4 + 19i}{13}$$

$$= -\frac{4}{13} + \frac{19}{13}i$$

Solutions to Chapter 8 "You Try It"

SECTION 8.1

You Try It 1

$$2x^2 = 7x - 3 \qquad \bullet \text{ Write in standard form.}$$

$$2x^2 - 7x + 3 = 0$$

$$(2x - 1)(x - 3) = 0 \qquad \bullet \text{ Solve by factoring.}$$

$$2x - 1 = 0 \qquad x - 3 = 0$$

$$2x = 1 \qquad x = 3$$

$$x = \frac{1}{2}$$

The solutions are $\frac{1}{2}$ and 3.

You Try It 2

$$x^2 - 3ax - 4a^2 = 0$$

$$(x + a)(x - 4a) = 0$$

$$x + a = 0 \qquad x - 4a = 0$$

$$x = -a \qquad x = 4a$$

The solutions are $-a$ and $4a$.

You Try It 3

$$(x - r_1)(x - r_2) = 0$$

$$(x - 3)\left[x - \left(-\frac{1}{2}\right)\right] = 0 \qquad \bullet \ r_1 = 3, r_2 = -\frac{1}{2}$$

$$(x - 3)\left(x + \frac{1}{2}\right) = 0$$

$$x^2 - \frac{5}{2}x - \frac{3}{2} = 0$$

$$2\left(x^2 - \frac{5}{2}x - \frac{3}{2}\right) = 2 \cdot 0$$

$$2x^2 - 5x - 3 = 0$$

You Try It 4

$$2(x + 1)^2 - 24 = 0$$

$$2(x + 1)^2 = 24 \qquad \bullet \text{ Solve for } (x + 1)^2.$$

$$(x + 1)^2 = 12$$

$$\sqrt{(x + 1)^2} = \sqrt{12} \qquad \bullet \text{ Take the square}$$

$$x + 1 = \pm\sqrt{12} \qquad \qquad \text{root of each side}$$

$$x + 1 = \pm 2\sqrt{3} \qquad \qquad \text{of the equation.}$$

$$x + 1 = 2\sqrt{3} \qquad x + 1 = -2\sqrt{3} \qquad \bullet \text{ Solve for } x.$$

$$x = -1 + 2\sqrt{3} \qquad x = -1 - 2\sqrt{3}$$

The solutions are $-1 + 2\sqrt{3}$ and $-1 - 2\sqrt{3}$.

SECTION 8.2

You Try It 1

$$4x^2 - 4x - 1 = 0 \qquad \bullet \text{ Write in the form}$$

$$4x^2 - 4x = 1 \qquad \qquad ax^2 + bx = -c.$$

$$\frac{1}{4}(4x^2 - 4x) = \frac{1}{4} \cdot 1 \qquad \bullet \text{ Multiply both sides by } \frac{1}{a}.$$

$$x^2 - x = \frac{1}{4}$$

$$x^2 - x + \frac{1}{4} = \frac{1}{4} + \frac{1}{4} \qquad \bullet \text{ Complete the square.}$$

$$\left(x - \frac{1}{2}\right)^2 = \frac{2}{4} \qquad \bullet \text{ Factor.}$$

$$\sqrt{\left(x - \frac{1}{2}\right)^2} = \sqrt{\frac{2}{4}} \qquad \bullet \text{ Take square roots.}$$

$$x - \frac{1}{2} = \pm\frac{\sqrt{2}}{2}$$

$$x - \frac{1}{2} = \frac{\sqrt{2}}{2} \qquad x - \frac{1}{2} = -\frac{\sqrt{2}}{2} \qquad \bullet \text{ Solve for } x.$$

$$x = \frac{1}{2} + \frac{\sqrt{2}}{2} \qquad x = \frac{1}{2} - \frac{\sqrt{2}}{2}$$

The solutions are $\frac{1 + \sqrt{2}}{2}$ and $\frac{1 - \sqrt{2}}{2}$.

You Try It 2

$x^2 + 4x + 8 = 0$

$x^2 + 4x = -8$

$x^2 + 4x + 4 = -8 + 4$ • **Complete the square.**

$(x + 2)^2 = -4$ • **Factor.**

$\sqrt{(x + 2)^2} = \sqrt{-4}$ • **Take square roots.**

$x + 2 = \pm 2i$

$x + 2 = 2i$	$x + 2 = -2i$ • **Solve for x.**
$x = -2 + 2i$	$x = -2 - 2i$

The solutions are $-2 + 2i$ and $-2 - 2i$.

SECTION 8.3

You Try It 1 $x^2 - 2x + 10 = 0$

$a = 1, b = -2, c = 10$

$x = \dfrac{-b \pm \sqrt{b^2 - 4ac}}{2a}$

$= \dfrac{-(-2) \pm \sqrt{(-2)^2 - 4(1)(10)}}{2 \cdot 1}$

$= \dfrac{2 \pm \sqrt{4 - 40}}{2} = \dfrac{2 \pm \sqrt{-36}}{2}$

$= \dfrac{2 \pm 6i}{2} = 1 \pm 3i$

The solutions are $1 + 3i$ and $1 - 3i$.

You Try It 2

$4x^2 = 4x - 1$

$4x^2 - 4x + 1 = 0$ • **Write in standard form.**

$a = 4, b = -4, c = 1$

$x = \dfrac{-b \pm \sqrt{b^2 - 4ac}}{2a}$

$= \dfrac{-(-4) \pm \sqrt{(-4)^2 - 4(4)(1)}}{2 \cdot 4}$

$= \dfrac{4 \pm \sqrt{16 - 16}}{8} = \dfrac{4 \pm \sqrt{0}}{8}$

$= \dfrac{4}{8} = \dfrac{1}{2}$

The solution is $\dfrac{1}{2}$.

You Try It 3 $3x^2 - x - 1 = 0$

$a = 3, b = -1, c = -1$

$b^2 - 4ac =$

$(-1)^2 - 4(3)(-1) = 1 + 12 = 13$

$13 > 0$

Because the discriminant is greater than zero, the equation has two real number solutions.

SECTION 8.4

You Try It 1 $x - 5x^{1/2} + 6 = 0$

$(x^{1/2})^2 - 5(x^{1/2}) + 6 = 0$

$u^2 - 5u + 6 = 0$

$(u - 2)(u - 3) = 0$

$u - 2 = 0$	$u - 3 = 0$
$u = 2$	$u = 3$

Replace u by $x^{1/2}$.

$x^{1/2} = 2$	$x^{1/2} = 3$
$\sqrt{x} = 2$	$\sqrt{x} = 3$
$(\sqrt{x})^2 = 2^2$	$(\sqrt{x})^2 = 3^2$
$x = 4$	$x = 9$

The solutions are 4 and 9.

You Try It 2

$\sqrt{2x + 1} + x = 7$ • **Solve for the radical.**

$\sqrt{2x + 1} = 7 - x$

$(\sqrt{2x + 1})^2 = (7 - x)^2$ • **Square each side.**

$2x + 1 = 49 - 14x + x^2$ • **Write in standard form.**

$0 = x^2 - 16x + 48$

$0 = (x - 4)(x - 12)$ • **Solve by factoring.**

$x - 4 = 0$	$x - 12 = 0$
$x = 4$	$x = 12$

4 checks as a solution.

12 does not check as a solution.

The solution is 4.

You Try It 3

$\sqrt{2x - 1} + \sqrt{x} = 2$

Solve for one of the radical expressions.

$\sqrt{2x - 1} = 2 - \sqrt{x}$

$(\sqrt{2x - 1})^2 = (2 - \sqrt{x})^2$ • **Square each side.**

$2x - 1 = 4 - 4\sqrt{x} + x$

$x - 5 = -4\sqrt{x}$

$(x - 5)^2 = (-4\sqrt{x})^2$ • **Square each side.**

$x^2 - 10x + 25 = 16x$

$x^2 - 26x + 25 = 0$

$(x - 1)(x - 25) = 0$

$x - 1 = 0$	$x - 25 = 0$
$x = 1$	$x = 25$

1 checks as a solution.

25 does not check as a solution.

The solution is 1.

You Try It 4

$$3y + \frac{25}{3y - 2} = -8$$

$$(3y - 2)\left(3y + \frac{25}{3y - 2}\right) = (3y - 2)(-8)$$

• Clear denominators.

$$(3y - 2)(3y) + (3y - 2)\left(\frac{25}{3y - 2}\right) = (3y - 2)(-8)$$

$$9y^2 - 6y + 25 = -24y + 16$$

• Write in standard form.

$$9y^2 + 18y + 9 = 0$$
$$9(y^2 + 2y + 1) = 0$$

• Solve by factoring.

$$9(y + 1)(y + 1) = 0$$

$$y + 1 = 0 \qquad y + 1 = 0$$
$$y = -1 \qquad y = -1$$

The solution is -1.

SECTION 8.5

You Try It 1
$$2x^2 - x - 10 \le 0$$
$$(2x - 5)(x + 2) \le 0$$

$$
\begin{array}{ll}
2x - 5 & \text{---} \mid \text{---------} \mid \text{+++} \\
x + 2 & \text{---} \mid \text{++++++++++} \mid \text{+++} \\
\end{array}
$$

$$-3 \;-2\; -1 \quad 0 \quad 1 \quad 2 \quad 3$$

$$-5\;-4\;-3\;-2\;-1\;\;0\;\;1\;\;2\;\;3\;\;4\;\;5$$

$$\left\{x \mid -2 \le x \le \frac{5}{2}\right\}$$

SECTION 8.6

You Try It 1

Strategy
• This is a geometry problem.
• Width of the rectangle: W
 Length of the rectangle: $W + 3$
• Use the equation $A = L \cdot W$.

Solution
$$A = L \cdot W$$
$$54 = (W + 3)(W)$$
$$54 = W^2 + 3W$$
$$0 = W^2 + 3W - 54$$
$$0 = (W + 9)(W - 6)$$

$$W + 9 = 0 \qquad W - 6 = 0$$
$$W = -9 \qquad W = 6$$

The solution -9 is not possible.

$$W + 3 = 6 + 3 = 9$$

The length is 9 m.

Solutions to Chapter 9 "You Try It"

SECTION 9.1

You Try It 1
x-coordinate of vertex:

$$-\frac{b}{2a} = -\frac{4}{2(4)} = -\frac{1}{2}$$

y-coordinate of vertex:

$$y = 4x^2 + 4x + 1$$
$$= 4\left(-\frac{1}{2}\right)^2 + 4\left(-\frac{1}{2}\right) + 1$$
$$= 1 - 2 + 1$$
$$= 0$$

Vertex: $\left(-\frac{1}{2}, 0\right)$

Axis of symmetry: $x = -\frac{1}{2}$

You Try It 2
x-coordinate of vertex:

$$-\frac{b}{2a} = -\frac{-2}{2(-1)} = -1$$

y-coordinate of vertex:

$$f(x) = -x^2 - 2x - 1$$
$$f(-1) = -(-1)^2 - 2(-1) - 1$$
$$= -1 + 2 - 1$$
$$= 0$$

Vertex: $(-1, 0)$

The domain is $\{x \mid x \in \text{real numbers}\}$. The range is $\{y \mid y \le 0\}$.

You Try It 3
$$y = x^2 + 3x + 4$$
$$0 = x^2 + 3x + 4$$

(Continued)

(Continued)

$$x = \frac{-b \pm \sqrt{b^2 - 4ac}}{2a}$$

$$= \frac{-3 \pm \sqrt{3^2 - 4(1)(4)}}{2 \cdot 1}$$ • $a = 1, b = 3, c = 4$

$$= \frac{-3 \pm \sqrt{-7}}{2}$$

$$= \frac{-3 \pm i\sqrt{7}}{2}$$

The equation has no real number solutions. There are no x-intercepts.

You Try It 4

$$g(x) = x^2 - x + 6$$
$$0 = x^2 - x + 6$$

$$x = \frac{-b \pm \sqrt{b^2 - 4ac}}{2a}$$

$$= \frac{-(-1) \pm \sqrt{(-1)^2 - 4(1)(6)}}{2(1)}$$ • $a = 1, b = -1, c = 6$

$$= \frac{1 \pm \sqrt{1 - 24}}{2}$$

$$= \frac{1 \pm \sqrt{-23}}{2} = \frac{1 \pm i\sqrt{23}}{2}$$

The zeros of the function are $\frac{1}{2} + \frac{\sqrt{23}}{2}i$ and $\frac{1}{2} - \frac{\sqrt{23}}{2}i$.

You Try It 5

$$y = x^2 - x - 6$$
$$a = 1, b = -1, c = -6$$

$$b^2 - 4ac$$
$$(-1)^2 - 4(1)(-6) = 1 + 24 = 25$$

Because the discriminant is greater than zero, the parabola has two x-intercepts.

You Try It 6

$$f(x) = -3x^2 + 4x - 1$$

$$x = -\frac{b}{2a} = -\frac{4}{2(-3)} = \frac{2}{3}$$ • The x-coordinate of the vertex

$$f(x) = -3x^2 + 4x - 1$$

$$f\left(\frac{2}{3}\right) = -3\left(\frac{2}{3}\right)^2 + 4\left(\frac{2}{3}\right) - 1$$ • $x = \frac{2}{3}$

$$= -\frac{4}{3} + \frac{8}{3} - 1 = \frac{1}{3}$$

Because a is negative, the function has a maximum value.

The maximum value of the function is $\frac{1}{3}$.

You Try It 7

Strategy • To find the time it takes the ball to reach its maximum height, find the t-coordinate of the vertex.

• To find the maximum height, evaluate the function at the t-coordinate of the vertex.

Solution

$$t = -\frac{b}{2a} = -\frac{64}{2(-16)} = 2$$ • The t-coordinate of the vertex

The ball reaches its maximum height in 2 s.

$$s(t) = -16t^2 + 64t$$
$$s(2) = -16(2)^2 + 64(2) = -64 + 128 = 64$$ • $t = 2$

The maximum height is 64 ft.

You Try It 8

Strategy
$$P = 2x + y$$
$$100 = 2x + y$$ • $P = 100$
$$100 - 2x = y$$ • Solve for y.

Express the area of the rectangle in terms of x.
$$A = xy$$
$$A = x(100 - 2x)$$ • $y = 100 - 2x$
$$A = -2x^2 + 100x$$
• To find the width, find the x-coordinate of the vertex of $f(x) = -2x^2 + 100x$.
• To find the length, replace x in $y = 100 - 2x$ by the x-coordinate of the vertex.

Solution $$x = -\frac{b}{2a} = -\frac{100}{2(-2)} = 25$$

The width is 25 ft.

$$100 - 2x = 100 - 2(25)$$
$$= 100 - 50 = 50$$

The length is 50 ft.

SECTION 9.2

You Try It 1

Any vertical line intersects the graph at most once. The graph is the graph of a function.

You Try It 2

Domain: $\{x \mid x \in \text{real numbers}\}$
Range: $\{y \mid y \in \text{real numbers}\}$

You Try It 3

Domain: $(-\infty, \infty)$
Range: $[0, \infty)$

You Try It 4

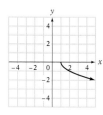

Domain: $\{x \mid x \geq 1\}$
Range: $\{y \mid y \leq 0\}$

SECTION 9.3

You Try It 1
$$(f + g)(-2) = f(-2) + g(-2)$$
$$= [(-2)^2 + 2(-2)] + [5(-2) - 2]$$
$$= (4 - 4) + (-10 - 2)$$
$$= -12$$

$(f + g)(-2) = -12$

You Try It 2 $(f \cdot g)(3) = f(3) \cdot g(3)$
$$= (4 - 3^2) \cdot [3(3) - 4]$$
$$= (4 - 9) \cdot (9 - 4)$$
$$= (-5)(5)$$
$$= -25$$

$(f \cdot g)(3) = -25$

You Try It 3 $\left(\dfrac{f}{g}\right)(4) = \dfrac{f(4)}{g(4)}$
$$= \frac{4^2 - 4}{4^2 + 2 \cdot 4 + 1}$$
$$= \frac{16 - 4}{16 + 8 + 1}$$
$$= \frac{12}{25}$$

$\left(\dfrac{f}{g}\right)(4) = \dfrac{12}{25}$

You Try It 4
$$g(x) = x^2$$
$$g(-1) = (-1)^2 = 1 \qquad \bullet \ \text{Evaluate } g \text{ at } -1.$$

$$f(x) = 1 - 2x \qquad \bullet \ \text{Evaluate } f \text{ at}$$
$$f[g(-1)] = f(1) = 1 - 2(1) = -1 \qquad g(-1) = 1.$$
$$f[g(-1)] = -1$$

You Try It 5
$$M(s) = s^3 + 1 \qquad \bullet \ \text{Evaluate } M \text{ at } L(s).$$

$$M[L(s)] = (s + 1)^3 + 1$$
$$= s^3 + 3s^2 + 3s + 1 + 1$$
$$= s^3 + 3s^2 + 3s + 2$$
$$M[L(s)] = s^3 + 3s^2 + 3s + 2$$

SECTION 9.4

You Try It 1 Because any horizontal line intersects the graph at most once, the graph is the graph of a 1–1 function.

You Try It 2
$$f(x) = \frac{1}{2}x + 4$$
$$y = \frac{1}{2}x + 4$$
$$x = \frac{1}{2}y + 4$$
$$x - 4 = \frac{1}{2}y$$
$$2x - 8 = y$$
$$f^{-1}(x) = 2x - 8$$

The inverse of the function is given by $f^{-1}(x) = 2x - 8$.

You Try It 3 $f[g(x)] = 2\left(\dfrac{1}{2}x - 3\right) - 6$
$$= x - 6 - 6 = x - 12$$

No, $g(x)$ is not the inverse of $f(x)$.

Solutions to Chapter 10 "You Try It"

SECTION 10.1

You Try It 1

$$f(x) = \left(\frac{2}{3}\right)^x$$

$$f(3) = \left(\frac{2}{3}\right)^3 = \frac{8}{27} \qquad \bullet \; x = 3$$

$$f(-2) = \left(\frac{2}{3}\right)^{-2} = \left(\frac{3}{2}\right)^2 = \frac{9}{4} \qquad \bullet \; x = -2$$

You Try It 2

$$f(x) = 2^{2x+1}$$

$$f(0) = 2^{2(0)+1} = 2^1 = 2 \qquad \bullet \; x = 0$$

$$f(-2) = 2^{2(-2)+1} = 2^{-3} = \frac{1}{2^3} = \frac{1}{8} \qquad \bullet \; x = -2$$

You Try It 3

$$f(x) = e^{2x-1}$$

$$f(2) = e^{2 \cdot 2 - 1} = e^3 \approx 20.0855 \qquad \bullet \; x = 2$$

$$f(-2) = e^{2(-2)-1} = e^{-5} \approx 0.0067 \qquad \bullet \; x = -2$$

You Try It 4

x	y
-4	4
-2	2
0	1
2	$\frac{1}{2}$
4	$\frac{1}{4}$

You Try It 5

x	y
-2	$\frac{5}{4}$
-1	$\frac{3}{2}$
0	2
1	3
2	5

You Try It 6

x	y
-2	6
-1	4
0	3
1	$\frac{5}{2}$
2	$\frac{9}{4}$

You Try It 7

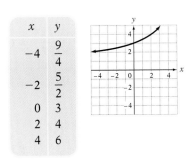

x	y
-4	$\frac{9}{4}$
-2	$\frac{5}{2}$
0	3
2	4
4	6

SECTION 10.2

You Try It 1

$$\log_4 64 = x \qquad \bullet \text{ Write the equivalent}$$
$$64 = 4^x \qquad \quad \text{exponential form.}$$
$$4^3 = 4^x \qquad \bullet \; 64 = 4^3$$
$$3 = x \qquad \bullet \text{ The bases are the same.}$$
$$\log_4 64 = 3 \qquad \quad \text{The exponents are equal.}$$

You Try It 2

$$\log_2 x = -4$$
$$2^{-4} = x \qquad \bullet \text{ Write the equivalent}$$
$$\frac{1}{2^4} = x \qquad \quad \text{exponential form.}$$
$$\frac{1}{16} = x$$

The solution is $\frac{1}{16}$.

You Try It 3

$$\ln x = 3$$
$$e^3 = x \qquad \bullet \text{ Write the equivalent}$$
$$20.0855 \approx x \qquad \text{exponential form.}$$

You Try It 4

$$\log_8 \sqrt[3]{xy^2} = \log_8(xy^2)^{1/3} = \frac{1}{3}\log_8(xy^2) \qquad \bullet \textbf{ Power Property}$$

$$= \frac{1}{3}(\log_8 x + \log_8 y^2) \qquad \bullet \textbf{ Product Property}$$

$$= \frac{1}{3}(\log_8 x + 2\log_8 y) \qquad \bullet \textbf{ Power Property}$$

$$= \frac{1}{3}\log_8 x + \frac{2}{3}\log_8 y \qquad \bullet \textbf{ Distributive Property}$$

You Try It 5

$$\frac{1}{3}(\log_4 x - 2\log_4 y + \log_4 z)$$

$$= \frac{1}{3}(\log_4 x - \log_4 y^2 + \log_4 z)$$

$$= \frac{1}{3}\left(\log_4 \frac{x}{y^2} + \log_4 z\right) = \frac{1}{3}\left(\log_4 \frac{xz}{y^2}\right)$$

$$= \log_4\left(\frac{xz}{y^2}\right)^{1/3} = \log_4 \sqrt[3]{\frac{xz}{y^2}}$$

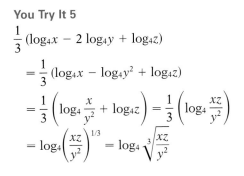

You Try It 6 Because $\log_b 1 = 0$, $\log_9 1 = 0$.

You Try It 7 $\log_3 0.834 = \dfrac{\ln 0.834}{\ln 3} \approx -0.16523$

You Try It 8 $\log_7 6.45 = \dfrac{\log 6.45}{\log 7} \approx 0.95795$

SECTION 10.3

You Try It 1

$$f(x) = \log_2(x - 1)$$
$$y = \log_2(x - 1) \quad \bullet \; f(x) = y$$
$$2^y = x - 1 \quad \bullet \; \text{Write the equivalent}$$
$$2^y + 1 = x \qquad \text{exponential equation.}$$

x	y
$\dfrac{5}{4}$	-2
$\dfrac{3}{2}$	-1
2	0
3	1
5	2

You Try It 2

$$f(x) = \log_3(2x)$$
$$y = \log_3(2x) \quad \bullet \; f(x) = y$$
$$3^y = 2x \quad \bullet \; \text{Write the equivalent}$$
$$\dfrac{3^y}{2} = x \qquad \text{exponential equation.}$$

x	y
$\dfrac{1}{18}$	-2
$\dfrac{1}{6}$	-1
$\dfrac{1}{2}$	0
$\dfrac{3}{2}$	1
$\dfrac{9}{2}$	2

You Try It 3

$$f(x) = -\log_3(x + 1)$$
$$y = -\log_3(x + 1) \quad \bullet \; f(x) = y$$
$$-y = \log_3(x + 1) \quad \bullet \; \text{Multiply both sides by } -1.$$
$$3^{-y} = x + 1 \quad \bullet \; \text{Write the equivalent}$$
$$3^{-y} - 1 = x \qquad \text{exponential equation.}$$

x	y
8	-2
2	-1
0	0
$-\dfrac{2}{3}$	1
$-\dfrac{8}{9}$	2

SECTION 10.4

You Try It 1

$$(1.06)^n = 1.5$$
$$\log(1.06)^n = \log 1.5 \qquad \bullet \; \text{Take the log of each side.}$$
$$n \log 1.06 = \log 1.5 \qquad \bullet \; \text{Power Property}$$
$$n = \dfrac{\log 1.5}{\log 1.06} \qquad \bullet \; \text{Divide both sides by log 1.06.}$$
$$n \approx 6.9585$$

The solution is 6.9585.

You Try It 2

$$4^{3x} = 25$$
$$\log 4^{3x} = \log 25 \qquad \bullet \; \text{Take the log of each side.}$$
$$3x \log 4 = \log 25 \qquad \bullet \; \text{Power Property}$$
$$3x = \dfrac{\log 25}{\log 4} \qquad \bullet \; \text{Divide both sides by log 4.}$$
$$3x \approx 2.3219$$
$$x \approx 0.7740 \qquad \bullet \; \text{Divide both sides by 3.}$$

The solution is 0.7740.

You Try It 3

$$\log_4(x^2 - 3x) = 1$$
$$4^1 = x^2 - 3x \qquad \bullet \; \text{Write in exponential form.}$$
$$4 = x^2 - 3x$$
$$0 = x^2 - 3x - 4$$
$$0 = (x + 1)(x - 4)$$
$$x + 1 = 0 \qquad x - 4 = 0$$
$$x = -1 \qquad x = 4$$

The solutions are -1 and 4.

You Try It 4

$$\log_3 x + \log_3(x + 3) = \log_3 4$$
$$\log_3[x(x + 3)] = \log_3 4 \qquad \bullet \; \text{Product Property}$$
$$x(x + 3) = 4 \qquad \bullet \; \text{1–1 Property of Logarithms}$$
$$x^2 + 3x = 4$$
$$x^2 + 3x - 4 = 0$$
$$(x + 4)(x - 1) = 0 \qquad \bullet \; \text{Solve by factoring.}$$
$$x + 4 = 0 \qquad x - 1 = 0$$
$$x = -4 \qquad x = 1$$

-4 does not check as a solution.
The solution is 1.

You Try It 5

$$\log_3 x + \log_3(x + 6) = 3$$
$$\log_3[x(x + 6)] = 3 \qquad \bullet \; \text{Product Property}$$
$$x(x + 6) = 3^3 \qquad \bullet \; \text{1–1 Property of Logarithms}$$
$$x^2 + 6x = 27$$
$$x^2 + 6x - 27 = 0$$
$$(x + 9)(x - 3) = 0 \qquad \bullet \; \text{Solve by factoring.}$$
$$x + 9 = 0 \qquad x - 3 = 0$$
$$x = -9 \qquad x = 3$$

-9 does not check as a solution.
The solution is 3.

SECTION 10.5

You Try It 1

Strategy To find the hydrogen ion concentration, replace pH by 2.9 in the equation pH = −log(H⁺) and solve for H⁺.

Solution

$$\text{pH} = -\log(\text{H}^+)$$
$$2.9 = -\log(\text{H}^+)$$
$$-2.9 = \log(\text{H}^+) \qquad \bullet \text{ Multiply by } -1.$$
$$10^{-2.9} = \text{H}^+ \qquad \bullet \text{ Write the equivalent}$$
$$0.00126 \approx \text{H}^+ \qquad\qquad \text{exponential equation.}$$

The hydrogen ion concentration is approximately 0.00126.

You Try It 2

Strategy To find the intensity, use the equation for the Richter scale magnitude of an earthquake, $M = \log \dfrac{I}{I_0}$. Replace M by 4.6 and solve for I.

Solution

$$M = \log \frac{I}{I_0}$$
$$4.6 = \log \frac{I}{I_0} \qquad \bullet \text{ Replace } M \text{ by 4.6.}$$
$$10^{4.6} = \frac{I}{I_0} \qquad \bullet \text{ Write in exponential form.}$$
$$10^{4.6}I_0 = I$$
$$39{,}811I_0 \approx I$$

The earthquake had an intensity that was approximately 39,811 times the intensity of a zero-level earthquake.

Solutions to Chapter 11 "You Try It"

SECTION 11.1

You Try It 1

$$y = x^2 + 2x + 1$$
$$-\frac{b}{2a} = -\frac{2}{2(1)} = -1$$

Axis of symmetry:
$$x = -1$$
$$y = (-1)^2 + 2(-1) + 1$$
$$= 0$$

Vertex: (−1, 0)

You Try It 2

$$x = -y^2 - 2y + 2$$
$$-\frac{b}{2a} = -\frac{-2}{2(-1)} = -1$$

Axis of symmetry:
$$y = -1$$
$$x = -(-1)^2 - 2(-1) + 2$$
$$= 3$$

Vertex: (3, −1)

You Try It 3

$$y = x^2 - 2x - 1$$
$$-\frac{b}{2a} = -\frac{-2}{2(1)} = 1$$

Axis of symmetry:
$$x = 1$$
$$y = 1^2 - 2(1) - 1$$
$$= -2$$

Vertex: (1, −2)

SECTION 11.2

You Try It 1

\bullet Center: (2, −3)
 Radius: $\sqrt{9} = 3$

You Try It 2

$$(x - h)^2 + (y - k)^2 = r^2$$
$$(x - 2)^2 + [y - (-3)]^2 = 4^2 \qquad \bullet \; h = 2,$$
$$(x - 2)^2 + (y + 3)^2 = 16 \qquad\quad k = -3,$$
$$\qquad\qquad\qquad\qquad\qquad\qquad\qquad r = 4$$

You Try It 3

$$x^2 + y^2 - 2x - 15 = 0$$
$$(x^2 - 2x) + y^2 = 15$$
$$(x^2 - 2x + 1) + y^2 = 15 + 1$$
$$(x - 1)^2 + y^2 = 16$$

Center: (1, 0)
Radius: 4

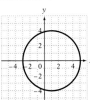

You Try It 4

$$x^2 + y^2 - 4x + 8y + 15 = 0$$
$$x^2 + y^2 - 4x + 8y = -15$$
$$(x^2 - 4x) + (y^2 + 8y) = -15$$
$$(x^2 - 4x + 4) + (y^2 + 8y + 16) = -15 + 4 + 16$$
$$(x - 2)^2 + (y + 4)^2 = 5$$

Center: $(2, -4)$
Radius: $\sqrt{5} \approx 2.2$

SECTION 11.3

You Try It 1
$a^2 = 4, b^2 = 25$
x-intercepts:
$(2, 0)$ and $(-2, 0)$

y-intercepts:
$(0, 5)$ and $(0, -5)$

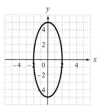

You Try It 2
$a^2 = 18, b^2 = 9$
x-intercepts:
$(3\sqrt{2}, 0)$ and $(-3\sqrt{2}, 0)$

y-intercepts:
$(0, 3)$ and $(0, -3)$

$$\left(3\sqrt{2} \approx 4\frac{1}{4}\right)$$

You Try It 3
$a^2 = 9, b^2 = 25$
Axis of symmetry:
x-axis

Vertices:
$(3, 0)$ and $(-3, 0)$

Asymptotes:
$y = \dfrac{5}{3}x$ and $y = -\dfrac{5}{3}x$

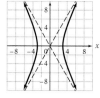

You Try It 4
$a^2 = 9, b^2 = 9$
Axis of symmetry:
y-axis

Vertices:
$(0, 3)$ and $(0, -3)$

Asymptotes:
$y = x$ and $y = -x$

SECTION 11.4

You Try It 1 (1) $y = 2x^2 + x - 3$
(2) $y = 2x^2 - 2x + 9$

Use the substitution method.

$$y = 2x^2 + x - 3$$
$$2x^2 - 2x + 9 = 2x^2 + x - 3$$
$$-3x + 9 = -3$$
$$-3x = -12$$
$$x = 4$$

Substitute into Equation (1).

$$y = 2x^2 + x - 3$$
$$y = 2(4)^2 + 4 - 3$$
$$y = 32 + 4 - 3$$
$$y = 33$$

The solution is $(4, 33)$.

You Try It 2
(1) $x^2 - y^2 = 10$
(2) $x^2 + y^2 = 8$

Use the addition method.

$$2x^2 = 18$$
$$x^2 = 9$$
$$x = \pm\sqrt{9} = \pm 3$$

Substitute into Equation (2).

$x^2 + y^2 = 8$	$x^2 + y^2 = 8$
$3^2 + y^2 = 8$	$(-3)^2 + y^2 = 8$
$9 + y^2 = 8$	$9 + y^2 = 8$
$y^2 = -1$	$y^2 = -1$
$y = \pm\sqrt{-1}$	$y = \pm\sqrt{-1}$

Because y is not a real number, the system of equations has no real number solution. The graphs do not intersect.

SECTION 11.5

You Try It 1 $\dfrac{x^2}{9} - \dfrac{y^2}{16} \leq 1$

Graph the hyperbola $\dfrac{x^2}{9} - \dfrac{y^2}{16} = 1$
as a solid curve. Shade the
region of the plane that includes
$(0, 0)$.

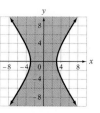

You Try It 2 $\dfrac{x^2}{16} + \dfrac{y^2}{4} > 1$

Graph the ellipse $\dfrac{x^2}{16} + \dfrac{y^2}{4} = 1$ as a dashed curve. Shade the region of the plane that does not include $(0, 0)$.

You Try It 3 $x^2 + y^2 < 16$
$y^2 > x$

Draw the circle $x^2 + y^2 = 16$ as a dashed curve. Shade inside the circle. Draw the parabola $y^2 = x$ as a dashed curve. Shade the region of the plane that includes $(-1, 0)$.

You Try It 4 $\dfrac{x^2}{4} + \dfrac{y^2}{9} \le 1$
$x > y^2 - 2$

Draw the ellipse $\dfrac{x^2}{4} + \dfrac{y^2}{9} = 1$ as a solid curve. Shade inside the ellipse. Draw the parabola $x = y^2 - 2$ as a dashed curve. Shade the region of the plane that includes $(0, 0)$.

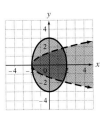

You Try It 5 $\dfrac{x^2}{16} + \dfrac{y^2}{25} \ge 1$
$x^2 + y^2 < 9$

Draw the ellipse $\dfrac{x^2}{16} + \dfrac{y^2}{15} = 1$ as a solid curve. Shade outside the ellipse. Draw the circle $x^2 + y^2 = 9$ as a dashed curve. Shade inside the circle.

The solution sets of the two inequalities do not intersect. This system of inequalities has no real number solution.

Solutions to Chapter 12 "You Try It"

SECTION 12.1

You Try It 1
$a_n = n(n + 1)$
$a_1 = 1(1 + 1) = 2$ The first term is 2.
$a_2 = 2(2 + 1) = 6$ The second term is 6.
$a_3 = 3(3 + 1) = 12$ The third term is 12.
$a_4 = 4(4 + 1) = 20$ The fourth term is 20.

You Try It 2
$a_n = \dfrac{1}{n(n + 2)}$

$a_6 = \dfrac{1}{6(6 + 2)} = \dfrac{1}{48}$ The sixth term is $\dfrac{1}{48}$.

$a_9 = \dfrac{1}{9(9 + 2)} = \dfrac{1}{99}$ The ninth term is $\dfrac{1}{99}$.

You Try It 3
$$\sum_{i=3}^{6} (i^2 - 2) = (3^2 - 2) + (4^2 - 2) + (5^2 - 2)$$
$$+ (6^2 - 2)$$
$$= 7 + 14 + 23 + 34 = 78$$

You Try It 4 $\displaystyle\sum_{n=1}^{4} \dfrac{x}{n} = \dfrac{x}{1} + \dfrac{x}{2} + \dfrac{x}{3} + \dfrac{x}{4}$
$$= x + \dfrac{x}{2} + \dfrac{x}{3} + \dfrac{x}{4}$$

SECTION 12.2

You Try It 1
$9, 3, -3, -9, \ldots$
$d = a_2 - a_1 = 3 - 9 = -6$ • Find d.
$a_n = a_1 + (n - 1)d$
$a_{15} = 9 + (15 - 1)(-6) = 9 + (14)(-6)$ • $n = 15$,
$= 9 - 84 = -75$ $d = -6$

You Try It 2

$7, 9, 11, \ldots, 59$

$d = a_2 - a_1 = 9 - 7 = 2$ • Find d.

$a_n = a_1 + (n - 1)d$
$59 = 7 + (n - 1)2$ • $a_n = 59, d = 2$
$59 = 7 + 2n - 2$
$59 = 5 + 2n$
$54 = 2n$
$27 = n$

There are 27 terms in the sequence.

You Try It 3

$-4, -2, 0, 2, 4, \ldots$

$d = a_2 - a_1 = -2 - (-4) = 2$ • Find d.

$a_n = a_1 + (n - 1)d$
$a_{25} = -4 + (25 - 1)2 = -4 + (24)2$ • $n = 25, d = 2$
$\quad = -4 + 48$
$a_{25} = 44$

$S_n = \dfrac{n}{2}(a_1 + a_n)$ • Use the formula.

$S_{25} = \dfrac{25}{2}(-4 + 44) = \dfrac{25}{2}(40) = 25(20)$

$S_{25} = 500$

You Try It 4 $\displaystyle\sum_{n=1}^{18}(3n - 2)$

$a_n = 3n - 2$
$a_1 = 3(1) - 2 = 1$ • Find a_1.
$a_{18} = 3(18) - 2 = 52$ • Find a_{18}.

$S_n = \dfrac{n}{2}(a_1 + a_n)$ • Use the formula.

$S_{18} = \dfrac{18}{2}(1 + 52) = 9(53) = 477$

You Try It 5

Strategy To find the value of the 20th-place prize:

• Write the equation for the nth-place prize.
• Find the 20th term of the sequence.

To find the total amount of prize money being awarded, use the Formula for the Sum of n Terms of an Arithmetic Sequence.

Solution

$10,000, 9700, \ldots$

$d = a_2 - a_1 = 9700 - 10,000 = -300$ • Find d.

$a_n = a_1 + (n - 1)d$ • Find the equation for
$\quad = 10,000 + (n - 1)(-300)$ the sequence.
$\quad = 10,000 - 300n + 300$
$\quad = -300n + 10,300$

$a_{20} = -300(20) + 10,300$ • Find the 20th term.
$\quad = -6000 + 10,300 = 4300$

The value of the 20th-place prize is \$4300.

$S_n = \dfrac{n}{2}(a_1 + a_n)$ • Use the formula.

$S_{20} = \dfrac{20}{2}(10,000 + 4300) = 10(14,300)$

$\quad = 143,000$

The total amount of prize money being awarded is \$143,000.

SECTION 12.3

You Try It 1

$5, 2, \dfrac{4}{5}, \ldots$

$r = \dfrac{a_2}{a_1} = \dfrac{2}{5}$ • Find r. $a_2 = 2, a_1 = 5$

$a_n = a_1 r^{n-1}$ • Use the formula.

$a_5 = 5\left(\dfrac{2}{5}\right)^{5-1} = 5\left(\dfrac{2}{5}\right)^4 = 5\left(\dfrac{16}{625}\right)$ • $n = 5, r = \dfrac{2}{5}$

$a_5 = \dfrac{16}{125}$

You Try It 2

$3, a_2, a_3, -192, \ldots$

$a_n = a_1 r^{n-1}$ • Use the formula.
$a_4 = 3r^{4-1}$ • $n = 4, a_1 = 3$
$-192 = 3r^{4-1}$ • $a_4 = -192$
$-192 = 3r^3$
$-64 = r^3$
$-4 = r$ • The common ratio is -4.

$a_n = a_1 r^{n-1}$
$a_3 = 3(-4)^{3-1} = 3(-4)^2 = 3(16) = 48$ • $n = 3, r = -4$

You Try It 3

$1, -\dfrac{1}{3}, \dfrac{1}{9}, -\dfrac{1}{27}$

$r = \dfrac{a_2}{a_1} = \dfrac{-\dfrac{1}{3}}{1} = -\dfrac{1}{3}$ • Find r. $a_2 = -\dfrac{1}{3}, a_1 = 1$

$S_n = \dfrac{a_1(1 - r^n)}{1 - r}$ • Use the formula.

$S_4 = \dfrac{1\left[1 - \left(-\dfrac{1}{3}\right)^4\right]}{1 - \left(-\dfrac{1}{3}\right)} = \dfrac{1 - \dfrac{1}{81}}{\dfrac{4}{3}} = \dfrac{\dfrac{80}{81}}{\dfrac{4}{3}}$

$\quad = \dfrac{80}{81} \cdot \dfrac{3}{4} = \dfrac{20}{27}$

You Try It 4 $\displaystyle\sum_{n=1}^{5}\left(\frac{1}{2}\right)^{n}$

$$a_n = \left(\frac{1}{2}\right)^n$$

$$a_1 = \left(\frac{1}{2}\right)^1 = \frac{1}{2} \qquad \text{• Find } a_1.$$

$$a_2 = \left(\frac{1}{2}\right)^2 = \frac{1}{4} \qquad \text{• Find } a_2.$$

$$r = \frac{a_2}{a_1} = \frac{\frac{1}{4}}{\frac{1}{2}} = \frac{1}{4} \cdot \frac{2}{1} = \frac{1}{2} \qquad \text{• Find } r.$$

$$S_n = \frac{a_1(1 - r^n)}{1 - r} \qquad \text{• Use the formula.}$$

$$S_5 = \frac{\frac{1}{2}\left[1 - \left(\frac{1}{2}\right)^5\right]}{1 - \frac{1}{2}} = \frac{\frac{1}{2}\left(1 - \frac{1}{32}\right)}{\frac{1}{2}} = \frac{\frac{1}{2}\left(\frac{31}{32}\right)}{\frac{1}{2}}$$

$$= \frac{31}{32}$$

You Try It 5 $3, -2, \dfrac{4}{3}, -\dfrac{8}{9}, \ldots$

$$r = \frac{a_2}{a_1} = \frac{-2}{3} = -\frac{2}{3} \qquad \text{• Find } r.\ a_2 = -2,\ a_1 = 3$$

$$S = \frac{a_1}{1 - r} = \frac{3}{1 - \left(-\frac{2}{3}\right)} \qquad \text{• Use the formula.}$$

$$= \frac{3}{1 + \frac{2}{3}} = \frac{3}{\frac{5}{3}} = \frac{9}{5}$$

You Try It 6 $0.\overline{66} = 0.6 + 0.06 + 0.006 + \cdots$

$$= \frac{6}{10} + \frac{6}{100} + \frac{6}{1000} + \cdots$$

$$S = \frac{a_1}{1 - r} = \frac{\frac{6}{10}}{1 - \frac{1}{10}} = \frac{\frac{6}{10}}{\frac{9}{10}} = \frac{6}{9} = \frac{2}{3}$$

An equivalent fraction is $\dfrac{2}{3}$.

You Try It 7
$0.3\overline{421} = 0.34 + 0.0021 + 0.000021 + \cdots$

$$= \frac{34}{100} + \frac{21}{10,000} + \frac{21}{1,000,000} + \cdots$$

$$S = \frac{a_1}{1 - r} = \frac{\frac{21}{10,000}}{1 - \frac{1}{100}} = \frac{\frac{21}{10,000}}{\frac{99}{100}} = \frac{21}{9900}$$

$$0.3\overline{421} = \frac{34}{100} + \frac{21}{9900} = \frac{3366}{9900} + \frac{21}{9900}$$

$$= \frac{3387}{9900} = \frac{1129}{3300}$$

An equivalent fraction is $\dfrac{1129}{3300}$.

You Try It 8

Strategy To find the total number of letters mailed, use the Formula for the Sum of n Terms of a Finite Geometric Series.

Solution $n = 6,\ a_1 = 3,\ r = 3$

$$S_n = \frac{a_1(1 - r^n)}{1 - r}$$

$$S_6 = \frac{3(1 - 3^6)}{1 - 3} = \frac{3(1 - 729)}{1 - 3}$$

$$= \frac{3(-728)}{-2} = \frac{-2184}{-2} = 1092$$

From the first through the sixth mailings, 1092 letters will have been mailed.

You Try It 9

Strategy To find the value of the investment, use the Formula for the nth Term of a Geometric Sequence, $a_n = a_1 r^{n-1}$.

Solution $n = 11,\ a_1 = 20{,}000,\ r = 1.08$

$$a_n = a_1 r^{n-1}$$

$$a_{11} = 20{,}000(1.08)^{11-1} = 20{,}000(1.08)^{10}$$

$$\approx 43{,}178.50$$

The value of the investment after 10 years is \$43,178.50.

SECTION 12.4

You Try It 1

$$(3m - n)^4 = \binom{4}{0}(3m)^4 + \binom{4}{1}(3m)^3(-n)$$

$$+ \binom{4}{2}(3m)^2(-n)^2 + \binom{4}{3}(3m)(-n)^3$$

$$+ \binom{4}{4}(-n)^4$$

$$= 1(81m^4) + 4(27m^3)(-n) + 6(9m^2)(n^2)$$

$$+ 4(3m)(-n^3) + 1(n^4)$$

$$= 81m^4 - 108m^3n + 54m^2n^2$$

$$- 12mn^3 + n^4$$

You Try It 2

$$(y - 2)^{10} = \binom{10}{0}y^{10} + \binom{10}{1}y^9(-2)$$

$$+ \binom{10}{2}y^8(-2)^2 + \cdots$$

$$= 1(y^{10}) + 10y^9(-2) + 45y^8(4) + \cdots$$

$$= y^{10} - 20y^9 + 180y^8 + \cdots$$

You Try It 3

$(t - 2s)^7$

$$\binom{n}{r - 1}a^{n-r+1}b^{r-1} \quad \bullet \; n = 7, a = t, b = -2s, r = 3$$

$$\binom{7}{3 - 1}(t)^{7-3+1}(-2s)^{3-1} = \binom{7}{2}(t)^5(-2s)^2$$

$$= 21t^5(4s^2) = 84t^5s^2$$

PREP TEST

1. $\frac{13}{20}$ **2.** $\frac{11}{60}$ **3.** $\frac{2}{9}$ **4.** $\frac{2}{3}$ **5.** 44.405 **6.** 73.63 **7.** 7.446 **8.** 54.06 **9. a.** Yes **b.** No **c.** Yes

d. Yes **10. a.** C **b.** D **c.** A **d.** B

SECTION 1.1

1. a. Integers: $0, -3$ **b.** Rational numbers: $-\frac{15}{2}, 0, -3, 2.\overline{33}$ **c.** Irrational numbers: $\pi, 4.232232223..., \frac{\sqrt{5}}{4}, \sqrt{7}$

d. Real numbers: all **3.** -27 **5.** $-\frac{3}{4}$ **7.** 0 **9.** $\sqrt{33}$ **11.** 91 **13.** 7 **15.** $-2, -1$ **17.** $9, 0, -9$

19. $4, 0, 4$ **21.** $\{-2, -1, 0, 1, 2, 3, 4\}$ **23.** $\{2, 4, 6, 8, 10, 12\}$ **25.** $\{3, 6, 9, 12, 15, 18, 21, 24, 27, 30\}$

27. $\{x \mid x > 4, x \in \text{integers}\}$ **29.** $\{x \mid x \geq -2\}$ **31.** $\{x \mid 0 < x < 1\}$ **33.** $\{x \mid 1 \leq x \leq 4\}$ **35.** $A \cup B = \{1, 2, 4, 6, 9\}$

37. $A \cup B = \{2, 3, 5, 8, 9, 10\}$ **39.** $A \cup B = \{-4, -2, 0, 2, 4, 8\}$ **41.** $A \cup B = \{1, 2, 3, 4, 5\}$ **43.** $A \cap B = \{6\}$

45. $A \cap B = \{5, 10, 20\}$ **47.** $A \cap B = \varnothing$ **49.** $A \cap B = \{4, 6\}$ **51.**

53. **55.** **57.**

59. **61.** **63.**

65. **67.** $\{x \mid 0 < x < 8\}$ **69.** $\{x \mid -5 \leq x \leq 7\}$ **71.** $\{x \mid -3 \leq x < 6\}$ **73.** $\{x \mid x \leq 4\}$

75. $\{x \mid x > 5\}$ **77.** $(-2, 4)$ **79.** $[-1, 5]$ **81.** $(-\infty, 1)$ **83.** $[-2, \infty)$ **85.**

87. **89.** **91.**

93. **95.** **97.** A **99.** B **101.** A **103.** R

105. R **107.** 0

SECTION 1.2

3. -30 **5.** -17 **7.** -96 **9.** -6 **11.** 5 **13.** 11,200 **15.** 20 **17.** 96 **19.** 23 **21.** 7 **23.** 17

25. -4 **27.** -62 **29.** -25 **31.** 7 **33.** 368 **35.** 26 **37.** -83 **41.** $\frac{43}{48}$ **43.** $-\frac{67}{45}$ **45.** $-\frac{13}{36}$

47. $\frac{11}{24}$ **49.** $\frac{13}{24}$ **51.** $-\frac{3}{56}$ **53.** $-\frac{2}{3}$ **55.** $-\frac{11}{14}$ **57.** $-\frac{1}{24}$ **59.** -12.974 **61.** -6.008 **63.** 1.9215

65. -6.02 **67.** -6.7 **69.** -1.11 **71.** -2030.44 **73.** 125 **75.** -8 **77.** -125 **79.** 324 **81.** -36

83. -72 **85.** 864 **87.** -160 **89.** 12 **91.** 2,654,208 **95.** -7 **97.** $\frac{177}{11}$ **99.** $\frac{16}{3}$ **101.** -40

103. 32 **105.** 24 **107.** $\frac{2}{15}$ **109.** $-\frac{1}{2}$ **111.** $-\frac{97}{72}$ **113.** 6.284 **115.** 5.4375 **117.** 1.891878 **119.** 0

121. No, the number zero has a multiplicative inverse that is undefined. **123.** 9 **125.** 625

127. First find b^c; then find $a^{(b^c)}$.

SECTION 1.3

1. 3 **3.** 3 **5.** 0 **7.** 6 **9.** 0 **11.** 1 **13.** $(2 \cdot 3)$ **15.** The Division Property of Zero

17. The Inverse Property of Multiplication **19.** The Addition Property of Zero **21.** The Division Property of Zero

23. The Distributive Property **25.** The Associative Property of Multiplication **27.** 10 **29.** 4 **31.** 0

33. $-\frac{1}{7}$ **35.** $\frac{9}{2}$ **37.** $\frac{1}{2}$ **39.** 2 **41.** 3 **43.** -12 **45.** 2 **47.** 6 **49.** -2 **51.** -14 **53.** 56

55. 256 **57.** $12x$ **59.** $-13ab$ **61.** $7x$ **63.** $-8a - 7b$ **65.** y **67.** $-5x + 45$ **69.** $-x - y$

71. $3a - 15$ **73.** $4x - 6y + 15$ **75.** $-7x + 14$ **77.** $48a - 75$ **79.** $-3x - 12y$ **81.** $-12x + 12y$

83. $33a - 25b$ **85.** $31x + 34y$ **87.** $4 - 8x - 5y$ **89.** $2x + 9$ **91.** Distributive Property

93. The statement is not correct; it mistakenly uses the Distributive Property. It is an irreducible statement. That is, the answer is $2 + 3x$. **95.** The statement is not correct; it mistakenly uses the Associative Property of Multiplication. The correct answer is $(2 \cdot 3)y = 6y$. **97.** Commutative Property of Addition

SECTION 1.4

1. $n - 8$ **3.** $\frac{4}{5}n$ **5.** $\frac{n}{14}$ **7.** $n - (n + 2) = -2$ **9.** $5(8n) = 40n$ **11.** $17n - 2n = 15n$

13. $n^2 - (12 + n^2) = -12$ **15.** $15n + (5n + 12) = 20n + 12$ **17.** $(15 - x + 2) + 2x = x + 17$ **19.** $\dfrac{5(34 - x)}{x - 3}$

21. Distance from Earth to the moon: d; $390d$ **23.** Amount of caramel in the mixture: c; $c + 3$

25. $10{,}000 - x$ **27.** Measure of angle B: x; measure of angle A: $2x$; measure of angle C: $4x$

29. The sum of twice x and 3 **31.** Twice the sum of x and 3 **33. a.** $\frac{1}{2}gt^2$ **b.** ma **c.** Av^2

CHAPTER 1 REVIEW EXERCISES*

1. $\{-2, -1, 0, 1, 2, 3\}$ [1.1B] **2.** $A \cap B = \{2, 3\}$ [1.1C] **3.** [1.1C]

4. The Associative Property of Multiplication [1.3A] **5.** -2.84 [1.2B] **6.** $\frac{7}{6}$ [1.3B] **7.** -288 [1.2C]

8. $16y - 15x + 18$ [1.3C] **9.** $\frac{3}{4}$ [1.1A] **10.** $\{x \mid x < -3\}$ [1.1B] **11.** [1.1C]

12. -15 [1.2A] **13.** $-\frac{7}{30}$ [1.2B] **14.** 4 [1.3A] **15.** $-\frac{5}{8}$ [1.2B] **16.** 0, 2 [1.1A]

17. 20 [1.3B] **18.** 14 [1.2A] **19.** 52 [1.2D] **20.** [1.1C]

21. $A \cup B = \{1, 2, 3, 4, 5, 6, 7, 8\}$ [1.1C] **22.** 12 [1.2A] **23.** $\{x \mid -2 \le x \le 3\}$ [1.1C] **24.** $\frac{2}{15}$ [1.2B]

25. 3 [1.3A] **26.** [1.1C] **27.** $-6x + 14$ [1.3C]

28. $-4, 0, -7$ [1.1A] **29.** The Inverse Property of Addition [1.3A] **30.** 3.1 [1.2B] **31.** 87 [1.1A]

32. $-1, 4$ [1.1A] **33.** $\{-3, -2, -1, 0, 1\}$ [1.1B] **34.** $\{x \mid x < 7\}$ [1.1B] **35.** $A \cup B = \{-4, -2, 0, 2, 4, 5, 10\}$ [1.1C] **36.** $A \cap B = \varnothing$ [1.1C] **37.** [1.1C] **38.** [1.1C]

39. 5 [1.2A] **40.** $-\frac{1}{2}$ [1.2B] **41.** 24 [1.2C] **42.** -107 [1.2D] **43.** -6 [1.3B] **44.** $5a + 9b$ [1.3C]

45. $4(x + 4)$; $4x + 16$ [1.4A] **46.** Flying time between San Diego and New York: t; $13 - t$ [1.4B]

47. Number of calories burned by walking at 4 mph for 1 h: C; $C + 396$ [1.4B] **48.** $2(x - 2) + 8$; $2x + 4$ [1.4A]

49. First integer: x; $4x + 5$ [1.4B] **50.** $12 - \dfrac{x + 3}{4} = \dfrac{-x + 45}{4}$ [1.4A] **51.** $2x + (40 - x) + 5 = x + 45$ [1.4A]

52. The width of the rectangle: w; $3w - 3$ [1.4B]

CHAPTER 1 TEST*

1. -30 [1.2A] **2.** $A \cap B = \{5, 7\}$ [1.1C] **3.** -72 [1.2C] **4.** [1.1C]

5. $A \cap B = \{-1, 0, 1\}$ [1.1C] **6.** -5 [1.3B] **7.** 2 [1.2A] **8.** $14x + 48y$ [1.3C] **9.** 12 [1.1A]

10. -100 [1.2C] **11.** [1.1C] **12.** 12 [1.2A] **13.** $\frac{25}{36}$ [1.2B]

14. 4 [1.3A] **15.** $-\frac{4}{27}$ [1.2B] **16.** -5 [1.1A] **17.** 2 [1.3B] **18.** -15 [1.2A] **19.** 10 [1.2D]

*Note: The numbers in brackets following the answers in the Chapter Review are a reference to the objective that corresponds to that problem. For example, the reference [1.2A] stands for Section 1.2, Objective A. This notation will be used for all Prep Tests, Chapter Reviews, Chapter Tests, and Cumulative Reviews throughout the text.

20. [1.1C] **21.** $A \cup B = \{1, 2, 3, 4, 5, 7\}$ [1.1C] **22.** $13x - y$ [1.3C]

23. 6 [1.2D] **24.** $\frac{2}{15}$ [1.2B] **25.** The Distributive Property [1.3A] **26.** [1.1C]

27. -1.41 [1.1C] **28.** $A \cup B = \{-2, -1, 0, 1, 2, 3\}$ [1.1C] **29.** $(x + 1) - 2(9 - x) = 3x - 17$ [1.4A]

30. Quantity of cocoa produced in Ghana: x; $3x$ [1.4B]

Answers to Chapter 2 Selected Exercises

PREP TEST

1. -4 [1.2A] **2.** -6 [1.2A] **3.** 3 [1.2A] **4.** 1 [1.2B] **5.** $-\frac{1}{2}$ [1.2B] **6.** $10x - 5$ [1.3C]

7. $6x - 9$ [1.3C] **8.** $3n + 6$ [1.3C] **9.** $0.03x + 20$ [1.3C] **10.** $20 - n$ [1.4B]

SECTION 2.1

5. Yes **7.** Yes **9.** 9 **11.** -10 **13.** -2 **15.** -15 **17.** 4 **19.** $-\frac{2}{3}$ **21.** $\frac{17}{6}$ **23.** $\frac{1}{6}$ **25.** $\frac{15}{2}$

27. $-\frac{32}{25}$ **29.** -21 **31.** $-\frac{2}{3}$ **33.** 1 **35.** $\frac{15}{16}$ **37.** -64 **39.** $\frac{5}{6}$ **41.** $\frac{3}{2}$ **43.** 0 **45.** 8 **47.** 6

49. 6 **51.** No solution **53.** -3 **55.** 1 **57.** $\frac{4}{3}$ **59.** 1 **61.** $-\frac{4}{3}$ **63.** -1.25 **65.** 2 **67.** $\frac{3}{2}$

69. -55 **71.** 6 **73.** $\frac{1}{2}$ **75.** -6 **77.** $\frac{5}{6}$ **79.** $-\frac{31}{7}$ **81.** $\frac{2}{3}$ **83.** $\frac{35}{12}$ **85.** -22 **87.** 6 **89.** $\frac{11}{2}$

91. $-\frac{4}{29}$ **93.** 9.4 **95.** 1500 **97.** $\frac{17}{8}$ **99.** 0 **101.** $r = \frac{C}{2\pi}$ **103.** $h = \frac{2A}{b}$ **105.** $M = \frac{IC}{100}$

107. $r = \frac{A - P}{Pt}$ **109.** $c = 2s - a - b$ **111.** $h = \frac{S - 2\pi r^2}{2\pi r}$ **113.** $R = Pn + C$ **115.** $H = \frac{S - 2WL}{2W + 2L}$

SECTION 2.2

1. The number is 5. **3.** The integers are 3 and 7. **5.** The integers are 21 and 29. **7.** The numbers are 21, 44, and 58. **9.** The integers are -20, -19, and -18. **11.** No solution **13.** The integers are 5, 7, and 9. **15.** There are 21 dimes in the collection. **17.** There are 12 dimes in the bank. **19.** There are eight 20¢ stamps and sixteen 15¢ stamps. **21.** There are five 3¢ stamps in the collection. **23.** There are seven 18¢ stamps in the collection. **25.** No solution

SECTION 2.3

1. The cost of the mixture is $2.95 per pound. **3.** 320 adult tickets were sold. **5.** The mixture must contain 22.5 L of imitation maple syrup. **7.** 20 oz of pure gold and 30 oz of the alloy were used. **9.** The cost of the mixture is $4.11 per pound. **11.** The mixture must contain 63.3 gal of cranberry juice. **13.** 200 lb of the 15% aluminum alloy must be used. **15.** The solution should contain 2.25 oz of rubbing alcohol and 1.25 oz of water. **17.** 45 oz of pure water must be added. **19.** 300 ml of alcohol must be added. **21.** The resulting drink is 6% fruit juice. **23.** 4 qt will have to be replaced with pure antifreeze. **25.** The helicopter will overtake the car 104 mi from the starting point. **27.** The speed of the first car is 58 mph. The speed of the second car is 66 mph. **29.** The distance between the two airports is 37.5 mi. **31.** The rate of the freight train is 30 mph. The rate of the passenger train is 48 mph.

33. The cyclist traveled 44 mi. **35.** 15 oz of the 12-karat gold should be used. **37.** The cars are $3.3\overline{3}$ (or $3\frac{1}{3}$ mi) apart 2 min before impact. **39.** The grass seed is 18 ft from the base of the 40-foot tower.

SECTION 2.4

3. a, c **5.** $\{x \mid x < 5\}$ (number line: $-5\,-4\,-3\,-2\,-1\ 0\ 1\ 2\ 3\ 4\ 5$) **7.** $\{x \mid x \le 2\}$ (number line: $-5\,-4\,-3\,-2\,-1\ 0\ 1\ 2\ 3\ 4\ 5$)

9. $\{x \mid x < -4\}$ (number line: $-5\,-4\,-3\,-2\,-1\ 0\ 1\ 2\ 3\ 4\ 5$) **11.** $\{x \mid x > 3\}$ **13.** $\{x \mid x > 4\}$ **15.** $\{x \mid x \le 2\}$ **17.** $\{x \mid x > -2\}$

19. $\{x \mid x \ge 2\}$ **21.** $\{x \mid x > -2\}$ **23.** $\{x \mid x \le 3\}$ **25.** $\{x \mid x < 2\}$ **27.** $\{x \mid x < -3\}$ **29.** $\{x \mid x \le 5\}$

31. $\left\{x \mid x \ge -\dfrac{1}{2}\right\}$ **33.** $\left(-\infty, \dfrac{23}{16}\right)$ **35.** $\left(-\infty, \dfrac{8}{3}\right)$ **37.** $(1, \infty)$ **39.** $\left(\dfrac{14}{11}, \infty\right)$ **41.** $(-\infty, 1]$ **43.** $\left(-\infty, \dfrac{7}{4}\right]$

45. $\left[-\dfrac{5}{4}, \infty\right)$ **47.** $(-\infty, 2]$ **51.** $[-2, 4]$ **53.** $(-\infty, 3) \cup (5, \infty)$ **55.** $(-4, 2)$ **57.** $(-\infty, -4) \cup (6, \infty)$

59. $(-\infty, -3)$ **61.** \varnothing **63.** $(-2, 1)$ **65.** $\{x \mid x < -2 \text{ or } x > 2\}$ **67.** $\{x \mid 2 < x < 6\}$ **69.** $\{x \mid -3 < x < -2\}$

71. $\left\{x \mid x > 5 \text{ or } x < -\dfrac{5}{3}\right\}$ **73.** \varnothing **75.** The set of real numbers. **77.** $\left\{x \mid \dfrac{17}{7} \le x \le \dfrac{45}{7}\right\}$ **79.** $\left\{x \mid -5 < x < \dfrac{17}{3}\right\}$

81. The set of real numbers. **83.** \varnothing **85.** The smallest number is -12. **87.** The maximum width of the rectangle is 11 cm. **89.** The TopPage plan is less expensive when service is for more than 460 pages. **91.** Paying with cash is less expensive when the call is 7 min or less. **93.** $32° < F < 86°$ **95.** George's amount of sales must be $44,000 or more. **97.** The first account is less expensive if more than 200 checks are written. **99.** $58 \le x \le 100$ **101.** The three even integers are 10, 12, and 14; or 12, 14, and 16; or 14, 16, and 18.

SECTION 2.5

1. Yes **3.** Yes **5.** 7 and -7 **7.** 4 and -4 **9.** 6 and -6 **11.** 7 and -7 **13.** No solution

15. No solution **17.** 1 and -5 **19.** 8 and 2 **21.** 2 **23.** No solution **25.** $-\dfrac{3}{2}$ and 3 **27.** $\dfrac{3}{2}$

29. No solution **31.** 7 and -3 **33.** 2 and $-\dfrac{10}{3}$ **35.** 1 and 3 **37.** $\dfrac{3}{2}$ **39.** No solution **41.** $\dfrac{11}{6}$ and $-\dfrac{1}{6}$

43. $-\dfrac{1}{3}$ and -1 **45.** No solution **47.** 3 and 0 **49.** No solution **51.** 1 and $\dfrac{13}{3}$ **53.** No solution

55. $\dfrac{7}{3}$ and $\dfrac{1}{3}$ **57.** $-\dfrac{1}{2}$ **59.** $-\dfrac{1}{2}$ and $-\dfrac{7}{2}$ **61.** $-\dfrac{8}{3}$ and $\dfrac{10}{3}$ **63.** No solution **65.** $\{x \mid x > 3 \text{ or } x < -3\}$

67. $\{x \mid x > 1 \text{ or } x < -3\}$ **69.** $\{x \mid 4 \le x \le 6\}$ **71.** $\{x \mid x \ge 5 \text{ or } x \le -1\}$ **73.** $\{x \mid -3 < x < 2\}$

75. $\left\{x \mid x > 2 \text{ or } x < -\dfrac{14}{5}\right\}$ **77.** \varnothing **79.** The set of real numbers. **81.** $\left\{x \mid x \le -\dfrac{1}{3} \text{ or } x \ge 3\right\}$

83. $\left\{x \mid -2 \le x \le \dfrac{9}{2}\right\}$ **85.** $\{x \mid x = 2\}$ **87.** $\left\{x \mid x < -2 \text{ or } x > \dfrac{22}{9}\right\}$ **89.** $\left\{x \mid -\dfrac{3}{2} < x < \dfrac{9}{2}\right\}$

91. $\left\{x \mid x < 0 \text{ or } x > \dfrac{4}{5}\right\}$ **93.** $\{x \mid x > 5 \text{ or } x < 0\}$ **95.** The lower and upper limits of the diameter of the bushing are 1.742 in. and 1.758 in. **97.** The lower and upper limits of the voltage of the electric motor are 195 volts and 245 volts.

99. The lower and upper limits of the length of the piston rod are $9\dfrac{19}{32}$ in. and $9\dfrac{21}{32}$ in. **101.** The lower and upper limits of the resistor are 28,420 ohms and 29,580 ohms. **103.** The lower and upper limits of the resistor are 23,750 ohms and 26,250 ohms. **105. a.** $\{x \mid x \ge -3\}$ **b.** $\{a \mid a \le 4\}$ **107. a.** \le **b.** \ge **c.** \ge **d.** $=$ **e.** $=$

CHAPTER 2 REVIEW EXERCISES

1. 6 [2.1B] **2.** $\left(\dfrac{5}{3}, \infty\right)$ [2.4A] **3.** $L = \dfrac{P - 2W}{2}$ [2.1D] **4.** -9 [2.1A] **5.** $\left\{x \mid -3 < x < \dfrac{4}{3}\right\}$ [2.4B]

6. $-\dfrac{40}{7}$ [2.1B] **7.** $-\dfrac{2}{3}$ [2.1A] **8.** -1 and 9 [2.5A] **9.** $\{x \mid 1 < x < 4\}$ [2.5B] **10.** $-\dfrac{9}{19}$ [2.1C]

11. $\dfrac{26}{17}$ [2.1C] **12.** $\{x \mid x > 2 \text{ or } x < -2\}$ [2.4B] **13.** $\left\{x \mid x \ge 2 \text{ or } x \le \dfrac{1}{2}\right\}$ [2.5B] **14.** $C = R - Pn$ [2.1D]

15. $-\dfrac{17}{2}$ [2.1C] **16.** No solution [2.5A] **17.** $(-\infty, \infty)$ [2.4B] **18.** $\dfrac{5}{2}$ [2.1C] **19.** -15 [2.1A]

20. $-\dfrac{1}{12}$ [2.1A] **21.** 7 [2.1A] **22.** $\dfrac{2}{3}$ [2.1A] **23.** $\dfrac{8}{5}$ [2.1B] **24.** 3 [2.1B] **25.** 9 [2.1C]

26. 8 [2.1C] **27.** $(-\infty, 1)$ [2.4A] **28.** $[-4, \infty)$ [2.4A] **29.** $\left\{x \mid x \le -\dfrac{39}{2}\right\}$ [2.4A] **30.** $\{x \mid x \ge 1\}$ [2.4A]

31. $(-1, 2)$ [2.4B] **32.** $-\dfrac{5}{2}, \dfrac{11}{2}$ [2.5A] **33.** $-\dfrac{8}{5}$ [2.5A] **34.** \varnothing [2.5B] **35.** The island is 16 mi from the dock. [2.3C] **36.** The mixture must contain 52 gal of apple juice. [2.3A] **37.** The executive's amount of sales must be \$55,000 or more. [2.4C] **38.** There are 7 quarters in the collection. [2.2B] **39.** The lower and upper limits of the diameter of the bushing are 2.747 in. and 2.753 in. [2.5C] **40.** The integers are 6 and 14. [2.2A] **41.** $82 \le x \le 100$ [2.4C] **42.** The speed of the first plane is 520 mph. The speed of the second plane is 440 mph. [2.3C] **43.** 375 lb of the 30% tin alloy and 125 lb of the 70% tin alloy were used. [2.3B] **44.** The lower and upper limits of the length of the piston are $10\dfrac{11}{32}$ in. and $10\dfrac{13}{32}$ in. [2.5C]

CHAPTER 2 TEST

1. -2 [2.1A] **2.** $-\dfrac{1}{8}$ [2.1A] **3.** $\dfrac{5}{6}$ [2.1A] **4.** 4 [2.1B] **5.** $\dfrac{32}{3}$ [2.1B] **6.** $-\dfrac{1}{5}$ [2.1B] **7.** 1 [2.1C]

8. $R = \dfrac{E - Ir}{I}$ [2.1D] **9.** $\dfrac{12}{7}$ [2.1C] **10.** $\{x \mid x \le -3\}$ [2.4A] **11.** $(-1, \infty)$ [2.4A] **12.** $\{x \mid x > -2\}$ [2.4B]

13. \varnothing [2.4B] **14.** $-\dfrac{9}{5}$ and 3 [2.5A] **15.** 7 and -2 [2.5A] **16.** $\left\{x \mid \dfrac{1}{3} \le x \le 3\right\}$ [2.5B]

17. $\left\{x \mid x < -\dfrac{1}{2} \text{ or } x > 2\right\}$ [2.5B] **18.** It costs less to rent from Gambelli agency if the car is driven less than 120 mi. [2.4C] **19.** The lower and upper limits of the diameter of the bushing are 2.648 in. and 2.652 in. [2.5C] **20.** The integers are 4 and 11. [2.2A] **21.** There are six 24¢ stamps. [2.2B] **22.** The price of the hamburger mixture is \$2.70 per pound. [2.3A] **23.** The jogger ran a total distance of 12 mi. [2.3C] **24.** The rate of the slower train is 60 mph. The rate of the faster train is 65 mph. [2.3C] **25.** It is necessary to add 100 oz of pure water. [2.3B]

CUMULATIVE REVIEW EXERCISES

1. -11 [1.2A] **2.** -108 [1.2C] **3.** 3 [1.2D] **4.** -64 [1.2D] **5.** -8 [1.3B] **6.** 1 [1.3B]

7. The Commutative Property of Addition [1.3A] **8.** $3n + (3n + 6) = 6n + 6$ [1.4A] **9.** $B = \dfrac{Fc}{ev}$ [2.1D]

10. $25y$ [1.3C] **11.** $\{-4, 0\}$ [1.1C] **12.** $\xleftarrow{\;\;\;\;}$ [1.1C] **13.** $y = \dfrac{-Ax - C}{B}$ [2.1D]

$\begin{array}{ccccccccccc} & & & & & (& & & &] & \\ \hline -5 & -4 & -3 & -2 & -1 & 0 & 1 & 2 & 3 & 4 & 5 \end{array}$

14. $\dfrac{1}{2}$ [2.1A] **15.** 1 [2.1B] **16.** 24 [2.1B] **17.** 2 [2.1C] **18.** 2 [2.1C] **19.** 1 [2.1C]

20. $-\dfrac{13}{5}$ [2.1C] **21.** $(-\infty, 1]$ [2.4A] **22.** $\{x \mid -4 < x \le 1\}$ [2.4B] **23.** -1 and 4 [2.5A]

24. 7 and -4 [2.5A] **25.** $\left\{x \mid x < -\dfrac{4}{3} \text{ or } x > 2\right\}$ [2.5B] **26.** $\{x \mid -2 < x < 6\}$ [2.5B] **27.** The second account is cheaper if the customer writes fewer than 50 checks. [2.4C] **28.** The first integer is -3. [2.2A] **29.** There are 15 dimes. [2.2B] **30.** 80 oz of pure silver were used in the mixture. [2.3A] **31.** The speed of the slower plane is 220 mph. [2.3C] **32.** The lower and upper limits of the diameter of the bushing are 2.449 in. and 2.451 in. [2.5C] **33.** 3 L of 12% acid solution must be used in the mixture. [2.3B]

Answers to Chapter 3 Selected Exercises

PREP TEST

1. $-4x + 12$ [1.3C] **2.** 10 [1.2D] **3.** -2 [1.2D] **4.** 11 [1.3B] **5.** 2.5 [1.3B] **6.** 5 [1.3B]
7. 1 [1.3B] **8.** 4 [2.1B] **9.** $y = 2x - 7$ [2.1D]

SECTION 3.1

1.

(graph)

3. $A(0, 3), B(1, 1), C(3, -4), D(-4, 4)$

5.

(graph)

7.

(graph)

9.

(graph)

11.

(graph)

13.

(graph)

15.

(graph)

17. Length: 4.47; midpoint: (4, 3) **19.** Length: 2.24; midpoint: $\left(-1, \dfrac{7}{2}\right)$ **21.** Length: 5.10; midpoint: $\left(-\dfrac{1}{2}, -\dfrac{9}{2}\right)$

23. Length: 9.90; midpoint: $\left(\dfrac{3}{2}, \dfrac{3}{2}\right)$ **25.** Length: 3; midpoint: $\left(\dfrac{7}{2}, -5\right)$ **27.** Length: 4.18; midpoint: $\left(\dfrac{1}{2}, \dfrac{1}{2}\right)$

29. a. After 20 min., the temperature is 280°F. **b.** After 50 min., the temperature is 160°F.

31.

33.

37.

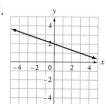

SECTION 3.2

3. Yes **5.** Yes **7.** No **9.** Function **11.** Function **13.** Function **15.** Not a function **17. a.** Yes

b. $y = \$34.75$ **19.** $f(3) = 11$ **21.** $f(0) = -4$ **23.** $G(0) = 4$ **25.** $G(-2) = 10$ **27.** $q(3) = 5$ **29.** $q(-2) = 0$

31. $F(4) = 24$ **33.** $F(-3) = -4$ **35.** $H(1) = 1$ **37.** $H(t) = \dfrac{3t}{t + 2}$ **39.** $s(-1) = 6$ **41.** $s(a) = a^3 - 3a + 4$

43. $4h$ **45. a.** \$4.75 per game **b.** \$4.00 per game **47. a.** \$3000 **b.** \$950 **49.** Domain = {1, 2, 3, 4, 5}

Range = {1, 4, 7, 10, 13}

51. Domain = {0, 2, 4, 6} **53.** Domain = {1, 3, 5, 7, 9} **55.** Domain = {-2, -1, 0, 1, 2}

Range = {1, 2, 3, 4} Range = {0} Range = {0, 1, 2}

57. Domain = {-2, -1, 0, 1, 2} **59.** 1 **61.** -8 **63.** None **65.** None **67.** 0 **69.** None **71.** None

Range = {-3, 3, 6, 7, 9}

73. -2 **75.** None **77.** Range = {-3, 1, 5, 9} **79.** Range = {-23, -13, -8, -3, 7} **81.** Range = {0, 1, 4}

83. Range = {2, 14, 26, 42} **85.** Range = $\left\{-5, \dfrac{5}{3}, 5\right\}$ **87.** Range = $\left\{-1, -\dfrac{1}{2}, -\dfrac{1}{3}, 1\right\}$ **89.** Range = {-38, -8, 2}

93. a. {(-2, -8), (-1, -1), (0, 0), (1, 1), (2, 8)}. **b.** Yes, the set of ordered pairs defines a function because each member of the domain is assigned to exactly one member of the range. **95.** The power produced will be 50.625 watts.

97. a. 22.5 ft/s **b.** 30 ft/s **99. a.** 68°F **b.** 52°F

SECTION 3.3

1.

(graph)

3.

(graph)

5.

(graph)

7.

(graph)

9.

11.

13.

15.

17.

19. *x*-intercept: (−4, 0); *y*-intercept: (0, 2)

21. *x*-intercept: $\left(\frac{9}{2}, 0\right)$; *y*-intercept: (0, −3)

23. *x*-intercept: $\left(\frac{3}{2}, 0\right)$; *y*-intercept: (0, 3)

25. *x*-intercept: $\left(\frac{4}{3}, 0\right)$; *y*-intercept: (0, 2)

27. *x*-intercept: $\left(\frac{3}{2}, 0\right)$; *y*-intercept: (0, −2)

29. Marlys receives $165 for tutoring 15 h.

31. The cost of receiving 32 messages is $14.40.

33. The cost of manufacturing 6000 compact discs is $33,000.

39.

SECTION 3.4

1. -1 **3.** $\dfrac{1}{3}$ **5.** $-\dfrac{2}{3}$ **7.** $-\dfrac{3}{4}$ **9.** Undefined **11.** $\dfrac{7}{5}$ **13.** 0 **15.** $-\dfrac{1}{2}$ **17.** Undefined **19.** $m = 40.$

The average speed of the motorist is 40 mph. **21.** $m = -5.$ The temperature of the oven decreases 5°/min.

23. $m = -0.05.$ For each mile the car is driven, approximately 0.05 gallon of fuel is used. **25.** $m \approx 343.9.$ The average

speed of the runner was 343.9 m/min. **27.** No **29.** $-3, 5, -3, (0, 5)$ **31.** $4, 0, 4, (0, 0)$

33. **35.** **37.** **39.**

41. **43.** **45.** **47.**

49. **51.** Increases by 2 **53.** Increases by 2 **55.** Decreases by $\dfrac{2}{3}$

57. i. D; ii. C; iii. B; iv. F; v. E; vi. A **61.** $k = -1$ **63.** $k = \dfrac{13}{2}$

SECTION 3.5

3. $y = 2x + 5$ **5.** $y = \dfrac{1}{2}x + 2$ **7.** $y = \dfrac{5}{4}x + \dfrac{21}{4}$ **9.** $y = -\dfrac{5}{3}x + 5$ **11.** $y = -3x + 9$ **13.** $y = -3x + 4$

15. $y = \dfrac{2}{3}x - \dfrac{7}{3}$ **17.** $y = \dfrac{1}{2}x$ **19.** $y = 3x - 9$ **21.** $y = -\dfrac{2}{3}x + 7$ **23.** $y = -x - 3$ **25.** $y = \dfrac{7}{5}x - \dfrac{27}{5}$

27. $y = -\dfrac{2}{5}x + \dfrac{3}{5}$ **29.** $x = 3$ **31.** $y = -\dfrac{5}{4}x - \dfrac{15}{2}$ **33.** $y = -3$ **35.** $y = -2x + 3$ **37.** $x = -5$

39. $y = x + 2$ **41.** $y = -2x - 3$ **43.** $y = \dfrac{2}{3}x + \dfrac{5}{3}$ **45.** $y = \dfrac{1}{3}x + \dfrac{10}{3}$ **47.** $y = \dfrac{3}{2}x - \dfrac{1}{2}$ **49.** $y = -\dfrac{3}{2}x + 3$

51. $y = -1$ **53.** $y = x - 1$ **55.** $y = -x + 1$ **57.** $y = -\dfrac{8}{3}x + \dfrac{25}{3}$ **59.** $y = \dfrac{1}{2}x - 1$ **61.** $y = -4$

63. $y = \dfrac{3}{4}x$ **65.** $y = -\dfrac{4}{3}x + \dfrac{5}{3}$ **67.** $x = -2$ **69.** $y = x - 1$ **71.** $y = \dfrac{4}{3}x + \dfrac{7}{3}$ **73.** $y = -x + 3$

75. $y = 1200x, 0 \le x \le 26\dfrac{2}{3}.$ The height of the plane 11 min after takeoff is 13,200 ft. **77. a.** $y = 85x + 30,000$

b. It will cost \$183,000 to build an 1800-ft² house. **79. a.** $y = -0.032x + 16, 0 \le x \le 500$ **b.** After 150 mi are

driven, 11.2 gal are left in the tank. **81. a.** $y = -20x + 230,000$ **b.** 60,000 cars would be sold at \$8500 each.

83. a. $y = 63x$ **b.** There are 315 Calories in a 5-ounce serving. **85.** $f(x) = x + 3$ **87.** 0 **89. a.** -10 **b.** 6

95. Answers will vary. The possible answers include $(0, 4), (3, 2),$ and $(9, -2).$ **97.** $y = -2x + 1$

SECTION 3.6

3. -5 **5.** $-\dfrac{1}{4}$ **7.** Yes **9.** No **11.** No **13.** Yes **15.** Yes **17.** Yes **19.** No **21.** Yes

23. $y = \dfrac{2}{3}x - \dfrac{8}{3}$ **25.** $y = \dfrac{1}{3}x - \dfrac{1}{3}$ **27.** $y = -\dfrac{5}{3}x - \dfrac{14}{3}$ **29.** $y = -2x + 15$ **31.** $\dfrac{A_1}{B_1} = \dfrac{A_2}{B_2}$ **33.** Any equation of

the form $y = 2x + b$ where $b \neq -13$ or of the form $y = -\dfrac{3}{2}x + c$, where $c \neq 8$.

SECTION 3.7

3. Yes **5.** No

7.

9.

11.

13.

15.

17.

19.

21.

23.

CHAPTER 3 REVIEW EXERCISES

1. $(4, 2)$ [3.1A] **2.** $P(-2) = -2$ [3.2A]
$P(a) = 3a + 4$

3. [3.1A]

4. [3.1A]

5. Range $= \{-1, 1, 5\}$ [3.2A] **6.** Domain $= \{-1, 0, 1, 5\}$ [3.2A] **7.** Midpoint: $\left(\dfrac{1}{2}, \dfrac{9}{2}\right)$; length: 5.10 [3.1B]
Range $= \{0, 2, 4\}$

8. -4 [3.2A] **9.** x-intercept: $(-3, 0)$ [3.3C]
y-intercept: $(0, -2)$

10. x-intercept: $(-2, 0)$ [3.3C]
y-intercept: $(0, -3)$

11. [3.3A]

12. [3.3B]

13. -1 [3.4A] **14.** $y = \dfrac{5}{2}x + \dfrac{23}{2}$ [3.5A] **15.** [3.1A] **16.** [3.4B]

17. $\{-2, -1, 2\}$ [3.2A] **18. a.** $y = -x + 295;\ 0 \le x \le 295$ **b.** When the rate is \$120, 175 rooms will be

occupied. [3.5C] **19.** $y = -4x - 5$ [3.6A] **20.** $y = \dfrac{5}{2}x + 8$ [3.6A] **21.** [3.3B]

22. [3.3B] **23.** $y = -\dfrac{2}{3}x + 1$ [3.5A] **24.** $y = \dfrac{1}{4}x + 4$ [3.5B] **25.** 10 [3.1B]

26. $(1, 3)$ [3.1B] **27.** [3.1C] **28.** [3.7A] **29.** [3.7A]

30. $y = -\dfrac{7}{6}x + \dfrac{5}{3}$ [3.5B] **31.** $y = 2x$ [3.6A] **32.** $y = -3x + 7$ [3.6A] **33.** $y = \dfrac{3}{2}x + 2$ [3.6A]

34. 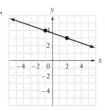 [3.4B] **35.** After 4 h, the car has traveled 220 mi. [3.3D]

36. The slope is 20. The manufacturing cost is \$20 per calculator. [3.4A] **37. a.** $y = 80x + 25,000$
b. The house will cost \$185,000 to build. [3.5C]

CHAPTER 3 TEST

1. 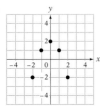 [3.1A] **2.** $(-3, 0)$ [3.1A] **3.** [3.3A] **4.** 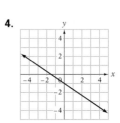 [3.3B]

5. $x = -2$ [3.5A] **6.** Length: 10.82; midpoint: $\left(-\dfrac{1}{2}, 5\right)$ [3.1B] **7.** $-\dfrac{1}{6}$ [3.4A] **8.** $P(2) = 9$ [3.2A]

9. [3.3C] **10.** [3.4B] **11.** $y = \frac{2}{5}x + 4$ [3.5A] **12.** 0 [3.2A]

13. $y = -\frac{7}{5}x + \frac{1}{5}$ [3.5B] **14.** $y = -3$ [3.5A] **15.** Domain $= \{-4, -2, 0, 3\}$ **16.** $y = -\frac{3}{2}x + \frac{7}{2}$ [3.6A]
Range $= \{0, 2, 5\}$ [3.2A]

17. $y = 2x + 1$ [3.6A] **18.** 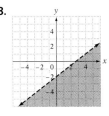 [3.7A] **19.** The slope is $-10,000$. The value of the house decreases by $10,000 each year. [3.4A]

20. a. $y = -\frac{3}{10}x + 175$ **b.** When the tuition is $300, 85 students will enroll. [3.5C]

CUMULATIVE REVIEW EXERCISES

1. The Commutative Property of Multiplication [1.3A] **2.** $\frac{9}{2}$ [2.1B] **3.** $\frac{8}{9}$ [2.1C] **4.** $-\frac{1}{14}$ [2.1C]

5. $\left\{ x \,\middle|\, x < -1 \text{ or } x > \frac{1}{2} \right\}$ [2.4B] **6.** $\frac{5}{2}$ and $-\frac{3}{2}$ [2.5A] **7.** $\left\{ x \,\middle|\, 0 < x < \frac{10}{3} \right\}$ [2.5B] **8.** 8 [1.2D]

9. -4.5 [1.3B] **10.** [1.1C] **11.** $C = R - Pn$ [2.1D] **12.** $x = 3 - \frac{3}{2}y$ [2.1D]

13. \varnothing [2.4B] **14.** 14 [3.2A] **15.** $(-8, 13)$ [3.1A] **16.** $-\frac{7}{4}$ [3.4A] **17.** $y = \frac{3}{2}x + \frac{13}{2}$ [3.5A]

18. $y = -\frac{5}{4}x + 3$ [3.5B] **19.** $y = -\frac{3}{2}x + 7$ [3.6A] **20.** $y = -\frac{2}{3}x + \frac{8}{3}$ [3.6A] **21.** There are 7 dimes in the

coin purse. [2.2B] **22.** The first plane is traveling at 200 mph and the second plane is traveling at 400 mph. [2.3C]

23. The mixture consists of 40 lb of $9.00 coffee and 20 lb of $6.00 coffee. [2.3A] **24.** [3.3C]

25. [3.4B] **26.** 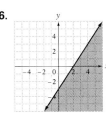 [3.7A]

27. a. $y = -5000x + 30,000$ **b.** The value of the truck decreases by $5000 per year. [3.5C]

Answers to Chapter 4 Selected Exercises

PREP TEST

1. $6x + 5y$ [1.3C] **2.** 7 [1.3B] **3.** 0 [2.1B] **4.** −3 [2.1C] **5.** 1000 [2.1C]

6. [3.3A] **7.** [3.3B] **8.** [3.7A]

SECTION 4.1

1. No **3.** Yes **5.** Independent **7.** Inconsistent **9.** **11.**

 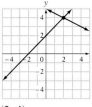

(3, −1) (2, 4)

13. **15.** **17.** **19.**

(4, 3) (4, −1) (2, −1) (4, −2)

21. **23.** **25.**

(3, −2) No solution $\left(x, \dfrac{2}{5}x - 2\right)$

27. (2, −1) **29.** (1, 2) **31.** (3, −2) **33.** $\left(-\dfrac{5}{2}, -4\right)$ **35.** (−1, 4) **37.** (−2, 5) **39.** (1, 5) **41.** (2, 0)

43. $\left(\dfrac{2}{3}, -1\right)$ **45.** No solution **47.** (−2, 5) **49.** (0, 0) **51.** $\left(x, \dfrac{1}{2}x - 4\right)$ **53.** (1, 5) **55.** $\left(\dfrac{2}{3}, 3\right)$

57. The amount invested at 4.2% is $5500. **59.** The amount invested at 6.5% must be $4000. **61.** The amount invested at 3.5% is $23,625. The amount invested at 4.5% is $18,375. **63.** There is $6000 invested in the mutual bond fund. **65.** (1.20, 1.40) **67.** (0.54, 1.03)

SECTION 4.2

1. (6, 1) **3.** (1, 1) **5.** (2, 1) **7.** (−2, 1) **9.** $(x, 3x - 4)$ **11.** $\left(-\dfrac{1}{2}, 2\right)$ **13.** No solution **15.** (−1, −2)

17. (−5, 4) **19.** (2, 5) **21.** $\left(\dfrac{1}{2}, \dfrac{3}{4}\right)$ **23.** (0, 0) **25.** (−1, 3) **27.** $\left(\dfrac{2}{3}, -\dfrac{2}{3}\right)$ **29.** (−2, 3) **31.** (2, −1)

33. (10, −5) **35.** $\left(-\dfrac{1}{2}, \dfrac{2}{3}\right)$ **37.** $\left(\dfrac{5}{3}, \dfrac{1}{3}\right)$ **39.** No solution **41.** (1, −1) **43.** (2, 1, 3) **45.** (1, −1, 2)

47. (1, 2, 4) **49.** (2, −1, −2) **51.** (−2, −1, 3) **53.** No solution **55.** (1, 4, 1) **57.** (1, 3, 2) **59.** (1, −1, 3)

61. $(0, 2, 0)$ **63.** $(1, 5, 2)$ **65.** $(-2, 1, 1)$ **69.** $(1, -1)$ **71.** $(1, -1)$ **73.** $y = \dfrac{2x}{1 - 3x}, \ y = \dfrac{-3x}{x + 2}$

SECTION 4.3

3. 11 **5.** 18 **7.** 0 **9.** 15 **11.** -30 **13.** 0 **15.** $(3, -4)$ **17.** $(4, -1)$ **19.** $\left(\dfrac{11}{14}, \dfrac{17}{21}\right)$ **21.** $\left(\dfrac{1}{2}, 1\right)$

23. Not independent **25.** $(-1, 0)$ **27.** $(1, -1, 2)$ **29.** $(2, -2, 3)$ **31.** Not independent **33a.** Sometimes true **b.** Always true **c.** Sometimes true **35.** The area is 239 ft².

SECTION 4.4

1. The rate of the motorboat in calm water is 15 mph. The rate of the current is 3 mph. **3.** The rate of the plane in calm air is 502.5 mph. The rate of the wind is 47.5 mph. **5.** The rate of the team in calm water is 8 km/h. The rate of the current is 2 km/h. **7.** The rate of the plane in calm air is 180 mph. The rate of the wind is 20 mph. **9.** The rate of the plane in calm air is 110 mph. The rate of the wind is 10 mph. **11.** There are 30 nickels in the bank. **13.** The cost of the pine is $1.10/ft. The cost of the redwood is $3.30/ft. **15.** The cost of the wool carpet is $52/yd. **17.** The company plans to manufacture 25 mountain bikes during the week. **19.** The chemist should use 480 g of the first alloy and 40 g of the second alloy. **21.** The Model VI computer costs $4000. **23.** There were 190 regular tickets, 350 member tickets, and 210 student tickets sold for the Saturday performance. **25.** The investor placed $10,400 in the 8% account, $5200 in the 6% account, and $9400 in the 4% account. **27.** The measures of the two angles are 35° and 145°. **29.** The age of the oil painting is 85 years, and the age of the watercolor is 50 years.

SECTION 4.5

1. **3.** **5.** **7.**

9. **11.** **13.** **15.**

17. **19.** **21.** **23.**

CHAPTER 4 REVIEW EXERCISES

1. $\left(6, -\dfrac{1}{2}\right)$ [4.1B] **2.** $(-4, 7)$ [4.2A] **3.** 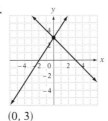 [4.1A] **4.** [4.1A]

$(0, 3)$

$(x, 2x - 4)$

5. $\left(x, -\dfrac{1}{4}x + \dfrac{3}{2}\right)$ [4.1B] **6.** $\left(x, \dfrac{1}{3}x - 2\right)$ [4.2A] **7.** $(3, -1, -2)$ [4.2B] **8.** $(5, -2, 3)$ [4.2B] **9.** 28 [4.3A]

10. 0 [4.3A] **11.** $(3, -1)$ [4.3B] **12.** $\left(\dfrac{110}{23}, \dfrac{25}{23}\right)$ [4.3B] **13.** $(-1, -3, 4)$ [4.3B] **14.** $(2, 3, -5)$ [4.3B]

15. [4.5A] **16.** [4.5A]

17. The rate of the cabin cruiser in calm water is 16 mph. The rate of the current is 4 mph. [4.4A]

18. The rate of the plane in calm air is 175 mph. The rate of the wind is 25 mph. [4.4A] **19.** On Friday, 100 children attended. [4.4B] **20.** The amount invested at 3% is $5000. The amount invested at 7% is $15,000. [4.1C]

CHAPTER 4 TEST

1. $(3, 4)$ [4.1A] **2.** $(-5, 0)$ [4.1A]

3. [4.5A] **4.** [4.5A] **5.** $\left(\dfrac{3}{4}, \dfrac{7}{8}\right)$ [4.1B]

6. $(-3, -4)$ [4.1B] **7.** $(2, -1)$ [4.1B] **8.** $(-2, 1)$ [4.2A] **9.** No solution [4.2A] **10.** $(1, 1)$ [4.2A]

11. No solution [4.2B] **12.** $(2, -1, -2)$ [4.2B] **13.** 10 [4.3A] **14.** -32 [4.3A] **15.** $\left(-\dfrac{1}{3}, -\dfrac{10}{3}\right)$ [4.3B]

16. $\left(\dfrac{59}{19}, -\dfrac{62}{19}\right)$ [4.3B] **17.** $\left(\dfrac{1}{5}, -\dfrac{6}{5}, \dfrac{3}{5}\right)$ [4.3B] **18.** The rate of the plane in calm air is 150 mph. The rate of the wind is 25 mph. [4.4A] **19.** The cost of cotton is $9/yd. The cost of wool is $14/yd. [4.4B]

20. The amount invested at 2.7% is $9000. The amount invested at 5.1% is $6000. [4.1C]

CUMULATIVE REVIEW EXERCISES

1. $-\dfrac{11}{28}$ [2.1C] **2.** $y = 5x - 11$ [3.5B] **3.** $3x - 24$ [1.3C] **4.** -4 [1.3B] **5.** $(-\infty, 6)$ [2.4B]

6. $\{x \mid -4 < x < 8\}$ [2.5B] **7.** $\{x \mid x < -1 \text{ or } x > 4\}$ [2.5B] **8.** $f(-3) = -98$ [3.2A]

9. The range is $\{0, 1, 5, 8, 16\}$. [3.2A] **10.** 1 [3.2A] **11.** $3h$ [3.2A] **12.** [1.1C]

13. $y = -\dfrac{2}{3}x + \dfrac{5}{3}$ [3.5A] **14.** $y = -\dfrac{3}{2}x + \dfrac{1}{2}$ [3.6A] **15.** 6.32 [3.1B] **16.** $\left(-\dfrac{1}{2}, 4\right)$ [3.1B]

17. [3.4B] **18.** [3.7A] **19.** $(2, 0)$ [4.1A]

20. [4.5A] **21.** $(1, 0, -1)$ [4.2B] **22.** 3 [4.3A] **23.** $(2, -3)$ [4.3B]

24. $(-5, -11)$ [4.1B] **25.** There are 16 nickels in the purse. [2.2B] **26.** The amount of water that should be added is 60 ml. [2.3B] **27.** The rate of the wind is 12.5 mph. [4.4A] **28.** The cost of 1 lb of steak is $6. [4.4B] **29.** The lower and upper limits of the resistor are 10,200 ohms and 13,800 ohms. [2.5C] **30.** The slope of the line is 40. The commission rate of the executive is $40 for every $1000 worth of sales. [3.4A]

Answers to Chapter 5 Selected Exercises

PREP TEST

1. $-12y$ [1.3C] **2.** -8 [1.2C] **3.** $3a - 8b$ [1.3C] **4.** $11x - 2y - 2$ [1.3C] **5.** $-x + y$ [1.3C]
6. $2 \cdot 2 \cdot 2 \cdot 5$ [1.2B] **7.** 4 [1.2B] **8.** -13 [1.3B] **9.** $-\dfrac{1}{3}$ [2.1B]

SECTION 5.1

1. a^4b^4 **3.** $-18x^3y^4$ **5.** x^8y^{16} **7.** $81x^8y^{12}$ **9.** $729a^{10}b^6$ **11.** x^5y^{11} **13.** $729x^6$ **15.** $a^{18}b^{18}$ **17.** $4096x^{12}y^{12}$
19. $64a^{24}b^{18}$ **21.** x^{2n+1} **23.** y^{6n-2} **25.** a^{2n^2-6n} **27.** x^{15n+10} **29.** $-6x^5y^5z^4$ **31.** $-12a^2b^9c^2$ **33.** $-6x^4y^4z^5$
35. $-432a^7b^{11}$ **37.** $54a^{13}b^{17}$ **39.** 243 **41.** y^3 **43.** $\dfrac{a^3b^2}{4}$ **45.** $\dfrac{x}{y^4}$ **47.** $\dfrac{1}{2}$ **49.** $-\dfrac{1}{9}$ **51.** $\dfrac{1}{y^8}$ **53.** $\dfrac{1}{x^6y^{10}}$
55. x^5y^5 **57.** $\dfrac{a^8}{b^9}$ **59.** $\dfrac{1}{2187a}$ **61.** $\dfrac{y^6}{x^3}$ **63.** $\dfrac{y^4}{x^3}$ **65.** $-\dfrac{1}{2x^2}$ **67.** $-\dfrac{1}{243a^5b^{10}}$ **69.** $\dfrac{16x^8}{81y^4z^{12}}$ **71.** $\dfrac{3a^4}{4b^3}$
73. $\dfrac{-9a}{8b^6}$ **75.** $\dfrac{16x^2}{y^6}$ **77.** $\dfrac{1}{b^{4n}}$ **79.** $-\dfrac{1}{y^{6n}}$ **81.** y^{n-2} **83.** $\dfrac{1}{y^{2n}}$ **85.** $\dfrac{x^{n-5}}{y^6}$ **87.** $\dfrac{8b^{15}}{3a^{18}}$ **89.** 4.67×10^{-6}
91. 1.7×10^{-10} **93.** 2×10^{11} **95.** 0.000000123 **97.** 8,200,000,000,000,000 **99.** 0.039 **101.** 150,000
103. 20,800,000 **105.** 0.000000015 **107.** 0.0000000000178 **109.** 140,000,000 **111.** 11,456,790
113. 0.000008 **115.** It would take a space ship 2.24×10^{15} h to cross the galaxy. **117.** It takes the satellite 8.86×10^3 h to reach Saturn. **119.** A proton is 1.83664508×10^3 times heavier than an electron. **121.** Light travels 8.64×10^{12} m in 8 h. **123.** The signals from Earth to Mars traveled $1.0\overline{81} \times 10^7$ mi/min. **125.** The centrifuge makes one revolution in 1.5×10^{-7} s. **127. a.** $\dfrac{8}{5}$ **b.** $\dfrac{5}{4}$

SECTION 5.2

1. $P(3) = 13$ **3.** $R(2) = 10$ **5.** $f(-1) = -11$ **7.** Polynomial: **a.** -1 **b.** 8 **c.** 2 **9.** Not a polynomial.
11. Not a polynomial. **13.** Polynomial: **a.** 3 **b.** π **c.** 5 **15.** Polynomial: **a.** -5 **b.** 2 **c.** 3
17. Polynomial: **a.** 14 **b.** 14 **c.** 0 **19.** **21.** **23.**

25. $6x^2 - 6x + 5$ **27.** $-x^2 + 1$ **29.** $5y^2 - 15y + 2$ **31.** $7a^2 - a + 2$ **33.** $3x^2 - 3xy - 2y^2$ **35.** $8x^2 - 2xy + 5y^2$
37. $S(x) = 3x^4 - 8x^2 + 2x$; $S(2) = 20$ **39. a.** $k = 8$ **b.** $k = -4$

SECTION 5.3

1. $2x^2 - 6x$ **3.** $6x^4 - 3x^3$ **5.** $6x^2y - 9xy^2$ **7.** $x^{n+1} + x^n$ **9.** $x^{2n} + x^ny^n$ **11.** $-4b^2 + 10b$
13. $-6a^4 + 4a^3 - 6a^2$ **15.** $-3y^4 - 4y^3 + 2y^2$ **17.** $-20x^2 + 15x^3 - 15x^4 - 20x^5$ **19.** $-2x^4y + 6x^3y^2 - 4x^2y^3$
21. $x^{3n} + x^{2n} + x^{n+1}$ **23.** $a^{2n+1} - 3a^{n+2} + 2a^{n+1}$ **25.** $5y^2 - 11y$ **27.** $6y^2 - 31y$ **29.** $9b^5 - 9b^3 + 24b$
31. $x^2 + 5x - 14$ **33.** $8y^2 + 2y - 21$ **35.** $8x^2 + 8xy - 30y^2$ **37.** $x^2y^2 + xy - 12$ **39.** $2x^4 - 15x^2 + 25$
41. $10x^4 - 15x^2y + 5y^2$ **43.** $x^{2n} - x^n - 6$ **45.** $6a^{2n} + a^n - 15$ **47.** $6a^{2n} + a^nb^n - 2b^{2n}$
49. $x^3 + 2x^2 - 11x + 20$ **51.** $10a^3 - 27a^2b + 26ab^2 - 12b^3$ **53.** $4x^5 - 16x^4 + 15x^3 - 4x^2 + 28x - 45$
55. $x^4 - 3x^3 - 6x^2 + 29x - 21$ **57.** $2a^3 + 7a^2 - 43a + 42$ **59.** $x^{3n} + 2x^{2n} + 2x^n + 1$
61. $x^{2n} - 2x^{2n}y^n + 4x^ny^n - 2x^ny^{2n} + 3y^{2n}$ **63.** $2y^5 - 10y^4 - y^3 - y^2 + 3$ **65.** $9x^2 - 4$ **67.** $36 - x^2$ **69.** $4a^2 - 9b^2$
71. $x^4 - 1$ **73.** $x^{2n} - 9$ **75.** $x^2 - 10x + 25$ **77.** $9a^2 + 30ab + 25b^2$ **79.** $x^4 - 6x^2 + 9$ **81.** $4x^4 - 12x^2y^2 + 9y^4$
83. $a^{2n} - 2a^nb^n + b^{2n}$ **85.** $-x^2 + 2xy$ **87.** $-4xy$ **89.** $(3x^2 + 10x - 8)$ ft^2 **91.** $(x^2 + 3x)$ m^2
93. $(x^3 + 9x^2 + 27x + 27)$ cm^3 **95.** $(2x^3)$ in^3 **97.** $(78.5x^2 + 125.6x + 50.24)$ in^2 **99. a.** $a^3 - b^3$ **b.** $x^3 + y^3$
101. a. $k = 5$ **b.** $k = 1$ **103.** $a^2 + b^2$

SECTION 5.4

1. $x - 2$ **3.** $-x + 2$ **5.** $xy + 2$ **7.** $x^2 + 3x - 5$ **9.** $3b^3 + 4b^2 + 2b$ **11.** $a^4 - 6a^2 + 1$
13. $x + 8$ **15.** $x^2 + 3x + 6 + \dfrac{20}{x - 3}$ **17.** $3x + 5 + \dfrac{3}{2x + 1}$ **19.** $5x + 7 + \dfrac{2}{2x - 1}$ **21.** $4x^2 + 6x + 9 + \dfrac{18}{2x - 3}$
23. $3x^2 + 1 + \dfrac{1}{2x^2 - 5}$ **25.** $x^2 - 3x - 10$ **27.** $x^2 - 2x + 1 - \dfrac{1}{x - 3}$ **29.** $2x^3 - 3x^2 + x - 4$ **31.** $2x + \dfrac{x + 2}{x^2 + 2x - 1}$
33. $x^2 + 4x + 6 + \dfrac{10x + 8}{x^2 - 2x - 1}$ **35.** $x^2 + 4 + \dfrac{3}{2x + 1}$ **37.** $2x - 8$ **39.** $3x - 8$ **41.** $3x + 3 - \dfrac{1}{x - 1}$
43. $2x^2 - 3x + 9$ **45.** $4x^2 + 8x + 15 + \dfrac{12}{x - 2}$ **47.** $2x^2 - 3x + 7 - \dfrac{8}{x + 4}$ **49.** $3x^3 + 2x^2 + 12x + 19 + \dfrac{33}{x - 2}$
51. $3x^3 - x + 4 - \dfrac{2}{x + 1}$ **53.** $3x + 1 + \dfrac{8}{x - 2}$ **55.** $P(3) = 8$ **57.** $R(4) = 43$ **59.** $P(-2) = -39$
61. $Z(-3) = -60$ **63.** $Q(2) = 31$ **65.** $F(-3) = 178$ **67.** $P(5) = 122$ **69.** $R(-3) = 302$ **71.** $Q(2) = 0$
73. a. $a^2 - ab + b^2$ **b.** $x^4 - x^3y + x^2y^2 - xy^3 + y^4$ **c.** $x^5 - x^4y + x^3y^2 - x^2y^3 + xy^4 - y^5$

SECTION 5.5

1. $3a(2a - 5)$ **3.** $x^2(4x - 3)$ **5.** Nonfactorable **7.** $x(x^4 - x^2 - 1)$ **9.** $4(4x^2 - 3x + 6)$ **11.** $5b^2(1 - 2b + 5b^2)$
13. $x^n(x^n - 1)$ **15.** $x^{2n}(x^n - 1)$ **17.** $a^2(a^{2n} + 1)$ **19.** $6x^2y(2y - 3x + 4)$ **21.** $4a^2b^2(6a - 1 - 4b^2)$
23. $y^2(y^{2n} + y^n - 1)$ **25.** $(a + 2)(x - 2)$ **27.** $(x - 2)(a + b)$ **29.** $(a - 2b)(x - y)$ **31.** $(x + 4)(y - 2)$
33. $(a + b)(x - y)$ **35.** $(y - 3)(x^2 - 2)$ **37.** $(3 + y)(2 + x^2)$ **39.** $(2a + b)(x^2 - 2y)$ **41.** $(2b + a)(3x - 2y)$
43. $(y - 5)(x^n + 1)$ **45.** $(2x - 1)(x^2 + 2)$ **47.** $(x + 10)(x + 2)$ **49.** $(a + 9)(a - 8)$ **51.** $(a + 6)(a + 1)$
53. $(y - 6)(y - 12)$ **55.** $(x + 12)(x - 11)$ **57.** $(x + 10)(x + 5)$ **59.** $(b + 2)(b - 8)$ **61.** $(a - b)(a - 2b)$
63. $(a + 11b)(a - 3b)$ **65.** $(x + 3y)(x + 2y)$ **67.** $-(x - 2)(x + 1)$ **69.** $-(x - 5)(x + 1)$ **71.** $(x - 3)(x - 2)$
73. $(2x + 1)(x + 3)$ **75.** $(2y + 3)(3y - 2)$ **77.** $(3b + 7)(2b - 5)$ **79.** $(3y - 13)(y - 3)$ **81.** Nonfactorable over
the integers **83.** $(a + 1)(4a - 5)$ **85.** $(5x - 2)(2x - 5)$ **87.** Nonfactorable over the integers
89. $(6x - y)(x + 7y)$ **91.** $(7a - 3b)(a + 7b)$ **93.** $(6x + 5y)(3x + 2y)$ **95.** $-(5x - 3)(x + 2)$
97. Nonfactorable over the integers **99.** $-(4b - 7)(2b + 5)$ **101.** $y^2(5y - 4)(y - 5)$ **103.** $-2x^2(5x - 2)(3x + 5)$
105. $y(4y - 3)(4y - 1)$ **107.** $f(x) = 2x - 3; g(x) = x + 6$ **109.** $F(x) = 2x + 1; G(x) = x - 5$
111. $f(b) = 4b - 5; g(b) = b - 3$ **113.** $f(x) = 2x + 1; h(x) = 2x - 7$ **115.** $f(x) = 4x + 3; g(x) = x + 5$
117. a. $-9, -6, 6,$ and 9 **b.** $-5, -1, 1,$ and 5 **c.** $-7, -5, 5,$ and 7 **d.** $-9, -3, 3,$ and 9 **e.** $-16, -8, 8,$ and 16
f. $-5, -1, 1,$ and 5

SECTION 5.6

1. $4; 25x^6; 100x^4y^4$ **3.** $4z^4$ **5.** $9a^2b^3$ **7.** $(x + 4)(x - 4)$ **9.** $(2x + 1)(2x - 1)$ **11.** $(4x + 11)(4x - 11)$
13. $(1 + 3a)(1 - 3a)$ **15.** $(xy + 10)(xy - 10)$ **17.** Nonfactorable over the integers **19.** $(5 + ab)(5 - ab)$

21. $(a^n + 1)(a^n - 1)$ **23.** $(x - 6)^2$ **25.** $(b - 1)^2$ **27.** $(4x - 5)^2$ **29.** Nonfactorable over the integers
31. Nonfactorable over the integers **33.** $(x + 3y)^2$ **35.** $(5a - 4b)^2$ **37.** $(x^n + 3)^2$ **39.** $(x - 7)(x - 1)$
41. $(x - y - a - b)(x - y + a + b)$ **43.** $8; x^9; 27c^{15}d^{18}$ **45.** $2x^3$ **47.** $4a^2b^6$ **49.** $(x - 3)(x^2 + 3x + 9)$
51. $(2x - 1)(4x^2 + 2x + 1)$ **53.** $(x - y)(x^2 + xy + y^2)$ **55.** $(m + n)(m^2 - mn + n^2)$ **57.** $(4x + 1)(16x^2 - 4x + 1)$
59. $(3x - 2y)(9x^2 + 6xy + 4y^2)$ **61.** $(xy + 4)(x^2y^2 - 4xy + 16)$ **63.** Nonfactorable over the integers
65. Nonfactorable over the integers **67.** $(a - 2b)(a^2 - ab + b^2)$ **69.** $(x^{2n} + y^n)(x^{4n} - x^{2n}y^n + y^{2n})$
71. $(x^n + 2)(x^{2n} - 2x^n + 4)$ **73.** $(xy - 5)(xy - 3)$ **75.** $(xy - 12)(xy - 5)$ **77.** $(x^2 - 3)(x^2 - 6)$
79. $(b^2 + 5)(b^2 - 18)$ **81.** $(x^2y^2 - 2)(x^2y^2 - 6)$ **83.** $(x^n + 1)(x^n + 2)$ **85.** $(3xy - 5)(xy - 3)$
87. $(2ab - 3)(3ab - 7)$ **89.** $(x^2 + 1)(2x^2 - 15)$ **91.** $(2x^n - 1)(x^n - 3)$ **93.** $(2a^n + 5)(3a^n + 2)$ **95.** $3(2x - 3)^2$
97. $a(3a - 1)(9a^2 + 3a + 1)$ **99.** $5(2x + 1)(2x - 1)$ **101.** $y^3(y + 11)(y - 5)$ **103.** $(4x^2 + 9)(2x + 3)(2x - 3)$
105. $2a(2 - a)(4 + 2a + a^2)$ **107.** $b^3(ab - 1)(a^2b^2 + ab + 1)$ **109.** $2x^2(2x - 5)^2$ **111.** $(x^2 + y^2)(x + y)(x - y)$
113. $(x^2 + y^2)(x^4 - x^2y^2 + y^4)$ **115.** Nonfactorable over the integers **117.** $2a(2a - 1)(4a^2 + 2a + 1)$
119. $a^2b^2(a + 4b)(a - 12b)$ **121.** $2b^2(3a + 5b)(4a - 9b)$ **123.** $(x - 2)^2(x + 2)$ **125.** $(2x + 1)(2x - 1)(x + y)(x - y)$
127. $x(x^n + 1)^2$ **129.** $b^n(b + 2)(3b - 2)$

SECTION 5.7

1. -3 and 5 **3.** -7 and 8 **5.** $-4, 0,$ and $\frac{2}{3}$ **7.** -5 and 3 **9.** 1 and 3 **11.** -2 and 5 **13.** $-\frac{1}{2}$ and $\frac{3}{2}$

15. $\frac{1}{2}$ **17.** -3 and 3 **19.** $-\frac{1}{2}$ and $\frac{1}{2}$ **21.** -3 and 5 **23.** -1 and $\frac{13}{2}$ **25.** -2 and $\frac{4}{3}$ **27.** $-5, 0,$ and 2

29. -2 and 2 **31.** $-3, -1,$ and 3 **33.** $-2, -\frac{2}{3},$ and 2 **35.** $-2, -\frac{2}{5},$ and 2 **37.** The number is -15 or 14.

39. The width of the rectangle is 10 cm and the length is 38 cm. **41.** The length of the hypotenuse is 13 ft.
43. The penny will hit the bottom of the wishing well in 2 s. **45.** The width of the cardboard is 8 in. and the length is 18 in.

CHAPTER 5 REVIEW EXERCISES

1. $6a^2b(3a^3b - 2ab^2 + 5)$ [5.5A] **2.** $5x + 4 + \frac{6}{3x - 2}$ [5.4B] **3.** $144x^2y^{10}z^{14}$ [5.1B]

4. $(a + 2b)(2x - 3y)$ [5.5B] **5.** $-(x - 4)(x + 3)$ [5.5C] **6.** $P(2) = 1$ [5.4D] **7.** $4x^2 - 8xy + 5y^2$ [5.2B]
8. $(6x + 5)(4x + 3)$ [5.5D] **9.** $(2x + 3y)^2$ [5.6A] **10.** $-6a^3b^6$ [5.1A] **11.** $(4a - 3b)(16a^2 + 12ab + 9b^2)$ [5.6B]
12. $4x^2 + 3x - 8 + \frac{50}{x + 6}$ [5.4C] **13.** $P(-2) = -7$ [5.2A] **14.** $(x - 8)(x + 5)$ [5.5C]

15. $(xy + 3)(xy - 3)$ [5.6A] **16.** $12x^5y^3 + 8x^3y^2 - 28x^2y^4$ [5.3A] **17.** $(x^n - 6)^2$ [5.6A] **18.** $\frac{5}{2}$ and 4 [5.7A]

19. $33x^2 + 24x$ [5.3A] **20.** $3a^2(a^2 - 6)(a^2 + 1)$ [5.6D] **21.** $16x^2 - 24xy + 9y^2$ [5.3C]

22. $x^3 + 4x^2 + 16x + 64 + \frac{252}{x - 4}$ [5.4C] **23.** $2x^2 - 5x - 2$ [5.2B] **24.** $70xy^2z^6$ [5.1B] **25.** $-\frac{x^3}{4y^2z^3}$ [5.1B]

26. 9.48×10^8 [5.1C] **27.** 2×10^{-6} [5.1C] **28.** 68 [5.4D] **29.** $4x^4 - 2x^2 + 5$ [5.4A]

30. $2x - 3 - \frac{4}{6x + 1}$ [5.4B] **31.** $a^{3n+3} - 5a^{2n+4} + 2a^{2n+3}$ [5.3A] **32.** $x^4 + 3x^3 - 23x^2 - 29x + 6$ [5.3B]

33. $5ab^2(2a^2b - 4ab^2 + 7)$ [5.5A] **34.** $x^2(5x^{n+3} + x^{n+1} + 4)$ [5.5A] **35.** $(y - 3)(x - 4)$ [5.5B]
36. $(x - 7)(x - 9)$ [5.5C] **37.** $(8x - 1)(3x + 8)$ [5.5D] **38.** $f(x) = 5x - 2; g(x) = x + 1$ [5.5D]
39. $(6 + a^n)(6 - a^n)$ [5.6A] **40.** $(2 - y^n)(4 + 2y^n + y^{2n})$ [5.6B] **41.** $(6x^4 - 5)(6x^4 - 1)$ [5.6C]
42. $3ab(a - b)(a^2 + ab + b^2)$ [5.6D] **43.** $(x^n + 2)^2(x^n - 2)^2$ [5.6D] **44.** $-2, 0, 3$ [5.7A] **45.** $-4, 0, 4$ [5.7A]
46. $-1, -6, 6$ [5.7A] **47.** $(3x^2 + 2)(5x^2 - 3)$ [5.6C] **48.** $\frac{c^{10}}{2b^{17}}$ [5.1B] **49.** $6x^3 - 29x^2 + 14x + 24$ [5.3B]
50. $(7x^2y^2 + 3)(3x^2y^2 + 2)$ [5.6C] **51.** $-4, 1,$ and 4 [5.7A] **52.** $25a^2 - 4b^2$ [5.3C] **53.** 0.00254 [5.1C]

54. $(3x - 2)(2x - 9)$ [5.5D] **55.** [5.2A] **56. a.** 3 **b.** 8 **c.** 5 [5.2A]

57. The mass of the moon is 8.103×10^{19} tons. [5.1D] **58.** The number is -8 or 7. [5.7B] **59.** The Great Galaxy of Andromeda is 1.291224×10^{19} miles from Earth. [5.1D] **60.** The area is $(10x^2 - 29x - 21)$ cm². [5.3D]

CHAPTER 5 TEST

1. $(4t + 3)^2$ [5.6A] **2.** $-18r^2s^2 + 12rs^3 + 18rs^2$ [5.3A] **3.** $P(2) = -3$ [5.2A] **4.** $(3x - 2)(9x^2 + 6x + 4)$ [5.6B]

5. $(4x + 5)(4x - 5)$ [5.6A] **6.** $6t^5 - 8t^4 - 15t^3 + 22t^2 - 5$ [5.3B] **7.** $35x^2 - 55x$ [5.3A]

8. $3x(2x - 3)(2x + 5)$ [5.6D] **9.** $-1, -\dfrac{1}{6},$ and 1 [5.7A] **10.** $2x^3 - 4x^2 + 6x - 14$ [5.2B]

11. 5.01×10^{-7} [5.1C] **12.** $2x + 1 + \dfrac{4}{7x - 3}$ [5.4B] **13.** $49 - 25x^2$ [5.3C] **14.** $(2a^2 - 5)(3a^2 + 1)$ [5.6C]

15. $6a^2 - 13ab - 28b^2$ [5.3B] **16.** $(3x^2 + 4)(x - 3)(x + 3)$ [5.6D] **17.** $64a^7b^7$ [5.1A] **18.** $-\dfrac{1}{3}$ and $\dfrac{1}{2}$ [5.7A]

19. $P(-2) = -8$ [5.4D] **20.** $\dfrac{2b^7}{a^{10}}$ [5.1B] **21.** $x^2 - 5x + 10 - \dfrac{23}{x + 3}$ [5.4C] **22.** $(2x - 3)(3x - 4)$ [5.5D]

23. $(3x - 2)(2x - a)$ [5.5B] **24.** There are 6.048×10^5 s in one week. [5.1D] **25.** The arrow will be 64 ft above the ground at 1 s and at 2 s after the arrow has been released. [5.7B] **26.** The area is $(10x^2 - 3x - 1)$ ft². [5.3D]

CUMULATIVE REVIEW EXERCISES

1. 4 [1.2D] **2.** $-\dfrac{5}{4}$ [1.3B] **3.** The Inverse Property of Addition [1.3A] **4.** $-18x + 8$ [1.3C]

5. $-\dfrac{1}{6}$ [2.1A] **6.** $-\dfrac{11}{4}$ [2.1B] **7.** $x^2 + 3x + 9 + \dfrac{24}{x - 3}$ [5.4C] **8.** -1 and $\dfrac{7}{3}$ [2.5A]

9. $P(-2) = 18$ [3.2A] **10.** $x = -2$ [3.2A] **11.** $\{-4, -1, 8\}$ [3.2A] **12.** The slope is $-\dfrac{1}{6}$. [3.4A]

13. $y = -\dfrac{3}{2}x + \dfrac{1}{2}$ [3.5A] **14.** $y = \dfrac{2}{3}x + \dfrac{16}{3}$ [3.6A] **15.** $\left(-\dfrac{7}{5}, -\dfrac{8}{5}\right)$ [4.3B] **16.** $\left(-\dfrac{9}{7}, \dfrac{2}{7}, \dfrac{11}{7}\right)$ [4.2B]

17. 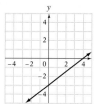 [3.3C] **18.** [3.7A] **19.** $(1, -1)$ [4.1A]

20. [4.5A] **21.** $\dfrac{b^5}{a^8}$ [5.1B] **22.** $\dfrac{y^2}{25x^6}$ [5.1B] **23.** $\dfrac{21}{8}$ [5.1B] **24.** $4x^3 - 7x + 3$ [5.3B]

25. $-2x(2x - 3)(x - 2)$ [5.5D] **26.** $(x - y)(a + b)$ [5.5B] **27.** $(x^2 + 4)(x + 2)(x - 2)$ [5.6D]
28. $2(x - 2)(x^2 + 2x + 4)$ [5.6D] **29.** The integers are 9 and 15. [2.2A] **30.** 40 oz of pure gold must be mixed with the alloy. [2.3A] **31.** The slower cyclist travels at 5 mph, the faster cyclist at 7.5 mph. [2.3C] **32.** The vehicle will reach the moon in 12 h. [5.1D] **33.** $m = 50$. The average speed is 50 mph. [3.4A]

Answers to Chapter 6 Selected Exercises

PREP TEST

1. 50 [1.2B] **2.** $-\dfrac{1}{6}$ [1.2B] **3.** $-\dfrac{3}{2}$ [1.2B] **4.** $\dfrac{1}{24}$ [1.2B] **5.** $\dfrac{5}{24}$ [1.2B] **6.** $-\dfrac{2}{9}$ [1.2D]

7. $\dfrac{1}{3}$ [1.3B] **8.** -2 [2.1C] **9.** $\dfrac{10}{7}$ [2.1C] **10.** The rates of the planes are 110 mph and 130 mph. [2.3C]

SECTION 6.1

3. 2 **5.** -2 **7.** $\dfrac{1}{9}$ **9.** $\dfrac{1}{35}$ **11.** $\dfrac{3}{4}$ **13.** The domain is $\{x\,|\,x \neq 3\}$. **15.** The domain is $\{x\,|\,x \neq -4\}$.

17. The domain is $\{x\,|\,x \neq -3\}$. **19.** The domain is $\left\{x\,\middle|\,x \neq \dfrac{2}{3}, 4\right\}$. **21.** The domain is $\{x\,|\,x \neq -3, 2\}$.

23. The domain is $\{x\,|\,x \in \text{real numbers}\}$. **27.** $1 - 2x$ **29.** $3x - 1$ **31.** $2x$ **33.** $-\dfrac{2}{a}$

35. The expression is in simplest form. **37.** $x^2y^2 - 4xy + 5$ **39.** $\dfrac{x^n}{x^n - y^n}$ **41.** $\dfrac{x - 3}{x - 5}$ **43.** $\dfrac{x + 4}{x - 4}$

45. $\dfrac{a - b}{a^2 - ab + b^2}$ **47.** $\dfrac{4x^2 + 2xy + y^2}{2x + y}$ **49.** $\dfrac{(a - 2)(x + 1)}{ax}$ **51.** $\dfrac{x^2 - 3}{x^2 + 1}$ **53.** $\dfrac{2xy + 1}{3xy - 1}$ **55.** $\dfrac{a^n + 4}{a^n + 1}$ **57.** $\dfrac{a^n + b^n}{a^n - b^n}$

59. $\dfrac{a + b}{(x + 1)(x - 1)}$ **61.** $\dfrac{abx}{2}$ **63.** $\dfrac{x}{2}$ **65.** $\dfrac{y(x - 1)}{x^2(x + 1)}$ **67.** $-\dfrac{x + 5}{x - 2}$ **69.** 1 **71.** $\dfrac{x^n + 4}{x^n - 1}$

73. $\dfrac{(x + 1)(x - 1)(x - 4)}{x - 2}$ **75.** $\dfrac{x + y}{3}$ **77.** $\dfrac{4by}{3ax}$ **79.** $\dfrac{4(x - y)^2}{9x^2y}$ **81.** $\dfrac{2x - 3y}{4y^2}$ **83.** $-\dfrac{2x + 5}{2x - 5}$ **85.** $x(x - 3)$

87. -1 **89.** $(x^n + 1)^2$ **91.** $\dfrac{(x + y)(x - y)}{x^3}$ **93.** $\dfrac{x + 3}{x - 3}$ **95.** $\dfrac{5(3y^2 + 2)}{y^2}$

SECTION 6.2

1. $\dfrac{9y^3}{12x^2y^4}, \dfrac{17x}{12x^2y^4}$ **3.** $\dfrac{2x^2 - 4x}{6x^2(x - 2)}, \dfrac{3x - 6}{6x^2(x - 2)}$ **5.** $\dfrac{3x - 1}{2x(x - 5)}, -\dfrac{6x^3 - 30x^2}{2x(x - 5)}$ **7.** $\dfrac{6x^2 + 9x}{(2x + 3)(2x - 3)}, \dfrac{10x^2 - 15x}{(2x + 3)(2x - 3)}$

9. $\dfrac{2x}{(x + 3)(x - 3)}, \dfrac{x^2 + 4x + 3}{(x + 3)(x - 3)}$ **11.** $\dfrac{6}{6(x + 2y)(x - 2y)}, \dfrac{5x + 10y}{6(x + 2y)(x - 2y)}$ **13.** $\dfrac{3x^2 - 3x}{(x + 1)(x - 1)^2}, \dfrac{5x^2 + 5x}{(x + 1)(x - 1)^2}$

15. $-\dfrac{x - 3}{(x - 2)(x^2 + 2x + 4)}, \dfrac{2x - 4}{(x - 2)(x^2 + 2x + 4)}$ **17.** $\dfrac{2x^2 + 6x}{(x - 1)(x + 3)^2}, -\dfrac{x^2 - x}{(x - 1)(x + 3)^2}$

19. $-\dfrac{12x^2 - 8x}{(2x - 3)(2x - 5)(3x - 2)}, \dfrac{6x^2 - 9x}{(2x - 3)(2x - 5)(3x - 2)}$ **21.** $\dfrac{5}{(3x - 4)(2x - 3)}, -\dfrac{4x^2 - 6x}{(3x - 4)(2x - 3)}, \dfrac{3x^2 - x - 4}{(3x - 4)(2x - 3)}$

23. $\dfrac{2x^2 + 10x}{(x + 5)(x - 3)}, -\dfrac{2x - 6}{(x + 5)(x - 3)}, -\dfrac{x - 1}{(x + 5)(x - 3)}$ **25.** $\dfrac{x - 5}{(x^n + 1)(x^n + 2)}, \dfrac{2x^{n+1} + 2x}{(x^n + 1)(x^n + 2)}$ **27.** $\dfrac{1}{2x^2}$ **29.** $\dfrac{1}{x + 2}$

31. $\dfrac{12ab - 9b + 8a}{30a^2b^2}$ **33.** $\dfrac{5 - 16b + 12a}{40ab}$ **35.** $\dfrac{7}{12x}$ **37.** $\dfrac{2xy - 8x + 3y}{10x^2y^2}$ **39.** $-\dfrac{a(2a - 13)}{(a + 1)(a - 2)}$ **41.** $\dfrac{5x^2 - 6x + 10}{(5x - 2)(2x - 5)}$

43. $\dfrac{a}{b(a - b)}$ **45.** $\dfrac{a^2 + 18a - 9}{a(a - 3)}$ **47.** $\dfrac{17x^2 + 20x - 25}{x(6x - 5)}$ **49.** $\dfrac{6}{(x - 3)^2(x + 3)}$ **51.** $-\dfrac{2(x - 1)}{(x + 2)^2}$ **53.** $-\dfrac{5x^2 - 17x + 8}{(x + 4)(x - 2)}$

55. $\dfrac{3x^n + 2}{(x^n + 1)(x^n - 1)}$ **57.** 1 **59.** $\dfrac{x^2 - 52x + 160}{4(x + 3)(x - 3)}$ **61.** $\dfrac{3x - 1}{4x + 1}$ **63.** $\dfrac{2(5x - 3)}{(x + 4)(x - 3)(x + 3)}$ **65.** $\dfrac{x - 2}{x + 3}$

67. 1 **69.** $\dfrac{x + 1}{2x - 1}$ **71.** $\dfrac{1}{2x - 1}$ **73.** $\dfrac{1}{x^2 + 4}$ **75.** $\dfrac{3 - a}{3a}$ **77.** $-\dfrac{2x^2 + 5x - 2}{(x + 2)(x + 1)}$ **79.** $\dfrac{b - a}{b + 2a}$ **81.** $\dfrac{2}{x + 2}$

83. a. $\dfrac{b + 6}{3(b - 2)}$ **b.** $\dfrac{6(x - 1)}{x(2x + 1)}$ **85. a.** $f(4) = \dfrac{2}{3}; g(4) = 4; S(4) = 4\dfrac{2}{3};$ Yes, $f(4) + g(4) = S(4)$ **b.** $S(a) = f(a) + g(a)$

SECTION 6.3

3. $\dfrac{5}{23}$ **5.** $\dfrac{2}{5}$ **7.** $\dfrac{x}{x-1}$ **9.** $-\dfrac{a}{a+2}$ **11.** $-\dfrac{a-1}{a+1}$ **13.** $\dfrac{2}{5}$ **15.** $-\dfrac{1}{2}$ **17.** $\dfrac{x^2-x-1}{x^2+x+1}$ **19.** $\dfrac{x+2}{x-1}$

21. $-\dfrac{x+4}{2x+3}$ **23.** $\dfrac{a(a^2+a+1)}{a^2+1}$ **25.** $\dfrac{x+1}{x-4}$ **27.** $\dfrac{(x-1)(x+1)}{x^2+1}$ **29.** $-\dfrac{1}{x(x+h)}$ **31.** $\dfrac{x-2}{x-3}$ **33.** $\dfrac{x-3}{x+4}$

35. $\dfrac{3x+1}{x-5}$ **37.** $-\dfrac{2a+2}{7a-4}$ **39.** $\dfrac{x+y}{x-y}$ **41.** $-\dfrac{2x}{x^2+1}$ **43.** $-\dfrac{a^2}{1-2a}$ **45.** $-\dfrac{3x+2}{x-2}$ **47.** $\dfrac{3n^2+12n+8}{n(n+2)(n+4)}$

49. a. $\dfrac{2}{2+x}$ **b.** -1 **c.** $\dfrac{x-3}{x-2}$

SECTION 6.4

3. 9 **5.** $\dfrac{15}{2}$ **7.** 3 **9.** $\dfrac{10}{3}$ **11.** -2 **13.** $\dfrac{2}{5}$ **15.** -4 **17.** $\dfrac{3}{4}$ **19.** $-\dfrac{6}{5}$ **21.** -2 **23.** There are

56.75 g of protein in a 454-gram box of the pasta. **25.** There would be 60,000 squawfish in the area. **27.** There would be 504 computers with defective CD-ROM drives. **29.** The dimensions of the room are 18 ft by 24 ft. **31.** 0.75 additional ounce of medicine is required. **33.** A person must walk 21.54 mi to lose 1 pound.

SECTION 6.5

1. -5 **3.** $\dfrac{5}{2}$ **5.** -1 **7.** 8 **9.** -3 **11.** -4 **13.** No solution **15.** $-\dfrac{5}{3}$ and -1 **17.** $-\dfrac{1}{3}$ and 5

19. $-\dfrac{11}{3}$ **21.** No solution **23.** $P_2=\dfrac{P_1V_1T_2}{V_2T_1}$ **25.** $b=\dfrac{af}{a-f}$ **27.** Working alone, the experienced

bricklayer can do the job in 14 h. **29.** The slower machine working alone would have taken 160 min to send the fax. **31.** When all three machines are working, it would take 4 h to fill the bottles. **33.** It would take 90 min to empty the tank. **35.** It will take 120 min to have 76 balloons. **37.** The two clerks, working together, can complete the task in 40 min (or 2400 s). **39.** The rate of the runner is 8 mph. **41.** The rate of the tortoise is 0.4 ft/s. The rate of the hare is 72 ft/s. **43.** The rate of the cyclist is 10 mph. **45.** The rate of the jet was 525 mph. **47.** The rate of the current is 3 mph. **49.** The rate of the wind is 60.80 mph. **51.** The rate of the current is 2 mph.

53. a. 3 **b.** 0 and y **c.** $3y$ and $\dfrac{3}{2}y$ **55.** The usual rate of the bus is 60 mph.

SECTION 6.6

1. The profit is \$80,000. **3.** The pressure is 6.75 lb/in². **5.** In 10s. the object will fall 1600 ft. **7.** In 3 s, the ball will roll 54 ft. **9.** When the width is 4 ft, the length is 10 ft. **11.** The gear that has 36 teeth will make 30 revolutions/min. **13.** The current is 7.5 amps. **15.** The intensity is 48 foot-candles when the distance is 5 ft. **17.** y is doubled. **19.** Inversely **21.** Inversely

CHAPTER 6 REVIEW EXERCISES

1. $\dfrac{1}{a-1}$ [6.1C] **2.** $-\dfrac{5x^2-17x-9}{(x-3)(x+2)}$ [6.2B] **3.** $P(4)=4$ [6.1A] **4.** 4 [6.4A] **5.** $-\dfrac{9x^2+16}{24x}$ [6.3A]

6. $\dfrac{16x^2+4x}{(4x+1)(4x-1)}$, $\dfrac{12x^2-7x+1}{(4x+1)(4x-1)}$ [6.2A] **7.** $r=\dfrac{S-a}{S}$ [6.5A] **8.** $P(-2)=\dfrac{2}{21}$ [6.1A] **9.** $\dfrac{2}{5}$ [6.4A]

10. The domain is $\{x\,|\,x\neq-3,2\}$. [6.1A] **11.** $\dfrac{3x^2-1}{3x^2+1}$ [6.1B] **12.** The domain is $\{x\,|\,x\neq3\}$. [6.1A]

13. x [6.1C] **14.** $\dfrac{4x-1}{x+2}$ [6.2B] **15.** $N=\dfrac{S}{1-Q}$ [6.5A] **16.** No solution [6.5A] **17.** $\dfrac{x}{x-3}$ [6.3A]

18. $\dfrac{x^2-7x+12}{(x-5)(x-4)}$, $\dfrac{x}{(x-5)(x-4)}$, $-\dfrac{x-5}{(x-5)(x-4)}$ [6.2A] **19.** 10 [6.5A] **20.** $\dfrac{1}{x}$ [6.1D] **21.** $\dfrac{3x+2}{x}$ [6.1D]

22. $\dfrac{9x^2 + x + 4}{(3x - 1)(x - 2)}$ [6.2B] **23.** $\dfrac{x^2 + 3x + 9}{x + 3}$ [6.1B] **24.** $\dfrac{21a + 40b}{24a^2b^4}$ [6.2B] **25.** The domain is

$\{x \,|\, x \in \text{real numbers}\}$. [6.1A] **26.** $4a^{2n} + 3a^n - 2$ [6.1B] **27.** $-\dfrac{x + 4}{x(x + 2)}$ [6.1B] **28.** $\dfrac{4x^2 + 6x + 9}{2x + 3}$ [6.1B]

29. $-\dfrac{(x + 4)(x + 3)}{6(x - 4)}$ [6.1C] **30.** $\dfrac{x^n + 6}{x^n - 3}$ [6.1D] **31.** $\dfrac{8(x - 2)(2x - 1)}{x(2x + 1)}$ [6.1D] **32.** $-\dfrac{x - 3}{x + 3}$ [6.1D]

33. $\dfrac{9 + 5ab}{6a^3b^2}$ [6.2B] **34.** $\dfrac{3x + 26}{(3x - 2)(3x + 2)}$ [6.2B] **35.** $\dfrac{x - 4}{x + 5}$ [6.3A] **36.** $\dfrac{x - 1}{4}$ [6.3A] **37.** 1, 13 [6.4A]

38. $\dfrac{15}{13}$ [6.5A] **39.** $R = \dfrac{V}{I}$ [6.5A] **40.** 6 [6.5A] **41.** It would take 40 min to empty the tub with both pipes

open. [6.5B] **42.** The rate of the bus is 45 mph. [6.5C] **43.** The rate of the helicopter is 180 mph. [6.5C]

44. To read 150 pages, it will take 375 min. [6.4B] **45.** The current is 2 amps. [6.6A] **46.** 48 mi would be

represented by 12 in. [6.4B] **47.** The stopping distance for a car traveling at 65 mph is 287.3 ft. [6.6A]

48. It would take the apprentice 104 min to complete the job alone. [6.5B]

CHAPTER 6 TEST

1. 2 [6.4A] **2.** $\dfrac{x + 2}{x - 1}$ [6.1D] **3.** $\dfrac{x^2 - 9x + 3}{(x - 3)(x + 2)}$ [6.2B] **4.** $\dfrac{x^2 - 2x - 3}{(x + 3)(x - 3)(x - 2)}, \dfrac{2x^2 - 4x}{(x + 3)(x - 3)(x - 2)}$ [6.2A]

5. No solution [6.5A] **6.** $\dfrac{v(v + 2)}{2v - 1}$ [6.1B] **7.** $\dfrac{6(x - 2)}{5}$ [6.1C] **8.** The domain is $\{x \,|\, x \neq 3, -3\}$. [6.1A]

9. $\dfrac{x - 4}{x + 3}$ [6.3A] **10.** $\dfrac{x + 4}{x + 2}$ [6.3A] **11.** $\dfrac{x + 1}{3x - 4}$ [6.1D] **12.** 1 and 2 [6.5A] **13.** $-\dfrac{2(a - 2)}{3a + 2}$ [6.1B]

14. $t = \dfrac{4r}{r - 2}$ [6.5A] **15.** $f(-1) = 2$ [6.1A] **16.** $\dfrac{-x^2 - 5x + 2}{(x + 1)(x + 4)(x - 1)}$ [6.2B] **17.** -3 and 5 [6.4A]

18. The rate of the cyclist is 10 mph. [6.5C] **19.** The resistance is 0.4 ohm. [6.6A] **20.** The office requires 14 rolls

of wallpaper. [6.4B] **21.** Working together, the landscapers can complete the task in 10 min. [6.5B]

22. The resistance is 1.6 ohms. [6.6A]

CUMULATIVE REVIEW EXERCISES

1. $\dfrac{36}{5}$ [1.2D] **2.** $\dfrac{15}{2}$ [2.1C] **3.** 7 and 1 [2.5A] **4.** The domain is $\{x \,|\, x \neq 3\}$. [6.1A] **5.** $P(-2) = \dfrac{3}{7}$ [6.1A]

6. 3.5×10^{-8} [5.1C] **7.** $\dfrac{a^5}{b^6}$ [5.1B] **8.** $(-\infty, 4]$ [2.4A] **9.** $-4a^4 + 6a^3 - 2a^2$ [5.3A]

10. $(2x^n - 1)(x^n + 2)$ [5.6C] **11.** $(xy - 3)(x^2y^2 + 3xy + 9)$ [5.6B] **12.** $x + 3y$ [6.1B] **13.** $y = \dfrac{3}{2}x + 2$ [3.6A]

14. $-\dfrac{3}{4}$ and $\dfrac{3}{2}$ [5.7A] **15.** $4x^2 + 10x + 10 + \dfrac{21}{x - 2}$ [5.4C] **16.** $\dfrac{3xy}{x - y}$ [6.1D]

17. $\dfrac{x^3y - 2x^2y}{2x^2(x + 1)(x - 2)}, \dfrac{2}{2x^2(x + 1)(x - 2)}$ [6.2A] **18.** $-\dfrac{x}{(3x + 2)(x + 1)}$ [6.2B] **19.** [3.3C]

20. [4.5A] **21.** -28 [4.3A] **22.** $\dfrac{x - 3}{x + 3}$ [6.3A] **23.** $(1, 1, 1)$ [4.2B]

24. $\left\{x \,\middle|\, x < -\dfrac{2}{3} \text{ or } x > 2\right\}$ [2.5B] **25.** 9 [6.4A] **26.** 13 [6.5A] **27.** $r = \dfrac{E - IR}{I}$ [2.1D] **28.** $\dfrac{x}{x - 1}$ [5.1B]

29. The smaller integer is 5 and the larger integer is 10. [2.2A] **30.** 50 lb of almonds must be used. [2.3A]

31. 75,000 people are expected to vote. [6.4B] **32.** It would take the new computer 14 min to do the job working alone. [6.5B] **33.** The rate of the wind is 60 mph. [6.5C] **34.** The first person has made 12 laps and is at the starting point. The second person has made 20 laps and is at the starting point. [2.3C]

Answers to Chapter 7 Selected Exercises

PREP TEST

1. 16 [1.2B] **2.** 32 [1.2C] **3.** 9 [1.2B] **4.** $\dfrac{1}{12}$ [1.2B] **5.** $-5x - 1$ [1.3C] **6.** $\dfrac{xy^5}{4}$ [5.1B]

7. $9x^2 - 12x + 4$ [5.3C] **8.** $-12x^2 + 14x + 10$ [5.3B] **9.** $36x^2 - 1$ [5.3C] **10.** -1 and 15 [5.7A]

SECTION 7.1

1. 2 **3.** 27 **5.** $\dfrac{1}{9}$ **7.** 4 **9.** Not a real number **11.** $\dfrac{343}{125}$ **13.** x **15.** $y^{1/2}$ **17.** $x^{1/12}$

19. $a^{7/12}$ **21.** $\dfrac{1}{a}$ **23.** $\dfrac{1}{y}$ **25.** $y^{3/2}$ **27.** $\dfrac{1}{x}$ **29.** $\dfrac{1}{x^4}$ **31.** a **33.** $x^{3/10}$ **35.** a^3 **37.** $\dfrac{1}{x^{1/2}}$ **39.** $y^{1/9}$

41. x^4y **43.** $x^6y^3z^9$ **45.** $\dfrac{x}{y^2}$ **47.** $\dfrac{x^{3/2}}{y^{1/4}}$ **49.** $\dfrac{x^2}{y^8}$ **51.** $\dfrac{1}{x^{11/12}}$ **53.** $\dfrac{1}{y^{5/2}}$ **55.** $\dfrac{1}{b^{7/8}}$ **57.** a^5b^{13} **59.** $\dfrac{m^2}{4n^{3/2}}$

61. $\dfrac{y^{17/2}}{x^3}$ **63.** $\dfrac{16b^2}{a^{1/3}}$ **65.** $y^2 - y$ **67.** $a - a^2$ **69.** x^{4n} **71.** $x^{3n/2}$ **73.** $y^{3n/2}$ **75.** x^{2n^2} **77.** $x^{2n}y^n$

79. $x^{4n}y^{2n}$ **81.** $\sqrt[4]{3}$ **83.** $\sqrt{a^3}$ **85.** $\sqrt{32t^5}$ **87.** $-2\sqrt[3]{x^2}$ **89.** $\sqrt[3]{a^4b^2}$ **91.** $\sqrt[5]{a^6b^{12}}$ **93.** $\sqrt[4]{(4x - 3)^3}$

95. $\dfrac{1}{\sqrt[3]{x^2}}$ **97.** $14^{1/2}$ **99.** $x^{1/3}$ **101.** $x^{4/3}$ **103.** $b^{3/5}$ **105.** $(2x^2)^{1/3}$ **107.** $-(3x^5)^{1/2}$ **109.** $3xy^{2/3}$

111. $(a^2 - 2)^{1/2}$ **113.** x^8 **115.** $-x^4$ **117.** xy^3 **119.** $-x^5y$ **121.** $4a^2b^6$ **123.** Not a real number

125. $3x^3$ **127.** $-4x^3y^4$ **129.** $-x^2y^3$ **131.** x^4y^2 **133.** $3xy^5$ **135.** $2ab^2$ **137. a.** False; 2 **b.** True

c. True **d.** False; $(a^n + b^n)^{1/n}$ **e.** False; $a + 2a^{1/2}b^{1/2} + b$ **f.** False; $a^{n/m}$

SECTION 7.2

1. $x^2yz^2\sqrt{yz}$ **3.** $2ab^4\sqrt{2a}$ **5.** $3xyz^2\sqrt{5yz}$ **7.** Not a real number **9.** $a^5b^2\sqrt[3]{ab^2}$ **11.** $-5y\sqrt[3]{x^2y}$

13. $abc^2\sqrt[3]{ab^2}$ **15.** $2x^2y\sqrt[4]{xy}$ **17.** $-6\sqrt{x}$ **19.** $-2\sqrt{2}$ **21.** $3\sqrt{2b} + 5\sqrt{3b}$ **23.** $-2xy\sqrt{2y}$

25. $6ab^2\sqrt{3ab} + 3ab\sqrt{3ab}$ **27.** $-\sqrt[3]{2}$ **29.** $8b\sqrt[3]{2b^2}$ **31.** $3a\sqrt[3]{2a}$ **33.** $17\sqrt{2} - 15\sqrt{5}$ **35.** $5b\sqrt{b}$

37. $-8xy\sqrt{2x} + 2xy\sqrt{xy}$ **39.** $2y\sqrt[3]{2x}$ **41.** $-4ab\sqrt[4]{2b}$ **43.** 16 **45.** $2\sqrt[3]{4}$ **47.** $xy^3\sqrt{x}$ **49.** $8xy\sqrt{x}$

51. $2x^2y\sqrt[3]{2}$ **53.** $2ab\sqrt[4]{3a^2b}$ **55.** 6 **57.** $x - \sqrt{2x}$ **59.** $4x - 8\sqrt{x}$ **61.** $x - 6\sqrt{x} + 9$ **63.** $84 + 16\sqrt{5}$

65. $672x^2y^2$ **67.** $4a^3b^3\sqrt[3]{a}$ **69.** $-8\sqrt{5}$ **71.** $x - y^2$ **73.** $12x - y$ **77.** $y\sqrt{5y}$ **79.** $b\sqrt{13b}$ **81.** $\dfrac{\sqrt{2}}{2}$

83. $\dfrac{2\sqrt{3y}}{3y}$ **85.** $\dfrac{3\sqrt{3a}}{a}$ **87.** $\dfrac{\sqrt{2y}}{2}$ **89.** $\dfrac{5\sqrt[3]{3}}{3}$ **91.** $\dfrac{5\sqrt[3]{9y^2}}{3y}$ **93.** $\dfrac{b\sqrt{2a}}{2a^2}$ **95.** $\dfrac{\sqrt{15x}}{5x}$ **97.** $2 + 2\sqrt{2}$

99. $-\dfrac{12 + 4\sqrt{2}}{7}$ **101.** $-\dfrac{10 + 5\sqrt{7}}{3}$ **103.** $-\dfrac{7\sqrt{x} + 21}{x - 9}$ **105.** $-\sqrt{6} + 3 - 2\sqrt{2} + 2\sqrt{3}$ **107.** $-\dfrac{17 + 5\sqrt{5}}{4}$

109. $\dfrac{8a - 10\sqrt{ab} + 3b}{16a - 9b}$ **111.** $\dfrac{3 - 7\sqrt{y} + 2y}{1 - 4y}$ **113. a.** False; $\sqrt[6]{432}$ **b.** True **c.** False; $x^{2/3}$ **d.** False; $\sqrt{x} + \sqrt{y}$

is in simplest form. **e.** False; $\sqrt[2]{2} + \sqrt[3]{3}$ is in simplest form. **f.** True **115.** $\sqrt[4]{a + b}$

SECTION 7.3

1. -2 **3.** 9 **5.** 7 **7.** $\dfrac{13}{3}$ **9.** 35 **11.** -7 **13.** -12 **15.** 9 **17.** 2 **19.** 2 **21.** 1 **23.** On the

moon, an object will fall 24.75 ft in 3 s. **25.** The HDTV is approximately 7.15 in. wider. **27.** The length of the

pendulum is 7.30 ft. **29.** 2 **31.** $r = \sqrt[3]{\dfrac{3V}{4\pi}}$

SECTION 7.4

3. $2i$ **5.** $7i\sqrt{2}$ **7.** $3i\sqrt{3}$ **9.** $4 + 2i$ **11.** $2\sqrt{3} - 3i\sqrt{2}$ **13.** $4\sqrt{10} - 7i\sqrt{3}$ **15.** $8 - i$ **17.** $-8 + 4i$
19. $6 - 6i$ **21.** $19 - 7i\sqrt{2}$ **23.** $6\sqrt{2} - 3i\sqrt{2}$ **25.** 63 **27.** -4 **29.** $-3\sqrt{2}$ **31.** $-4 + 12i$ **33.** $-2 + 4i$

35. $17 - i$ **37.** $8 + 27i$ **39.** 1 **41.** 1 **43.** $-3i$ **45.** $\frac{3}{4} + \frac{1}{2}i$ **47.** $\frac{10}{13} - \frac{2}{13}i$ **49.** $\frac{4}{5} + \frac{2}{5}i$ **51.** $-i$

53. $-\frac{\sqrt{5}}{5} + \frac{2\sqrt{5}}{5}i$ **55.** $\frac{3}{10} - \frac{11}{10}i$ **57.** $\frac{6}{5} + \frac{7}{5}i$ **59. a.** Yes **b.** Yes

CHAPTER 7 REVIEW EXERCISES

1. $20x^2y^2$ [7.1A] **2.** 7 [7.3A] **3.** $39 - 2i$ [7.4C] **4.** $7x^{2/3}y$ [7.1B] **5.** $6\sqrt{3} - 13$ [7.2C]

6. -2 [7.3A] **7.** $\frac{1}{x^5}$ [7.1A] **8.** $\frac{8\sqrt{3y}}{3y}$ [7.2D] **9.** $-2a^2b^4$ [7.1C] **10.** $2a^2b\sqrt{2b}$ [7.2B]

11. $\frac{x\sqrt{x} - x\sqrt{2} + 2\sqrt{x} - 2\sqrt{2}}{x - 2}$ [7.2D] **12.** $\frac{2}{3} - \frac{5}{3}i$ [7.4D] **13.** $3ab^3\sqrt{2a}$ [7.2A] **14.** $-4\sqrt{2} + 8i\sqrt{2}$ [7.4B]

15. $5x^3y^3\sqrt[3]{2x^2y}$ [7.2B] **16.** $4xy^2\sqrt[3]{x^2}$ [7.2C] **17.** $7 + 3i$ [7.4C] **18.** $3\sqrt[4]{x^3}$ [7.1B] **19.** $-2ab^2\sqrt[5]{2a^3b^2}$ [7.2A]

20. $-2 + 7i$ [7.4D] **21.** $-6\sqrt{2}$ [7.4C] **22.** 30 [7.3A] **23.** $3a^2b^3$ [7.1C] **24.** $5i\sqrt{2}$ [7.4A]

25. $-12 + 10i$ [7.4B] **26.** $31 - 10\sqrt{6}$ [7.2C] **27.** $6x^2\sqrt{3y}$ [7.2B] **28.** $\frac{1}{3}$ [7.1A] **29.** $\frac{1}{a^{10}}$ [7.1A]

30. $-7x^3y^8$ [7.1C] **31.** $4\sqrt[3]{a^2}$ [7.1B] **32.** $b^{5/8}$ [7.1B] **33.** $xy^2z^2\sqrt[4]{x^2z^2}$ [7.2A] **34.** $5\sqrt{6}$ [7.2B]

35. $4x^2\sqrt{3xy} - 4x^2\sqrt{5y}$ [7.2B] **36.** 40 [7.2C] **37.** $3x + 3\sqrt{3x}$ [7.2C] **38.** $5x\sqrt{x}$ [7.2D]

39. $\frac{x + 2\sqrt{xy} + y}{x - y}$ [7.2D] **40.** $6i$ [7.4A] **41.** $7 - 4i$ [7.4B] **42.** $10\sqrt{2} + 2i\sqrt{3}$ [7.4B] **43.** $9 - i$ [7.4B]

44. $14 + 2i$ [7.4B] **45.** $-4 - 9i$ [7.4B] **46.** -16 [7.4C] **47.** $6i$ [7.4D] **48.** $\frac{14}{5} + \frac{7}{5}i$ [7.4D]

49. $1 + i$ [7.4D] **50.** -24 [7.3A] **51.** The width is 5 in. [7.3B] **52.** The amount of power is 120 watts. [7.3B] **53.** The distance required is 242 ft. [7.3B] **54.** The distance is 6.63 ft. [7.3B]

CHAPTER 7 TEST

1. $\frac{1}{2}x^{3/4}$ [7.1B] **2.** $-2x^2y\sqrt[3]{2x}$ [7.2B] **3.** $3\sqrt[5]{y^2}$ [7.1B] **4.** $18 + 16i$ [7.4C] **5.** $4x + 4\sqrt{xy} + y$ [7.2C]

6. $r^{1/6}$ [7.1A] **7.** 4 [7.3A] **8.** $2xy^2$ [7.1C] **9.** $-4x\sqrt{3}$ [7.2C] **10.** $-3 + 2i$ [7.4B] **11.** $4x^2y^3\sqrt{2y}$ [7.2A]
12. $14 + 10\sqrt{3}$ [7.2C] **13.** $10 + 2i$ [7.4C] **14.** 2 [7.2D] **15.** $8a\sqrt{2a}$ [7.2B] **16.** $2a - \sqrt{ab} - 15b$ [7.2C]

17. $\frac{64x^3}{y^6}$ [7.1A] **18.** $\frac{x + \sqrt{xy}}{x - y}$ [7.2D] **19.** $-\frac{4}{5} + \frac{7}{5}i$ [7.4D] **20.** -3 [7.3A] **21.** $\frac{b^3}{8a^6}$ [7.1A]

22. $3abc^2\sqrt[3]{ac}$ [7.2A] **23.** $\frac{4x^2}{y}$ [7.2D] **24.** -4 [7.4C] **25.** The distance is 576 ft. [7.3B]

CUMULATIVE REVIEW EXERCISES

1. The Distributive Property [1.3A] **2.** $f(-3) = 34$ [3.2A] **3.** $\frac{3}{2}$ [2.1B] **4.** $\frac{2}{3}$ [2.1C]

5. $\frac{1}{3}$ and $\frac{7}{3}$ [2.5A] **6.** $\{x \mid x > 1\}$ [2.4A] **7.** $\{x \mid -6 \le x \le 3\}$ [2.5B] **8.** $(9x + y)(9x - y)$ [5.6A]

9. $x(x^2 + 3)(x + 1)(x - 1)$ [5.6D] **10.** $y = \frac{1}{3}x + \frac{7}{3}$ [3.5B] **11.** 3 [4.3A] **12.** $C = R - nP$ [2.1D]

13. $2x^2y^2$ [5.1B] **14.** $\frac{x(2x - 3)}{y(x + 4)}$ [6.1C] **15.** $-x\sqrt{10x}$ [7.2B] **16.** $-\frac{1}{2}$ and 3 [6.5A]

17. The slope is $\frac{3}{2}$, and the y-intercept is $(0, 3)$. [3.3B] **18.** [3.7A]

19. $-\dfrac{1}{5} + \dfrac{3}{5}i$ [7.4D] **20.** -20 [7.3A] **21.** $\dfrac{x^2 + 12x - 12}{(2x - 3)(x + 4)}$ [6.2B] **22.** $\left(\dfrac{17}{4}, \dfrac{9}{2}\right)$ [4.3B] **23.** There are nineteen 18¢ stamps. [2.2B] **24.** The rate of the plane is 250 mph. [2.3C] **25.** The time is 1.25 s. [5.1D] **26.** The height of the periscope is 32.7 ft. [7.3B] **27.** $m = 0.08$. The annual income is 8% of the investment. [3.4A]

Answers to Chapter 8 Selected Exercises

PREP TEST

1. $3\sqrt{2}$ [7.2A] **2.** Not a real number [7.2A] **3.** $\dfrac{2x - 1}{x - 1}$ [6.2B] **4.** 8 [1.3B] **5.** Yes [5.6A] **6.** $(2x - 1)^2$ [5.6A] **7.** $(3x + 2)(3x - 2)$ [5.6A] **8.** [1.1C] **9.** -3 and 5 [5.7A] **10.** 4 [6.4A]

SECTION 8.1

3. $2x^2 - 4x - 5 = 0$; $a = 2, b = -4, c = -5$ **5.** $4x^2 - 5x + 6 = 0$; $a = 4, b = -5, c = 6$ **7.** 0 and 4 **9.** -5 and 5 **11.** -2 and 3 **13.** 3 **15.** 0 and 2 **17.** -2 and 5 **19.** 2 and 5 **21.** $-\dfrac{3}{2}$ and 6 **23.** $\dfrac{1}{4}$ and 2 **25.** -4 and $\dfrac{1}{3}$ **27.** $-\dfrac{2}{3}$ and $\dfrac{9}{2}$ **29.** -4 and $\dfrac{1}{4}$ **31.** -2 and 9 **33.** -2 and $-\dfrac{3}{4}$ **35.** -5 and 2 **37.** -4 and $-\dfrac{3}{2}$ **39.** $2b$ and $7b$ **41.** $-c$ and $7c$ **43.** $-b$ and $-\dfrac{b}{2}$ **45.** $\dfrac{2a}{3}$ and $4a$ **47.** $-\dfrac{a}{3}$ and $3a$ **49.** $-\dfrac{3y}{2}$ and $-\dfrac{y}{2}$ **51.** $-\dfrac{4a}{3}$ and $-\dfrac{a}{2}$ **53.** $x^2 - 7x + 10 = 0$ **55.** $x^2 + 6x + 8 = 0$ **57.** $x^2 - 5x - 6 = 0$ **59.** $x^2 - 9 = 0$ **61.** $x^2 - 8x + 16 = 0$ **63.** $x^2 - 5x = 0$ **65.** $x^2 - 3x = 0$ **67.** $2x^2 - 7x + 3 = 0$ **69.** $4x^2 - 5x - 6 = 0$ **71.** $3x^2 + 11x + 10 = 0$ **73.** $9x^2 - 4 = 0$ **75.** $6x^2 - 5x + 1 = 0$ **77.** $10x^2 - 7x - 6 = 0$ **79.** $8x^2 + 6x + 1 = 0$ **81.** $50x^2 - 25x - 3 = 0$ **83.** -7 and 7 **85.** $-2i$ and $2i$ **87.** -2 and 2 **89.** $-\dfrac{9}{2}$ and $\dfrac{9}{2}$ **91.** $-7i$ and $7i$ **93.** $-4\sqrt{3}$ and $4\sqrt{3}$ **95.** $-5\sqrt{3}$ and $5\sqrt{3}$ **97.** $-3i\sqrt{2}$ and $3i\sqrt{2}$ **99.** -5 and 7 **101.** -6 and 0 **103.** -7 and 3 **105.** 0 and 1 **107.** $-5 - \sqrt{6}$ and $-5 + \sqrt{6}$ **109.** $3 - 3i\sqrt{5}$ and $3 + 3i\sqrt{5}$ **111.** $-\dfrac{2 - 9\sqrt{2}}{3}$ and $-\dfrac{2 + 9\sqrt{2}}{3}$ **113.** $x^2 - 2 = 0$ **115.** $x^2 + 1 = 0$ **117.** $x^2 - 8 = 0$ **119.** $x^2 + 2 = 0$ **121.** $-\dfrac{3b}{a}$ and $\dfrac{3b}{a}$ **123.** $-a - 2$ and $-a + 2$ **125.** $-\dfrac{1}{2}$

SECTION 8.2

1. -1 and 5 **3.** -9 and 1 **5.** 3 **7.** $-2 - \sqrt{11}$ and $-2 + \sqrt{11}$ **9.** $3 - \sqrt{2}$ and $3 + \sqrt{2}$ **11.** $1 - i$ and $1 + i$ **13.** -3 and 8 **15.** -9 and 4 **17.** $\dfrac{3 - \sqrt{5}}{2}$ and $\dfrac{3 + \sqrt{5}}{2}$ **19.** $\dfrac{1 - \sqrt{5}}{2}$ and $\dfrac{1 + \sqrt{5}}{2}$ **21.** $3 - \sqrt{13}$ and $3 + \sqrt{13}$ **23.** 3 and 5 **25.** $2 - 3i$ and $2 + 3i$ **27.** $-3 - 2i$ and $-3 + 2i$ **29.** $1 - 3\sqrt{2}$ and $1 + 3\sqrt{2}$ **31.** $\dfrac{1 - \sqrt{17}}{2}$ and $\dfrac{1 + \sqrt{17}}{2}$ **33.** $1 - 2i\sqrt{3}$ and $1 + 2i\sqrt{3}$ **35.** $\dfrac{1}{2} - i$ and $\dfrac{1}{2} + i$ **37.** $\dfrac{1}{3} - \dfrac{1}{3}i$ and $\dfrac{1}{3} + \dfrac{1}{3}i$ **39.** $\dfrac{2 - \sqrt{14}}{2}$ and $\dfrac{2 + \sqrt{14}}{2}$ **41.** $-\dfrac{3}{2}$ and 1 **43.** $1 - \sqrt{5}$ and $1 + \sqrt{5}$ **45.** $\dfrac{1}{2}$ and 5 **47.** $2 - \sqrt{5}$ and $2 + \sqrt{5}$ **49.** -3.236 and 1.236 **51.** 0.293 and 1.707 **53.** -0.809 and 0.309 **55.** $-a$ and $2a$ **57.** $-5a$ and $2a$ **59.** No, the ball will have gone only 197.2 ft when it hits the ground.

SECTION 8.3

3. -2 and 5 **5.** -9 and 4 **7.** $4 - 2\sqrt{22}$ and $4 + 2\sqrt{22}$ **9.** -8 and 3 **11.** $\dfrac{-5 - \sqrt{33}}{4}$ and $\dfrac{-5 + \sqrt{33}}{4}$ **13.** $-\dfrac{1}{4}$ and $\dfrac{3}{2}$ **15.** $7 - 3\sqrt{5}$ and $7 + 3\sqrt{5}$ **17.** $\dfrac{1 - \sqrt{3}}{2}$ and $\dfrac{1 + \sqrt{3}}{2}$ **19.** $-1 - i$ and $-1 + i$ **21.** $1 - 2i$ and $1 + 2i$ **23.** $2 - 3i$ and $2 + 3i$ **25.** $\dfrac{1 - \sqrt{11}}{2}$ and $\dfrac{1 + \sqrt{11}}{2}$ **27.** $-\dfrac{3}{2} - \dfrac{1}{2}i$ and $-\dfrac{3}{2} + \dfrac{1}{2}i$

29. $\frac{3}{4} - \frac{3\sqrt{3}}{4}i$ and $\frac{3}{4} + \frac{3\sqrt{3}}{4}i$ **31.** 0.394 and 7.606 **33.** -4.236 and 0.236 **35.** -1.351 and 1.851

37. Two complex number solutions **39.** One real number solution **41.** Two real number solutions

43. No. The arrow does not reach a height of 275 ft. (The discriminant is less than zero.) **45.** $\{p \,|\, p < 9\}$

47. $(1, \infty)$ **49.** $-2i$ and i

SECTION 8.4

1. $-2, 2, -3$, and 3 **3.** $-2, 2, -\sqrt{2}$, and $\sqrt{2}$ **5.** 1 and 4 **7.** 16 **9.** $-2i, 2i, -1$, and 1 **11.** $-4i, 4i, -2$, and 2

13. 16 **15.** 1 and 512 **17.** $-\frac{2}{3}, \frac{2}{3}, -1$, and 1 **19.** 3 **21.** 9 **23.** -1 and 2 **25.** 0 and 2 **27.** $-\frac{1}{2}$ and 2

29. -2 **31.** 1 **33.** 1 **35.** -3 **37.** -1 and 10 **39.** $-\frac{1}{2} - \frac{\sqrt{7}}{2}i$ and $-\frac{1}{2} + \frac{\sqrt{7}}{2}i$ **41.** -3 and 1

43. -1 and 0 **45.** $-\frac{1}{3}$ and $\frac{1}{2}$ **47.** $-\frac{2}{3}$ and 6 **49.** $\frac{4}{3}$ and 3 **51.** $-\frac{1}{4}$ and 3 **53.** 9 and 36 **55.** $-\sqrt{5}$ or $\sqrt{5}$

SECTION 8.5

3. $\{x \,|\, x < -2 \text{ or } x > 4\}$ **5.** $\{x \,|\, x \le 1 \text{ or } x \ge 2\}$

7. $\{x \,|\, -3 < x < 4\}$ **9.** $\{x \,|\, x < -2 \text{ or } 1 < x < 3\}$

11. $\{x \,|\, -4 \le x \le 1 \text{ or } x \ge 2\}$ **13.** $\{x \,|\, x < -2 \text{ or } x > 4\}$

15. $\{x \,|\, -1 < x \le 3\}$ **17.** $\{x \,|\, x \le -2 \text{ or } 1 \le x < 3\}$

19. $\{x \,|\, x > 4 \text{ or } x < -4\}$ **21.** $\{x \,|\, -3 \le x \le 12\}$ **23.** $\left\{x \,\middle|\, \frac{1}{2} < x < \frac{3}{2}\right\}$ **25.** $\left\{x \,\middle|\, x < 1 \text{ or } x > \frac{5}{2}\right\}$

27. $\{x \,|\, x < -1 \text{ or } 1 < x \le 2\}$ **29.** $\left\{x \,\middle|\, \frac{1}{2} < x \le 1\right\}$ **31.** $\{x \,|\, 2 < x \le 3\}$ **33.** $\{x \,|\, x > 5 \text{ or } -4 < x < -1\}$

35. **37.** **39.**

SECTION 8.6

1. The height is 3 cm. The base is 14 cm. **3.** The dimensions of Colorado are 272 mi by 383 mi. **5.** The maximum speed is 33 mph. **7.** The rocket takes 12.5 s to return to Earth. **9.** A driver can be going approximately 72.5 km/h and still be able to stop within 150 m. **11.** It would take the larger pipe 6 min to fill the tank. It would take the smaller pipe 12 min to fill the tank. **13.** The rate of the wind was approximately 108 mph. **15.** The rowing rate of the guide is 6 mph. **17.** The radius of the cone is 1.5 in.

CHAPTER 8 REVIEW EXERCISES

1. 0 and $\frac{3}{2}$ [8.1A] **2.** $-2c$ and $\frac{c}{2}$ [8.1A] **3.** $-4\sqrt{3}$ and $4\sqrt{3}$ [8.1C] **4.** $-\frac{1}{2} - 2i$ and $-\frac{1}{2} + 2i$ [8.1C]

5. -3 and -1 [8.2A] **6.** $\frac{7 - 2\sqrt{7}}{7}$ and $\frac{7 + 2\sqrt{7}}{7}$ [8.2A] **7.** $\frac{3}{4}$ and $\frac{4}{3}$ [8.3A] **8.** $\frac{1}{2} - \frac{\sqrt{31}}{2}i$ and $\frac{1}{2} + \frac{\sqrt{31}}{2}i$ [8.3A]

9. $x^2 + 3x = 0$ [8.1B] **10.** $12x^2 - x - 6 = 0$ [8.1B] **11.** $1 - i\sqrt{7}$ and $1 + i\sqrt{7}$ [8.2A] **12.** $-2i$ and $2i$ [8.2A]

13. $\frac{11 - \sqrt{73}}{6}$ and $\frac{11 + \sqrt{73}}{6}$ [8.3A] **14.** Two real number solutions [8.3A] **15.** $\left\{x \,\middle|\, -3 < x < \frac{5}{2}\right\}$ [8.5A]

16. $\left\{x \,\middle|\, x \le -4 \text{ or } -\frac{3}{2} \le x \le 2\right\}$ [8.5A] **17.** -64 and 27 [8.4A] **18.** $\frac{5}{4}$ [8.4B] **19.** -1 and 3 [8.4C]

20. -1 [8.4C] **21.** $\left\{x \,\middle|\, x < \frac{3}{2} \text{ or } x \ge 2\right\}$ [8.5A]

22. $\left\{x \,\middle|\, x \le -3 \text{ or } \frac{1}{2} \le x < 4\right\}$ [8.5A] **23.** 4 [8.4B] **24.** 5 [8.4B]

25. $\dfrac{-3 - \sqrt{249}}{10}$ and $\dfrac{-3 + \sqrt{249}}{10}$ [8.4C] **26.** $\dfrac{-11 - \sqrt{129}}{2}$ and $\dfrac{-11 + \sqrt{129}}{2}$ [8.4C] **27.** $3x^2 + 8x - 3 = 0$ [8.1B]

28. -5 and $\dfrac{1}{2}$ [8.1B] **29.** $-1 + 3\sqrt{2}$ and $-1 - 3\sqrt{2}$ [8.1C] **30.** $-3 + i$ and $-3 - i$ [8.3A] **31.** 2 and 3 [8.4C]

32. $-5, 5, -\sqrt{3}$, and $\sqrt{3}$ [8.4A] **33.** $\dfrac{25}{18}$ [8.4B] **34.** -8 and $\dfrac{1}{8}$ [8.4A] **35.** 2 [8.4B] **36.** $5 + 3\sqrt{2}$ and

$5 - 3\sqrt{2}$ [8.3A] **37.** 5 and $-\dfrac{4}{9}$ [8.4C] **38.** $\dfrac{1 + \sqrt{5}}{6}$ and $\dfrac{1 - \sqrt{5}}{6}$ [8.3A] **39.** $\dfrac{5}{4}$ [8.4B] **40.** $\dfrac{-13 + \sqrt{217}}{2}$

and $\dfrac{-13 - \sqrt{217}}{2}$ [8.4C] **41.** Two real number solutions [8.3A] **42.** $\{x \mid -2 \le x \le 5\}$ [8.5A] **43.** The sculling

crew's rate of rowing in calm water is 6 mph. [8.6A] **44.** The width of the rectangle is 5 cm. The length of the

rectangle is 12 cm. [8.6A] **45.** The integers are 2, 4, and 6 or -6, -4, and -2. [8.6A] **46.** Working alone, the new

computer can print the payroll in 12 min. [8.6A] **47.** The rate of the first car is 40 mph. The rate of the second car is

50 mph. [8.6A]

CHAPTER 8 TEST

1. -4 and $\dfrac{2}{3}$ [8.1A] **2.** $-\dfrac{2}{3}$ and $\dfrac{3}{2}$ [8.1A] **3.** $x^2 - 9 = 0$ [8.1B] **4.** $2x^2 + 7x - 4 = 0$ [8.1B]

5. $2 - 2\sqrt{2}$ and $2 + 2\sqrt{2}$ [8.1C] **6.** $3 - \sqrt{11}$ and $3 + \sqrt{11}$ [8.2A] **7.** $\dfrac{3 - \sqrt{15}}{3}$ and $\dfrac{3 + \sqrt{15}}{3}$ [8.2A]

8. $\dfrac{1 - \sqrt{3}}{2}$ and $\dfrac{1 + \sqrt{3}}{2}$ [8.3A] **9.** $-2 - 2i\sqrt{2}$ and $-2 + 2i\sqrt{2}$ [8.3A] **10.** Two real number solutions [8.3A]

11. Two complex number solutions [8.3A] **12.** $\dfrac{1}{4}$ [8.4A] **13.** $\sqrt{3}, -\sqrt{3}, 1$, and -1 [8.4A] **14.** 4 [8.4B]

15. No solution [8.4B] **16.** -9 and 2 [8.4C] **17.** $\{x \mid x < -4 \text{ or } 2 < x < 4\}$ [8.5A]

18. $\left\{x \mid -4 < x \le -\dfrac{3}{2}\right\}$ [8.5A] **19.** The ball hits the basket 1.88 s after it is

released. [8.6A] **20.** The rate of the canoe in calm water is 4 mph. [8.6A]

CUMULATIVE REVIEW EXERCISES

1. 14 [1.3B] **2.** -28 [2.1C] **3.** $-\dfrac{3}{2}$ [3.4A] **4.** $y = x + 1$ [3.6A] **5.** $-3xy(x^2 - 2xy + 3y^2)$ [5.5A]

6. $(2x - 5)(3x + 4)$ [5.5D] **7.** $(x + y)(a^n - 2)$ [5.5B] **8.** $x^2 - 3x - 4 - \dfrac{6}{3x - 4}$ [5.4B] **9.** $\dfrac{x}{2}$ [6.1C]

10. $2\sqrt{5}$ [3.1B] **11.** $b = \dfrac{2S - an}{n}$ [2.1D] **12.** $-8 - 14i$ [7.4C] **13.** $1 - a$ [7.1A] **14.** $\dfrac{x\sqrt[3]{4y^2}}{2y}$ [7.2D]

15. $-\dfrac{3}{2}$ and -1 [8.4C] **16.** $-\dfrac{1}{2}$ [6.5A] **17.** $2, -2, \sqrt{2}$, and $-\sqrt{2}$ [8.4A] **18.** 0 and 1 [8.4B]

19. $\left\{x \mid -2 < x < \dfrac{10}{3}\right\}$ [2.5B] **20.** $\left(\dfrac{5}{2}, 0\right)$ and $(0, -3)$ [3.3C] **21.** [4.5A]

22. $(-1, 1, 2)$ [4.2B] **23.** $-\dfrac{7}{3}$ [6.1A] **24.** $\{x \mid x \ne -3, 5\}$ [6.1A] **25.** $\{x \mid x < -3 \text{ or } 0 < x < 2\}$

[8.5A]

26. $\{x \mid -3 < x \le 1 \text{ or } x \ge 5\}$ **27.** The lower limit is $9\dfrac{23}{64}$ in. The upper limit is $9\dfrac{25}{64}$ in. [2.5C]

[8.5A]

28. The area is $(x^2 + 6x - 16)$ ft². [5.3D] **29.** Two complex number solutions [8.3A]

30. $m = -\dfrac{25,000}{3}$. The building depreciates $\dfrac{\$25,000}{3}$, or about $8333, each year. [3.4A]

Answers to Chapter 9 Selected Exercises

PREP TEST

1. 1 [1.3B] **2.** -7 [1.3B] **3.** 30 [3.2A] **4.** $h^2 + 4h - 1$ [3.2A] **5.** $-\dfrac{2}{3}$ and 3 [8.1A]

6. $2 - \sqrt{3}$ and $2 + \sqrt{3}$ [8.3A] **7.** $y = \dfrac{1}{2}x - 2$ [2.1D] **8.** Domain: $\{-2, 3, 4, 6\}$. Range: $\{4, 5, 6\}$.

The relation is a function. [3.2A] **9.** 8 [3.2A] **10.** [3.3B]

SECTION 9.1

5. -5 **7.** $x = 7$ **9.** **11.** **13.**

Vertex: $(1, -5)$
Axis of symmetry: $x = 1$

Vertex: $(1, -2)$
Axis of symmetry: $x = 1$

Vertex: $\left(\dfrac{1}{2}, -\dfrac{25}{4}\right)$

Axis of symmetry: $x = \dfrac{1}{2}$

15. **17.** **19.** **21.**

Vertex: $\left(\dfrac{3}{2}, -\dfrac{1}{4}\right)$

Axis of symmetry: $x = \dfrac{3}{2}$

Vertex: $\left(\dfrac{3}{2}, \dfrac{9}{2}\right)$

Axis of symmetry: $x = \dfrac{3}{2}$

Vertex: $(0, -1)$
Axis of symmetry: $x = 0$

Vertex: $(2, -1)$
Axis of symmetry: $x = 2$

23. 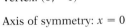 **25.** Domain: $\{x \mid x \in \text{real numbers}\}$ **27.** Domain: $\{x \mid x \in \text{real numbers}\}$

Range: $\{y \mid y \geq -5\}$.

Range: $\{y \mid y \leq 0\}$

Vertex: $\left(1, -\dfrac{5}{2}\right)$

Axis of symmetry: $x = 1$

29. Domain: $\{x \mid x \in \text{real numbers}\}$ **33.** $(3, 0)$ and $(-3, 0)$ **35.** $(0, 0)$ and $(-2, 0)$ **37.** $(4, 0)$ and $(-2, 0)$
Range: $\{y \mid y \geq -7\}$

39. $\left(-\dfrac{1}{2}, 0\right)$ and $(3, 0)$ **41.** $\left(-2 + \sqrt{7}, 0\right)$ and $\left(-2 - \sqrt{7}, 0\right)$ **43.** No x-intercepts **45.** 3 **47.** $\dfrac{3 + \sqrt{41}}{2}$ and

$\dfrac{3 - \sqrt{41}}{2}$ **49.** 0 and $\dfrac{4}{3}$ **51.** $i\sqrt{2}$ and $-i\sqrt{2}$ **53.** $\dfrac{1}{6} + \dfrac{\sqrt{47}}{6}i$ and $\dfrac{1}{6} - \dfrac{\sqrt{47}}{6}i$ **55.** $\dfrac{1 + \sqrt{41}}{4}$ and $\dfrac{1 - \sqrt{41}}{4}$

57. Two **59.** One **61.** No x-intercepts **63.** Two **65.** No x-intercepts **67.** Two **69.** $(-4, 0)$ and $(5, 0)$

73. a. Minimum **b.** Maximum **c.** Minimum **75.** Minimum value: -2 **77.** Maximum value: -3

79. Maximum value: $\dfrac{9}{8}$ **81.** Minimum value: $-\dfrac{11}{4}$ **83.** Maximum value: $\dfrac{9}{4}$ **85.** Minimum value: $-\dfrac{1}{12}$

87. Parabola **b** has the greatest maximum value. **89.** The maximum value of the function is -5. **91.** A price of \$250 will give the maximum revenue. **93.** The diver reaches a height of 13.1 m above the water. **95.** The minimum height is 24.36 ft. **97.** The water will land at a height of $31\dfrac{2}{3}$ ft. **99.** The car can be traveling 20 mph and still stop at a stop sign 44 ft away. **101.** The two numbers are 10 and 10. **103.** The length is 100 ft and the width is 50 ft.
105. $k = 16$

SECTION 9.2

1. Yes **3.** No **5.** Yes

7. Domain: $\{x \mid x \in \text{real numbers}\}$
Range: $\{y \mid y \geq 0\}$

9. Domain: $\{x \mid x \in \text{real numbers}\}$
Range: $\{y \mid y \in \text{real numbers}\}$

11. Domain: $\{x \mid x \leq 4\}$
Range: $\{y \mid y \geq 0\}$

13. Domain: $\{x \mid x \in \text{real numbers}\}$
Range: $\{y \mid y \in \text{real numbers}\}$

15. Domain: $\{x \mid x \geq -2\}$
Range: $\{y \mid y \leq 0\}$

17. Domain: $\{x \mid x \in \text{real numbers}\}$
Range: $\{y \mid y \geq 0\}$

19. $a = 18$ **21.** 17 **23.** $f(14) = 8$ **25.** $\{x \mid -2 < x < 2\}$ **27.** $x = 1$

SECTION 9.3

1. 5 **3.** 1 **5.** 0 **7.** $-\dfrac{29}{4}$ **9.** $\dfrac{2}{3}$ **11.** 2 **13.** 39 **15.** -8 **17.** $-\dfrac{4}{5}$ **19.** -2 **21.** 7

25. -13 **27.** -29 **29.** $8x - 13$ **31.** 4 **33.** 3 **35.** $x + 4$ **37.** 1 **39.** 5 **41.** $x^2 + 1$
43. 5 **45.** 11 **47.** $3x^2 + 3x + 5$ **49.** -3 **51.** -27 **53.** $x^3 - 6x^2 + 12x - 8$

55. a. $S(M(x)) = 80 + \dfrac{16000}{x}$ **b.** \$83.20 **c.** When 5000 digital cameras are manufactured, the camera store sells each camera for \$83.20. **57. a.** $d[r(p)] = 0.90p - 1350$ **b.** $r[d(p)] = 0.90p - 1500$ **c.** $r[d(p)]$ (The cost is less.) **59.** 0 **61.** -2 **63.** 7 **65.** $h^2 + 6h$ **67.** $2 + h$ **69.** $2a + h$ **71.** -1 **73.** -6 **75.** $6x - 13$

SECTION 9.4

3. Yes **5.** No **7.** Yes **9.** No **11.** No **13.** No **17.** $\{(0, 1), (3, 2), (8, 3), (15, 4)\}$ **19.** No inverse **21.** $\{(-2, 0), (5, -1), (3, 3), (6, -4)\}$ **23.** No inverse **25.** $f^{-1}(x) = \dfrac{1}{4}x + 2$ **27.** $f^{-1}(x) = \dfrac{1}{2}x - 2$ **29.** $f^{-1}(x) = 2x + 2$ **31.** $f^{-1}(x) = -\dfrac{1}{2}x + 1$ **33.** $f^{-1}(x) = \dfrac{3}{2}x - 6$ **35.** $f^{-1}(x) = -3x + 3$ **37.** $f^{-1}(x) = \dfrac{1}{2}x + \dfrac{5}{2}$ **39.** $f^{-1}(x) = \dfrac{1}{5}x + \dfrac{2}{5}$ **41.** $f^{-1}(x) = \dfrac{1}{6}x + \dfrac{1}{2}$ **43.** $\dfrac{5}{3}$ **45.** 3 **47.** Yes; Yes **49.** Yes **51.** No **53.** Yes **55.** Yes **57.**

59.

61.

65. 5 **67.** 0 **69.** -4

CHAPTER 9 REVIEW EXERCISES

1. Yes [9.2A] **2.** Yes [9.4A] **3.**

Domain: $\{x | x \in \text{real numbers}\}$
Range: $\{y | y \in \text{real numbers}\}$ [9.2A]

4.

Domain: $\{x | x \geq -4\}$
Range: $\{y | y \geq 0\}$ [9.2A]

5. Two [9.1B] **6.** $(0, 0)$ and $(-3, 0)$ [9.1B] **7.** $-\dfrac{1}{3} + \dfrac{\sqrt{5}}{3}i$ and $-\dfrac{1}{3} - \dfrac{\sqrt{5}}{3}i$ [9.1B] **8.** 3 [9.1C]

9. $-\dfrac{17}{4}$ [9.1C] **10.** 5 [9.3B] **11.** 10 [9.3B] **12.** $12x^2 + 12x - 1$ [9.3B] **13.** Yes [9.4B]

14.

Domain: $\{x | x \in \text{real numbers}\}$
Range: $\{y | y \geq -5\}$ [9.1A]

15.

Vertex: $(1, 2)$
Axis of symmetry: $x = 1$ [9.1A]

16.

Domain: $\{x | x \in \text{real numbers}\}$
Range: $\{y | y \geq -3\}$ [9.2A]

17. No [9.4A] **18.** 7 [9.3A] **19.** -9 [9.3A] **20.** 70 [9.3A] **21.** $\dfrac{12}{7}$ [9.3A] **22.** $6x^2 + 3x - 16$ [9.3B]

23. $f^{-1}(x) = -\dfrac{1}{6}x + \dfrac{2}{3}$ [9.4B] **24.** $f^{-1}(x) = \dfrac{3}{2}x + 18$ [9.4B] **25.** $f^{-1}(x) = 2x - 16$ [9.4B] **26.** The dimensions are 7 ft by 7 ft. [9.1D]

CHAPTER 9 TEST

1. $\frac{3}{4} + \frac{\sqrt{23}}{4}i$ and $\frac{3}{4} - \frac{\sqrt{23}}{4}i$ [9.1B] **2.** $\frac{-3 + \sqrt{41}}{2}$ and $\frac{-3 - \sqrt{41}}{2}$ [9.1B] **3.** Two real zeros [9.1B]

4. -2 [9.3A] **5.** 234 [9.3A] **6.** $-\frac{13}{2}$ [9.3A] **7.** -5 [9.3A] **8.** 5 [9.3B]

9. $f[g(x)] = 2x^2 - 4x - 5$ [9.3B] **10.** 9 [9.1C] **11.** $f^{-1}(x) = \frac{1}{4}x + \frac{1}{2}$ [9.4B] **12.** $f^{-1}(x) = 4x + 16$ [9.4B]

13. $\{(6, 2), (5, 3), (4, 4), (3, 5)\}$ [9.4B] **14.** Yes [9.4B] **15.** 100 [9.1D]

16.

Domain: $\{x \mid x \le 3\}$
Range: $\{y \mid y \le 0\}$ [9.2A]

17.

Domain: $\{x \mid x \in \text{real numbers}\}$
Range: $\{y \mid y \ge -2\}$ [9.2A]

18.

Domain: $\{x \mid x \in \text{real numbers}\}$
Range: $\{y \mid y \in \text{real numbers}\}$ [9.2A]

19. No [9.4A] **20.** Dimensions of 50 cm by 50 cm would give a maximum area of 2500 cm². [9.1D]

CUMULATIVE REVIEW EXERCISES

1. $-\frac{23}{4}$ [1.3B] **2.** [number line] [1.1C] **3.** 3 [2.1C] **4.** $\{x \mid x < -2 \text{ or } x > 3\}$ [2.4B]

5. The set of all real numbers [2.5B] **6.** $-\frac{a^{10}}{12b^4}$ [5.1B] **7.** $2x^3 - 4x^2 - 17x + 4$ [5.3B] **8.** $\left(\frac{1}{2}, 3\right)$ [4.2A]

9. $xy(x + 3y)(x - 2y)$ [5.6D] **10.** -3 and 8 [5.7A] **11.** $\{x \mid x < -3 \text{ or } x > 5\}$ [8.5A]

12. $\frac{5}{2x - 1}$ [6.2B] **13.** -2 [6.5A] **14.** $-3 - 2i$ [7.4D] **15.**

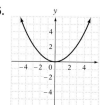

Vertex: $(0, 0)$ [9.1A]
Axis of symmetry: $x = 0$

16. [graph] [3.7A] **17.** $y = -2x - 2$ [3.5B] **18.** $y = -\frac{3}{2}x - \frac{7}{2}$ [3.6A]

19. $\frac{1}{2} + \frac{\sqrt{3}}{6}i$ and $\frac{1}{2} - \frac{\sqrt{3}}{6}i$ [8.3A] **20.** 3 [8.4B] **21.** -3 [9.1C] **22.** $\{1, 2, 4, 5\}$ [3.2A] **23.** Yes [3.2A]

24. 2 [7.3A] **25.** 10 [9.3B] **26.** $f^{-1}(x) = -\frac{1}{3}x + 3$ [9.4B] **27.** The cost per pound of the mixture is $3.96. [2.3A] **28.** 25 lb of the 80% copper alloy must be used. [2.3B] **29.** An additional 4.5 oz of insecticide are required. [6.4B] **30.** It would take the larger pipe 4 min to fill the tank. [8.6A] **31.** A force of 40 lb will stretch the string 24 in. [6.6A] **32.** The frequency is 80 vibrations/min. [6.6A]

Answers to Chapter 10 Selected Exercises

PREP TEST

1. $\frac{1}{9}$ [5.1B] **2.** 16 [5.1B] **3.** -3 [5.1B] **4.** 0; 108 [3.2A] **5.** -6 [2.1B] **6.** -2 and 8 [8.1A]

7. 6326.60 [1.3B] **8.** [9.1A]

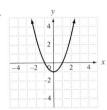

SECTION 10.1

3. c **5. a.** 9 **b.** 1 **c.** $\frac{1}{9}$ **7. a.** 16 **b.** 4 **c.** $\frac{1}{4}$ **9. a.** 1 **b.** $\frac{1}{8}$ **c.** 16 **11. a.** 7.3891

b. 0.3679 **c.** 1.2840 **13. a.** 54.5982 **b.** 1 **c.** 0.1353 **15. a.** 16 **b.** 16 **c.** 1.4768

17. a. 0.1353 **b.** 0.1353 **c.** 0.0111 **19.** **21.**

23. **25.** **27.** **29.**

31. b and **d** **33.** (0, 1) **35.** No x-intercept; (0, 1) **37.** **39.**

41. a. **b.** The point (2, 27.7) means that after 2 s, the object is falling at a speed of 27.7 ft/s.

SECTION 10.2

3. $\log_5 25 = 2$ **5.** $\log_4 \frac{1}{16} = -2$ **7.** $\log_{10} x = y$ **9.** $\log_a w = x$ **11.** $3^2 = 9$ **13.** $10^{-2} = 0.01$ **15.** $e^y = x$

17. $b^y = u$ **19.** 4 **21.** 7 **23.** 2 **25.** 3 **27.** 0 **29.** 4 **31.** 9 **33.** 64 **35.** $\frac{1}{7}$ **37.** 1

39. 316.23 **41.** 0.02 **43.** 7.39 **45.** 0.61 **49.** $\log_3(x^3 y^2)$ **51.** $\ln \frac{x^4}{y^2}$ **53.** $\log_7 x^3$ **55.** $\ln(x^3 y^4)$

57. $\log_4(x^2y^2)$ **59.** $\log_3 \dfrac{x^2z^2}{y}$ **61.** $\ln \dfrac{x}{y^2z}$ **63.** $\log_6 \sqrt{\dfrac{x}{y}}$ **65.** $\log_4 \dfrac{s^2r^2}{t^4}$ **67.** $\ln \dfrac{x}{y^2z^2}$ **69.** $\log_2 \dfrac{t^3v^2}{r^2}$

71. $\log_4 \sqrt{\dfrac{x^3z}{y^2}}$ **73.** $\ln \sqrt{\dfrac{x}{y^3}}$ **75.** $\log_2 \dfrac{\sqrt{xz}}{\sqrt[3]{y^2}}$ **77.** $\log_8 x + \log_8 z$ **79.** $5 \log_3 x$ **81.** $\log_b r - \log_b s$

83. $2 \log_3 x + 6 \log_3 y$ **85.** $3 \log_7 u - 4 \log_7 v$ **87.** $2 \log_2 r + 2 \log_2 s$ **89.** $2 \ln x + \ln y + \ln z$

91. $\log_5 x + 2 \log_5 y - 4 \log_5 z$ **93.** $2 \log_8 x - \log_8 y - 2 \log_8 z$ **95.** $\dfrac{3}{2} \log_4 x + \dfrac{1}{2} \log_4 y$ **97.** $\dfrac{3}{2} \log_7 x - \dfrac{1}{2} \log_7 y$

99. $\log_3 t - \dfrac{1}{2} \log_3 x$ **101.** 0.8451 **103.** −0.2218 **105.** 1.3863 **107.** 1.0415 **109.** 0.8617 **111.** 2.1133

113. −0.6309 **115.** 0.2727 **117.** 1.6826 **119.** 1.9266 **121.** 0.6752 **123.** 2.6125 **125. a.** False
b. True **c.** False **d.** True **e.** False **f.** True

SECTION 10.3

5. [graph] **7.** [graph] **9.** [graph] **11.** [graph]

13. [graph] **15.** [graph] **17.** [graph] **19.** [graph]

21. [graph] **23. a.** [graph] **b.** The point (25.1, 2) means that a star that is 25.1 parsecs from Earth has a distance modulus of 2.

SECTION 10.4

3. $-\dfrac{1}{3}$ **5.** −3 **7.** 1 **9.** 6 **11.** $\dfrac{4}{7}$ **13.** $\dfrac{9}{8}$ **15.** 1.1133 **17.** 1.0986 **19.** 1.3222 **21.** −2.8074

23. 3.5850 **25.** 1.1309 **29.** $\dfrac{11}{2}$ **31.** −4 and 2 **33.** $\dfrac{5}{3}$ **35.** $\dfrac{1}{2}$ **37.** $\dfrac{9}{2}$ **39.** 17.5327 **41.** 3 **43.** 2
45. No solution **47.** 1.7233 **49.** 0.8060 **51.** 5.6910 **53. a.**

b. It will take 2.64 s for the object to travel 100 ft.

SECTION 10.5

1. The value of the investment after 2 years is $1172. **3.** The investment will be worth $15,000 in approximately 18 years.
5. a. After 3 hours, the technetium level will be 21.2 mg. **b.** The technetium level will reach 20 mg after 3.5 h.
7. The half-life is 2.5 years. **9. a.** At 40 km above Earth's surface, the atmospheric pressure is approximately
0.098 newtons/cm². **b.** On Earth's surface, the atmospheric pressure is approximately 10.13 newtons/cm².
c. The atmospheric pressure decreases as you rise above Earth's surface. **11.** The pH of milk is 6.4.

13. The depth is 2.5 m. **15.** Normal conversation emits 65 decibels. **17.** The average time of a Major League baseball game increased by 22.5 min. **19.** The thickness of the copper is 0.4 cm. **21.** The Richter scale magnitude of the earthquake was 6.8. **23.** The intensity of the earthquake was $794,328,235I_0$. **25.** The magnitude of the earthquake for the seismogram given is 5.3. **27.** The magnitude of the earthquake for the seismogram given is 5.6. **29.** The investment has a value of $3210.06 after 5 years.

CHAPTER 10 REVIEW EXERCISES

1. 1 [10.1A] **2.** $5^2 = 25$ [10.2A] **3.** [10.1B] **4.** [10.3A]

5. $\frac{2}{5}\log_3 x + \frac{4}{5}\log_3 y$ [10.2B] **6.** $\log_3 \frac{x^2}{y^5}$ [10.2B] **7.** -12 [10.4A] **8.** $\frac{1}{4}$ [10.4B] **9.** 1.7251 [10.2C]

10. 32 [10.2A] **11.** 79 [10.4B] **12.** 1000 [10.2A] **13.** $\log_7 \sqrt[3]{xy^4}$ [10.2B] **14.** $\frac{1}{2}(5\log_8 x - 3\log_8 y)$ [10.2B]

15. $\log_2 32 = 5$ [10.2A] **16.** 0.4278 [10.2C] **17.** -0.535 [10.4A] **18.** $\frac{3}{2}$ [10.1A] **19.** $\frac{2}{3}$ [10.4B]

20. 2 [10.4B] **21.** [10.1B] **22.** 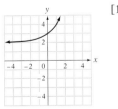 [10.3A] **23.** 2 [10.2A]

24. $\log_2 \sqrt[3]{\frac{x}{y}}$ [10.2B] **25.** 1 [10.4A] **26.** 1 [10.1A] **27.** $\frac{1}{5}$ [10.2A] **28.** $\log_3 81 = 4$ [10.2A]

29. 5 [10.4B] **30.** $\frac{2}{3}\log_5 x + \frac{1}{3}\log_5 y$ [10.2B] **31.** 2 [10.4A] **32.** $\frac{1}{7}$ [10.1A] **33.** 4 [10.2A]

34. 2 [10.4B] **35.** 2.3219 [10.2C] **36.** 3 [10.4A] **37.** 625 [10.2A] **38.** $\log_b \frac{x^2}{y^7}$ [10.2B]

39. 5 [10.1A] **40.** 3.2091 [10.4A] **41.** The value of the investment in 2 years is $4692. [10.5A]

42. The Richter scale magnitude of the earthquake is 8.3. [10.5A] **43.** The half-life is 27 days. [10.5A]

44. The sound emitted from a busy street corner is 107 decibels. [10.5A]

CHAPTER 10 TEST

1. $f(0) = 1$ [10.1A] **2.** $f(-2) = \frac{1}{3}$ [10.1A] **3.** [10.1B] **4.** [10.1B]

5. 2 [10.2A] **6.** $\frac{1}{9}$ [10.2A] **7.** [10.3A] **8.** [10.3A]

9. $\frac{1}{2}(\log_6 x + 3\log_6 y)$ [10.2B] **10.** $\log_3 \sqrt{\frac{x}{y}}$ [10.2B] **11.** $\ln x - \frac{1}{2}\ln z$ [10.2B] **12.** $\ln \frac{x^3}{y\sqrt{z}}$ [10.2B]

13. -2 [10.4A] **14.** -3 [10.4A] **15.** 2.5789 [10.4A] **16.** 6 [10.4B] **17.** 3 [10.4B]

18. 1.3652 [10.2C] **19.** 2.6801 [10.2C] **20.** The half-life is 33 h. [10.5A]

CUMULATIVE REVIEW EXERCISES

1. $\frac{8}{7}$ [2.1C] **2.** $y = 2x - 6$ [3.6A] **3.** $(4x^n + 3)(x^n + 1)$ [5.6C] **4.** $\frac{x - 3}{x + 3}$ [6.3A] **5.** $\frac{x\sqrt{y} + y\sqrt{x}}{x - y}$ [7.2D]

6. $2 + \sqrt{10}$ and $2 - \sqrt{10}$ [8.2A] **7.** $3x^2 + 8x - 3 = 0$ [8.1B] **8.** [4.5A] **9.** $(0, -1, 2)$ [4.2B]

10. $-\frac{2x^2 - 17x + 13}{(x - 2)(2x - 3)}$ [6.2B] **11.** $\{x \mid -5 \le x \le 1\}$ [8.5A] **12.** $\{x \mid 1 \le x \le 4\}$ [2.5B]

13. [10.1B] **14.** [10.3A] **15.** 4 [10.1A]

16. 125 [10.2A] **17.** $\log_b \frac{x^3}{y^5}$ [10.2B] **18.** 1.7712 [10.2C] **19.** 2 [10.4A] **20.** $\frac{1}{2}$ [10.4B]

21. The customer can write at most 49 checks. [2.4C] **22.** The cost per pound of the mixture is $3.10. [2.3A]

23. The rate of the wind is 25 mph. [6.5C] **24.** The spring will stretch 10.2 in. [6.6A] **25.** The cost of redwood is $.40 per foot. The cost of fir is $.25 per foot. [4.4B] **26.** In approximately 10 years, the investment will be worth $10,000. [10.5A]

Answers to Chapter 11 Selected Exercises

PREP TEST

1. 7.21 [3.1B] **2.** $x^2 - 8x + 16; (x - 4)^2$ [8.2A] **3.** $0, -4,$ and 4 [8.1C] **4.** $(1, -1)$ [4.1B]

5. $(1, -5)$ [4.2A] **6.** $x = 2; (2, -2)$ [9.1A] **7.** [9.1A] **8.** [3.7A]

9. 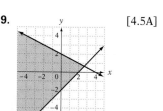 [4.5A]

SECTION 11.1

1. a. vertical line **b.** The parabola opens up. **3. a.** horizontal line **b.** The parabola opens right.

5. a. horizontal line **b.** The parabola opens left.

7.

Axis of symmetry: $y = \dfrac{3}{2}$

Vertex: $\left(-\dfrac{25}{4}, \dfrac{3}{2}\right)$

9.

Axis of symmetry: $x = 0$

Vertex: $(0, 2)$

11.

Axis of symmetry: $y = 0$

Vertex: $(-1, 0)$

13.

Axis of symmetry: $y = 2$

Vertex: $(-1, 2)$

15.

Axis of symmetry: $x = -1$

Vertex: $\left(-1, -\dfrac{7}{2}\right)$

17.

Axis of symmetry: $y = \dfrac{1}{2}$

Vertex: $\left(-\dfrac{25}{4}, \dfrac{1}{2}\right)$

19.

Axis of symmetry: $x = -1$

Vertex: $(-1, -7)$

21.

Axis of symmetry: $x = \dfrac{1}{4}$

Vertex: $\left(\dfrac{1}{4}, -\dfrac{25}{8}\right)$

23.

Axis of symmetry: $x = -\dfrac{5}{2}$

Vertex: $\left(-\dfrac{5}{2}, -\dfrac{1}{4}\right)$

25. $\left(0, \dfrac{1}{8}\right)$

SECTION 11.2

1.

3.

5.

7. $(x - 2)^2 + (y + 1)^2 = 4$

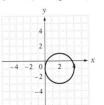

9. $(x + 1)^2 + (y - 1)^2 = 5$ **11.** $(x - 1)^2 + (y + 2)^2 = 25$ **13.** $(x + 3)^2 + (y + 4)^2 = 16$ **15.** $(x - 1)^2 + (y + 1)^2 = 25$

17. $(x - 4)^2 + y^2 = 16$ **19.** The radius is $6\sqrt{3}$ in.

SECTION 11.3

1.

3.

5.

7.

9.

11.

13.

15.

17.

19.

21.

23. a. $\dfrac{x^2}{25} + \dfrac{y^2}{16} = 1$, ellipse

b.

25. a. $\dfrac{x^2}{25} - \dfrac{y^2}{4} = 1$, hyperbola **b.**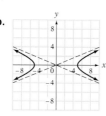

27. a. $\dfrac{y^2}{9} - \dfrac{x^2}{36} = 1$, hyperbola **b.**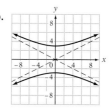

SECTION 11.4

1. $(5, 19)$ and $(-2, 5)$ **3.** $(-1, -2)$ and $(2, 1)$ **5.** $(2, -2)$ **7.** $(2, 2)$ and $\left(-\dfrac{2}{9}, -\dfrac{22}{9}\right)$ **9.** $(3, 2)$ and $(2, 3)$

11. $\left(\dfrac{\sqrt{3}}{2}, 3\right)$ and $\left(-\dfrac{\sqrt{3}}{2}, 3\right)$ **13.** No real number solution **15.** $(1, 2), (1, -2), (-1, 2),$ and $(-1, -2)$

17. $(3, 2), (3, -2), (-3, 2),$ and $(-3, -2)$ **19.** $(2, 7)$ and $(-2, 11)$ **21.** $(3, \sqrt{2}), (3, -\sqrt{2}), (-3, \sqrt{2}),$ and $(-3, -\sqrt{2})$

23. No real number solution **25.** $(\sqrt{2}, 3), (\sqrt{2}, -3), (-\sqrt{2}, 3),$ and $(-\sqrt{2}, -3)$ **27.** $(2, -1)$ and $(8, 11)$ **29.** $(1, 0)$

31. $(2, -1)$ and $(1, -2)$ **33.** $\left(-\dfrac{1}{2}, \dfrac{9}{4}\right)$ and $(-4, 4)$ **35.** $(1.000, 2.000)$ **37.** $(1.755, 0.811)$ and $(0.505, -0.986)$

39. $(5.562, -1.562)$ and $(0.013, 3.987)$

SECTION 11.5

3.

5.

7.

9.

11.

13.

15.

17.

19.

21.

23.

25.

27.

29.

31.

33.

35.

37.

39.

41.

43.

45.

47.

CHAPTER 11 REVIEW EXERCISES

1. [11.1A]

2. [11.3A]

3. [11.5A]

4. [11.2A]

5. Axis of symmetry: $x = 2$; vertex: $(2, 4)$ [11.1A]

6. $(1, \sqrt{5}), (1, -\sqrt{5})$ [11.4A]

7. [11.5B]

8. [11.3B]

9. $(x + 1)^2 + (y - 2)^2 = 18$ [11.2A]

10. Axis of symmetry: $x = \dfrac{7}{2}$; vertex: $\left(\dfrac{7}{2}, \dfrac{17}{4}\right)$ [11.1A] **11.** $\left(1, \dfrac{3}{2}\right)$ and $\left(\dfrac{2}{9}, \dfrac{1}{3}\right)$ [11.4A]

12. $x^2 + (y + 3)^2 = 97$ [11.2A] **13.** $(x + 1)^2 + (y - 5)^2 = 36$ [11.2A]

14. $(\sqrt{5}, 3), (\sqrt{5}, -3), (-\sqrt{5}, 3),$ and $(-\sqrt{5}, -3)$ [11.4A] **15.** $(x + 2)^2 + (y - 1)^2 = 9$ [11.2B]

16. $(-2, -12)$ [11.4A]

17. [11.5A]

18. [11.3B]

19. [11.5A]

20. [11.2A]

21. [11.3A]

22. [11.5A]

23. [11.5B]

24. [11.5B]

25. [11.1A]

26. [11.5B]

27. [11.1A]

28. [11.1A]

29. $(x - 3)^2 + (y + 4)^2 = 25$ [11.2A] **30.** $(x - 3)^2 + (y + 2)^2 = 36$ [11.2B]

31. [11.2A]

32. [11.2B]

33. [11.3A]

34. [11.3A] **35.** [11.3B] **36.** [11.3B]

37 $(-1, 0), (5, 6)$ [11.4A] **38.** [11.5A]

CHAPTER 11 TEST

1. $(x + 3)^2 + (y + 3)^2 = 16$ [11.2A] **2.** $(x + 2)^2 + (y - 1)^2 = 32$ [11.2A] **3.** [11.3B]

4. [11.5B] **5.** $x = 3$ [11.1A] **6.** $(5, 1), (-5, 1), (5, -1),$ and $(-5, -1)$ [11.4A]

7. $\left(\dfrac{3}{2}, \dfrac{1}{4}\right)$ [11.1A] **8.** [11.1A] **9.** [11.3A]

10. 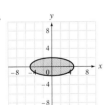 [11.5A] **11.** $(x + 2)^2 + (y - 4)^2 = 9$ [11.2A] **12.** $\left(-\dfrac{9}{4}, \dfrac{1}{2}\right)$ and $(36, -4)$ [11.4A]

13. $\left(\dfrac{2}{3}, \dfrac{4}{3}\right)$ and $(2, 0)$ [11.4A] **14.** $(x + 1)^2 + (y + 3)^2 = 58$ [11.2A] **15.** \varnothing 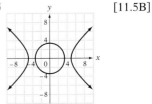 [11.5B]

16. $(x - 2)^2 + (y + 1)^2 = 4$ [11.2B] **17.** [11.1A]

18. [11.2A] **19.** [11.3B] **20.** [11.5A]

CUMULATIVE REVIEW EXERCISES

1. ... [1.1C] **2.** $\frac{38}{53}$ [2.1C] **3.** $\left\{ x \mid -\frac{4}{3} < x < 0 \right\}$ [2.5B] **4.** $y = -\frac{3}{2}x$ [3.5A]

5. $y = x - 6$ [3.6A] **6.** $x^{4n} + 2x^{3n} - 3x^{2n+1}$ [5.3A] **7.** $(x - y - 1)(x^2 - 2x + 1 + xy - y + y^2)$ [5.6B]

8. $\{x \mid -4 < x \le 3\}$ [8.5A] **9.** $\frac{x}{x + y}$ [6.1B] **10.** $\frac{x - 5}{3x - 2}$ [6.2B] **11.** $\frac{5}{2}$ [6.5A]

12. 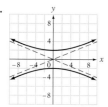 [3.7A] **13.** $\frac{4a}{3b^{12}}$ [5.1B] **14.** $2x^{3/4}$ [7.1B] **15.** $3\sqrt{2} - 5i$ [7.4A]

16. $\dfrac{-1 + \sqrt{7}}{2}$ and $\dfrac{-1 - \sqrt{7}}{2}$ [8.3A] **17.** $(4, 2), (4, -2), (-4, 2),$ and $(-4, -2)$ [11.4A] **18.** 6 [8.4B]

19. -20 [3.2A] **20.** $f^{-1}(x) = \frac{1}{4}x - 2$ [9.4B] **21.** 0 [9.1C] **22.** 400 [9.1D] **23.** 5 [3.1B]

24. $(x + 1)^2 + (y - 2)^2 = 17$ [11.2A] **25.** [11.1A] **26.** [11.3A]

27. [11.3B] **28.** [11.5B]

29. There were 82 adult tickets sold. [2.3A] **30.** The rate of the motorcycle is 60 mph. [6.5C]

31. The rowing rate of the crew is 4.5 mph. [6.5C] **32.** The gear will make 18 revolutions/min. [6.6A]

Answers to Chapter 12 Selected Exercises

PREP TEST

1. 12 [1.2D] **2.** $\frac{3}{4}$ [3.2A] **3.** 18 [1.3B] **4.** 96 [1.3B] **5.** -410 [1.3B] **6.** $\frac{4}{9}$ [1.2D]

7. $x^2 + 2xy + y^2$ [5.3C] **8.** $x^3 + 3x^2y + 3xy^2 + y^3$ [5.3C]

SECTION 12.1

3. 8 **5.** 2, 3, 4, 5 **7.** 3, 5, 7, 9 **9.** 0, -2, -4, -6 **11.** 2, 4, 8, 16 **13.** 2, 5, 10, 17 **15.** 0, $\frac{7}{2}$, $\frac{26}{3}$, $\frac{63}{4}$

17. 40 **19.** 110 **21.** 225 **23.** $\frac{1}{256}$ **25.** 380 **27.** $-\frac{1}{36}$ **29.** 45 **31.** 20 **33.** 91 **35.** 0 **37.** 432

39. $\frac{5}{6}$ **41.** $2x + 2x^2 + 2x^3 + 2x^4 + 2x^5$ **43.** $x + \frac{x^2}{2} + \frac{x^3}{3} + \frac{x^4}{4} + \frac{x^5}{5}$ **45.** $x^2 + x^4 + x^6 + x^8 + x^{10}$

47. $x + \frac{x^2}{4} + \frac{x^3}{9} + \frac{x^4}{16}$ **49.** $\frac{1}{x} + \frac{1}{x^2} + \frac{1}{x^3} + \frac{1}{x^4} + \frac{1}{x^5}$ **51.** $\sum_{i=1}^{n} \frac{1}{i}$

SECTION 12.2

1. 141 **3.** 50 **5.** 17 **7.** $12\frac{1}{2}$ **9.** -109 **11.** $a_n = 3n - 2$ **13.** $a_n = -3n + 6$ **15.** $a_n = -2.5n + 9.5$

17. 20 **19.** 20 **21.** 13 **23.** 20 **25.** 2500 **27.** -760 **29.** 189 **31.** 345 **33.** $\frac{187}{2}$ **35.** -180

37. In 11 weeks, the person will walk 60 min per day. **39.** The object will drop 144 ft during the 5th second.
41. The salary for the eighth month is \$3600. The total salary for eight months is \$20,400. **43.** log 8

SECTION 12.3

3. 131,072 **5.** $\frac{128}{243}$ **7.** 32 **9.** 6 and 4 **11.** -2 and $\frac{4}{3}$ **13.** 12 and -48 **15.** 2186 **17.** $\frac{2343}{64}$ **19.** 62

21. $\frac{121}{243}$ **23.** 9 **25.** $\frac{7}{9}$ **27.** $\frac{8}{9}$ **29.** $\frac{2}{9}$ **31.** $\frac{5}{11}$ **33.** $\frac{1}{6}$ **35.** There will be 50 mg of radioactive material

in the sample at the beginning of the fourth hour. **37.** The ball bounces to a height of 1.8 ft on the sixth bounce.
39. The total amount earned in 30 days is 1,073,741,823 cents, or \$10,737,418.23. **41.** 27

SECTION 12.4

1. 6 **3.** 40,320 **5.** 1 **7.** 10 **9.** 1 **11.** 84 **13.** 21 **15.** 45 **17.** 1 **19.** 20 **21.** 11 **23.** 6
25. $x^4 + 4x^3y + 6x^2y^2 + 4xy^3 + y^4$ **27.** $x^5 - 5x^4y + 10x^3y^2 - 10x^2y^3 + 5xy^4 - y^5$ **29.** $16m^4 + 32m^3 + 24m^2 + 8m + 1$
31. $32r^5 - 240r^4 + 720r^3 - 1080r^2 + 810r - 243$ **33.** $a^{10} + 10a^9b + 45a^8b^2 + \cdots$ **35.** $a^{11} - 11a^{10}b + 55a^9b^2 + \cdots$
37. $256x^8 + 1024x^7y + 1792x^6y^2 + \cdots$ **39.** $65,536x^8 - 393,216x^7y + 1,032,192x^6y^2 + \cdots$ **41.** $x^7 + 7x^5 + 21x^3 + \cdots$
43. $-560x^4$ **45.** $-6x^{10}y^2$ **47.** $126y^5$ **49.** $5n^3$ **51. a.** False **b.** False **c.** False **d.** False **e.** False
f. True

CHAPTER 12 REVIEW EXERCISES

1. $2 + 2x + 2x^2 + 2x^3$ [12.1B] **2.** 28 [12.1A] **3.** $\frac{19}{30}$ [12.3C] **4.** $\frac{1}{3}$ [12.3C] **5.** 1830 [12.2B]

6. 35 [12.2B] **7.** $\frac{23}{99}$ [12.3C] **8.** 84 [12.4A] **9.** 726 [12.3B] **10.** 22 terms [12.2A] **11.** 0.996 [12.3B]

12. $5670x^4y^4$ [12.4A] **13.** $-\frac{5}{27}$ [12.1A] **14.** 1127 [12.3B] **aa15.** $a_n = 5n - 12$ [12.2A]

16. $-42,240x^4y^7$ [12.4A] **17.** $243\sqrt{3}$ [12.3A] **18.** 2000 [12.2B] **19.** 220 [12.4A] **20.** $-\frac{13}{60}$ [12.1B]

21. $2x + 2x^2 + \dfrac{8}{3}x^3 + 4x^4 + \dfrac{32}{5}x^5$ [12.1B] **22.** $\dfrac{1}{729}$ [12.3A]

23. $x^5 - 15x^4y^2 + 90x^3y^4 - 270x^2y^6 + 405xy^8 - 243y^{10}$ [12.4A] **24.** $\dfrac{16}{5}$ [12.3C] **25.** 6 [12.3C] **26.** 99 [12.4A]

27. 77 [12.2A] **28.** $4x + 4x^2 + 4x^3 + 4x^4 + 4x^5 + 4x^6$ [12.1B] **29.** 15 [12.2A] **30.** 135 [12.3A]

31. 9 [12.3C] **32.** $\dfrac{1}{2}$ [12.1A] **33.** 44 [12.2A] **34.** 729 [12.2B] **35.** -168 [12.3B] **36.** 21 [12.4A]

37. $2268x^3y^6$ [12.4A] **38.** 45 [12.2B] **39.** $\dfrac{5}{8}$ [12.1A] **40.** $a_n = 4n - 8$ [12.2A] **41.** $\dfrac{2}{125}$ [12.3A]

42. $\dfrac{11}{30}$ [12.3C] **43.** 143 [12.2A] **44.** 341 [12.3B] **45.** A person will be walking 60 min each day in 16 weeks. [12.2C] **46.** The temperature of the spa after 8 h is 67.7°F. [12.3D] **47.** The total salary for the nine-month period is $24,480. [12.2C] **48.** At the beginning of the seventh hour, the amount of radioactive material in the sample is 3.125 mg. [12.3D]

CHAPTER 12 TEST

1. 34 [12.2B] **2.** 84 [12.4A] **3.** 32 [12.3A] **4.** $\dfrac{1}{2}$ [12.1A] **5.** $3x + 3x^2 + 3x^3 + 3x^4$ [12.1B]

6. 468 [12.2B] **7.** 16 [12.3C] **8.** 16 [12.2A] **9.** $\dfrac{7}{30}$ [12.3C] **10.** 70 [12.4A] **11.** $\dfrac{2}{27}$ [12.3A]

12. $-280x^4y^3$ [12.4A] **13.** -115 [12.2A] **14.** $a_n = -3n + 15$ [12.2A] **15.** -66 [12.3B] **16.** $\dfrac{7}{6}, \dfrac{8}{7}$ [12.1A]

17. $\dfrac{665}{32}$ [12.3B] **18.** 1575 [12.2B] **19.** The inventory after the October 1 shipment was 2550 skeins. [12.2C]

20. There will be 20 mg of radioactive material in the sample at the beginning of the fifth day. [12.3D]

CUMULATIVE REVIEW EXERCISES

1. [3.3B] **2.** $2(x^2 + 2)(x^4 - 2x^2 + 4)$ [5.6D] **3.** $\dfrac{x^2 + 5x - 2}{(x + 2)(x - 1)}$ [6.2B] **4.** $f(-2) = 14$ [3.2A]

5. $4y\sqrt{x} - y\sqrt{2}$ [7.2C] **6.** $\dfrac{1}{4} + \dfrac{\sqrt{55}}{4}i$ and $\dfrac{1}{4} - \dfrac{\sqrt{55}}{4}i$ [8.3A] **7.** 4 [7.3A] **8.** $(x + 1)^2 + (y + 1)^2 = 34$ [11.2A]

9. $\left(\dfrac{7}{6}, \dfrac{1}{2}\right)$ [4.2A] **10.** -10 [4.3A] **11.** $\{x \mid x < -2 \text{ or } x > 2\}$ [2.4B] **12.** [3.7A]

13. $\dfrac{1}{2}\log_5 x - \dfrac{1}{2}\log_5 y$ [10.2B] **14.** 3 [10.4A] **15.** $a_5 = 20; a_6 = 30$ [12.1A] **16.** 6 [12.1B]

17. -103 [12.2A] **18.** $\dfrac{9}{5}$ [12.3C] **19.** $\dfrac{7}{15}$ [12.3C] **20.** $12xy^5$ [12.4A] **21.** The amount of water that must be added is 120 oz. [2.3B] **22.** The new computer takes 24 min to complete the payroll. The old computer takes 40 min to complete the payroll. [6.5B] **23.** The boat was traveling 6.25 mph and the current was traveling 1.25 mph. [6.5C] **24.** The half-life is 55 days. [10.5A] **25.** There are 1536 seats in the theater. [12.2C] **26.** The height of the fifth bounce is 2.6 ft. [12.3D]

FINAL EXAM

1. -31 [1.2D] **2.** -1 [1.3B] **3.** $33 - 10x$ [1.3C] **4.** 8 [2.1B] **5.** $\dfrac{2}{3}$ [2.1C] **6.** $-\dfrac{2}{3}$ and 4 [2.5A]

7. [3.3C] **8.** $y = -3x + 7$ [3.5B] **9.** $y = -\dfrac{2}{3}x - \dfrac{1}{3}$ [3.6A] **10.** $6a^3 - 5a^2 + 10a$ [5.3A]

x-intercept: $\left(\dfrac{9}{2}, 0\right)$

y-intercept: $(0, -3)$

11. $(2 - xy)(4 + 2xy + x^2y^2)$ [5.6B] **12.** $(x - y)(1 - x)(1 + x)$ [5.6D] **13.** $x^2 - 2x - 3 - \dfrac{5}{2x - 3}$ [5.4B]

14. $\dfrac{x(x - 1)}{2x - 5}$ [6.1D] **15.** $\dfrac{-10x}{(x + 2)(x - 3)}$ [6.2B] **16.** $\dfrac{x + 3}{x + 1}$ [6.3A] **17.** $-\dfrac{7}{4}$ [6.5A] **18.** $d = \dfrac{a_n - a_1}{n - 1}$ [2.1D]

19. $\dfrac{y^4}{162x^3}$ [5.1B] **20.** $\dfrac{1}{64x^8y^5}$ [7.1A] **21.** $-2x^2y\sqrt{2y}$ [7.2B] **22.** $\dfrac{x^2\sqrt{2y}}{2y^2}$ [7.2D] **23.** $\dfrac{6}{5} - \dfrac{3}{5}i$ [7.4D]

24. $2x^2 - 3x - 2 = 0$ [8.1B] **25.** $\dfrac{3 + \sqrt{17}}{4}$ and $\dfrac{3 - \sqrt{17}}{4}$ [8.3A] **26.** 27 and -8 [8.4A] **27.** $\dfrac{3}{2}$ and -2 [8.4C]

28. [11.1A] **29.** [11.3A] **30.** $f^{-1}(x) = \dfrac{3}{2}x + 6$ [9.4B] **31.** $(3, 4)$ [4.2A]

32. 10 [4.3A] **33.** $\left(\dfrac{5}{2}, -\dfrac{3}{2}\right)$ [11.4A] **34.** $\left(\dfrac{3}{2}, \infty\right)$ [2.4B] **35.** $\{x \mid -4 < x < -1\}$ [2.5B]

36. [3.7A] **37.** [10.3A] **38.** $\log_2 \dfrac{a^2}{b^2}$ [10.2B] **39.** 6 [10.4B]

40. $2y + 2y^2 + 2y^3 + 2y^4 + 2y^5$ [12.1B] **41.** $\dfrac{23}{45}$ [12.3C] **42.** $144x^7y^2$ [12.4A] **43.** The range of scores is $69 \le x \le 100$. [2.4C] **44.** The cyclist traveled 40 mi. [2.3C] **45.** The amount invested at 8.5% is \$8000. The amount invested at 6.4% is \$4000. [4.1C] **46.** The width is 7 ft. The length is 20 ft. [8.6A] **47.** 200 additional shares would need to be purchased. [6.4B] **48.** The rate of the plane is 420 mph. [4.4B] **49.** The distance traveled is 88 ft. [7.3B] **50.** The rate of the plane for the first 360 mi is 120 mph. [8.6A] **51.** The intensity is 200 foot-candles. [6.6A] **52.** The rate of the boat in calm water is 12.5 mph. The rate of the current is 2.5 mph. [4.4A] **53.** The value of the investment is \$4785.65. [10.5A] **54.** The value of the house will be \$256,571. [12.3D]

Glossary

abscissa The first number of an ordered pair; it measures a horizontal distance and is also called the first coordinate of an ordered pair. [3.1]

absolute value of a number The distance of the number from zero on the number line. [1.1, 2.5]

absolute value equation An equation containing the absolute-value symbol. [2.5]

absolute value function A function containing an absolute value symbol. [9.2]

addition method An algebraic method of finding an exact solution of a system of linear equations. [4.2]

additive inverse of a polynomial The polynomial with the sign of every term changed. [5.2]

additive inverses Numbers that are the same distance from zero on the number line but lie on different sides of zero; also called opposites. [1.1]

analytic geometry Geometry in which a coordinate system is used to study relationships between variables. [3.1]

arithmetic progression A sequence in which the difference between any two consecutive terms is constant; also called an arithmetic sequence. [12.2]

arithmetic sequence A sequence in which the difference between any two consecutive terms is constant; also called an arithmetic progression. [12.2]

arithmetic series The indicated sum of the terms of an arithmetic sequence. [12.2]

asymptotes The two straight lines that a hyperbola "approaches." [11.3]

axes The two number lines that form a rectangular coordinate system; also called coordinate axes. [3.1]

axes of symmetry of a hyperbola The line that passes through the vertices of the hyperbola. [11.3B]

axis of symmetry of a parabola A line of symmetry that passes through the vertex of the parabola and is parallel to the y-axis for an equation of the form $y = ax^2 + bx + c$ or parallel to the x-axis for an equation of the form $x = ay^2 + by + c$. [9.1, 11.1]

base In an exponential expression, the number that is taken as a factor as many times as indicated by the exponent. [1.2]

binomial A polynomial of two terms. [5.2]

center of a circle The central point that is equidistant from all the points that make up a circle. [11.2]

center of an ellipse The intersection of the two axes of symmetry of the ellipse. [11.3]

circle The set of all points (x, y) in the plane that are a fixed distance from a given point (h, k) called the center. [11.2]

clearing denominators Removing denominators from an equation that contains fractions by multiplying each side of the equation by the LCM of the denominators. [6.5]

closed interval In interval notation, an interval that contains its endpoints. [1.1]

coefficient The number part of a variable term. [1.3]

cofactor of an element of a matrix $(-1)^{i+j}$ times the minor of that element, where i is the row number of the element and j is its column number. [4.3]

combined variation A variation in which two or more types of variation occur at the same time. [6.6]

combining like terms Using the Distributive Property to add the coefficients of like variable terms; adding like terms of a variable expression. [1.3]

common difference of an arithmetic sequence The difference between any two consecutive terms in an arithmetic sequence. [12.2]

common logarithms Logarithms to the base 10. [10.2]

common monomial factor A monomial factor that is a factor of the terms in a polynomial. [5.5]

common ratio of a sequence In a geometric sequence, each successive term of the sequence is the same nonzero constant multiple of the preceding term. This common multiple is called the common ratio of the sequence. [12.3]

completing the square Adding to a binomial the constant term that makes it a perfect-square trinomial. [8.2]

complex fraction A fraction whose numerator or denominator contains one or more fractions. [1.2, 6.3]

complex number A number of the form $a + bi$, where a and b are real numbers and $i = \sqrt{-1}$. [7.4]

composite number A number that has whole-number factors besides 1 and itself. For example, 12 is a composite number. [1.1]

composition of two functions The operation on two functions f and g denoted by $f \circ g$. The value of the composition of f and g is given by $(f \circ g)(x) = f[g(x)]$. [9.3]

compound inequality Two inequalities joined with a connective word such as *and* or *or*. [2.4]

compound interest Interest that is computed not only on the original principal but also on the interest already earned. [10.5]

compound interest formula $A = P(1 + i)$, where P is the original value of an investment, i is the interest rate per compounding period, n is the total number of compounding periods, and A is the value of the investment after n periods. [10.5]

conditional equation An equation that is true if the variable it contains is replaced by the proper value. $x + 2 = 5$ is a conditional equation. [2.1]

conic section A curve that can be constructed from the intersection of a plane and a right circular cone. The four conic sections are the parabola, hyperbola, ellipse, and circle. [11.1]

conjugates Binomial expressions that differ only in the sign of a term. The expressions $a + b$ and $a - b$ are conjugates. [7.2]

consecutive even integers Even integers that follow one another in order. [2.2]

consecutive integers Integers that follow one another in order. [2.2]

consecutive odd integers Odd integers that follow one another in order. [2.2]

constant function A function given by $f(x) = b$, where b is a constant. Its graph is a horizontal line passing through $(0, b)$. [3.3]

constant of proportionality k in a variation equation; also called the constant of variation. [6.6]

constant of variation k in a variation equation; also called the constant of proportionality. [6.6]

constant term A term that contains no variable part. [1.3, 5.2]

contradiction An equation in which any replacement for the variable will result in a false equation. $x = x + 1$ is a contradiction. [2.1]

coordinate axes The two number lines that form a rectangular coordinate system; also called axes. [3.1]

coordinates of a point The numbers in the ordered pair that is associated with the point. [3.1]

cube root of a perfect cube One of the three equal factors of the perfect cube. [5.6]

cubic function A third-degree polynomial function. [5.2]

degree of a monomial The sum of the exponents of the variables. [5.1]

degree of a polynomial The greatest of the degrees of any of its terms. [5.2]

dependent system of equations A system of equations whose graphs coincide. [4.1]

dependent variable A variable whose value depends on that of another variable known as the independent variable. [3.2]

descending order The terms of a polynomial in one variable are arranged in descending order when the exponents of the variable decrease from left to right. [5.2]

determinant A number associated with a square matrix. [4.3]

difference of two perfect squares A polynomial in the form of $a^2 - b^2$. [5.6]

difference of two perfect cubes A polynomial in the form $a^3 - b^3$. [5.6]

direct variation A special function that can be expressed as the equation $y = kx$, where k is a constant called the constant of variation or the constant of proportionality. [6.6]

discriminant For an equation of the form $ax^2 + bx + c = 0$, the quantity $b^2 - 4ac$ is called the discriminant. [8.3]

domain The set of the first coordinates of all the ordered pairs of a function. [3.2]

double function A function that pairs a number with twice that number. [3.2]

double root When a quadratic equation has two solutions that are the same number, the solution is called a double root of the equation. [8.1]

element of a matrix A number in a matrix. [4.3]

elements of a set The objects in the set. [1.1]

ellipse An oval shape that is one of the conic sections. [11.3]

empty set The set that contains no elements. [1.1]

endpoints of an interval In interval notation, the values that mark the interval's beginning and end, whether or not either or both of those values are included in the interval. [1.1]

equation A statement of the equality of two mathematical expressions. [2.1]

equation in two variables An equation in which two different variables appear. [3.1]

equivalent equations Equations that have the same solution. [2.1]

evaluating a function Determining $f(x)$ for a given value of x. [3.2]

evaluating a variable expression Replacing the variables in a variable expression by numerical values and then simplifying the resulting expression. [1.3]

even integer An integer that is divisible by 2. [2.2]

expanding by cofactors A technique for finding the value of a 3×3 or larger determinant. [4.3]

expansion of the binomial $(a + b)^n$ To write the expression as the sum of its terms. [12.4]

exponent In an exponential expression, the raised number that indicates how many times the factor, or base, occurs in the multiplication. [1.2]

exponential decay equation Any equation that can be written in the form $A = A_0 b^{kt}$, where A is the size at time t, A_0 is the initial size, b is between 0 and 1, and k is a positive real number. [10.5]

exponential equation An equation in which a variable occurs in the exponent. [10.4]

exponential form The expression 2^6 is in exponential form. Compare *factored form*. [1.2]

exponential function The exponential function with base b is defined by $f(x) = b^x$, where b is a positive real number not equal to 1. [10.1]

exponential growth equation Any equation that can be written in the form $A = A_0 b^{kt}$, where A is the size at time t, A_0 is the initial size, b is greater than 1, and k is a positive real number. [10.5]

extraneous solution When each side of an equation is raised to an even power, the resulting equation may have a solution that is not a solution of the original equation. Such a solution is called an extraneous solution. [7.3]

factor One number is a factor of another when it can be divided into that other number with a remainder of zero. [5.5]

factored form The multiplication $2 \cdot 2 \cdot 2 \cdot 2 \cdot 2 \cdot 2$ is in factored form. Compare *exponential form*. [1.2]

factoring a polynomial Writing the polynomial as a product of other polynomials. [5.5]

factoring a quadratic trinomial Expressing the trinomial as the product of two binomials. [5.5]

factoring by grouping Grouping the terms of the polynomial in such a way that a common binomial factor is found. [5.5]

finite sequence A sequence that contains a finite number of terms. [12.1]

finite set A set for which all the elements can be listed. [1.1]

first coordinate of an ordered pair The first number of the ordered pair; it measures a horizontal distance and is also called the abscissa. [3.1]

first-degree equation in one variable An equation in which all variables have an exponent of 1 and all variable terms include one and the same variable. [2.1]

FOIL A method of finding the product of two binomials. The letters stand for First, Outer, Inner, and Last. [5.3]

formula A literal equation that states a rule about measurement. [2.1]

function A relation in which no two ordered pairs that have the same first coordinate have different second coordinates. [3.2]

functional notation Notation used for those equations that define functions. The letter f is commonly used to name a function. [3.2]

general term of a sequence In the sequence $a_1, a_2, a_3, \ldots, a_n, \ldots$ the general term of the sequence is a_n. [12.1]

geometric progression A sequence in which each successive term of the sequence is the same nonzero constant multiple of the preceding term; also called a geometric sequence. [12.3]

geometric sequence A sequence in which each successive term of the sequence is the same nonzero constant multiple of the preceding term; also called a geometric progression. [12.3]

geometric series The indicated sum of the terms of a geometric sequence. [12.3]

graph of a function A graph of the ordered pairs that belong to the function. [3.3]

graph of a quadratic inequality in two variables A region of the plane that is bounded by one of the conic sections. [11.5]

graph of a real number A heavy dot placed directly above the number on the number line. [1.1]

graph of an ordered pair The dot drawn at the coordinates of the point in the plane. [3.1]

graph of $x = a$ A vertical line passing through the point $(a, 0)$. [3.3]

graph of $y = b$ A horizontal line passing through the point $(0, b)$. [3.3]

graphing a point in the plane Placing a dot at the location given by the ordered pair; also called plotting a point in the plane. [3.1]

graphing, or plotting, an ordered pair Placing a dot, on a rectangular coordinate system, at the location given by the ordered pair. [3.1]

greater than A number that lies to the right of another number on the number line is said to be greater than that number. [1.1]

greatest common factor (GCF) The GCF of two or more numbers is the largest integer that divides evenly into all of them. [1.2]

grouping symbols Parentheses (), brackets [], braces { }, the absolute value symbol, and the fraction bar. [1.2]

half-open interval In set-builder notation, an interval that contains one of its endpoints. [1.1]

half-plane The solution set of a linear inequality in two variables. [3.7]

horizontal-line test A graph of a function represents the graph of a one-to-one function if any horizontal line intersects the graph at no more than one point. [9.4]

hyperbola A conic section formed by the intersection of a cone and a plane perpendicular to the base of the cone. [11.3]

hypotenuse In a right triangle, the side opposite the 90° angle. [3.1, 7.3]

identity An equation in which any replacement for the variable will result in a true equation. $x + 2 = x + 2$ is an identity. [2.1]

imaginary number A number of the form ai, where a is a real number and $i = \sqrt{-1}$. [7.4]

imaginary part of a complex number For the complex number $a + bi$, b is the imaginary part. [7.4]

inconsistent system of equations A system of equations that has no solution. [4.1]

independent system of equations A system of equations whose graphs intersect at only one point. [4.1]

independent variable A variable whose value determines that of another variable known as the dependent variable. [3.2]

index In the expression $\sqrt[n]{a}$, n is the index of the radical. [7.1]

index of a summation The variable used in summation notation. [12.1]

infinite geometric series The indicated sum of the terms of an infinite geometric sequence. [12.3]

infinite sequence A sequence that contains an infinite number of terms. [12.1]

infinite set A set in which the list of elements continues without end. [1.1]

input In a function, the value of the independent variable. [9.1]

input/output table A table that shows the results of evaluating a function for various values of the independent variable. [9.1]

integers The numbers $\ldots, -3, -2, -1, 0, 1, 2, 3, \ldots$. [1.1]

intersection of two sets The set that contains all elements that are common to both of the sets. [1.1]

interval notation A type of set notation in which the property that distinguishes the elements of the set is their location within a specified interval. [1.1]

inverse of a function The set of ordered pairs formed by reversing the coordinates of each ordered pair of the function. [9.4]

inverse variation A function that can be expressed as the equation $y = \frac{k}{x}$, where k is a constant. [6.6]

irrational number The decimal representation of an irrational number never terminates or repeats and can only be approximated. [1.1]

joint variation A variation in which a variable varies directly as the product of two or more variables. A joint variation can be expressed as the equation $z = kxy$, where k is a constant. [6.6]

leading coefficient In a polynomial, the coefficient of the variable with the largest exponent. [5.2]

least common multiple (LCM) The LCM of two or more numbers is the smallest number that is a multiple of each of those numbers. [1.2]

least common multiple (LCM) of two or more polynomials The simplest polynomial of least degree that contains the factors of each polynomial. [6.2]

leg In a right triangle, one of the two sides that are not opposite the 90° angle. [3.1, 7.3]

less than A number that lies to the left of another number on the number line is said to be less than that number. [1.1]

like terms Terms of a variable expression that have the same variable part. Having no variable part, constant terms are like terms. [1.3]

linear equation in three variables An equation of the form $Ax + By + Cz = D$, where A, B, and C are coefficients of the variables and D is a constant. [4.2]

linear equation in two variables An equation of the form $y = mx + b$ or $Ax + By = C$. [3.3]

linear function A function that can be expressed in the form $f(x) = mx + b$. Its graph is a straight line. [3.3, 5.2]

linear inequality in two variables An inequality of the form $y > mx + b$ or $Ax + By > C$. (The symbol $>$ could be replaced by \geq, $<$, or \leq.) [3.7]

literal equation An equation that contains more than one variable. [2.1]

logarithm For b greater than zero and not equal to 1, the statement $y = \log_b x$ (the logarithm of x to the base b) is equivalent to $x = b^y$. [10.2]

lower limit In a tolerance, the lowest acceptable value. [2.5]

main fraction bar The fraction bar that is placed between the numerator and denominator of a complex fraction. [1.2]

mantissa The decimal part of a common logarithm. [10.2]

matrix A rectangular array of numbers. [4.3]

maximum value of a function The greatest value that the function can take on. [9.1]

midpoint of a line segment The point on a line segment that is equidistant from its endpoints. [3.1]

minimum value of a function The least value that the function can take on. [9.1]

minor of an element The minor of an element in a 3×3 determinant is the 2×2 determinant obtained by eliminating the row and column that contain that element. [4.3]

monomial A number, a variable, or a product of a number and variables; a polynomial of one term. [5.1, 5.2]

multiplicative inverse The multiplicative inverse of a nonzero real number a is $\frac{1}{a}$; also called the reciprocal. [1.2]

n factorial The product of the first n natural numbers; n factorial is written $n!$ [12.4]

natural exponential function The function defined by $f(x) = e^x$, where $e \approx 2.71828$. [10.1]

natural logarithm When e (the base of the natural exponential function) is used as the base of a logarithm, the logarithm is referred to as the natural logarithm and is abbreviated $\ln x$. [10.2]

natural numbers The numbers $1, 2, 3, \ldots$; also called the positive integers. [1.1]

negative integers The numbers $\ldots, -3, -2, -1$. [1.1]

negative reciprocal The negative reciprocal of a nonzero real number a is $-1/a$. [3.6]

negative slope The slope of a line that slants downward to the right. [3.4]

nonfactorable over the integers A polynomial is nonfactorable over the integers if it does not factor using only integers. [5.5]

nonlinear system of equations A system of equations in which one or more of the equations are not linear equations. [11.4]

nth root of a A number b such that $b^n = a$. The nth root of a can be written $a^{1/n}$ or $\sqrt[n]{a}$. [7.1]

null set The set that contains no elements. [1.1]

numerical coefficient The number part of a variable term. [1.3]

odd integer An integer that is not divisible by 2. [2.2]

one-to-one function In a one-to-one function, given any y, there is only one x that can be paired with the given y. [9.4]

open interval In set-builder notation, an interval that does not contain its endpoints. [1.1]

opposites Numbers that are the same distance from zero on the number line but lie on different sides of zero; also called additive inverses. [1.1]

order $m \times n$ A matrix of m rows and n columns is of order $m \times n$. [4.3]

Order of Operations Agreement Rules that specify the order in which we perform operations in simplifying numerical expressions. [1.2]

ordered pair A pair of numbers expressed in the form (a, b) and used to locate a point in the plane determined by a rectangular coordinate system. [3.1]

ordered triple Three numbers expressed in the form (x, y, z) and used to locate a point in the xyz-coordinate system. [4.2]

ordinate The second number of an ordered pair; it measures a vertical distance and is also called the second coordinate of an ordered pair. [3.1]

origin The point of intersection of the two number lines that form a rectangular coordinate system. [3.1]

output The result of evaluating a function. [9.1]

parabola The graph of a quadratic function is called a parabola. [9.1]

parallel lines Lines that have the same slope and thus do not intersect. [3.6]

Pascal's Triangle A pattern for the coefficients of the terms of the expansion of the binomial $(a + b)^n$ that can be formed by writing the coefficients in a triangular array known as Pascal's Triangle. [12.4]

percent mixture problem A problem that involves combining two solutions or alloys that have different concentrations of the same substance. [2.3]

perfect cube The product of the same three factors. [5.6]

perfect square The product of a term and itself. [5.6]

perfect-square trinomial The square of a binomial. [5.6]

perpendicular lines Lines that intersect at right angles. The slopes of perpendicular lines are negative reciprocals of each other. [3.6]

plane The infinitely extending, two-dimensional space in which a rectangular coordinate system lies; may be pictured as a large, flat piece of paper. [3.1]

plotting a point in the plane Placing a dot at the location given by the ordered pair; also called graphing a point in the plane. [3.1]

point-slope formula The equation $y - y_1 = m(x - x_1)$, where m is the slope of a line and (x_1, y_1) is a point on the line. [3.5]

polynomial A variable expression in which the terms are monomials. [5.2]

polynomial function An expression whose terms are monomials. [5.2]

positive integers The numbers 1, 2, 3, . . . ; also called the natural numbers. [1.1]

positive slope The slope of a line that slants upward to the right. [3.4]

power In an exponential expression, the number of times (indicated by the exponent) that the factor, or base, occurs in the multiplication. [1.2]

prime number A number whose only whole-number factors are 1 and itself. For example, 17 is a prime number. [1.1]

prime polynomial A polynomial that is nonfactorable over the integers. [5.5]

principal square root The positive square root of a number. [7.1]

Principle of Zero Products If the product of two factors is zero, then at least one of the factors must be zero. [5.7]

product of the sum and difference of two terms A polynomial that can be expressed in the form $(a + b)(a - b)$. [5.3, 5.6]

proportion An equation that states the equality of two ratios or rates. [6.4]

Pythagorean Theorem The square of the hypotenuse of a right triangle is equal to the sum of the squares of the two legs. [3.1]

quadrant One of the four regions into which a rectangular coordinate system divides the plane. [3.1]

quadratic equation An equation of the form $ax^2 + bx + c = 0$, where a and b are coefficients, c is a constant, and $a \neq 0$; also called a second-degree equation. [5.7, 8.1]

quadratic equation in standard form A quadratic equation written in descending order and set equal to zero. [5.7]

quadratic formula A general formula, derived by applying the method of completing the square to the standard form of a quadratic equation, used to solve quadratic equations. [8.3]

quadratic function A function that can be expressed by the equation $f(x) = ax^2 + bx + c$, where a is not equal to zero. [5.2, 9.1]

quadratic inequality An inequality that can be written in the form $ax^2 + bx + c < 0$ or $ax^2 + bx + c > 0$, where a is not equal to zero. The symbols \leq and \geq can also be used. [8.5]

quadratic in form A trinomial is quadratic in form if it can be written as $au^2 + bu + c$. [5.6, 8.4]

quadratic trinomial A trinomial of the form $ax^2 + bx + c$, where a and b are nonzero coefficients and c is a nonzero constant. [5.5]

radical In a radical expression, the symbol $\sqrt{}$. [7.1]

radical equation An equation that contains a variable expression in a radicand. [7.3]

radical function A function containing a radical. [9.2]

radicand The expression under a radical sign. [7.1]

radius of a circle The fixed distance, from the center of a circle, of all points that make up the circle. [11.2]

range The set of the second coordinates of all the ordered pairs of a function. [3.2]

rate The quotient of two quantities that have different units. [6.4]

rate of work That part of a task that is completed in one unit of time. [6.5]

ratio The quotient of two quantities that have the same unit. [6.4]

rational expression A fraction in which the numerator or denominator is a polynomial. [6.1]

rational function A function that is written in terms of a rational expression. [6.1]

rational number A number of the form $\frac{a}{b}$, where a and b are integers and b is not equal to zero. [1.1]

rationalizing the denominator The procedure used to remove a radical from the denominator of a fraction. [7.2]

real numbers The rational numbers and the irrational numbers taken together. [1.1]

real part of a complex number For the complex number $a + bi$, a is the real part. [7.4]

reciprocal The reciprocal of a nonzero real number a is $\frac{1}{a}$; also called the multiplicative inverse. [1.2]

reciprocal of a rational expression The rational expression with the numerator and denominator interchanged. [6.1]

rectangular coordinate system A coordinate system formed by two number lines, one horizontal and one vertical, that intersect at the zero point of each line. [3.1]

relation A set of ordered pairs. [3.2]

Remainder Theorem If the polynoimal $P(x)$ is divided by $x - a$, the remainder is $P(a)$. [5.4]

repeating decimal A decimal formed when dividing the numerator of its fractional counterpart by the denominator results in a decimal part wherein one or more digits repeat infinitely. [1.1]

right triangle A triangle that contains one 90° angle. [3.1]

root(s) of an equation The replacement value(s) of the variable that will make the equation true; also called the solution(s) of the equation. [2.1]

roster method A method of designating a set by enclosing a list of its elements in braces. [1.1]

scatter diagram A graph of ordered-pair data. [3.1]

scientific notation Notation in which a number is expressed as the product of a number between 1 and 10 and a power of 10. [5.1]

second coordinate of an ordered pair The second number of the ordered pair; it measures a vertical distance and is also called the ordinate. [3.1]

second-degree equation An equation of the form $ax^2 + bx + c = 0$, where a and b are coefficients, c is a constant, and $a \neq 0$; also called a quadratic equation. [8.1]

sequence An ordered list of numbers. [12.1]

series The indicated sum of the terms of a sequence. [12.1]

set A collection of objects. [1.1]

set-builder notation A method of designating a set that makes use of a variable and a certain property that only elements of that set possess. [1.1]

sigma notation Notation used to represent a series in a compact form; also called summation notation. [12.1]

simplest form of a rational expression A rational expression is in simplest form when the numerator and denominator have no common factors. [6.1]

slope A measure of the slant, or tilt, of a line. The symbol for slope is m. [3.4]

slope-intercept form of a straight line The equation $y = mx + b$, where m is the slope of the line and $(0, b)$ is the y-intercept. [3.4]

solution of a system of equations in three variables An ordered triple that is a solution of each equation of the system. [4.2]

solution of a system of equations in two variables An ordered pair that is a solution of each equation of the system. [4.1]

solution of an equation in three variables An ordered triple (x, y, z) whose coordinates make the equation a true statement. [4.2]

solution of an equation in two variables An ordered pair whose coordinates make the equation a true statement. [3.1]

solution set of a system of inequalities The intersection of the solution sets of the individual inequalities. [4.5]

solution set of an inequality A set of numbers, each element of which, when substituted for the variable, results in a true inequality. [2.4]

solution(s) of an equation The replacement value(s) of the variable that will make the equation true; also called the root(s) of the equation. [2.1]

solving an equation Finding a root, or solution, of the equation. [2.1]

square function A function that pairs a number with its square. [3.2]

square matrix A matrix that has the same number of rows as columns. [4.3]

square of a binomial A polynomial that can be expressed in the form $(a + b)^2$. [5.3]

square root of a perfect square One of the two equal factors of the perfect square. [5.6]

standard form of a quadratic equation A quadratic equation is in standard form when the polynomial is in descending order and equal to zero. [8.1]

substitution method An algebraic method of finding an exact solution of a system of linear equations. [4.1]

summation notation Notation used to represent a series in a compact form; also called sigma notation. [12.1]

sum of two perfect cubes A polynomial that can be written in the form $a^3 + b^3$. [5.6]

sum of two perfect squares A binomial in the form $a^2 + b^2$, which is nonfactorable over the integers. [5.6]

synthetic division A shorter method of dividing a polynomial by a binomial of the form $x - a$. This method uses only the coefficients of the variable terms. [5.4]

system of equations Two or more equations considered together. [4.1]

system of inequalities Two or more inequalities considered together. [4.5]

term of a sequence A number in a sequence. [2.1]

terminating decimal A decimal formed when dividing the numerator of its fractional counterpart by the denominator results in a remainder of zero. [1.1]

terms of a variable expression The addends of the expression. [1.3]

tolerance of a component The amount by which it is acceptable for the component to vary from a given measurement. [2.5]

trinomial A polynomial of three terms. [5.2]

undefined slope The slope of a vertical line is undefined. [3.4]

uniform motion The motion of an object whose speed and direction do not change. [2.3, 6.5]

uniform motion problem A problem that involves the motion of an object whose speed and direction do not change. [2.3]

union of two sets The set that contains all elements that belong to either of the sets. [1.1]

upper limit In a tolerance, the greatest acceptable value. [2.5]

value mixture problem A problem that involves combining two ingredients that have different prices into a single blend. [2.3]

value of a function The value of the dependent variable for a given value of the independent variable. [3.2]

variable A letter used to stand for a quantity that is unknown or that can change. [1.1]

variable expression An expression that contains one or more variables. [1.3]

variable part of a variable term The variables, taken together, that occur in a variable term. [1.3]

variable term A term composed of a numerical coefficient and a variable part. When the numerical coefficient is 1 or -1, the 1 is usually not written. [1.3]

vertex of a parabola The point on the parabola with the smallest y-coordinate or the largest y-coordinate. [9.1]

vertical-line test A graph defines a function if any vertical line intersects the graph at no more than one point. [9.2]

whole numbers The numbers 0, 1, 2, 3, 4, 5, ... [1.1]

x-coordinate The abscissa in an xy-coordinate system. [3.1]

x-intercept The point at which a graph crosses the x-axis. [3.3]

xy-coordinate system A rectangular coordinate system in which the horizontal axis is labeled x and the vertical axis is labeled y. [3.1]

xyz-coordinate system A three-dimensional coordinate system formed when a third coordinate axis (the z-axis) is located perpendicular to the xy-plane. [4.2]

y-coordinate The ordinate in an xy-coordinate system. [3.1]

y-intercept The point at which a graph crosses the y-axis. [3.3]

zero of a function A value of x for which $f(x) = 0$. [9.1]

zero slope The slope of a horizontal line. [3.4]

TI-30X IIS

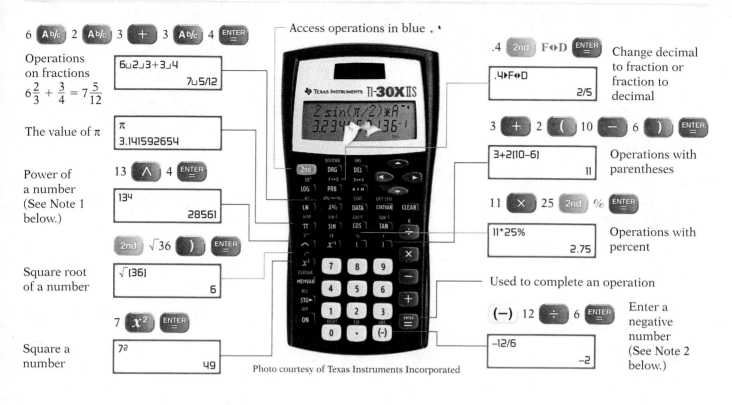

Access operations in blue

Operations on fractions
6 [Ab/c] 2 [Ab/c] 3 [+] 3 [Ab/c] 4 [ENTER]
$6\frac{2}{3} + \frac{3}{4} = 7\frac{5}{12}$

6⎵2⎵3+3⎵4
7⎵5/12

The value of π
π
3.141592654

Power of a number (See Note 1 below.)
13 [∧] 4 [ENTER =]
13⁴
28561

Square root of a number
[2nd] √ 36 [)] [ENTER =]
√(36)
6

Square a number
7 [x²] [ENTER =]
7²
49

Change decimal to fraction or fraction to decimal
.4 [2nd] [F◊D] [ENTER]
.4▸F◊D
2/5

Operations with parentheses
3 [+] 2 [(] 10 [−] 6 [)] [ENTER]
3+2(10−6)
11

Operations with percent
11 [×] 25 [2nd] [%] [ENTER]
11*25%
2.75

Used to complete an operation

Enter a negative number (See Note 2 below.)
[(−)] 12 [÷] 6 [ENTER]
−12/6
−2

Photo courtesy of Texas Instruments Incorporated

fx-300MS

Access operations in gold

Square root of a number
[√] 36 [=]
√ 36
6

Operations on fractions
6 [ab/c] 2 [ab/c] 3 [+] 3 [ab/c] 4 [=]
$6\frac{2}{3} + \frac{3}{4} = 7\frac{5}{12}$

6⎵2⎵3+3⎵4
7⎵5⎵12

Square a number
7 [x²] [=]
7²
49

Enter a negative number (See Note 2 below.)
[(−)] 12 [÷] 6 [=]
−12÷6
−2

Change decimal to fraction
.4 [=] [=]
.4
2⎵5

Power of a number (See Note 1 below.)
13 [∧] 4 [=]
13⁴
28561

Operations with parentheses
3 [+] 2 [(] 10 [−] 6 [)] [=]
3+2(10−6)
11

Operations with percent
11 [×] 25 [=]
11x25%
2.75

Used to complete an operation

[=]
π
3.141592654

The value of π

Photo courtesy of Casio, Inc.

NOTE 1: Some calculators use the [yˣ] key to calculate a power. For those calculators, enter 13 [yˣ] 4 [=] to evaluate 13^4.

NOTE 2: Some calculators use the [+/−] key to enter a negative number. For those calculators, enter 12 [+/−] [÷] 6 [=] to calculate $-12 \div 6$.

Math Study Skills Workbook

Second Edition

Your Guide to Reducing Test Anxiety
and Improving Study Strategies

Paul D. Nolting, Ph.D.
Learning Specialist

Preface

Wouldn't it be nice if all we had to do was listen to a lecture on math and read the textbook in order to learn it? That would be paradise. However, most math courses take place on Earth, and students have to do much more than just take notes and read a textbook. They need a system of study skills that help them understand and master mathematics.

Many students in all levels of math courses have difficulty learning math because it is one of the most difficult subjects in college. First, many students who have struggled with math before going to college continue to struggle when they take their first college math courses, even developmental math courses. Second, some students who have done well in the past begin to struggle when they take upper level math like college algebra or calculus. They made A's and B's in previous math courses, but all of a sudden they are failing upper level math courses. These students probably lived off their intelligence until they took a math course that challenged them. Since they never had to study for math, they did not know where to begin to study. Students in all levels of math courses benefit from designing a system of study skills.

Does this sound far fetched? Not really. I have watched calculus students read this workbook and improve their grades in order to be more competitive candidates for engineering programs. For example, I asked two calculus II students why they were taking my math study skills course. I thought they needed a one credit hour course to allow them to graduate, but I was wrong. Both students made a C in calculus I and said that they needed to make an A or B in calculus II because they were engineering majors. In most cases, students who make C's in calculus are not admitted to engineering schools. Both students made an A in my course and a B in their calculus II course and went on to engineering school. Success for them!

Another success story belongs to a student who failed a beginning algebra course three times. Yes, three times. She came to my office ready to quit college all together. I convinced her to focus on designing a system of study skills before taking the course one more time. She did. With her new system of study skills, she passed with a B and stayed in college. Success for her!

Earning better math grades doesn't have to be the only benefit from spending time developing a system for studying math. According to many students, they were able to take the math study skills and adapt them to their other courses. In fact, some students claim it is easier than adapting general study skills to learning math. They reported that their other course grades also went up. What a good experience!

So, what kind of assistance do most students want? Students want tips and procedures they can use to help them improve their math grades. The math study suggestions, however, have to be based on research and be statistically proven to improve student learning and grades. **Math Study Skills Workbook** is based on **Winning at Math: Your Guide to Learning Mathematics Through Successful Study Skills** (2002), which is the only math study skills text that can boast statistical evidence demonstrating an improvement in students' ability to learn math and make better grades. Learning and using these study skills techniques will improve your grades.

Math Study Skills Workbook is designed to supplement your math course or study skills course or to be used as part of a math anxiety workshop. The workbook is designed for independent work or to be supplemented by lectures. To gain the most benefit, the workbook needs to be completed by midterm.

This workbook is designed to enhance learning by teaching math learning skills in small "chunks." After each section, you are required to recall the most important information by writing it down. The writing exercises are especially designed for you to personalize learning techniques, such as the ten steps to better test taking. Each chapter review is designed to reinforce your learning and to give you the opportunity to immediately select and use the best learning strategy from that chapter. Lecture Notes pages have been added, where possible, to allow you to include your notes in this workbook.

As you adapt and practice each study skill, place them into a larger system for studying math. Decide which study skills will help with each of the following tasks you must do when learning math:

1. Taking and reworking lecture notes
2. Learning vocabulary
3. Completing homework in a way that helps you learn and master the math

4. Mastering and memorizing the concepts
5. Preparing for tests
6. Taking tests
7. Managing any anxiety whether math or text anxiety

Remember, to reap the most benefit, you need to complete this workbook and be using your system of study skills by midterm. Practice and master! Then, maybe we can use your success story in our next edition of the workbook!

What You Need to Know to Study Math

Mathematics courses are not like other courses. Because they are different, they require different study procedures. Passing most of your other courses requires only that you read, understand, and recall the subject material. To pass math, however, an extra step is required: You must use the information you have learned to correctly solve math problems.

Learning general study skills can help you pass most of your courses, but special math study skills are needed to help you learn more and get better grades in math. In this chapter, you will find out

- why learning math is different from learning other subjects,
- what the differences are between high school and college math, and
- why your first math test is very important.

Why learning math is different from learning other subjects

In a math course, you must be able to do four things:

1. *Understand* the material
2. *Process* the material
3. *Apply* what you have learned to correctly solve a problem, and
4. *Remember* what you have learned to learn new material.

Of these four tasks, applying what you have learned to correctly solve a problem is the hardest.

Examples: Political science courses require that you learn about politics and public service. But your instructor will not make you run for

continued

governor to pass the course. Psychology courses require you to understand the concepts of different psychology theories. But you will not have to help a patient overcome depression to pass the course. In math, however, you must be able to correctly solve problems to pass the course.

Sequential Learning Pattern

Another reason learning math is different than learning other subjects is that it follows a *sequential learning pattern*, which simply means that the material learned on one day is used the next day and the next day, and so forth. This building-block approach to learning math is the reason it is difficult to catch up when you fall behind. *All* building blocks must be included to be successful in learning math.

You can compare learning math to building a house. Like a house, which must be built foundation first, walls second, and roof last, math must be learned in a specific order. Just as you cannot build a house roof first, you cannot learn to solve complex problems without first learning to solve simple ones.

Example: In a history class, if you study Chapters 1 and 2, do not understand Chapter 3, and end up studying and having a test on Chapter 4, you *could* pass. Understanding Chapter 4 in history is not totally based

continued

on comprehending Chapters 1, 2, and 3. To succeed in math, however, each previous chapter has to be completely understood before you can continue to the next chapter.

Sequential learning affects your ability to study for math tests, as well. If you study Chapter 1 and understand it, study Chapter 2 and understand it, and study Chapter 3 and *do not* understand it, then when you study for a test on Chapter 4, you are not going to understand it either, and you probably will not do well on the test.

REMEMBER

To learn the new math material for the test on Chapter 5, you must first go back and learn the material in Chapter 4. This means you will have to go back and learn Chapter 4 while learning Chapter 5. (The best of us can fall behind under these circumstances.) However, if you do not understand the material in Chapter 4, you will not understand the material in Chapter 5 either, and you will fail the test on Chapter 5. This is why the sequential learning of math concepts if so important.

The sequential learning pattern is also affected by

- your previous math course grade,
- your math placement test scores, and
- the time elapsed since your last math course.

Sequential learning is influenced by how much math knowledge you have at the beginning of your course. Students who forgot or never acquired the necessary skills from their previous math course will have difficulty with their current math course. If you do not remember what you learned in your last math course, you will have to relearn the math concepts from the previous course as well as the new material for the current course. In most other courses, such as the humanities, previous course knowledge is not required. However, in math you must remember what the last course taught you so that you are prepared for the current course. Measuring previous course knowledge will be explained in Chapter 2, "How to Discover Your Math-Learning Strengths and Weaknesses."

Math placement scores also affect sequential learning. If you barely scored high enough to be placed into a math course, then you will have math-learning gaps. Learning problems will occur when new math material is based on one of your learning gaps. The age of the placement test score also affects sequential learning. Placement test scores are designed to measure your *current* math knowledge and are to be used immediately.

Sequential learning is interrupted if math courses are taken irregularly. Math courses are designed to be taken one after another. By taking math courses each semester, without semester breaks in between courses, you are less likely to forget the concepts required for the next course. Research has shown that the more time between math courses, the more likely that a student will fail the math course.

Now that you understand the "building block" nature of math, think about your math history.

- What are your previous math grades?
- How well did you do on the math placement test at your college?

- How long has it been since you took a math course?
- When you look at your math history, are there semesters when you did not take math?

These questions are important because if there was too much time in between your different math courses, you may have forgotten important math concepts that you need in your current class. To use the building block analogy, the blocks may not be as strong any more.

Now that you understand how learning math is a building experience, what should you do? *First,* don't get anxious. Stay calm. *Second,* if your college has a diagnostic math inventory in the tutoring center or math lab, take it to see what math concepts you have forgotten. Then ask your instructor where you can go to relearn these math concepts. *Third,* take the time to follow through. Many students just give up too easily, or they think they will catch up a little at a time. Don't think that and don't give up. The energy put into your class at the beginning of the semester will be more productive than energy put into class at the end of the semester when you try to learn everything during the last week before the final exam. *Fourth,* study and really learn the math; don't practice mimicking it. *Finally,* when it is time to register for the next semester, register immediately so you will be able to get into the math class you need. Why do all this? Because math is sequential!

Math as a Foreign Language

Another helpful technique for studying math is to consider it a foreign language. Looking at math as a foreign language can improve your study procedures. If you do not practice a foreign language, what happens? You forget it. If you do not practice math, what happens? You are likely to forget it, too.

Students who excel in a foreign language must practice it *at least* every other day. The same study skills apply to math, because it is considered a foreign language.

Like a foreign language, math has unfamiliar vocabulary words or terms that must be put in sentences called *equations*. Understanding and solving a math equation is similar to speaking and understanding a sentence in a foreign language.

> **Example:** Math sentences use symbols (which stand for spoken words), such as
>
> = (for which you *say*, "equal"),
> - (for which you *say*, "less"), and
> *a* (for which you *say*, "unknown").

Learning *how* to speak math as a language is the key to math success. Currently, most universities consider computer and statistics (a form of math) courses as foreign languages. Some universities have now gone so far as to actually classify math as a foreign language.

Math as a Skill Subject

Math is a *skill subject*, which means you have to practice actively the skills involved to master it. Learning math is similar to learning to play a sport, learning to play a musical instrument, or learning auto mechanics skills. You can listen and watch your coach or instructor all day, but unless you *practice* those skills yourself, you will not learn.

Examples: In basketball, the way to improve your free throw is to *see and understand* the correct shooting form and then *practice* the shots yourself. Practicing the shots improves your free-throwing percentage. However, if you simply listen to your coach describe the correct form and see him demonstrate it, but you do not practice the correct form yourself, you will not increase your shooting percentage.

Suppose you want to play the piano, and you hired the best available piano instructor. You sit next to your instructor on the piano bench and watch the instructor demonstrate beginning piano-playing techniques. You see and understand how to place your hand on the keys and play. But what does it take to learn to play the piano? You have to place your hands on the keys and *practice*.

Math works the same way. You can go to class, listen to your instructor, watch the instructor demonstrate skills, and understand everything that is said (and feel that you are quite capable of solving the problems). However, if you leave the class *and do not practice* — by working and successfully solving the problems — you will not learn math.

Many of your other courses can be learned by methods other than practicing. In social studies, for example, information can be learned by listening to your instructor, taking good notes, and participating in class discussions. Many students mistakenly believe that math can be learned the same way.

REMEMBER

Math is different. If you want to learn math, you must practice. Practice not only means doing your homework but also means spending the time it takes to understand the reasons for each step in each problem.

Math as a Speed Subject

Math is a *speed subject,* which means, in most cases, it is taught faster than your other subjects. Math instructors have a certain amount of material that must be covered each semester. They have to finish certain chapters because the next math course is based on the information taught in their courses. In many cases a common math department final exam is given to make sure you know the material for the next course. Instructors are under pressure to make sure you are ready for the final exam because it demonstrates how prepared you are for the next level in math. This is different from, let's say, a sociology course where if an instructor doesn't teach the last chapter it will not cause students too many problems in the next sociology or social science course. So don't complain to the math instructor about the speed of the course. Instead, improve your study skills so you can keep up!

Another way math is a speed subject is that most of the tests are timed, and many students think

that they will run out of time. THIS CAUSES PANIC AND FEAR! This is different than most of your other courses where you generally have enough time to complete your tests or in your other courses that have multiple choice tests where you can start bubbling the responses on the scantron sheet if you start running out of time. Students must not only understand how to do the math problems but also must learn the math well enough to complete the problems with enough speed to finish the test.

What makes me curious is, if students feel like they don't have enough time to complete the math test, why are most of them gone before the test time is over? Sure, students who have learned the math thoroughly may complete the test early. That makes sense. Some students leave, however, because they either don't know the material, want to leave the anxious environment, or carelessly work through the test.

So, since speed is an issue in learning math, what should you do? *First,* to use an analogy, start a daily workout program to stay in shape. Review, review, and review as you learn new material. *Second,* practice doing problems within a time constraint. Give yourself practice tests.

A Bad Math "Attitude"

Students' attitudes about learning math are different from their attitudes about learning their other subjects. Many students who have had bad experiences with math do not like math and have a bad attitude about learning it. In fact, some students actually *hate* math, even though these same students have positive experiences with their other subjects and look forward to going to those classes.

Society, as a whole, reinforces students' negative attitudes about math. It has become socially acceptable not to do well in math. This negative attitude has become evident even in popular comic

strips, such as *Peanuts*. The underlying message is that math should be feared and hated and that it is all right not to learn math.

This "popular" attitude toward math may reinforce your belief that it is all right to fail math. Such a belief is constantly being reinforced by others.

The bad math attitude is not a major problem, however. Many students who hate math pass it anyway, just as many students who hate history still pass it. The major problem concerning the bad math attitude is how you use this attitude. If a bad math attitude leads to poor class attendance, poor concentration, and poor study skills, then you have a bad math attitude *problem*.

REMEMBER

Passing math is your goal, regardless of your attitude.

Section Review

1. How does a sequential learning pattern affect math learning?

Give two examples:

Example 1: _____

Example 2: _____

2. List two examples of how learning math is similar to learning a foreign language.

Example 1: _____

Example 2: _____

3. How is math similar to a skills subject?

4. List three ways a bad math attitude could affect your mathematics learning.

First Way: _____

Second Way: _____

Third Way: _____

5. Why is math study skills important at all levels of math (see Preface)?

6. Math study skills can help you in other subjects but general study skills usually cannot improve math learning. Why is this statement important? (see Preface)

The differences between high school and college math

Math, as a college-level course, is almost two to three times as difficult as high school-level math courses. There are many reasons for the increased difficulty: course class time allowance, the amount of material covered in a course, the length of a course, and the college grading system.

The first important difference between high school and college math courses is the length of time devoted to instruction each week. Most college math

instruction, for the fall and spring semesters, has been cut to three hours per week; high school math instruction is provided five hours per week. Additionally, college courses cover twice the material in the same time frame as do high school courses. What is learned in one year of high school math is learned in one semester (four months) of college math.

Simply put, in college math courses you are receiving less instructional time per week and covering twice the ground per course as you were in high school math courses. The responsibility for learning in college is the student's. As a result, most of your learning (and *practicing*) will have to occur outside of the college classroom.

College Summer Semester Versus Fall or Spring Semester and the Difference Between Night and Day

College math courses taught during summer semesters are more difficult than those taught during fall or spring. Further, math taught during night courses is more difficult than math taught during day courses.

Students attending a six-week summer math session must learn the information — and master the skills — two and a half times as fast as students attending regular, full-semester math sessions. Though you receive the same amount of instructional classroom time, there is less time to understand and *practice the skills* between class sessions.

Summer classes are usually two hours per day, four days per week (nighttime summer classes are four hours per night, two nights per week).

Example: If you do not understand the lecture on Monday, then you have only Monday night to learn the material before progressing to more difficult material on Tuesday. During a night course, you have to learn and understand the material before the break; after the break, you will move on to the more difficult material — *that night.*

Because math is a sequential learning experience, where every building block must be understood before proceeding to the next block, you can quickly fall behind, and you may never catch up. In fact, some students become lost during the first half of a math lecture and never understand the rest of the lecture (this can happen during just one session of night class). This is called "kamikaze" math because most students do not survive summer courses.

If you *must* take a summer math course, take a 10- or 12-week *daytime* session so that you will have more time to process the material between classes.

Course Grading System

The grading system for math is different in college than in high school.

Example: While in high school, if you make a D or borderline D/F, the teacher more than likely will give you

continued

a D, and you may continue to the next course. However, in some college math courses, a D is not considered a passing grade, or if a D is made, the course will not count toward graduation.

College instructors are more likely to give the grade of N (no grade), W (withdrawal from class), or F for barely knowing the material. This is because the instructor knows that you will be unable to pass the next course if you barely know the current one.

Most colleges require students to pass two college-level algebra courses to graduate. In most high schools, you may graduate by passing one to three math courses. In some college-degree programs, you may even have to take four math courses and make a C in all of them to graduate.

The grading systems for math courses are very precise compared to the grading systems for humanities courses.

Example: In a math course, if you have a 79 percent average and you need 80 percent to get a B, you will get a C in the course. But if you make a 79 percent in English class, you may be able to talk to your instructor and do extra credit work to earn a B.

Because math is an exact science and not as subjective as English, do not expect your math instructor to let you do extra work to earn a better grade. In college, there usually is not a grade given for "daily work," as is often offered in high school.

In fact, *your test scores may be the only grades that will count toward your final grade.* Therefore, you should not assume that you will be able to "make up" for a bad test score.

The Ordering of College Math Courses

College math courses should be taken, *in order,* from the fall semester to the spring semester. If at all possible, avoid taking math courses from the spring to fall semesters. There is less time between the fall and spring semester for you to forget the information. During the summer break, you are more likely to forget important concepts required for the next course and therefore experience greater difficulty.

Section Review

1. Compare the amount of college class time to high school class time.

2. How do college summer math courses differ from courses offered in the spring or fall?

3. How can the order of taking math courses affect your learning?

Why your first math test is very important

Making a high grade on the first major math test is more important than making a high grade on the first major test in other subjects. The first major math test is the easiest and, most often, the one the student is least prepared for.

Students often feel that the first major math test is mainly a review and that they can make a B or C without much study. These students are overlooking an excellent opportunity to make an A on the easiest major math test of the semester. (Do not forget that this test counts the same as the more difficult remaining math tests.)

At the end of the semester, these students sometimes do not pass the math course or do not make an A because of their first major test grade. In other words, the first math test score was not high enough to "pull up" a low test score on one of the remaining major tests.

Studying hard for the first major math test and obtaining an A offers you several advantages:

• A high score on the first test can compensate for a low score on a more difficult fourth or fifth math test. All major tests have equal value in the final grade calculations.

• A high score on the first test can provide assurance that you have learned the basic math skills required to pass the course. This means you will not have to spend time relearning the misunderstood material covered on the first major test while learning new material for the next test.

• A high score on the first test can motivate you to do well. Improved motivation can cause you to increase your math study time, which will allow you to master the material.

• A high score on the first test can improve your confidence for higher test scores. With more confidence, you are more likely to work harder on the difficult math homework assignments, which will increase your chances of doing well in the course.

What happens if after all your studying, you make a low score on your first math test? You can

still use this test experience to help you improve your next grade or to help determine if you are in the right math course. Your first math test, no matter what you make on it, can be used as a diagnostic test. Your teacher can review your test with you to see which type of math problems you got right and which ones you need to learn how to solve. It may be that you missed only a few concepts that caused the low score, and you can learn how to do these problems by getting help from the teacher, a learning resource center, or the math lab. However, you need to learn how to do these problems immediately so that you don't fall behind in the course. After meeting with your teacher, ask for the best way you can learn the concepts that are the bases of the missed problems and how to prepare for the next test. Even students who made a B on the first math test can benefit by seeing the teacher.

If you made below a 50 on your first math test, I suggest that it might be a good idea to drop back to a lower level math course. Even though it might be beyond the first week of drop and add, many colleges/universities will let you drop back to a lower math class after the first major math test. Students who drop back get a good foundation in mathematics that helps them in their next math courses. On the other hand, I have seen students who insisted on staying in the math course and repeated it several times before passing. Some of the students stopped taking the math course and dropped out of college. Dropping back to a lower level math course and passing it is the smartest move. These students went on to become more successful in their math courses.

REMEMBER

It does not matter where you start as long as you graduate. Discuss this option with your teacher.

Section Review

1. Why is your first math test supposed to be the easiest?

2. Give three reasons why your first math test is so important.

First Reason: _____

Second Reason: _____

continued

Third Reason: _____

3. What do you do if you fail your first test?

Option One: _____

Option Two: _____

Option Three: _____

Chapter 1 Review

1. In a math course, you must be able to do three things. You must _____ the material, _____ the material, and _____ what you have learned to solve the problem.

2. Math requires _____ _____, which means one concept builds on the next concept.

3. Placement test scores are designed to measure your _____ math knowledge and are to be used _____.

4. Like a foreign language, math has unfamiliar vocabulary words or terms that must be put in sentences called _____.

5. Learning how to speak math as a _____ _____ is one key to math success.

6. Keeping a _____ attitude about math will help you study more efficiently.

7. College math courses are _____ to _____ times as difficult as high school math courses.

8. Math course are even more difficult than other courses because a grade of _____ or better is usually required to take the next course.

9. The math grading is exact; in many cases, you cannot do _____ _____ to improve your grade.

10. If you fail your first math test, you need to make an appointment with your _____ to review your _____. Then you can decide if you should _____ to a lower math course.

What is the most important information you learned from this chapter?

How can you immediately use it?

Lecture Notes

How to Discover Your
Math-Learning Strengths and Weaknesses

Math-learning strengths and weaknesses affect students' grades. You need to understand your strengths and weaknesses to improve your math-learning skills.

Just as a mechanic does a diagnostic test on a car before repairing it, you need to do your own testing to learn what you need to improve upon. You do not want the mechanic to charge you for something that does not need repairing, nor do you want to work on learning areas that do not need improvement. You want to identify learning areas you need to improve or better understand.

In this chapter, you will learn

- how what you know about math affects your grades,
- how quality of math instruction affects your grades,
- how affective student characteristics affect your math grades,
- how to determine your learning style,
- how to develop a math-learning profile of your strengths and weaknesses, and
- how to improve your math knowledge.

Areas of math strengths and weaknesses include math knowledge, level of test anxiety, study skills, study attitudes, motivation, and test-taking skills. Before we start identifying your math strengths

and weaknesses, you need to understand what contributes to math academic success.

Dr. Benjamin Bloom, a famous researcher in the field of educational learning, discovered that your IQ (intelligence) and your cognitive entry skills account for 50 percent of your course grade.

Figure 1
Variables Contributing to
Student Academic Achievement

Cognitive Entry Skills
(how much math you know before
entering a new math course)
and
IQ
(how fast you can learn old and new math concepts)
50%

Quality of Instruction
(Effectiveness of math instruction: textbook, teaching style, extra teaching aids, etc.)
25%

Affective Characteristics
(Personality, self-concept, locus of control, motivation, attitudes, anxiety, study habits)
25%

*Bloom (1976)

Quality of instruction represents 25 percent of your course grade, while affective student characteristics reflect the remaining 25 percent of your grade.

- *Intelligence* may be considered, for our purpose, how fast you can learn or relearn math concepts.
- *Cognitive entry skills* refer to how much math you knew before entering your current math course. For example, previous math knowledge can be based on previous math grades, the length of time since the last math course, the type of math courses completed (algebra I, algebra II, intermediate algebra), and placement test scores.
- *Quality of instruction* is concerned with the effectiveness of math instructors when presenting material to students in the classroom and math lab. This effectiveness depends on the course textbook, curriculum, teaching style, extra teaching aids (videos, audiocassettes), and other assistance. Quality of instruction may also include math lab resources, quality of tutoring based on tutor training and Supplemental Instruction.
- *Affective student characteristics* are characteristics you possess that affect your course grades — excluding how much math you knew before entering your math course. Some of these affective characteristics are anxiety, study skills, study attitudes, self-concept, motivation, and test-taking skills.

How what you know about math affects your grades

Poor math knowledge can cause low grades. A student placed in a math course that requires a more extensive math background than the student possesses will probably fail that course. Without the correct math background, you will fall behind and never catch up.

Placement Tests and Previous Course Grades

The math you need to know to enroll in a particular math course can be measured by a placement test (ACT, SAT) or by the grade received in the prerequisite math course. Some students are incorrectly placed in math courses by placement tests.

If, by the second class meeting, everything looks like Greek and you do not understand what is being explained, move to a lower-level course. In the lower-level math course you will have a better chance to understand the material and to pass the course. A false evaluation of math ability and knowledge can only lead to frustration, anxiety, and failure.

Requests by Students for Higher Placement

Some students try to encourage their instructors to move them to a higher-level math course because they believe they have received an inaccurate placement score. Many students try to avoid noncredit math courses, while other students do not want to repeat a course that they have previously failed. These moves can also lead to failure.

Some older students imagine that their math skills are just as good as when they completed their last math course, which was five to ten years ago. If they have not been practicing their math skills, they are just fooling themselves. Still other students believe they do not need the math skills obtained in a prerequisite math course to pass the next course. This is also incorrect thinking. Research indicates that students who were

placed correctly in their prerequisite math course, and who subsequently failed it, will not pass the next math course.

What My Research Shows

I have conducted research on thousands of students who have either convinced their instructors to place them in higher-level math courses or have placed themselves in higher-level math courses. The results? These students failed their math courses many times before realizing they did not possess the prerequisite math knowledge needed to pass the course. Students who, without good reason, talk their instructors into moving them up a course level are setting themselves up to fail.

To be successful in a math course, you must have the appropriate math knowledge. If you think you may have difficulty passing a higher-level math course, you probably do not have an adequate math background. Even if you do pass the math course with a D or C, research indicates that you will most likely fail the next higher math course.

It is better to be conservative and pass a lower-level math course with an A or B instead of making a C or D in a higher-level math course and failing the next course at a higher level.

This is evident when many students repeat a higher-level math course up to five times before repeating the lower-level math course that was barely passed. After repeating the lower-level math course with an A or B, these students passed their higher-level math course.

How quality of math instruction affects your grades

Quality of instruction accounts for 25 percent of your grade. Quality of instruction includes such things as classroom atmosphere, instructor's teaching style, lab instruction, and textbook content and format. All of these "quality" factors can affect your ability to learn in the math classroom.

Interestingly enough, probably the most important "quality" variable is the compatibility of an instructor's teaching style with your learning style. You need to discover your learning style and compare it to the instructional style. Noncompatibility can be best solved by finding an instructor who better matches your learning style. However, if you cannot find an instructor to match your learning style, improving your math study skills and using the math lab or learning resource center (LRC) can compensate for most of the mismatch.

Use of the math lab or learning resource center (LRC) can dramatically improve the quality of instruction. With today's new technologies, students are able to select their best learning aids. These learning aids could be videotapes, CD-ROMs, computer programs, math study skills computer programs, math texts, and websites.

The quality of tutors is also a major factor in the math lab or learning resource center. A low student-to-tutor ratio and trained tutors are essential for good tutorial instruction. Otherwise, the result is just a math study hall with a few helpers.

The math textbook should be up to date with good examples and a solutions manual. This increases the amount of learning from the textbook compared to older, poorly designed texts.

The curriculum design affects the sequence of math courses, which could cause learning problems.

Some math courses have gaps between them that cause learning problems for all students.

How affective student characteristics affect your math grades

Variable 2: _____

Affective student characteristics account for about 25 percent of your grade. These affective characteristics include math study skills, test anxiety, motivation, locus of control, learning style, and other variables that affect your personal ability to learn math.

Most students do not have this 25 percent of the grade in their favor. In fact, most students have never been taught *any* study skills, let alone how to study math, specifically. Students also do not know their best learning style, which means they may study ineffectively by using their least effective learning style. However, it is not your fault that you were not taught math study skills or made aware of your learning style.

Variable 3: _____

By working to improve your affective characteristics, you will reap the benefits by learning math more effectively and receiving higher grades. Thousands of students have improved their math affective characteristics and thereby their grades by using this workbook.

Section Review

2. How does your placement test or previous math course grade affect your math learning?

1. List and explain the three variables that contribute to your academic success.

Variable 1: _____

3. How does the quality of math instructors affect your grade?

4. How do your affective characteristics affect your grade?

How to determine your learning style

Research has shown that matching a student's best learning style with the instructional style improves learning. Research has also shown that students who understand their learning style can improve their learning

effectiveness. A learning disadvantage will occur for students who do not know or do not understand their learning style. Students should talk to their instructor or counselor about taking one or more learning style inventories.

Taking Stock of Your Learning Style

There are different types of learning styles assessments. One type of learning style assessment focuses on learning modalities, whereas other assessments focus on cognitive or environmental learning styles. We will focus on learning modalities and cognitive learning styles.

Learning modalities
(using your senses)

If it is available to you, take a learning style inventory that measures _learning modalities_. Learning modalities focus on the best way your brain receives information — that is, learning _visually_ (seeing), _auditorially_ (hearing), or _kinesthetically_ (touching).

If possible, take an inventory that offers math-learning modalities, such as _Learning Styles Inventory_ (Brown and Cooper, 1978). Other learning style inventories can measure learning modalities, but they focus on English- or reading-learning modes. Sometimes students' math-learning modes may be different from their language-learning modes. Your instructor can go to www.academicsuccess.com to learn more about using the Learning Styles Inventory.

If you cannot locate a modality-type inventory, then you may want to do a self-assessment on your math-learning modality. Do you learn math best by seeing it, by hearing it, or by having hands-on learning experiences? Hands-on learning means you need to touch or feel things to best learn about them. Group

learning is when two to four students get together to discuss math problems and talk about the solutions. Individual learning is studying by yourself, using your book, CD or web resources. Do you learn best in groups or by yourself?

This is not the most accurate way to determine your learning modality; however, you may use this informal assessment until you have an opportunity to take a more formal modality learning style inventory.

Cognitive learning styles
(processing what you sense)

The second type of learning-styles inventory focuses on *cognitive learning styles*. Cognitive learning styles describe how you process information, once you have heard, seen or felt the information. Based on McCarthy (1981), you can process information four different ways: Innovative, Analytic, Common sense, or Dynamic. You can find information on the McCarthy inventory by going to www.AboutLearning.com.

If you do not have the opportunity to take a cognitive learning-style inventory, you may want to do a self-assessment of your cognitive learning style. Which of the following best describes you? Remember that you may not have all the characteristics mentioned for each learning style. Look for the style that best matches you.

Innovative learners solve problems or make decisions by personally relating to them and using their feelings. They are interested in people and like cultural events. They want to work towards peace and harmony. They work towards self-improvement. Some of their careers are in counseling, social work, personnel, humanities, social sciences, English and most fields dealing with people. They function through life by using their values.

They learn best by listening and discussing their ideas with others, while searching for the meaning. Their best learning activities are motivational stories, simulations, group discussion, journal writing, integrative

lectures, group projects, and discussion lectures. They usually like subjective tests. They are students with good ideas, and they find unique solutions to problems. They ask the question, "Why do I need to learn this?" (McCarthy, 1981). In the math course they may ask for the underlying value of learning the course material as it relates to them. Their favorite question is, "Why?" Is this your cognitive learning style?

Analytic learners look for facts and ask experts for advice when solving problems. They want to know what the experts think and the research to back up the facts. In some cases they are more interested in ideas and concepts than people. They want to collect information before making logical decisions. If they are having difficulty with a decision, they will re-examine the facts or go to experts. They are very uncomfortable with subjective decisions. They generally major in math and the hard sciences; they enjoy careers involving research (McCarthey, 1981).

They learn by thinking about the concepts and solving problems by using logic. They learn best in the traditional lecture classroom by instructors giving them facts. Their best learning activities are lecture with visual aids, textbook reading assignments, problem solving and demonstration by the instructor (expert), independent research and gathering data. They usually like objective tests. They want to know, "What exactly is the content or skill?" (McCarthey, 1981). They learn best by taking exact and well organized notes and consider class discussion an interruption in learning. They usually do not like group learning and prefer studying individually. They usually major in areas like math, physics, chemistry and accounting. Their favorite question is "What?" Is this your learning style?

Common-sense learners solve problems by knowing how things work. They want to know how to use the material today and how it can be used in real life. They want to use facts to build concepts and then test the practicality of the concepts. In their discussions they get right to the point. They are skills oriented and like to tinker with

things. They use their sensory experiences to help make decisions (McCarthy, 1981).

They learn by testing concepts in a practical way and make decisions by applying theories. They want to know practical ways to use theories and concepts. They need hands on experiences as part of their learning such as labs and manipulatives. Some additional learning activities are homework problems, field trips, computer-aided instruction, and lecture with demonstration, individual reports and computer simulations. They usually like objective tests (Harb, Terry, Hurt, & Williamson, 1981). They major in applied sciences, health fields, computer science, engineering and technologies. Their favorite question is "How?" Is this your learning style?

Dynamic learners solve problems by looking at hidden possibilities and processing information concretely. They like self-discovery. They like new things and can adapt to change. They can be very flexible. They can also reach correct conclusions and answers without knowing how they did it. They need to know what else can be done with this information. They seek to answer this question, "If I do this, what possibilities will it create?" (McCarthy, 1981).

They learn best by independent self-instruction and trial-and-error practice. They are flexible and will take risks. They function in life by acting out and testing experiences. Some of their best learning activities are Socratic lecture, role playing, open-ended laboratories, student lectures, outside projects, simulations and group problem solving (Harb, Terry, Hurt, & Williamson, 1981). They major in business, marketing and sales occupations. Their favorite question is, "What can this become?" Is this your learning style?

By now you may have identified your cognitive learning style. If not, take time to examine how you are learning and then revisit these descriptions. Also you may have recognized that innovative learners are the opposite of common sense learners and analytic learners are the opposite of dynamic learners. This does not mean that you cannot learn from someone teaching in the opposite learning style, but it will take more concentration and better study skills.

REMEMBER

It is important to recognize that students who do not have learning styles that fit easily with learning math can still learn math. Good study skills and awareness on the part of the math instructor and student can compensate a mismatch of modality/ learning styles. Working together to find strategies to help improve the teaching and learning is most productive for everyone.

Section Review

1. My learning modality is _____

 because I learn math best by _____

2. My cognitive learning style is _____

 because I process math information by _____

Assessing your math strengths and weaknesses

Students have learning strengths and weaknesses that affect math learning and their course grades. They must identify and understand these strengths and weaknesses to know which strengths support them and which weaknesses to improve upon. After reading this section, you will be asked to complete the Math-Learning Profile Sheet (Figure 2).

Examples: Having good math study skills is a *positive* math-learning characteristic, whereas having high test anxiety is a *negative* math-learning characteristic.

Measuring Your Math Study Skills

Math study skills is the main part of the affective characteristics that represents about 25% of your math grade. You can determine your math study skills expertise by taking the free Math Study Skills Evaluation at www.academicsuccess.com, or you can do a self determination about your math study skills. If you take the online Math Study Skills Evaluation, the results will indicate your score. If you score below 70, you need to really focus on developing a set of study skills. A score between 70 and 80 means you have average study skills and a score above 90 means that you have good study skills. A self determination can be made by asking yourself if you have ever been taught math study skills. If you

have, then you probably have either good or excellent math study skills. If you have not, then you most likely need to develop good math study skills.

Scoring low on the study skills inventory is not the end of the world. In fact, it might indicate that your problems with learning math may be due to your study habits instead of math intelligence.

A low score on the math study skills evaluation means that most of your learning problems are because you have not been taught math study skills. This workbook can teach you math study skills. If you had previous poor math grades, it was probably due to lack of math study skills, and this is not your fault.

You can improve your math study skills and become a successful math student. If you have good math study skills, you can still improve them to make A's in math. If you have excellent math study skills, you can still learn some math study skills that can help you in future courses.

Measuring Your Math Anxiety Level

It is also important to assess any math anxiety you might have, so that you can start managing it immediately. If you took the online Math Study Skills Evaluation, your answer should indicate if you "seldom", "often" or "almost always" become anxious and forget important concepts during a test.

If you have medium or high math test anxiety, it can be reduced. This problem can be fixed if you follow the suggestions in the next chapter. This also means that some of your problems learning math (if you have any) are due to anxiety instead of your intellectual ability to learn math.

If you have high text anxiety, then you will probably need to use the *How to Reduce Test Anxiety* audio cassette tape. You can order this tape by going to www.academicsuccess.com. It is

Figure 2
Math-Learning Profile Sheet

After completing this profile, use your answers to select which areas you would like to improve first. If you have a serious problem with math or test anxiety, get help in this area immediately. Circle the X that best describes your belief.

	Strongly Disagree	Disagree	Agree	Strongly Agree
1. I have good math study skills.	X	X	X	X
2. Math tests don't make me anxious.	X	X	X	X
3. I have a strong basic knowledge of math to pass this class.	X	X	X	X
4. I have a positive attitude about learning math.	X	X	X	X
5. It has been less than a year since I took my last math class.	X	X	X	X
6. I have taken a math study skills course before this class.	X	X	X	X
7. I have good reading skills to be successful in this class.	X	X	X	X
8. I know if I improve my study habits, I can improve my math grades.	X	X	X	X

This informal survey is to help you decide what kind of shape you are in to do well in your course.

Which areas are strong and help you do well in a math class?

Which areas keep you from doing the best you can in a math class? How can you change them?

important that you find support and guidance in managing this anxiety immediately!

You can reduce and learn to manage test anxiety! It just takes work!

Measuring Previous Math Knowledge

Now it's time to measure your previous math knowledge. Your previous math knowledge is an excellent predictor of future math success. In fact, the best predictor of math success is your previous math grade (if taken within one year of your current course).

The measurement of your math knowledge is based on your previous math grade or your math placement test. If your college does not require a placement test, contact the chairperson of the math department to help you with your course placement. Selecting a math course yourself is like playing Russian roulette with five bullets in the chamber. Without help, you will probably commit mathematical suicide.

Visit your counselor or instructor to discuss your math placement test score. Ask your counselor or instructor how your placement score relates to the scores for the math course below and above your course.

Find out if your score barely placed you into your course, if your score was in the middle, or if your score almost put you into the next course. If your placement score barely put you into your current math course, you will have difficulty learning the math required in that course.

A barely passing placement score means that you are missing some of the math knowledge you require to be successful, and you will really struggle to pass the course.

A middle placement score indicates that you probably have most of the math knowledge required for the course; however, you still may have difficulty passing the course. A high placement score means you have most of the math knowledge needed to pass

and may have a good chance of passing the course. (Still, if you have poor math study skills and high test anxiety, you may not pass the course.)

Students who have low placement test scores need to improve their math knowledge as soon as possible. Students with middle placement scores should also consider improving their math knowledge.

You need to improve your math skills within the first few weeks of the semester. If you wait too long, you will become hopelessly lost. Suggestions for improving your math knowledge are included in a later section in this chapter.

If you are taking your second math course, use your previous math course grade to determine your math knowledge. Your previous course grade is the best predictor of success in your next math course. (However, this is only true if your last math course was taken within the last year.)

Students who made an A or B in their previous math course have a good chance of passing their next math course. However, students who made a C in their previous math course have a poor chance of passing their next math course. Students who made below a C or withdrew from the previous math course have virtually no chance of passing their next math course.

What level of math knowledge preparation do you possess right now for this course? If you have poor or good previous math knowledge, you need to build up those math concepts and skills as soon as possible. Go to your instructor and discuss how to build up your skills. Some suggestions are to review the chapters of the previous math text or ask your instructor about some websites that help students review math. Students with poor math knowledge have about three to four weeks to improve their math before it can become a major problem. Don't wait. Start now to review your math skills in order to improve your mathematics learning and grades. Additional information on improving your math knowledge is in the next section.

Determining Your Math Attitude

Math attitude can play a major part in being successful in your course. Some students who have a poor attitude toward math may not attend class as often as they should or may procrastinate in doing their homework. Other students avoid math until they *have* to take it. What is your math attitude — good, neutral, or negative? Remember, students with negative math attitudes can pass math. It depends on what you do with your attitude that counts. Don't let your negative attitude cause problems in your math course. Also, if you have a good math attitude, don't forget to study. A good math attitude can only get you so far!

The Amount of Time Since Your Last Math Course

Another indicator of math success can be the length of time since your last math course. The longer the time since your last math course, the less likely you will be to pass the current course. The exception to this rule is if you were practicing math while you were not taking math courses. Math is very easily forgotten if not practiced.

How long has it been since your last math course? If it has been two or more years since your last math class, your math may be similar to someone who has a low placement score or made a C or lower in the last math class. You can improve your math knowledge by following the same suggestions made for those students and applying the information in the next section (How to improve your math knowledge).

Previous Study Skills Courses or Training

General study skills training that does not focus on math may help some students in time management, reading techniques, learning styles, and overcoming procrastination. Even though math requires specific study techniques, you can learn many other helpful general study skills and personal habits that will help you in math classes. If you have had no study skills training, then this workbook will help you in your math and other courses. Students with no past study skills training should also take advantage of any other opportunities to learn about college success skills in addition to learning from this workbook. Contact your learning centers, instructors or counselors about study skills workshops and when they will be offered.

Previous/Current Required Reading Courses

College reading skills are also needed for success in math courses. This is especially true when you are trying to solve story or word problems.

At community colleges and some universities, there are three levels of reading courses. Usually, the first two levels are noncredit reading courses, and the third level is for college credit.

If you are enrolled in the lowest reading level, you will have difficulty reading and understanding the math text. Students who are enrolled in the second level may also have difficulty reading the text. Students who are not required to take a reading class or who are reading at grade level may still have some difficulty reading the math text.

REMEMBER

Math texts are not written like English or history texts. Even students with college-level reading skills may experience difficulty understanding their math text.

Students not at college level reading can pass math courses; however, if you need help in reading, get that help as soon as possible. Or, many learning or study skills centers help students with reading skills for specific courses. Work with a learning specialist to design a system for reading your math textbook. If you do not know your reading level, ask to take a reading test. If in doubt about your reading level, take a reading class.

Your Locus of Control

"Locus of control" is a concept that describes how much control you feel you have over both your life and your course grades. Some students feel they have a lot of control over their lives and learning, while other students feel they have very little control over their lives or grades. For example, when a student with external locus of control does poorly on a test, the blame is put on the instructor, when in reality the student did not study well enough. On the other hand, the student with internal locus of control admits that studying more and paying more attention in class would have helped. Students can improve their internal locus of control and be successful in mathematics at the same time.

Your Best Learning Modality

Learning modalities are best described by your preference in how you receive lecture information, either visually, auditory or kinesthetically (hands-on). As discussed, some of these modalities match your math instructor and some do not. Most math instructors teach through a visual mode, with the distant second modality being auditory.

Even if your learning style is different from your math instructor's, there is no excuse for not learning math. The awareness of your learning modality and cognitive style can help you become a better learner in and outside of the classroom. This information will be discussed in Chapter 6, "How to Remember What You Have Learned."

Your Cognitive Learning Style

Cognitive learning styles are best described as how information is processed once it is received by your brain. Most math instructors process information *analytically*, and they therefore teach math the same way. Using *common sense* is the second most popular way math instructors cognitively process information.

If you do not process information analytically or through common sense, then you may be at a learning disadvantage. However, improved math study skills can compensate for a mismatch in learning modes/cognitive styles. Chapter 6 suggests different study skills that can help you learn math through your preferred learning modality and cognitive style.

Evaluating Your Findings

Now that you understand some of the general characteristics that help students be successful, complete the informal Math-Learning Profile Sheet (Figure 2).

When you look at your profile, think of it as similar to going to a fitness center for the first time. A professional trainer asks questions, runs you through equipment, analyzing your strengths and weaknesses, and then designs a program for you. You are doing the same thing here.

Section Review

1. List your strengths for learning math.

2. List three areas you will start improving right now.

 Area One: _____

 Area Two: _____

 Area Three: _____

3. How will you improve these areas?

 Area One: _____

Area Two: _____

Area Three: _____

How to improve your math knowledge

Instructors always operate on the premise that you finished your previous math course just last week; they do not wait for you to catch up on current material. It does not matter if your previous math course was a month ago or five years ago. Instructors expect you to know the previous course material — period.

Review Your Previous Math Course Material and Tests

There are several ways to improve your math knowledge. Review your previous math course material before attending your present math course. Look closely at your final exam to determine your

weak areas. Work on your weak areas as soon as possible so they can become building blocks (instead of stumbling blocks) for your current course.

If it has been some time since your last math course, visit the math lab or learning resource center to locate review material. Ask the instructor if there are any computer programs that will assess your math skills to determine your strengths and weaknesses for your course. Review math videotapes on the math course below your level. Also review any computer software designed for the previous math course.

Another way to enhance learning is to review the previous math course text by taking all of the chapter review tests. If you score above 80 percent on one chapter review test, move on to the next chapter. A score below 80 percent means you need to work on that chapter before moving on to the next chapter. Get a tutor to help you with those chapters if you have trouble. Make sure you review all the chapters required in the previous course as soon as possible. If you wait more than two weeks to conclude this exercise, it may be too late to catch up (while learning new material at the same time).

Employ a Tutor

One last way to improve your cognitive entry skills is to employ a private tutor. If you have a history of not doing well in math courses, you may need to start tutorial sessions *the same week class begins.* This will give the tutor a better chance of helping you regain those old math skills.

You still need to work hard to relearn old math skills while continuing to learn the new material. If you wait four to five weeks to employ a tutor, it will probably be too late to catch up and do well or even pass the course.

REMEMBER

Tutorial sessions work best when the sessions begin during the first two weeks of a math course.

Schedule Math Courses "Back to Back"

Another way to maintain appropriate math knowledge is to take your math courses "back to back." It is better to take math courses every semester — even if you do not like math — so that you can maintain sequential (linear) learning.

I have known students who have made B's or C's in a math class and who then waited six months to a year to take the next math course. Inevitably, many failed. These students did not complete any preparatory math work before the math course and were lost after the second chapter. This is similar to having one semester of Spanish, not speaking it for a year, then visiting Spain and not understanding what is being said.

The only exception to taking math courses "back to back" is taking a six-week "kamikaze" math course (an ultracondensed version of a regular course), which should be avoided.

If you are one of the unfortunate many who are currently failing a math course, you need to ask yourself, "Am I currently learning any math or just becoming more confused?" If you are learning some math, stay in the course. If you are getting more confused, withdraw from the course. Improve your math knowledge prior to enrolling in a math course during the next semester.

Example: You have withdrawn from a math course after midterm due to low grades. Instead of waiting until next semester, attend a math lab or seek a tutor and learn Chapters 1, 2, and 3 *to perfection*. Also use this time to improve your math study skills. You will enter the same math course next semester with excellent math knowledge and study skills. In fact, you can make an A on the first test and complete the course with a high grade. Does this sound farfetched? It may, but I know hundreds of students who have used this learning procedure and passed their math course instead of failing it again and again.

Finding Your Best Instructor

Finding an instructor who best matches your learning style can be a difficult task. Your learning style is important; your learning style is how you best acquire information.

Example: Auditory learners do better when hearing the information over and over again instead of carefully reading the information. If an auditory learner is taught by a visual-style instructor who prefers that students read materials on their own and who prefers working problems instead of

continued

describing them, the mismatch could cause the student to do worse than if the student were taught by an auditory instructor.

Most students are placed in their first math course by an academic advisor. Usually, academic advisors know who are the most popular and least popular math instructors. However, advisors can be reluctant to discuss teacher popularity. And unfortunately, students may want the counselor to devise a course schedule based on the student's time limits instead of teacher selection.

To learn who are the best math instructors, ask the academic advisor which math instructor's classes fill up first. This does not place the academic advisor in the position of making a value judgment; neither does it guarantee the best instructor. But it will increase the odds in your favor.

Another way to acquire a good math instructor is to ask your friends about their current and previous math instructors. However, if another student says that an instructor is excellent, make sure your learning style matches your friend's learning style. Ask you friend, "Exactly what makes the instructor so good?" Then compare the answer with how you learn best. If you have a different learning style than your friend, look for another instructor, or ask another friend whose learning style more closely matches your own.

To obtain the most from an instructor, discover your best learning style and match it to the instructor's teaching style. Most learning centers or student personnel offices will have counselors who can measure and explain your learning style. Interview or observe the instructor while the instructor is teaching. This process is time consuming, but it is well worth the effort!

Once you have found your best instructor, *do not* change. Remain with the same instructor for every math class whenever possible.

REMEMBER

The first step in becoming a better math student is knowing your learning strengths and weaknesses. Now you can focus on what you need to improve.

Whether your learning styles are a natural fit for learning math or not, you still own the responsibility to be smart in selecting your study system for math class. You can be successful!

Section Review

1. Give four examples of how you can review math course materials.

 Example 1: _____

 Example 2: _____

 Example 3: _____

 Example 4: _____

2. What is the advantage of taking math courses back to back?

3. Give two examples of how to find your best math instructor.

 Example 1: _____

 Example 2: _____

Chapter 2 Review

1. _____ _____ _____
 refers to how much math you knew before entering your current math course.

2. _____ _____ _____
 are the learning skills you possess that affect your grades.

3. Students who talk their instructors into moving up a course level usually _____ that course.

4. The _____ of tutors is a major part of the math lab or learning resource center.

5. Affective student characteristics account for about _____ percent of your grade.

6. _____ learners look for facts and ask experts for their advice.

7. Hands-on learners are called _____ learners.

8. Assessing your math-learning _____ and _____ will help you improve your studying and learning.

9. Students having previous difficulty in mathematics should start tutorial sessions the _____ week class begins.

10. Once you find your best math instructor _____ _____ change instructors for your next math course.

What is the most important information you learned from this chapter?

How can you immediately use it?

Lecture Notes

3

How to Reduce Math Test Anxiety

In this chapter you will first learn about math anxiety and then explore test anxiety. They are different issues that can affect each other. It is best to learn about them as separate entities first.

Math anxiety is a common problem for many high school, college and university students. It is especially difficult for students in developmental courses who normally have more math anxiety than other students. However, there are students who are in higher level math courses that also struggle with this problem. It is very common for students to have anxiety only about math and not in their other subjects.

Most students think that math anxiety only affects them when taking a test, but it also affects other areas. It can affect the way you do your homework, learn in the classroom or through distance learning courses, and the way you choose a career. Students who have math anxiety may procrastinate in doing their homework or put off sitting down and completing an online lesson. This can lead to math failure. Students also select a major based on the amount of math that is required, which could lead to lower paying or dissatisfying careers. However, most students with math anxiety meet it face to face during the test, experiencing test anxiety as well.

Mild test anxiety can be a motivational factor in that it can make students properly prepare for a test. However, high test anxiety can cause major problems in both learning and test taking, as students avoid studying for the test when anxiety begins to afflict their thought processes. Reducing test anxiety is the key for many students to become successful in math. Such students need to learn the causes of test anxiety and how to reduce the test anxiety that affects their learning and grades.

Several techniques have proven helpful in reducing both math anxiety and math test anxiety. However, reducing them does not guarantee good math grades. It must be coupled with effective study skills and a desire to do well in math.

In this chapter, you will learn about math anxiety:

- definitions of math anxiety,
- types of math anxiety,
- causes of math anxiety, and
- effects of math anxiety on learning.

And you will learn about test anxiety:

- definitions of test anxiety

- why math tests create anxiety
- causes of test anxiety
- different types of test anxiety
- how to reduce test anxiety

Understanding math anxiety

Definition of Math Anxiety

Math anxiety is a relatively new concept in education. During the 1970s, certain educators began using the terms "mathophobia" and "mathemaphobia" as a possible cause for children's unwillingness to learn math. Math anxiety is an extreme emotional and/or physical reaction to a very negative attitude toward math. There is a strong relationship between low math confidence and high math test anxiety (Fox, 1977).

Math anxiety is the feeling of tension and anxiety that interferes with the manipulation of numbers and the solving of math problems during tests. (Richardson and Sulnn, 1972). Math anxiety is a state of panic, helplessness, paralysis, and mental disorganization that occurs in some students when they are required to solve math problems. This discomfort varies in intensity and is the outcome of numerous previous experiences students have had in their past learning situations (Tobias, 1986).

Today, math anxiety is accepted as one of the major problems students have in completing their math courses. It is real, but it can be overcome.

Types of Math Anxiety

Math anxiety can be divided into three separate anxieties: Math Test Anxiety, Numerical Anxiety and Abstraction Anxiety. Math Test Anxiety involves anticipation, completion and feedback of math tests. Numerical Anxiety refers to everyday situations requiring working with numbers and performing arithmetic calculations. Numerical anxiety can also include students who are trying to figure out the amount for a tip, thinking about mathematics, doing math homework or listening/seeing math instruction. Abstraction Anxiety deals with working with variables and mathematical concepts used to solve equations. Students can have all three math anxieties or only one anxiety. Most often, the students I have worked with have Math Test Anxiety and Abstraction Anxiety. These students don't have any anxiety working with numbers, but once they start learning algebra, they develop both conditions. This may have happened in high school or college.

The Causes of Math Anxiety

Since math anxiety is a learned condition, its causes are unique with each student, but they are all rooted in individuals' past experiences. Bad experiences in elementary school are one of the most common sources for students' math anxiety: coming in last in math races at the blackboard, watching a classmate next to them finish a problem twice as fast as they do, teachers saying, "That's okay. You just aren't good in math. You are better in English," or classmates and teachers calling them stupid. These words and experiences remain with people; they can still hear the words and eventually begin telling themselves the same thing. When these students walk into the classroom or open a math book, or take a test, these "mental tapes" play in their minds. When asked, many students indicated that they were made fun of when trying to solve math problems at the chalkboard. When they could not solve the problem, the teacher and/or students would call them "stupid."

Teacher and peer embarrassment and humiliation become the conditioning experience that causes some students' math anxiety. Over the years, this math anxiety is reinforced and even increases in magnitude. In fact, many math anxious students — now 30, 40 and 50 years old — *still* have extreme fear about working math problems on the board. One 56 year old indicated that he had a great deal of fear that the instructor would call him to the board. Even if he knew how to do the problem, displaying that knowledge to his peers was very difficult. Some students have said that they absolutely refused to go to the board.

Being embarrassed by family members can also cause math anxiety. According to students who have been interviewed on the matter, their parents tried to help them with math and this sometimes led to serious trauma. These students claim that the tutoring from their guardians, mainly their fathers, often resulted in scolding when they were not able to complete the problems. One student reported that his father hit him *every* time he got a multiplication problem wrong. Brothers and sisters can also tease one another about being dumb in math. This is particularly true of boys telling girls that they cannot do math. When people hear these statements enough times, they may start to believe them and associate these bad feelings with the word math. So, for students many years later just hearing the word "math" triggers a response of anxiety, consciously or unconsciously recalling the bad feelings, and becoming uneasy.

A good example of this is a student who I worked with who had completed her BS degree fifteen years ago at a college that did not require much math. She was returning to college to be an elementary school teacher, which required her to take math and a math placement test. As soon as I mentioned that she had to take math, she said, "I can't do math and I will have to wait a few days to get psychologically ready to take the math placement test." She indicated her old feeling of not being able

to do math rushed through her and she almost had an anxiety attack. This is an extreme case but a true example of math anxiety. In most cases math anxiety is not this bad, but it is disruptive enough to cause learning and testing problems.

If you have math anxiety, try to remember the first time you had anxiety about learning math. This doesn't include anxiety when taking a math test, which will be discussed later in this chapter. Was it in elementary, middle or high school? Can you recall a specific incident? Was it being called on in class or getting a poor grade on a homework assignment? Was it a parent saying he/she cannot do math and neither can you? If you can't remember a specific incident, then when was the last time you told yourself that you couldn't learn math? Today? Yesterday? A year ago?

Some students with math anxiety also have had positive experiences with learning math. Now, try to remember the first positive experience with math. Was it a teacher, parent, friend, or other person who praised you? Was it getting back a good grade on a homework assignment or test? Was it making a good grade in a math course? Think back to your last positive experience with math one more time. How did it make you feel? Those who don't have math anxiety still need to understand it, so they can help classmates that might have it and need support. Also, it is possible that they will develop it in the future.

One way to overcome math anxiety is to try to find out when and why it first occurred and how it is still affecting you. Right now, reread the questions in the previous two paragraphs and on separate paper, brainstorm your answers to the questions. You do not have to write complete sentences. Just record the information. If you think of other memories as you brainstorm, write those down too.

If some of the memories are tough ones, just make a short note of them right now. Don't dwell on them because they are in the past. This can be frustrating for some students, but it is worth working through it. You will realize that math anxiety is

usually a result of events, not because of being "dumb" or other personal flaws.

If you have had a very positive experience in the past when you were studying math, brainstorm answers to the appropriate questions. You may want to keep in mind friends you know who struggle with math and think about the other questions.

This brainstorming will help you write your math autobiography assignment at the end of this chapter.

How Math Anxiety Affects Learning

Math Anxiety can cause learning problems in several ways. It can affect how you do your homework and your participation in the classroom and study groups. Let's first start by looking at how math anxiety could affect your homework. Students with high math anxiety may have difficulty starting or completing their homework. Doing homework reminds some students of their learning problems in math. More specifically, it reminds them of their previous math failures, which causes further anxiety. This anxiety can lead to total avoidance of homework or "approach-avoidance" behavior.

Total homework avoidance is called procrastination. The very thought of doing their homework causes these students anxiety, which causes them to put off tackling their homework. This makes them feel better for a short amount of time — *until test day.*

Math anxiety can also affect your classroom participation and learning. Usually students with math anxiety are afraid to speak out in class and ask questions. They remember that in the past they were made fun of when giving the wrong answer. They are also afraid of asking a question that others, including the teacher, will consider dumb. So they sit in class fearful of being asked a question, looking like they understand the lecture so they will not be

called on. They also take a lot of notes even though they don't understand them, in order to give the illusion of comprehension. If you are one of these students, these are hard habits to break. However, these habits may cause you to be unsuccessful in your math class. Here are some suggestions to break these habits:

1. Make an appointment to talk to your math instructor. Math instructors want you to talk to them. When I do my consulting around the country one of the major complaints I get from math instructors is that the students don't come and see them. Make an appointment to see your math instructor before the first major test to discuss your math history and to ask for learning suggestions. In most cases it is easier to talk to the instructor before you get your first grade.

2. Before class, ask the instructor to work one homework problem. You might want to write the problem on the board before the instructor arrives. This is less stressful because you are not asking the question in front of the whole class. In fact, one of my good friends, Dr. Mike Hamm, suggests that his students put the problems they do not know on the board before class. Other students go to the board and solve the problems. Dr. Hamm then solves the ones the students cannot do.

3. Prepare one question from your homework and ask it within the first 15 minutes of class. Instructors are more likely to answer questions in the first part of class when they have time instead of the end of class when time is running out.

4. Ask a question that you already know the answer. That way if the instructor starts to ask you a question about the problem, you will know the answer. This is good practice for asking those questions to which you don't know the answers.

5. Use email to send questions to your instructor. This way you can still get the answer with very little anxiety.

By using these suggestions you can reduce your math anxiety and learn more mathematics. A question unanswered could be a missed test question.

REMEMBER

The instructor's job is to answer your questions and you are paying for the course.

Math anxious students sometimes avoid doing additional math outside of the classroom. They avoid study groups and Supplemental Instruction. It is like asking a person with aquaphobia (fear of water) to take vacation at the beach. However, a person with aquaphobia can go to the beach and enjoy himself or herself and not get wet. In other words, students can still attend study groups and Supplemental Instruction and just listen. When they are ready to get their feet wet, they can ask a few questions. Don't let these great opportunities go by.

Math anxiety can affect how you learn mathematics. It can be overcome with your effort. You don't have to live in the past with your math fears. Today is a new day and you can change how math affects you. The next step is to understand how test anxiety can affect your demonstration of math knowledge.

Section Review

1. List and explain the three types of math anxiety.

First Type: _____

Second Type: _____

Third Type: _____

2. List and explain two causes of math anxiety.

First Cause: _____

Second Cause: _____

continued

3. List three ways math anxiety can cause learning problems.

First Way: _____

Second Way: _____

Third Way: _____

4. List and explain three ways students can manage or avoid the affects of math anxiety.

First Way: _____

Second Way: _____

Third Way: _____

How to recognize test anxiety

Test anxiety has existed for as long as tests have been used to evaluate student performance. Because it is so common and because it has survived the test of time, test anxiety has been carefully studied over the last 50 years. Pioneering studies indicate that test anxiety generally leads to low test scores.

At the University of South Florida (Tampa), Dr. Charles Spielberger investigated the relationship

between test anxiety and intellectual ability. The study results suggested that anxiety coupled with high ability can improve academic performance; but anxiety coupled with low or average ability can interfere with academic performance. That is:

Anxiety + high ability = improvement
Anxiety + low or average ability = no improvement

> **Example:** Students with average ability and low test anxiety had better performance and higher grades than did students with average ability and high test anxiety. However, there are students who make good grades, take calculus, and still have test anxiety.

Test anxiety is a *learned* response; a person is not born with it. An environmental situation brings about test anxiety. The good news is that because it is a learned response, it can be *unlearned*. Test anxiety is a special kind of general stress. General stress is considered "strained exertion," which can lead to physical and psychological problems.

Defining Test Anxiety

There are several definitions of test anxiety. One definition states, "Test anxiety is a conditioned emotional habit to either a single terrifying experience, recurring experience of high anxiety, or a continuous condition of anxiety" (Wolpe, 1958).

Another definition of test anxiety relates to the educational system. The educational system develops evaluations that measure one's mental performance, and this creates test anxiety. This definition suggests that test anxiety is the *anticipation* of some realistic or nonrealistic situational threat (Cattell, 1966). The "test" can be a research paper, an oral report, work at the chalkboard, a multiple-choice exam, a written essay, or a math test.

Math test anxiety is a relatively new concept in education. *Ms.* magazine published "Math Anxiety: Why Is a Smart Girl Like You Counting on Your Fingers?" (Tobias, 1976) and coined the phrase *math anxiety*. During the 1970s, other educators began referring to *mathophobia* and *mathemaphobia* as a possible cause for children's unwillingness to learn math. Additional studies on the graduate level discovered that math anxiety was common among adults as well as children.

One of my students once described math test anxiety as "being in a burning house with no way out." No matter how you define it, math test anxiety is real, and it affects millions of students.

Why Math Tests Create Anxiety

It has been shown that math anxiety exists among many students who usually do not suffer from other tensions. Counselors at a major university reported that one-third of the students who enrolled in behavior therapy programs offered through counseling centers had problems with math anxiety (Suinn, 1988).

Educators know that math anxiety is common among college students and is more prevalent in women than in men. They also know that math anxiety frequently occurs in students with a poor high school math background. These students were found to have the greatest amount of anxiety.

Approximately half of the students in college prep math courses (designed for students with inadequate high school math background or low placement scores) could be considered to have math anxiety. However, math anxiety also occurs in

students in high-level math courses, such as college algebra and calculus.

Educators investigating the relationship between anxiety and math have indicated that anxiety contributes to poor grades in math. They also found that simply *reducing* math test anxiety does not guarantee higher math grades. Students often have other problems that affect their math grades, such as poor study skills, poor test-taking skills, or poor class attendance.

Section Review

1. What is your personal definition of test anxiety?

2. What type of student does math anxiety affect most?

The causes of test anxiety

The causes of test anxiety can be different for each student, but they can be explained by seven basic concepts. See Figure 3 on the next page.

The causes of math test anxiety can be different for each student. It could possibly have first occurred in middle or high school. However, for many students it first occurs in college when passing tests is the only way to pass the course. Homework and extra credit in most college courses don't count toward your grade. Now students must have a passing average and in some cases must pass the departmental final exam. Additional pressure also exists because not passing algebra means you won't graduate and you might not get the job you want. As you can see, there are more reasons to have math test anxiety in college than in high school.

If you have math test anxiety, I want you to try to remember the first time it surfaced. Was it in middle school, high school or college? Can you recall a specific incident? Was it your first algebra test? Was it your first math test after being out of school for a long time? Was it after you decided to get serious about college? Was it a college algebra course that was required for your major? Was it an instructor who told you that if you don't pass the next math test you would fail the course? Was it when you needed to pass the next math test to pass the course so you could keep your financial aid? Was it your children asking you why you failed your last math test? If you cannot remember a specific incident when you had test anxiety, do you expect to have any major anxiety on the next test?

Students with math test anxiety also have had positive experiences with taking math tests. Now try to remember your first positive experience with taking a math test. Was it in middle school, high

school or in college? Was it after studying many hours for the test? Now think back to your last positive experience with a math test. How did it make you feel?

Since we have already explored your experiences with taking math tests, let's look at some of the direct causes of your math test anxiety. If you do have test anxiety, what is the main cause? If you don't know, then review the seven Causes of Test Anxiety in Figure 3. Does one of these reasons fit you? If you don't have test anxiety, what would be a reason that could cause test anxiety?

Like you did before, brainstorm answers to the questions listed in the previous three paragraphs. You will also use this brainstorming to help you write your autobiography.

REMEMBER

For some students, just writing about their previous math history helps them.

Section Review

1. List and describe five causes of test anxiety.

 First Cause: _____

Figure 3
Causes of Test Anxiety

- Test anxiety can be a learned behavior resulting from the expectations of parents, teachers, or other significant people in the student's life.

- Test anxiety can be caused by the association between grades and a student's personal worth.

- Test anxiety develops from fear of alienating parents, family, or friends due to poor grades.

- Test anxiety can stem from a feeling of lack of control and an inability to change one's life situation.

- Test anxiety can be caused by a student's being embarrassed by the teacher or other students when trying to do math problems.

- Test anxiety can be caused by timed tests and the fear of not finishing the test, even if one can do all the problems.

- Test anxiety can be caused by being put in math courses above one's level of competence.

What are the cause(s) of *your* test anxiety?

Second Cause: _____

Third Cause: _____

Fourth Cause: _____

Fifth Cause: _____

2. Put into your own words your cause(s) of test anxiety. If you don't have test anxiety, list what you think is the major cause of test anxiety.

The different types of test anxiety

The two basic types of test anxiety are emotional (educators call this *somatic*) and worry (educators call this *cognitive*). Students with high test anxiety have *both* emotional and worry anxiety.

Signs of emotional anxiety are upset stomach, nausea, sweaty palms, pain in the neck, stiff shoulders, high blood pressure, rapid, shallow breathing, rapid heartbeat, or general feelings of nervousness. As anxiety increases, these feelings intensify. Some students even run to the bathroom to throw up or have diarrhea.

Even though these *feelings* are caused by anxiety, the physical response is real. These feelings and physical inconveniences can affect your concentration and your testing speed, and they can cause you to completely "draw a blank."

Worry anxiety causes the student to think about failing the test. These negative thoughts can happen either before or during the test. This negative "self-talk" causes students to focus on their anxiety instead of recalling math concepts.

The effects of test anxiety range from a "mental block" on a test to avoiding homework. One of the most common side effects of test anxiety is getting the test and immediately forgetting information that you know. Some students describe this event as having a "mental block," "going blank," or finding that the test looks like a foreign language.

After five or ten minutes into the test, some of these students can refocus on the test and start working the problems. They have, however, lost valuable time. For other students, anxiety persists throughout the test and they cannot recall the needed math information. It is only after they walk out the door that they can remember how to work the problems.

Sometimes math anxiety does not cause students to "go blank" but slows down their mental processing speed. This means it takes longer to recall formulas and concepts and to work problems. The result is frustration and loss of time, leading to *more* anxiety. Since, in most cases, math tests are *speed* tests (those in which you have a certain amount of time to complete the test), you may not have enough time to work all the problems or to check the answers if you have mentally slowed down. The result is a lower test score, because even though you know the material, you do not complete all of the questions before test time runs out.

Not using all of the time allotted for the test is another problem caused by test anxiety. Students know that they should use all of the test time to check their answers. In fact, math is one of the few subjects in which you can check test problems to find out if your work is correct. However, most students do not use all of the test time, and this results in lower test scores. Why does this happen?

Students with high test anxiety do not want to stay in the classroom. This is especially true of students whose test anxiety increases as the test progresses. The test anxiety gets so bad that they would rather leave early and receive a lower grade than stay in that "burning house."

Figure 4
The 12 Myths About Test Anxiety

Students are born with test anxiety.

Test anxiety is a mental illness.

Test anxiety cannot be reduced.

Any level of test anxiety is bad.

All students who are not prepared have test anxiety.

Students with test anxiety cannot learn math.

Students who are well prepared will not have test anxiety.

Very intelligent students and students taking high-level courses, such as calculus, do not have test anxiety.

Attending class and doing homework should reduce all test anxiety.

Being told to relax during a test will make a student relaxed.

Doing nothing about test anxiety will make it go away.

Reducing test anxiety will guarantee better grades.

Students have another reason for leaving the test early: the fear of what the instructor and other students will think about them for being the last one to hand in the test. These students refuse to be the last ones to finish the test because they think that the instructor or other students will think they are dumb. This is middle-school thinking, but the feelings are still real — no matter what the age of the student. These students do not realize that some students who turn in their tests first fail, whereas many students who turn in their tests last make A's and B's.

Another effect of test anxiety relates to completing homework assignments. Students who have high test anxiety may have difficulty starting or completing their math homework. Doing homework reminds some students of their learning problems in math. More specifically, it reminds them of their previous math failures, which causes further anxiety. This anxiety can lead to total homework avoidance or "approach-avoidance" behavior.

Total homework avoidance is called procrastination. The very thought of doing their homework causes these students anxiety, which causes them to put off tackling their homework. This makes them feel better for a short amount of time — *until test day.*

Example: Some students begin their homework and work some problems successfully. They then get stuck on a problem that causes them anxiety, so they take a break. During their break the anxiety disappears until they start doing their homework again. Doing their homework causes more anxiety, which leads to another break. The breaks become more frequent.

continued

Finally, the student ends up taking one long break and not doing the homework. Quitting, to them, means *no more anxiety* until the next homework assignment.

The effects of math test anxiety can be different for each student. Students can have several of the mentioned characteristics that can interfere with math learning and test taking. However, there are certain myths about math that each student needs to know. Review Figure 4, The 12 Myths About Test Anxiety, to see which ones you believe. If you have test anxiety, which of the mentioned characteristics are true for you?

Section Review

1. List and describe the two basic types of test anxiety.

 First Type: _____

 Second Type: _____

2. List two reasons students leave the test room early instead of checking their answers.

First Reason: _____

Second Reason: _____

3. List 6 of the 12 myths about test anxiety that you most believed.

First Myth: _____

Second Myth: _____

Third Myth: _____

Fourth Myth: _____

Fifth Myth: _____

Sixth Myth: _____

How to reduce test anxiety

To reduce math test anxiety, you need to understand both the relaxation response and how negative self-talk undermines your abilities.

Relaxation Techniques

The relaxation response is any technique or procedure that helps you to become relaxed. It will take the place of an anxiety response. Someone simply telling you to relax or even telling yourself to relax, however, does little to reduce your test anxiety. There are both short-term and long-term relaxation response techniques that help control emotional (somatic) math test anxiety. These techniques will also help reduce worry (cognitive) anxiety. Effective *short-term* techniques include the tensing and differential relaxation method, the palming method, and deep breathing.

Short-Term Relaxation Techniques

The Tensing and Differential Relaxation Method

The tensing and differential relaxation method helps you relax by tensing and relaxing your muscles all at once. Follow these procedures while you are sitting at your desk before taking a test:

1. Put your feet flat on the floor.
2. With your hands, grab underneath the chair.
3. Push down with your feet and pull up on your chair at the same time for about five seconds.
4. Relax for five to ten seconds.
5. Repeat the procedure two or three times.
6. Relax all your muscles except the ones that are actually used to take the test.

The Palming Method

The palming method is a visualization procedure used to reduce test anxiety. While you are at your desk before or during a test, follow these procedures:

1. Close and cover your eyes using the center of the palms of your hands.
2. Prevent your hands from touching your eyes by resting the lower parts of your palms on your cheekbones and placing your fingers on your forehead. Your eyeballs must not be touched, rubbed, or handled in any way.
3. Think of some real or imaginary relaxing scene. Mentally visualize this scene. Picture the scene as if you were actually there, looking through your own eyes.
4. Visualize this relaxing scene for one to two minutes.
5. Open your eyes and wait about two minutes and repeat the visualization scene again. This time also imagine any sounds or smells that can enhance your scene. For example, if you are imagining a beach scene, hear waves on the beach and smell the salty air. This technique can also be completed without having your hands over your eyes by just closing your eyes.

Practice visualizing this scene several days before taking a test and the effectiveness of this relaxation procedure will improve.

Deep Breathing

Deep breathing is another short-term relaxation technique that can help you relax. Proper breathing is a way to reduce stress and decrease test anxiety. When breathing properly, enough oxygen gets into your bloodstream to nourish your body and mind. A lack of oxygen in your blood contributes to an anxiety state that makes it more difficult to react to stress. Proper deep breathing can help you control your test anxiety.

Deep breathing can replace the rapid, shallow breathing that sometimes accompanies test anxiety, or it can prevent test anxiety. Here are the steps to deep breathing:

1. Sit straight up in your chair in a good posture position.
2. Slowly inhale through your nose.
3. As you inhale, first fill the lower section of your lungs and work your way up to the upper part of your lungs.
4. Hold your breath for a few seconds.
5. Exhale slowly through your mouth.
6. Wait a few seconds and repeat the cycle.
7. Keep doing this exercise for four or five minutes. This should involve going through about ten breathing cycles. Remember to take two normal breaths between each cycle. If you start to feel light-headed during this exercise, stop for thirty – forty-five seconds and then start again.
8. Throughout the entire exercise, make sure you keep breathing smoothly and in a regular rhythm without gulping air or suddenly exhaling.
9. As an extra way to improve your relaxation, say "relax" or "be calm" to yourself as you exhale. This can start a conditioned response that can trigger relaxation when you repeat the words during anxious situations. As you keep practicing, this conditioned response will strengthen. Practice is the key to success.

You need to practice this breathing exercise for several weeks before using the technique during tests. If you don't practice this technique, then it will not work. After taking your first test keep doing the exercise several times a week to strengthen the relaxation response.

Side one of the audiocassette *How to Reduce Test Anxiety* (Nolting, 1986) further explains test anxiety and discusses these and other short-term relaxation response techniques. Short-term relaxation techniques can be learned quickly but are not as successful as the long-term relaxation technique. Short-term techniques are intended to be used while learning the long-term technique.

Long-Term Relaxation Techniques

The cue-controlled relaxation response technique is the best long-term relaxation technique. It is presented on side two of the audiocassette *How To Reduce Test Anxiety* (Nolting, 1986). Cue-controlled relaxation means you can induce your own relaxation based on repeating certain cue words to yourself. In essence, you are taught to relax and then silently repeat cue words, such as "I am relaxed."

After enough practice, you can relax during math tests. The cue-controlled relaxation technique has worked with thousands of students. For a better understanding of test anxiety and how to reduce it, listen to *How to Reduce Test Anxiety* (Nolting, 1986).

Negative Self-Talk

According to cognitive psychologists, self-talk is what we say to ourselves as a response to an event or situation. These dialogues determine our feelings about that event or situation. Sometimes we say it so quickly and automatically that we don't even hear ourselves. We then think it is the situation that causes the feeling, but in reality it is our interactions or thoughts about the experience that are controlling our emotions. This sequence of events can be represented by this timeline:

External Events
(math test)
↓
Interpretation of Events and Self-Talk
(how you feel about the test
and what you are telling yourself)
↓
Feelings and Emotions
(happy, glad, angry, mad, upset)

In most cases this means that you are responsible for how and what you feel. You can have positive self-talk or negative self-talk about taking a math test. Yes, some students have positive self-talk about math tests and see it as a challenge and something to accomplish that makes them feel good, while others see it as an upsetting event that leads to anger and anxiety. In other words, you are what you tell yourself.

Negative self-talk is a form of worry (cognitive) anxiety. This type of worrying can interfere with your test preparation and can keep you from concentrating on the test. Worrying can motivate you to study, but too much worrying may prevent you from studying at all.

Negative self-talk is defined as the negative statements you tell yourself before and during tests. Negative self-talk causes students to lose confidence and to give up on tests. Further, it can give you an inappropriate excuse for failing math and cause you to give up on learning math.

Examples of Negative Self-Talk

"No matter what I do, I will not pass this course."

"I failed this course last semester, and I will fail it again."

"I am no good at math, so why should I try?"

"I cannot do it; I cannot do the problems, and I am going to fail this test."

"I have forgotten how to do the problems, and I am going to fail."

"I am going to fail this test and never graduate."

"If I can't pass this test, I am too dumb to learn math and will flunk out."

Students who have test anxiety are prone to negative self talk. Test anxiety can be generated or heightened by repeatedly making statements to yourself that usually begin with "what if." For example, "What if I fail the test?" or "What if I fail this class again?" These "what if" statements generate more anxiety that can cause students to feel sick. These statements tell them to be anxious. Some other aspects of self-talk are:

- Self–talk can be in telegraphic form with short words or images.
- Self-talk can be illogical but at the time the person believes it.
- Negative self-talk can lead to avoidance like not taking the test or skipping classes.
- Negative self-talk can cause more anxiety.
- Negative self-talk can lead to depression and a feeling of helplessness.
- Negative self-talk is a bad habit that can be changed.

There are different types of negative self-talk. If you have negative self-talk, then review the types below and see which one fits you best. You may use a combination of them:

1. *The Critic* is the person inside us who is always trying to put us down. It constantly judges behaviors and finds fault even if it is not there. It jumps on any mistake and exaggerates it to cause more anxiety. The Critic puts us down for mistakes on the test and blames us for not controlling the anxiety. The Critic reminds us of previous comments from real people who have criticized us. It compares us to other students who are doing better in the class. It loves to say, "That was a stupid mistake!" or "You are a total disappointment. You can't pass this math class like everyone else can!" The Critic's goal is to promote low self-esteem.

2. *The Worrier* is the person inside us who looks at the worst-case scenario. It wants to scare us with the ideas of disasters and complete failure. When it sees the first sign of anxiety, it "blows it out of proportion" to the extent that we will not remember anything and totally fail the test. The Worrier creates more anxiety than normal. The Worrier anticipates the worst, underestimates our ability, and sees us not only failing the test but "failing life." The Worrier loves to ask "What if?" For example, "What if I fail the math test and don't graduate?" or "What if I can't control my anxiety and throw up in the math class?" The goal of the Worrier is to cause more anxiety so we will quit.

3. *The Victim* is the person inside us who wants us to feel helpless or hopeless. It wants us to believe that no matter what we do, we will not be successful in math. The Victim does not blame other events (poor schooling) or people (bad teachers) for math failures. It blames us. It dooms us and puts us into a learned helpless mode,

meaning that if we try to learn math, we will fail or if we don't try to learn math we will fail. So why try? The Victim likes to say, "I can't learn math." The goal of the Victim is to cause depression and make us not try.

4. *The Perfectionist* is similar to the Critic, but is the opposite of the Victim. It wants us to do the best we can and will guide us into doing better. It tells us that we are not studying enough for the math test and that a B is not good enough and that we must make an A. In fact, sometimes an A is not good enough unless it is a 100%. So, if we make a B on the test, the Perfectionist says, "A or B is just like making an F." The Perfectionist is the hard-driving part of us that wants the best but cannot stand mistakes or poor grades. It can drive us to mental and physical exhaustion to make that perfect grade. It is not interested in self-worth, just perfect grades. Students driven by the Perfectionist often drop a math course because they only have a B average. The Perfectionist loves to repeat, "I should have . . ." or "I must . . ." The goal of the Perfectionist is to cause chronic anxiety that leads to burnout.

Review these types of personalities to see which one may fit you best. We may have a little of each one in us, but what is the dominant one for you in math? Now we can look at how to stop these negative thoughts.

Managing Negative Self-Talk

Students need to change their negative self-talk to positive self-talk without making unrealistic statements.

Positive self-statements can improve your studying and test preparation. During tests, positive self-talk can build confidence and decrease your test anxiety. These positive statements (see examples) can help reduce your test anxiety and improve your grades. Some more examples of positive self-statements are on the cassette *How to Reduce Test Anxiety* (Nolting, 1986). Before the test, make up some positive statements to tell yourself.

There are several ways to counter and control negative self-talk. Students can replace the negative self-talk with positive statements and questions that make them think in a realistic way about the situation. Another way is to develop thought stopping techniques to reduce or eliminate the negative thoughts. Try each way or a combination to see what works best for you.

Countering self-talk involves writing down and practicing positive statements that can replace negative statements. Students can develop their own positive statements. Some rules for writing positive statements are:

1. Use the first person present tense, for example, "I can control my anxiety and relax myself."

2. Don't use negatives in the statement. For example, don't say, "I am not going to have anxiety." Instead say, "I will be calm during the test."

3. Have a positive and realistic belief in the statement. For example say, "I am going to pass this test," instead of saying, "I am going to make an A on this test" when you have not made an A on any of the tests.

The statements used to counter the negative thoughts can be based on the type of negative self-talk. The Critic who puts down by saying, "Your test anxiety is going to cause you to fail the test" can be countered with, "I have test anxiety, but I am learning to control it." The Worrier who says, "What if I fail the test" can be countered with "So what? I will do better on the next test." The Victim who thinks things are hopeless and says, "I will never be

able to pass math" can be countered with, "I have studied differently for this test, and I can pass the math course." The Perfectionist who says, "I need to make an A on the test or I will drop out of school" can be countered with, " I don't need to make an A to please anyone. All I need is to pass the math course to get the career I want." These are some of the examples of counter statements that can control anxiety.

Examples of Positive Self-Talk

"I failed the course last semester, but I can now use my math study skills to pass this course."

"I went blank on the last test, but I now know how to reduce my test anxiety."

"I know my poor math skills are due to poor study skills, not my own ability, and since I am working on my study skills, my math skills will improve."

"I know that with hard work, I will pass math."

"I prepared for this test and will do the best I can. I will reduce my test anxiety and use the best test-taking procedures. I expect some problems will be difficult, but I will not get discouraged."

"I am solving problems and feel good about myself. I am not going to worry about that difficult problem; I am going to work the problems that I can do. I am going to use all the test time and check for careless errors. Even if I do not get the grade I want on this test, it is not the end of the world."

Thought-Stopping Techniques

Some students have difficulty stopping their negative self-talk. These students cannot just tell themselves to eliminate those thoughts. These students need to use a thought-stopping technique to overcome their worry and become relaxed.

Thought stopping involves focusing on the unwanted thoughts and, after a few seconds, suddenly stopping those thoughts by emptying your mind. Using the command "Stop!" or a loud noise like clapping your hands can effectively interrupt the negative self-talk. In a homework situation, you may be able to use a loud noise to stop your thoughts, but don't use it during a test.

To stop your thoughts in the classroom or during a test, silently shout to yourself "Stop!" or "Stop thinking about that." After your silent shout, either relax yourself or repeat one of your positive self-talk statements. You may have to shout to yourself several times during a test or while doing homework to control negative self-talk. After every shout, use a different relaxation technique/scene or positive self-talk statement.

Thought stopping works because it interrupts the worry response before it can cause high negative emotions. During that interruption, you can replace the negative self-talk with positive statements or relaxation. However, students with high worry anxiety should practice this technique three days to one week before taking a test. Contact your counselor if you have additional questions about the thought-stopping techniques.

Writing the Math Autobiography

An autobiography relates how you remember and feel about your past experiences. In addition, many autobiographies explore how these past

feelings and experiences shape current life. While some people write autobiographies for others to learn about their lives, many people write private autobiographies in order to understand what is going on in their lives.

Use the Appendix as a guide to write your Math Autobiography. Your final product will be a typed (or handwritten if your instructor approves) paper that summarizes your experiences learning math, how you feel about these experiences, and how they shape your current perspective on learning math.

Section Review

1. Describe your best short-term relaxation technique.

2. You can use the palming method by closing your eyes and visualizing a scene without putting your hands to your face. Describe a very relaxing scene that you could visualize. Make sure to include some sounds and visual images in your scene.

3. Practice your relaxation scene for three to five minutes for the next five days. List the times and dates you practiced your scene.

Date: _____ Time: _____
Date: _____ Time: _____
Date: _____ Time: _____
Date: _____ Time: _____
Date: _____ Time: _____

4. How does negative self-talk cause you to have text anxiety?

5. Make up three positive self-talk statements that are not listed in this text.

Statement 1: _____

Statement 2: _____

Statement 3: _____

6. Describe the thought-stopping technique in your
 own words.

7. What words will you use as your silent shout?

8. What will you do after your silent shout?

9. How does the thought-stopping technique work?

Chapter 3 Review

1. Reducing math test anxiety _____ _____ guarantee good math grades.

2. Test anxiety is a _____ response; a person is not _____ with it.

3. Math test anxiety involves _____, _____, and _____ of math tests.

4. One cause of test anxiety is that a student goes to the board to work a problem but is called _____ when he or she cannot work the problem.

5. The two basic types of test anxiety are _____ and _____.

6. The effects of test anxiety range from a " _____ _____ " on a test to _____ homework.

7. The _____ and _____ _____ method helps you relax by tensing and relaxing your muscles all at once.

8. The palming method is a _____ procedure used to reduce text anxiety.

9. Negative self-talk is a form of _____ anxiety.

10. Thought stopping works because it interrupts the _____ response before it can cause _____ emotions.

What is the most important information you learned from this chapter?

How can you immediately use it?

Lecture Notes

How to Improve Your Listening and Note-Taking Skills

Listening and note-taking skills in a math class are very important, since most students either do not read the math text or have difficulty understanding it. In most of your other classes, if you do not understand the lecture, you can read the text and get almost all the information. In math class, however, the instructor can usually explain the textbook better than the students can read and understand it.

Students who do not have good listening skills or note-taking skills will be at a disadvantage in learning math. Most math understanding takes place in the classroom. Students must learn how to take advantage of learning math in the classroom by becoming effective listeners, calculator users, and notetakers.

In this chapter you will learn

- how to become an effective listener,
- how to become a good notetaker,
- when to take notes,
- the seven steps to math note-taking, and
- how to rework your notes.

How to become an effective listener

Becoming an effective listener is the foundation for good note-taking. You can become an effective listener using a set of skills that you can learn and practice. To become an effective listener, you must prepare yourself both physically and mentally.

Sitting in the Golden Triangle

The physical preparation for becoming an effective listener involves *where you sit* in the classroom. Sit in the best area to obtain high grades, that is, in "the golden triangle of success." The golden triangle of success begins with seats in the front row facing the instructor's desk (see Figure 5 on the next page).

Students seated in this area (especially in the front row) directly face the teacher and are most likely to pay attention to the lecture. This is a great seating location for visual learners. There is also less tendency

for them to be distracted by activities outside the classroom or by students making noise within the classroom.

The middle seat in the back row is another point in the golden triangle for students to sit, especially those who are auditory (hearing) learners. You can hear the instructor better because the instructor's voice is projected to that point. This means that there is less chance of misunderstanding the instructor, and you can hear well enough to ask appropriate questions.

By sitting in the golden triangle of success, you can force yourself to pay more attention during class and be less distracted by other students. This is very important for math students because math instructors usually go over a point once and continue on to the next point. If you miss that point in the lesson, then you could be lost for the remainder of the class.

This may seem obvious, but I know too many students who do the following if they have early morning classes. Roll out of bed, put on the closest clothes to them, brush their hair, maybe their teeth, and race to class. Don't do this. You won't wake up until the end of your class! Train yourself to be somewhere on campus to review your notes at least fifteen minutes before class starts and do some of the other strategies suggested in the next section. So, another physical way to be ready is to be awake and nourished!

Warming Up for Math Class

The mental preparation for note-taking involves "warming up" before class begins and becoming an active listener. Just as an athlete must warm up before a game begins, you must warm up before taking notes. Warm up by

- reviewing the previous day's notes,
- reviewing the reading material,
- reviewing the homework,
- preparing questions, and
- working one or two unassigned homework problems.

This mental warm up before the lecture allows you to refresh your memory and prepare pertinent questions, making it easier to learn the new lecture material.

Figure 5
The Golden Triangle of Success

Instructor

Visual Learners

Auditory Learners

Back of the Classroom

How to Become an Active Listener

Becoming an active listener is the second part of the mental preparation for note-taking.

Do not anticipate what the instructor is going to say or immediately judge the instructor's information before the point is made. This will distract you from learning the information.

> **Examples:** Watch the speaker, listen for main ideas, and nod your head or say to yourself, "I understand," when agreeing with the instructor.

Expend energy looking for interesting topics in the lecture. When the instructor discusses information that you need to know, immediately repeat it to yourself to begin the learning process.

You can practice this exercise by viewing math videotapes and repeating important information. This is an especially good learning technique for auditory learners.

Ask appropriate questions when you have them. Sometimes if you are so confused you do not know what question to ask, still let the instructor know you are confused. Also, mark your notes where you got lost so that if the instructor doesn't have time to answer your question, you can visit during office hours to get an answer.

REMEMBER

Class time is an intense study period that should not be wasted.

Listening and Learning

Some students think that listening to the instructor and taking notes is a waste of valuable time. Students too often sit in class and use only a fraction of their learning ability. Class time should be considered a valuable study period where you can listen, take notes, and learn at the same time. One way to do this is by memorizing important facts when the instructor is talking about material you already know. Another technique is to repeat back to yourself the important concepts right after the instructor says them in class. Using class time to learn math is an efficient learning system.

Section Review

1. Explain how sitting in the golden triangle of success makes you a better listener.

2. List five ways you can warm up before math class begins.

First Way: _____

Second Way: _____

Third Way: _____

Fourth Way: _____

Fifth Way: _____

3. Select your best warm-up process and try it before your next math class. What was your warm-up process, and how did it help you?

4. How can you listen and learn at the same time?

How to become a good notetaker

Becoming a good notetaker requires two basic strategies. One strategy is to be specific in detail. In other words, *copy* the problems down, step by step. The second strategy is to *understand* the general principles, general concepts, and general ideas.

Copying from the Board

While taking math notes, you need to copy each and every step of the problem even though you may already know every step of the problem. While in the classroom, you might understand each step, but a week later you might not remember how to do the problem unless all the steps were written down. In addition, as you write down each step, you are memorizing it. Make sure to copy every step for each problem written on the board.

There will be times when you will get lost while listening to the lecture. Nevertheless, you should keep taking notes, even though you do not understand the problem. This will provide you with a reference point for further study. Put a question mark (?) by those steps that you do not understand;

then, after class, review the steps you did not understand with the instructor, your tutor, or another student.

If you seem to think many other students are lost too, ask questions or let the instructor know.

Taking Notes

The goal of note-taking is to take the least amount of notes and get the greatest amount of information on your paper. This could be the opposite of what most instructors have told you. Some instructors tell you to take down everything. This is not necessarily a good note-taking system, since it is very difficult to take precise, specific notes while trying to understand the instructor.

What you need to develop is a note-taking system in which you write the least amount possible and get the most information down while still understanding what the instructor is saying.

Develop an Abbreviation List

To reduce the amount of written notes, an abbreviation system is needed. An abbreviation system is your own system of reducing long words to shorter versions that you still can understand. By writing less, you can listen more and have a better understanding of the material.

Example: When the instructor starts explaining the commutative property, you need to write "commutative property" out the first time. After that, use "COM." You should develop abbreviations for all the most commonly used words in math.

Figure 6
Abbreviations

E.G.	(for example)
CF.	(compare, remember in context)
N.B.	(note well, this is important)
\	(therefore)
∴	(because)
É	(implies, it follows from this)
>	(greater than)
<	(less than)
=	(equals, is the same)
¹	(does not equal, is not the same)
()	(parentheses in the margin, around a sentence or group of sentences, indicates an important idea)
?	(used to indicate you do not understand the material)
O	(a circle around a word may indicate that you are not familiar with it; look it up)
E	(marks important materials likely to be used in an exam)
1, 2, 3,	(to indicate a series of facts)
D	(shows disagreement with statement or passage)
REF	(reference)
et al.	(and others)
bk	(book)
p	(page)
etc.	(and so forth)
V	(see)
VS	(see above)
SC	(namely)
SQ	(the following)
Com.	(commutative)
Dis.	(distributive)
APA	(associative property of addition)
AI	(additive inverse)
IPM	(identity property of multiplication)

Figure 6 on the previous page provides a list of abbreviations. Add your own abbreviations to this list. By using abbreviations as much as possible, you can obtain the same meaning from your notes and have more time to listen to the instructor.

When to take notes

To become a better notetaker, you must know when to take notes and when not to take notes. The instructor will give cues that indicate what material is important. These cues include

- presenting usual facts or ideas,
- writing on the board,
- summarizing,
- pausing,
- repeating statements,
- enumerating, such as "1, 2, 3" or "A, B, C,"
- working several examples of the same type of problem on the chalkboard,
- saying, "This is a tricky problem." Most students will miss it (for example, 5/0 is "undefined" instead of "zero").
- saying, "This is the most difficult step in the problem,"
- indicating that certain types of problems will be on the test, such as coin or age problems, and
- explaining bold-face words.

You must learn the cues your instructor gives to indicate important material. If you are in doubt about the importance of the class material, do not hesitate to ask the instructor about its importance.

While taking notes, you may become confused about math material. At that point, take as many notes as possible, and do not give up on note-taking. One thing you shouldn't do—bother the student next to you for an explanation. This can be annoying, and both of you will miss important information while you are talking.

As you take notes on confusing problem steps, leave extra space; then go back and fill in information that clarifies your misunderstanding of the steps in question. Ask your tutor or instructor for help with the uncompleted problem steps, and write down the reasons for each step in the space provided.

Another procedure to save time while taking notes is to stop writing complete sentences. Write your main thoughts in phrases. Phrases are easier to jot down and easier to memorize.

Section Review

1. Note-taking requires two basic strategies:

 _____ and

 _____ .

2. List five abbreviations (and their meanings) that you use in your math class (one abbreviation and meaning per line).

3. List five cues that your instructor gives that indicate what material is important (one cue per line).

The seven steps to math note-taking

No matter what format you use to take notes, a good set of notes contains the following:

- examples of problems
- explanations of how to complete the problems
- appropriate math vocabulary and rules that are used for each problem

In addition, it is wise to make them as organized as possible. It may get on your nerves for a while, but when you see that organized notes save time when studying, you will be more willing to take the time to organize them.

Many successful math students, whether in developmental math courses or calculus, use a form of the three column note-taking system. It organizes the information, making it easy to review. While creating the three column system, students are also learning the math, so it is not busy work. It does take a few weeks to feel comfortable with it, but it is worth sticking with it. Many students have adopted this system and improved their studying time.

The seven steps lead you through several important stages in the learning process. Steps one through three help you *record the information* accurately, completely, and in an organized fashion. Steps four through seven help you *rehearse and understand the information*. These steps also help you learn to *recall the information*, helping you prepare for the test.

To prepare for taking notes this way, get a good chunk of paper ready. Here are some choices:

- Use regular paper, portrait or landscape style, and draw two lines to create three columns. They should be proportioned as follows:

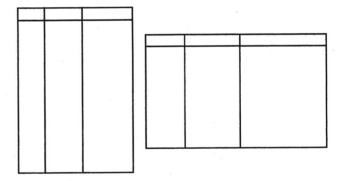

- Some students purchase graph paper because it helps them keep numbers lined up correctly.
- Other students design the page on the computer and print out copies.
- Choose two columns next to each other to be the "Examples" and "Explanations" columns. Label them as such. These are the two columns you will use all the time during class.
- Label the other column "Key Words and Rules." You might fill this column in during class but will probably complete it outside of class.

If you do this ahead of time, you do not have to frantically draw lines while the instructor starts talking. If you forget or run out of paper, it might be quicker to fold the paper into three columns.

Figure 7

Three Column Note-taking Method

Key Words/Rules	Example	Explanation
Solve linear equation	$4(x+4) + 3(x-4) = 2(x-1)$	Have to get x on one side of the = and numbers on other side of the =
Distributive Property	$4x + 16 + 3x -12 = 2x -2$	Multiply numbers to the left of the () with each variable and number in the ().
Community Property	$4x + 3x + 16 -12 = 2x -2$	Regroup numbers and variables.
Combine like terms	$7x + 4 = 2x -2$	
Additive inverse	$7x -2x + 4 = 2x -2x -2$ $5x + 4 = -2$	Subtract $2x$ from both sides in order to get variables all on the right of the =
Additive inverse	$5x + 4 -4 = -2 -4$ $5x = -6$	Subtract 4 from both sides to get numbers all on the right of the =
Mulitiplicative inverse	$\dfrac{5x}{5} = \dfrac{-6}{5}$	Divide both sides by 5 to get x by itself on the left side of the =
Simplify	$x = -1\ 1/5$	Put as a whole number and fraction.
	Insert new problem	

Draw a line dividing the end of one problem from the beginning of the next. Or, do one problem per page to give plenty of room for notes. Graph paper can help students keep numbers and words lined up. Turn the paper landscape.

During class you will use two columns most of the time—the "Example" and the "Explanations" columns. Sometimes, when instructors really want their students to use this method, they will organize their board work to reflect what the notes should look like. Most of the time, however, this does not happen.

The "Examples" column is where you will place the examples the instructor uses during class.

The "Explanations" column is where you put the instructor's verbal explanations of the examples. If the instructor tends to explain by just doing the problem without talking about it, ask questions that will force the instructor to explain the problem verbally.

When the instructor simply explains a vocabulary word or rule, write the vocabulary word or a phrase naming the rule in the 'Key Words and Rules" column. Write the explanation in the "Explanation" column. If the instructor provides an example, obviously it goes in the "Example" column. If no example is provided, ask for one. Then if your wish is not granted, look one up in the textbook after class.

Here are the "Seven Steps to Taking Math Notes." Remember, these are the basic steps. You can adapt them to work for you.

Step 1 *Record the instructor's example problems in the "Example" column.*

Step 2 *Record the reasons for each step in the "Explanations" column.* After completing a problem, draw a horizontal line, separating it from the next problem.

 If you get lost during the lecture, place a question mark in this column and move on.

Step 3 *If possible, record the key words and rules in the other column.* Here is an opportunity for you to ask what rules are being used if the instructor is assuming the class knows them and you do not. Otherwise, you will need to refer to the textbook, work with a tutor, or use online resources after class to complete this column.

 It is important to understand the vocabulary and rules –not just know how to do them!

Step 4 *In order to keep everything in your heads, not letting anything slip out, use the notes to review.* Cover up the "Examples" and "Explanations" columns and quiz yourself on the vocabulary and rules. Or, fold back the "Key Words and Rules" column, look at the examples and explanations and choose the correct key words and rules used to complete the examples.

Step 5 *Place a check mark by the key words and concepts that you did not know.*

Step 6 *Review the information that you checked until you understand and remember it.*

Step 7 *Develop a glossary of the math terms and rules that are difficult for you to remember.*

For a while you might spend more time organizing this system outside of class. Some students have to still depend on their current process, concentrating on getting the information down. Then they transfer the information into the three column method. However, it is best to learn how to do most of the process during math class.

Remember the learning modalities and learning styles? This process can use all of them!

- *Kinesthetic:* Writing out the notes
- *Auditory:* Saying information out loud
- *Visual:* Looking at the notes and thinking; color code each column with highlighters, making sure you do not smear the pencil writing
- *Group:* Working on the notes with one or two other people to make sure they are complete and accurate

• *Individual:* Reading them alone while waiting to meet someone.

Current research is discovering that using language (vocabulary, discussion) to learn math helps students to understand the concepts more deeply. Then completing many problems helps master the skills and increase the speed which is important for preparing for tests. These notes will help you understand the math so that it will stick with you. You will remember more information. Homework will help you master the skills and increase the speed at which you can complete problems.

A Math Glossary

The math glossary is created to define a math vocabulary in your own words. Since math is considered a foreign language, understanding the math vocabulary becomes the key to comprehending math. Creating a glossary for each chapter of your textbook will help you understand math.

Your glossary should include all words printed in bold face in the text and any words you do not understand. If you cannot explain the math vocabulary in your own words, ask your instructor or tutor for help. You may want to use the last pages in your notebook to develop a math glossary for each chapter in your textbook. Some students use note cards. Review your math glossary before each test.

Section Review

1. List and describe the seven steps to note-taking.

First Step: _____

Second Step: _____

Third Step: _____

Fourth Step: _____

Fifth Step: _____

Sixth Step: _____

Seventh Step: _____

2. Make several copies of the modified three-column note-taking system, and take notes using this system in your next math class. Or develop the modified three-column note-taking system in your own notebook.

3. List two benefits of a math glossary.

Benefit 1: _____

Benefit 2: _____

4. From your current math chapter, list and define five words that you can put in your math glossary.

Word 1: _____

Word 2: _____

Word 3: _____

Word 4: _____

Word 5: _____

How to rework your notes

The note-taking system does not stop when you leave the classroom. As soon as possible after class, rework your notes. You can rework the notes between classes or as soon as you get home. By reworking your notes as soon as possible, you can decrease the amount of forgetting. This is an excellent procedure to transfer math information from short-term memory to long-term memory.

REMEMBER

Most forgetting occurs right after learning the material. You need to rework the notes as soon as possible. Waiting means that you probably will not understand what was written.

Here are important steps in reworking your notes:

Step 1 *Rewrite the material you cannot read or will not be able to understand a few weeks later.* If you do not rework your notes, you will be frustrated when studying for a test if you come across notes you cannot read. Another benefit of rewriting the notes is that you immediately learn the new material. Waiting means it will take more time to learn the material.

Step 2 *Fill in the gaps.* Most of the time, when you are listening to the lecture, you cannot write down everything. It is almost impossible to write down everything, even if you know shorthand. Locate the portions of your notes that are incomplete. Fill in the concepts that were left out. In the future, skip two or three lines in your notebook page for anticipated lecture gaps.

Step 3 *Add additional key words and rules in the keywords and rules column.* These key words or ideas were the ones not recorded during the lecture.

Example: You did not know that you should add the *opposite* of 18 to solve a particular problem, and you incorrectly added 18. Put additional important key words and ideas (such as "opposite" and "negative of") in the notes; these are the words that will improve your understanding of math.

Step 4 *Make a problem log on those problems that the teacher worked in class.* The problem log is a separate section of your notebook that contains a listing of the problems (without explanations — just problems) that your teacher worked in class. If your teacher chose those problems to work in class, you can bet that they are considered important. The problems in this log can be used as a practice test for the next exam. Your regular class notes will not only contain the solutions but also all the steps involved in arriving at those solutions and can be used as a reference when you take your practice test.

Write the problem on the left half of the page. Then save some room on the right side for writing out what the test instructions might be for the problem. Sometimes we forget what the test instructions are, and we miss points on the test by following the directions wrong. Skip several lines to complete the problem when you use the problem log as a practice test.

Step 5 *Make a calculator handbook and put in it your keystroke sequences.* The calculator handbook can be a spiral-bound set of note cards or a separate section of your notebook that holds only calculator-related information. Your handbook should also include an explanation of when that particular set of keystrokes is to be used.

Step 6 *Reflection and synthesis.* Once you have finished going over your notes, review the major points in your mind. Combine your new notes with your previous knowledge to have a better understanding of what you have learned today.

If you are a group learner and even if you are not, it helps to review and reflect with other students.

Use a Tape Recorder

If you have a math class during which you cannot get all the information down while listening to the lecture, ask your instructor about using a tape recorder. To ensure success, the tape recorder must have a tape counter and must be voice activated.

The tape counter displays a number indicating the amount of tape to which you have listened. When you find you are in an area of confusing information, write the beginning and ending tape counter number in the left margin of your notes. When reviewing your notes, the tape count number will be a reference point for obtaining information to work the problem. You can also reduce the time it takes to listen to the tape by using the pause button to stop the recording of unnecessary material.

Ask Questions

To obtain the most from a lecture, you must ask questions in class. By asking questions, you improve your understanding of the material and decrease your homework time. By *not* asking questions, you create for yourself unnecessary confusion during the remainder of the class period. Also, it is much easier to ask questions in class about potential homework problems than it is to spend hours trying to figure out the problems on your own at a later time.

If you are shy about asking questions in class, write down the questions and read them to your instructor. If the instructor seems confused about the questions, tell him or her that you will discuss the problem after class. To encourage yourself to ask questions, remember

- you have paid for the instructor's help,
- five other students probably have the same question,
- the instructor needs feedback on his or her teaching to help the class learn the material, and
- there is no such thing as a "stupid" question.

Record Each Problem Step

The final suggestion in note-taking is to record each step of every problem written or verbally explained. By recording each problem step, you begin *overlearning* how to work the problems. This will increase your problem-solving speed during future tests. If you get stuck on the homework, you will also have complete examples to review.

The major reason for recording every step of a problem is to understand how to do the problems while the instructor is explaining them instead of trying to remember unwritten steps. Although it may seem time consuming, it pays off during homework and test time.

Section Review

1. List and describe the six steps to reworking your notes.

First Step: _____

Second Step: _____

Third Step: _____

Fourth Step: _____

Fifth Step: _____

Sixth Step: _____

2. How can using a tape recorder in class improve your learning?

3. List three reasons for asking questions in class.

First Reason: _____

Second Reason: _____

Third Reason: _____

4. What is the major reason for recording each problem step?

Chapter 4 Review

1. Sitting in the golden triangle of success can help you pay more _____ during class.

2. Class time is an intense _____ _____ that should not be wasted.

3. Auditory learners should review the math videotape and _____ back the important information.

4. The goal of note-taking is to take the _____ amount of notes and get the _____ amount of information.

5. Using abbreviations can help you write _____ and listen _____.

6. Your _____ should include any words you don't know and their explanations.

7. Rework your notes to transfer information from _____ - _____ memory to _____ - _____ memory.

8. A _____ _____ is a separate section in your notebook that contains problems that your teacher worked in class.

9. Asking _____ in class will improve your understanding and decrease your homework time.

10. _____ how to work problems will increase your problem-solving speed and memory.

What is the most important information you learned from this chapter?

How can you immediately use it?

Lecture Notes

5

How to Improve Your
Reading, Homework, and Study Techniques

Reading a math textbook is more difficult than reading other textbooks. Math textbooks are written differently than your English or social science textbooks. Math textbooks contain condensed material and therefore take longer to read.

Mathematicians can reduce a page of writing to one paragraph using math formulas and symbols. To make sure you understood that same information, an English instructor would take that original page of writing and expand it into two pages. Mathematicians pride themselves on how little they can write and still cover the concept. This is one reason it may take you two to three times as long to read your math text as it would any other text.

REMEMBER

Reading your math text will take longer than reading your other texts.

Math students are expected to know how to do their homework; however, most math students do not have a homework system. Most students begin their homework by going directly to the problems and trying to work them. When they get stuck, they usually quit. This is not a good homework system.

A good homework system will improve your homework success and math learning at the same time.

In this chapter, you will learn

- how to read a math textbook,
- how to do your homework,
- how to solve word problems,
- how to work with a study buddy.

How to read
a math textbook

The way you read a math textbook is different from the traditional way students are taught to read textbooks in high school or college. Students are taught to read quickly or skim the material. If you do not understand a word, you are supposed to keep on reading. Instructors of other courses want students to continue to read so they can pick up the unknown words and their meanings from context.

This reading technique may work with your other classes, but using it in your math course will be

totally confusing. By skipping some major concept words or bold-face words, you will not understand the math textbook or be able to do the homework. Reading a math textbook takes more time and concentration than reading your other textbooks.

If you have a reading problem, it would be wise to take a developmental reading course before taking math. This is especially true with reform math courses, where reading and writing are emphasized.

Reform math classes deal more with word problems than traditional math courses do. If you cannot take the developmental reading course before taking math, then take it during the same semester as the math course.

Ten Steps to Understanding Reading Materials

There are several appropriate steps in reading a math textbook:

Step 1 *Skim the assigned reading material.* Skim the material to get the general idea about the major topics. Read the chapter introduction and each section summary. You do not want to learn the material at this time; you simply want to get an overview of the assignment. Then think about similar math topics that you already know.

> **Example:** Skimming will allow you to see if problems presented in one chapter section are further explained in later chapter sections.

Step 2 *As you skim the chapter, circle (using*

pencil) the new words that you do not understand. If you do not understand these new words after reading the assignment, then ask the instructor for help. Skimming the reading assignments should take only five to ten minutes.

Step 3 *Put all your concentration into reading.* While reading the textbook, highlight the material that is important to you. However, do not highlight more than 50 percent of a page because the material is not being narrowed down enough for future study. Especially highlight the material that is also discussed in the lecture. Material discussed in both the textbook and lecture usually appears on the test. The purpose of highlighting is to emphasize the important material for future study. Do not skip reading assignments.

REMEMBER

> *Reading a math textbook is very difficult. It might take you half an hour to read and understand just one page.*

Step 4 *When you get to the examples, go through each step.* If the example skips any steps, make sure you write down each one of those skipped steps in the textbook for better understanding. Later on, when you go back and review, the steps are already filled in. You will understand how each step was completed. Also, by filling in the extra steps, you are starting to overlearn the material for better recall on future tests.

Step 5 *Mark the concepts and words that you do not know.* Maybe you marked them the first time while skimming. If you

understand them now, erase the marks. If you do not understand the words or concepts, then reread the page or look them up in the glossary. Try not to read any further until you understand all the words and concepts.

Step 6 *If you do not clearly understand some words or concepts, add these words to the note-taking glossary in the back of your notebook.* Your glossary will contain the bold-face words that you do not understand. If you have difficulty understanding the bold-face words, ask the instructor for a better explanation. You should know all the words and concepts in your notebook's glossary before taking the test.

Step 7 *If you do not understand the material, follow these eight points, one after the other, until you do understand the material:*

Point 1 — Go back to the previous page and reread the information to maintain a train of thought.

Point 2 — Read ahead to the next page to discover if any additional information better explains the misunderstood material.

Point 3 — Locate and review any diagrams, examples, or rules that explain the misunderstood material.

Point 4 — Read the misunderstood paragraph(s) several times aloud to better understand their meaning.

Point 5 — Refer to your math notes for a better explanation of the misunderstood material.

Point 6 — Refer to another math textbook, computer software program, or videotape that expands the explanation of the misunderstood material.

Point 7 — Define exactly what you do not understand and call your study buddy for help.

Point 8 — Contact your math tutor or math instructor for help in understanding the material.

Step 8 *Recall important material.* Try to recall some of the important concepts and rules. If you have difficulty recalling this information, look at the highlighted areas of the text, bold print words or formulas. If you are a visual learner, write down these words several times on a separate sheet of paper. If you are an auditory learner, repeat these word several times out loud.

If you are a kinesthetic (hands on) learner, write out the formulas twice on two separate cards. Cut one of them up into puzzle pieces. Mix up the parts of the puzzle. Turn over the other formula card so you cannot see it. Put the puzzle back together. Do this several times.

For bold print words, write the bold print word on one card and its explanation twice on two other separate cards. Make another puzzle. Mix up the pieces of the explanation and put them back together.

Then add these words and formulas to your note-taking glossary. Recalling important material right after you read it improves you ability to remember that information.

Step 9 *Reflect on what you have read.* Combine what you already know with the new information that you just read. Think about how this new information enhances your math knowledge. Prepare questions for your instructor on the confusing information. Ask those questions at the next class meeting.

Step 10 *Write anticipated test questions.* Research has noted that students have about 80 percent accuracy in predicting test questions. Think about what is the most important concept you just read and what problems the instructor could give you that would test the knowledge of that concept. Make up three to five problems and add them to your problem log (this was part of your note-taking system). Indicate in the log that these are your questions, not the instructors.

By using this reading technique, you have narrowed down the important material to be learned. You have skimmed the textbook to get an overview of the assignment. You have carefully read the material and highlighted the important parts. You then added to your note-taking glossary unknown words or concepts.

REMEMBER

The highlighted material should be reviewed before doing the homework problems, and the glossary has to be learned 100 percent before taking the test.

How Reading Ahead Can Help

Reading ahead is another way to improve learning. If you read ahead, do not expect to understand everything. Read ahead two or three sections and put question marks (in pencil) by the material you do not understand.

When the instructor starts discussing that material, have your questions prepared and take good notes. Also, if the lecture is about to end, ask the instructor to explain the confusing material in the textbook. Reading ahead will take more time and effort, but it will better prepare you for the lectures.

How to Establish Study Period Goals

Before beginning your homework, establish goals for the study period. Do not just do the homework problems.

Ask yourself this question: "What am I going to do tonight to become more successful in math?"

By setting up short-term homework goals and reaching them, you will feel more confident about math. This also improves your self-esteem and helps you become a more internally motivated student. Set up homework tasks that you can complete. Be realistic.

Study period goals are set up either on a time-line basis or an item-line basis. Studying on a time-line basis is studying math for a certain amount of time.

Example: You may want to study math for an hour, then switch to another subject. You will study on a time-line basis.

Studying on an item-line basis means you will study your math until you have completed a certain number of homework problems.

Example: You might set a goal to study math until you have completed all the odd-numbered problems in the chapter review. The odd-numbered problems are the most important problems to work. These, in most texts, are answered in the answer section in the back of the book. Such problems provide the opportunity to recheck your work if you do not get the answer correct. Once you have completed these problems, do the even-numbered problems.

No matter what homework system you use, remember this important rule: Always finish a homework session by understanding a concept or doing a homework problem correctly.

Do not end a homework session with a problem you cannot complete. You will lose confidence, since all you will think about is the last problem you could not solve instead of the 50 problems you correctly solved. If you did quit on a problem you could not solve, return and rework problems you have done correctly.

REMEMBER

Do not end your study period with a problem you could not complete.

Section Review

1. List and describe the ten steps to reading a math textbook.

First Step: _____

Second Step: _____

Third Step: _____

Fourth Step: _____

Fifth Step: _____

Sixth Step: _____

Seventh Step: _____

Eighth Step: _____

Ninth Step: _____

Tenth Step: _____

2. After trying each of the ten steps to reading your math text, make up, list, and try your own condensed version.

First Step: _____

Second Step: _____

Third Step: _____

Fourth Step: _____

Fifth Step: _____

Sixth Step: _____

Seventh Step: _____

Eighth Step: _____

Ninth Step: _____

Tenth Step: _____

3. List two reasons why reading ahead can improve your learning.

First Reason: _____

Second Reason: _____

4. Describe the two types of study period goals. Which one do you use?

First Goal: _____

Second Goal: _____

How to do your homework

Doing your homework can be frustrating or rewarding. Most students jump right into their homework, become frustrated, and stop studying. These students usually go directly to the math problems and start working them without any preparation. When they get stuck on one problem, they flip to the back of the text for the answer. Then they either try to work the problem backward to understand the problem steps or they just copy down the answer.

Other students go to the solution guide and just copy the steps. After getting stuck several times, these students will inevitably quit doing their homework assignment. Their homework becomes a frustrating experience, and they may even quit doing their math homework altogether.

10 Steps to Doing Your Homework

To improve your homework success and learning, refer to these ten steps:

Step 1 *Review the textbook material that relates to the homework.* A proper review will increase the chances of successfully completing your homework. If you get stuck on a problem, you will have a better chance of remembering the location of similar problems. If you do not review prior to doing your homework, you could get stuck and not know where to find help in the textbook.

REMEMBER

To be successful in learning the material and in completing homework assignments, you must first review your textbook.

Step 2 *Review your lecture notes that relate to the homework.* If you could not understand

the explanation in the textbook on how to complete the homework assignment, then review your notes.

REMEMBER

Reviewing your notes will give you a better idea about how to complete your homework assignment.

Step 3 *Do your homework as neatly as possible.* Doing your homework neatly has several benefits. When approaching your instructor about problems with your homework, he or she will be able to understand your previous attempts to solve the problem. The instructor will easily locate the mistakes and show you how to correct the steps without having to decipher your handwriting. Another benefit is that when you review for midterm or final exams, you can quickly relearn the homework material without having to decipher your own writing.

REMEMBER

Neatly prepared homework can help you now and in the future.

Step 4 *When doing your homework, write down every step of the problem.* Even if you can do the step in your head, write it down anyway. This will increase the amount of homework time, but you are overlearning how to solve problems, which improves your memory. Doing every step is an easy way to memorize and understand the material. Another

advantage is that when you rework the problems you did wrong, it is easy to review each step to find the mistake.

REMEMBER

In the long run, doing every step of the homework will save you time and frustration.

Step 5 *Understand the reasons for each problem step and check your answers.* Do not get into the bad habit of memorizing how to do problems without knowing the reasons for each step. Many students are smart enough to memorize procedures required to complete a set of homework problems. However, when similar homework problems are presented on a test, the student cannot solve the problems. To avoid this dilemma, keep reminding yourself about the rules, laws, or properties used to solve problems.

Example: *Problem:* $2(a + 5) = 0$. What property allows you to change the equation to $2a + 10 = 0$? *Answer:* The distributive property.

Once you know the correct reason for going from one step to another in solving a math problem, you can solve any problem requiring that property. Students who simply memorize how to do problems instead of understanding the reasons for correctly working the steps will eventually fail their math course.

How to Check Your Answers

Checking your homework answers should be part of your homework duties. Checking your answers can improve your learning and help you prepare for tests.

Check the answers of the problems for which solutions are not given. This may be the even-numbered or odd-numbered problems or the problems not answered in the solutions manual.

First, check your answer by estimating the correct answer.

> **Example:** If you are multiplying 2.234 by 5.102, remember that 2 times 5 is 10. The answer should be a little over 10.

You can also check your answers by substituting the answer back into the equation or doing the opposite function required to answer the question. The more answers you check, the faster you will become. This is very important because increasing your answer-checking speed can help you catch more careless errors on future tests.

Step 6 *If you do not understand how to do a problem, refer to the following points:*

Point 1 — Review the textbook material that relates to the problem.

Point 2 — Review the lecture notes that relate to the problem.

Point 3 — Review any similar problems, diagrams, examples, or rules that explain the misunderstood material.

Point 4 — Refer to another math textbook, solutions guide, math computer program software, or videotape to obtain a better understanding of the material.

Point 5 — Call your study buddy.

Point 6 — Skip the problem and contact your tutor or math instructor as soon as possible for help.

Step 7 *Always finish your homework by successfully completing problems.* Even if you get stuck, go back and successfully complete previous problems before quitting. You want to end your homework assignment with feelings of success.

Step 8 *After finishing your homework assignment, recall to yourself or write down the most important learned concepts.* Recalling this information will increase your ability to learn these new concepts. Additional information about step 8 will be presented later in this chapter.

Step 9 *Make up note cards containing hard-to-remember problems or concepts.* Note cards are an excellent way to review material for a test. More information on the use of note cards as learning tools is presented later in this chapter.

Step 10 *Do not fall behind.* As mentioned in Chapter 1, math is a sequential learning process. If you get behind, it is difficult to catch up because each topic builds on the next. It would be like going to Spanish class without learning the last set of vocabulary words. The teacher would be talking to you using the new vocabulary,

but you would not understand what was being said.

Do Not Fall Behind

To keep up with your homework, it is necessary to complete the homework every school day and even on weekends. Doing your homework one-half hour each day for two days in a row is better than one hour every other day.

If you have to get behind in one of your courses, *make sure it is not math.* Fall behind in a course that does not have a sequential learning process, such as psychology or history. After using the ten steps to doing your homework, you may be able to combine two steps into one. Find your best combination of homework steps and use them.

REMEMBER

Getting behind in math homework is the fastest way to fail the course.

Section Review

1. List the ten steps to doing your math homework.

First Step: _____

Second Step: _____

Third Step: _____

Fourth Step: _____

Fifth Step: _____

Sixth Step: _____

Seventh Step: _____

Eighth Step: _____

Ninth Step: _____

Tenth Step: _____

2. After trying the ten steps to doing your math homework, make up, list, and try your own condensed version.

First Step: _____

Second Step: _____

Third Step: _____

Fourth Step: _____

Fifth Step: _____

Sixth Step: _____

Seventh Step: _____

Eighth Step: _____

Ninth Step: _____

Tenth Step: _____

3. How will reviewing your notes or math text help you complete your homework more successfully?

4. List two ways you can check your homework answers.

First Way: _____

Second Way: _____

5. What happens to students who fall behind in their homework?

How to solve word problems

The most difficult homework assignment for most math students is working story/word problems. Solving word problems requires excellent reading comprehension and translating skills.

Students often have difficulty substituting algebraic symbols and equations for English terms. But once an equation is written, it is usually easily solved. To help you solve word problems, follow these ten steps:

Step 1 *Read the problem three times.* Read the problem quickly the first time as a scanning procedure. As you are reading the problem the second time, answer these three questions:

1. *What is the problem asking me?* (This is usually at the end of the problem.)
2. *What is the problem telling me that is useful?* (Cross out unneeded information.)
3. *What is the problem implying?* (This is usually something you have been told to remember.) Read the problem a third time

to check that you fully understand its meaning.

Step 2 *Draw a simple picture of the problem to make it more real to you* (e.g., a circle with an arrow can represent travel in any form — by train, by boat, by plane, by car, or by foot).

Step 3 *Make a table of information and leave a blank space for the information you are not told.*

Step 4 *Use as few unknowns in your table as possible.* If you can represent all the unknown information in terms of a single letter, do so! When using more than one unknown, use a letter that reminds you of that unknown. Then write down what your unknowns represent. This eliminates the problem of assigning the right answer to the wrong unknown. Remember that you have to create as many separate equations as you have unknowns.

Step 5 *Translate the English terms into an algebraic equation using the list of terms in Figures 8 and 9 on the next page.* Remember, the English terms are sometimes stated in a different order than the algebraic terms.

Step 6 *Immediately retranslate the equation, as you now have it, back into English.* The translation will not sound like a normal English phrase, but the meaning should be the same as the original problem. If the meaning is not the same, the equation is incorrect and needs to be rewritten. Rewrite the equation until it means the same as the English phrase.

Step 7 *Review the equation to see if it is similar to equations from your homework and if it makes sense.* Some formulas dealing with specific word problems may need to

Figure 8

Translating English Terms
into Algebraic Symbols

Sum	+
Add	+
In addition	+
More than	+
Increased	+
In excess	+
Greater	+
Decreased by	-
Less than	-
Subtract	-
Difference	-
Diminished	-
Reduce	-
Remainder	-
Times as much	x
Percent of	x
Product	x
Interest on	x
Per	/
Divide	/
Quotient	/
Quantity	()
Is	=
Was	=
Equal	=
Will be	=
Results	=
Greater than	>
Greater than or equal to	≥
Less than	<
Less than or equal to	≤

Figure 9

Translating English Words
into Algebraic Expressions

English Words	Algebraic Expressions
Ten more than x	$x + 10$
A number added to 5	$5 + x$
A number increased by 13	$x + 13$
5 less than 10	$10 - 5$
A number decreased by 7	$x - 7$
Difference between x and 3	$x - 3$
Difference between 3 and x	$3 - x$
Twice a number	$2x$
Ten percent of x	$0.10x$
Ten times x	$10x$
Quotient of x and 3	$x/3$
Quotient of 3 and x	$3/x$
Five is three more than a number	$5 = x + 3$
The product of 2 and a number is 10	$2x = 10$
One-half a number is 10	$x/2 = 10$
Five times the sum of x and 2	$5(x + 2)$
Seven is greater than x	$7 > x$
Five times the difference of a number and 4	$5(x - 4)$
Ten subtracted from 10 times a number is that number plus 5	$10x - 10 = x + 5$
The sum of $5x$ and 10 is equal to the product of x and 15	$5x + 10 = 15x$
The sum of two consecutive integers	$(x) + (x + 1)$
The sum of two consecutive even integers	$(x) + (x + 2)$
The sum of two consecutive odd integers	$(x) + (x + 2)$

be rewritten. Distance problems, for example, may need to be written solving for each of the other variables in the formula. Distance = rate x time; therefore, time = distance/rate, and rate = distance/time. Usually, a distance problem will identify the specific variable to be solved.

Step 8 *Solve the equation using the rules of algebra.*

REMEMBER

> *Whatever is done to one side of the equation must be done to the other side of the equation. The unknown must end up on one side of the equation, by itself. If you have more than one unknown, then use the substitution or elimination method to solve the equations.*

Step 9 *Look at your answer to see if it makes sense.*

> **Example:** If tax was added to an item, it should cost more; if a discount was applied to an item, it should cost less. Is there more than one answer? Does your answer match the original question? Does your answer have the correct units?

Step 10 *Put your answer back into the original equation to see if it is correct.* If one side of the equation equals the other side of the equation, then you have the correct answer. If you do not have the correct answer, go back to step 5.

The most difficult part of solving word problems is translating part of a sentence into algebraic symbols and then into algebraic expressions. Review Figure 8 (Translating English Terms into Algebraic Symbols) and Figure 9 (Translating English Words into Algebraic Expressions), both on the previous page.

How to Recall What You Have Learned

After completing your homework problems, a good visual learning technique is to make note cards. Note cards are 3" by 5" index cards on which you place information that is difficult to learn or material you think will be on the test.

On the front of the note cards write a math problem or information that you need to know. Color-code the important information in red or blue. On the back of the note card, write how to work the problem or explain important information.

> **Example:** If you are having difficulty remembering the rules for multiplying positive and negative numbers, you would write some examples on the front of the note card with the answers on the back.

Make note cards on important information you might forget. Every time you have five spare minutes, pull out your note cards and review them. You can glance at the front of the card, repeat to yourself the answer, and check yourself with the back of the card. If you are correct and know the information on a card, do not put it back in the deck. Mix up the cards you do not know and pick another card to test yourself on the information. Keep doing

this until there are no cards left that you do not know.

If you are an auditory learner, then use the tape recorder like the note cards. Record the important information as you would on the front of the note card. Then leave a blank space on the recording before recording the answer. Play the tape back. When you hear the silence, put the tape on pause. Then say the answer out loud to yourself. Take the tape player off pause and see if you were correct.

Review What You Have Learned

After finishing your homework, close the textbook and try to remember what you have learned. Ask yourself these questions: "What major concepts did I learn tonight?" and "What test questions might the instructor ask on this material?"

Recall for about three to four minutes the major points of the assignment, especially the areas you had difficulty understanding. Write down questions for the instructor or tutor. Since most forgetting occurs right after learning the material, this short review will help you retain the new material.

Section Review

1. List the ten steps to solving word problems.

 First Step: _____

 Second Step: _____

 Third Step: _____

 Fourth Step: _____

 Fifth Step: _____

 Sixth Step: _____

 Seventh Step: _____

 Eighth Step: _____

 Ninth Step: _____

 Tenth Step: _____

2. What is your best way to recall what you have learned?

3. How does reviewing right after doing your homework help you remember?

problems and to discuss math. If you miss class, get the notes from your study buddy so you will not get behind.

Call your study buddy when you get stuck on your homework. You can solve math problems over the phone. Do not sit for half an hour or an hour trying to work one problem; that will destroy your confidence, waste valuable time, and possibly alienate your study buddy. Think how much you could have learned by trying the problem for 15 minutes and then calling your study buddy for help. Spend, at the maximum, 15 minutes on one problem before going on to the next problem or calling your study buddy.

REMEMBER

A study buddy can improve your learning while helping you complete the homework assignment. Just do not overuse your study buddy or expect that person to do your homework for you.

How to work with a study buddy

You need to have a study buddy when you miss class or when you do your homework. A study buddy is a friend or classmate who is taking the same course. You can find a study buddy by talking to your classmates or making friends in the math lab.

Try to find a study buddy who knows more about math than you do. Tell the class instructor that you are trying to find a study buddy and ask which students make the best grades. Meet with your study buddy several times a week to work on

Section Review

1. How can you select your study buddy?

2. How can a study buddy help you learn math?

3. Who is your study buddy?

The benefits of study breaks

Psychologists have discovered that learning decreases if you do not take study breaks. Therefore, take a break after studying math for 45 minutes to one hour.

If you have studied for only 15 or 20 minutes and feel you are not retaining the information or your mind is wandering, take a break. If you continue to force yourself to study, you will not learn the material. After taking a break, return to studying.

If you still cannot study after taking a break, review your purpose for studying and your educational goals. Think about what is required to graduate. It will probably come down to the fact that you will have to pass math. Think about how studying math today will help you pass the next test; this will increase your chances of passing the course and of graduating.

Write on an index card three positive statements about yourself and three positive statements about studying. Look at this index card every time you have a study problem. Use every opportunity available to reinforce your study habits.

Section Review

1. How can study breaks improve your learning?

2. What do you do when you don't want to study? How do you get started?

Chapter 5 Review

1. Reading one page of a math text might take you _____ _____ _____.

2. After reading your textbooks, _____ important material can improve your ability to remember that information.

3. Write anticipated tests question in your _____ _____ and indicate that they are your questions.

4. Checking your homework answers can improve your _____ and help prepare you for a _____.

5. Matching:

(for students in pre-algebra class)

_____ Product	A. +
_____ Increased	B. -
_____ Sum	C. x
_____ Per	D. /
_____ Product	E. =
_____ Is	
_____ Reduced	
_____ Remainder	

(for students taking elementary algebra or higher-level math course)

_____ Difference between x and 3	A. $10x$
_____ Quotient of x and 3	B. $x/3$
_____ Five times the sum of x and 2	C. $5(x-4)$
	D. $5(x+2)$
	E. $5 = x+3$
_____ A number added to 5	F. $5+x$

_____ Five less than 10	G. $10-5$
_____ Ten times x	H. $x-3$
_____ Five is three more than a number	
_____ Five times the difference of a number and four	

6. Make sure you finish your homework by working problems you _____ do.

7. Using _____ _____ is a visual way to recall what you learned in math.

8. Using a_____ _____ is an auditory way to recall what you learned in math.

9. Most_____ occurs right after learning; a short review will help you _____ the new material.

10. A _____ _____ can improve your learning while helping you complete the homework assignment.

What is the most important information you learned from this chapter?

How can you immediately use it?

6

How to Remember What You Have Learned

To understand the learning process, you must understand how your memory works. You learn by conditioning and thinking. But memorization is different from learning. For memorization, your brain must perform several tasks, including receiving the information, storing the information, and recalling the information.

By understanding how your memory works, you will be better able to learn at which point your memory is failing you. Most students usually experience memory trouble between the time their brain receives the information and the time the information is stored.

There are many techniques for learning information that can help you receive and store information without losing it in the process. Some of these techniques may be more successful than others, depending on how you learn best.

In this chapter, you will discover

- how you learn,
- how short-term memory affects what you remember,
- how working memory affects what you remember,
- how long-term memory/reasoning affects what you remember,
- how to use learning styles to improve memory,
- how to use memory techniques,

- how to develop practice tests, and
- how to use number sense.

How you learn

Educators tell us that learning is the process of "achieving competency." More simply put, it is how you become good at something. The three ways of learning are by conditioning, by thinking, and by a combination of conditioning and thinking.

Learning by Conditioning and Thinking

Conditioning is learning things with a maximum of physical and emotional reaction and a minimum of thinking.

Examples: Repeating the word *pi* to yourself and practicing where the

continued

symbol is found on a calculator are two forms of conditioned learning. You are learning using your voice and your eye-hand coordination (physical activities), and you are doing very little thinking.

Thinking is defined as learning with a maximum of thought and a minimum of emotional and physical reaction.

Example: Learning about *pi* by thinking is different from learning about it by conditioning. To learn "pi" by thinking, you would have to do the calculations necessary to result in the numeric value that the word *pi* represents. You are learning *using your mind* (thought activities), and you are using very little emotional or physical energy to learn *pi* in this way.

The most successful way to learn is to combine thinking and conditioning. The best learning combination is to learn by thinking first and conditioning second.

Learning by thinking means that you learn by

- *observing,*
- *processing,* and
- *understanding* the information.

How Your Memory Works

Memory is different from learning; it requires reception, storage, and retrieval of information. You

receive information through your five senses (*sensory input*): what you see, feel, hear, smell, and taste.

Examples: In math class, you will use your sense of *vision* both to watch the instructor demonstrate problems on the chalkboard and to read printed materials. You will use your sense of *hearing* to listen to the instructor and other students discuss the problems. Your sense of *touch* will be used to operate your calculator and to appreciate geometric shapes. In chemistry and other classes, however, you may additionally use your senses of *smell* and *taste* to identify substances.

The sensory register briefly holds an exact image or sound of each sensory experience until it can be processed. If the information is not processed immediately, it is forgotten. The sensory register helps us go from one situation to the next without cluttering up our minds with trivial information.

Processing the information involves placing it in either short-term memory, working memory, or long-term memory.

How short-term memory affects what you remember

Information that passes through the sensory register is stored in short-term memory. Remembering something for a short time is not hard to do for most students. By conscious effort, you can remember

math laws, facts, and formulas received by the sensory register (your five senses) and put them in short-term memory. You can recognize them and register them in your mind as something to remember for a short time.

> **Example:** When you are studying math, you can tell yourself the distributive property is illustrated by $a(b + c) = ab + ac$. By deliberately telling yourself to remember that fact (by using conditioning — repeating or writing it again and again), you can remember it, at least for a while, because you have put it in short-term memory.

Psychologists have found that short-term memory cannot hold an unlimited amount of information. You may be able to use short-term memory to remember one phone number or a few formulas but not five phone numbers or ten formulas.

Items placed into short-term memory usually fade fast, as the name suggests.

> **Examples:** Looking up a telephone number in the directory, remembering it long enough to dial, then forgetting it immediately; learning the name of a person at a large party or in a class but forgetting it completely within a few seconds; cramming for a test and forgetting most of it before taking the test.

Short-term memory involves the ability to recall information immediately after it is given (without any interruptions). It is useful in helping you to concentrate on a few concepts at a time, but it is not the best way to learn. This is especially true of students who have attention problems or problems with short-term memory.

Students with short-term memory problems forget a math step as soon as the instructor explains it. To remember more facts or ideas and keep them in memory a longer period of time — especially at test time — use of a better system than short-term memory is required.

Changes in the way math is taught include the idea that conditioning (the most often used teaching form) leads only to short-term memory. Conditioning is often used for short-term recall of concept skills. The information learned in this way can only be stored for longer periods of time by thinking. Thinking takes place when applying the skills to working the problems.

How working memory affects what you remember

Working memory (or long-term retrieval) is that process in the brain that works on problems for a longer period of time than short-term memory. Working memory, then, offers an increase in the amount of *time* information is held in memory. (An increase in the *volume* of information that can be held requires long-term memory.)

Working memory is like the amount of RAM in a computer. Working memory uses the information (such as multiplication tables) recalled from long-term memory, along with new information to learn new concepts. It is the ability to think about and use many pieces of information at the same time. For instance, when you solve a linear equation you must use all

the math that you learned in elementary school like addition, subtraction, and multiplication. You then have to add this to the new rules for linear equations.

Working memory can be compared to a mental workspace or an internal chalkboard. Just like a chalkboard, working memory has limited space, which can cause a "bottleneck" in learning. It involves the ability to recall information after learning has been consistently interrupted over a period of several minutes. Students with working memory problems may listen to a math lecture and understand each step as it is explained. When the instructor goes back to a previous step discussed several minutes prior, however, the student has difficulty explaining or remembering the reasons for the steps. These students have difficulty remembering series of steps long enough to understand the concept.

Working memory can go both ways in the memory process. First, it can go into long-term memory and abstract reasoning. Second, working memory can bring information out from long-term memory and abstract memory to use in learning new concepts. When learning a mathematical concept, working memory goes into abstract memory. When learning just facts such as the multiplication tables or definition of words, information goes directly into long-term memory. When solving math problems, in most cases, information is brought in from abstract reasoning and long-term memory into working memory. Students use their working memory to do their homework and when answering test questions. The amount of space in working memory is critical to answering test questions just like the amount of RAM is critical to running computer programs.

Example: In solving 26 x 32, you would put the intermediate products

continued

52 (*from 2 x 26*) and 780 (*from 30 x 26* — remember, 3 is in the 10s place, so make it 30) into working memory and add them together. The more automatic the multiplication, the less working memory you use. If you cannot remember your multiplication, you use up working memory trying to solve the multiplication problem. This leaves you with less working memory to solve the resulting addition problem.

How long-term memory/reasoning affects what you remember

Long-term memory is a storehouse of material that is retained for long periods of time. Working memory places information in long-term memory and long-term memory is recalled into working memory to solve problems. It is *not* a matter of trying harder and harder to remember more and more unrelated facts or ideas; it is a matter of organizing your short-term memories and working memories into meaningful information.

In most cases long-term memory is immeasurable. Students have so much room in their long-term memory that no one has measured its total capacity. Long-term memory also relates more to language skills rather than abstract skills. Students with good long-term memory and poor abstract skills can sometimes do well in every subject except math and the physical sciences. These students can use

their long-term memory language skills by learning the vocabulary. By understanding the language of mathematics they can put into words how to solve math problems and recall these words during the test instead of depending mainly on their abstract memory. This information must be reviewed many times for it to get into long-term memory. This concept will be explored more in later chapters.

Reasoning or abstract memory is thinking about memories, comprehending their meanings and understanding their concepts. Abstract reasoning involves learning how the rules and laws apply to solving math problems. Students who understand the principals of a certain mathematical concept can use these principles to solve any similar problem. That is why the focus in mathematics is learning how to solve the problem based on concepts instead of just memorizing how to do a problem. In most cases you cannot memorize enough problems and hope they appear on the test. Without understanding the concept, the information cannot be transferred into abstract reasoning. Many students with excellent abstract reasoning major in math related careers.

The main problem most students face is converting information from working memory to long-term and/or abstract reasoning – and understanding it. To place information into long-term memory students must understand math vocabulary and practice problems. To place information into abstract reasoning students must understand the concept and remember it. In most cases student must use part of the long-term memory and abstract reasoning to solve math problems. It will depend on the skills in each area as to which one they use the best to solve math problems.

REMEMBER

Securing math information into long-term memory is not accomplished by just doing the homework — you must also understand it.

The Role of "Memory Output" in Testing

Memory output is what educators call a "retrieving process." It is necessary for verbal or written examinations. The retrieving process is used when answering questions in class, doing math homework or doing math tests. It is the method by which you recall information stored into long-term memory and by using abstract reasoning are able to place it into words or on paper. This retrieval process can come directly from long-term memory. For example, "What are whole numbers?" This is a fact question that comes from long-term memory. The retrieval process can also come from abstract reasoning. For example, "Write down the distributive property and substitute numbers to prove that it works." However, most math problems are solved through working memory by using information from both long-term memory and abstract reasoning. For example, solve: $3y - 10 = 9y + 21$. For this problem you are pulling in your number facts from long-term memory and the rules for solving equations from abstract reasoning.

Memory output can be blocked by three things:

1. insufficient processing of information into long-term memory,
2. test anxiety, or
3. poor test-taking skills.

If you did not completely place all of the information you learned into long-term memory, you may not be able to give complete answers on tests.

Test anxiety can decrease your ability to recall important information or it can cause you to block out information totally. Ways to decrease test anxiety were discussed in Chapter 3. Students who work on their test-taking skills can improve their memory output. Techniques to improve both your memory output and your test-taking skills will be explained in Chapter 7.

Understanding the Stages of Memory

Understanding the stages of memory will help you answer this common question about learning math: "How can I understand the procedures to solve a math problem one day and forget how to solve a similar problem two days later?"

There are three good answers to this question. First, after initially learning how to solve the problem, you did not *rehearse* the solving process enough for it to enter your long-term memory. Second, you did get the information into long-term memory, but the information was not *reviewed* frequently enough and was therefore forgotten. Third, you memorized how to work the problem but did not *understand* the concept.

REMEMBER

Locating where your memory breaks down and compensating for those weaknesses will improve your math learning.

Section Review

1. Label the boxes in Figure 10 on the facing page with stages of memory.

2. List, define and give examples for each stage of memory by explaining and giving examples of the following terms:

Stage One: _____

Example: _____

Stage Two: _____

Example: _____

Stage Three: _____

continued on page 102

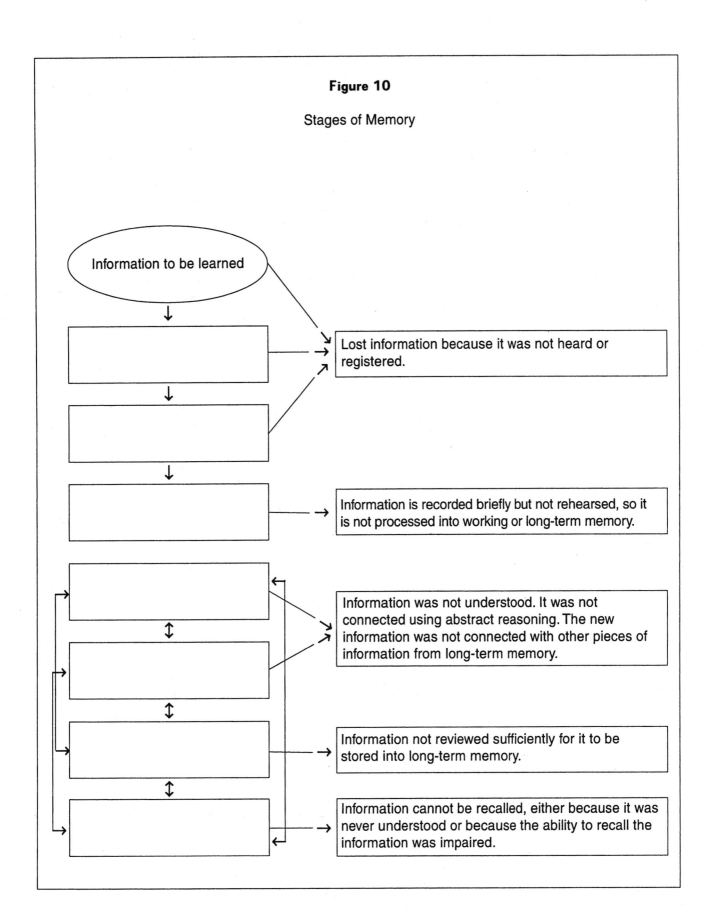

Figure 10

Stages of Memory

Example: _____

Stage Four: _____

Example: _____

Stage Five: _____

Example: _____

Stage Six: _____

Example: _____

Stage Seven: _____

Example: _____

3. What are three conditions that can block your memory output?

 First Condition: _____

 Second Condition: _____

 Third Condition: _____

4. Review the stages of memory and list the stages at which *your* memory breaks down. For each stage of breakdown, say how you can prevent it.

 Stage: _____

 Prevention: _____

Stage: _____

Prevention: _____

Stage: _____

Prevention: _____

5. What is the stage of memory you do well? Why?

Stage: _____

I do well because _____

How to use learning styles to improve memory

There are many different techniques, which can help you store information in your long-term memory and reasoning. Using your learning sense or learning style and decreasing distraction while studying are very efficient ways to learn. Using your best *learning sense* (what educators call your "predominate learning modality") can improve how

well you learn and enhance the transfer of knowledge into long-term memory/reasoning. Your learning senses are vision, hearing, touching, etc. Ask yourself if you learn best by watching (vision), listening (hearing), or touching (feeling).

Another helpful tool is the Learning Styles Inventory. It is the only inventory that has separate math learning style information. You can go to www.AcademicSuccess.com to learn more about the Learning Styles Inventory. Based on your preferred learning style practice those learning suggestions first.

REMEMBER

Learning styles are neither good nor bad and are based on genetics and environment. Knowing your best learning styles and using them effectively can dramatically improve your math learning and grades.

Visual (watching) Learner

Knowing that you are a visual math learner can help you select the memory technique that will work best for you. Repeatedly reading and writing down math materials being studied is the best way for a visual learner to study.

Based on the Learning Styles Inventory, students who learn math best by seeing it written are Visual Numerical Learners. If you are a visual numerical learner you will learn best by following the 10 suggestions in Figure 11, on the next page. Try as many of these suggestions as possible and then select and practice those that are most helpful.

A visual way to decrease distractions is by using the "my mind is full" concept. Imagine that your mind is completely filled with thoughts of

Figure 11
Visual Numerical Learners

These students learn math best by seeing it written. If you are a visual numerical learner, you may learn best by following these suggestions:

1. Studying a variety of written materials, such as additional handouts and math texts.

2. Playing games with and being involved in activities with visible printed number problems.

3. Using visually orientated computer programs, CD-ROMs, and math websites.

4. Reworking your notes using suggestions from this workbook.

5. Visualizing numbers and formulas, in detail.

6. Checking out videocassette tapes from the math lab or LRC.

7. Making 3x5 note (flash) cards, in color.

8. Using different colors of ink to emphasize different parts of the math formula.

9. Asking your tutor to *show you* how to do the problems instead of *telling you* how to do the problems.

10. Writing down each problem step the tutor tells you to do. Highlight the important steps or concepts, which cause you difficulty.

Auditory (hearing) Learner

If you are an *auditory learner* (one who learns best by hearing the information) then learning formulas is best accomplished by repeating them back to yourself, or recording them on a tape recorder and listening to them. Reading out loud is one of the best auditory ways to get important information into long-term memory. Stating facts and ideas out loud improves your ability to think and remember. If you cannot recite out loud, recite the material to yourself, emphasizing the key words.

Based on the Learning Styles Inventory students who learn math best by hearing it are Auditory Numerical Learners. If you are an auditory numerical learner you may learn best by following the 10 suggestions in Figure 12 on the next page. Try as many of these suggestions as possible and then select and practice those that are most helpful.

An auditory way to improve your concentration is by becoming aware of your distractions and telling yourself to concentrate. If you are in a location where talking out loud will cause a disturbance, mouth the words "start concentrating" as you say them in your mind. Your concentration periods should increase.

Tactile/Concrete (touching) Learner

A *tactile/concrete learner* needs to feel and touch the material to learn it. Tactile concrete learners, who are also called *kinesthetic* learners, tend to learn best when they can concretely manipulate the information to be learned. Unfortunately, most math instructors do not use this learning sense. As a result, students who depend heavily upon feeling and touching for learning will usually have the most difficulty developing effective

learning math, and other distracting thoughts cannot enter. Your mind has one-way input and output, which only responds to thinking about math when you are doing homework or studying.

Figure 12
Auditory Numerical Learners

1. Say numbers to yourself or moving your lips as you read problems.

2. Tape record your class and play it back while reading your notes.

3. Read aloud any written explanations.

4. Explain to your tutor how to work the math problem.

5. Make sure all important facts are spoken aloud with auditory repetition.

6. Remember important facts by auditory repetition.

7. Read math problems aloud and try solutions verbally and sub-verbally as you talk yourself through the problem.

8. Record directions to difficult math problems on audiotape and refer to them when solving a specific type of problem.

9. Have your tutor explain how to work the problems instead of just showing you how to solve them.

10. Record math laws and rules in your own words, by chapters, and listen to them every other day (auditory highlighting).

math learning techniques. This learning style creates a problem with math learning because math is more abstract than concrete. Also, most math instructors are visual abstract learners and have difficulty teaching math tactilely. Ask for the math instructors and tutors who give the most practical examples and who may even "act out" the math problems.

As mentioned before a tactile concrete learner will probably learn most efficiently by hands-on learning. For example, if you want to learn the FOIL method, you would take your fingers and trace the "face" to remember the steps. See Figure 14 (The FOIL Method). Also, learning is most effective when physical involvement with manipulation is combined with sight and sound. For example, as you trace the face you also say the words out loud.

Based on the Learning Styles Inventory, Tactile/Concrete Learners best learn math by manipulating the information that is to be taught. If you are a tactile/concrete learner, you may learn best by following the 10 suggestions in Figure 13. Try as many of these suggestions as possible and then select and practice the best suggestions that help. If you do not have these manipulatives or don't know how to use them, ask the math lab supervisor or instructor if they have any manipulative materials or models. If the math lab does not have any manipulative materials, you may have to ask for help to develop your own.

Tactile/concrete learners can also use graphing calculators to improve their learning. By entering the keystrokes it is easier to remember how to solve the problems. This practice is also an excellent way to remember how to solve the problem when using a calculator while taking a test.

Another way tactile/concrete learners can learn is to trace the graph with their fingers when it appears on the calculator. They should say out loud and trace every equation to "feel" how the graph changes when using different equations. For example, if you add 2 to one side of the equation, move your finger to where the graph changes and say out loud how much it moved.

A tactile/concrete way to improve your study concentration is by counting the number of distractions for each study session. Place a sheet of paper by your book when doing homework. When you catch yourself not concentrating put the letter

Figure 13
Tactile Concrete Learners

1. Cut up a paper plate to represent a fraction of a whole.

2. Fold up a piece of paper several times and cut along the fold marks to represent a fraction of a whole.

3. In order to understand math concepts, ask to be shown how to use qusinar rods or algebra tiles as manipulatives.

4. Try to use your hands and body to "act out" a solution. For example, you may "become" the car in a rate-and-distance word problem.

5. Obtain diagrams, objects or manipulatives and incorporate activities such as drawing and writing into your study time. You may also enhance your learning by doing some type of physical activity such as walking.

6. Try to get involved with at least one other student, tutor or instructor that uses manipulatives to help you learn math.

7. Ask to use the Hands-on Equations Learning System using manipulatives to learn basic algebra. You can go to their Web site (www.Borenson.com) to learn more about this system and other systems to help you learn math.

8. Go to one of the "learning stores" usually in your local mall to see if they have manipulatives.

9. Go to a K-12 learning resource store to see if they have manipulatives, such as magnetic boards with letters and numbers.

10. Talk to the coordinator of students with disabilities to see if they use manipulatives when tutoring their students.

Figure 14
The FOIL Method

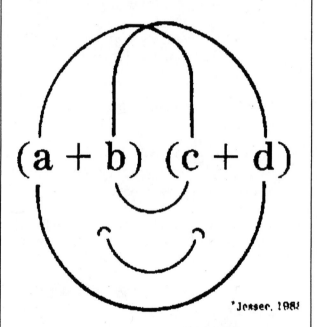

FOIL is used to remember the procedure to multiply two binomials. The letters in FOIL stand for First, Outside, Inside and Last. To use FOIL, multiply the following

- the First terms ((a) (c)),
- the Outside terms ((a) (d)),
- the Inside terms ((b) (c)),
- the Last terms ((b) (d)).

To learn FOIL, trace your finger along the FOIL route.

"C" on the sheet of paper. This will remind you to concentrate and get back to work. After each study period, count up the number of "C's" and watch the number decrease.

Social Individual Learner

If you are a social individual learner, learning math may best be done individually. You may learn best by yourself, working with computer programs and being individually tutored. In some cases, social individuals may have to meet in groups to develop practice tests but leave socializing to a minimum. If you are a social individual learner and visual learner, using the computer may be one of the best learning tools available. If you are a social individual learner, based on the Learning Styles Inventory, you may learn best by following the eight suggestions in Figure 15. Try as many of these suggestions as possible and then select those that are most helpful.

A problem that a social individual learner may encounter is working too long on a problem for which they could have received help. Social individual learners must understand that getting help is okay, especially if it saves study time and makes them more study efficient.

Social Group Learners

If you are a social group learner (one who best learns in a group) then learning math may best be done in study groups and in math classes that have collaborative learning (group learning). Social group learners may learn best by discussing information. They can usually develop their own study groups and discuss how to solve problems over the phone. If you are a social group learner and an auditory learner then you definitely learn best by talking to people. If you are a social group learner,

Figure 15
Social Individual Learners

1. Study math, English or other subjects alone.

2. Utilize videocassette tapes or auditory tapes to learn by yourself.

3. Prepare individual questions for your tutor or instructor.

4. Obtain individual help from the math lab or hire your own tutor.

5. Set up a study schedule and study area so other people will not bother you.

6. Study in the library or in some other private, quiet place.

7. Use group study times only as a way to ask questions, obtain information and take pre-tests on your subject material.

8. Use math websites to help you learn.

based on the Learning Styles Inventory, you may learn best by following the eight suggestions in Figure 16, on the following page. Try as many of these suggestions as possible and then select and practice those that are most helpful.

A learning problem that a social group learner may have is talking too much about other subjects when in a study group. This is called being off task. You may want to have a student serve as a discussion monitor to let the students know when they need to get back on task. Also, social group learners need to know that they still must study math individually to be successful. During this individual study session prepare questions for the group.

Figure 16
Social Group Learners

1. Study math, English or your other subjects in a study group.

2. Sign up for math course sections which use cooperative learning (learning in small groups).

3. Review your notes with a small group.

4. Obtain help in the math lab or other labs where you can work in group situations.

5. Watch math videocassette tapes with a group and discuss the subject matter.

6. Listen to audiocassette tapes on the lecture and discuss them with the group.

7. Obtain several "study buddies" so you can discuss with them the steps to solving math problems.

8. Form a study group. Each member brings 10 test questions with explanations on the back. The group completes all the test questions.

Multiple Senses

If you have difficulty learning material from one sense (learning style), you might want to try learning material through two or three senses. Involving two or more senses in learning improves your learning and remembering. Review the figures in this section on using your learning styles and whenever possible combine the learning styles.

If your primary sense is visual and your secondary sense is auditory, you may want to write down equations while saying them out loud to yourself. Writing and reciting the material at the same time combines visual, auditory, and some tactile/concrete styles of learning.

Studying with a pen or highlighter is a visual as well as a tactile/concrete way to improve your concentration. Placing the pen or highlighter in your hand and using it will force you to concentrate more on what you are reading. After you write and recite the material back to yourself, do it five or ten more times to over learn it.

Section Review

1. List five ways visual learners can improve their memory:

 First Way: _____

 Second Way: _____

 Third Way: _____

 Fourth Way: _____

 Fifth Way: _____

2. List five ways auditory learners can improve their memory:

 First Way: _____

Second Way: _____

Third Way: _____

Fourth Way: _____

Fifth Way: _____

3. List five ways tactile/concrete learners can improve their memory:

First Way: _____

Second Way: _____

Third Way: _____

Fourth Way: _____

Fifth Way: _____

4. List three ways social individual learners can improve their memory:

First Way: _____

Second Way: _____

Third Way: _____

5. List three ways social group learners can improve their memory:

First Way: _____

Second Way: _____

Third Way: _____

6. What is your best modality learning style for math? Visual, auditory or tactile/concrete?

7. What is your second best modality learning style for math? _____

8. Are you more of an individual or group learner?

Why? _____

9. How can you use multiple senses to improve your memory?

10. What is your best combination of learning styles to learn math? _____

Give an example of how you can combine your learning styles to learn a math concept (if you cannot think of a concept ask your instructor for one).

How to use memory techniques

There are many different techniques that can help store information in your long-term memory: a positive attitude about studying, selective learning, organization, visual imagery, mnemonic devices, and acronyms.

A Good Study/Math Attitude

A positive attitude about studying will help you concentrate and improve your retention of information. This means you should have at least a neutral math attitude (you neither like nor dislike it) and reserve the right to actually learn to like math. If you still don't like math, just pretend that you do while studying it. View studying as an opportunity to learn rather than an unpleasant task. Tell yourself that you can learn the material, and it will help you pass the course and graduate.

Be a Selective Learner

Being selective in your math learning will improve your memory. Prioritize the materials you are studying; decide which facts you need to know and which ones you can ignore. Narrow down information into laws and principles that can be generalized. Learn the laws and principles 100 percent.

> **Example:** If you have been given a list of math principles and laws to learn for a test, put each one on an index card. As you go through them, create two piles: a "I already know this" pile and a "I don't know this" pile. Then study *only* the "I don't know this" pile. Study the "I don't know this" pile until it is completely memorized and understood.

Become an Organizer

Organizing your math material into idea/fact clusters or groups will help you learn and memorize it. Grouping similar material in a problem log or calculator log is an example of categorizing information. Do not learn isolated facts; always try to connect them to other similar material.

Use Visual Imagery

Using mental pictures or diagrams to help you learn the material is especially helpful for the visual learners and those who are right-hemisphere dominant (who tend to learn best by visual and spatial methods). Mental pictures and actual diagrams involve 100 percent of your brain power. Picture the steps to solve difficult math problems in your mind.

Example: Use the FOIL method (see Figure X, on the facing page) to visually learn how to multiply binomials. Memorize the face until you can sketch it from memory. If you need to use it during a test, you can then sketch the face onto your scratch paper and refer to it.

Examples: When learning the *commutative property*, remember that the word *commutative* sounds like the word *community*. A community is made up of different types of people who could be labeled *as* an *a* group and a *b* group. However, in a community of *a* people and *b* people, it does not matter if we count the *a* people first or the *b* people first; we still have the same total number of people in the community. Thus, $a + b = b + a$.

When learning the *distributive law of multiplication over addition*, such as $a(b + c)$, remember that "distributive" sounds like "distributor," which is associated with giving out a product. The distributor *a* is giving its products to *b* and *c*.

Make Associations

Association learning can help you remember better. Find a link between new facts and some well-established old facts and study them together. The recalling of old facts will help you remember the new facts and strengthen a mental connection between the two. Make up your own associations to remember math properties and laws.

REMEMBER

The more ridiculous the association, the more likely you are to remember it.

Use Mnemonic Devices

The use of mnemonic devices is another way to help you remember. Mnemonic devices are easily remembered words, phrases, or rhymes associated with difficult-to-remember principles or facts.

Example: Many students become confused with the *order of operations*. These students mix up the order of the steps in solving a problem, such as dividing before adding the numbers

continued

in the parentheses. A mnemonic device to remember the order of operations is "Please Excuse My Dear Aunt Sally." The first letter in each of the words represents the math function to be completed from the first to the last. Thus, the order of operations is **P**arentheses (*Please*), **E**xponents (*Excuse*), **M**ultiplication (*My*), **D**ivision (*Dear*), **A**ddition (*Aunt*), and **S**ubtraction (*Sally*). Remember to multiply and/or divide whatever comes first, from left to right. Also, add or subtract whatever comes first, from left to right.

expression) and 7 (in the second expression). The **I**nside quantities are 3 (in the first expression) and x (in the second expression). The **L**ast quantities are 3 (in the first expression) and 7 (in the second expression). This results in

F $(2x)(x)$ + **O** $(2x)(7)$ + **I** $(3)(x)$ + **L** $(3)(7)$.

Do the multiplication to get $2x^2 + 14x + 3x + 21$, which adds up to $2x^2 + 17x + 21$. See Figure 13 (The FOIL Method).

Use Acronyms

Acronyms are another memory device to help you learn math. Acronyms are word forms created from the first letters of a series of words.

Example: FOIL is one of the most common math acronyms. FOIL is used to remember the procedure for multiplying two binomials. Each letter in the word *FOIL* represents a math operation. **FOIL** stands for **F**irst, **O**utside, **I**nside, and **L**ast, as it applies to multiplying two binomials such as $(2x + 3)(x + 7)$. The **F**irst quantities are $2x$ (in the first expression) and x (in the second expression). The **O**utside quantities are $2x$ (in the first

continued

Section Review

1. How can a good math attitude help you learn?

2. Give an example of being a selective learner in your math class.

3. From your current math lessons, make up and explain one association remembering device that is not in this workbook.

4. For your next major math test, make up and explain one mnemonic device that is not in this workbook.

5. For your next major math test, make up and explain one acronym that is not in this workbook.

How to develop practice tests

Developing a practice test is one of the best ways to evaluate your memory and math skills before taking a real test. You want to find out what you do not know *before* the real test instead of *during* the test. Practice tests should be as real as possible and should include time constraints.

You can create a practice test by reworking all the problems since your last test that you have recorded in your problem log. Another practice test can be developed using every other problem in the textbook chapter tests. Further, you can use the solutions manual to generate other problems with which to test yourself. You can also use old exams from the previous semester. Check to see if the math lab/LRC or library has tests on file from previous semesters, or ask your instructor for other tests. For some students, a better way to prepare for a test is the group method.

Example: Hold a group study session several days before the test. Have each student prepare a test with ten questions. On the back of the test, have the answers listed, worked out step by step. Have each member of the study group exchange his or her test with another member of the group. Once all the tests have been completed, have the author of each test discuss with the group the procedures used to solve those problems.

If group work improves your learning, you may want to hold a group study session at least once a week. Make sure the individual or group test is completed at least three days before the real test.

Completing practice math tests will help you increase your testing skills. It will also reveal your test problem weaknesses in enough time for you to

learn how to solve the problem before the real test. If you have difficulty with any of the problems during class or after taking the practice test, be sure to see your tutor or instructor.

After taking the practice test(s), you should know what parts you do not understand (and need to study) and what is likely to be on the test. Put this valuable information on one sheet of paper. This information needs to be understood and memorized. It may include formulas, rules, or steps to solving a problem.

Use the learning strategies discussed in this chapter to remember this information. A good example of how this information should look is what students might call a mental "cheat sheet." Obviously, you cannot use the written form of this sheet during the real test.

If you cannot take a practice test, put down on your mental cheat sheet the valuable information you will need for the test. Work to understand and memorize your mental cheat sheet. Chapter 7, "How to Improve Your Math Test-Taking Skills," will discuss how to use the information on the mental cheat sheet — *without cheating.*

Section Review

1. List three different ways you can make up a practice test.

First Way: _____

Second Way: _____

Third Way: _____

2. List ten different problems you would put on your practice test for the next exam.

First Problem: _____

Second Problem: _____

Third Problem: _____

Fourth Problem: _____

Fifth Problem: _____

Sixth Problem: _____

Seventh Problem: _____

Eighth Problem: _____

Ninth Problem: _____

Tenth Problem: _____

3. Give these ten problems to your study buddy and get ten problems from him or her. Compare your answers with those of your study buddy. How many did you get right? _____ How can you correct the wrong answers?

How to use number sense

When taking your practice tests, you should use "number sense," or estimations, to make sure your answer is reasonable. Number sense is like common sense but it applies to math. Number sense is the ability to see if your answer makes sense without using algorithms. (Algorithms are the sequential math steps used to solve problems.) These following examples demonstrate solving a math problem (from a national math test given to high school students) using algorithms and number sense.

Example: Solve 3.04 x 5.3. Students use algorithms to solve this problem by multiplying the number 3.04 by 5.3, in sequence. Seventy-two percent of the students answered the problem

continued

correctly using algorithms.

Example: Estimate the product 3.04 x 5.3, and choose from the following answers.

A) 1.6 C) 160
B) 16 D) 1600

Only 15 percent of the students chose B, which is the correct answer. Twenty-eight percent of the students chose A. Using *estimating* to solve the answer, a whopping 85 percent of the students got the problem wrong.

These students were incorrectly using their "mental chalkboard" instead of using number sense. In using number sense to answer, you would multiply the numbers to the left of the decimal in each number to get an estimate of the answer. To estimate the answer, then, you would multiply 3 (the number to the left of the decimal in 3.04) by 5 (the number to the left of the decimal in 5.3) and expect the answer to be a little larger than 15.

It appears that the students' procedural processing (the use of algorithms) was good, but when asked to solve a non-routine problem using estimating (which is easier than using algorithms), the results were disappointing.

Another example of using number sense, or estimating, is in "rounding off."

Taking the time to estimate the answer to a math problem is a good way to check your answer. Another way to check your answer is to see if it is reasonable. Many students forget this important step and get the answer wrong. This is especially true of word or story problems.

Example: Solve 48 + 48 by rounding off. Rounding off means mentally changing the number (up or down) to make it more manageable to you, without using algorithms. By rounding off, 48 becomes 50 (easier to work with). 50 + 50 = 100. If the choices for answers were 104, 100, 98 and 96, you would then subtract 4 from the 100 (since each number was rounded up by 2) and you would get 96.

Example: In solving the equation $x + 3 = 9$, you calculated that $x = 5$. To check your answer, substitute x with 5 and see if the problem works out correctly. 5 + 3 does not equal 9, so you know you have made a mistake and need to recalculate the problem. The correct answer, by the way, is $x = 6$.

Also, when dealing with an equation, make sure to put the answer back into the equation to see if one side of the equation equals the other. If the two sides are not equal, you have the wrong answer. If you have extra time left over after you have completed a test, you should check answers using this method.

REMEMBER

Number sense is a way to get more math problems correct by estimating your answer to determine if it is reasonable.

Examples: When solving a rate-and-distance problem, use your common sense to realize that one car cannot go 500 miles per hour to catch the other car. However, the car can go 50 miles per hour.

The same common sense rule applies to age-word problems, where the age of a person cannot be 150 years, but it can be 15 years.

Further, in solving equations, x is *usually* a number that is less than 20. When you solve a problem for x and get 50, then, this isn't reasonable, and you should recheck your calculations.

continued

Section Review

1. Give an explanation of number sense.

2. How does number sense compare to common
 sense?

3. Give an example of a recent math problem you
 missed due to poor number sense.

4. List two ways you can improve your number
 sense.

 First Way: _____

 Second Way: _____

Name: _____ Date: _____

Chapter 6 Review

1. The way you receive information is through your five senses, which are _____, _____, _____, _____, and _____.

2. The main problem most students have is converting learned material into _____ _____ and _____ it.

3. Repeatedly reading and writing math material is one of the best ways for a _____ learner to study.

4. Reading _____ _____ is one of the best auditory ways to learn material.

5. A _____/_____ learner needs to feel and touch material to learn it.

6. Being a selective learner means _____ the material to study and learning the laws and principles _____ percent.

7. Mnemonic devices are easy-to-remember _____, _____, or _____ associated with difficult-to-remember principles or facts.

8. _____ are word forms created from the first letters of a series of words.

9. The reason to develop a practice test is to find out what you do not know _____ the test instead of _____ the test.

10. Social learners benefit from forming _____ _____.

What is the most important information you learned from this chapter?

How can you immediately use it?

How to Improve Your Math Test-Taking Skills

Taking a math test is different from taking tests in other subjects. Math tests require not only that you recall the information, but that you apply the information. Multiple-choice tests, for example, usually test you on recall, and if you do not know the answer, you can guess.

Math tests build on each other, whereas history tests often do not test you on previous material. Most math tests are speed tests, where the faster your are, the better grade you can receive; most social science tests are designed for everyone to finish.

Math test preparation and test-taking skills are different from preparation and skills needed for other tests. You need to have a test-taking plan and a test analysis plan to demonstrate your total knowledge on math tests. Students with these plans make better grades compared to students without them. Math instructors want to measure your math knowledge, not your poor test-taking skills.

In this chapter, you will learn

- why attending class and doing your homework may not be enough to pass,
- the general pretest rules,
- the ten steps to better test taking,
- the six types of test-taking errors, and
- how to prepare for the final exam.

Why attending class and doing your homework may not be enough to pass

Most students and some instructors believe that attending class and doing all the homework ensures an A or B on tests. This is far from true. Doing all the homework and getting the correct answers is very different in many ways from taking tests:

1. While doing homework, there is little anxiety. A test situation is just the opposite.

2. You are not under a time constraint while doing your homework; you may have to complete a test in 55 minutes or less.

3. If you get stuck on a homework problem, your textbook and notes are there to assist you. This is not true for most tests.

4. Once you learn how to do several problems in a homework assignment, the rest are similar. In a test, the problems may be in random order.

5. In doing homework, you have the answers to at least half the problems in the back of the text and answers to all the problems in the solutions guide. This is not true for tests.

6. While doing homework, you have time to figure out how to correctly use your calculator. During the test, you can waste valuable time figuring out how to use your calculator.

7. When doing homework, you can call your study buddy or ask the tutor for help, something you cannot do on the test.

8. While doing your homework, you can go to websites for online tutoring to help you solve problems you do not understand.

Do not develop a false sense of security by believing you can make an A or B by doing just your homework. Tests measure more than just your math knowledge.

The general pretest rules

General rules are important when taking any type of test:

1. *Get a good night's sleep before taking a test.* This is true for the ACT, the SAT, and your math tests. If you imagine you are going to cram all night and perform well on your test with three to four hours of sleep, you are wrong. It would be better to get seven or eight hours of sleep and be fresh enough to use your memory to recall information needed to answer the questions.

2. *Start studying for the test at least three days ahead of time.* Make sure you take a practice test to find out, before the test, what you do not know. Review your problem log and work the problems. Review the concept errors you made on the last test. (How to identify and correct your concept errors will be discussed later in this chapter.) Meet with the instructor or tutor for help on those questions you cannot solve.

3. *Review only already-learned material the night before a test.*

4. *Make sure you know all the information on your mental cheat sheet.* Review your notebook and glossary to make sure you understand the concepts. Work a few problems and recall the information on your mental cheat sheet right before you go to bed. Go directly to bed; do not watch television, listen to the radio, or party. While you are asleep, your mind will work on and remember the last thing you did before going to sleep.

5. *Get up in the morning at your usual time and review your notes and problem log.* Do not do any new problems, but make sure your calculator is working.

Section Review

1. List three reasons why only attending class and doing your homework may not be enough to pass your math course.

First Reason: _____

Second Reason: _____

Third Reason: _____

2. List and explain three general pretest rules that best apply to you.

First Rule: _____

Second Rule: _____

Third Rule: _____

The ten steps to better test taking

You need to have a game plan to take a math test. This plan is different from taking history, English, humanities and some science tests. The game plan is to get the most points in the least amount of time. Many students lose test points because they use the wrong test-taking strategies for math. By following these ten steps you can demonstrate more knowledge on the test and get more problems right.

Step 1 *Use a memory data dump.* When you get your test, turn it over and write down the information that you might forget. Remember, this is your mental cheat sheet that you should have already made while preparing for the test. Your mental cheat sheet has now turned into a mental list, and writing down this information is not cheating. Do not put your name on the test, do not skim it, just turn it over and write down those facts, figures, and formulas from your mental cheat sheet or other information you might not remember during the test. This is called your *first memory data dump.* The data dump provides memory cues for test questions.

Example: It might take you a while to remember how to do a coin problem. However, if you had immediately turned your test over and written down different ways of solving coin problems, it would be easier to solve the coin problem.

Step 2 *Preview the test.* Put your name on the test and start previewing. Previewing the test requires you to look through the entire test to find different types of problems and their point values. Put a mark by the questions that you can do without thinking. These are the questions that you will solve first.

Step 3 *Do a second memory data dump.* The secondary data dump is for writing down material that was jarred from your memory while previewing the test. Write this information on the back of the test.

Step 4 *Develop a test progress schedule.* When you begin setting up a test schedule, determine the point value for each question. Some test questions might be worth more points than others.

In some tests, word problems are worth 5 points and other questions might be worth 2 or 3 points. You must decide the best way to get the most points in the least amount of time. This might mean working the questions worth 2 or 3 points first and leaving the more difficult word problems for last.

Decide how many problems should be completed halfway though the test. You should have more than half the problems completed by that time.

Step 5 *Answer the easiest problems first.* Solve, in order, the problems you marked while previewing the test. Then review the answers to see if they make sense. Start working through the test as fast as you can while being accurate. Answers should be reasonable.

Example: The answer to a problem asking you to find the area of a rectangle cannot be negative, and the answer to a land-rate-distance problem cannot be 1000 miles per hour.

Clearly write down each step to get partial credit. Even if you end up missing the problem, you might get some credit. In most math tests, the easier problems are near the beginning of the first page; you need to answer them efficiently and quickly. This will give you both more time for the harder problems and time to review.

Step 6 *Skip difficult problems.* If you find a problem that you do not know how to work, read it twice and automatically skip it. Reading it twice will help you understand the problem and put it into your working memory. While you are solving other problems, your mind is still working on that problem. Difficult problems could be the type of problem you have never seen before or a problem in which you get stuck on the second or third step. In either case, skip the problem and go on to the next one.

Step 7 *Review the skipped problems.* When working the skipped problems, think how you have solved other, similar problems as a cue to solving the skipped ones. Also try to remember how the instructor solved that type of problem on the board.

While reviewing skipped problems, or at any other time, you may have the "Aha!" response. The "Aha!" response is your remembering how to do a skipped problem. Do not wait to finish your current problem. Go to the problem on which you had the "Aha!" and finish that problem. If you wait to finish your current problem, your "Aha!" response could turn into an "Oh, no!" response.

Step 8 *Guess at the remaining problems.* Do as much work as you can on each problem, even if it is just writing down the first step. If you cannot write down the first step, rewrite the problem. Sometimes rewriting the problem can jar your memory enough to do the first step or the entire problem. This is true of students, particularly of tactile/concrete learners where writing can trigger the

memory process of solving the problem. Remember, you did not learn how to solve math problems with your hands tied behind your back. If you leave the problem blank, you will get a zero. Do not waste too much time on guessing or trying to work the problems you cannot do.

Step 9 *Review the test.* Look for careless errors or other errors you may have made. Students usually lose 2 to 5 test points on errors that could have been caught in review. Do not talk yourself out of an answer just because it may not look right. This often happens when an answer does not come out even. It is possible for the answer to be a fraction or decimal.

REMEMBER

Answers in math do not have "dress codes." Research reveals that the odds of changing a right answer to a wrong answer are greater than the odds of changing a wrong answer to a right one.

Step 10 *Use all the allowed test time.* Review each problem by substituting the answer back into the equation or doing the opposite function required to answer the question. If you cannot check the problem by the two ways mentioned, rework the problem on a separate sheet of paper and compare the answers. Do not leave the test room unless you have reviewed each problem two times or until the bell rings.

Even though we encourage students to work until the end of the test period most students leave the classroom before the end of the period. These students state that even though they know they should use all the test time, they cannot stay in the room until the end of the test time. These students also know that their grades would probably improve if they kept checking their answers or kept working the problems they are having difficulty with. After talking to hundreds of these students, I discovered two different themes for leaving the classroom early. First, test anxiety gets so overwhelming that they cannot stay in the room. The relief from the test anxiety (leaving the room) is worth more than getting a better grade. If you are one of these students, you must learn how to decrease your test anxiety by following the suggestions in Chapter 4 or by using the *How to Reduce Test Anxiety* cassette tape. Don't let the anxiety control your grades!

The other reason for leaving the test early is that they do not want to be the last or one of the last few students to turn in their tests. They still believe that students who turn their tests in last are "dumb and stupid." These students also believe that students who turn their tests in first make "A's" and "B's" and those students who turn their tests in last make "D's" and "F's." If you are one of these students, you don't need to care about what other students think about you (it's usually wrong anyway). YOU need to fight the urge to leave early and use all the test time. Remember, passing mathematics is the best way to get a high paying job and support yourself or your family. DO IT NOW!

REMEMBER

There is no prize for handing your test in first, and students who turn their papers in last do make A's.

Stapling your scratch paper to the math test when handing it in has several advantages:

- If you miscopied the answer from the scratch paper, you will probably get credit for the answer.
- If you get the answer incorrect due to a careless error, your work on the scratch paper could give you a few points.
- If you do get the problem wrong, it will be easier to locate the errors when the instructor reviews the test. This will prevent you from making the same mistakes on the next math test.

REMEMBER

Handing in your scratch paper may get you extra points or improve your next test score.

Section Review

1. List and explain the ten steps to test taking.

First Step: _____

Second Step: _____

Third Step: _____

Fourth Step: _____

Fifth Step: _____

Sixth Step: _____

Seventh Step: _____

Eighth Step: _____

Ninth Step: _____

Tenth Step: _____

2. After trying the ten steps to test taking, develop your own personalized test-taking steps.

First Step: _____

Second Step: _____

Third Step: _____

Fourth Step: _____

Fifth Step: _____

Sixth Step: _____

Seventh Step: _____

Eighth Step: _____

Ninth Step: _____

Tenth Step: _____

Additional steps

The six types of test-taking errors

To improve future test scores, you must conduct a test analysis of previous tests. In analyzing your tests, you should look for the following kinds of errors:

1. misread-directions errors

2. careless errors
3. concept errors
4. application errors
5. test-taking errors
6. study errors

Students who conduct math test analyses will improve their total test scores.

Misread-directions errors occur when you skip directions or misunderstand directions, but do the problem anyway.

Examples: Suppose you have this type of problem to solve:

$$(x + 1)(x + 1)$$

Some students will try to solve for x, but the problem only calls for multiplication. You would solve for x only if you have an equation such as $(x + 1)(x + 1) = 0$.

Another common mistake is not reading the directions before doing several word problems or statistical problems. All too often, when a test is returned, you find only three out of the five problems had to be completed. Even if you did get all five of them correct, it cost you valuable time that could have been used obtaining additional test points.

To avoid misread-direction errors, carefully read and interpret all the directions. Look for anything that is unusual, or if the directions have two parts. If you do not understand the directions, ask the instructor for clarification. If you feel uneasy

about asking the instructor for interpretation of the question, remember the instructor in most cases does not want to test you on the interpretation of the question but how you answer it. Also, you don't want to make the mistake of assuming that the instructor will not interpret the question. Let the instructor make the decision to interpret the question, not you.

Careless errors are mistakes that you can catch automatically upon reviewing the test. Both good and poor math students make careless errors. Such errors can cost a student a higher letter grade on a test.

Examples: *Dropping the sign:* $-3(2x) = 6x$, instead of $-6x$ which is the correct answer.

Not simplifying your answer: Leaving $(3x - 12)/3$ as your answer instead of simplifying it to $x - 4$.

Adding fractions: $1/2 + 1/3 = 2/5$, instead of $5/6$, which is the correct answer.

Word problems: $x = 15$ instead of "The student had 15 tickets."

However, many students want all their errors to be careless errors. This means that the students *did* know the math, but simply made silly mistakes. In such cases, I ask the student to solve the problem immediately, while I watch.

If the student can solve the problem or point out his or her mistake in a few seconds, it is a careless error. If the student cannot solve the problem immediately, it is not a careless error and is probably a concept error.

When working with students who make careless errors, I ask them two questions: First,

"How many points did you lose due to careless errors?" Then I follow with, "How much time was left in the class period when you handed in your test?" Students who lose test points to careless errors are giving away points if they hand in their test papers before the test period ends.

To reduce careless errors, you must realize the types of careless errors made and recognize them when reviewing your test. If you cannot solve the missed problem immediately, it is not a careless error. If your major error is not simplifying the answer, review each answer as if it were a new problem and try to reduce it.

Concept errors are mistakes made when you do not understand the properties or principles required to work the problem. Concept errors, if not corrected, will follow you from test to test, causing a loss of test points.

Examples: Some common concept errors are not knowing:
$(-)(-)x = x$, *not -x*
$-1(2) > x(-1) = 2 < x$, *not 2 > x*
$5/0$ is undefined, *not 0*
$(a + x) / x$ is *not* reduced to *a*
the order of operations

Concept errors must be corrected to improve your next math test score. Students who have numerous concept test errors will fail the next test and the course if concepts are not understood. Just going back to rework the concept error problems is not good enough. You must go back to your textbook or notes and learn why you missed those types of problems, not just the one problem itself.

The best way to learn how to work those types of problems is to set up a concept problem error page in the back of your notebook. Label the

first page "Test One Concept Errors." Write down all your concept errors and how to solve the problems. Then, work five more problems that use the same concept. Now, in your own words, write the concepts that you are using to solve these problems.

If you cannot write the concept in your own words, you do not understand it. Get assistance from your instructor if you need help finding similar problems using the same concept or cannot understand the concept. Do this for every test.

Application errors occur when you know the concept but cannot apply it to the problem. Application errors usually are found in word problems, deducing formulas (such as the quadratic equation), and graphing. Even some better students become frustrated with application errors; they understand the material but cannot apply it to the problem.

To reduce application errors, you must predict the type of application problems that will be on the test. You must then think through and practice solving those types of problems using the concepts.

> **Example:** If you must derive the quadratic formula, you should practice doing that backward and forward while telling yourself the concept used to move from one step to the next.

Application errors are common with word problems. After completing the word problem, reread the question to make sure you have applied the answer to the intended question. Application errors can be avoided with appropriate practice and insight.

Test-taking errors apply to the specific way you take tests. Some students consistently make

the same types of test-taking errors. Through recognition, these bad test-taking habits can be replaced by good test-taking habits. The result will be higher test scores. The list that follows includes the test-taking errors that can cause you to lose many points on an exam:

1. *Missing more questions in the first third, second third, or last third of a test* is considered a test-taking error.

 Missing more questions in the first third of a test can be due to carelessness when doing easy problems or due to test anxiety.

 Missing questions in the last part of the test can be due to the fact that the last problems are more difficult than the earlier questions or due to increasing your test speed to finish the test.

 If you consistently miss more questions in a certain part of the test, use your remaining test time to review that section of the test first. This means you may review the last part of your test first.

2. *Not completing a problem to its last step* is another test-taking error. If you have this bad habit, review the last step of the test problem first, before doing an in-depth test review.

3. *Changing test answers from correct ones to incorrect ones* is a problem for some students. Find out if you are a good or bad answer-changer by comparing the number of answers changed to correct and to incorrect answers. If you are a bad answer-changer, write on your test, "Don't change answers." Change answers only if you can prove to yourself or the instructor that the changed answer is correct.

4. *Getting stuck on one problem and spending too much time on it* is another test-taking error. You need to set a time limit on each

problem before moving to the next problem. Working too long on a problem without success will increase your test anxiety and waste valuable time that could be used in solving other problems or in reviewing your test.

5. *Rushing through the easiest part of the test and making careless errors* is a common test-taking error for the better student. If you have the bad habit of getting more points taken off for the easy problems than for the hard problems, first review the easy problems, and later review the hard problems.

6. *Miscopying an answer from your scratch work to the test* is an uncommon test-taking error, but it does cost some students points. To avoid these kinds of errors, systematically compare your last problem step on scratch paper with the answer written on the test. In addition, always hand in your scratch work with your test.

7. *Leaving answers blank* will get you zero points. If you look at a problem and cannot figure out how to solve it, do not leave it blank. Write down some information about the problem, rewrite the problem, or try to do at least the first step.

REMEMBER

Writing down the first step of a problem is the key to solving the problem and obtaining partial credit.

8. *Solving only the first step of a two-step problem* causes problems for some students. These students get so excited when answering the first step of the problem that they forget about the second part. This is especially true on two-step word problems. To correct this test-taking error, write "two" in the margin of the problem. That will remind you that there are two steps or two answers to this problem.

9. *Not understanding all the functions of your calculator* can cause major testing problems. Some students only barely learn how to use the critical calculator functions. They then forget or have to relearn how to use their calculator, which costs test points and test time. Do not wait to learn how to use your calculator on the test. Overlearn the use of your calculator *before* the test.

10. *Leaving the test early without checking all your answers is a costly habit.* Do not worry about the first person who finishes the test and leaves. Many students start to get nervous when students start to leave after finishing the test. This can lead to test anxiety, mental blocks, and loss of recall.

According to research, the first students finishing the test do not always get the best grades. It sometimes is the exact opposite. Ignore the exiting students, and always use the full time allowed.

Make sure you follow the ten steps to better test taking. Review your test taking procedures for discrepancies in following the ten steps to better test taking. Deviating from these proven ten steps will cost you points.

Study errors, the last type of mistake to look for in test analysis, occur when you study the wrong type of material or do not spend enough time on pertinent material. Review your test to find out if you missed problems because you did not practice that type of problem or because you did practice it but forgot how to do it during the test. Study errors

Figure 17
Math Test for Prealgebra

The answers are listed in bold. The correct answer on missed questions is shaded. Identify the type of error based on the "6 Types of Test-taking Errors." The student's test score is 70.

1. Write in words: 32.685

 Thirty-two and six hundred eighty-five thousandths

2. Write as a fraction and simplify: 0.078

 $\dfrac{78}{1000}$ **$\dfrac{39}{500}$**

3. Round to the nearest hundredth: 64.8653

 64.865 64.87

4. Combine like terms:

 6.78x -3.21 + 7.23x -6.19

 = 6.78x + 7.23x + (-3.21) + (-6.19)
 = 14.01x -9.4

5. Divide and round to the nearest hundredth:
 68.1357 2.1

 32.445 → 32.45

6. Write as a decimal: $\dfrac{5}{16}$

 0.3125

7. Insert < or > to make a true statement:
 $\dfrac{3}{8}$ **<** $\dfrac{6}{13}$

8. Solve: $\dfrac{3}{x} = \dfrac{9}{12}$

 9x = 3(12)

 $\dfrac{9x}{9} = \dfrac{36}{9}$

 x = 5 x = 4

9. What number is 35% of 60?

 2100 **21.00**

10. 20.8 is 40% of what number? **52**

11. 567 is what percent of 756?

 $\dfrac{756}{567} = \dfrac{9}{100}$

 = 133.3% 75%

12. Multiply:

 (-6.03) (-2.31) = 13.9 **13.9293**

continued

Answer Key for Exercise

1. Correct
2. *Misread directions* — Forgot to simplify by reducing the fraction.
3. *Concept error* — Did not know that hundredths is two places to the right of the decimal.
4. Correct
5. Correct
6. Correct
7. Correct
8. *Careless error* — Divided incorrectly in the last step.
9. *Test-taking error* — Not following step 5 in test-taking steps: Reviewing answers to see if they make sense. The number that equals 35% can't be larger than 60.
10. *Test-taking error* — Did not follow step 7 in test taking steps: Don't leave an answer blank
11. *Application error* — Solved the equation correctly but did not apply the correct equation.
12. *Concept error* — Did not know that when you multiply with one number in the hundredths, the answer must include the hundredths column.

will take some time to track down. But correcting study errors will help you on future tests.

Most students, after analyzing one or several tests, will recognize at least one major, common test-taking error. Understanding the effects of this test-taking error should change your study techniques or test-taking strategy.

Example: If there are seven minutes left in the test, should you review for careless errors or try to answer those two problems you could not totally solve? This is a trick question. The real question is, Do you miss more points due to careless errors or concept errors, or are the missed points about even? The answer to this question should determine how you will spend the last minutes of the test. If you miss more points due to careless errors or miss about the same number of points due to careless/concept errors, review for careless errors.

Careless errors are easier to correct than concept errors. However, if you make very few or no careless errors, you should be working on those last two problems to get the greatest number of test points. Knowing your test-taking errors can add more points to your test by changing your test-taking procedure.

Section Review

1. List the six types of test taking errors.

 Error One: _____

 Error Two: _____

 Error Three: _____

 Error Four: _____

 Error Five: _____

 Error Six: _____

2. Which of these errors have you made in the past?

3. How can you avoid these types of errors on the next test?

4. Review the Math Test for Prealgebra if you are in a prealgebra course. Review for type of errors, the points lost for each error, and example of each error.

 Error Type: _____

 Points Lost for Error: _____

 Example of Error: _____

 Error Type: _____

 Points Lost for Error: _____

 Example of Error: _____

 Error Type: _____

 Points Lost for Error: _____

 Example of Error: _____

 Error Type: _____

 Points Lost for Error: _____

 Example of Error: _____

 Error Type: _____

 Points Lost for Error: _____

 Example of Error: _____

 Error Type: _____

 Points Lost for Error: _____

 Example of Error: _____

Error Type: _____

Points Lost for Error: _____

Example of Error: _____

5. Review your last test to list the type of errors, the points lost for each error, and examples of the errors. You may have more errors than allowed for in the list below. Use a separate sheet of paper if you need to list more. (Remember: Careless errors are problems you can solve immediately.)

Error Type: _____

Points Lost for Error: _____

Example of Error: _____

Error Type: _____

Points Lost for Error: _____

Example of Error: _____

Error Type: _____

Points Lost for Error: _____

Example of Error: _____

Error Type: _____

Points Lost for Error: _____

Example of Error: _____

6. List any additional concept error problems that were not included in question 2.

Points Lost for Error: _____

Example of Error: _____

Points Lost for Error: _____

Example of Error: _____

Using the pages in the back of your math notebook, rework the concept error problems and do five math problems just like them. For each concept error, write why you could not solve the problem on the test and what new information you learned to solve the problem. Do this for every major test. Give an example of one of your concept error(s) and the reasons you can now solve it.

Concept Error Problem(s):

Reasons I Can Now Solve It:

How to prepare for the final exam

The first day of class is when you start preparing for the final exam. Look at the syllabus or ask the instructor if the final exam is cumulative. A cumulative exam covers everything from the first chapter to the last chapter. Most math final exams are cumulative.

The second question you should ask is if the final exam is a departmental exam or if it is made up by your instructor. In most cases, departmental exams are more difficult and need a little different preparation. If you have a departmental final, you need to ask for last year's test and ask other students what their instructors say will be on the test.

The third question is, How much will the final exam count? Does it carry the same weight as a regular test or, as in some cases, will it count a third of your grade? If the latter is true, the final exam will usually make a letter grade difference on your final grade. The final exam could also determine if you pass or fail the course. Knowing this information before the final exam will help you prepare.

Preparing for the final exam is similar to preparing for each chapter test. You must create a pretest to discover what you have forgotten. You can use questions from the textbook chapter tests or questions from your study group.

Review the concept errors that you recorded in the back of your notebook labeled "Test One," "Test Two," and so on. Review your problem log for questions you consistently miss. Even review material that you knew for the first and second test but which was not used on any other tests. Students forget how to work some of these problems.

If you do not have a concept error page for each chapter and did not keep a problem log you need to develop a pre-test before taking the final exam. If you are individually preparing for the final, then copy each chapter test and do every fourth problem to see what errors you may make. However, it is better to have a study group where each of the four members brings in ten problems with the answers worked out on a separate page. Then each group member can take the 30-question test to find out what they need to study. You can refer to the answers to help you solve the problems that you miss. Remember you want to find out what you don't know before the test not during the test. Developing and using these methods before each test will improve your grades.

Make sure to use the 10 steps to better test taking and the information gained from your test analysis. Use all the time on the final exam because you could improve your final grade by a full letter if you make an "A" on the final.

Section Review

1. What are the three qustions you need to ask about your final exam?

 First Question: _____

 Second Question: _____

 Third Question: _____

Chapter 7 Review

1. You should start studying for a test at least _____ days ahead of time.

2. You should be reviewing only already-_____ material the night before the test.

3. When previewing the test, put a _____ by the questions you can solve without thinking, and do those problems _____.

4. When taking a test, you must decide the best way to get the _____ points in the _____ amount of time.

5. Halfway through the test, you should have _____ than half of the problems completed.

6. If you cannot write down the first step of a problem, you should _____ the problem to help you remember how to solve it.

7. Do not leave the test room until you have reviewed each problem _____ times or until the bell rings.

8. In reviewing the test, if you cannot solve the problem immediately (within a few seconds), it is not a _____ error and is probably a _____ error.

9. Students who have numerous _____ errors will fail the next test and the course if the _____ are not understood.

10. The _____ day of class is when you start preparing for the final exam.

What is the most important information you learned from this chapter?

How can you immediately use it?

Appendix: Math Autobiography

This is the purpose for your math autobiography—to reflect on your past experiences when learning math, how they made you feel then, and how these experiences and feelings shape how you feel about what you are doing in your current math classes. It will close with solutions you believe will help you improve your experiences in math classes.

The following writing process will simplify this task while improving the final paper.

1. Use the questions in the sections on types of math anxiety and causes of math anxiety in Chapter 3 to brainstorm ideas for your paper. Record these answers on paper.

2. By each experience you record, write how the experience made you feel.

3. To further your brainstorming, complete the following sentences:

 My favorite math instructor was

 _____,

 because he/she

_____,

I always felt

_____.

One math instructor who bothered me was

_____,

because he/she

_____,

I always felt

_____.

continued

When I hear the word math, I think of

_____.

I did well in math until

_____.

The first time I experienced math anxiety was when

_____.

The first time I was anxious before or during a math test was _____

_____.

The most recent time I was anxious about math class was when

_____.

The most recent time I was anxious before or during a test was when

_____.

4. Since an autobiography is in chronological order, identify the important information in your brainstorming and place into the correct time category in the tables on the following pages. These tables will be the outline for your paper.

5. You will write the introduction last, after you have thought through the meaning of your experiences. Then, when you type or write the final paper, you will place the introduction paragraph first.

Elementary School
Main Paragraph One

Experiences	Feelings

As a result of these experiences, now when I am either in math class, stydying for math, or taking a math test, I . . .

Middle School
Main Paragraph Two

Experiences	Feelings

As a result of these experiences, now when I am either in math class, studying for math, or taking a math test, I . . .

High School
Main Paragraph Three

Experiences	Feelings

As a result of these experiences, now when I am either in math class, studying for math, or taking a math test, I . . .

College and/or Work Experience
Main Paragraph Four (If this applies)

Experiences	Feelings

As a result of these experiences, now when I am either in math class, studying for math, or taking a math test, I . . .

Conclusion Paragraph

When I look back on my experiences learning or using math, I understand why I have developed attitudes like

_____ .

To develop a more positive perspective toward math, I need to improve certain behaviors.

Behaviors	Solutions for Change
1.	
2.	
3.	
4.	
As a result, I will become . . .	

Introduction Paragraph

(When you write your final paper, this paragraph will be first. This paragraph should introduce the main lesson you have learned through this activity. Look at what you have written for your conclusion paragraph for some ideas.)

6. Now, using the tables as your outline, either handwrite (if allowed by the instructor) or type the final paper to be turned in to your instructor.

Bibliography

Bloom, B. 1976. *Human Characteristics and School Learning.* New York: McGraw-Hill Book Company.

Brown, J. F., and R. M. Cooper. 1978. *Learning Styles Inventory.* Freeport, NY: Educational Activities, Inc.

Cattell, R. B. 1 966. "The Screen Test for the Number of Factors." *Multivariate Behavioral Research*, no. 1, pp. 245-276.

Fox, L. H. 1977. *Women and Mathematics: Research Perspective for Change.* Washington, DC: National Institute of Education.

Kolb, D. 1985. *Learning-Styles Inventory.* Boston, MA: McBer and Company.

McCarthy, B. 1981. *The 4-Mat System: Teaching to Learning Styles with Right/Left Mode Techniques.* Barrington, IL: Excel.

Nolting, P. D. 1987. *How to Reduce Test Anxiety.* An audiocassette tape. Bradenton, FL: Academic Success Press. (www.academicsuccess.com)

Richardson. F. C., and R. M. Suinn. 1973. "A comparison of Traditional Systematic Desensitization, Accelerated Mass Desensitization, and Mathematics Anxiety." *Behavior Therapy*, no. 4, pp. 212-218.

Suinn, R. M. 1988. *Mathematics Anxiety Rating Scale (MARS).* Fort Collins, CO: RMBSI, Inc.

Tobias, S. 1978. "Who's Afraid of Math and Why?" *Atlantic Monthly*, September, pp. 63-65.

Wolpe, J. 1958. *Psychotherapy by Reciprocal Inhibition.* Stanford, CA: Stanford University Press.

About the Author

Over the past 15 years, Learning Specialist Dr. Paul Nolting has helped thousands of students improve their math learning and obtain better grades. Dr. Nolting is a national expert in assessing math learning problems — from study skills to learning disabilities — and developing effective learning strategies and testing accommodations.

Dr. Nolting is also a nationally recognized consultant and trainer of math study skills and of learning and testing accommodations for students with learning disabilities. He has conducted national training grant workshops on math learning for the Association on Higher Education and Disabilities and for a two-year University of Wyoming 1991 Training Grant of Basic Skills.

Dr. Nolting has conducted numerous national conference workshops on math learning for the National Developmental Education Association, the National Council of Educational Opportunity Association, and the American Mathematical Association of Two-Year Colleges. He is a consultant for the American College Test (ACT) and the Texas Higher Education Coordinating Board. He is also a consultant for Houghton Mifflin faculty development programs, conducting workshops on math study skills for math faculty. His text *Winning at Math: Your Guide to Learning Mathematics Through Successful Study Skills* is used throughout the United States, Canada, and the world as the definitive text for math study skills.

Dr. Nolting has consulted with numerous universities and colleges. Some of the universities with which Dr. Nolting has consulted are the University of Massachusetts, the University of Colorado-Boulder, Texas Tech. University, Black Hills State University, Tennessee Tech. University, and the University of Connecticut.

Some of the colleges with which he has consulted are San Antonio College, St. Louis Community College, J. Sargeant Reynolds College, Montgomery College, Broward Community College, Miami-Dade Community College, Northeast State Technical Community College, Landmark College, Denver Community College, Valencia Community College, and Pensacola Junior College. Dr. Nolting has consulted with over 75 colleges, universities, and high schools over the last 15 years.

Dr. Nolting holds a Ph.D. in education from the University of South Florida. His Ph.D. dissertation was "The Effects of Counseling and Study Skills Training on Mathematics Academic Achievement." He is an adjunct instructor for the University of South

Florida, teaching assessment and appraisal courses at the graduate level.

His book *Winning at Math: Your Guide to Learning Mathematics Through Successful Study Skills* was selected Book of the Year by the National Association of Independent Publishers. "The strength of the book is the way the writer leads a reluctant student through a course from choosing a teacher to preparing for the final examination," says *Mathematics Teacher*, a publication of the National Council of Teachers of Mathematics. *Winning at Math* and Dr. Nolting's other titles are available at http://www.academicsuccess.com.

His two audiocassettes, *How to Reduce Test Anxiety* and *How to Ace Tests*, were also winners of awards in the National Association of Independent Publishers competition. "Dr. Nolting," says *Publisher's Report*, "is an innovative and outstanding educator and learning specialist."

A key speaker at numerous regional and national education conferences and conventions, Dr. Nolting has been widely acclaimed for his ability to communicate with faculty and students on the subject of improving math learning.

Books and Materials by Academic Success Press and Dr. Paul Nolting

How to Ace Tests. Academic Success Press, 1987.
How to Reduce Test Anxiety. Academic Success Press, 1987. (audiocassette)

How to Develop Your Own Math Study Skills Workshop and Course. Academic Success Press, 1991.

Improving Mathematics Studying and Test-Taking Skills. Boston, MA: Houghton Mifflin Company, 1994. (videotape, ISBN 0-669-33291-7)
Math and the Learning Disabled Student: A Practical Guide for Accommodations. Academic Success Press, 1991.
Math and Students with Learning Disabilities: A Practical Guide to Course Substitutions. Academic Success Press, 1993.
Strategy Cards for Higher Grades. Academic Success Press, 1989. Twenty 3-inch by 5-inch cards. (card deck)
Winning at Math — Computer Evaluation Software. Academic Success Press.
Winning at Math: Your Guide to Learning Mathematics Through Successful Study Skills. Academic Success Press, 1997.

Ordering Information

For ordering information on *Winning at Math* and other books and learning materials listed above, call toll free: 1-800-247-6553 or visit www.academicsuccess.com. Credit card, checks, and purchase orders are accepted. Shipping charges are $6.00 for the first item and $2.00 for any additional items. Academic Success Press can be contacted at 6023 26th Street West, PMB 132, Bradenton, FL 34207. For a free catalog, call (941) 359-2819.

To order Houghton Mifflin videos, contact customer service toll free at 1-800-225-1464.